S0-BLL-310

DIRECTIONS IN ELECTROMAGNETIC WAVE MODELING

DIRECTIONS IN ELECTROMAGNETIC WAVE MODELING

Edited by
Henry L. Bertoni
Weber Research Institute
Polytechnic University
Brooklyn, New York

and
Leopold B. Felsen
Weber Research Institute
Polytechnic University
Farmingdale, New York

PLENUM PRESS • NEW YORK AND LONDON

Library of Congress Cataloging-in-Publication Data

Directions in electromagnetic wave modeling / edited by Henry L. Bertoni and Leopold B. Felsen.
p. cm.
Proceedings of a conference held Oct. 22-24, 1990, in New York City.
Includes bibliographical references and index.
ISBN 0-306-44023-7
1. Electromagnetic waves--Mathematical models--Congresses.
I. Bertoni, Henry L. II. Felsen, Leopold B.
QC669.D57 1991
530.1'41--dc20 91-26500
CIP

Proceedings of an international conference on Directions in Electromagnetic Wave Modeling, held October 22–24, 1990, in New York, New York

ISBN 0-306-44023-7

Printed in the United States of America

PREFACE

In 1945, Dr. Ernst Weber founded, and was the first director of, the **Microwave Research Institute (MRI)** at POLYTECHNIC UNIVERSITY (at that time named the Polytechnic Institute of Brooklyn). MRI gained world-wide recognition in the 50's and 60's for its research in electromagnetic theory, antennas and radiation, network theory and microwave networks, microwave components and devices. It was also known through its series of topical symposia and the widely distributed hard bound MRI Symposium Proceedings. Rededicated as the **Weber Research Institute (WRI)** in 1986, the research focus today is on such areas as electromagnetic propagation and antennas, ultra-broadband electromagnetics, pulse power, acoustics, gaseous electronics, plasma physics, solid state materials, quantum electronics, electromagnetic launchers, and networks. Following the MRI tradition, WRI intends to organize its own series of in-depth topical conferences with published proceedings.

This volume constitutes the Proceedings of the first WRI International Conference on Directions in Electromagnetic Wave Modeling. The conference was held October 22-24, 1990, in New York, New York, in cooperation with the Antennas and Propagation Society of the IEEE, the Microwave Theory and Techniques Society of the IEEE, and the U. S. National Committee of URSI.

Henry L. Bertoni
Leopold B. Felsen

March 1991

CONTENTS

INTRODUCTION

Scanning the Conference 3
H. L. Bertoni and L. B. Felsen

Trends in Electromagnetic Research 7
A. K. Jordan (Invited)

ANALYTICALLY BASED TECHNIQUES

Wave Objects, Spectra, and Their Role in Analytic Modeling 11
L. B. Felsen (Invited)

Phase-Space Sampling in Asymptotic Diffraction Theory 23
J. M. Arnold

Derivation of Extended Parabolic Theories for Vector Electromagnetic Wave Propagation 31
R. I. Brent and L. Fishman

Recent Advances in Diffraction Analysis of Reflector Antennas 39
Y. Rahmat-Samii (Invited)

High Frequency EM Scattering by Non-Uniform Open Waveguide Cavities Containing an Interior Obstacle 67
P. H. Pathak, P. H. Law and R. J. Burkholder (Invited)

Waves and Beams in Biological Media 79
J. Bach Andersen (Invited)

Complex Source Pulsed Beams: Propagation, Scattering and Applications 87
E. Heyman (Invited)

Combinations of Local Scattering Operators and Global Propagators 101
I-T. Lu and B-L. Ma (Invited)

Wave Theory Modeling in Underwater Acoustics 111
H. Schmidt (Invited)

Seismic Waveform Modelling of High-Frequency Body Waves 113
C. H. Chapman and R. T. Coates (Invited)

Current-Based Hybrid Analysis in Surface-Wave Hybrid Formulation for Electromagnetics Modeling 123
D-S. Wang and L. N. Medgyesi-Mitschang

Transient Plane Wave Scattering from a Circular Cylinder Backed by a Slit-Coupled Coaxial Cavity 133
G. Vecchi and L. B. Felsen

Transient Acoustic Radiation in a Continuously Layered Fluid - An Analysis Based on the Cagniard Method 145
A. T. de Hoop and M. D. Verweij

Diffracted Microwave Fields Near Dielectric Shells: Computation, Measurement, and Decomposition 153
R. A. Hayward, E. L. Rope and G. Tricoles

Green's Function of Wave Equation for Inhomogeneous Medium 161
E. G. Saltykov

NUMERICAL SOLUTION OF MAXWELL'S EQUATIONS

Direct Maxwell's Equation Solvers in Time and Frequency Domains - A Review 171
R. Mittra and J-F. Lee (Invited)

Assessing the Impact of Large Scale Computing on the Size and Complexity of First-Principles Electromagnetic Models 185
E. K. Miller (Invited)

State of the Art and Future Directions in Finite-Difference and Related Techniques in Supercomputing Computational Electromagnetics 197
A. Taflove (Invited)

Application of Conjugate Gradient Method for the Solution of Large Matrix Problems 215
T. K. Sarkar (Invited), S. Ponnapalli and P. Petre

Electromagnetic Modeling in Accelerator Designs 229
R. K. Cooper and K. C. D. Chan (Invited)

Application of Adaptive Remeshing Techniques to the Finite Element Analysis of Nonlinear Optical Waveguides 239
R. D. Ettinger, F. A. Fernandez and J. B. Davies

MODELING OF PRINTED ELEMENT STRUCTURES

Numerical Modeling of Passive Network and Circuit Components in Monolithic Microwave Integrated Circuits (MMIC) 249
D. C. Chang (Invited), D. I. Wu and J. X. Zheng

On the Modeling of Printed Circuits and Antennas on Curved Substrates 265
A. Nakatani and N. G. Alexopoulos (Invited)

A General, Full-Wave Approach for Modeling Signal Lines and Discontinuities in Computer Packages 273
B. J. Rubin (Invited)

Numerical Modeling of Frequency Dispersive Boundaries in the Time Domain Using Johns Matrix Techniques 287
W. J. R. Hoefer (Invited)

Wave Interactions in Planar Active Circuit Structures 299
J. Birkeland, S. El-Ghazaly and T. Itoh (Invited)

High Frequency Numerical Modeling of Passive Monolithic Circuits and Antenna Feed Networks 307
L. P. B. Katehi (Invited)

Leaky Waves on Multiconductor Microstrip Transmission Lines 319
L. Carin (Invited)

Closed-Form Asymptotic Representations for the Grounded Planar Single and Double Layer Material Slab Green's Functions and their Applications in the Efficient Analysis of Arbitrary Microstrip Geometries 329
S. Barkeshli and P. H. Pathak

Improvements of Spectral Domain Analysis Techniques for Arbitrary Planar Circuits 339
T. Becks and I. Wolff

Integral Equation Analysis of Microwave Integrated Circuits 347
K. A. Michalski

Semi-Discrete Finite Element Method Analysis of Microstrip Structures 355
M. Davidovitz and Z. Wu

Spectral-Domain Modeling of Radiation and Guided Wave Leakage in a Printed Transmission Line 363
N. K. Das

On the Modeling of Traveling Wave Currents for Microstrip Analysis 371
J. S. Herd

S-Parameter Calculation of Arbitrary Three-Dimensional Lossy Structures by Finite Difference Method 379
D. Hollmann, S. Haffa, and W. Wiesbeck

Modeling of Electromagnetic Fields in Layered Media by the Simulated Image Technique 387
Y. L. Chow and J. J. Yang

MODELING INTERACTIONS WITH MATERIALS

Modeling for Waves in Random Media - A Need for Analytical, Numerical, and Experimental Investigations 397
A. Ishimaru (Invited)

Spectral Changes Induced by Scattering from Space-Time Fluctuations 405
E. Wolf (Invited)

Rough Surface Scattering 407
J. A. DeSanto and R. J. Wombell (Invited)

Domain of Validity of the Wiener-Hermite Functional Expansion Approach to Rough Surface Scattering 417
C. Eftimiu

Wave Intensity Fluctuations in a One Dimensional Discrete Random Medium 425
S. S. Saatchi and R. H. Lang

Fractal Electrodynamics and Modeling 435
D. L. Jaggard (Invited)

Feynman-Diagram Approach to Wave Diffraction by Media Having Multiple Periodicities 447
T. Tamir, K-Y. Tu and H. Lee (Invited)

Electromagnetic Bandgap Engineering in Three-Dimensional Periodic Dielectric Structures 457
K. M. Leung (Invited)

Modeling of Superconductivity for EM Boundary Value Problems 467
R. Pous, G. C. Liang and K.K. Mei (Invited)

Derivation and Application of Approximate Boundary Conditions 477
T. B. A. Senior (Invited)

Chirality in Electrodynamics: Modeling and Applications 485
D. L. Jaggard and N. Engheta

Electromagnetic Wave Modeling for Remote Sensing 495
S. V. Nghiem, J. A. Kong and T. Le. Toan (Invited)

Approximate Scattering Models in Inverse Scattering: Past, Present and Future 507
A. J. Devaney (Invited)

Wave Modeling for Inverse Problems with Acoustic, Electromagnetic, and Elastic Waves 517
K. J. Langenberg (Invited)

Self-Focusing Revisited: Spatial Solitons, Light Bullets and Optical Pulse Collapse 529
Y. Silberberg, J. S. Aitchison, A. M. Weiner, D. E. Leaird, M. K. Oliver, J. L. Jackel, and P. W. E. Smith (Invited)

Parametric Excitation of Whistler Waves by Circularly Polarized Electromagnetic Pumps in a Nonuniform Magnetoplasma: Modeling and Analysis 531
S. P. Kuo (Invited) and R. J. Barker

Nonlinear Modulational Instability of an Electromagnetic Pulse in a Neutral Plasma 541
D. J. Kaup (Invited)

Index 557

INTRODUCTION

SCANNING THE CONFERENCE

Henry L. Bertoni and Leopold B. Felsen

Weber Research Institute
Polytechnic University
Farmingdale, NY

GENERAL OVERVIEW

The **WRI** International Conference on **Directions in Electromagnetic Wave Modeling** was intended to be a forum for critical in-depth assessment of the state-of-the-art, new directions and challenges in the modeling of electromagnetic wave radiation, guiding, propagation, and scattering in diverse environments. The focus was on analytic, numerical and combined techniques with emphasis on their scope, limitations, and anticipated role in future research and applications.

Modeling of electromagnetic wave interactions is an essential ingredient in applications ranging from remote sensing to integrated circuits, and has traditionally been addressed within the context of various applications. In contrast, this Conference was devoted to the techniques of electromagnetic wave modeling as such, in order to integrate specialized approaches into a broader perspective.

Analytically based asymptotic techniques for representing propagation in large scale structures continues to be an active research area. Development of ray type algorithms, and the introduction of Gabor beam expansion methods, make possible accurate modeling of ducting and scattering in complex environments, as demonstrated by various papers presented at the Conference.

One important aspect of electromagnetic wave modeling is the ubiquitous presence of digital computers. Hardly a theoretical paper on electromagnetics appears in the technical literature whose contents do not depend in some way on computer usage. However, the stage of the mathematical development at which a computer is employed covers the broad range from the direct numerical approximation of Maxwell's equations to the mere numerical evaluation of a final expression. One purpose of this Conference was to highlight this range of computer implementation.

While direct implementation of Maxwell's equations via finite difference and finite element methods is feasible for structures of limited size, techniques that analytically account for propagation are required when the problem size is much larger than the wavelength, especially in three dimensions. In one approach to large structures, local numerical solutions are married to global propagators. Alternatively, representations for unknown currents and/or polarization together with Green's function representations in either the spatial or spectral domain can be used to find solutions by imposing self consistency. These latter methods find

Directions in Electromagnetic Wave Modeling
Edited by H.L. Bertoni and L.B. Felsen, Plenum Press, New York, 1991

particular application to printed element structures where layering of the structure admits of spectral representations for the Green's function. All of these methods were explored in various papers presented at the Conference, together with their strengths and limitations.

Another aspect of electromagnetic wave modeling involves the choice of efficient and physically insightful approaches for representing wave interactions in different classes of media. This aspect was covered in the Conference through papers treating random, fractal, periodic and nonlinear media, including scattering at rough surfaces and boundary conditions to be used with superconducting and layered materials.

Several papers dealt with ultra wide-band wave phenomena and the challenges for treating these problems directly in the time domain. It is anticipated that this subject area will receive increased attention in the future.

In order to foster a critical in-depth assessment of the stat-of-the-art in electromagnetic modeling, the Conference was organized into twelve non-overlapping sessions for oral presentation of papers, and one poster session. This format of a well defined topic covered in non-overlapping sessions, which permitted immersion in all aspects of the Conference, was praised by the attendees.

The Opening Secession was primarily devoted to an overview by program managers from various U.S. Government agencies as to where electromagnetic wave modeling fits into their current objectives. The remaining 53 papers presented at the Conference, and covering the topics cited above, were organized into the twelve sessions under a variety of topic headings. For presentation in this volume, they have been reorganized into the following four broad areas:

I. Analytically Based Techniques
II. Numerical Solution of Maxwell's Equations
III. Modeling of Printed Element Structures
IV. Modeling Interactions With Materials

A few authors did not provide a written manuscript of their presentation. For completeness, the abstracts of their papers have been included in this volume. Also included are several papers substantially over the assigned page length that were submitted too late to be revised.

PERSPECTIVES OF U.S. GOVERNMENT AGENCIES

The presentations by program managers of U.S. Government agencies was intended to connect the conference topics with problem priorities within the relevant organizations charge with promoting research. Presentations were made by:

Lawrence S. Goldberg, National Science Foundation
Arthur K. Jordan, Naval Research Laboratory
James W. Mink, U.S. Army Research Office
Arje Nachman, Air Force Office of Scientific Research.

While specific priorities differed between the various agencies, there was a commonality of emphasis on the importance of the subject areas in the conference program, and on the need for strong coupling between analysis, computation and experiment.

Listed specifically by Dr. Mink as focus items in the U.S. Army research program were conformal, dielectric and printed antennas; antennas for use near air-medium interfaces; numerical modeling techniques; active quasi-optical systems; target imaging; and the investigation of background clutter at millimeter wavelengths. These focus items find application to communications, surveillance and target acquisition systems; detection of buried objects; operation of antennas on composite vehicles; diakoptic theory; quasi-optical techniques of portable power transmission; and target classification and identification. Dr. Nachman identified linear FM wave propagation through dispersive and/or stochastic media, nonlinear wave propagation through Kerr media, inverse scattering and the exploitation of wavelets as focus items.

Dr. Jordan stressed broadband applications of electromagnetic waves and the integration of optics and electronics into components, which will require augmenting classical electromagnetics by quantum theory. Dr. Jordan supplemented his conference presentation with a written version of Trends in Electromagnetic Research, which is included as the next section of this Introduction. Although this material addresses specific electromagnetic problem areas of current interest within the Office of Naval Research, his summary also includes broadly applicable concepts.

TRENDS IN ELECTROMAGNETIC RESEARCH

Arthur K. Jordan

Naval Research Laboratory, Code 4210
Washington, DC

Electromagnetic wave research has been advancing along three basic trends that characterize scientific research in general.

First, electromagnetic wave research has become more diverse and has more comprehensive objectives and applications.

- For example, broadband and multiband integrated circuits are being developed for both optical and electronic, that is optoelectronic, applications.
- These devices require novel materials and complex designs.
- Users of these devices have different and changing system requirements.
- The sponsoring organizations have economic and administrative objectives.
- These are only a few of the possible examples.

Second, we now have a greater and more flexible theoretical and experimental capabilities to meet these research objectives.

- More computing power and digital technology are available to facilitate the translation of mathematical analysis to engineering results.
- For example, symbolic computation and large-scale numerical techniques such as finite-element and finite-difference methods in both the frequency and time domains are used to model electromagnetic wave scattering from complex objects.
- These capabilities together with high-resolution graphics have led to greatly enhanced imaging and modeling techniques.
- More fundamental physical theories have been applied to analyze electromagnetic waves.

Directions in Electromagnetic Wave Modeling
Edited by H.L. Bertoni and L.B. Felsen, Plenum Press, New York, 1991

- One example is simplifying the analysis of radiation from layered structures by applying gauge theoretical concepts.

- A second example is generalizing the design of optical waveguides by applying inverse scattering theory.

- A third example is analyzing nonlinear waves and their interactions by using non-relativistic and relativistic models.

- Experimental techniques are now possible that bridge the gap between classical and quantum theories.

- For example, by using very short pulse (pico- and femto-second) techniques, it is possible to probe atomic and molecular structures.

- We can control material properties to the extent that semiconductors can be engineered on the atomic scale.

- Thin-film superconductors have been developed for microwave applications.

Third, these research trends have nonlinear dependence on individual researchers.

- This nonlinear dependence can be partially attributed to two factors:

  - the continual education of researchers by feedback from new results in their own and related disciplines.

  - the accelerated feedback from individual researchers in the form of rapid publications, electronic mail and the availability of scientific meetings.

- Since this research feedback has become faster and more detailed, the influence of individual researchers on research trends has great importance.

- The individual researchers create their specific research plans and produce results.

- Sponsors identify general research objectives and endeavor to obtain funding to meet these objectives.

These research trends will affect electromagnetics in several ways:

- Research objectives will become more flexible, reflecting changing national and international priorities.

- Individual scientists will have more freedom to pursue independent lines of research.

- With this increased freedom and flexibility there will be increased responsibility to be sensitive to societal and economic priorities.

Both the speed and variety of research results are increasing dramatically. In order to continue productive research it is essential to be guided by fundamental physical theories and sound scientific methods.

ANALYTICALLY BASED TECHNIQUES

WAVE OBJECTS, SPECTRA, AND THEIR ROLE IN ANALYTIC MODELING

L.B. Felsen

Polytechnic University
Weber Research Institute
Farmingdale, NY

INTRODUCTION

Modeling of complex wave propagation and scattering relies most generally on a combination of analytical and numerical techniques. How to apportion that combination is problem dependent, with analytical modeling most effective for relatively simple but numerically large portions of a still larger composite. For example, in a large multireflector antenna system, the feed and shaped reflectors constitute complex portions that are separated by long (in terms of wavelengths) propagation paths in free space. While purely numerical treatment of the entire configuration would be prohibitive, a self-consistent parametrization that blends analytical free space propagators with numerical solutions for the feed and reflector portions can render the problem tractable. The most effective analytical models are structured around "observable-based" wave objects tied to the physical phenomena that are operative in establishing the final field. For the propagation links that are most amenable to analytical modeling, "good" wave objects in the frequency domain include ray fields, beams, paraxial propagators, as well as global and local guided mode fields, either separately or in hybrid combinations, that seek to optimize the utility of each for any given propagation environment and input signal. These wave objects and their combinations have direct counterparts under transient conditions. Success in this endeavor requires integrated treatment of input signal, propagation channel, and processing at the receiver so as to *unify* the entire source-medium-detection process. This is done effectively in a configuration-spectrum phase space representation that treats space-time and wavenumber-frequency simultaneously. It should be noted that the "source" fields here may be actual (directly generated) or virtual (induced on a scatterer surface or an equivalent fictitious surface enclosing a scatterer).

The concepts described above are discussed below and illustrated on selected examples. Emphasis is restricted to the frequency domain. Some relevant time domain aspects are summarized in references 1-4.

Directions in Electromagnetic Wave Modeling
Edited by H.L. Bertoni and L.B. Felsen, Plenum Press, New York, 1991

Rays and Beams

Rays: Ray fields constitute the most versatile propagators for high frequency waves in general propagation and scattering environments. The ray trajectories trace the progress of local plane waves, which represent constructive interference maxima of progressing global wave fields and thereby systematize the localization of high frequency wave phenomena. The total field at any observation point is synthesized by the sum of all ray fields passing through that point. The classes of ray fields that describe various propagation and diffraction processes are encompassed by the geometrical theory of diffraction (GTD).[5] Ray fields, being representative of local plane waves, have minimal spectral content -- the *point* spectrum which defines the *direction* of the ray. Local plane waves are inadequate to describe high frequency wave phenomena in transition regions, in which the fields undergo rapid variation transverse to the local propagation (ray) direction; examples are transitions through light-shadow boundaries, total reflection boundaries, caustics and foci, etc., formed by the ensemble of rays. To accommodate these transitions, the skeletal ray field spectra must be fleshed out by wavelength-dependent uniform transition functions, which are incorporated in various uniform theories of diffraction.[5,6-8] The GTD and uniformized ray field methodology has been well established and constitutes one of the most frequently used modeling tools for high frequency wave propagation and diffraction.

Beams: The need for keeping track of transition regions, and for introducing uniformizing corrections in these regions, when tracing ray fields through complicated environments has more recently stimulated interest in the use of beam fields, which have more spectral flesh than the skeletal ray fields and therefore remain valid in ray field transition regions. In fact, because of the availability of an additional parameter, the beam width w_o at the waist relative to the wavelength λ and the corresponding Fresnel length $b=w_o^2/\lambda$ defining the interval with good collimation (Fig. 1a), it is possible to span the entire spectral domain from the *point spectrum* of a *spatially extended* plane wave ($w_o \rightarrow \infty$) to the *wide spectrum* of a *spatially localized* point source ($w_o \rightarrow 0$) (Fig. 1b). Gaussian beams are strongly favored because of their self-similar spatial and spectral properties.[9] Thus, in the waist plane $z=0$ of a sheet beam in the two-dimensional (x,z) domain, the field $u(x,0)$ and its spatial wavenumber spectrum $\hat{u}(\xi,0)$, with ξ representing the normalized spatial wavenumber, are given by

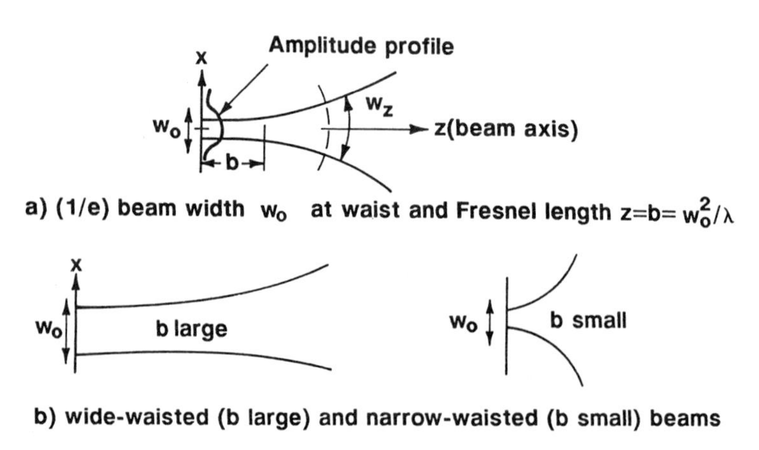

Fig. 1 Gaussian beams

$$u(x,0) = \left[\sqrt{2}/w_o\right]^{1/2} \exp\left\{-\pi\,[x/w_o]^2\right\} \ ; \ \hat{u}(\xi,0) = \left[\sqrt{2}w_o\right]^{1/2} \exp\left\{-\pi[(\xi w_o/2\pi]^2\right\} \tag{1}$$

Moreover, a Gaussian beam can be modeled by a point source at a *complex* location,[10,11] thereby making the machinery of complex ray tracing available for describing its progress through a propagation and scattering environment.[8,11] Here, the "axis" of the beam, whereon the field attains its maximum exponential value, follows a real ray trajectory, whereas off-axis points are reached by real-space intersections of complex rays (Fig. 2). Complex ray field tracing is more involved than real ray field tracing. However, in the paraxial region near the beam axis, the complex ray fields are almost real and their value can be ascertained by perturbation with respect to the on-axis fields. In the far zone well beyond the Fresnel length, the complex ray fields can likewise be expressed by simplifying asymptotic approximations.

Narrow-waisted beam fields have short Fresnel lengths (Fig. 1b) and reach their far zone within a few wavelengths. They behave almost like line or point source fields and can therefore be tracked essentially like real ray fields. Furthermore, as noted earlier, they have the advantage of retaining validity in ray field transition regions, such as caustic or focal regions, and require no special treatment there. When this feature is coupled with the fact that linearly phased beamlike windowed wave objects can be employed as discrete or continuous basis functions in a configuration-spectrum phase space for *rigorous* analysis and synthesis of general wave fields,[12,13a] a narrow-waisted Gaussian beam basis set constitutes an attractive tool for accurate description of high frequency wavefields in certain complex environments.

The discretized phase space strategy has been applied to wave transmission from extended nonfocused and focused plane aperture distributions through a plane dielectric layer[13b] and through a curved layer[13c] (as encountered, for example, in a plane aperture antenna covered by a shaped dielectric radome). The implementation is carried out in the discretized phase-space (Gabor lattice) schematized in Fig. 3. First, the aperture field is decomposed into narrow-waisted basis beams with amplitudes A_{mn}, whose distribution in the phase space is shown in Fig. 4 for a uniform truncated nonphased aperture illumination. Clearly favored in the bulk $-d \leq x \leq d$ of the aperture region are the $n=0$ (forward propagating) beams, whereas the abrupt truncation of the field at the aperture edges $x = \pm d$ produces edge centered tilted beams ($m = \pm d/L_x$, $|n| \geq 0$) which establish the diffraction side lobes as well as the

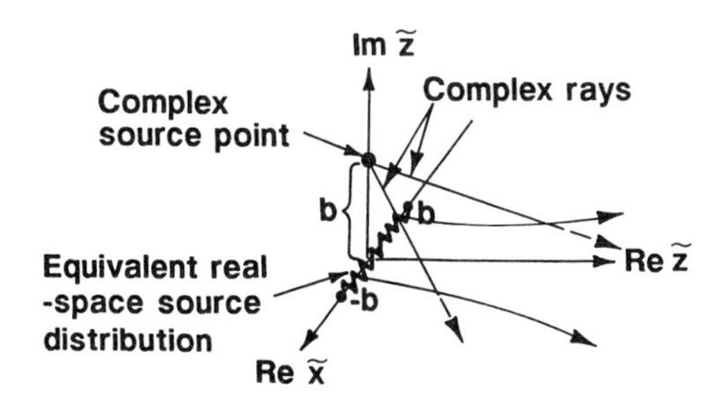

Fig. 2 Complex source point modeling of two-dimensional Gaussian beams. Complex source point: $\tilde{\underline{r}}' = (\tilde{x}', \tilde{z}') = (0, 0 + ib)$, $b > 0$. Beam profile in real coordinate space ($\mathrm{Re}\,\tilde{x}$, $\mathrm{Re}\,\tilde{z}$) (see Fig. 1) is established by real-space intersections of complex rays.

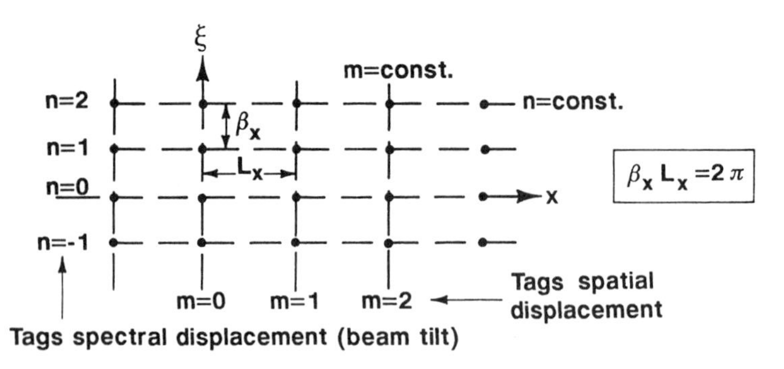

Fig. 3 Discretized phase space lattice. L_x and β_x denote the lattice spacing along the configurational (x) and spectral (ξ) coordinates, respectively. Spatial displacements: $F(x - mL_x x)$, where F is a Gaussian envelope function; spectral displacement: $\exp(i\, n\, \beta_x x)$.

enhancement in the edge truncation (light-shadow) transition region. Next, the fields produced by the basis beams away from the aperture plane in free space are modeled by the complex source point method (Fig. 2) and, therefore, can be tracked through the layers by complex ray tracing via the formula[13b,c]

$$u(\underline{r}) \propto \sum_{m,n} A_{mn} e^{-kb}\, G_{mn}(x,z) \tag{2}$$

where

$$G_{mn}(x,z) = \sum_{q} \left[G_o(\tilde{L}_\alpha)\tilde{t}_1 \tilde{t}_2 (-\tilde{\Gamma}_1 \tilde{\Gamma}_2)^q \tilde{D}_q \exp(ik\tilde{\psi}_q) \right]_{m,n} \tag{2a}$$

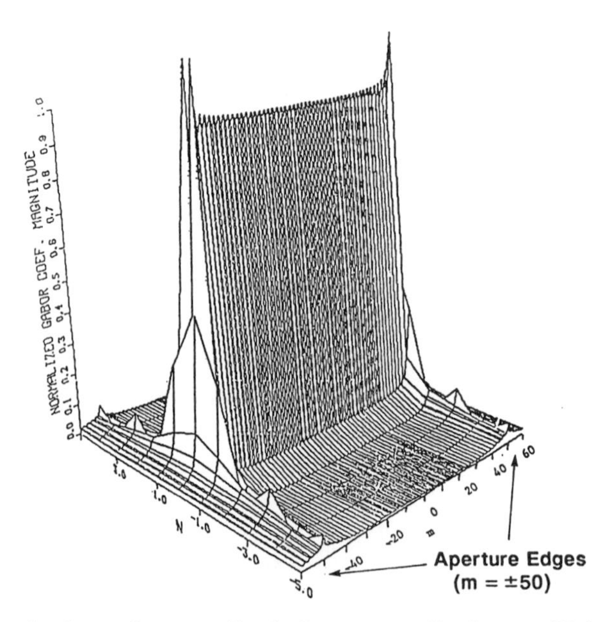

Fig. 4 Magnitudes of normalized beam amplitude coefficients A_{mn}/A_{oo}, corresponding to the lattice of Fig. 3, for a nonphased uniform aperture distribution. Aperture width $2d = 25\,\lambda$, λ = wavelength; $L_x = 0.25\,\lambda$.

The geometrical parameters appearing in G_{mn} in (2a), which tracks a single beam (m,n), are identified in Fig. 5 for transmission through a plane layer,[13b] with q denoting the number of multiple reflections along the complex ray inside the layer; analogous definitions apply to circular layer. $G_0(\tilde{L}_\alpha)$ is the line source Green's function in free space, observed along $\tilde{L}_\alpha$. Here, $\tilde{t}_{1,2}$ are plane wave transmission coefficients, and $\tilde{\Gamma}_{1,2}$ are plane wave reflection coefficients, at each of the layer interfaces, while $\tilde{D}_{\tilde{q}}$ is the ray tube divergence, which determines the complex ray field amplitude, and $\tilde{\psi}_q$ is the accumulated phase along the ray. In the absence of the layer, one puts $\tilde{t}_{1,2}=1$, $q=0$, to obtain the radiated field in free space. The tilde $\tilde{}$ identifies quantities defined along complex ray paths and propagation angles.

The beam algorithm for *various* narrow beam basic elements was tested first on a plane layer, for which comparison with an independently devised exact spectral integral reference solution confirmed that the various beam arrangements *stabilize* accurately around the *same* solution, which *is* the *reference* solution (Fig. 6).[13b] The algorithm has then been applied to a cylindrically curved layer[13c] and also to the three-dimensional problem of a planar square aperture distribution covered by an ogive-shaped layer;[13d] here, the stability of results produced by alternative beam stackings in the phase space is taken as evidence of their correctness because no independent reference data are available.

The continuously windowed phase space strategy has been applied to nonfocused and focused plane aperture distributions in free space, both in the time-harmonic and transient regimes, and it has been compared critically with the corresponding discretized phase space formulation. These results are presented in references.[2a,b,14]

Guided Modes

Guided modes are important basic elements for representing general electromagnetic wavefields in environments that confine the energy transversely along

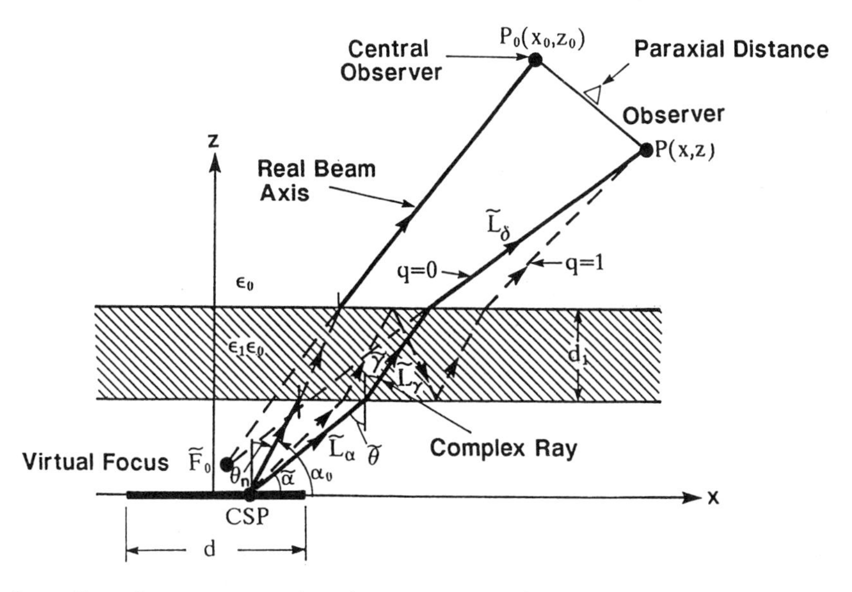

Fig. 5 Complex source point (CSP), launching complex ray from complex extension of aperture domain. Complex ray trajectory through plane layer, undergoing q reflections. Real-space observer at x,z.

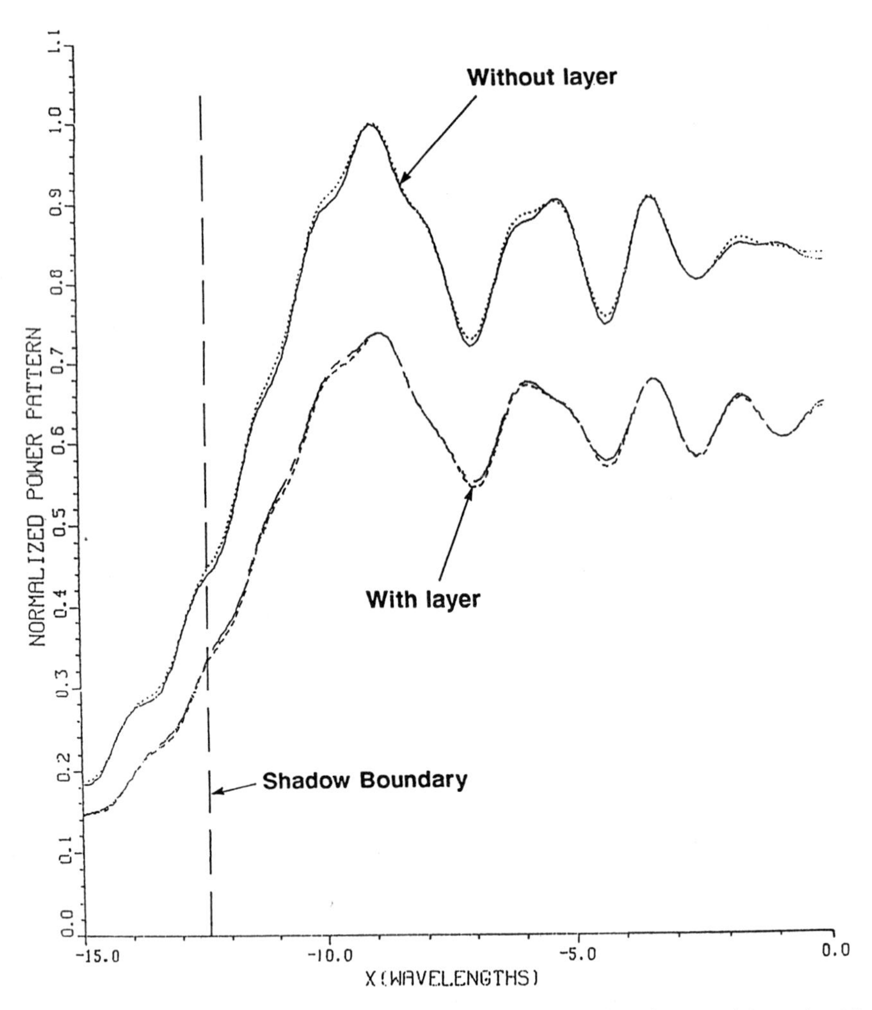

Fig. 6 Near zone field at $z=17\lambda$ from the aperture in Fig. 4, with and without plane layer. Solid lines: independently computed reference solution. Dots, dashes: beam synthesis. The shadow boundary is located at the aperture edge.

channels formed either by boundaries and interfaces, by continuous refraction in an inhomogeneous medium, or both. A guided mode is a source-free solution of the field equations, and under ideal (coordinate separable) conditions, it satisfies the boundary conditions in the transverse domain exactly, has its distinct transverse periodicity (wavenumber spectral parameter), and propagates longitudinally along the channel direction with its distinct propagation coefficient. Ideal modes are decoupled from, and propagate independently of, other modes in the set. They may be "proper" (completely trapped) or "improper" (leaky) through continuous shedding to the exterior.[15,16] When the channel does not conform with separability in one of the basic coordinate systems, ideal modes defined for separable conditions become coupled. If the departure from separability is weak, one may define adiabatic modes (AM) that adapt smoothly, without coupling to other AM, to the slow nonseparable longitudinal variations, retaining their distinct spectral identity throughout. Each AM fails in its own cutoff region, across which its wavefield changes from being propagating to becoming evanescent. This deficiency, attributable to the sparse AM *point* spectrum, can be overcome by synthesizing a new wave object, an intrinsic mode (IM), as a spectral *continuum* of AM. For a summary of the AM-IM connection, see reference 17.

Both the AM and IM perform best in a coordinate frame that comes closest to satisfying separability. Therefore, the systematic search for a "good" coordinate system (good parametrization) is of prime importance. Asymptotic approximations that decompose the modal wavefields into local plane wave constituents permit tracking of these fields along modal ray trajectories, thereby charting the evolution of the local mode under rather general environmental conditions. These conditions may include those that transform an initially well-guided mode into a radiating beam, while retaining the distinct modal (spectral) footprint throughout. Viewed in this manner establishes a "guided" (i.e., transversely confined) AM or IM as a very general spectral wave object.

Implementation of the scenario outlined above for a longitudinally changing guiding environment depends on whether the AM or IM in question remain confined between physical boundaries (walls or interfaces) or whether they are confined by continuous refraction. In the former case, the boundaries parametrize the width of the local waveguide for *all* modes. In the latter instance, each refractively confined mode has its own "duct width" defined by the caustics of its modal ray system. How to search for coordinate frames with "good" local separability in each case can be explored on canonical test problems which are exactly separable in "correct" coordinate systems but nonseparable in "incorrect" coordinate systems. Examples involving a wedge waveguide and a graded index waveguide with special transverse and longitudinal variation have been discussed elsewhere.[18] Here, cylindrical polar coordinates are correct but local rectangular coordinates are not; the latter must be updated so as to approach the correct system.

The understanding gained from test environments can point the way toward attacking more general problems that are not separable in any of the standard coordinate frames. For example, in a longitudinally homogeneous surface duct in the region

$$x > 0, \ -\infty < z < \infty ,$$

with transversely inhomogeneous refractive index

$$n^2(x,z) = 1 - a_o x, \ a_o > 0 \tag{3}$$

that decays monotonically away from the bounding surface at $x=0$, the upgoing and downgoing ray (local plane wave) congruences that synthesize the high frequency asymptotic trapped mode field are schematized in Fig. 7. Upgoing ray fields for mode m in this coordinate separable environment are converted into downgoing ray fields at the mode-dependent caustic x_m, whereas downgoing ray fields are converted into upgoing ray fields at the fixed (mode-independent) boundary $x=0$. The width x_m of the modal duct is established by invoking self-consistent closure on the phase accumulation of the global upgoing-downgoing ray field congruences. Because of the longitudinal invariance, the ray congruences remain self-similar, and the mode field remains trapped throughout. It can be shown that by modal ray tracing, with enforcement of self-consistency, one may construct the asymptotic approximation of the exact modal field.[19]

Let the canonical profile above now be changed to the longitudinally varying nonseparable form $n^2(x,z)=1+(a_o z)x$, which is guiding for $z<0$. As in the canonical

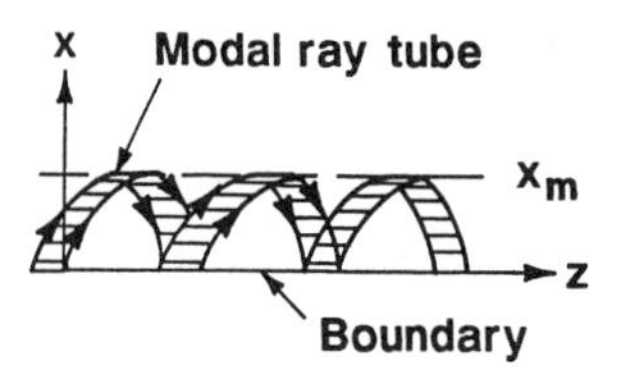

Fig. 7 Modal ray congruences for the surface duct in (3).

case above, it is suggestive to establish the evolution of an initially well trapped mode from the guiding region $z<<0$ through the transition $z\approx 0$ into the antiguiding region $z>0$ by local tracking along the modal rays.[20] The resulting ray plot reveals that the mode trapped in the surface duct gradually detaches from the boundary and is converted into a radiated beam (Fig. 8). Throughout the detachment, the modal beam retains the transverse periodicity imprint from the initial trapping region.

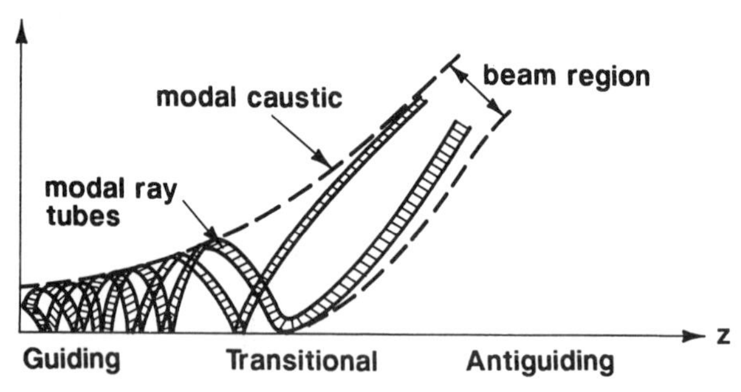

Fig. 8 Modal ray congruences for guiding to antiguiding transition.

SUMMARY

This brief overview has been intended to outline a strategy that involves analytical modeling for wave propagation and scattering in complex environments, under conditions that make the problem scale numerically large, thereby rendering purely numerical approaches inefficient or even intractable. Typically, such large problems are in the short-wavelength regime relative to the problem dimensions. In many applications, the complex portions of the overall environment are localized and are connected within the global configuration by relatively regular portions. These more regular interconnects may be capable of sustaining propagating wave objects of relatively simple form, thereby allowing analytical modeling of interactions over long ranges in terms of wavelengths. To render such modeling efficient, it is desirable that the wave parametrization is *observable-based,* i.e., the wave objects in the model should be able to reproduce observed features in measured or numerically produced reference data in as direct a manner as possible. Even apart from possible computational advantages, such an observable-based parametrization (OBP) furnishes the *physical insight* into complex wave phenomena, which is essential for intelligent *interrogation* of the environment and for *interpretation* of data. Thus, shaping of input signals and signal processing at the receiver can be closely tied to the selected OBP wave objects to structure a *global* strategy. Under these conditions, synthesis of total response is established by *self-consistent interaction*, via propagators, between complex scattering regions. The latter would be treated either numerically or, if the complexity has sufficient regularity, by an OBP model adapted to that complexity.

The first step in pursuit of these objectives is the assembling of wave objects (propagators) that are well suited for various quasi-canonical conditions, thereby permitting their construction from canonical environments, which are amenable to rigorous analysis. Some examples have been given here, with emphasis on how the *detailed understanding* of a *canonical* problem within the *context of OBP* may lead to extension to noncanonical conditions. Detailed understanding requires balanced consideration of configurational and spectral aspects, implemented most effectively in a configuration-spectrum phase space.

Clearly, the OBP strategy is problem dependent. The user must decide whether to avail himself of this option for a problem under consideration. If an alternative is sought to a purely numerical approach, the considerations introduced here become relevant. Even if one follows the purely numerical route, OBP treatment, though possibly less accurate, can clarify the *phenomenology* and thereby provide clues for better numerical construction. These various aspects should be kept in mind when devising problem strategies for wave interaction with complex environments.

ACKNOWLEDGMENT

The research pertaining to this paper has been sponsored by the U.S. Army Research Office under Contract No. DAAG29-85-K-0180, the Office of Naval Research under Grant No. N00014-90-J-1120, and the Joint Services Electronics Program under Contract No. F49620-88-C-0075.

REFERENCES

1. G. Vecchi and L.B. Felsen, "Transient Plane Wave Scattering from a Circular Cylinder Backed by a Slit-Coupled Coaxial Cavity," this volume.

2(a). B.Z. Steinberg, E. Heyman and L.B. Felsen, "Phase Space Beam Summation for Time-Dependent Radiation from Large Apertures: Continuous Parametrization," to be published in J. Opt. Soc. Am. A.

2(b). B.Z. Steinberg and E. Heyman, "Phase Space Beam Summation for Time-Dependent Radiation from Large Apertures: Discretized Parametrization," to be published in J. Opt. Soc. Am. A.

3. E. Heyman and L.B. Felsen, "Complex Source Pulsed Beams," J. Opt. Soc. Am. A6, p. 806-817, 1989.

4. E. Heyman, "Complex Source Pulsed Beams: Propagation and Scattering," this volume.

5. R. C. Hansen, Ed., *Geometrical Theory of Diffraction*, IEEE Press, New York (1981).

6. J.M. Arnold and L.B. Felsen, "Spectral Reconstruction of Uniformized Wavefields from Nonuniform Ray or Adiabatic Mode Forms for Acoustic Propagation and Diffraction," J. Acoust. Soc. Am., 87, 587-600, 1990.

7. Yu. A. Kravtsov and Yu. I. Orlov, "Geometrical Optics of Inhomogeneous Media," Springer Verlag, New York, 1990.

8. L.B. Felsen, "Novel Ways for Tracking Rays," J. Opt. Soc. Am. A2, 954-963, 1985.

9. J.A. Arnaud, *Beam and Fiber Optics,* Academic Press, New York 1976.

10. G.A. Deschamps, "Gaussian Beam as a Bundle of Complex Rays," Elec. Letts., Vol. 7, pp. 684-685, 1971.

11. L.B. Felsen, "Complex-Source-Point Solutions of the Field Equations and their Relation to the Propagation and Scattering of Gaussian Beams," Symposia Matematica, Istituto Nazionale di Alta Matematica, Vol. XVIII, Acad. Press, London and New York, 40-56, 1976.

12. P.D. Einziger, S. Raz and M. Shapira, "Gabor Representation and Aperture Theory," J. Opt. Soc. Am., A3, p. 508, 1986.

13a. J.J. Maciel and L.B. Felsen, "Systematic Study of Fields Due to Extended Apertures by Gaussian Discretization," IEEE Transactions Ant. and Propagation, AP-37, pp. 884-892 (1989).

13b. J.J. Maciel and L.B. Felsen, "Gaussian Beam Analysis of Propagation from an Extended Plane Aperture Distribution Through Dielectric Layers: I-Plane Layer," IEEE Trans. on Antennas and Propagation, AP-38, pp. 1607-1617, 1990.

13c. J.J. Maciel and L.B. Felsen, "Gaussian Beam Analysis of Propagation from an Extended Plane Aperture Distribution Through Dielectric Layers: II-Circular Cylindrical Layer," IEEE Trans. on Antennas and Propagation, AP-38, pp. 1618-1624, 1990.

13d. J.J. Maciel and L.B. Felsen, "Gaussian Beam Algorithm for Transmission of Electromagnetic Fields from a Two-Dimensional Plane Aperture Through a Rotationally Symmetric Shaped Radome," presented at the 1990 IEEE Antenna and Propagat. Soc./URSI Meeting, Dallas, TX.

14. B.Z. Steinberg, E. Heyman and L.B. Felsen, "Phase Space Beam Summation for Time-Harmonic Radiation from Large Apertures," J. Opt. Soc. Am. A8, pp. 41-59, 1991.

15. T. Tamir and A.A. Oliner, "Guided Complex Waves. I. Fields at an Interface. II. Relation to Radiation Patterns," Proc. IEE (London) 110, p. 107, 1963.

16. L.B. Felsen and N. Marcuvitz, "Radiation and Scattering Waves," Prentice Hall, Englewood Cliffs, NJ, 1973, Chapter 5.

17. J.M. Arnold and L.B. Felsen, "Local Intrinsic Modes -- Layer with Nonplanar Interface," Wave Motion 8, pp. 1-14, 1986.

18. L.B. Felsen and L. Sevgi, "Adiabatic and Intrinsic Modes for Wave Propagation in Guiding Environments with Transverse and Longitudinal Variation: Formulation and Canonical Test," to be published in IEEE Trans. Antennas and Propagat.

19. S.J. Maurer and L.B. Felsen, "Ray Methods for Trapped and Slightly Leaky Mode in Multilayered or Multiwave Regions," IEEE Trans. Microwave Theory and Tech., MTT-18, No. 9, 393-584, Sept. 1970.

20. L.B. Felsen and L. Sevgi, "Adiabatic and Intrinsic Modes for Wave Propagation in Guiding Environments with Transverse and Longitudinal Variation: Continuously Refracting Media," to be published in IEEE Trans. Antennas and Propagat.

PHASE-SPACE SAMPLING IN ASYMPTOTIC DIFFRACTION THEORY

J. M. Arnold

Department of Electronics and Electrical Engineering
University of Glasgow
Scotland

INTRODUCTION

In this paper, the discretisation of 2-dimensional wavefields is studied from two points of view: discrete Fourier synthesis and discrete Gaussian synthesis (the Gabor representation). In high-frequency fields localisation of the wave occurs, summarised *in extremis* by the laws of geometrical optics and GTD. Efficient numerical methods should incorporate this localisation if at all possible. In the case of Fourier synthesis, a wave is represented as a superposition of plane waves in arbitrary directions; high-frequency localisation occurs through the principle of stationary phase, wherein at a given point only a small part of the spectrum makes the dominant contribution, the remainder of the spectrum being self-cancelling in superposition due to destructive interference. The directions of dominant plane waves are those closest to the GO ray directions through the point of observation. In the Gaussian synthesis case, localisation occurs in the spectral elements themselves. Under certain circumstances, which can be determined rigorously, both the plane wave and Gaussian syntheses can be discretised, and comparison between them becomes feasible. In order to carry this out, consistent definitions of localisation are required. We show that spatial localisation is most conveniently expressed, in both cases, in terms of the unit of length $L=(2\pi|R|/k)^{1/2}$ and $\Omega=2\pi/L=(2\pi k/|R|)^{1/2}$, where R is the radius of curvature of the wavefront at the observation point of interest. In ray terms, L is the width of the spatial zone around any given ray which influences strongly the field on that ray, and for that reason it is called the *spatial influence zone.*; the parameter Ω plays the same role in the spectral domain and is called the *spectral influence zone.* The concept of the influence zone is then extended to show that the field over a connected portion of wavefront is closely approximated by a function which is nearly-bandlimited both spectrally and spatially. The Landau-Pollak Theorem [1] then states that this function lives in a finite-dimensional space of N dimensions, where $N=KD/2\pi$, and D and K are the spatial and spectral bandwidths respectively. Thus, N arithmetic data are required to specify the field accurately; these may be N spatial samples, N spectral samples, or some linear admixture of the two sets.

The Gabor representation utilises Gaussian elementary functions having a prescribed spatial width and linear phase [2,3,4]. We show that the optimum value for the width is one spatial influence zone, a choice which permits further localisation in phase-space. Intuitively, one expects the Lagrange manifold in phase-space to play a dominant role in localisation, and this role is most efficiently exploited by the matching of the beam parameters to the influence zone. However, we are able to show finally that the explicit localisation inherent in the Gabor representation does not necessarily lead to a numerical advantage for this representation over the Fourier representation; wherever the FFT algorithm is appropriate, this will out-perform Gaussian synthesis in terms of the number of computations required when $k\to\infty$.

THE INFLUENCE ZONES

Consider a diverging or converging wave crossing the plane $z=0$. We assume that one ray in this wave is along the z-axis. In the Fresnel approximation, which is adequate for our purposes here, we have

$$u(x) = u_0 e^{i\left(\frac{k}{2R}\right)x^2}, \qquad u_0 = e^{-i\pi/4}\left(\frac{k}{2\pi R}\right)^{-1/2} \tag{1}$$

$$\tilde{u}(\beta) = e^{-i\left(\frac{R}{2k}\right)\beta^2} \tag{2}$$

for the field u and its spectrum over the plane $z=0$. The Fourier representation for $u(x)$ is

$$u(x) = \frac{1}{2\pi}\int_{-\infty}^{\infty}\tilde{u}(\beta)e^{i\beta x}d\beta \tag{3}$$

When the spectral range is reduced to $|\beta|<K/2$, we have

$$u_K(x) = \frac{1}{2\pi}\int_{-K/2}^{K/2} e^{-i\left(\frac{R}{2k}\right)\beta^2} e^{i\beta x}d\beta \tag{4}$$

and we seek the minimum value of K which permits u_K to be regarded as a good approximation to u near $x=0$. It is easily shown by simple asymptotic methods that

$$u_K(0) = u(0)\left\{1 - \left(\frac{8k}{\pi K^2 R}\right)^{1/2} e^{-i\left(\frac{K^2 R}{8k}\right)} e^{-i\pi/4} + O(K^{-2})\right\} \tag{5}$$

Clearly as $K \to \infty$, the error $\varepsilon = |u_K(0) - u(0)|/u(0)$ diminishes as $O(K^{-1})$. It is convenient to define a spectral domain zone-width Ω by

$$\frac{R\Omega^2}{2k} = \pi \tag{6}$$

and a number m by $m = K/\Omega$. Then (6) becomes

$$\varepsilon = \frac{2}{m\pi} + O(m^{-2}) \tag{7}$$

Obviously, m can be chosen large enough to make ε arbitrarily small. We call the zone $|\beta| < \Omega/2$ the *spectral influence zone*; if the window K is larger than a few zone widths, the error ε can be neglected, and $u_K(0)$ is a good approximation to $u(0)$.

A ray whose transverse wavenumber is $\Omega/2$ intercepts $z=0$ at a spatial point $x = L/2 = \Omega R/2k$. Therefore, all rays with $|\beta| < \Omega/2$ intercept $z=0$ in the zone $|x| < L/2$. Since each ray 'carries' a spectral plane wave with the same β, the *spectral* zone of width Ω is equivalent to a *spatial* zone of width L. From the definitions we have

$$\frac{R\Omega^2}{2k} = \frac{kL^2}{2R} = \pi \tag{8}$$

and therefore $L = 2\pi/\Omega$. Similarly, a *spectral* window of width K is equivalent to a *spatial* window of width D, where

$$\frac{K^2R}{2k} = \frac{kD^2}{2R} = \pi \tag{9}$$

and $D = mL$ if $K = m\Omega$. Thus

$$KD = m^2\Omega L = 2\pi m^2 \tag{10}$$

The significance of the influence zones is as follows. Each ray in a GO bundle carries along with it a surrounding sub-bundle of rays which significantly influence the value of the field at the central ray [5]; rays which intercept the observation space $z=0$ at distances further from the central ray than a few influence zones (say $m=10$) have only a small effect on the field on the central ray. It follows that the spectrum may be truncated outside the spectral region of width $K = m\Omega$ for some suitable m with negligible effect on the field on the central ray.

The decay of the error in (7) is not sufficiently rapid for most applications in diffraction theory. This is because the width of the spatial influence zone decreases as $O(k^{-1/2})$, according to (8), and so for fixed window width D the error ε behaves like $k^{-1/2}$, which is comparable with edge diffraction effects. Faster decay can be obtained by smoothing the truncation window, as in classical signal processing theory. Consider the general windowed transform

$$u_w(x) = \frac{1}{2\pi}\int_{-\infty}^{\infty}\tilde{u}(\beta)w(\beta)e^{i\beta x}d\beta \tag{11}$$

where $w(\beta)$ indicates some general windowing function such that $|w|=0$ if $|\beta|>K/2$; equation (4) refers to a particular choice of $w=1$ for $|\beta|<K/2$. If instead of the rectangular window function, which is discontinuous at $\beta=\pm K/2$, we choose a function which has r continuous derivatives, it then follows from standard Fourier theory that $\varepsilon \rightarrow O(m^{-r-2})$ as $m\rightarrow\infty$. In particular the window

$$\begin{aligned} w&=0, \quad |\beta|>K/2 \\ &=\cos^2\theta, \quad |\beta|<K/2; \quad \theta=\pi\beta/K \end{aligned} \tag{12}$$

is sufficiently smooth for most purposes.

BANDLIMITED REPRESENTATIONS

Consider now a section of wavefront of width D_0, consisting of those rays with transverse wavenumbers $|\beta|<K_0/2$, where $K_0/k=D_0/R$. The union of all the zones of width mL around each of these rays occupies the spatial range $|x|<(D_0+mL)/2$, and includes all rays of wavenumbers $|\beta|<(K_0+m\Omega)/2$. The spectrum can be truncated (with a smoothing window if required) beyond the limits $|\beta|<(K_0+m\Omega)/2$ without significantly affecting the values of u(x) in the primary cross-section $|x|<D_0/2$; hence, for this range of x-values, the field u is effectively spectrally bandlimited. Furthermore, once the observation point is moved a further m influence zones outside the range $|x|<(D_0+mL)/2$, the bandlimited transform falls away to negligible values. Therefore the field u is simultaneously nearly-bandlimited both *spatially* and *spectrally*, with effective bandlimits

$$\begin{aligned} |\beta| &< (K_0 + m\Omega)/2 = K \\ |x| &< (D_0 + 2mL)/2 = D \end{aligned} \tag{13}$$

This situation satisfies the conditions of the Landau-Pollak Theorem (LP) [1], which asserts that if a function u is nearly-bandlimited spatially and spectrally, it can be approximated accurately by a function which lives in a finite-dimensional space. The number of dimensions of this space, and hence the number of degrees of freedom of u, is the Landau-Pollak dimension

$$N = KD / 2\pi \tag{14}$$

If the cross-section of the wave D_0 is measured in influence zones, so that D_0=ML and K_0=MΩ, then N=(M+m)(M+2m)ΩL/2π=(M+m)(M+2m). Consequently, the number of arithmetic data values needed to accurately specify the wave over the primary cross-section $|x|<D_0/2$ is determined by the number M of influence zones contained in that cross-section, and by the width m of the guard-bands supplemented to the primary cross-section. In asymptotic theories one would like the relative truncation errors in this process to diminsh at least as fast as $O(k^{-1})$ when $k\rightarrow\infty$; since $k\rightarrow\infty$ implies $L=O(k^{-1/2})$ from (8), and hence $M=O(k^{1/2})$ and $N_0=M^2=K_0D_0/2\pi=O(k)$, it is necessary to take m=O(M)=κM, where κ=O(1) as $k\rightarrow\infty$, and to apply smoothing windows which are at least continuous (r=0 above).

The LP Theorem does not specify how these arithmetic values should be constructed, only that no more or less than N are needed. Different constraints on the residual truncation errors may lead to different choices for the optimal data set; they may be N spatial samples, N spectral samples, or N coefficients of some linear representation such as the Gabor representation. If N uniformly-spaced spectral samples are given, then N uniformly-spaced spatial samples are recoverable using the discrete Fourier Transform. If this is executed by the FFT algorithm, then $O(N \log_2 N)$ computations are required when $N\rightarrow\infty$.

THE LAGRANGE MANIFOLD

The use of a Fourier representation fails to take full account of the localisation inherent in high-frequency wavefields. The spectral intregral (4) is dominated by a stationary phase contribution arising from spectral influence zones around the single GO wavenumber β=kx/R. In the composite (x,β)-space (phase-space) the dominant stationary phase points form a submanifold called the Lagrange manifold, Λ. The stationary points on Λ are also the transverse wavenumbers of GO rays through the corresponding observation points. In the elementary case considered here (converging or diverging wave in the Fresnel approximation) the Lagrange manifold is a straight line in (x,β)-space, with slope k/R. In more general cases, where R is a function of x, Λ acquires curvature and may even fold over, indicating the presence of caustics. It is desirable to construct a representation for the wavefield which explicitly localises around Λ, as opposed to the Fourier representation which sums over whole ranges of β-values for each x-value.

THE GABOR REPRESENTATION

The Gabor representation is [2,3,4]

$$u(x) = \sum_{m,n} U_{mn} e^{-\pi(\xi - m)^2} e^{2\pi i n\xi} \tag{15a}$$

$$\tilde{u}(\beta) = L \sum_{m,n} U_{mn} e^{-\pi(\nu - n)^2} e^{-2\pi i m\nu} \tag{15b}$$

where ξ=x/L, ν=β/Ω, $\Omega L=2\pi$ and $\{U_{mn}\}$ are coefficients given by

$$U_{mn} = C \sum_{q=0}^{\infty} (-1)^q e^{-\pi(q+\frac{1}{2})^2} \int_{-(q+\frac{1}{2})}^{(q+\frac{1}{2})} u((\xi + m)L) e^{\pi\xi^2} e^{-2\pi i n\xi} d\xi \tag{16}$$

with $C=2^{-1/2}(\zeta/\pi)^{-3/2}$ and ζ=1.8541. The Gaussian basis functions in this representation are centred on the phase-space points x=mL, $\beta=n\Omega$, which form a rectangular grid in phase-space. The choice of L and Ω is arbitrary, consistent with $\Omega L=2\pi$. Because of the localisation of the Gaussian functions it is possible to choose Ω so that the Gabor grid points closest to the Lagrange manifold contribute the dominant terms to the phase-space summations in (15). For dual-bandlimited functions of the Landau-Pollak type, N Gabor functions each occupying a phase-space area of $\Omega L=2\pi$ units will occupy a total phase-space area of $2N\pi$ units; this exactly fills the area required by the LP Theorem of $KD=2N\pi$ units.

It is easy to show from (15), by considering the representation for $u(\xi+1)$, that for the Fresnel wave

$$u(\xi) = u_0 e^{i\alpha\pi\xi^2} = u_0 e^{i\frac{k}{2R}x^2} \tag{17}$$

then for integer α

$$U_{m+1,n+\alpha} = U_{mn} e^{i\alpha\pi} \tag{18}$$

The parameter α measures the number of spatial influence zones occupied by the elementary Gaussian basis functions ($\alpha\pi=kL^2/2R$). Choosing L so that α=1 has the effect of matching the width of the Gaussian basis functions to the spatial influence zone-width of the wavefront. For any α, the Lagrange manifold Λ is given by

$$\beta = \frac{kx}{R} \Rightarrow v = \frac{kL}{\Omega R}\xi = \alpha\xi \tag{19}$$

so the Gabor grid points lying on Λ are given by $n=\alpha m$. Therefore the translation $(m,n)\rightarrow(m+1,n+\alpha)$ is translation parallel to Λ, and the difference formula (18) indicates how the coefficients U_{mn} are generated over the whole phase-space grid when only the subset $\{U_{0n}\}$ is given at m=0. The most natural choice for the Gaussian basis parameters is $\alpha=1$, but the theory can be adapted for any α.

To study the convergence of the Gabor representation away from the discrete set of points on Λ, it is necessary to know the behaviour of the U_{0n} as $|n|\rightarrow\infty$. It can be shown by direct asymptotic methods applied to (16) that for *integer* α

$$U_{0n} \rightarrow \frac{Cu_0}{1+i\alpha} e^{n\pi i} e^{i\alpha\pi/4} e^{\frac{-\pi}{4(1+i\alpha)}} e^{-\lambda|n|} \tag{20}$$

as $|n|\rightarrow\infty$, where $\lambda=\pi/(1+i\alpha)$. Clearly, $\mathrm{Re}(\lambda)>0$, and the fastest decay of U_{0n} away from Λ occurs when $\alpha=1$ (excluding the case $\alpha=0$ corresponding to a plane wave for u), which is also the matched case where all the dominant basis functions lie at grid points on the Lagrange manifold. Consequently, it follows from the translation property (18) that the U_{mn} decay exponentially away from the Lagrange manifold Λ, and the greatest localisation near Λ occurs for $\alpha=1$. This behaviour represents much stronger localisation than the Fourier representation allows. However, it does not necessarily lead to any numerical advantage over the FFT. To reconstruct the field u(x) from (15a) to the same order of accuracy as the FFT, which produces a relative error of ε due to aliasing, it is necessary to sum (15a) over all n-terms for which $|\exp(-\lambda|n|)|>\varepsilon$, and for $\alpha=1$ this requires that all terms for which $|n|<2\pi^{-1}\ln\varepsilon^{-1}$ be included. This must be done for each of the N spatial points required by the Landau-Pollak Theorem in order to specify the field accurately. Thus the m-summation must be evaluated $2N\pi^{-1}\ln\varepsilon^{-1}$ times. The localisation of the Gaussian functions reduces the number of terms contributing significantly to this sum to those for which $\exp(-\pi(\xi-m)^2)>\varepsilon$, and there are $\pi^{-1/2}(\ln\varepsilon^{-1})^{1/2}$ of these in number. Hence the total number of computational steps required is

$$N_{comp} = O(N\,(\ln\varepsilon^{-1})^{3/2}) \tag{23}$$

Assuming that $\varepsilon=O(N^{-\mu})$, which is reasonable for a smoothing-windowed transform, (23) becomes

$$N_{comp} = O(N\,(\ln N)^{3/2}) \tag{24}$$

CONCLUSIONS

Although the Gabor representation is more compact in phase space than the Fourier representation for a Fresnel wave, this does not nessarily lead to a significant compression of the computational requirements. If N is the Landau-Pollak dimension of the restricted-aperture Fresnel wave ($N=kD^2/2\pi R$), the discrete FFT calculation of N spatial points requires $O(N\log_2 N)$ computational steps, whereas the Gabor representation requires $O(N(\ln N)^{3/2})$ steps to compute the values of the field at the same N spatial sample points. Actually this estimate overstates the requirements of the Gabor representation, since only ND_0/D of these samples actually lie in the window of interest, $|x|<D_0/2$, but this does not affect the O-estimate of (23) for $N\rightarrow\infty$ if $D=O(1)$ in this limit. These calculations also assume that both the Fourier and Gabor coefficients are given *a priori*. In practice, for more general fields than the Fresnel wave, these coefficients will have to be determined, which may or may not add computational overheads. For example, the Fourier plane wave spectrum of a converging wave with caustics is easily constructed using asymptotic methods such as spectral synthesis or Maslov's method [6], with negligible additional computational work since no integration is required, only O(N) function evaluations. The Gabor coefficients are usually harder to obtain because of the complexity of (16), which involves integration and a slowly convergent sum.

REFERENCES

1. Slepian, D.: 'On bandwidth', Proc. IEEE, **64**, 292-300, 1976
2. Einziger, P. D., Raz, S. and Shapira, M.: 'Gabor representation and aperture theory', J. Opt. Soc. Am., **3A**, 508-522, 1986
3. Steinberg, B. Z., Heyman, E. and Felsen, L. B.: 'Phase space beam summation for the time-harmonic radiation from large apertures', to be published in J. Opt. Soc. Am., 1990
4. Bastiaans, M. J.: 'A sampling theorem for the complex spectrogram and Gabor's expansion of a signal in Gaussian elementary signals', Opt. Eng., **20**, 594-598, 1981
5. Kravtsov, Yu. A.: 'Rays and caustics as physical objects', Progress in Optics XXVI, Elsevier Science Publishers, 1988
6. Arnold, J. M.: 'Spectral synthesis of uniform wavefunctions', Wave Motion, **8**, 135-150, 1986

DERIVATION OF EXTENDED PARABOLIC THEORIES FOR VECTOR ELECTROMAGNETIC WAVE PROPAGATION

Ronald I. Brent

Department of Mathematics
University of Lowell
Lowell, MA 01854

Louis Fishman

Department of Mathematical and Computer Sciences
Colorado School of Mines
Golden, CO 80401

ABSTRACT

The recent application of pseudo-differential operator and functional integral methods to the factored scalar Helmholtz equation has yielded extended parabolic wave theories and corresponding path integral solutions for a large variety of acoustic wave propagation problems. These known techniques are applied here to the full vector electromagnetic problem resulting in a new first-order Weyl pseudo-differential equation, which is recognized as an extended parabolic wave equation. Perturbation treatments of the Weyl composition equation for the operator symbol matrix yield high-frequency and other asymptotic wave theories. Unlike the scalar Helmholtz equation case, the one-way vector EM equation (and a scalar analogue provided by the Klein-Gordon equation of relativistic physics) require the solution of a generalized quadratic operator equation. While these operator solutions do not have a simple formal representation as in the straightforward square root case, they are conveniently constructed in the Weyl pseudo-differential operator calculus.

INTRODUCTION

We propose to apply the phase space and path integral methods [1] to vector propagation problems. In particular, Maxwell's equations provide the appropriate starting point to model atmospheric EM problems [2-4] as well as integrated optics problems [4]. These modern, " microscopic " methods can be motivated by the parabolic approximation method [1]. The solution of realistic electromagnetic propagation problems has been greatly aided by the derivation of parabolic, or one-way, wave equations. These equations can be solved easily using range-stepping algorithms, thus eliminating the difficulties experienced with shooting or relaxation methods. While originally proposed by Leontovich and Fock [2] for ground wave propagation in the 1940's, computer limitations of the time did not allow for much study of complicated problems. A more complete derivation was given by Lax et. al. [3] in 1975, with additional corrections for higher-order fields. Sudarshan et. al. [4], and Mukunda et. al. [4] first analyzed the scalar problem using the "front form" and then expanded upon the method to examine the full vector problem. EM propagation in anisotropic media has also been modeled using parabolic approximations [4]. By beginning with the full vector equations, a mathematical scaling and asymptotic analysis was presented which resulted in the derivation of a vector system of parabolic equations. An extension of the split-step method [1] was presented which was used to solve the coupled partial differential equations, with the resulting solutions displaying the effects of such an external field on atmospheric propagation.

Quite frequently the somewhat stringent requirements necessary for the derivation of parabolic equations might not be valid, thereby leading to the desirability of more general propagation models. Based on the experience with the scalar (acoustic) propagation problem [1], the phase space and path integral approach should lead to wide-angle, one-way EM wave equations, corresponding rapid marching algorithms, the basis for inverse methods, and a physically-oriented approach to the full two-way problem. The initial step in applying known scalar results to the vector problem is the formal decomposition of the EM field into forward and backward propagating fields. Pseudo-differential operator (ΨDO) theory may then be applied to explicitly construct the resulting (matrix) operator equation. The construction of the symbol matrix associated with the one-way propagation operator can be achieved through the analysis of the Weyl composition equation for the operator symbol matrix, ultimately leading to approximate parabolic theories valid in different regimes. This paper presents the results of these steps and sets the framework for the future research goals of path integral representations and corresponding numerical algorithms for both the one-way and the full two-way vector problems.

Unlike the scalar Helmholtz equation case, the one-way EM equation requires the solution of a generalized quadratic operator equation. The Klein-Gordon (K-G) equation of relativistic physics [5] provides a scalar analogue of this complication. Thus, in Section 2, we formulate and motivate our problem by first discussing the scalar Klein-Gordon equation and the difficulties in diagonalizing the K-G operator into positive- and negative- frequency solutions for time-independent scalar and vector potentials. Beginning with Maxwell's equations, we then derive propagation equations for the components of the electric field. The derived vector equations are then formally split into two equations governing forward and backward propagating waves using the insight gained from the scalar K-G problem. For this preliminary study, we shall assume that the medium dielectric properties are transversely inhomogeneous, and boundaries are horizontal (that is, parallel to the propagation direction). We proceed in Section 3 by explicitly representing the up-to-now formal one-way wave equations using pseudo-differential operator theory. In Section 4, we present the derivation of the symbol matrix for a homogeneous half space and the nonuniform high-frequency limit. The symbol matrix corresponding to typical parabolic regimes, incorporating narrow-angle, weak-inhomogeneity, and weak-gradient conditions, is also presented. This reproduces known parabolic approximation results obtained from typical asymptotic scaling arguments. Finally, there is a discussion of numerical applications and future directions in Section 5.

FORMULATION

The scalar Helmholtz operator, $\partial_x^2 + \nabla_T^2 + k^2(\tilde{x}_T)$, formally factors as $(\partial_x + iC_1)(\partial_x + iC_2)$ in terms of the two operator solutions of the simple quadratic operator equation

$$C^2 - (\nabla_T^2 + k^2(\tilde{x}_T)) = 0 . \tag{1}$$

The physical wave splitting in this case is thus formally given by

$$\partial_x^2 + \nabla_T^2 + k^2(\tilde{x}_T) = (\partial_x + i\sqrt{\nabla_T^2 + k^2(\tilde{x}_T)})(\partial_x - i\sqrt{\nabla_T^2 + k^2(\tilde{x}_T)}) . \tag{2}$$

For the vector EM problem, this simple type of decomposition is no longer appropriate. Before proceeding directly to the vector formulation, however, it is insightful to discuss the Klein-Gordon equation of relativistic physics, which provides a scalar model for the subtleties of the vector decomposition problem.

The K-G equation for time-independent scalar and vector potentials in one time and three spatial dimensions is given by

$$\left((i\partial_t - e\phi(\tilde{x}))^2 - (-i\nabla - e\tilde{A}(\tilde{x}))^2 - m^2 \right)\psi(t,\tilde{x}) = 0, \tag{3}$$

where $\phi(\tilde{x})$ is the scalar potential, $\tilde{A}(\tilde{x})$ is the vector potential, and the units have been chosen such that $h/2\pi = c = 1$ [5]. The K-G equation is known to decouple into positive- and negative-frequency solutions when the potentials are time independent [5]. For $\phi(\tilde{x})$=constant, this result is clearly seen by factoring the K-G operator as

$$\left((i\partial_t - e\phi)^2 - (-i\nabla - e\tilde{A})^2 - m^2 \right) = \left((i\partial_t - e\phi) - \sqrt{(-i\nabla - e\tilde{A})^2 + m^2} \right) \bullet \left((i\partial_t - e\phi) + \sqrt{(-i\nabla - e\tilde{A})^2 + m^2} \right) . \tag{4}$$

Each of the two operators on the right hand side of Eq.(4) corresponds to one of the uncoupled solutions. For spatially-dependent scalar potentials, however, this factorization is not correct because of the noncommuting operators $\phi(\tilde{x})$ and ∇. The form of the factors in Eq.(4) suggests that the diagonalization of the K-G operator into positive- and negative-frequency solutions should probably require a generalized quadratic operator equation. Indeed,

$$\partial_t \begin{pmatrix} \psi^+ \\ \psi^- \end{pmatrix} = \begin{pmatrix} iC_1 & 0 \\ 0 & iC_2 \end{pmatrix} \begin{pmatrix} \psi^+ \\ \psi^- \end{pmatrix}, \tag{5}$$

where

$$\begin{pmatrix} \psi \\ \partial_t \psi \end{pmatrix} = \begin{pmatrix} 1 & 1 \\ iC_1 & iC_2 \end{pmatrix} \begin{pmatrix} \psi^+ \\ \psi^- \end{pmatrix}, \tag{6}$$

and C_1 and C_2 are the two operator solutions of the generalized quadratic operator equation

$$C^2 + 2e\phi(\tilde{x})C - \left((-i\nabla - e\tilde{A}(\tilde{x}))^2 + m^2 - e^2\phi^2(\tilde{x}) \right) = 0 . \tag{7}$$

It is straightforward to show that Eqs.(3),(6), and (7) imply Eq.(5) and, conversely, that Eqs.(5) - (7) imply Eq.(3). Moreover, Eqs.(5) and (6) are consistent. Choosing C_1 to correspond to the positive-frequency solution (ψ^+) and C_2 to correspond to the negative-frequency solution (ψ^-) completes the identification.

Returning to the vector EM problem, we begin with a reduced form of Maxwell's equations in three dimensions governing the propagation of an electric field $\tilde{E}$ and a magnetic field $\tilde{H}$. It is assumed that time enters the problem through a harmonic dependence $\exp(-i\omega t)$, and that magnetic inductance $\tilde{B} = \mu_0\tilde{H}$ and displacement $\tilde{D} = \varepsilon\tilde{E}$. These equations are [4]

$$\nabla \times \tilde{E} = i\omega\mu_o\tilde{H} \;, \text{ and } \quad \nabla \times \tilde{H} = -i\omega\varepsilon_o\kappa\tilde{E} \;, \tag{8a,b}$$

where $\kappa = \varepsilon/\varepsilon_o + i(\sigma/\omega\varepsilon_o)$ represents the dielectric variations in the medium. In these equations, ε_o and μ_o are the free-space permittivity and permeability, while ε and σ are the spatially-dependent permittivity and conductivity. Using Eqs.(8a) and (8b), one can derive

$$\partial_x^2\tilde{E} + \nabla_T^2\tilde{E} + \nabla(\kappa^{-1}\nabla\kappa \bullet \tilde{E}) + k_o^2\kappa\tilde{E} = 0, \tag{9}$$

where $k_o = \omega/c_o$ is a reference wave number. Equation (9) can be written as

$$\partial_x^2\tilde{E} + \tilde{\tilde{\mathbf{B}}}\bullet\partial_x\tilde{E} + \tilde{\tilde{\mathbf{A}}}^2\bullet\tilde{E} = 0 \;, \tag{10a}$$

where under transversely inhomogeneous conditions

$$\tilde{\tilde{\mathbf{A}}}^2 = \begin{pmatrix} \nabla_T^2 + k_o^2\kappa & 0 & 0 \\ 0 & \nabla_T^2 + k_o^2(\kappa + a_{yy}) & k_o^2 a_{yz} \\ 0 & k_o^2 a_{zy} & \nabla_T^2 + k_o^2(\kappa + a_{zz}) \end{pmatrix}, \tag{10b}$$

$$\tilde{\tilde{\mathbf{B}}} = \begin{pmatrix} 0 & \kappa^{-1}\left(\dfrac{\partial\kappa}{\partial y}\right) & \kappa^{-1}\left(\dfrac{\partial\kappa}{\partial z}\right) \\ 0 & 0 & 0 \\ 0 & 0 & 0 \end{pmatrix}, \text{ and } \quad k_o^2 a_{\tau\upsilon} = \partial_\tau\left(\kappa^{-1}\left(\frac{\partial\kappa}{\partial\upsilon}\right)\right) \;. \tag{10c,d}$$

In Eq's.(10c) and (10d), $\partial/\partial\upsilon$ terms are isolated derivatives while the τ-derivative operator applies to both the coefficient and the electric field. Equation (10b) is derived under isotropic conditions, although modifications could include anisotropic cases. However, this adds many difficulties and will not be addressed in this paper. The difficulties in factoring the operator in Eq.(10a) are those encountered with the K-G operator since the operators $\tilde{\tilde{\mathbf{A}}}^2$ and $\tilde{\tilde{\mathbf{B}}}$ do not commute. That is, noting that $\tilde{\tilde{\mathbf{B}}}^2 = 0$, writing $\partial_x^2 + \tilde{\tilde{\mathbf{B}}}\partial_x + \tilde{\tilde{\mathbf{A}}}^2 = (\partial_x + \tilde{\tilde{\mathbf{B}}}/2)^2 + \tilde{\tilde{\mathbf{A}}}^2 = ((\partial_x + \tilde{\tilde{\mathbf{B}}}/2) + i\tilde{\tilde{\mathbf{A}}})((\partial_x + \tilde{\tilde{\mathbf{B}}}/2) - i\tilde{\tilde{\mathbf{A}}})$ is not correct in general.

For a transversely inhomogeneous environment, the electric field vector ($\tilde{E}$) can be split into forward ($\tilde{E}^+$) and backward ($\tilde{E}^-$) propagating wave field components. In analogy with the scalar K-G problem,

$$\partial_x\begin{pmatrix} \tilde{E}^+ \\ \tilde{E}^- \end{pmatrix} = \begin{pmatrix} i\tilde{\tilde{C}}_1 & 0 \\ 0 & i\tilde{\tilde{C}}_2 \end{pmatrix}\begin{pmatrix} \tilde{E}^+ \\ \tilde{E}^- \end{pmatrix}, \tag{11}$$

where

$$\begin{pmatrix} \tilde{E} \\ \partial_x\tilde{E} \end{pmatrix} = \begin{pmatrix} \tilde{\tilde{I}} & \tilde{\tilde{I}} \\ i\tilde{\tilde{C}}_1 & i\tilde{\tilde{C}}_2 \end{pmatrix}\begin{pmatrix} \tilde{E}^+ \\ \tilde{E}^- \end{pmatrix}, \tag{12}$$

$\tilde{\tilde{I}}$ is the 3x3 Identity matrix, and $\tilde{\tilde{C}}_1$ and $\tilde{\tilde{C}}_2$ are the two matrix operator solutions of the generalized quadratic operator equation

$$\tilde{\tilde{C}}^2 - i\tilde{\tilde{\mathbf{B}}}\bullet\tilde{\tilde{C}} - \tilde{\tilde{\mathbf{A}}}^2 = 0 \;. \tag{13}$$

Again, Eqs.(10), (12), and (13) imply Eq.(11), while, conversely, Eqs.(11)-(13) imply Eq.(10). Eqs.(11) and (12) are consistent and $\tilde{\tilde{C}}_1$ must be chosen to correspond to the forward propagating wave field with $\tilde{\tilde{C}}_2$ corresponding to the backward propagating wave field.

PSEUDO-DIFFERENTIAL OPERATOR REPRESENTATION

Equations (11) and (13) formally represent the decoupled Maxwell's equations for a transversely inhomogeneous environment. Unlike the scalar Helmholtz case in Eq.(2), the operator $\tilde{\tilde{C}}_1$ does not admit a simple formal representation in terms of square root operators in general. However, $\tilde{\tilde{C}}_1$ can be explicitly constructed within the Weyl pseudo-differential operator calculus [1]. Using this construction, the forward one-way wave equation can be written as [1]

$$\partial_x \tilde{E}^+(x,\tilde{x}_T) = i\left(\frac{k_o}{2\pi}\right)^2 \int_{R^4} d\tilde{p}_T\, d\tilde{x}_T'\, \exp\left(ik_o\tilde{p}_T \bullet (\tilde{x}_T - \tilde{x}_T')\right)\, \tilde{\tilde{\Omega}}_{\tilde{\tilde{C}}_1}\left(\tilde{p}_T, \frac{\tilde{x}_T + \tilde{x}_T'}{2}\right) \bullet \tilde{E}^+(x,\tilde{x}_T'), \qquad (14)$$

in terms of the operator symbol matrix $\tilde{\tilde{\Omega}}_{\tilde{\tilde{C}}_1}(\tilde{p},\tilde{q})$. Equation (13) then takes the form

$$\tilde{\tilde{\Omega}}_{\tilde{\tilde{A}}^2}(\tilde{p},\tilde{q}) = \left(\frac{k_o}{\pi}\right)^4 \int_{R^8} d\tilde{t}\, d\tilde{x}\, d\tilde{y}\, d\tilde{z} \left(\tilde{\tilde{\Omega}}_{\tilde{\tilde{C}}}(\tilde{t}+\tilde{p}, \tilde{x}+\tilde{q}) - i\tilde{\tilde{B}}(\tilde{x}+\tilde{q})\right)$$
$$\bullet\, \tilde{\tilde{\Omega}}_{\tilde{\tilde{C}}}(\tilde{y}+\tilde{p}, \tilde{z}+\tilde{q}) \exp\left(2ik_o(\tilde{x}\bullet\tilde{y} - \tilde{t}\bullet\tilde{z})\right), \qquad (15)$$

which relates the symbol for the operator $\tilde{\tilde{C}}$ to the symbols for the operators $\tilde{\tilde{A}}^2$ and $\tilde{\tilde{B}}$. This equation can be written in a formal manner appropriate for high-frequency analysis as [1]

$$\tilde{\tilde{\Omega}}_{\tilde{\tilde{A}}^2}(\tilde{p},\tilde{q}) = \lim_{\substack{\tilde{\eta}\to\tilde{p} \\ \tilde{y}\to\tilde{q}}} \left(\exp\left(\frac{i}{2k_o}(\nabla_{\tilde{\eta}}\bullet\nabla_{\tilde{q}} - \nabla_{\tilde{p}}\bullet\nabla_{\tilde{y}})\right)\left(\tilde{\tilde{\Omega}}_{\tilde{\tilde{C}}}(\tilde{p},\tilde{q}) - i\tilde{\tilde{B}}(\tilde{q})\right)\bullet\tilde{\tilde{\Omega}}_{\tilde{\tilde{C}}}(\tilde{\eta},\tilde{y})\right). \qquad (16)$$

We reiterate, that when solving the composition equation (15), we choose the appropriate root that corresponds to the physically forward propagating wave.

We note that by associating $x \leftrightarrow t$, $\tilde{\tilde{A}}^2 \leftrightarrow \left((-i\nabla - e\tilde{A})^2 + m^2 - e^2\phi^2\right)$, $\tilde{x}_T \leftrightarrow \tilde{x}$, and $-i\tilde{\tilde{B}} \leftrightarrow 2e\phi$, one can use the scalar equivalents of Eqs.(14)-(16) to explicitly construct the set of equations in the scalar K-G case.

SYMBOL CONSTRUCTIONS

The approximate solution of the composition equation (15) or (16) begins with the construction of $\tilde{\tilde{\Omega}}_{\tilde{\tilde{A}}^2}(\tilde{p},\tilde{q})$. Referring to Eq.(10b) and applying the Weyl pseudo-differential operator calculus [1] result in

$$\tilde{\tilde{\Omega}}_{\tilde{\tilde{A}}^2}(\tilde{p},\tilde{q}) = k_o^2\, \tilde{\tilde{\Omega}}^{(0)}_{\tilde{\tilde{A}}^2}(\tilde{p},\tilde{q}) + k_o \tilde{\tilde{\Omega}}^{(1)}_{\tilde{\tilde{A}}^2}(\tilde{p},\tilde{q}), \qquad (17a)$$

where

$$\tilde{\tilde{\Omega}}^{(0)}_{\tilde{\tilde{A}}^2}(\tilde{p},\tilde{q}) = \left(\kappa(\tilde{q}) - \tilde{p}^2\right)\tilde{\tilde{I}}, \qquad (17b)$$

$$\tilde{\tilde{\Omega}}^{(1)}_{\tilde{\tilde{A}}^2}(\tilde{p},\tilde{q}) = \frac{2i}{\kappa(\tilde{q})}\begin{pmatrix} 0 & 0 & 0 \\ 0 & \frac{\partial\kappa}{\partial q_1}p_1 & \frac{\partial\kappa}{\partial q_2}p_1 \\ 0 & \frac{\partial\kappa}{\partial q_1}p_2 & \frac{\partial\kappa}{\partial q_2}p_2 \end{pmatrix}, \tag{17c}$$

and $\tilde{p} = \langle p_1, p_2 \rangle$ and $\tilde{q} = \langle q_1, q_2 \rangle$.

In the homogeneous medium limit, $\kappa(\tilde{q}) = \kappa$, $\tilde{\tilde{B}}(\tilde{q}) = 0$, $\tilde{\tilde{\Omega}}^{(1)}_{\tilde{\tilde{A}}^2}(\tilde{p},\tilde{q}) = 0$, and Eq.(15) reduces to

$$\tilde{\tilde{\Omega}}_{\tilde{\tilde{A}}^2}(\tilde{p}) = k_o^2(\kappa - \tilde{p}^2)\tilde{\tilde{I}} = \tilde{\tilde{\Omega}}^2_{\tilde{\tilde{C}}}(\tilde{p}) \quad . \tag{18}$$

The symbol matrix corresponding to the forward propagating wave condition is then given by

$$\tilde{\tilde{\Omega}}_{\tilde{\tilde{C}}_1}(\tilde{p},\tilde{q}) = k_o \sqrt{\kappa - \tilde{p}^2}\;\tilde{\tilde{I}} \quad . \tag{19}$$

In the high-frequency ($k_o \rightarrow \infty$) limit, we assume an expansion of the form

$$\tilde{\tilde{\Omega}}_{\tilde{\tilde{C}}}(\tilde{p},\tilde{q}) = k_o^{\alpha}\left(\tilde{\tilde{\Omega}}^{(0)}_{\tilde{\tilde{C}}}(\tilde{p},\tilde{q}) + \frac{1}{k_o}\tilde{\tilde{\Omega}}^{(1)}_{\tilde{\tilde{C}}}(\tilde{p},\tilde{q}) + \frac{1}{k_o^2}\tilde{\tilde{\Omega}}^{(2)}_{\tilde{\tilde{C}}}(\tilde{p},\tilde{q}) + \cdots\right) . \tag{20}$$

Applying the composition equation (16) and Eqs.(10c) and (17), one determines that $\alpha = 1$ gives the proper balancing, leading to

$$\tilde{\tilde{\Omega}}^{(0)}_{\tilde{\tilde{C}}_1}(\tilde{p},\tilde{q}) = \sqrt{\kappa(\tilde{q}) - \tilde{p}^2}\;\tilde{\tilde{I}} \quad . \tag{21}$$

Thus, to lowest order, homogeneous results generalize by letting $\kappa \rightarrow \kappa(\tilde{q})$. The first higher-order correction is

$$\tilde{\tilde{\Omega}}^{(1)}_{\tilde{\tilde{C}}_1}(\tilde{p},\tilde{q}) = \frac{i}{\kappa(\tilde{q})\sqrt{\kappa(\tilde{q}) - \tilde{p}^2}}\begin{pmatrix} 0 & 0 & 0 \\ 0 & \frac{\partial\kappa}{\partial q_1}p_1 & \frac{\partial\kappa}{\partial q_2}p_1 \\ 0 & \frac{\partial\kappa}{\partial q_1}p_2 & \frac{\partial\kappa}{\partial q_2}p_2 \end{pmatrix} + \frac{i}{2\kappa(\tilde{q})}\begin{pmatrix} 0 & \frac{\partial\kappa}{\partial q_1} & \frac{\partial\kappa}{\partial q_2} \\ 0 & 0 & 0 \\ 0 & 0 & 0 \end{pmatrix} . \tag{22}$$

Equations (20)-(22) provide a two-term, nonuniform high-frequency approximation to the symbol matrix.

To examine the typical parabolic regimes, we begin by letting $\kappa - 1 \rightarrow \Delta(\kappa - 1)$, $\tilde{p}^2 \rightarrow \Delta\tilde{p}^2$, and $\frac{\partial\kappa}{\partial q_{1,2}} \rightarrow \Delta\frac{\partial\kappa}{\partial q_{1,2}}$ such that Δ represents the small scale size of the terms. These assumptions correspond to weak inhomogeneity, narrow angle, and weak gradient, the typical parabolic regime. We can then write

$$\tilde{\tilde{\Omega}}_{\tilde{\tilde{A}}^2}(\tilde{p},\tilde{q}) = k_o^2\tilde{\tilde{I}} + \Delta\left(k_o^2((\kappa-1)-\tilde{p}^2)\right)\tilde{\tilde{I}} + \Delta^{3/2}\,2ik_o\begin{pmatrix} 0 & 0 & 0 \\ 0 & \frac{\partial\kappa}{\partial q_1}p_1 & \frac{\partial\kappa}{\partial q_2}p_1 \\ 0 & \frac{\partial\kappa}{\partial q_1}p_2 & \frac{\partial\kappa}{\partial q_2}p_2 \end{pmatrix}, \quad (23)$$

where we have neglected terms of $O(\Delta^2)$. To similar accuracy

$$\tilde{\tilde{\Omega}}_{\tilde{\tilde{B}}}(\tilde{q}) = \tilde{\tilde{B}}(\tilde{q}) = \Delta\begin{pmatrix} 0 & \frac{\partial\kappa}{\partial q_1} & \frac{\partial\kappa}{\partial q_2} \\ 0 & 0 & 0 \\ 0 & 0 & 0 \end{pmatrix}. \quad (24)$$

By assuming an asymptotic series solution for the symbol $\tilde{\tilde{\Omega}}_{\tilde{\tilde{C}}}$ in powers of $\Delta^{1/2}$, it is possible to derive the result

$$\tilde{\tilde{\Omega}}_{\tilde{\tilde{C}}_1}(\tilde{p},\tilde{q}) \cong k_o\left(\frac{1}{2}((\kappa+1)-\tilde{p}^2)\right)\tilde{\tilde{I}} + \frac{i}{2}\begin{pmatrix} 0 & \frac{\partial\kappa}{\partial q_1} & \frac{\partial\kappa}{\partial q_2} \\ 0 & 0 & 0 \\ 0 & 0 & 0 \end{pmatrix} + i\begin{pmatrix} 0 & 0 & 0 \\ 0 & \frac{\partial\kappa}{\partial q_1}p_1 & \frac{\partial\kappa}{\partial q_2}p_1 \\ 0 & \frac{\partial\kappa}{\partial q_1}p_2 & \frac{\partial\kappa}{\partial q_2}p_2 \end{pmatrix}, \quad (25)$$

where we have let $\Delta \to 1$ to revert to unscaled notation. Equation (25) also follows from the high-frequency result in Eqs.(20)-(22) in the narrow-angle and weak-inhomogeneity limits. Using Eqs.(14), (25), and known Weyl calculus results, one derives the parabolic equations

$$\nabla_T^2\tilde{E}_T^+ + 2ik_o\partial_x\tilde{E}_T^+ + k_o^2(\kappa(\tilde{x}_T)+1)\tilde{E}_T^+ + \nabla_T(\nabla_T\kappa\bullet\tilde{E}_T^+) = 0, \quad (26a)$$

and

$$\nabla_T^2 E_x^+ + 2ik_o\partial_x E_x^+ + k_o^2(\kappa(\tilde{x}_T)+1)E_x^+ + ik_o(\nabla_T\kappa\bullet\tilde{E}_T^+) = 0, \quad (26b)$$

where $\tilde{E}_T^+$ is the transverse electric field vector. Using the substitution $\tilde{E}^+ = e^{ik_o x}\tilde{F}$ and the assumption $\tilde{F} = \tilde{F}^{(0)} + \Delta^{1/2}\tilde{F}^{(1)} + \Delta\tilde{F}^{(2)} + \ldots$ yield the results

$$F_x^{(0)} = 0 \quad \text{and} \quad F_x^{(1)} = \frac{i}{k_o}\nabla_T\bullet\tilde{F}_T^{(0)}, \quad (27a,b)$$

and

$$\nabla_T^2\tilde{F}_T^{(0)} + 2ik_o\partial_x\tilde{F}_T^{(0)} + k_o^2(\kappa(\tilde{x}_T)-1)\tilde{F}_T^{(0)} = 0, \quad (27c)$$

where we have made use of the equation $\nabla\bullet(\kappa\tilde{E}) = 0$. This result is consistent with [3] as well as [4]. Equation (27) states that when seeking asymptotic solutions for the components of the electric field, the leading terms for the transverse and longitudinal components are of different order. Zeroth-order transverse components satisfy a parabolic equation, while the longitudinal component starts with a first-order term that is related to the zeroth-order transverse components. Higher-order corrections can be derived by obtaining higher-order solutions to the composition equation.

By examining different asymptotic regimes and using scalar results from [1], it is possible to derive parabolic approximations for the transverse components under the assumptions of (1) arbitrary angle, weak inhomogeneity, weak gradient, and (2) arbitrary inhomogeneity, narrow angle, weak gradient.

DISCUSSION

The framework for the development of extended parabolic (one-way) wave theories for the vector EM case has been formulated. As opposed to the scalar Helmholtz equation case, the one-way propagation operator satisfies a generalized quadratic operator equation. This equation is analyzed within the Weyl pseudo-differential operator calculus through an appropriate composition equation. The singular high-frequency result in Eqs.(20)-(22) demonstrates that high-frequency, wide-angle wave theories fully accounting for the mode coupling will require uniform high-frequency asymptotic analysis [1]. The generalization of the phase space path integral involves an exponentiated symbol matrix. The corresponding numerical marching algorithm is complicated by the appearance of this symbol matrix as opposed to just a scalar symbol, hence a generalization of the split-step method involves the diagonalization of a matrix for a particular medium. This idea has been used in anisotropic EM propagation [4], where the matrix there accounts for electric-field component coupling resulting from medium anisotropy. Had there been slow range dependence in the medium, our formulation could be used by neglecting the small range derivatives of the operators $\tilde{\tilde{\mathbf{A}}}^2$ and $\tilde{\tilde{\mathbf{B}}}$. This would lead to a generalized split-step algorithm involving a matrix diagonalization at every range step which could seemingly be numerically prohibitive. Path integral representations and corresponding numerical algorithms based on the underlying physics of Maxwell's equations are currently under study [1]. Preliminary discussions of source modeling and initial-field construction and boundary conditions have been given in [4].

ACKNOWLEDGMENTS

The support of NSF, AFOSR, and ONR is gratefully acknowledged.

REFERENCES

1) L. Fishman, J.J. McCoy, and S.C. Wales, Factorization and path integration of the Helmholtz equation: numerical algorithms, J. Acoust. Soc. Am. 81:1355 (1987).

2) M. A. Leontovich and V.A. Fock, Solution of the problem of propagation of electromagnetic waves along the earth's surface by the method of parabolic equations, J. Phys. USSR 10:13 (1946).

3) M. Lax, W. H. Louisell, and W. B. McKnight, From Maxwell to paraxial wave optics, Phys. Rev. 11:1365 (1975).

4) R. I. Brent, W. L. Siegmann, and M. J. Jacobson, Parabolic approximations for atmospheric propagation of EM waves, including the terrestrial magnetic field, to appear in Radio Science 25 (1990).

5) J. D. Bjorken and S. D. Drell, Relativistic Quantum Mechanics (McGraw-Hill Book Company, New York, 1964).

RECENT ADVANCES IN DIFFRACTION ANALYSIS OF REFLECTOR ANTENNAS

Y. Rahmat-Samii

Electrical Engineering Department
University of California, Los Angeles
Los Angeles, California 90024-1594

1. INTRODUCTION

Reflector antenna configurations have evolved considerably in the last decade. For example, a reflector antenna configuration envisioned for future satellite communications is shown in Fig. 1. This antenna system utilizes a dual offet reflector antenna with the following characteristics: (a) the main reflector is mesh deployable, (b) the subreflector is made of multi-layered frequency selective surfaces (FSS), (c) a deployable mast is shown to support the subreflector, (d) there are complex conformal array feeds operating at different frequency bands, and (e) adaptive beam forming networks (BFN) with reconfigurable functionality are used. This antenna system architecture is envisioned to fulfill the demanding requirements imposed upon the future -generation antenna systems. Some of the key design aspects of these future-generation antenna systems are high gain, low sidelobes, low cross polarization, multi-frequency operation and scanning beam capabilities. Clearly, sophisticated and advanced analytical/numerical and measurement techniques are required to accurately and properly assess the performance characteristics of this type of an antenna configuration.

In this chapter, an overview of some of the recent developments in the analysis of reflector antennas is presented. The focus primarily will be on the analysis/numerical techniques. The reader is referenced to [1] for the advanced measurement concepts. The proper evaluation of the reflector antenna configuration shown in Fig. 1 requires a multidisciplinary treatment of different segments of this antenna. We start our discussion by presenting a generalized diagram, shown in Fig. 2, which summarizes some of the most current ideas in applying analytical/numerical techniques. The degree of maturity in applying different techniques will be discussed later. For each segment, references to the most recent publications are made with the understanding that the reader will refer to these references, which in turn provide additional references in each area.

In the next several sections we present representative results for each antenna segment. This we believe should provide a good insight as to the recent advances in applying analytical/numerical techniques for the modeling of complex reflector antenna

Directions in Electromagnetic Wave Modeling
Edited by H.L. Bertoni and L.B. Felsen, Plenum Press, New York, 1991

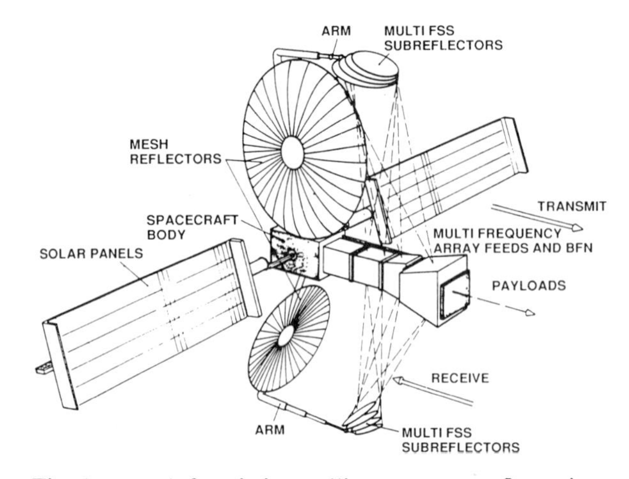

Fig. 1. A futuristic satellite antenna configuration.

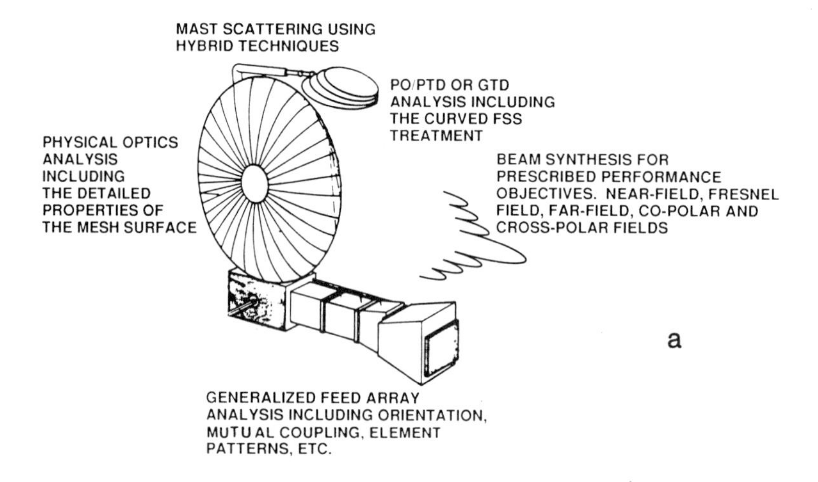

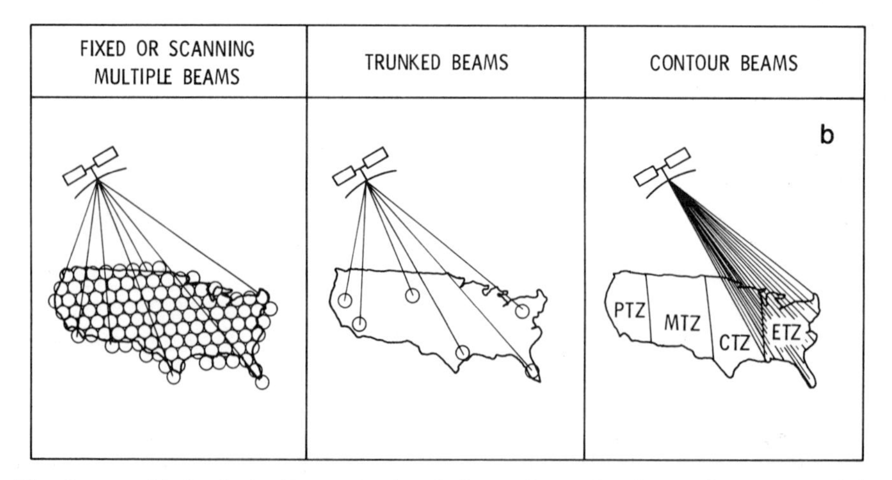

Fig. 2. (a) Analytical/numerical techniques for evaluating each segment of the antenna system. (b) Typical beam performance requirements.

configurations. Due to the page limitations, topics such as reflector feeds, multi-reflector performances, reflector tolerance studies and blockage effects are not addressed in this chapter.

2. REFLECTOR ANALYSIS METHODOLOGY

2.1 Coordinate Systems

It has been our experience that in dealing with multi reflector/feed systems, it is the best to erect a local coordinate system for each segment of the antenna where the description of the radiation or scattering phenomenon can be most readily presented and then utilize the Eulerian angle coordinate transformation construction to relate the location and the fields from one coordinate system into the next one. A detailed description of such a construction can be found in [2]. Even though this is a simple step, it requires careful attention in its proper implementation. Note that these local coordinates are used to describe both the points and the vector fields.

2.2 Near, Fresnel and Far Fields

Due to the potential proximity of the array feed to the subreflector and the subreflector to the main reflector, one must consider analytical/numerical techniques that properly account for different distances. Additionally, recent interests in applying antenna radiating systems for High- Power Microwave (HPM) applications have necessitated the near-field computations from the reflector surfaces [3]. This can be an important consideration when breakdown issues and safety measures become important. In applying reflector antennas as compact ranges, it is very important to characterize the near-field amplitude and phase when assessing the quiet-zone regions. Also, for many applications, it is essential that both the co-pol and cross-pol fields be determined. Obviously, the determination of both the amplitude and phase is very desirable.

2.3 Analysis of Mesh Reflectors

Mesh reflectors are considered to be an integral part of future large deployable antennas [4]. These mesh surfaces could potentially be used at frequencies up to 35 GHz. It is very important that the detailed characteristics of mesh geometry be properly incorporated in the diffraction analysis formulations. The reader is referred to [5-8] for details. Since the Physical Optics (PO)/Physical Theory of Diffraction (PTD) formulations are the preferred choice for the main reflector analysis, one must be able to effectively modify the physical optics induced current by the mesh topology and performance.

In the past, simple mesh models were used to study the mesh effects. However, it has been noticed that due to the complexity of the mesh structure (typically, knit wires), one must give special attention to the contact and no contact effects among the wires, etc. Block diagram in Fig. 3 shows the steps required to implement the PO/mesh -model diffraction analysis. As shown in Fig. 4, in order to incorporate the effect of the mesh, one needs to modify the PO induced current by properly taking into account the local mesh transmission/reflection matrix in the model. The most recent model utilizes the combination of the method of moments and the Floquet expansion formulation [8]. The steps of this construction are shown in Fig. 5. Once the reflection/transmission coefficients are constructed, the modified physical optics current can be determined.

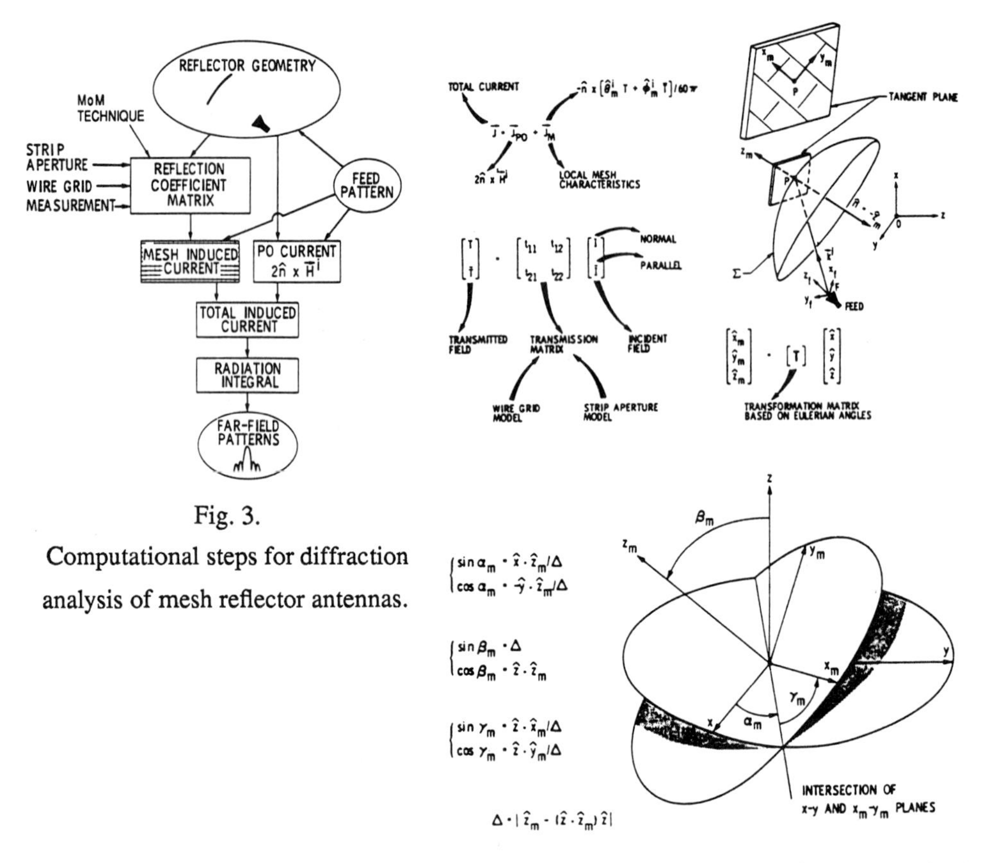

Fig. 3. Computational steps for diffraction analysis of mesh reflector antennas.

Fig. 4. Incorporation of the local mesh transmission/reflection matrix.

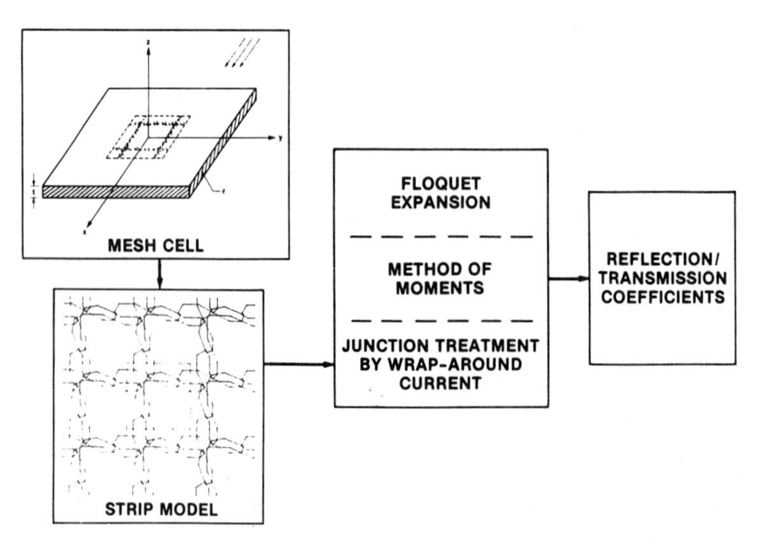

Fig. 5. Computational steps for a complex mesh structure.

The formulation steps are depicted in the block diagram shown in Fig. 5. First, the complex mesh structure in a given cell is broken into many linear segments by identifying the "node" points for the initial and end points of the linear segment. If there is an intersect on the contact point, the same node point is used for all the branches intersecting at this point and making electrical contact there. Each linear segment is then divided into several subdivisions to define the current basis element for the implementation of the method of moments. The method of moments formulation is developed utilizing the Floquet expansion technique and keeping the total tangential electric field at zero on the strip segment. Since the strip width is very narrow in terms of the wavelength, only current components in the strip direction are considered.

The "wrap-around" current is introduced to give special attention to junction treatment. This has been found to be essential to obtain converging results. Additionally, to make the formulation of the problem more complete and realistic, the effects of the mesh surface conductivity have been added to the formulation by modifying appropriate terms of the matrix equation. The precise nature of this modification depends on the current basis function and the models used to introduce the surface conductivity. The problem has been formulated for the cases where the strip structure is backed by a dielectric. Inclusion of the dielectric allows utilization of the alorithm developed to analyze frequency selective surfaces (FSS). The output of the computer program are the induced currents, reflection and transmission coefficients.

As an example, Fig. 6 shows the layout of two mesh configurations used for measurements and computations. These layouts clearly demonstrate the simple contact and no-contact cases. Experimental and numerical results have been obtained for the dielectric slab with a thickness of 70-mil and a dielectric constant of 4.8. Results are presented in terms of transmission coefficient amplitudes. Good agreement has been achieved, demonstrated in Fig. 6 [8]. In particular, it is noticed that the no-contact (broken connection) case does demonstrate significantly different transmission loss characteristics than the contact case at the measured frequencies.

The physical optics radiation integral in Fig. 3 may be evaluated in a variety of ways. Some of the most popular approaches are summarized in [9,10]. The advantage and disadvantage of each technique strongly depend on the particularities of the application. For example, for far-field computations in the forward direction, the Jacobi-Bessel [11], and Fourier-Bessel [12] techniques are very powerful and efficient. These techniques allow one to represent the far-field in terms of summation of known functions but with unknown coefficients. Once these coefficients are determined, they can be used for many observation angles. These expansions are, in particular, very useful for the focusing-type beam patterns that typically result from main reflectors. For the Jacobi-Bessel and the Fourier-Bessel, these expansions take the following general forms:

$$\vec{T}(\theta,\phi) = 2\pi \sum_{p=0}^{P} ab e^{jkz_c(w-w_B)} \frac{1}{p!} (jk)^p (w-w_B)^p \sum_{m=0}^{M} \sum_{n=0}^{N} (j)^n$$

$$\text{x } [{}_p\vec{C}_{mn} \cos n\Phi + {}_p\vec{D}_{mn} \sin n\Phi] \sqrt{2(n+2m+1)} \frac{J_{n+2m+1}(kB)}{kB} \tag{1a}$$

$$B = \sqrt{a^2(u-u_0)^2 + b^2(\upsilon-\upsilon_0)^2} \tag{1b}$$

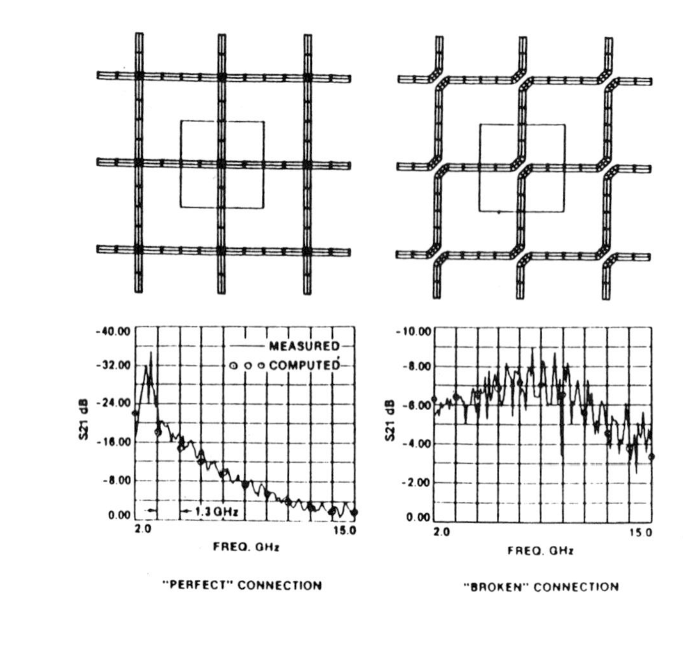

a

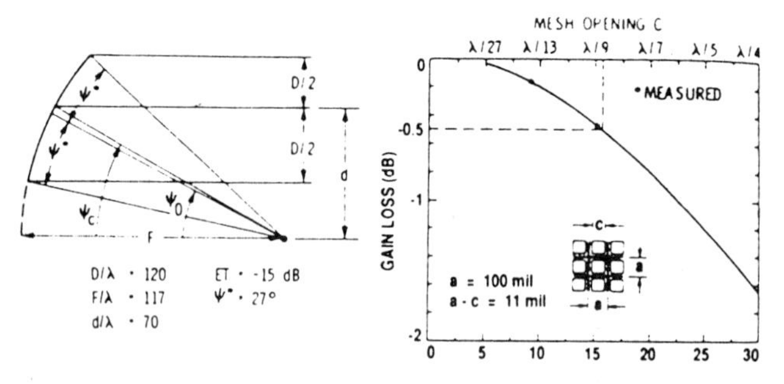

• REFLECTOR GAIN LOSS IS DIRECTLY RELATED TO THE MESH REFLECTIVITY LOSS

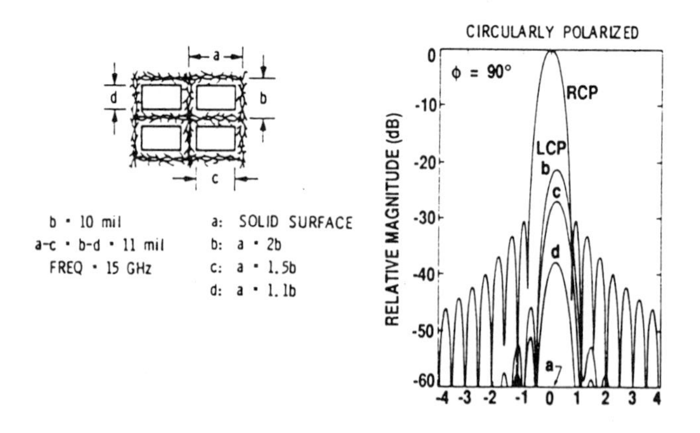

b

• CO-POLAR PATTERN DOES NOT CHANGE APPRECIABLY
• CROSS-POLAR PATTERN CHANGES DRASTICALLY

Fig. 6. (a) Transmission loss for "perfect" and "broken" connections with the dielectric slab backing. (b) Reflector antenna directivity and pattern performance as a function of mesh cell configuration.

and

$$\vec{T}(\theta,\phi) = \sum_{p=0}^{P} \frac{1}{p!} \left(\frac{jk}{4F}\right)^p (\cos\theta - \cos\theta_0)^p \, e^{jk(\frac{H^2}{4F} - F)\cos\theta}$$

$$\times \sum_{m=-M/2}^{M/2-1} \sum_{n=-N/2}^{N/2-1} {}_p\vec{g}_{mn} R_{mn} \tag{2a}$$

$$R_{mn} = \frac{2\pi ab J_1(k\sqrt{u_{mn}^2 a^2 + \upsilon_{mn}^2 b^2})}{k\sqrt{u_{mn}^2 a^2 + \upsilon_{mn}^2 b^2}} \tag{2b}$$

$$\begin{cases} u_{mn} = B_{mn}\cos\Phi_{mn} \\ \upsilon_{mn} = B_{mn}\sin\Phi_{mn} \end{cases} \tag{2c}$$

$$\begin{cases} B_{mn} = \sqrt{[u - u_0 + \frac{H}{2F}(\cos\theta - \cos\theta_0) - \frac{m\pi}{ka}]^2 + (\upsilon - \upsilon_0 - \frac{n\pi}{kb})^2} \\ \Phi_{mn} = \tan^{-1}\left[\dfrac{\upsilon - \upsilon_0 - \frac{n\pi}{kb}}{u - u_0 + \frac{H}{2F}(\cos\theta - \cos\theta_0) - \frac{m\pi}{ka}}\right] \end{cases} \tag{2d}$$

All the parameters appearing in the above equations are defined in [13]. Fig. 7 shows an example of the results of the far-field computations using both the Jacobi-Bessel and Fourier-Bessel expansion techniqus applied to an offset paraboloidal reflector with an elliptical projected aperture [13].

However, if one's interest lies with an application that necessitates the determination of the near-field or Fresnel fields of the main reflector, one must give considerable attention to the accurate evaluation of the radiation integral by using very tight integration grids. In these cases, the brute-force integration method with specialized phase terms is typicaly applied. Also, application of improved integration techniques could become very effective in reducing the computational time [14]. This may become necessary for constructing the near field of compact range reflector antennas and for HPM applications.

2.4 Subreflector Analysis

Among all the different analysis techniques applicable to subreflectors, GTD and PO+PTD have received the most attention. It should be realized that in contrast to the main reflector, the subreflector typically produces broad radiating patterns and that, in most cases, the near-field of the subreflector must be constructed because the main reflector could be residing in the near-field of the subreflector. Also, it becomes very

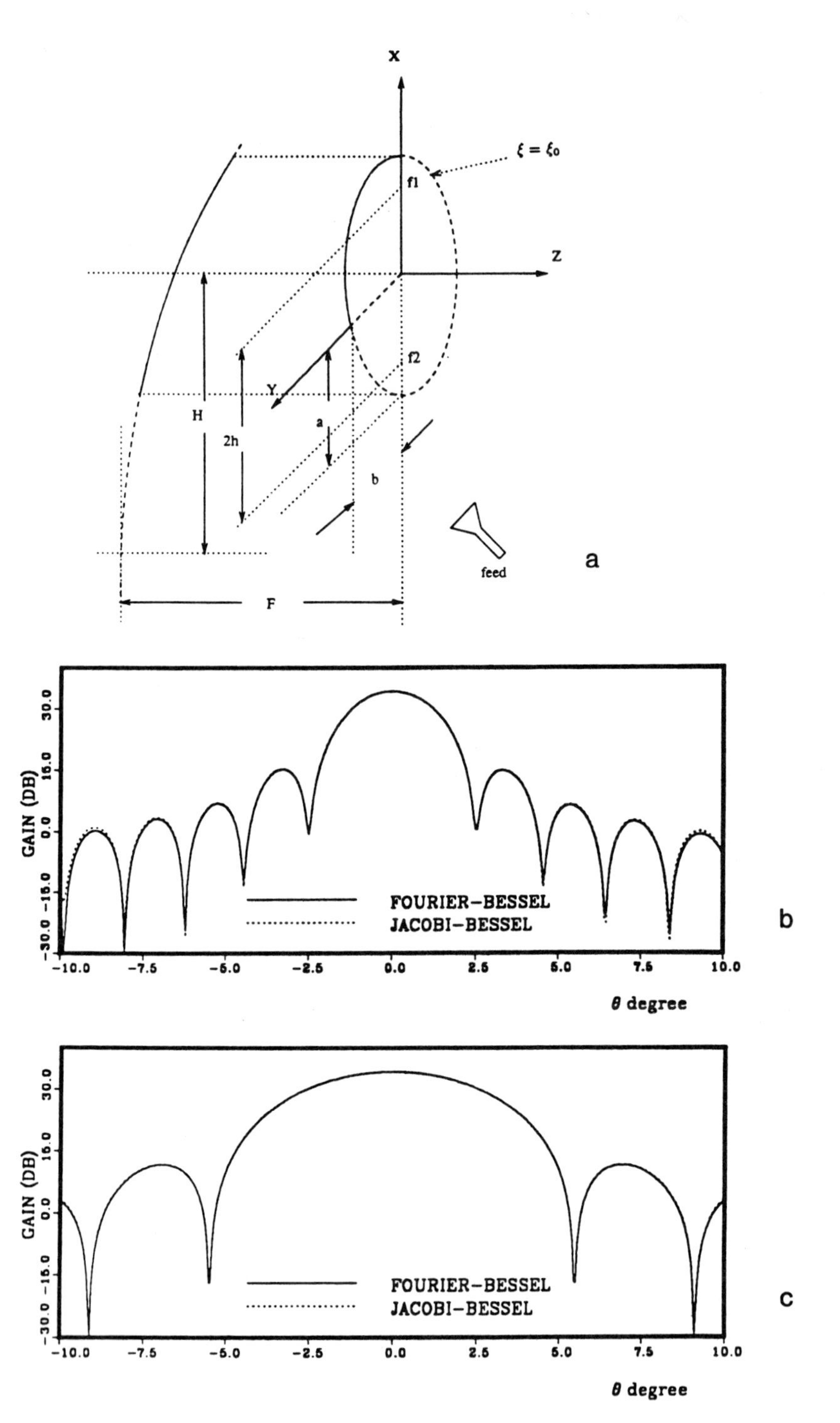

Fig. 7. (a) Geometry of an offset parabolic reflector antenna with an elliptical projected aperture. (b-c) Comparative study between the Fourier-Bessel and Jacobi-Bessel algorithms as applied to an offset parabolic antenna with an elliptical aperture, $a=15\lambda$, $a/b=2.0$, $F=30\lambda$, $H=18\lambda$, $q_1=7.14$, $q_2=44.9$, $t=(-18\lambda,0,0)$, P,M,N=(1,16,8) and (1,7,7) for the Fourier-Bessel and Jacobi-Bessel algorithms respectively; (b) $\phi=0°$, (c)$\phi=90°$.

important that both the co-pol and cross-polarized field components be computed correctly. To fully appreciate the effects of different analysis models in the prediction of the scattered field from subreflector type geometries, the scattering patterns of several representative geometries are presented next, along with a comparison of several popular techniques.

3. COMPARATIVE STUDY AMONG LEADING DIFFRACTION TECHNIQUES

The purpose of this section is to review several of the numerous diffraction techniques (see, for example, [9,10,15]) for the analysis of reflector antennas. These techniques include those that have been used extensively for reflector analysis such as (i) Physical Optics (PO) [9], (ii) Geometrical Theory of Diffraction (GTD) [16], (iii) Uniform Asymptotic Theory (UAT) [17,18], (iv) Uniform Geometrical Theory of Diffraction (UGTD) [19], and (v) Theory of Gaussian Beams (TGB) [20]. A class of Physical Theory of Diffraction (PTD) techniques, which are modifications of Ufimtsev's PTD [21], is also studied: (vi) Mitzner's Incremental Length Diffraction Coefficients (ILDC) [22], (vii) Michaeli's Equivalent Edge Currents (EEC) [23], and (viii) Ando's Modified Physical Theory of Diffraction (MPTD) [24]. Comparative results are presented for flat circular discs (an unfocused system), offset ellipsoidal reflectors, and offset parabolic reflectors (focused systems).

3.1 Circular Discs

The PTD and GTD techniques will be studied in this section by analyzing the scattering from a conducting circular disc of radius a (Fig. 8). The feed is a short dipole located a distance d in front of the disc center. This problem is chosen because it facilitates effective comparisons among various techniques. The PTD techniques that will be studied are modifications of Ufimtsev's physical theory of diffraction [21]. In these techniques, the total scattered field consists of two components, the PO field and the fringe field:

$$\vec{E}^{PTD} = \vec{E}^{PO} + \vec{E}^{fr}. \tag{3}$$

In the far field region, the PO field can be constructed by

$$\vec{E}^{PO} = -j\omega\mu \frac{e^{-jkr}}{4\pi r}(\hat{\hat{1}} - \hat{r}\hat{r})\cdot \int\int_{\Sigma} (2\hat{n} \times \vec{H}^{i}) e^{j k \hat{r}\cdot\vec{r}_{\sigma}} d\sigma, \tag{4}$$

where Σ is the scatterer surface, $\hat{n}$ is a unit normal of the surface, $2\hat{n} \times \vec{H}^{i}$ is the PO current, and $\hat{\hat{1}}$ is the unit dyad. The dyad $(\hat{\hat{1}} - \hat{r}\hat{r})$ serves to extract the transverse (to $\hat{r}$) component of the surface integral. The fringe field is determined by the electric and magnetic equivalent currents along the scatterer edge:

$$\vec{E}^{fr} = -j\omega(\hat{1} - \hat{r}\hat{r})\cdot\vec{A} + j\ k\hat{r} \times \vec{F} \tag{5}$$

$$\vec{A} = \frac{\mu}{4\pi}\int_{\Gamma} \hat{z}' I^{eq} \frac{e^{-jk|\vec{r}-\vec{r}'|}}{|\vec{r}-\vec{r}'|} dl' \tag{6}$$

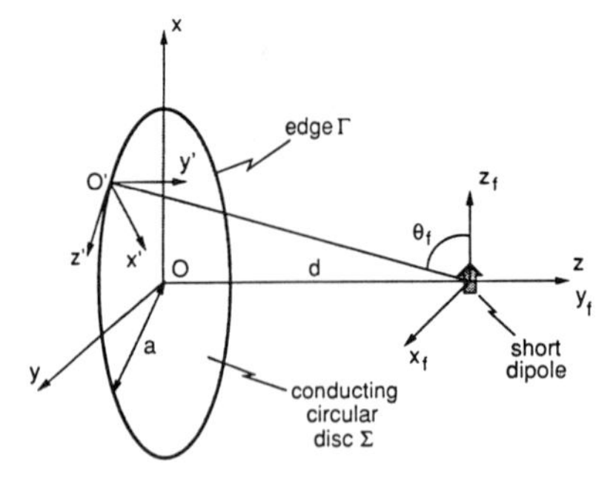

Fig. 8.

Geometry of the disc-dipole radiation system.

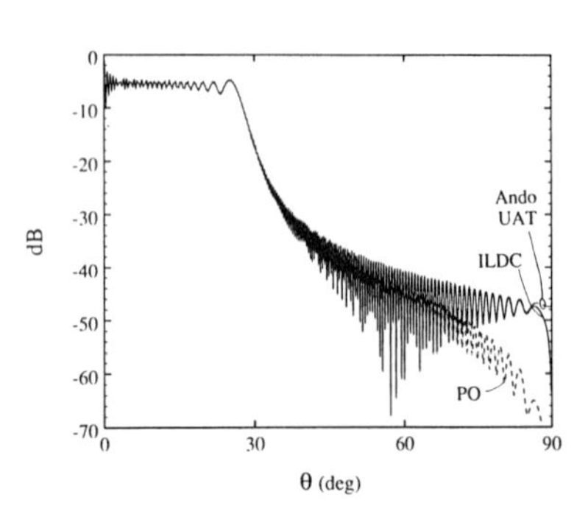

Fig. 9.

Co-polar far-field patterns ($\phi = 0°$) for a circular disc of radius $a = 96\lambda$ and a disc-dipole distance $d = 176\lambda$. (All curves have been normalized to the boresight PO field.)

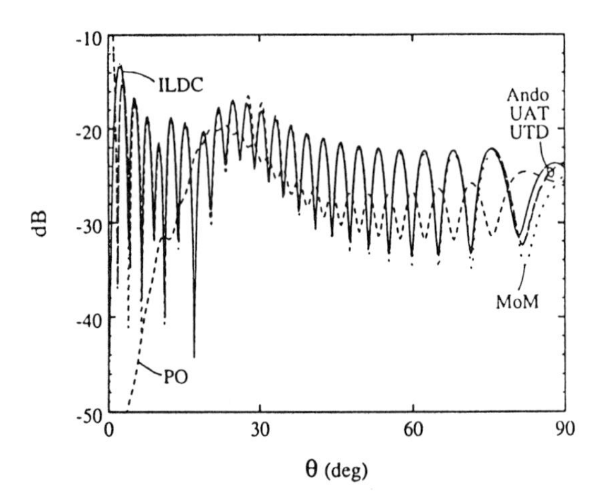

Fig. 10.

Cross-polar far-field patterns ($\phi = 45°$) for a circular disc of radius $a = 12\lambda$ and a disc-dipole distance $d = 22\lambda$. (All curves have been normalized to the boresight co-polar PO field).

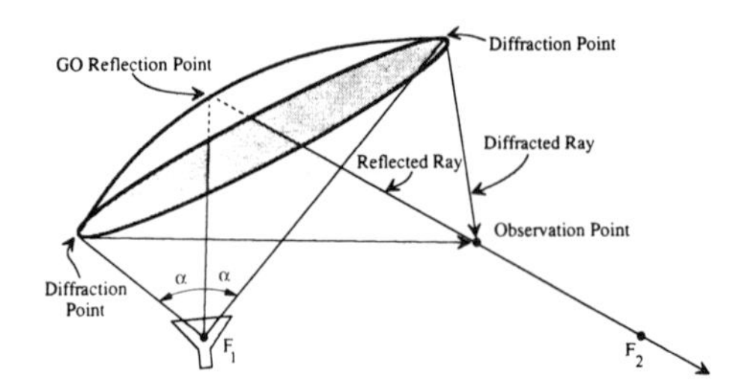

Fig. 11. Offset ellipsoidal reflector geometry.

$$\vec{F} = \frac{1}{4\pi} \int_{\Gamma} \hat{z}' M^{eq} \frac{e^{-jk|\vec{r}-\vec{r}'|}}{|\vec{r}-\vec{r}'|} dl', \tag{7}$$

where dl' is a differential path length along the scatterer edge Γ. For every point on the scatterer edge a local coordinate system $\hat{x}' - \hat{y}' - \hat{z}'$ is defined (Fig. 8) in a manner such that $\hat{z}'$ is along the tangential direction of the edge. Therefore, the vector $\hat{z}'$ in equations (6) and (7) varies along the integration path. The equivalent edge currents I^{eq} and M^{eq} in equations (6) and (7) can be written in terms of the tangential (to the scatterer edge) components of the incident field and the diffraction coefficients D_e, D_x, and D_m as

$$I^{eq} = \frac{j2E^i_{z'}}{\omega\mu \sin^2\theta^i} D_e + \frac{j2H^i_{z'}}{k \sin^2\theta^i} D_x \tag{8}$$

$$M^{eq} = \frac{-j2H^i_{z'}}{\omega\varepsilon \sin\theta' \sin\theta^i} D_m, \tag{9}$$

where θ^i is associated with the incident direction and θ' is associated with the observation direction, both with respect to the local coordinate system. Mitzner's [22], Michaeli's [23], and Ando's [24] formulations differ in the detailed expressions of the diffraction coefficients.

The total scattered field in Keller's GTD [16] is divided into a geometrical optics (GO) part and a diffracted part. The detailed expressions for these field components can be found in [25, 26]. The singularities of GTD at the shadow boundary (SB) and reflection boundary (RB) are cured by its uniform versions, UTD and UAT. In UTD, Fresnel integrals are incorporated to smooth out the singularities in the diffracted field. Similar Fresnel functions were introduced in UAT to modify the GO field so that the singularities in the diffracted field can be cancelled.

In Fig. 9, the far-field patterns are plotted for a large disc ($a = 96\lambda$, $d = 176\lambda$) where the method of moments is not easily applicable. As can be seen, the PTD and GTD techniques agree well except in the last sidelobe region. The PO solution works well up until 30 degrees, but starts to deviate from all other solutions thereafter.

The cross-polar fields obtained by the various techniques have also been examined. Generally, the cross-polar fields are very weak around $\phi = 0°$ and $\phi = 90°$ for symmetric antenna configurations, but become significant at $\phi = 45°$. Most of the observations for the co-polar fields discussed above are true for the cross-polar fields, except that PO predicts very different patterns from those predicted by MoM. This is illustrated in Fig. 10 using $a = 6\lambda$, $d = 11\lambda$. Note that in the PO cross-polar field, the first few lobes are not predicted, and the behavior of the pattern outside the main beam region is ''out of phase'' with the results. Nevertheless, the envelope of the field pattern outside the main beam region predicted by PO is not disparate from others for this large disc.

3.2 Ellipsoidal Reflector

We will next consider an offset ellipsoidal reflector. This reflector is of special interest since it gives us a second focal point. The GTD group has a singularity at a focal point and it is of interest to find how close to the focal point these techniques will still yield meaningful results. The reflector will be defined by the intersection of an ellipsoid of revolution with a circular cone emanating from one of the ellipsoid's foci, as shown in Fig. 11. This particular way of defining the reflector will lead to the reflector having a planar edge and an elliptical aperture [9].

The GTD solution is an approximate high-frequency solution that is useful in determining the scattered fields of objects that are large compared to a wavelength. For the given incident field produced by the feed, the total reflected field at the observation point (P_f) is found through the asymptotic solution given by Keller's Geometrical Theory of Diffraction [4]. The GTD solution can be written in the form

$$\vec{E}(P_f) = \vec{E}_g + \vec{E}_d + 0(k^{-1}) \tag{10}$$

where the g-subscript represents the geometrical optics field, which is of the order k^0. The d-subscript represents the edge-diffracted field, which is of the order $k^{-1/2}$. These fields can be thought of as "ray fields" that are locally plane waves.

The geometrical optics field at the reflection point is found through the application of Snell's law. In terms of the incident field unit vector, the reflected field unit vector is given as

$$\hat{e}_r = -\hat{e}_i + 2(\hat{n}_\Sigma \cdot \hat{e}_i)\hat{n}_\Sigma \tag{11}$$

The unit vector $\hat{e}_r$ gives the direction of the electric field for the reflected ray. The vector $\hat{n}_\Sigma$ is the unit normal to the surface at the reflection point. The reflected field can, for our geometry, be written as follows

$$\vec{E}_g(P_f) = \frac{A\eta \exp[-jk(d_1 + d_2)]}{1 - d_2/R}\hat{e}_r \tag{12}$$

where d_1 and d_2 are the distances from the feed to the reflection point and from the reflection point to the observation point, respectively. R is the radii of curvature of the reflected field at the reflection point. Notice that the magnitude of the field will become infinite when the observation point is at the focus ($d_2 = R$). Further, notice the phase shift of 180° that occurs when the ray passes through the focus (i.e., $d_2 > R$). This phase shift has been discussed extensively in the literature and is known as the Phase Anomaly or Gouy's phase shift, after the French scientist who first discovered it.

The edge-diffracted field is due to the contribution from diffraction points lying on the edge of the reflector. These diffraction points are found through the application of the law of edge diffraction. The total diffracted field is found by summing up the contribution from all of the diffraction points. The diffracted field is given by the expression

$$\vec{E}(x,y,z) = g(kd_4)DF\frac{1}{\sin\theta_i}[D_s E^i_\theta \hat{\theta} + D_h E^i_\phi \hat{\phi}] \tag{13}$$

in which $g(kd_4)$ is a cylindrical wave factor, DF is the divergence factor, and $D_{s,h}$ are the soft and hard diffraction coefficients. Detailed expressions for each of these terms can be found in the literature [25, 26].

In a scattering problem we can find the scattered *near-field* from the integral representations

$$\vec{E} = -j\omega\mu \int_{\Sigma} \left[\vec{J}g + \frac{1}{k^2} (\vec{J} \cdot \nabla)\nabla g \right] dS' \tag{14}$$

$$\vec{H} = \int_{\Sigma} [\vec{J} \times \nabla g] \, dS' \tag{15}$$

where $\vec{J}$ is the induced current on the scatterer and g is the free-space Green's function given by

$$g = \frac{e^{-jkR}}{4\pi R}, \quad R = |\vec{r} - \vec{r}'| \tag{16}$$

and Σ is the surface of the scatterer. The PO assumes that the scatterer is an infinite ground plane and thus assumes $\vec{J}$ to be given by

$$\vec{J} = 2\hat{n} \times \vec{H}^i \tag{17}$$

where $\hat{n}$ is the normal to the surface and $\vec{H}^i$ is the incident H-field. This approximation for $\vec{J}$ has been shown to be useful for relatively large and smooth scatterers.

A method frequently used for the solution of beam waveguide and lens problems is the theory of Gaussian beams. Gaussian beams are a result of an approximate analytic solution to the wave equation. We assume that the beam consists of field components purely transverse to the direction of propagation and that the phase varies essentially as a plane wave. That is, we look for solutions to the scalar wave equation of the form

$$E(x,y,z) = E_o \psi(x,y,z) e^{-jkz} \tag{18}$$

The factor $\psi(x,y,z)$ describes how the beam deviates from a plane wave. Substituting Eq. (18) in the wave equation and assuming a paraxial beam, we find the fundamental Gaussian mode, which is given in cylindrical coordinates by

$$\psi(r,z) = \sqrt{\frac{2}{\pi}} \frac{1}{w(z)} \exp\left[\frac{-r^2}{w^2(z)} - j\left[\frac{-kr^2}{2R^2(z)} - \Phi(z) \right] \right] \tag{19}$$

where R(z), w(z) and $\Phi(z)$ are defined as

$$R(z) = z\left[1 + \left(\frac{\pi w_0^2}{\lambda z}\right)^2\right] \qquad (20)$$

$$w^2(z) = w_0^2\left[1 + \left(\frac{\pi w_0^2}{\lambda z}\right)^2\right] \qquad (21)$$

$$\Phi(z) = \arctan\left[\frac{\lambda z}{\pi w_0^2}\right] \qquad (22)$$

The first term in the exponent is a radial amplitude factor, the second term is a radial phase factor, and thc last two terms are longitudinal phase factors. The term w(z) is called the beam width and gives the radius at which the field has fallen off to a value of e^{-1} times the value on the axis. The term R(z) is the radius of curvature of the phase front.

Let us choose an ellipsoid given by $a = 40\lambda$ and $b = 50\lambda$. This ellipsoid has focal length, $f = 20\lambda$ and eccentricity, $e = 0.6$. Consider a reflector produced by specifying $\beta = 5\pi/6$, $\alpha = \pi/12$. Referring to the geometry of Fig. 11, these specifications lead to a reflector with an elliptical aperture described by $a' = 19.73\lambda$, and $b' = 16.85\lambda$. Such a reflector could conceivably be used in a beam waveguide or a Gregorian antenna system. Having specified the reflector, we also need to define the feed of this system. We choose to investigate the case of a feed pointing towards the center of the reflector as seen from the feed. The taper of the feed is determined by the choice of q_x and q_y. For the particular value of α chosen, a value of $q_x = q_y = 50$ will lead to a feed edge taper (ET) of -20 log $[\cos^{50}(\alpha)] = 15.06$ dB.

Figs. 12a-b show the magnitudes and phases for the dominant components of the E-field along the central reflected ray, as obtained using the GTD, PO and Gaussian Beam methods. We see that away from the focal point in the main beam direction, the GTD and Gaussian Beam solutions both give results that are very close to the PO solution. Near the focal point both the Gaussian Beam and the GTD methods predict erroneous results for the magnitude. The Gaussian Beam method predicts a phase that is very close to that predicted by PO, even though the magnitude differs. Notice also that the GTD solution displays the same sort of ripple as the PO solution in both its magnitude and phase.

3.3 Parabolic Reflector

The scattered near-field, far-field, and Fresnel-field obtained by these techniques have been compared to obtain some information about the relative accuracy of the various techniques for the parabolic reflector geometry [27]. It has been found that as was the case for the circular disc, the GTD technique fails drastically along the boresight direction but gives accurate patterns past the first few sidelobes. In the case of the parabolic reflector, the failure of the GTD technique is three-fold, since in the boresight direction we have caustics for both the GO field and the diffracted field, and additionally, this is also the direction of the shadow boundaries. Notice that the GTD solution fails on the shadow boundaries. The use of the uniform theories (UAT, UTD) would eliminate this problem. The fundamental order Gaussian beam solution gives a reasonable value for the peak of the main beam, but does not predict any sidelobes.

For certain applications such as high-power microwave (HPM) applications, the near-field power density must be known, in order to predict the maximum power levels limited by air breakdown. Fig. 13 shows a plot of the total (scattered plus incident) near-field power density of an offset parabolic reflector that has a feed radiating a total average power of 1 GW at 1 GHz. Notice the standing wave that occurs in front of the reflector. It can be seen from Fig. 13 that if we assume that the air breakdown occurs at a field strength of 10^6 V/m, then the total field in front of the reflector could exceed the breakdown field strength.

4. ANTENNA SURFACE DISTORTION COMPENSATION

Reflector surface distortion degrades antenna performance by lowering the gain and increasing the sidelobe levels [28]. The distortion may result from thermal, gravitational or other effects. It is the purpose of this section to present a novel distortion compensation concept based on the application of properly matched array feeds. Both numerical simulations and measured data are presented to demonstrate the utility of this concept. In particular, it is shown that the concept is most useful for overcoming the deterioration effects of slowly varying surface distortions. The application of this concept could prove very useful for future large space and ground antennas. The demand for using large antennas in space has been highlighted in a recent review paper [4]. It is anticipated that these antennas must exhibit high gain and low sidelobes for future multiple-beam satellites. For example, these characteristics are very important to future generations of Mobile Satellite System (MSS) and Personal Access Satellite System (PASS) spacecraft antennas, since they directly impact system capacity by increasing frequency reusability. Additionally, there are potentially several applications for large antennas for space science (e.g., orbiting VLBI). Among the designers of large-antenna configurations, reflectors still enjoy more acceptance than any other antenna concept. Most probably these antennas will be in the category of deployable, erectable, or inflatable [4]. In the space environment, the surface of these large reflectors will be distorted, resulting in degraded antenna performance. Fig. 14 shows a schematic of potential applications of these large reflectors.

The mechanism that causes this antenna performance degradation is principally aperture phase errors introduced by distortion of the reflector surface (Fig. 15a). Figure 15b summarizes how a corrective apparatus can be used to compensate for the distorted radiated phase front. As observed, the corrective apparatus may be positioned at different locations in accordance with its compensating characteristics. In this section, attention is focused on the application of an array feed compensating technique. This technique is particularly attractive because of recent developments in monolithic technology.

4.1 Compensation Algorithm

The array-feed excitation coefficients may be determined in different ways, as discussed schematically in [29]. In this study, the concept of the focal-plane field match is used to develop the required compensation algorithm. In general, this concept may be implemented in several ways. This focal-plane field can be constructed by illuminating the dis-

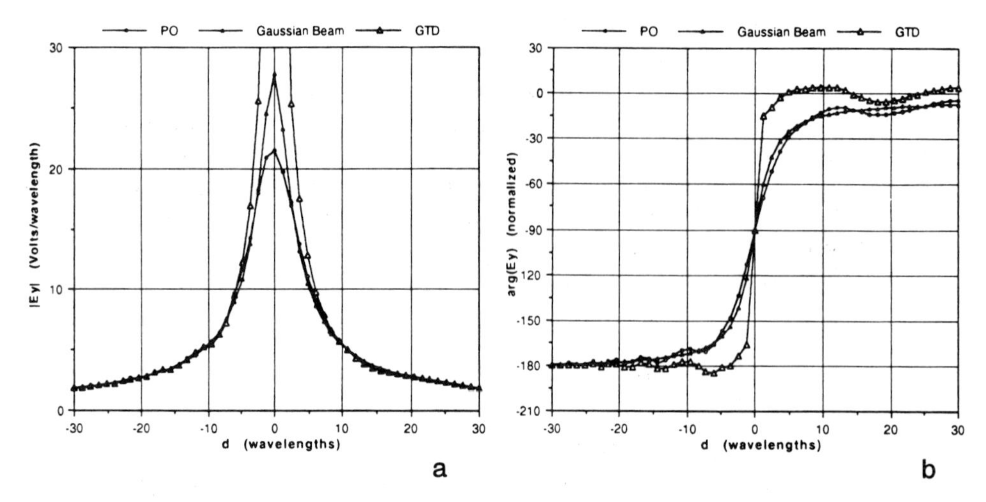

Fig. 12. (a) $|E_y|$ along the central reflected ray for an offset ellipsoidal reflector with $b = 50\lambda$, $a = 40\lambda$, $\beta = 5\pi/6$, $\alpha = \pi/12$, and a y-polarized feed with ET = 15 dB. (b) Phase of E_y (in degrees) normalized to $\exp[-jk(d_1 + d_2)]$ for a reflector with $b = 50\lambda$, $a = 40\lambda$, $\beta = 5\pi/6$, $\alpha = \pi/12$, and a y-polarized feed with ET = 15 dB.

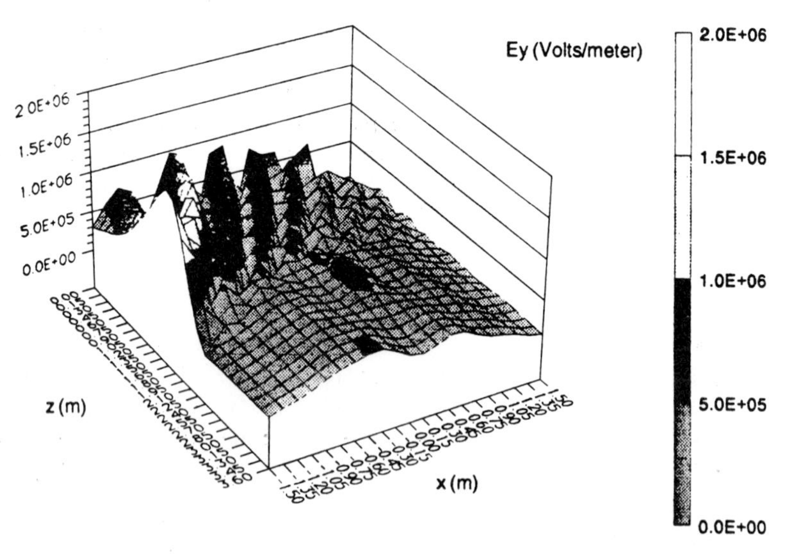

Fig. 13. Total (scattered plus incident) near-field of an offset parabolic reflector with F/D = 1 and D = 3 m and with a feed radiating 1 GW of power at 1 GHz, having an edge taper of 10 dB.

torted reflector by a plane wave and then constructing the focal-plane field using physical optics or GTD constructions. This approach requires the development of new computer programs for the determination of the focal-plane fields. Once the focal-plane fields are constructed, the array-feed excitation can be conjugate matched to this focal plane.

If we assume that the focal-plane field can be estimated from the far-field pattern by invoking the reciprocity theorem, then the far-field is computed in the desired direction (say the boresight) for the expected location of each array-feed element and the conjugate value of this computed far field is used to determine the excitation coefficients of the element. The advantage of this approach is the use of a modified version of an existing far-field computer program. This is the approach used in this chapter. A previously developed computer program has been modified to perform the computation for implementation of the compensation algorithm.

To apply the surface compensation algorithm, the surface distortion profile must be known either in terms of a functional description or at discrete points; this information can be obtained by using optical, contact, microwave holographic [30,31], photogrammetric [32], or other metrological techniques. A comprehensive review of modern measurement techniques applicable to large reflector antennas is given in [1]. Figure 16 depicts the steps necessary to implement this compensating diffraction algorithm. As will be discussed later, it is possible to perform the measurements required to compensate for the reflector distortion without actual knowledge of the surface distortion. This is very important in the actual implementation of the surface-compensation techniques in space applications.

The concept of conjugate-field matching is used subsequently to determine the complex excitation coefficients of the array-feed elements. This approach can provide the array excitation coefficients that either maximize reflector gain in the desired direction or control sidelobe levels [29]. Note that for sidelobe control a fictitious directive feed pattern is used to effectively taper the induced current distribution for the desired sidelobe level.

This fictitious feed pattern is used only during the array excitation coefficient determination by the computer program. The actual feed pattern is used finally to determine the resulting compensated antenna pattern.

4.2 Numerical Simulations

To verify the accuracy of the developed computer programs, many simulations have been performed. Simulations were based on both the closed and discrete form representations of the surface distortion. For numerical simulation purposes, an offset reflector with the dimensions shown in Fig. 17 is considered. This reflector geometry was also used to build an antenna for the experimental verification, as will be discussed in the next section. The figure describes the simulated distortion that is a representation of a typical dominant thermal or gravitational distortion. Notice that this distortion represents an 11-mm surface deviation of a parabolic antenna at its rim. In terms of the wavelength, it is equal to about 1/3 of the wavelength at the operating frequency of 8.45 GHz. This, of course, is a substantial amount of distortion.

Many numerical studies have been performed using a different number of feed elements both with and without compensation. Figure 18 summarizes the results of these numeri-

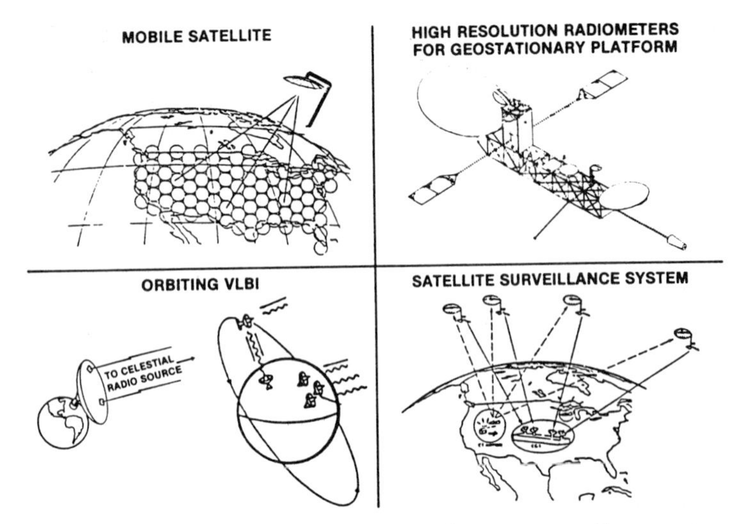

Fig. 14. Future mission applications of large space reflector antennas.

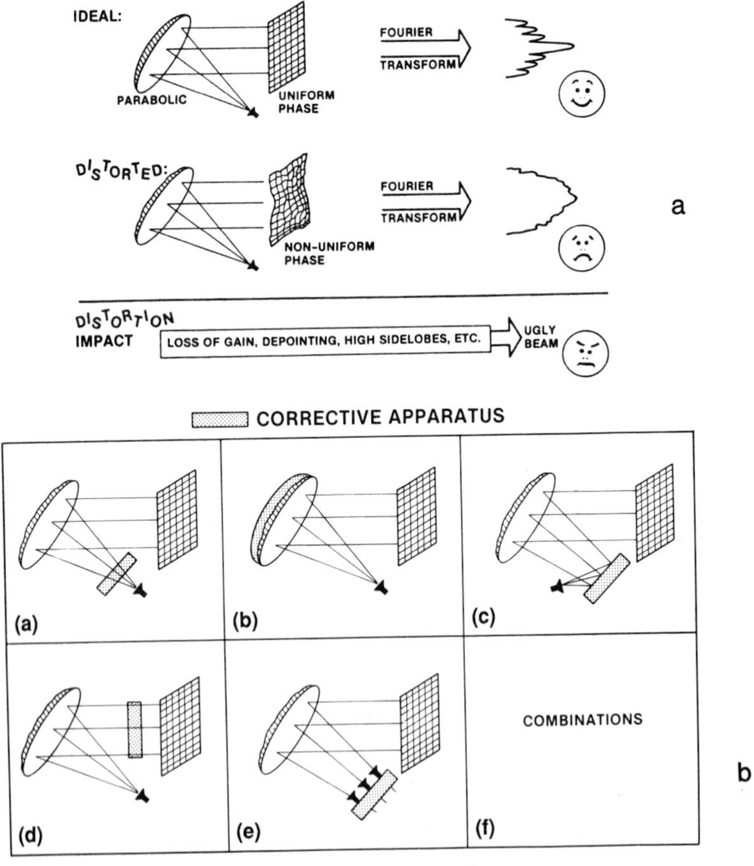

Fig. 15. (a) A simple interpretation of why the reflector surface distortions result in unwanted pattern degradations. (b) A conceptual presentation of the use of a corrective apparatus to produce a uniform aperture phase in front of a distorted reflector antenna surface. Notice that, in general, the corrective apparatus (shown as a rectangular box with dots) may be placed at different locations.

cal simulations and provides appropriate references to the resulting far-field patterns. Also identified in this figure is the computed antenna directivity for the simulated cases. As examples for the 19-element, 16-element, and 7-element array-feed configurations, the complex excitation coefficients have been computed and typical excitation coefficient data are shown in Table 1 for a 19-element array feed.

4.3 Experimental Study

For the experimental study, it has been assumed that the reflector surface is distorted in the fashion shown in Fig. 17, which describes a dominant term in a typical thermal or gravitational distortion. The functional expression in Fig. 17 has been used to fabricate a test reflector to demonstrate the utility of the compensation technique via measured data. Both 16-element and 19-element array feeds using cigar elements were used to perform the experiment (in practice, other array elements such as microstrip elements can also be used). The reader may refer to [29] for the details of the experimental steps.

Array design was done for a 19-element array with a hexagonal arrangement of elements (see Fig. 19). This configuration was chosen for a number of reasons: it allows a single element to be activated at the center of the array, yielding the narrowest possible symmetrical overall antenna pattern, hence making boresighting easier; also, the 19-element configuration allows high-density element packing, which in turn allows more effective compensation since the average distance from the center of the array to the individual elements is smaller.

Feed-pattern measurements of the 19-element configuration were carried out to deduce the array-element pattern in the array environment. The measurements were performed by setting the element excitations to amplitude = 0 dB, phase = 0 deg. for active elements, and amplitude = -35 dB, phase = 0 deg. for inactive elements in the cases where 1, 7, and 19 elements were used. Azimuth patterns were taken for various polarization angles of the feed to provide phi = 0-, 45- and 90-deg. cuts. The results shown in Fig. 19 demonstrate a fair comparison with the computed data. The computed result used the appropriate E- and H-plane COS**Q models for the array-element patterns.

It was decided to undertake the construction of a distorted reflector for purposes of demonstrating the practical application of the method of compensation to an actual reflector. Analytical calculations and computer simulations were carried out, and an appropriate surface with a distortion of the form $\rho^3 \cos 2\phi$ in aperture coordinates was used (see Fig. 17). As mentioned earlier, this distribution is a proper manifestation of the dominant-distortion mode that results from thermal or gravitational effects. The Optical Science Laboratory (OSL) at the University of Arizona constructed the reflector. The final specifications consisted of the surface shape, the surface tolerance, and the general requirement that the reflector be as lightweight as possible while retaining high rigidity [29].

Once the beam-forming network (BFN) was assembled, it was necessary to determine the attenuator amplitude and phase characteristics (since the phase shift through the attenuators is a function of attenuation) as well as the phase-shifter phase characteristics. The

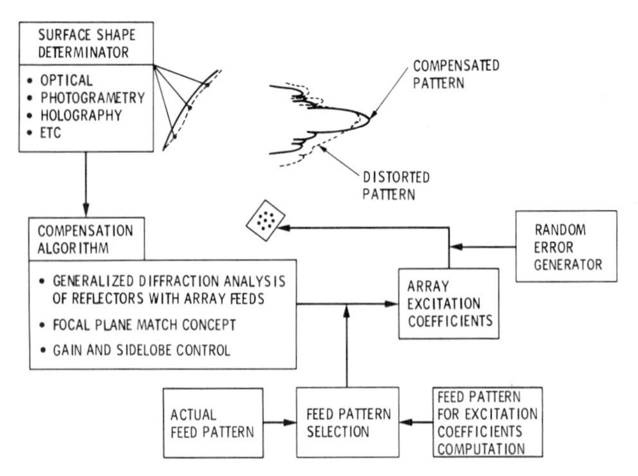

Fig. 16. A block diagram of the computer program development.

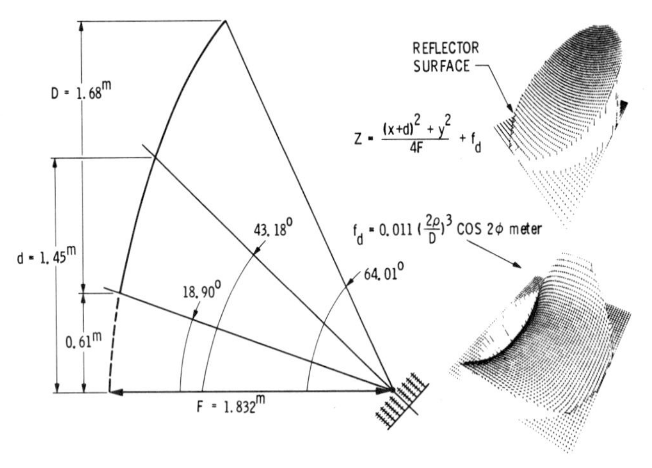

Fig. 17. The reflector surface configuration and its functional representation incorporating slowly varying distortion.

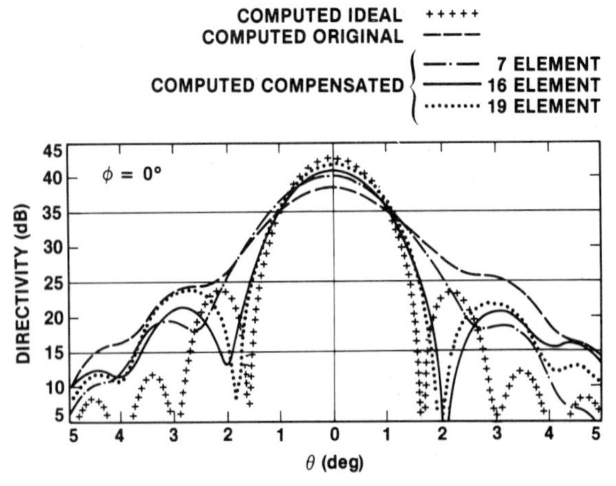

Fig. 18. Compensated and non-compensated far-field patterns generated by the diffraction analysis computer program for different numbers of array elements.

TABLE I

Complex (Amplitude & Phase) Excitation Coefficients of a 19-Element Array Feed to Compensate for the Reflector Surface Distortion.

(Element positions in the feed plane: x_f upward, y_f to the right; columns listed left to right, elements within each column top to bottom.)

Column	Element (top to bottom)	MAXIMUM GAIN	SIDELOBE CONTROL
1	1	-10.6 dB, -67.0°	-12.1 dB, -58.8°
1	2	-5.8 dB, -83.5°	-7.4 dB, -69.1°
1	3	-12.1 dB, -40.1°	-12.6 dB, -35.5°
2	1	-9.1 dB, -303.4°	-10.6 dB, -319.8°
2	2	-3.0 dB, -19.2°	-3.1 dB, -13.7°
2	3	-3.8 dB, -11.2°	-3.5 dB, -8.5°
2	4	-8.7 dB, -300.8°	-9.9 dB, -313.7°
3	1	-6.5 dB, -226.9°	-11.2 dB, -247.4°
3	2	-2.6 dB, -325.0°	-3.4 dB, -335.9°
3	3	0.0 dB, 0°	0.0 dB, 0°
3	4	-3.0 dB, -320.8°	-3.6 dB, -332.0°
3	5	-8.8 dB, -225.3°	-12.3 dB, -246.7°
4	1	-9.1 dB, -303.4°	-10.6 dB, -319.8°
4	2	-3.0 dB, -19.2°	-3.1 dB, -13.7°
4	3	-3.8 dB, -11.2°	-3.5 dB, -8.5°
4	4	-8.7 dB, -300.8°	-9.9 dB, -313.7°
5	1	-10.6 dB, -67.0°	-12.1 dB, -58.8°
5	2	-5.8 dB, -83.5°	-7.4 dB, -69.1°
5	3	-12.1 dB, -40.1°	-12.6 dB, -35.5°

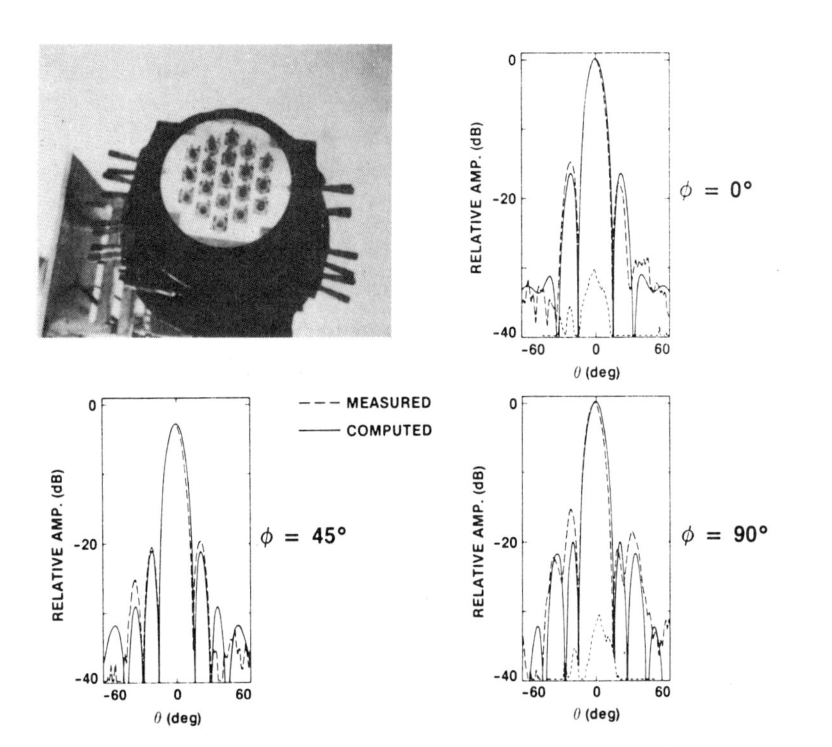

Fig. 19. A comparison between the measured and computed patterns for the 19-element array (frequency = 8.45 GHz).

attenuator phase characteristic model was derived empirically, based upon graphical analysis of several attenuator phase characteristics. To perform these measurements, the array/BFN assembly was mounted on the antenna structure (Fig. 20). It was discovered that setting individual elements to the corresponding measured excitation rather than to the complex conjugate as in the computer simulation was effective in compensating for the reflector distortion. This is because the antenna was operating in a receive mode. It is very important to notice that in this procedure there was no need to know what the actual surface distortion was; the procedure compensated for the effects of the distortion directly as described next. The compensation procedure for both the 16- and 19-element array feeds began by setting all elements to amplitude = 0 dB, phase = 0 deg. Patterns were then taken for comparisons to the computer simulations of the same configuration as well as for before/after comparisons of the patterns. Next, individual elements were set one at a time to amplitude = 0 dB, phase = 0 degrees while all other elements were attenuated to -35 dB; the amplitude and phase of the resulting output were then measured at the boresight. These field values constitute the excitation coefficients for the array elements. Based upon these excitation coefficients, appropriate dial settings for the variable attenuators and phase shifters were computed for the individual elements. Finally, these dial settings were applied to the BFN and the compensated patterns were measured.

Figure 21 shows the measured antenna patterns for different feed-element configurations. Notice that in this figure, the "original" measured pattern refers to the case where the distorted reflector is illuminated by a 7-element array, producing an almost 10-dB taper at the reflector edge without any compensation invoked.

It is worthwhile to mention that the power dividing network used suffered from loss because it was not tailored for the number of elements of interest. Therefore, the on-axis improvement is referred to only the directivity improvement. This is clearly observed by comparing the radiated far-field patterns as shown in Fig. 21. In actual implementation of the technique for overall gain improvement, one should use a power amplifier behind each element to compensate for the unwanted power losses due to the power dividing network. For this reason, employing monolithic technology would be useful and effective.

4.4 Related Studies

As was mentioned earlier, the array feed can be used to overcome the unwanted pattern degradation due to wide-angle scan in reflector antennas [33]. An example is shown in Fig. 22.

Finally, an integrated RF, control, and structural experiment, aboard the space shuttle or space station, has recently been proposed, to properly assess the effectiveness of an adaptive compensation concept in the actual space environment. An artist's rendition of such a concept is shown in Fig. 23. Use of the non-uniform sampling technique [34] has been considered to measure the RF performance of the antenna in space.

5. SOLUTION METHODOLOGY FOR FSS REFLECTORS

The geometry of an FSS subreflector is shown in Fig. 1. It is assumed that this subreflector is illuminated by a feed whose radiation characteristic is known. In general,

Fig. 20. The reflector antenna and the array feed in position in the far-field range.

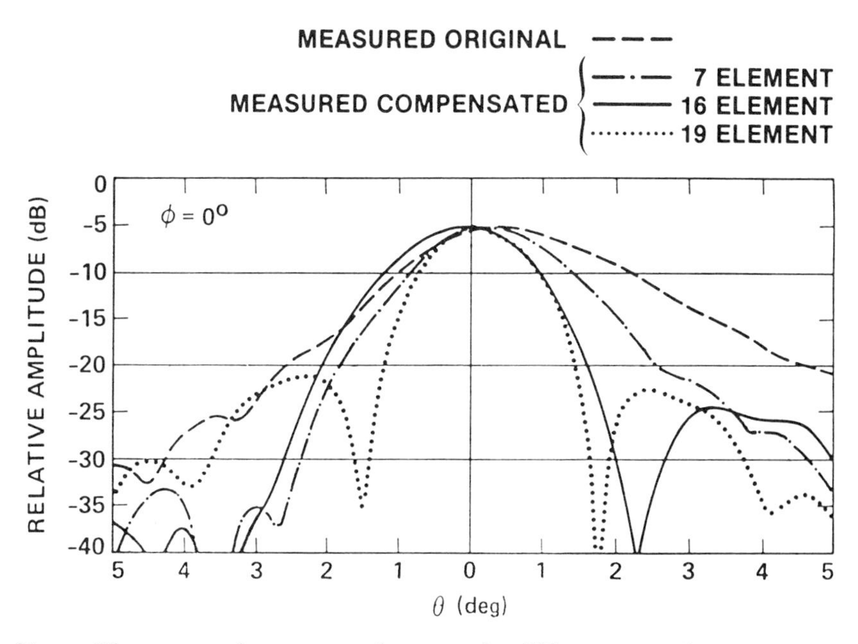

Fig. 21. The measured compensated patterns for different array element configurations. The dashed curve is the measured pattern with no compensation. The patterns are normalized with respect to their peak values.

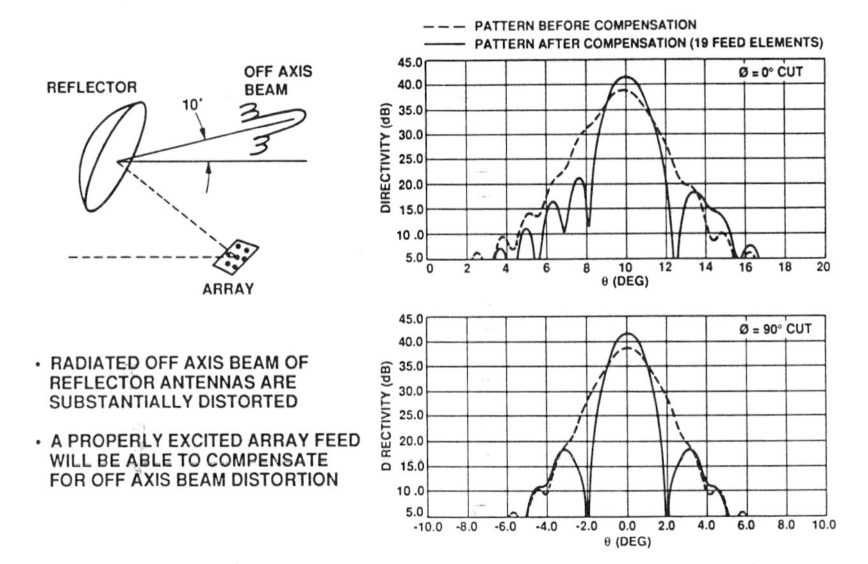

Fig. 22. An example to demonstrate the applicability of the array feed to overcome the wide-angle scan degradation characteristics of reflector antennas.

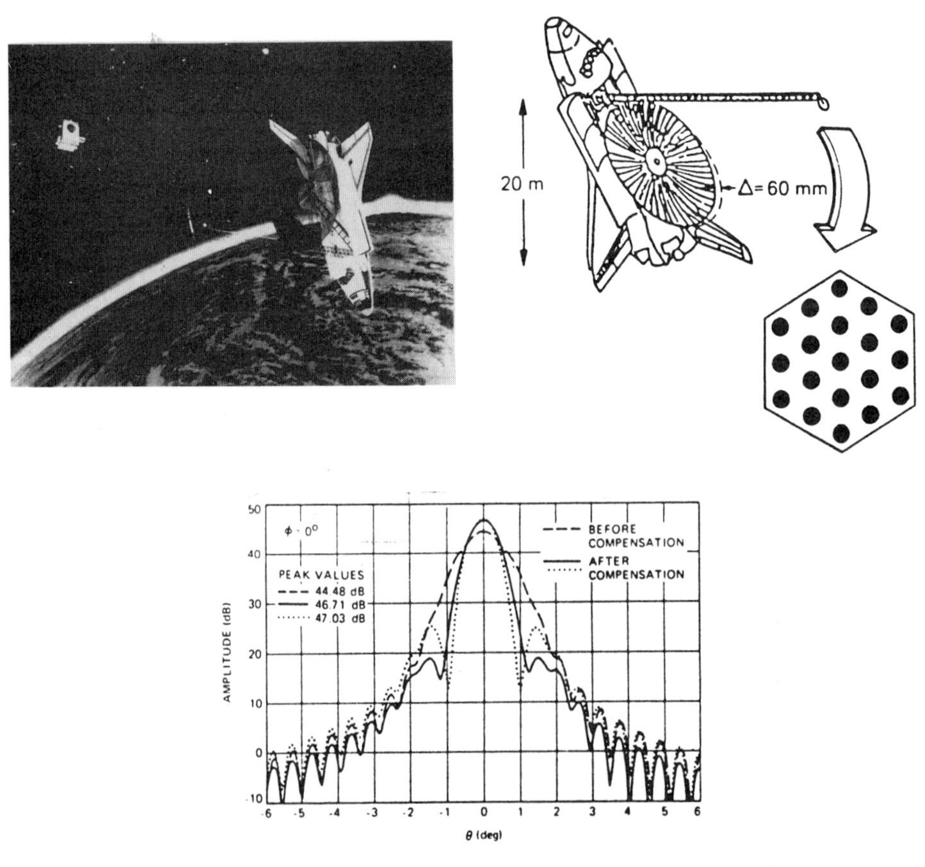

Fig. 23. An artist's rendition of the proposed large-antenna integrated RF/Control/Structure experiments in a configuration attached to a shuttle.

the feed coordinate system may be tilted and displaced with respect to the antenna coordinate system.

A local coordinate system at an arbitrary point on the subreflector surface is defined with respect to the surface normal direction. The Eulerian angle coordinate representation is used to ordinate the local coordinate with respect to the normal direction. This allows us to align the surface coordinate system with the local geometry of the FSS subreflector. With respect to this local coordinate system, the transmission/reflection coefficient matrix of the FSS is defined. This matrix may be given based on actual curved FSS analysis, local flat FSS analysis or measured data. In most cases, one uses the local flat FSS analysis to perform some numerical simulations.

Next, the physical optics current is defined, which incorporates the effects of the FSS by properly incorporating the reflection coefficient matrix of FSS. It is noticed at any location on the subreflector the Eulerian angles are used to relate the local FSS coordinate to the overall subreflector coordinate system. Once these modified physical optics currents are obtained, they are then used in the computation of the physical optics integral for the determination of the scattered field from the subreflector. This scattered field is then used in the computation of the induced current on the main reflector, which will finally result into the determination of the dual reflector radiated pattern and directivity. There are obviously different ways to numerically compute the physical optics integral as discussed earlier.

6. ACKNOWLEDGEMENT

Some of the work reported in this chapter was originally supported in part by the Jet Propulsion Laboratory, California Institute of Technology, under contract with the National Aeronautics and Space Administration. The discussion related to HPM was supported by a U.S. Army research grant. The author would like to thank his students at UCLA and his colleagues at JPL for their assistance in the development of the topics discussed in this chapter. The author would also like to thank Ms. C. Swiggett of UCLA for her expert typing of this manuscript.

References

[1] Y. Rahmat-Samii, "Large Antenna Measurement and Compensation Techniques," 11th ESTEC Antenna Workshop on Antenna Measurements, pp. 57-68, Gothenburg, Sweden, June 1988.

[2] Y. Rahmat-Samii, "Useful Coordinate Transformations for Antenna Applications," *IEEE Trans. Antennas Propagat.*, vol AP-27, pp. 571-574, July 1979.

[3] D. W. Duan and Y. Rahmat-Samii, "Antennas for High Power Microwave Applications--Part I," UCLA Report No. Eng-90-21, Sept. 1990.

[4] A. G. Roederer and Y. Rahmat-Samii, "Unfurable Satellite Antennas," *Annales des Telecommunications,* vol. 44, pp. 475-483, Nov. 1989.

[5] M. I. Astrakhan, "Reflection and Screening Properties of Plane Wire Grids," Radio Eng., (Moscow), vol. 23, pp. 76-83, 1968.

[6] Y. Rahmat-Samii and S. W. Lee, "Vector Diffraction Analysis of Reflector Antennas with Mesh Surfaces," *IEEE Trans. Antennas Propagat.,* vol. AP-33, pp. 76-90, Jan., 1985.

[7] J. R. Wait, Electromagnetic Scattering, P. L. E. Uslenghi (editor), New York: Academic Press, 1987.

[8] Y. Rahmat-Samii, W. Imbriale and V. Galindo, "Reflector Antennas with Mesh Surfaces: Contact or no Contact at Wire Intersections?," 1989 IEEE AP-S Int. Symp., San Jose, CA, June 26-30, 1989.

[9] Y. Rahmat-Samii, "Reflector Antennas,". Y. T. Lo and S-W Lee (editors), Antenna Handbook, chapter 15. New York: Van Nostrand Reinhold Company, 1988.

[10] W.V.T. Rusch, "The Current State of the Reflector Antenna Art," *Antennas Propagat.,* vol. AP-32, pp. 313-329, April 1984.

[11] Y. Rahmat-Samii, "Jacobi-Bessel Analysis of Reflector Antennas with Elliptical Apertures," *IEEE Trans. Antennas Propagat.,* vol. AP-35, pp. 1070-1073, Sept. 1987.

[12] C. S. Hung and R. Mittra, "Secondary Pattern and Focal Region Distribution of Reflector Antennas Under Wide-Angle Scanning," *IEEE Trans. Antennas Propagat.,* vol. AP-31, pp. 99-103, Sept. 1983.

[13] C. S. Kim and Y. Rahmat-Samii, "Analysis of Antennas with Elliptical Apertures using Fourier-Bessel Expansion: A Comparative Study," S. A. Kong (editor), Pro gress in Electromagnetics Research, vol. 4, New York: Elsevier Science Publishing, 1990.

[14] D. W. Duan and Y. Rahmat-Samii, "Comments on Numerical Evaluation of Radiation Integrals for Reflector Antenna Analysis Including a New Measure of Accuracy," *IEEE Trans. Antennas Propagat.,* to be published, May 1991.

[15] A. W. Love, (editor), Reflector Antennas, New York: IEEE Press, 1978.

[16] J. B. Keller,"Geometrical Theory of Diffraction," *J. Opt. Soc. of America,* vol. 52, pp. 116-130, Feb. 1962.

[17] D. S. Ahluwalia, R. M. Lewis and J. Boersma, "Uniform Asymptotic Theory of Diffraction by a Plane Screen," *SIAM J. Appl. Math.,* vol. 16, pp. 783-807, 1968.

[18] S.-W. Lee and G. A. Deschamps, "A Uniform Asymptotic Theory of Electromagnetic Diffraction by a Curved Wedge," *IEEE Trans. Antennas Propagat.,* vol. AP-24, pp. 25-34, Jan. 1976.

[19] R. G. Kouyoumjian and P. H. Pathak, "A Uniform Geometrical Theory of Diffraction for an Edge in a Perfectly Conducting Surface," *Proc. IEEE,* vol. 62, pp. 1448-1461, Nov, 1974.

[20] T. S. Chu, "An Imaging Beam Waveguide Feed," *IEEE Trans. Antennas Propagat.,* vol. AP-31, pp. 614-619, July 1983.

[21] P. Y. Ufimtsev, "Method of Edge Waves in the Physical Theory of Diffraction," *Izd-Vo Sovyetskoye Radio,* pages 1-243, 1962. Translation prepared by the U. S. Air Force Foreign Technology Division Wright-Patterson, AFB, Ohio (1971), available from NTIS, Springfield, VA 22161, AD733203.

[22] K. M. Mitzner, "Incremental Length Diffraction Coefficients," Technical Report AFAL-TR-73-296, Aircraft Division Northrop Corp., April 1974.

[23] A. Michaeli, "Elimination of Infinities in Equivalent Edge Currents, part I: Fringe Current Components". *IEEE Trans. Antennas Propagat.,* vol. AP-34, pp. 912-918, July 1986.

[24] M. Ando, ''Radiation Pattern Analysis of Reflector Antennas,'' *Electronics and Communications in Japan, Part 1,* vol. 68, pp. 93-102, 1985.

[25] S.-W. Lee, ''Uniform Asymptotic Theory of Electromagnetic Edge Diffraction: A Review,'' P. L. E. Uslenghi, (editor), Electromagnetic Scattering, pp. 67-119, New York: Academic Press, 1978.

[26] Y. Rahmat-Samii and R. Mittra, ''Spectral Analysis of High Frequency Diffraction of an Arbitrary Incident Field by a Half Plane - Comparison With Four Asymptotic Techniques,'' *Radio Science,* vol. 13, pp. 31-48, Jan.-Feb. 1978.

[27] Y. Rahmat-Samii, P. O. Iverson and D. W. Duan, ''GTD, PO, PTD and Gaussian Beam Diffraction Analysis Techniques Applied to Reflector Antennas,'' to be published in ACES Journal, in 1991.

[28] Y. Rahmat-Samii, ''Effects of Deterministic Surface Distortions on Reflector Antenna Performance,'' *Annales Des Telecommunications*, Vol. 40, No. 7-8, pp. 350-360, August 1985.

[29] Y. Rahmat-Samii, ''Array Feed for Reflector Surface Distortion Compensation: Concepts and Implementation,'' *IEEE AP-S magazine*, vol. 32, pp. 20-26, Aug. 1990.

[30] Y. Rahmat-Samii, ''Surface Diagnosis of Large Reflector Antennas Using Microwave Holographic Metrology: An Iterative Approach,'' *Radio Science*, vol. 19, pp. 1205-1217, Sept.-Oct. 1984.

[31] Y. Rahmat-Samii, ''Communicating From Space,'' *IEEE Potentials*, pp. 31-35, October 1988.

[32] C. F. Fraser, ''Photogrammetric Measurement of Antenna Reflectors,'' Antenna Measurement Techniques Association Meeting, pp. 374-379, Seattle, Washington, September 1987.

[33] Y. Rahmat-Samii, ''Improved Reflector Antenna Performance Using Optimized Feed Arrays,'' Proc. of the 1989 URSI International Symposium on Electromagnetic Theory, Stockholm, Sweden, August 14-17, 1989, pp. 559-561.

[34] Y. Rahmat-Samii and D. Chueng, ''Nonuniform Sampling Techniques for Antenna Applications,'' *IEEE Trans. Antennas Propagat.,* vol. 35, pp. 268-279, March 1987.

HIGH FREQUENCY EM SCATTERING BY NON-UNIFORM OPEN WAVEGUIDE CAVITIES CONTAINING AN INTERIOR OBSTACLE[1]

P.H. Pathak, P.H. Law and R.J. Burkholder

The Ohio State University ElectroScience Laboratory
1320 Kinnear Road, Columbus, Ohio 43212

1. Introduction

An analysis of the high frequency (HF) electromagnetic scattering by an open waveguide cavity containing a complex interior obstacle is of interest in scattered field and electromagnetic coupling predictions. A typical cavity-obstacle geometry is shown in Figure 1; this geometry is illuminated by an external source. The interior walls of the cavity are assumed to be perfectly conducting but may contain a thin material coating. The HF scattering by the configuration in Figure 1 is analyzed here primarily for the case when the observation point lies on the same side of the scattering geometry as the source, and for incidence and scattering angles within about 75° from the axis of the waveguide at the open front end. Outside of this region the scattering by external features of the cavity are generally more dominant than the scattering from the cavity interior. The method of analysis described here is based on a hybrid combination of HF ray/beam techniques, or

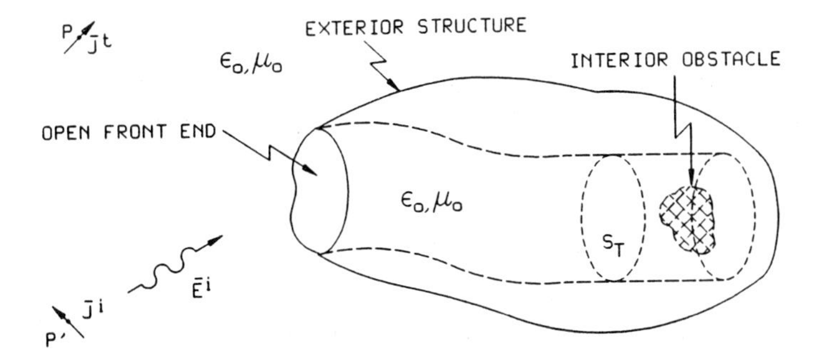

Figure 1. Open-ended cavity with an interior obstacle.

[1]The work reported here was supported in part by the General Electric Company, the Joint Services Electronics Program (Contract N00014-89-J-1007), and the NASA/Lewis Research Center (Grant NAG3-476).

modal techniques (mostly for cavities built up by connecting separable waveguide sections), with other methods. The HF techniques, or the modal techniques, are used to track the fields coupled into the waveguide cavity; whereas, other (numerical or experimental) methods are required in general to deal with the effect of the complex interior obstacle.

In the high frequency approximation, the total electric field scattered from the cavity-obstacle configuration in Figure 1 may be viewed as a superposition of separate first order contributions. One such contribution results from the field ($\bar{E}_d^s$) that is directly scattered by the edge of the aperture S_A defined at the open front end which is illuminated. There also exists a contribution to the external scattered fields that is produced by the outer structure which houses the open cavity. This contribution is not of direct interest here and it is thus ignored; however, it may be found via HF methods such as the GTD [1,2] and its uniform version [2,3], or the PTD [3,4]. The main contribution of interest here results from the field ($\bar{E}_c^s$) scattered from the interior cavity region containing the obstacle. Thus, the scattered field $\bar{E}^s(P)$ of interest here, which arrives at an external observation point P, is given by

$$\bar{E}^s(P) \quad = \quad \bar{E}_d^s(P) + \bar{E}_c^s(P). \tag{1}$$

It is noted that the effects of all higher order or multiple wave interactions are also ignored in (1) since the interaction of waves across the aperture S_A can be neglected as they are not significant for an electrically large aperture. Furthermore, the interior waves reflected from the obstacle are assumed to be much stronger than the interior waves reflected from the aperture S_A at the open end; this assumption is valid for an electrically large obstacle as well as for a large aperture S_A, and under this assumption the effects of multiple wave interactions between the aperture S_A and the interior obstacle can be ignored. Since one is interested in the HF regime and large obstacles in this work, and in incidence and aspect angles within about 75° of the waveguide axis, the above assumptions are easily satisfied. It is noted that $\bar{E}_c^s(P)$ is generally the dominant contributor to $\bar{E}^s(P)$ in (1).

A summary of the analytical formulation is described in the next section for estimating the contributions $\bar{E}_d^s$ and $\bar{E}_c^s$ to the scattered field $\bar{E}^s$ in (1). Included are some numerical results based on this analytical formulation which are compared with results based on other independent approaches where possible to establish confidence in the analysis. An $e^{j\omega t}$ time dependence is assumed and suppressed in the following.

2. Summary of Analytical Development

Let the composite cavity-obstacle geometry of Figure 1 be illuminated by an electric current source $\bar{J}^i$ at P′ in the exterior region which is assumed to be free space. $\bar{J}^i$ generates the incident fields ($\bar{E}^i, \bar{H}^i$) which exist in the absence of the cavity-obstacle configuration. The ($\bar{E}^i, \bar{H}^i$) in the presence of the cavity-obstacle combination produces the scattered fields ($\bar{E}^s, \bar{H}^s$) of interest at the point P. In the following, an analytical formulation is summarized initially for $\bar{E}_d^s$ in this section and then for $\bar{E}_c^s$; a superposition of these as in (1) provides the desired $\bar{E}^s$. Finally, methods for finding the field quantities used in the formulation for $\bar{E}_c^s$ are discussed. It was mentioned earlier that the contribution $\bar{E}_c^s$ to $\bar{E}^s$ is generally the strongest for a large cavity, while $\bar{E}_d^s$ is usually much weaker but not always negligible.

Expression for $\bar{E}_d^s$

The field $\bar{E}_d^s$ at P can be obtained via a GTD based equivalent current method (ECM) as [3]:

$$\bar{E}_d^s(P) \approx \frac{jkZ_o}{4\pi}\frac{e^{-jkr}}{r}\left\{ \oint_{\substack{\text{rim}\\ \text{(edge)}}} \left[\hat{r}\times\hat{r}\times\hat{t}' I_{eq}(\bar{r}') + Y_o\hat{r}\times\hat{t}' M_{eq}(\bar{r}')\right] e^{jk\hat{r}\cdot\bar{r}'} dt' \right\} \quad (2)$$

where the equivalent currents I_{eq} and M_{eq} in (2) are located on the rim (or edge) of the aperture S_A but with the cavity-obstacle configuration removed. The I_{eq} and M_{eq} depend on the GTD diffraction coefficients for the aperture edge. Thus, I_{eq} and M_{eq} asymptotically produce the same fields in external free space as those diffracted by the edge of the aperture S_A. The $\hat{t}'$ is a unit vector tangent to the edge, $\bar{r}'$ is the position vector from the coordinate origin to the point on the rim at t', and $\bar{r}$ is the vector from the origin to the observation point P. Z_o and Y_o are the free space impedance and admittance, respectively. One may find expressions for (I_{eq}, M_{eq}) in [5] for a perfectly conducting edge; they are not listed here due to space limitations.

Formulation for evaluating $\bar{E}_c^s$

Consider a test electric current point source $\bar{J}^t$ which is placed at the observation point at P and which generates the fields $(\bar{E}_t^{ig}, \bar{H}_t^{ig})$ inside the waveguide cavity but in the *absence* of the obstacle. Also, let the original source $\bar{J}^i$ at P' generate the fields $(\bar{E}^{ig}, \bar{H}^{ig})$ inside the waveguide cavity but with the obstacle *absent*. Next, consider a planar cross-sectional surface S_T within the waveguide cavity; it is chosen conveniently but close to the interior obstacle as shown in Figure 1. If $(\bar{E}^{sg}, \bar{H}^{sg})$ are the interior fields scattered by the obstacle when it is illuminated within the interior by $(\bar{E}^{ig}, \bar{H}^{ig})$ due to the original source $\bar{J}^i$, then the desired obstacle scattered field $\bar{E}_c^s(P)$ at a point P in the exterior region of interest can be expressed in terms of the fields $(\bar{E}_t^{ig}, \bar{H}_t^{ig})$ and $(\bar{E}^{sg}, \bar{H}^{sg})$ evaluated on S_T by:

$$\bar{E}_c^s(P)\cdot\bar{J}^t \approx \iint_{S_T} \left(\bar{E}^{sg}\times\bar{H}_t^{ig} - \bar{E}_t^{ig}\times\bar{H}^{sg}\right)\cdot d\bar{S}. \quad (3)$$

The fields $(\bar{E}^{ig}, \bar{H}^{ig})$ and $(\bar{E}_t^{ig}, \bar{H}_t^{ig})$ can be evaluated within the cavity (in the absence of the obstacle) using, for example, ray or beam techniques if the cavity has an arbitrary shape, or by modal techniques if the cavity is built up of separable waveguide sections and the modes in the different sections can be found easily. It then remains to find the fields $(\bar{E}^{sg}, \bar{H}^{sg})$ at S_T which are scattered by the interior obstacle when illuminated by $(\bar{E}^{ig}, \bar{H}^{ig})$.

Evaluation of $(\bar{E}^{ig}, \bar{H}^{ig})$ and $(\bar{E}_t^{ig}, \bar{H}_t^{ig})$

If the cavity interior is made up of piecewise separable waveguide sections for which the waveguide modal fields can be found easily, then modal or hybrid-modal methods can be used to find the coupling of the externally applied fields into the cavity and the subsequent propagation of these fields from the open end to the

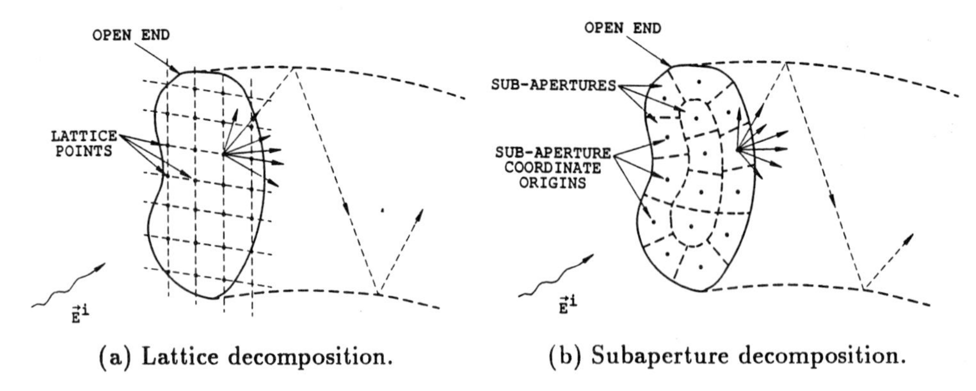

(a) Lattice decomposition.

(b) Subaperture decomposition.

Figure 2. Ray or beam launching into the cavity from an array of points in the open end.

plane S_T [5-9]. For non-uniform cavities for which modes cannot be defined in the conventional sense, numerical approaches such an integral equation or finite element method could be used. However, the cavities under consideration are electrically large so that numerical methods become poorly convergent and hence are generally unsuitable. Furthermore, such numerical approaches do not provide much physical insight into the important scattering mechanisms associated with cavities. Therefore, it is suggested that ray or beam tracking methods be employed for arbitrarily shaped non-uniform cavities, possibly with material coated walls.

In ray and beam tracking methods, the external fields incident on the open end of the cavity are expanded into a set of rays or beams which are then launched into the cavity and tracked within via multiple reflections from the interior cavity walls using prescribed laws of ray or beam propagation and reflection. The rays/beams bounce around inside the cavity until they reach either the plane S_T or they "turnaround" and exit back out the open end S_A without ever reaching S_T (this commonly occurs if the waveguide cavity tapers in cross-section away from the open end). The majority of rays/beams will reach the plane S_T where they collectively provide $(\bar{E}^{ig}, \bar{H}^{ig})$ and $(\bar{E}_t^{ig}, \bar{H}_t^{ig})$ via superposition. The effects of "turnaround" rays/beams must be added separately to $\bar{E}_c^s(P)$ in addition to the expression given in (3). This contribution is easily found by performing an aperture integration over the equivalent sources defined by the intersection of each ray/beam with the plane of the open end S_A as the ray/beam exits the cavity [10].

The most basic ray tracking approach is the geometrical optics (GO) ray shooting method [5,6], sometimes referred to as the shooting and bouncing ray (SBR) method [10,11]. In this method, the portion of the incident plane wave intercepted by the aperture at the open end of the cavity is broken up into a dense grid of parallel ray tubes which are tracked inside the cavity using the laws of GO. The effects of a thin material lining the otherwise perfectly conducting walls are easily included by using Fresnel reflection coefficients at each ray tube reflection point. This method has the advantage of being conceptually simple; however, the fields diffracted into the cavity by the rim edges at the open end are neglected because only the incident GO field entering the cavity is tracked. Furthermore, a new set of ray tubes must be tracked for each new plane wave incidence angle.

A more rigorous approach to ray or beam tracking within non-uniform cavities is to use a discrete phase-space like expansion of the fields coupled into the cavity in terms of a set of high frequency ray or Gaussian beam type basis elements.

These basis elements, which are launched from an array of points in the aperture, evolve within the cavity via multiple reflections at the interior cavity walls as they propagate toward (or away from, in the case of "turnaround") the obstacle region, as illustrated in Figure 2. An important property of this phase-space like expansion of the interior cavity fields is that the initial ray or beam launching directions remain independent of the excitation or the incident wave direction. Thus, the ray/beam paths do not change with excitation and hence need to be tracked only once through the cavity via the rules of ray/beam optics. Only the initial launching amplitudes of the rays/beam need to be changed according to the excitation; these initial ray/beam amplitudes and directions are chosen via certain rules based on physical considerations. Furthermore, the discrete phase-space like interior field expansion also intrinsically includes, to within the Kirchhoff approximation, the effects of fields coupled into the cavity via diffraction of the incident field by the edges at the open end. The Kirchhoff approximation is assumed to be valid here as the aperture is electrically large and the incident wave directions are chosen to exclude very wide angles with respect to the normal to the aperture.

The rays/beams may be launched from a uniformly spaced array of lattice points, as in Figure 2(a), or from an array of phase centers of subapertures, as in Figure 2(b). In the lattice method, the incident fields in the plane of the aperture are expanded into a set of basis functions which can be conveniently tracked from the lattice points throughout the cavity. The Gabor expansion can be used to expand the fields of an aperture into a set of linearly phased Gaussian functions centered around each of the lattice points [12]. These functions give rise to or launch Gaussian beams (i.e., bi-laterally symmetric beams with Gaussian amplitude distributions) which can then be tracked, approximately like real rays along their axis, via the laws of beam optics [5,6] or more rigorously as complex rays emanating from a source located in complex space because the paraxial fields of such a source are equivalent to a Gaussian beam [13]. Since the the Gabor expansion gives rise to a set of Gaussian beams which are non-uniformly shifted in angle and which have non-identical beam widths, some of these beams may not be well focussed enough to be tracked like real rays using beam optics, because they diverge and become distorted or non-Gaussian after a small number of reflections from curved surfaces (the laws of beam optics assume that a beam remains Gaussian after multiple reflections). On the other hand, it is very cumbersome to track complex rays through multiple reflections from arbitrarily shaped cavity walls, hence, this method although feasable in principle has not yet been applied to the problem being considered here. Instead, a Gaussian beam shooting method similar to the Gabor-complex ray method has been developed wherein sets of identical Gaussian beams with constant interbeam angular spacing are launched radially from the phase centers of subaperture domains [5,6] as in Figure 2(b). In this method, the original aperture at the open end of the cavity is subdivided into a small number of conveniently chosen subapertures, and the incident fields in the aperture are replaced with equivalent currents or sources based on the Kirchhoff approximation. The Gaussian beams launched from with each subaperture are then appropriately weighted according to the far field radiation pattern of the equivalent currents existing over that subaperture with the cavity walls absent. The subaperture sizes and the initial widths of the Gaussian beams are chosen such that the beams remain relatively well focussed after undergoing multiple reflections from the interior cavity walls; this allows the beams to be tracked like real rays using the well known laws of beam optics. Nevertheless, this method is limited to cavities which are less than approximately four times as long as they are wide, due to beam distortion effects [5,6].

To overcome the above limitation of the Gaussian beam shooting method due to distortion effects, a generalized ray expansion (GRE) method has been developed which launches GO ray tubes rather than beams from the phase centers of subapertures (as in Figure 2(b)) [6,14]. Unlike pure Gaussian beams which are wide, ray tubes can be made arbitrarily narrow and overcome the problems associated with beam distortion upon reflection. In the GRE method, a dense cone-shaped grid of GO ray tubes is launched radially from the phase centers of each subaperture, initially with a spherical wave spread factor. Each ray tube is weighted according to the far field pattern of its subaperture, similar to the Gaussian beam method. These ray tubes are then tracked via the laws of GO in exactly the same manner as ray tubes of the SBR method; the only difference is in the way the ray tubes are launched initialy. However, unlike the SBR method, but like the Gaussian beam method, the GRE method intrinsically includes the fields diffracted into the cavity by the edges at the open end (to within the Kirchhoff approximation) and the ray tubes need to be tracked only once independent of the excitation. It is noted that in the ray based SBR and GRE methods, many more ray tubes would generally need to be tracked than Gaussian beams for a given cavity geometry. Furthermore, many more ray tubes would generally need to be tracked in the GRE method than in the SBR method *for a single incidence angle*; however, far fewer ray tubes would generally need to be tracked in the GRE than the total number of ray tubes which need to be tracked in the SBR method for a large number incidence angles.

Figure 3 shows plots of backscatter vs. incidence angle (with the $1/\sqrt{\rho}$ dependence suppressed) for a perfectly conducting 2-D S-shaped cavity with a simple planar termination, calculated using the hybrid modal method (reference solution), the Gaussian beam shooting method and the GRE method. Figure 4 shows the echo area patterns (backscatter, in decibels relative to a square wavelength) of a 3-D S-shaped cavity which transitions from rectangular at the open end to elliptical at the planar termination, calculated using the GRE and SBR methods[2] (no reference solution available). In Figure 4, the 6 subaperture GRE result is expected to be more accurate than the 4 subaperture GRE result, and both the GRE results are expected to be more accurate than the SBR result because the GRE method includes the fields diffracted into the cavity by the rim edges at the open end.

Evaluation of $(\bar{E}^{sg}, \bar{H}^{sg})$

It may be possible that the interior reflection from some types of obstacles can be analyzed using ray/beam methods; in this case, the ray/beam representation for the fields $(\bar{E}^{ig}, \bar{H}^{ig})$ which enter the cavity (after being excited by the original source $\bar{J}^i$) continue beyond S_T into the obstacle region to then reflect back from the obstacle to S_T again as the ray/beam fields $(\bar{E}^{sg}, \bar{H}^{sg})$. If this is not the case, then it may be possible to use numerical or experimental approaches to determine $(\bar{E}^{sg}, \bar{H}^{sg})$. While there appear to be a few different possibilities for determining

[2]The ray tracing subroutines used to compute both the GRE and SBR results using a super-ellipse model for the cavity were provided by S.W. Lee of the University of Illinois, Champaign.

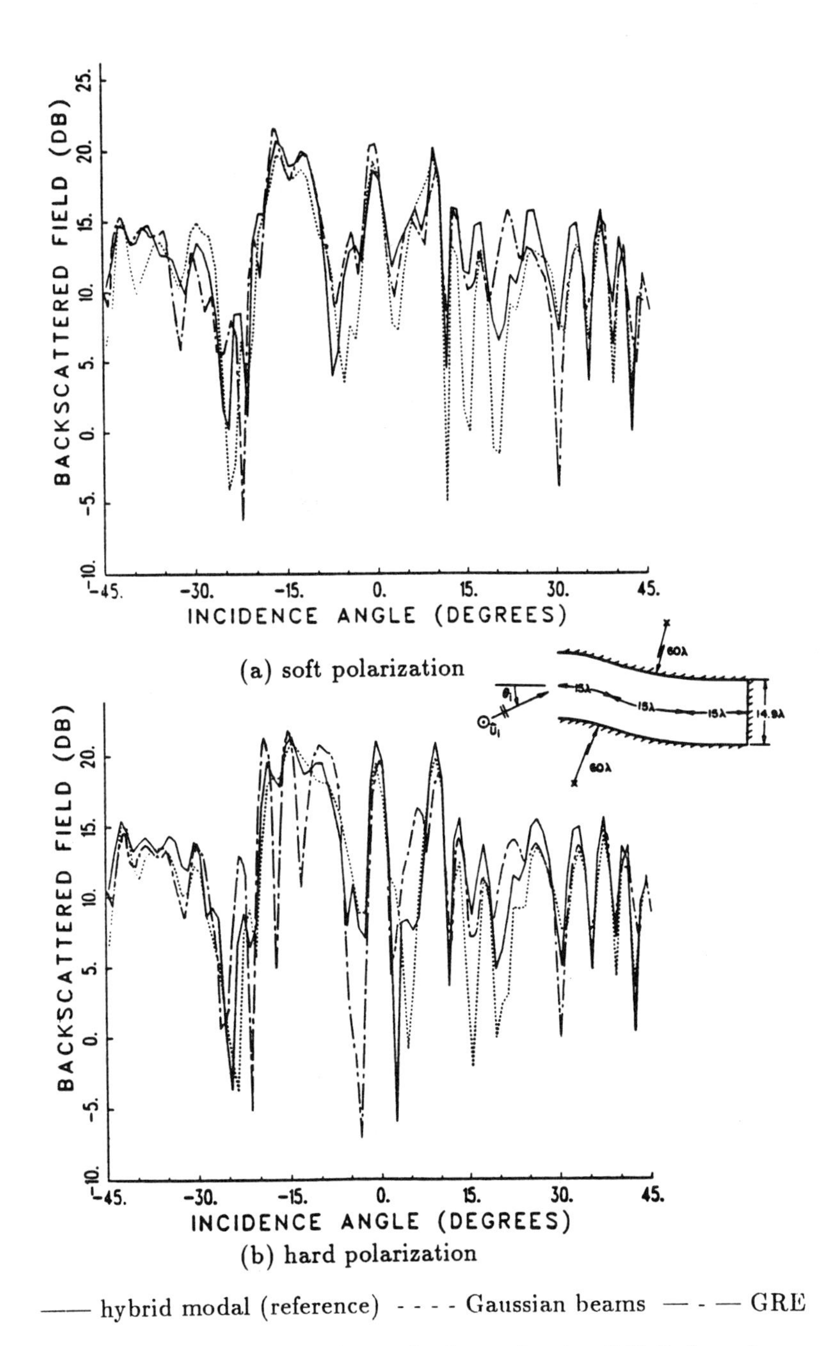

Figure 3. Backscatter patterns of a perfectly conducting 2-D S-shaped waveguide cavity with a planar termination The Gaussian beam and GRE calculations each used 7 uniform subapertures.

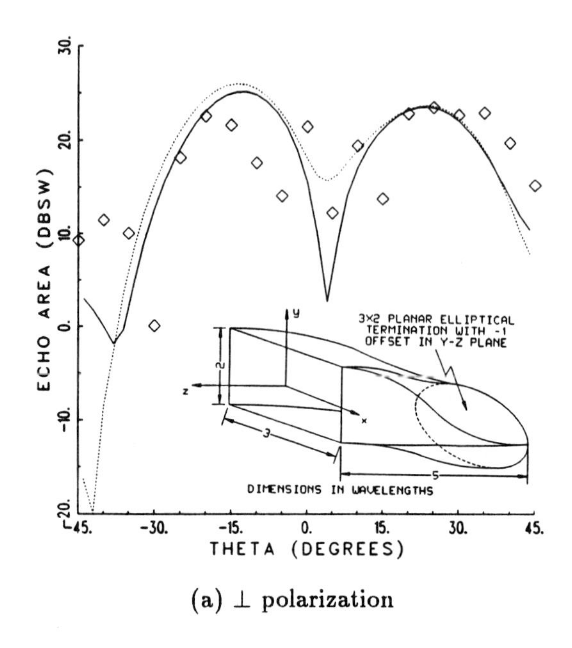

(a) ⊥ polarization

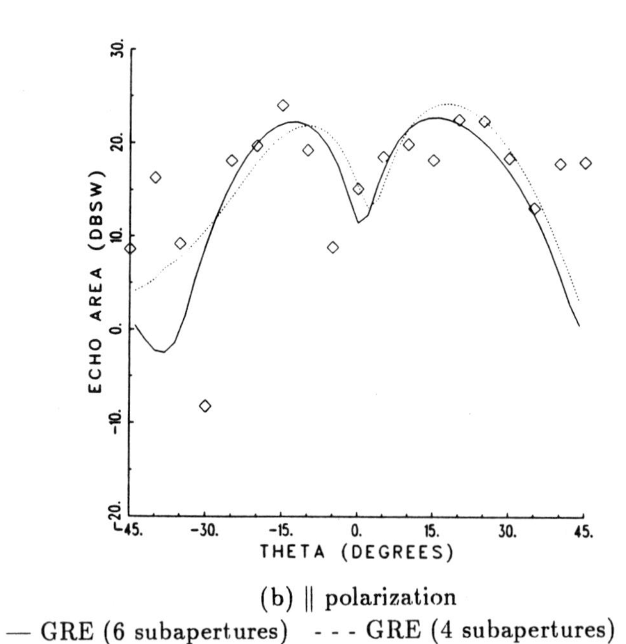

(b) ∥ polarization

— GRE (6 subapertures) - - - GRE (4 subapertures) ◇ SBR

Figure 4. Electromagnetic echo area (backscatter) patterns in the y-z plane of an S-shaped rectangular-to-elliptical cavity with a planar termination. The GRE calculations used uniform rectangular subapertures.

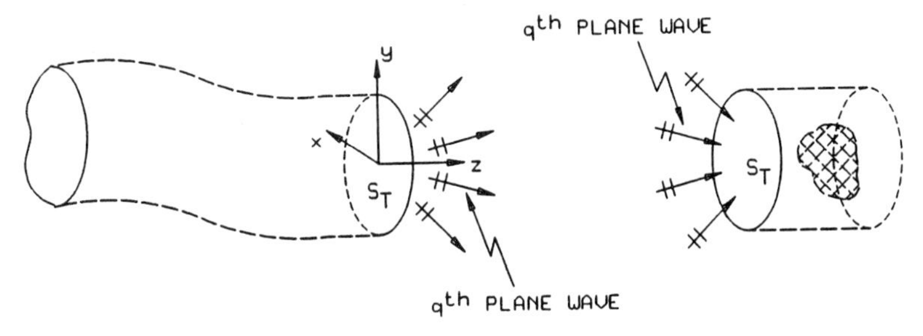

(a) Open waveguide section. (b) obstacle-termination section.

Figure 5. Original cavity geometry decomposed into two separate problems.

$(\bar{E}^{sg}, \bar{H}^{sg})$ at S_T using analytical/numerical/experimental procedures, only one such possible approach is discussed below.

In the first step, one begins by breaking up the original geometry of Figure 1 into two parts as shown in Figure 5. The cavity-obstacle region beyond the plane S_T constitutes one part as in Figure 5(b) and the remaining part is formed by the generally longer section of the open waveguide cavity between S_A and S_T as in Figure 5(a). In the second step, the field $\bar{E}^r$ radiated from S_T when $\bar{E}^{ig}$ is incident there, as shown in Figure 5(a), is expressed in its plane wave spectral form as

$$\bar{E}^r(x,y,z) = \int_{-\infty}^{\infty}\int_{-\infty}^{\infty} \bar{\mathcal{E}}^{ig}(k_x, k_y)e^{-j(k_x x + k_y y + k_z z)}dk_x dk_y \tag{4}$$

where $k_z = \sqrt{k^2 - k_x^2 - k_y^2}$. The x, y, z coordinate system used in (4) is shown in Figure 5(a). This continuous plane wave spectrum is approximated by a sufficient number of discrete *propagating* plane waves:

$$\bar{E}^r \approx \sum_{q=1}^{Q} \bar{\mathcal{E}}^{ig}(k_{xq}, k_{yq})e^{-j(k_{xq}x + k_{yq}y + k_{zq}z)}\Delta k_{xq}\Delta k_{yq} \equiv \sum_{q=1}^{Q} \bar{E}_q^r \tag{5}$$

where $k_{zq} = \sqrt{k^2 - k_{xq}^2 - k_{yq}^2}$ and $k^2 > k_{xq}^2 + k_{yq}^2$. Δk_{xq} and Δk_{yq} are chosen small enough so that (5) adequately reproduces $\bar{E}^{ig}$ at S_T within the Kirchhoff approximation. Equation (5) is simply a superposition of Q plane waves, each propagating in the direction $\hat{k}_q = \hat{x}k_{xq} + \hat{y}k_{yq} + \hat{z}k_{zq}$, so $\bar{H}^{ig}$ at S_T has the same form as (5) with $\bar{\mathcal{E}}^{ig}(k_{xq}, k_{yq})$ replaced by $\bar{\mathcal{H}}^{ig}$ where

$$\bar{\mathcal{H}}^{ig}(k_{xq}, k_{yq}) = \hat{k}_q \times \bar{\mathcal{E}}^{ig}(k_{xq}, k_{yq})/Z_o.$$

In the third step, each of these Q plane waves is allowed to illuminate the obstacle region from the directions $\hat{k}_q$, as shown in Figure 5(b). The problem of finding $(\bar{E}^{sg}, \bar{H}^{sg})$ from $(\bar{E}^{ig}, \bar{H}^{ig})$ at S_T is now reduced to one of finding $(\bar{E}_q^{sg}, \bar{H}_q^{sg})$ due to the qth plane wave of (5) illuminating the obstacle region of Figure 5(b). Thus,

$$(\bar{E}^{sg}, \bar{H}^{sg}) = \sum_{q=1}^{Q}(\bar{E}_q^{sg}, \bar{H}_q^{sg}). \tag{6}$$

The $(\bar{E}_q^{sg}, \bar{H}_q^{sg})$ can be found numerically via an integral equation or finite element method, for example, or analytically if the obstacle region is characterized by some

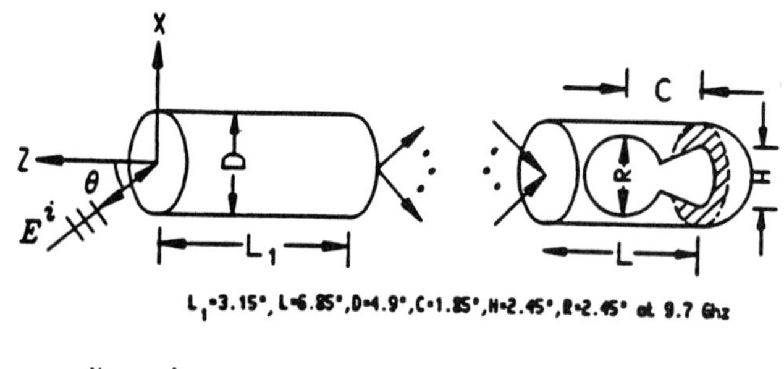

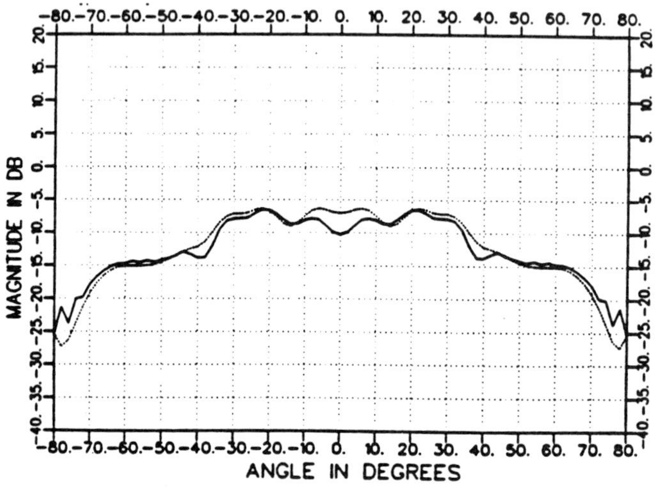

(a) ⊥ polarization

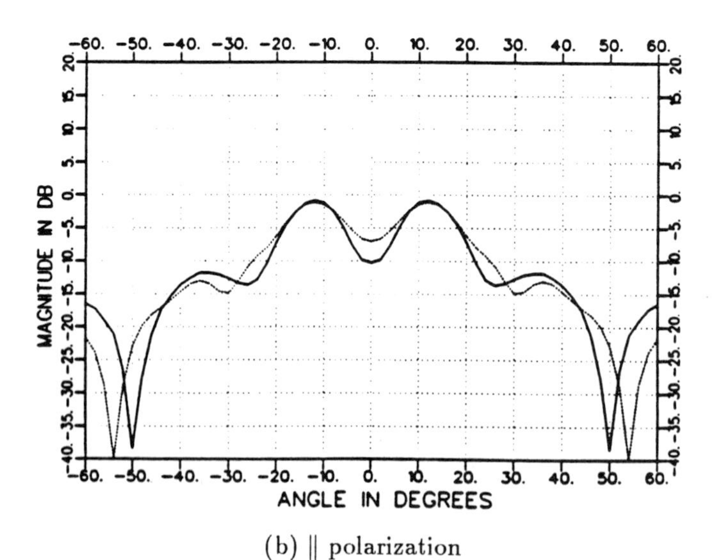

(b) || polarization

Figure 6. Backscatter patterns of a perfectly conducting circular cavity with a cone-sphere obstacle mounted on a planar termination.

sufficiently simple canonical geometry. In principle, $(\bar{E}_q^{sg}, \bar{H}_q^{sg})$ could also be found experimentally via bistatic scattering measurments. In this scenario, a plane wave incident from the qth direction with the same polarization as $\bar{E}_q^r$ (with $\bar{E}_q^r$ = value of $\bar{E}^r$ in q^{th} direction) would illuminate the geometry of Figure 5(b) and the far field bistatic scattering would be measured in the half-space to the left of S_T in Figure 5(b). Next, this far field data would be weighted by the amplitude of $\bar{\mathcal{E}}^{ig}(k_{xq}, k_{yq})$ and transformed (via a fast Fourier transform) to give the fields $(\bar{E}_q^{sg}, \bar{H}_q^{sg})$ back at the plane S_T. However, a facility for obtaining such extensive bistatic scattering measurements is not readily available at the present time, so numerical and analytical models are currently being investigated. Finally, (6) is incorporated into (3) to yield the desired $\bar{E}_c^s$.

It is noted that the main advantage of the decomposition shown in Figure 5 is that the two sub-problems can be analyzed independently and combined via (3), (5) and (6). Therefore, different front end waveguides like the one in Figure 5(a) can be combined with different obstacle-termination geometries like the one in Figure 5(b), as long as they share a common S_T. Figure 6 shows the backscatter patterns of a perfectly conducting circular waveguide cavity with a cone-sphere obstacle mounted on a planar termination. The solid line in these plots is a moment method calculation for the complete geometry of length $L+L_1$ and the dashed line is calculated using the decomposition method described above. In the latter case, the short obstacle-termination section of length L was analyzed using the method of moments for each of the plane waves in the expansion of (5), and the front section of length L_1 was analyzed using the hybrid modal method to find the coupling of the incident plane wave through the open end and the propagation down the guide to S_T. For more general shapes, the ray/beam approach is useful for tracking fields through L_1.

References

[1] J.B. Keller, Geometrical theory of diffraction, *Journal Opt. Soc. Am.* 52:116-130 (1962).

[2] R.G. Kouyoumjian and P.H. Pathak, "A UTD Approach to EM Scattering and Radiation," Chapter 3 in Low and High Frequency Asymptotics, V.K. Varadan and V.V. Varadan, eds., North-Holland, Elsevier Science Publishers, B.V. Netherlands (1986).

[3] P.H. Pathak, "Techniques for high frequency problems," in Antenna Handbook – Theory, Applications, and Design, Van Nostrand Reinhold, New York (1988).

[4] P. Ya. Ufimtsev, Method of edge waves in the physical theory of diffraction, (from the Russian, Method krayevykh voin v fizicheskoy teoril difraktsii), Izd-Vo Sov. Radio 1-243 (1962). Translation prepared by the U.S. Air Force Foreign Technology Division, Wright-Patterson Air Force Base, OH; released for public distribution September 7, 1971.

[5] P.H. Pathak and R.J. Burkholder, "Modal, Ray and Beam Techniques for Analyzing the EM Scattering by Open-ended Waveguide Cavities, *IEEE Trans. Antennas Prop.* Vol. AP-37, No. 5, pp. 635-647, May 1989.

[6] R.J. Burkholder, "High-Frequency Asymptotic Methods for Analyzing the EM Scattering by Open-Ended Waveguide Cavities," Ph.D. Dissertation, The Ohio State University, Columbus, June 1989.

[7] A. Altintas, P.H. Pathak, and M.C. Liang, "A Selective Modal Scheme for the Analysis of EM Coupling into or Radiation from Large Open-Ended Waveguides," *IEEE Trans. Antennas Prop.* Vol. AP-36, No. 1, pp. 84-96, January 1988.

[8] H. Shirai and L.B. Felsen, "Rays, Modes and Beams for Plane Wave Coupling into a Wide Open-Ended Parallel-Plane Waveguide," *Wave Motion,* 9, pp. 301-317, 1987.

[9] H. Shirai and L.B. Felsen, "Rays and Modes for Plane Wave Coupling into a Large Open-Ended Circular Waveguide," *Wave Motion,* 9, pp. 461-482, 1987.

[10] S.W. Lee, H. Ling and R.-C. Chou, "Ray-Tube Integration in Shooting and Bouncing Ray Method," *Microwave and Optical Tech. Letters,* Vol. 1, No. 8, October 1988.

[11] H. Ling, R.-C. Chou, and S.W. Lee, "Shooting and Bouncing Rays: Calculating the RCS of an Arbitray Cavity," *IEEE Trans. Antennas and Prop.* Vol. AP-37, No. 2, pp. 194-205, February 1989.

[12] J. Maciel and L.B. Felsen, "Systematic Study of Fields Due to Extended Apertures by Gaussian Beam Discretization," *IEEE Trans. Antennas and Prop.* Vol. AP-37, No. 7, pp. 884-892, July 1989.

[13] J. Maciel and L.B. Felsen, "Gaussian Beam Analysis of Propagation from an Extended Plane Aperture Distribution Through Dielectric Layers, Part I – Plane Layer, Part II – Circular Cylindrical Layer," IEEE Trans. Antennas and Prop. Vol. AP-38, No. 10, pp. 1607-1624, October 1990.

[14] P.H. Pathak and R.J. Burkholder, "High Frequency EM Scattering by Open-Ended Waveguide Cavities," accepted for publication in *J. Radio Science,* Jan.-Feb. 1991.

WAVES AND BEAMS IN BIOLOGICAL MEDIA

J. Bach Andersen

Aalborg University
Denmark

1. INTRODUCTION

There is a need for understanding and utilizing propagation of electromagnetic waves in the human body. This need stems from various applications of a diagnostic or therapeutic nature. One application is in hyperthermia treatment for cancer, where electromagnetic energy is used as a heat source with the goal of heating tumor tissue without overheating healthy tissue. Another application is in dosimetry, where the goal is to prevent inadvertent heating or illumination by electromagnetic sources. Still another application is in high frequency imaging of internal organs.

The basic problems derive from the fact that the human body is a very inhomogeneous structure consisting of different tissues with high losses. In terms of modelling this seemingly very complicated situation has led to the use of strictly numerical techniques, such as finite-elements (FE), finite-difference-time-domain (FDTD) and frequency domain integral equation techniques (IE). For an overview of these methods, the reader is referred to Paulsen (1990).

In interpreting physical phenomena and as a guideline for further investigations simple analytical models are indispensable. Especially in lossy media the well-known Gaussian beam with complex source points has proven of great value (1,2,3). In the applications the antennas or applicators are placed near the tissue in question, so we need models that can carry from near-field to far-field and still be simple. The reason for the success of the beam model in lossy media is the fact that discontinuities in aperture distributions are smeared out by the losses a short distance from the aperture. Thus waveguide or horn apertures or other sources look smooth, and a Gaussian model is often an excellent fit both in amplitude and phase all the way from near to far. The difficulties around using Gaussian beams in free space disappear with the high losses of biological media.

Directions in Electromagnetic Wave Modeling
Edited by H.L. Bertoni and L.B. Felsen, Plenum Press, New York, 1991

2. BASIC THEORY AND APERTURE MODELLING

The basic scalar theory of Gaussian beams in lossless media is well-known (Kogelnik and Li (1966), Deschamps (1971)). The easiest way to explain the beam equation is to use a complex source point in the scalar Green's function,

$$E = \exp(-jkr) / r \tag{1}$$

assuming rotational symmetry,

$$r^2 = (z-z_0)^2 + (\rho)^2, \tag{2}$$

assuming the paraxial approximation , $\rho < |z-z_o|$,

$$r \approx z - z_0 + \frac{\rho^2}{2(z-z_0)} \tag{3}$$

leading to

$$E \approx \frac{1}{z-z_0} e^{-j k z -j k \rho^2/2 (z-z_0)} \tag{4}$$

Since k is complex, we have a Gaussian distribution in ρ even for a real source point, but in general a complex source point will allow us sufficient complexity to model a variety of different phenomena. The approximation in (3) should be kept in mind, however, so in some cases it may be necessary to check the validity by normal aperture integrations. An alternative explanation of the model comes from the wave equation, assuming that the slowly varying part ψ in

$$E = \exp(-jkz)\ \psi(x,y,z) \tag{5}$$

satisfies

$$\left(\frac{\delta^2}{\delta x^2} + \frac{\delta^2}{\delta y^2} + k^2\right) \psi(x,y,z) = 0 \tag{6}$$

i.e the variation in the axial variation of ψ may be neglected compared with the variations in the transverse direction and relative to the wavelength.

The three-dimensional solution to eqs. 5 and 6 is the general astigmatic beam equation with an elliptical cross-section

$$E = \frac{\exp(-jkz)}{\sqrt{(z-z_1)(z-z_2)}} \exp\left(-j \frac{kx^2}{2(z-z_1)} - j \frac{ky^2}{2(z-z_2)}\right). \tag{7}$$

Thus, in the general case the beam may be characterized by two complex numbers, z_1 and z_2 , and the field may be determined everywhere in three-dimensional space. The simplicity is of course only justified, if the model is a good model of the real world. The justification has been done by comparing with experimental result (Lumori et al, 1990).

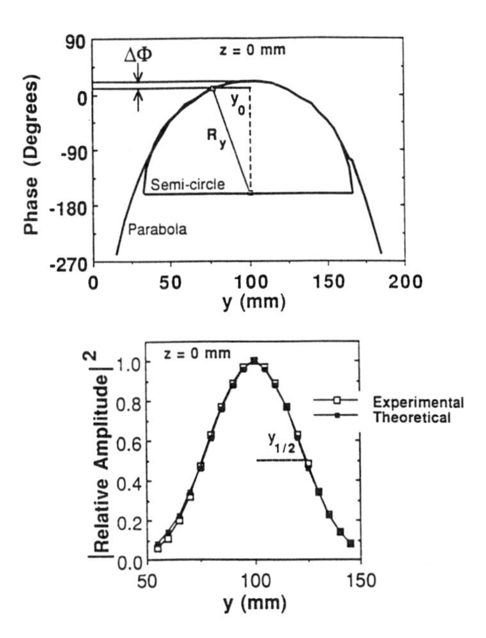

Fig.1 Experimental results for a horn antenna in a saline solution, compared with a fitted Gaussian beam.

Fig. 1 shows some experimental results for one polarization in the y-direction for a horn antenna a few mm in front of the aperture in a saline solution equivalent to muscle tissue (Lumori et al (1990)). A Gaussian function is fitted to the experimental results in such a way that the phase curvature and the half-power points agree. From this experiment the complex number z_2 may be found. Similarly by making phase and amplitude measurements in the x-direction, the complex source point z_1 is obtained. Eq. 7 may then be used to find the fields everywhere, all the way from the radiating aperture to the far field. The experimental agreement with the model for other values of z is excellent. The model has also been used to study phased array systems, where the fields from each antenna are modelled by Gaussian beams.

The Gaussian shape of the aperture distribution has another interesting feature. It may be shown that the aperture distribution which gives maximum gain at a point near the aperture, is a focused, tapered distribution (Andersen, J.B. (1985)). The tapering will have a Gaussian-like shape, since it can be conceived as the conjugate of the fields at the aperture from a source at the point in question. The losses in the medium will automatically give a Gaussian distribution. That the case is a maximum gain case can be understood from the fact, that for a large constant aperture the fields near the edges will contribute relatively little to the point gain, whereas they will contribute fully to the total losses.

3. INTERPRETATION OF THE MODEL

The simple rotationally symmetric case (eq. 4) may be used to study how the normal beam parameters, such as width of waist, focal point, point of maximum power density etc. depend on the complex source point parameters.

The power attenuation along the axis is given by

$$P(z) = e^{-2\alpha z} \frac{a^2 + b^2}{(z-a)^2 + b^2} \tag{8}$$

where a and b are the real and imaginary part of the complex source point coordinates, respectively. From eq. 8 we can derive that there may be a local maximum at depth for a focused beam, provided

$$|b| < \frac{1}{2\alpha} \tag{9}$$

i.e the imaginary space coordinate should be less than half the penetration depth of a plane wave.

In Fig. 2 is shown a plot of eq. 8 for various cases. The focused beam (a) has an optical focus (phase convergence) at z=4, the maximum is however at z=3, so we have a focal shift due to the losses. Note that the decay rate after the focus is much higher than for a plane wave (c), very similar to the case of the defocused case (b), or rather the case of a diverging wavefront at the aperture, corresponding to a focus behind the aperture. It is a common experience in hyperthermia applications that a finite aperture gives a higher attenuation than an infinite aperture (a plane wave), this feature is well explained by the Gaussian beam.

Since both the wavenumber and coordinates are complex it is difficult to overview the relationships. To this purpose a graphical plot in the complex source plane is helpful. The lateral variation of the beam in the aperture (z=0) is given by

$$k\,\rho^2/(2\,z_0) = (u+jv)\,\rho^2/2 \tag{10}$$

so u=constant means constant amplitude (or constant beamwidth) and v=constant means constant phase. In the lossless case (k real) it is easy to see, that

a > 0	corresponds to	converging beam
a < 0		diverging beam
b < 0		transverse decay
b > 0		transverse growth

so the four quadrants can be designated accordingly (Fig. 3). In the lossy case these axes are turned an angle to the right equal to the phase angle, ϕ of the wavenumber,

$$k = |k| e^{-j\phi} . \tag{11}$$

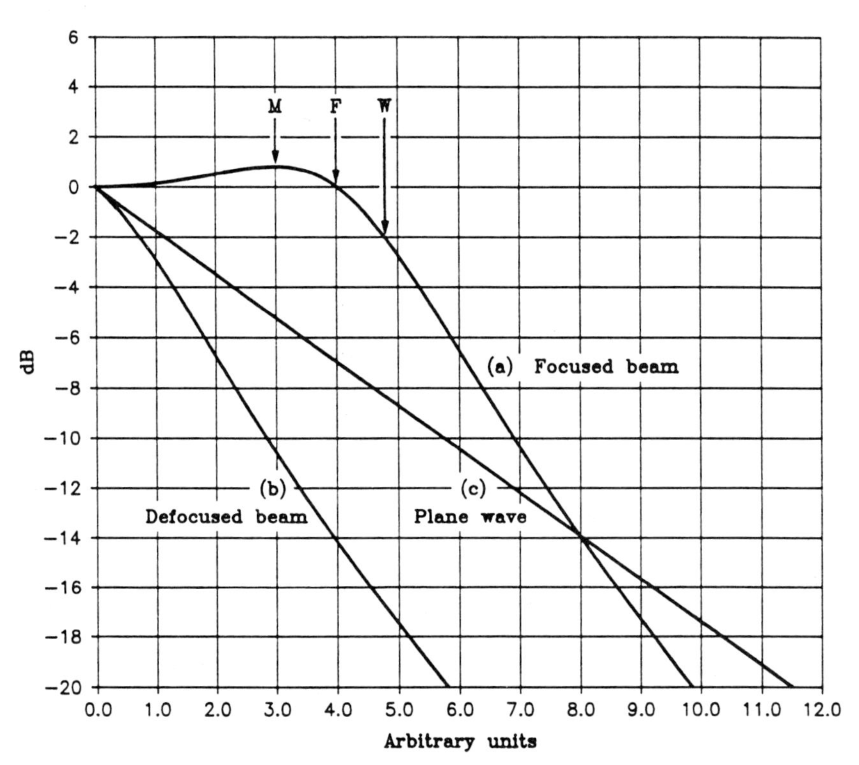

Fig 2. Axial variation of intensity for a) a focused beam with a local maximum, b) a beam with a diverging wavefront, and c) a plane wave.

The evolution of the beam parameters as a function of z is easily followed in the diagram by starting with the source point S (a complex number) and follow a straight line parallel to the a-axis , but in the negative direction. In the case chosen in Fig 3 the imaginary part of z_0 is negative, the absolute value less than half the penetration depth. The transverse decay rate is zero, corresponding to a large aperture with constant amplitude. Following the beam for increasing values of z we first arrive at the point M of maximum power density (this point may also be constructed geometrically). Later, at point F on the imaginary axis, the focal point if there were no losses. The point of minimum waist, W, is reached where the line is tangent to a constant u-circle, and finally point D, where the beam makes a

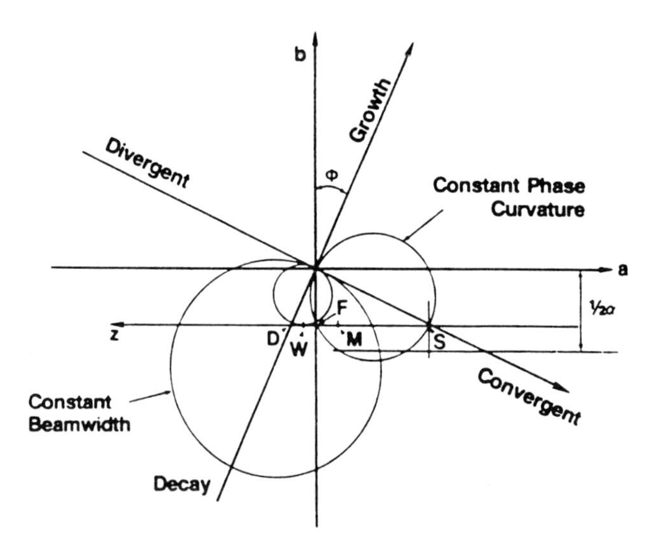

Fig.3 Complex source point plane. S is at aperture for the case of a constant magnitude, focused illumination. F is lossless focus, M is point of maximum intensity, W is point of minimum waist, D point of phase divergence.

transition from a convergent to a divergent wavefront. It is noteworthy, that in the lossless case all these points coalesce to a single point. The points are also indicated on Fig. 2.

It may also be seen that in order to obtain a maximum at larger depths in the general case, the beam at the surface should have laterally growing amplitudes, which of course will lead to a very small gain due to the losses near the surface. A maximum gain situation requires laterally decreasing amplitudes. In contrast to lossless media the simultaneous requirements of high gain and high resolution are not possible in lossy media (Andersen (1985)).

Inclusion of curved interfaces between different media is simple, so many practical situations are readily covered with a minimum of computer power.

4. CONCLUSION

Modelling of finite apertures near lossy media by means of Gaussian beams is accurate and straightforward, and gives a good insight into the various propagation mechanisms and possibilities of increasing the penetration. In terms of computations the results are almost instantaneous, even for the case of multilayered, curved interfaces, where multiple reflections are taken into account.

More general inhomogeneities need numerical algorithms, but a hybrid technique should prove useful, where the overall features are covered by the beam model, and detailed calculations around a three-dimensional inhomogeneity could be solved by other means with the beam as the incident field.

REFERENCES

Andersen, J. Bach, 1985, "Theoretical limitations on radiation into muscle tissue", Intern. Journal of Hyperthermia, vol 1, no 1, pp 45-55.

Andersen,J. Bach, 1987, "Electromagnetic power deposition: Inhomogeneous media, applicators, and phased arrays", ch.7 in "Physics and Technology of Hyperthermia", eds. Field and Franconi, NATO ASI Series E: Appl. Sciences-No 127, Martinus Nijhoff Publishers, 30 pages.

Andersen, J. Bach, Gross, E., 1986, "Gaussian beams in lossy media", URSI Symposium on Electromagnetic Theory, Budapest, Hungary, pp 134-136.

Deschamps, G.A., 1971, "Gaussian beam as a bundle of complex rays", Electronics Letters, vol 7, no 23, pp 684-685.

Kogelnik, H. and Li, T., 1966, "Laser beams and resonators", Proc. IEEE, vol 54, no 10, pp 1312-1328.

Lumori, M.L.D., Andersen, J.Bach, Gopal, M.K., Cetas, T.C., 1990, "Gaussian beam representation of aperture fields: simulation and experiment", accepted by IEEE Transact. MTT.

Paulsen, K.D., 1990, "Power deposition models for hyperthermia applicators", ch. 12 in "An introduction to Practical Aspects of Clinical Hyperthermia" (eds. Field and Hand), Taylor and Francis.

COMPLEX SOURCE PULSED BEAMS:

PROPAGATION, SCATTERING AND APPLICATIONS

Ehud Heyman*

Department of Electrical Engineering
Tel-Aviv University
Tel-Aviv, 69978, Israel

1. INTRODUCTION

The excitation, propagation and diffraction of short pulse signals is receiving increased attention in a variety of fields. Due to the broad frequency spectra of such signals, the conventional route of inversion from the frequency domain is often less convenient, and physically less transparent, than direct treatment of time domain events. Analysis and synthesis of such wave solutions require use of source functions (e.g., time-dependent Green's functions or plane waves). Important among source functions, for a variety of applications (some examples will be given below), are those which establish well focused fields. In the frequency domain, such fields take the form of transversely confined beams, the most favored among which have Gaussian-like profiles. In the time domain, such fields take the form of pulsed teams (PB), i.e. highly directed space-time wave packets. Several types of PB solutions of the wave equation have recently been introduced,[1-3] in particular in connection with long range focused energy transfer. Particular interest has been given to the focus wave modes (FWM)[1] and the related wave functions,[2] which are non diffracting or slowly diffracting free space wave packets. However, since these exact solutions comprise backward propagating spectral components, their role in modeling *physical* radiation is yet to be clarified.[4-6] Efforts have been made to minimize the backward propagating part in these exact solutions,[7] but the effect of eliminating them completely has not yet been fully established.

This work is concerned with another family of exact PB solutions, the so-called complex source pulsed beams (CSPB).[8-10] Unlike the above mentioned solutions, these PBs may be synthesized in terms of *causal* radiation from time-dependent source distribution of *finite* spatial support. Furthermore, analytical expressions for these waves can be generated by displacing a pulsed point source appropriately into a complex coordinate space (as an extension of the time

* This work is supported by the United States - Israel Binational Science Foundation, Jerusalem, Israel, under Grant No. 88-00204.

harmonic complex source technique[11,12]). Thus, by the complex source coordinate substitution approach one may convert conventional Green's function solutions for pulsed point source excited spherical wave inputs in an environment, into solutions for PB inputs in that environment. The PB's properties (e.g. direction and collimation) are controlled by the complex extension of the source coordinates.

CSPBs are of importance per se for a variety of applications, e.g. modeling of *scalar*[5] or *electromagnetic*[13] focused pulsed fields (wave packets) or for local probing and exploration of an environment.[14] CSPBs are also important as they form a *complete* set of basis functions for *exact* angular spectrum expansions of time-dependent fields.[15,16] These new alternatives to the Sommerfeld-Weyl plane wave spectral representations have several advantages, not the least of which are the spectral compactization achieved and the use of local beam propagators that can be tracked paraxially in an inhomogeneous environment.

The aim of this paper is to review the time-dependent complex source method and some of the above mentioned applications. The free space properties of the CSPB are described in Sec.2. In Sec. 3 we examine three alternative schemes for utilizing CSPBs as basis functions in spectral expansions of transient radiation. Finally the procedure for constructing exact solutions for PB scattering problems is described in Sec. 4, with a numerical example for reflections at a plane dielectric interface. Concluding remarks are made in Sec. 5.

2. PULSED BEAMS IN FREE SPACE

We consider the scalar field due to a time-dependent complex point source with time history $\overset{+}{f}(t-t')$ located in a homogeneous medium with wave speed v. The complex source coordinates $\mathbf{r}'$ and the time reference t′ are expressed in the form

$$\mathbf{r}' = (x',y',z') = \mathbf{r}_0 + ib\hat{\mathbf{b}} \quad , \quad |\hat{\mathbf{b}}| = 1 \quad , \quad b > 0 \tag{1}$$

$$t' = it_1 \quad , \quad 0 < b \le t_1 \tag{2}$$

where $\mathbf{r}_0$, $\hat{\mathbf{b}}$ and b will be identified as the *real beam origin*, *direction* and *Fresnel length*, respectively. Without loss of generality let $\mathbf{r}_0 \equiv (0,0,0)$ and $\hat{\mathbf{b}} \equiv \hat{\mathbf{z}}$. The second condition in (2) is a sufficient *convergence* requirement for the solutions. In order to cope with the complex space-time coordinate domain we utilize analytic signals, defined via

$$\overset{+}{f}(t) = \frac{1}{\pi}\int_0^{\infty} d\omega\, e^{-i\omega t}\, \hat{f}(\omega) \quad , \quad \text{Im } t \le 0 \, . \tag{3}$$

where $\hat{f}(\omega)$ is the frequency spectrum corresponding to the *real* time signal f(t). Clearly, $\overset{+}{f}(t)$ is analytic in the lower half of the complex t-plane; the physical (real) signal f(t) is recovered from the real limit

$$\overset{+}{f}(t) = f(t) + i\mathbf{H}f(t) \quad , \quad \mathbf{H}f = -\frac{1}{\pi}\int_{-\infty}^{\infty} \mathbf{P}\, \frac{f(\tau)}{t-\tau}\, d\tau \quad , \tag{4}$$

H being the Hilbert transform and **P** is Cauchy's principal value.

For real $\mathbf{r}$ and t, the analytic field due to the complex source in (1) and (2) is given by

$$\overset{+}{F}(\mathbf{r},t;\mathbf{r}',t') = \overset{+}{f}\,(t-t'-s/v)/4\pi s \tag{5}$$

where

$$s \equiv \sqrt{(x-x')^2 + (y-y')^2 + (z-z')^2} = \sqrt{\rho^2 + (z-ib)^2} \quad , \quad \text{Re } s > 0 \tag{6}$$

and $\rho = \sqrt{x^2+y^2}$. Both Re $\overset{+}{F}$ and Im $\overset{+}{F}$ = H Re $\overset{+}{F}$ are exact real solutions. For real initial data, Re $\overset{+}{F}$ is the conventional retarded point source field. For complex initial data, Eq. (5) yields exact beam-type solutions of the time-dependent wave equation. The following properties of these waves follow from the behavior of s($\mathbf{r}$) as a function of the real observation point (Fig. 1) and from the properties of the analytic signal $\overset{+}{f}$ (t) which, from (3), *decays monotonically* as the imaginary part of its argument becomes more negative:[5,13,15] *(1)* $\overset{+}{F}(\mathbf{r},t)$ is continuous everywhere in the *real* coordinate space, except across a disk $\rho < b$ in the $z = 0$ plane, which is therefore referred to as the *source disk* since it represents the real source distribution that gives rise to $\overset{+}{F}$ (Fig. 1). *(2)* $\overset{+}{F}$ is an *outward* propagating (causal) wave solution with respect to this disk; the waveforms propagate along phase paths (propagation lines, see Fig. 1) whereby they are gradually Hilbert transformed (i.e, the far field waveform is a Hilbert transform of the near field waveform along the same phase path). They also exhibit a $1/r$ decay along the phase path for $r<b$. *(3)* The waveform is strongest on the positive z-axis and decreases monotonically along a phase front to a minimum along the negative axis. The beamwidth (i.e., the rate of decay transverse to the positive z-axis) depends on the decay rate of $\overset{+}{f}$ in the lower half of the complex t-plane, which is determined by the frequency bandwidth of f(t); the wider the frequency band, the faster the decay of $\overset{+}{f}$ (t) and the narrower is the beamwidth. *(4)* When well collimated, the paraxial field near the positive z-axis may be approximated by using $s \simeq z-ib+\rho^2/2(z-ib)$, giving

$$\overset{+}{F} \simeq \overset{+}{f}\,[t-z/v-i\beta/v-\rho^2/2v(z-ib)]/4\pi(z-ib), \tag{7}$$

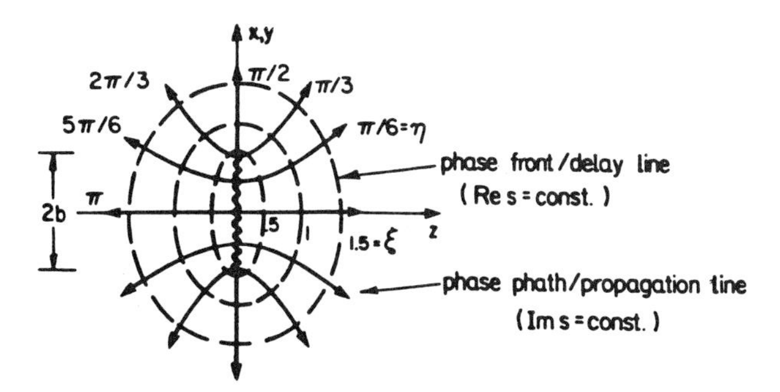

Fig. 1. Oblate spheroidal coordinates for complex source field $(x, y, z) = b(\cos\phi \sin\eta \cosh\xi, \sin\phi \sin\eta \cosh\xi, \cos\eta \sinh\xi)$, $\xi>0$, $-\pi<\eta<\pi$. In this system $s=b(\sinh\xi-i\cos\eta)$ hence the phase paths and phase fronts coincide with the ξ and η coordinates, respectively.

with an *exact* equality *on* the beam axis. The positive parameter $\beta \equiv vt_1 - b$ (see (2)) may actually be considered as part of $\overset{+}{f}$. *(5)* b is recognized as the Fresnel (or diffraction length); it affects both the transverse spatial decay in (7), via the negative imaginary part of $-\rho^2/(z-ib)$ in the argument of $\overset{+}{f}$ in (7), and the axial decay of the amplitude for $z > b$. *(6)* The axial energy decay of the CSPB can be determined, without loss of generality, from the real field $F = \mathrm{Re}\, \overset{+}{F}$ with $\beta = 0$ (recall that $\mathrm{Im}\, \overset{+}{F}$ is also a solution while $\beta \neq 0$ can be considered as part of $\overset{+}{f}$). From (7) we obtain *on* the positive beam axis

$$F(\mathbf{r},t)\Big|_{\rho=0} = [4\pi(z^2+b^2)]^{-1}(z-bH)\, f(t-z/v) \quad , \quad f = \mathrm{Re}\, \overset{+}{f} \tag{8}$$

Calculating the pulse energy $E = \langle F,F\rangle$, using $\langle f,Hf\rangle = 0$ and $\langle Hf,Hf\rangle = \langle f,f\rangle$, we obtain

$$E(z)\Big|_{\rho=0} = (4\pi)^{-2} E_0 (z^2+b^2)^{-1} \quad , \quad E_0 = \langle f,f\rangle \tag{9}$$

Thus, E remains essentially constant for $z \ll b$, whereas for $z \gg b$ it decays like z^{-2}. An important feature in the CSPB is that all its temporal-frequency components have the same diffraction length b.[5] The CSPB therefore determines the aperture distribution for an *efficient energy transfer*, where all frequency components stay collimated essentially up to the same distance b. For $z \gg b$ the PB diverges with a frequency dependent diffraction angle $\theta \sim O(kb)^{-1/2}$. The *actual* diffraction angle of the *time-dependent* wave packet depends on the choice of f (see example in (12b).

An important example is the analytic time-dependent Green's function $\overset{+}{G}$ (Fig.2), defined by (4) with $\overset{+}{f} = \overset{+}{\delta}$ where

$$\overset{+}{\delta}(t) = \begin{cases} 1/\pi i t & , \ \mathrm{Im}\, t < 0 \quad (10a) \\ \delta(t) + P/\pi i t & , \ \mathrm{Im}\, t = 0 \quad (10b) \end{cases}$$

is the analytic δ function; (10b) being the distributional limit (4) of (10a). From (7) with (10a), the time-dependent field near the positive z-axis and its spectrum are given by

$$\overset{+}{G} \simeq [4\pi(z-ib)]^{-1} (\pi i)^{-1} [t - z/v - i\beta - \rho^2/2v(z-ib)]^{-1} \tag{11a}$$

$$\hat{G} \simeq [4\pi(z-ib)]^{-1} \exp[ik(z+\rho^2/2(z-ib))] e^{-k\beta} \tag{11b}$$

Thus, the transverse Gaussian width of the frequency spectrum in (11b) is given by

$$W(z) = W(0)\sqrt{1+(z/b)^2} \quad , \quad W(0) = \sqrt{b/k} \tag{12a}$$

For the time-dependent solution in (11b), the PB's beamwidth W_G in the $z = vt$ plane, the pulsewidth T_G on the z-axis, and the diffraction angle θ_G are given, respectively, by

$$W_G(z) = W_G(0)\sqrt{1+(z/b)^2} \quad , \quad W_G(0) = \sqrt{\beta b}\, \sqrt{2(\sqrt{2}-1)}\, ,$$

$$T_G = 2\beta/v \quad , \quad \theta_G = \sqrt{\beta/b} \tag{12b}$$

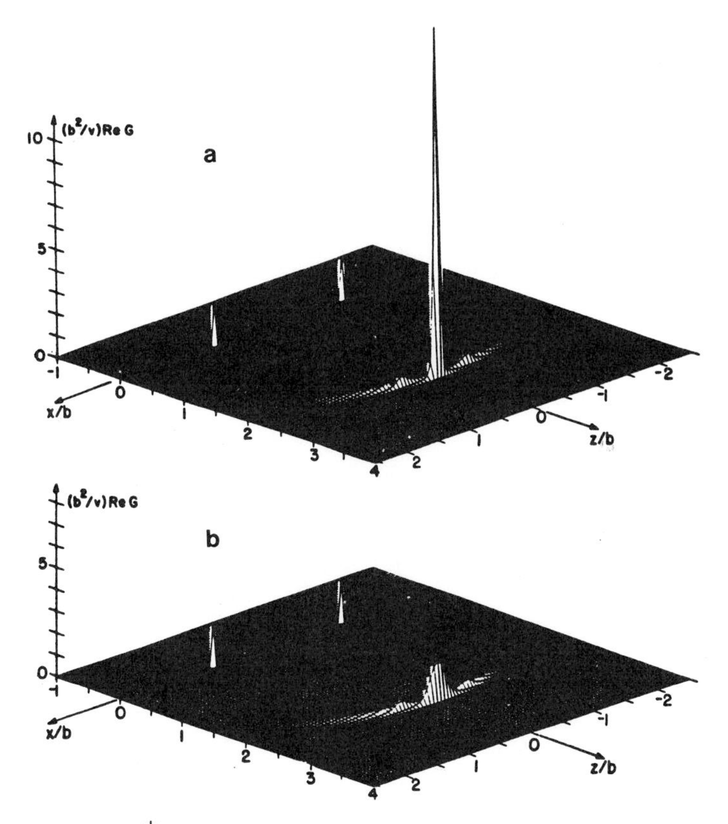

Fig. 2. The CSPB Re $\overset{+}{G}$. The plot depicts an axial cross sectional cut in the (x,z) plane through the rotationally symmetric 3D-space at time $vt=2.5b$. Beam parameters: (a) $\beta=0.0005b$; (b) $\beta=0.005b$. All axes are normalized with respect to b and v. The wavepacket is localized around $z=vt$, $\rho=0$. The field spikes at $z=0$, $\rho=b$ are stationary and are attributable to singularities of the equivalent *real* source distribution on the source disk corresponding to the complex source location. They do not contribute to the propagating wavepacket[5].

As $\beta\to 0$, the PB becomes sharper (see Fig. 2); the limit $\beta=0$ renders the PB impulsive *on* the positive z-axis (note that the axial frequency decay in (11b) is $\exp(-k\beta)$). More examples and properties are given in references 5, 13 and 15.

3. PB EXPANSION OF TRANSIENT RADIATION

As mentioned in the Introduction PBs can be utilized as basis functions in an angular spectral expansion of transient radiation, propagation and diffraction problems. These new alternatives to the conventional Sommerfeld-Weyle plane wave integral (in the time domain) has many favorable properties: PB fields can be tracked locally in complicated inhomogeneous environment, accounting for local interactions only. Having finite spectral width they are insensitive to caustic and focii. Furthermore, the PB spectral integrals are localized *a priori* since only those PBs that pass near the observer actually contribute. Plane wave spectral integrals, on the other hand, comprise *global* wave functions, and are therefore distributed over *all* spectral angles so that localization can be affected only via asymptotic evaluation. Thus, the PB expansion approach combines the algorithmical ease of the ray methods with the uniform features of the spectral representation. Similar strategy has been employed in the Gaussian beam summation method, which has been introduced as an efficient tool for computing *time-harmonic* wavefields in inhomogeneous medium.[17] For the transient case considered here, the spectral propagators are PBs. Unlike the Gaussian beams, CSPBs are *exact* solutions that furnish a *complete* set for spectral expansion, thereby providing means to examine the effects on the spectral representation of paraxial and other approximations in the PB propagation model. The various localization mechanisms discussed above are demonstrated in three expansion schemes in subsections, 1, 2 and 3 below.

3.1. Radiation From Localized Sources

The CSPB spectral representation for the field u(r,t) due to a *real* point source with time history f(t) located at the origin is given by

$$\overset{+}{u}(\mathbf{r},t) = \overset{+}{f}(t - r/v)/4\pi r \simeq \frac{ib}{2\pi v} \iint_{\Theta} d\boldsymbol{\theta} \, \frac{d}{dt} \overset{+}{F}[\mathbf{r},t; ib\hat{\mathbf{b}}(\boldsymbol{\theta}), ib/v] \quad , \quad r > b \tag{13}$$

where we use the analytic signal representation of the field; the real field is obtained by taking the real part of (13). The first equality in (13), with $r=|\mathbf{r}|$, is the conventional expression for the retarded radiated field. Inside the integral, $\overset{+}{F}$ is the CSPB (5) with $\hat{\mathbf{b}}(\boldsymbol{\theta})$ being the unit vector in the direction $\boldsymbol{\theta} = (\theta, \phi)$ and Im $t' = b/v$ (see (2)). Thus, the integral involves an angular spectrum of collimated PBs with Fresnel length b, emanating radially from the source in direction $\boldsymbol{\theta}$ (Fig. 3). Unlike the conventional Sommerfeld plane wave integrals, the PB integral in (13) is localized *a priori*, since only those PBs passing in a certain angular range Θ near the observer yield nonnegligible contributions. As shown in Ref. 15, the approximation in (13) converges to an identity by increasing the angular integration domain Θ. The convergence rate depends on the problem parameters (e.g. the distance r and the pulsewidth of f(t)) andon the expansion parameter b. A proof of (13) and a detailed analytical and numerical study of the convergence rate and of the truncation errors are given in Ref. 15.

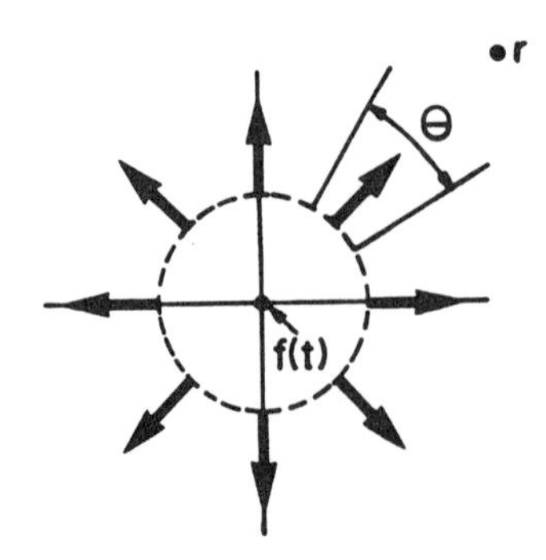

Fig.3. CSPB expansion for radiation from a point source: The source launches an angular continuum of PBs, schematized by the arrows.

3.2. Local PB analysis of Radiation From Extended Source Distribution

The integral representation (13) is localized since only those PBs that pass near the observer actually contribute there. For distributed sources, the PB approach can further localize the *radiation* process since, unlike the conventional point source (Green's function) superposition approach, each PB basis element accounts for the bulk effect of radiation from a finite region in the source domain. To demonstrate this property, we consider radiation from large extended aperture distribution $u_0(\mathbf{r}_0,t)$ in the z=0 plane, $\mathbf{r}_0 \equiv (x_0,y_0,0)$. The radiating field is expressed in the form

$$\overset{+}{u}(\mathbf{r},t) = \overset{+}{O}(t) * \int d^2\bar{r}\; d^2\bar{\theta}\; d\bar{t}\; \overset{+}{U}_0(\bar{r},\bar{\theta},\bar{t})\, \overset{+}{P}(\bar{r},t;\bar{r},\bar{\theta},\bar{t}) \tag{14}$$

where the operation outside the integral is a convolution $*$ with a normalizing function $\overset{+}{O}(t)$ which described essentially differentiation. The integral (14) comprises a continuous superposition of shifted, tilted and delayed PBs $\overset{+}{P}$ which are related to, but not identical with, the CSPB. The departure points, departure directions and departure times of these PBs from the z=0 plane are specified, respectively, by $\bar{r} = (\bar{x},\bar{y})$, $\bar{\theta} = (\theta_x,\theta_y)$ and $\bar{t}$ (see Fig. 4(a)). This integral representation has therefore been termed *phase-space PB summation* (PS-PB-S). The $(\bar{r},\bar{\theta},t)$ phase-space distribution of PBs is matched to the aperture distribution $u_0(\mathbf{r}_0,t)$ via the PB amplitude function $\overset{+}{U}_0$, which is calculated by the projection in the z=0 plane

$$\overset{+}{U}_0(\bar{r},\bar{\theta},\bar{t}) = \int d^2\mathbf{r}_0\; dt\; \overset{+}{u}_0(\mathbf{r}_0,t)\, \overset{+}{P}^*(\mathbf{r}_0,t;\bar{r},\bar{\theta},\bar{t}) \tag{15}$$

where the raised asterisk denotes a complex conjugate. Recalling the physical interpretation of $\overset{+}{P}$, the window kernel in (15) samples $\overset{+}{u}_0(\mathbf{r}_0,t)$ inside a window which is centered at $(\mathbf{r}_0,t) = (\bar{r},\bar{t})$ and tilted along a surface $(x_0-\bar{x})\sin\theta_x + (y_0-\bar{y})\sin\theta_y = t-\bar{t}$ (see two dimensional schematization in Fig. 4(c)). This operation, which defines the *local spatial spectrum* of the time-dependent field, has been termed by us *local Radon transform*, as an extension of the *global* Radon transform in the $(\mathbf{r}_0,t)$ domain, that defines the plane wave spectrum of a time-dependent field.[18] This is analogue to the *local* Fourier transform that defines the local spatial spectrum in the time harmonic case.[19]

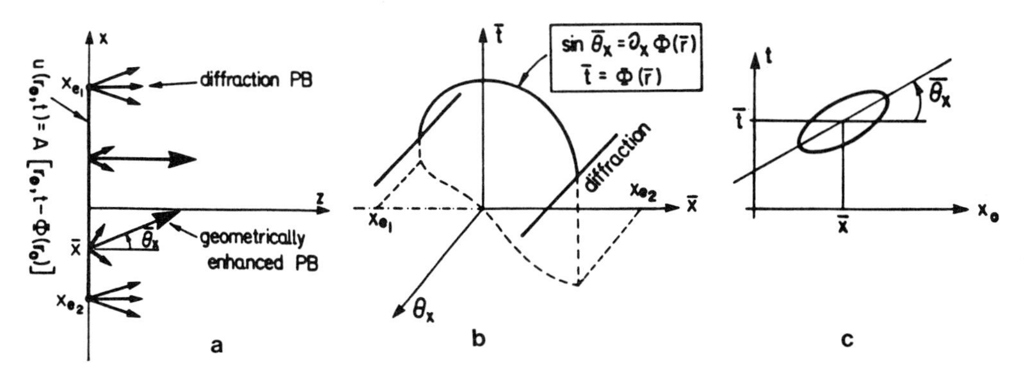

Fig.4. 2D schematization for phase-space pulsed beam expansion for radiation from distributed aperture. (a) Configurational domain interpretation. (b) phase-space interpretation. (c) Local Random transform.

In (15) spatial, spectral and temporal widths of the PB window in the z=0 plane can be chosen so that each PB senses, via (15), the local radiation properties of u_0 at $(\bar{r},\bar{t})$. As an example, let $u_0(r_0,t) = A[r_0, t-\Phi(r_0)]$ where $\Phi(r_0)$ is a delay function and $A(r_0,t)$ is a short pulse on the scale of the spatial variation of A and of Φ. Then, for each departure point $\bar{r}$ in the aperture plane, $\overset{+}{U}_0$ favors *a priori* the PB with initiation time $\bar{t}(\bar{r}) = \Phi(\bar{r})$, that propagates along the local preferred (geometrical) direction of radiation $\sin\bar{\theta}(\bar{r}) = \nabla\Phi(\bar{r})$, thereby establishing localization of $\overset{+}{U}_0$ in the $(\bar{r},\bar{\theta},\bar{t})$ domain around the so-called Lagrangian manifold of u(r,t) (see Fig. 4(b)). Similarly, if u_0 has end point discontinuities at r_e, $\overset{+}{U}_0$ also favors diffraction PBs, whose phase-space parameters are defined by $\bar{r} = r_e$ and $\bar{t} = \Phi(r_e)$ (Fig. 4(b)). Further localization of the PS-PB-S integral (14) is due to the fact that only those PBs which pass near the observer at time t actually contribute. A thorough analytical, asymptotical and numerical study of this representation is presented in Ref. 16.

3.3. Global PB Analysis of Transient Radiation from Well Collimated Aperture Distribution

The representation in Sec. 3.2 expresses the field in terms of the *local* properties of u_0. It is therefore most efficient if the source distribution (or the data) has a wide spectral spread as schematized in Fig. 4(a). For well collimated pulsed aperture distribution, on the other hand, one may wish to expand the field in terms of wider PBs that cover the *entire* aperture. Each of these PBs basis functions than manifests the well collimation properties of the entire aperture distribution, i.e. it has a long diffraction length and a narrow diffraction angle. The basis functions for this type of expansion are generalization of the CSPB $\overset{+}{G}$ in (10)-(12), and are defined by

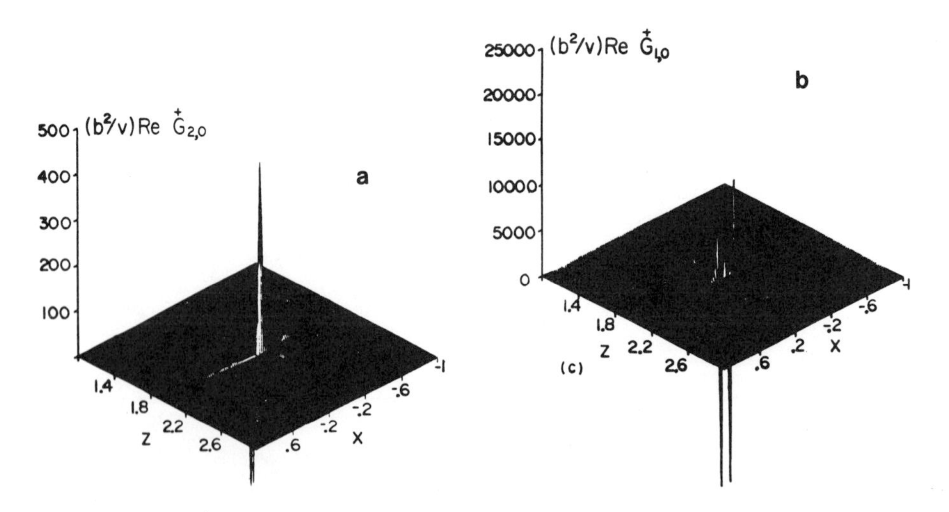

Fig.5. CMPB Re $\overset{+}{G}_{m,n}$. The plot depicts a cross-sectional cut in the (x,z) plane at vt=2.5b for: (a) (m,n)=(1,0). (b) (m,n)=(2,0). The (m,n)=(0,0) element is shown in Fig. 2.

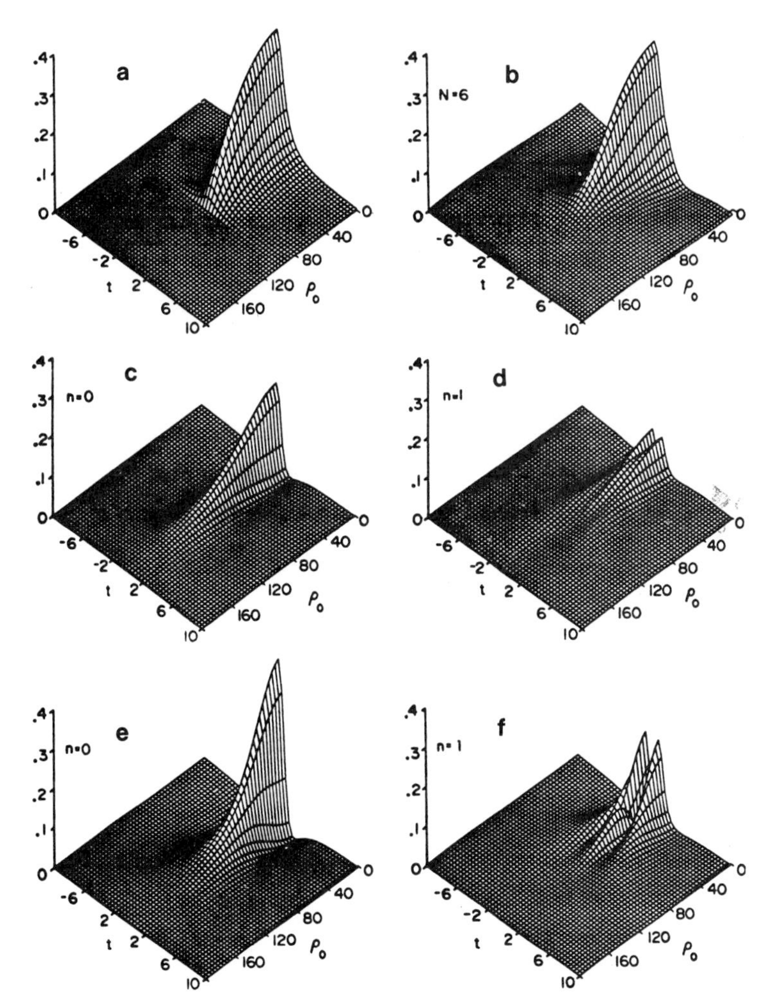

Fig.6. CMPB expansion of transient radiation from well collimated aperture. (a) Radially symmetrical aperture distribution $\overset{+}{U}_0(\rho_0,t)=\overset{+}{\delta}(t-iT)[1-(\rho_0/d)^2]H(d-\rho_0)$, H is the Heaviside function, T and d=100vt are the pulse width and aperture width, respectively. The aperture diffraction length is $D \simeq (2d)^2/vT = 4000vT$. (b) Reconstructed aperture distribution using 6 elements in (14). The expansion is matched to the aperture distribution using b=4000vT. (c),(d) Contribution to (b) of the zero and first order elements in (16), respectively. (e),(f) As in (c),(d) but for non-matched PBs with b=2000vT. In this case the convergence is poorer.

$$\overset{+}{G}_{m,n}(\mathbf{r},t;\ \mathbf{r};t') = \partial_x^m\, \partial_y^n\, \overset{+}{G}(\mathbf{r},t;\ \mathbf{r};t'), \quad m,n=0,1,... \tag{16}$$

Following Ref. 20 they are denoted as complex multiple pulsed beams (CMPB).[21] Their oscillatory behavior transverse to the beam axis (see Fig. 5) makes them a complete set for an expansion of well collimated radiation. The radiated field is then expressed in the form

$$\overset{+}{u}(\mathbf{r},t) = \sum_{m,n} \overset{+}{a}_{m,n}(t) * \overset{+}{G}_{m,n}(\mathbf{r},t) \tag{17}$$

where the amplitude functions $\overset{+}{a}_{m,n}$ are found via the projection

$$\overset{+}{a}_{m,n}(t) = \int d^2\mathbf{r}_0\, dt_0\, \overset{+}{u}_0(\mathbf{r}_0,t_0)\overset{+}{G}{}^{\dagger}_{m,n}(\mathbf{r}_0,t-t_0) \quad , \mathbf{r}_0 = (x_0,y_0,0) \tag{18}$$

where $\overset{+}{G}{}^{\dagger}_{m,n}$ is a biorthogonal function that can be found.[21] Furthermore, using the fact that the function $\overset{+}{G}_{m,n}$ are defined in terms of an analytic δ functions (see (11a) and (16)), the t_0-integrals in (17) and (18) can be evaluated in closed form. A thorough study of the properties of the expansion (17) has been carried out in Ref. 21. As expected, the expansion is efficient *only* if the collimation parameter of the PBs (i.e. b) is matched to the collimation properties of the *entire* aperture distribution. A numerical example is shown in Fig. 6.

4. PULSED BEAM INTERACTION WITH PROPAGATION ENVIRONMENT

The propagation of PB through an ambient environment and its scattering by structural features therein can be analyzed by substitution of complex space-time source coordinates in the conventional real source transient Green's function for that environment, which is given, in general, in a spectral integral form. This opens up a whole class of *canonical* propagation and diffraction problems, through which one may systematically explore how PBs are affected by environmental features and develop local (paraxial) models for the interaction processes. Since the PB is localized, information from canonical problem will be relevant directly for more general structures that can be modeled by the canonical prototype.

Exact, closed form expressions for the interaction of the CSPB with a scattering environment may be derived via the spectral theory of transients (STT). The method has been introduced originally to deal with propagation and scattering of pulsed fields in complicated environment,[22-24] but has been modified subsequently to analyze CSPBs.[9,14] Within this framework, the analytic source excited time-dependent fields can be derived from the canonical STT integral

$$\overset{+}{G}(\mathbf{r},t;\ \mathbf{r}',t') = \frac{1}{2\pi^2}\int_C \frac{A(\mathbf{r},\mathbf{r}';\ \xi)}{t-t'-\tau(\mathbf{r},\mathbf{r}';\ \xi)}\, d\xi \ , \tag{19}$$

which encompasses a distribution of source excited, analytic time-dependent plane waves with amplitude function A and travel time τ, tagged by the spatial spectral parameter ξ (the plane wave's angle). The amplitude may describe either the incident field or the reflected/diffracted field that reaches the observer at $\mathbf{r}$ after being emitted by a localized source at $\mathbf{r}'$. The behavior of $\overset{+}{G}$ is determined by the singularities (poles and branch points) of the integrand in the complex ξ-plane. The analytic properties of the integrand have been discussed previously.[22-24] The function $\tau(\xi)$ usually has real branch points at $\xi=\pm\xi_c$ that separate the plane wave spectral range into propagating (visible) and evanescent portions $|\xi|\lessgtr\xi_c$, respectively. The integration contour C follows the real axis but is chosen so as to avoid these singularities, as well as possible singularities of A, in a manner consistent with the radiation condition. The integrand in (19) also contains time-dependent poles $\xi(t)$, determined by

$$\tau[\xi(t)] = t - t' , \tag{20}$$

that define the transient plane waves arriving at the observer at time t. The integral in (10) may be evaluated by contour deformation in terms of the above singularities. By adding to C a semicircle at infinity in either the upper or lower half planes, one obtains, assuming only the time-dependent pole singularities defined in (20),

$$\overset{+}{G} = \overset{+}{G}{}_p^{+} + I^{+} = \overset{+-}{G}_p + I^{-} , \tag{21}$$

respectively. Here (with $\tau' \equiv d\tau/d\xi$)

$$\overset{+}{G}_p(t) = \sum_{\hat{\xi}(t)} -\frac{i}{\pi} \frac{A[\hat{\xi}(t)]}{\tau'[\hat{\xi}(t)]} , \quad \overset{+-}{G}_p(t) = \sum_{\check{\xi}(t)} \frac{i}{\pi} \frac{A[\check{\xi}(t)]}{\tau'[\check{\xi}(t)]} . \tag{22}$$

denote the contributions at time t from poles $\hat{\xi}(t)$ and $\check{\xi}(t)$ in the upper or lower half planes, respectively, while $I^{\pm}$ are integrals about the right or left branch cut, respectively. The criteria for choosing lower or upper half-plane closure are problem dependent and are motivated by de-emphasizing the contribution of the branch cut integral, thereby leaving the field described *only* by the pole contribution in (26).[14]

As an example, we have analyzed reflection of two-dimensional CSPB from a plane interface separating two homogeneous half spaces with wave propagation speed v_0 and v_1, that occupy the regions $z > 0$ and $z < 0$, respectively, in the two-dimensional coordinate frame $\rho=(x,z)$ (Fig. 7). The PB is incident from $z > 0$ at an angle θ_0. It is generated by a pulsed line source located at

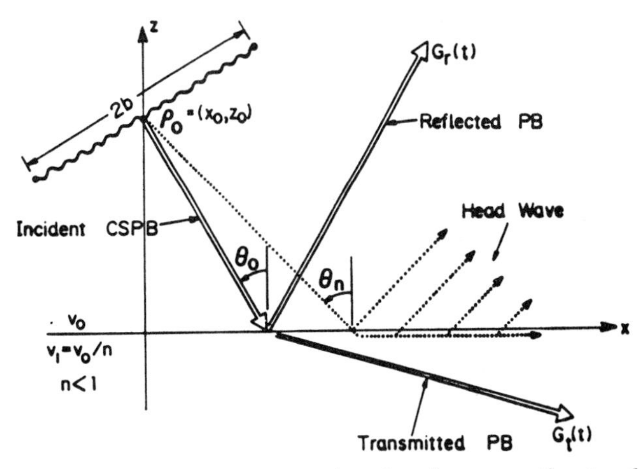

Fig. 7. CSPB reflection and transmission at a plane interface separating two homogeneous half spaces with wave speed v_0 and $v_1 > v_0$. $\rho_0=(0,z_0)$, locates the center of the equivalent *real* source distribution (wiggly line). Double lined arrow: beam axes. Dotted lines: critically incident ray from ρ_0 and trajectories of the lateral (head) wave.

the complex coordinates point (1) with $\rho_0=(0,z_0)$ and $\hat{\mathbf{b}}=(\sin\theta_0,-\cos\theta_0)$, $0\leq\theta_0<\pi/2$. We assume $v_0 < v_1$, thereby admitting conditions of total reflection beyond the critical angle $\theta_n=\sin^{-1}n$, $n\equiv v_0/v_1<1$, and the consequence generation of lateral (head) waves along the interface (see Fig.7). The total field G is assumed to be continuous across the interface. Figure 8 depicts the transient incident and reflected fields, Re $\overset{+}{G}_i$ and Re $\overset{+}{G}_r$ respectively, computed at various points along the interface when the PB is incident at the critical angle as described above. Here $n=1/2$ so that $\theta_n=30^0=\theta_0$. With $\rho_0=(0,\sqrt{3}/2)$, the axis of the incident PB intersects the interface at $x=1/2$. The symbol * identifies the PB geometrical arrival time $t^{go}(x)=\sqrt{3/4+x^2}/v_0$ along the interface, based on assuming a real source location ρ_0, and the symbol + the head wave arrival time $t^h(x)=1/v_0 + (x-0.5)/v_1$ in the total reflection domain ($x>0.5$). One observes that the incident PB peaks at $t=t^{go}$, the strongest peak occurs on the PB axis at $x=0.5$. The relatively slow *temporal* rise and decay of the peak are attributable to the two-dimensionality of the problem; they are deemphasized for a 3D PB (Fig.2). The 3D PB also has a higher degree of *spatial* collimation. The reflected waveforms essentially replicate the incident waveforms except for the amplitude modification due to the local reflection coefficient. After the onset of total reflection ($x>0.5$) one may observe a weak lateral wave contribution; again it is anticipated that this phenomenon will be more pronounced in the 3D case. A detailed parametric study, including an analysis of the transmitted field can be found in Ref. 14.

The exact STT solution for canonical problems can be used to derive expressions for the paraxial interaction of the PB.[14] Analysis of several canonical configurations have revealed that, unless the PB hits at a critical direction (e.g., close to the critical angle in Fig. 7), simple paraxial models of the type used in beam optics or Geophysical exploration are quite accurate. For criti-

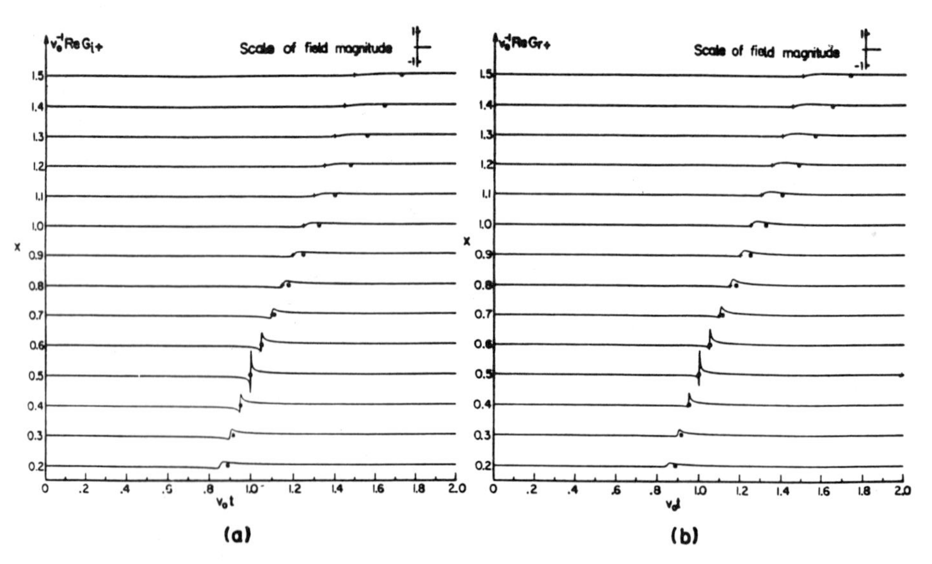

Fig. 8. Temporal evolution of CSPB field at various points (x) along the interfaces for critically incident beam ($\theta_0=\theta_n$). The symbols * and + identify the ray and the head wave (for $x>0.5$) arrival times. Beam parameters: $b=3$, $\beta=10^{-4}b$. (a) Re $\overset{+}{G}_i$. (b) Re $\overset{+}{G}_r$.

cal incidence conditions, one may still use the exact STT solution for the canonical prototype in order to derive more sophisticated paraxial models.

5. SUMMARY

The motivation in this presentation was to review some applications and research directions associated with the CSPB. They include: *a)* Synthesis of PB solutions with desirable properties; *b)* Three alternative strategies for using CSPBs as basis functions for spectral expansion of time-dependent fields. Each expansion scheme represents a different approach to spectral compactization and therefore is applicable to different types of source configuration. *c)* Analysis of the interaction of CSPB inputs with a scattering environment. The analytic procedure presented permits a systematic study of a new class of prototype (canonical) problems through which one may assess how environmental conditions affect these highly localized wave packets and how they may be utilized for local environmental exploration. Since PBs are localized in space-time, these prototype structures can be extended to more general configurations. Together with the PB expansion schemes discussed in Sec. 3, these solutions provide efficient computation tools that may accommodate time-dependent fields in complicated environments. It is hoped that the presentation has been sufficiently convincing to establish the CSPB as a basic element for constructing *new phenomenological and computational theories* of transient propagation and scattering.

REFERENCES

1. J. B. Brittingham, Focus wave modes in homogeneous Maxwell's equation, *J. Appl. Phys., 54*, 1179-1189 (1983).
2. R. W. Ziolkowsky, Localized transmission of electromagnetic energy, *Phys. Rev. A, 39(4)*, 2005-2033 (1989).
3. H. E. Moses and R. T. Prosser, Initial conditions, sources and currents for prescribed time-dependent acoustic and electromagnetic fields in three dimensions. Part I: the inverse initial value problem. Acoustic and electromagnetic 'bullets', expanding waves and imploding waves, *IEEE Trans. Antennas Propagat., AP-34*, 188-196 (1986).
4. E. Heyman, B. Z. Steinberg and L. B. Felsen, Spectral analysis of focus wave modes (FWM), *J. Opt. Soc. Am. A, 4*, 2084-2091 (1987).
5. E. Heyman and L. B. Felsen, Complex source pulsed beam fields, *J. Opt. Soc. Am. A, 6*, 806-817 (1989).
6. E. Heyman, The focus wave mode: A dilemma with causality, *IEEE Trans. Antennas Propagat., AP-37*, 1604-1608 (1989).
7. I. M. Besieries, A. M. Shaarawi and R. W. Zialkowski, A bidirectional ravelling plane wave representation of exact solutions of the scalar wave equation, *J. Math. Phys., 30*, 1254-1267 (1989).
8. E. Heyman and L. B. Felsen, Propagating pulsed beam solutions by complex source parameter substitution, *IEEE Trans. Antennas Propagat., AP -34*, 1062-1065 (1986).
9. E. Heyman and B. Z. Steinberg, Spectral analysis of complex source pulsed beams, *J. Opt. Soc. Am. A, 4*, 473-480 (1987).
10. P. D. Einziger. and S. Raz, Wave solutions under complex space-time shifts, *J. Opt. Soc. Am. A, 4*, 3-10 (1987).
11. G. A. Deschamps, Gaussian beams as a bundle of complex rays, *Electron. Lett., 7*, 684-685 (1971).
12. L. B. Felsen, Complex-source-point solutions of the field equations and their relation to the propagation and scattering of Gaussian beams, in *Symp. Matemat. Instituto Nazionale di Alta Matematica, XVIII*, 40-56 Academic Press, London (1976).
13. E. Heyman, B. Z. Steinberg and R. Ianconescu, Electromagnetic complex source pulsed beam, *IEEE Trans. Antennas Propagat., AP-38*, 957-963, (1990).

14. E. Heyman and R. Ianconescu, Pulsed beam reflection and transmission at a dielectric interface: two dimensional fields, *IEEE Trans, Antennas Propagat., AP-39*, Nov. (1990).
15. E. Heyman, Pulsed beam expansion of transient radiation, *Wave Motion, 11*, 337-349 (1989).
16. B. Z. Steinberg, E. Heyman and L. B. Felsen , phase space beam summation for time-dependent radiation from large apertures: Continuous parametrization, *J. Opt. Soc. Am. A*, in press (1991).
17. V. Cerveny, M. M. Popov and I. Psencik, Computation of wave fields in inhomogeneous media - Gaussian beam approach, *Geophys. J. R, astr. Soc., 70*, 109-128 (1982).
318. C. H. Chapman, A new method for computing synthetic seismograms, *Geophys. J. R. astr. Soc., 54*, 481-518 (1978).
19. B. Z. Steinberg. E. Heyman and L. B. Felsen. phase space beam summation for time harmonic radiation from large apertures, *J. Opt. Soc. Am. A, 7*, Nov. (1990).
20. S. Y. Shin and L. B. Felsen, Gaussian beam modes by multiples with complex source points, *J. Opt. Soc. Am., 67*, 699-700 (1977).
21. E. Heyman and I. Beracha, Complex multiple pulsed beam expansion of transient radiation from well collimated apertures, in preparation.
22. E. Heyman and L. B. Felsen, Weakly dispersive spectral theory of transients. Part I: Formulation and interpretation, *IEEE Trans. Antennas Propagat., AP-35*, 80-86 (1987).
23. E. Heyman and L. B. Felsen, Weakly dispersive spectral theory of transients. Part II: Evaluation of the spectral integral, *IEEE Trans. Antennas Propagat., AP-35*, 574-580 (1987).
24. E. Heyman, Weakly dispersive spectral theory of transients. Part III: Applications, *IEEE Trans. Antennas Propagat., AP-35*, 1258-1266 (1987).

COMBINATIONS OF LOCAL SCATTERING OPERATORS AND GLOBAL PROPAGATORS

I-Tai Lu and Bai-Lin Ma

Department of Electrical Engineering
Weber Research Institute
Polytechnic University, Farmingdale, NY 11735

ABSTRACT

This paper addresses hybrid methods which employ analytic or asymptotic approaches as global propagators and employ numerical algorithms as local scattering operators for studying wave propagation and scattering in complex environments. Specifically, a ray - mode - moments method for wave scattering from an aperture coupled system is shown as an example.

I. INTRODUCTION

For wave propagation and scattering in complex environments, no one single solution method can work satisfactorily for a broad range of parameters. Analytic methods, such as spectral integration, mode expansion, generalized ray expansion, and ray-mode combination, are efficient global propagators for large structures but work only for separable geometries. Approximate approaches are more flexible than analytic techniques but are restricted to some asymptotic regimes and are difficult to assess the accuracy of the computed results. For example, high frequency methods, such as asymptotic ray theory and Gaussian beam method, are flexible enough to be adapted to non-planar and inhomogeneous layered media, but fail when media properties vary rapidly. Numerical algorithms are flexible operators for modeling any geometries but are inefficient global propagators. For example, low frequency methods, such as method of moments, T-matrix, finite elements, boundary elements, and finite difference, are convenient for small size structures, but are too computer intensive for large media. Therefore, it may be advantageous to combine various approaches into a single framework, which will provide numerical efficiency and physical insight. Moreover, comparing the results from different hybrid combinations may provide validity checks for a complex problem.

In the wave equation, we have three configurational coordinate and one temporal coordinate. The most general hybrid method can be constructed by partitioning the four-dimensional space into subregions. In every subregion the temporal coordinate defines a time spectrum, and each configurational coordinate defines a local spectral domain. The solution in the subregion can then be solved in the original coordinate system, transformed domain, or mixed (phase) domain. Furthermore, there are various ways of partitioning spectra and of performing inverse transforms. Therefore, there exist usually many solution methods available for each subregion, and proper options must be chosen to provide physical insight and to ensure numerical efficiency and accuracy. (Hybrid methods may be desired even in

Directions in Electromagnetic Wave Modeling
Edited by H.L. Bertoni and L.B. Felsen, Plenum Press, New York, 1991

some subregions.) To integrate various methods in these subregions, we impose boundary conditions at the interfaces between subregions. The field variables along these interfaces are then formulated in terms of system equations which are to be solved numerically. Efficient iterative schemes for solving the system equations are available when the coupling among subsystems is weak. To illustrate the scenario we consider a scatterer or an aperture embedded in a global environment. Typical global propagators of the environment, local scattering operators of the scatterer or the aperture, and a systematic way of combining them are discussed in Section II.

A specific example employing ray-mode-moment method to analyze an aperture coupled system is shown in Section III. The method of moments serves as a local coupling operator, and the hybrid ray-mode method serves as a global propagator. Based on the equivalence principle, the aperture is replaced by an electric conducting surface with unknown magnetic currents on both sides of this surface, and the system is then divided into subregions. The response of each subregion is governed by its Green's function. By invoking the boundary conditions, we obtain a system equation which may be solved by the method of moments. By expressing these unknown distributions in terms of appropriate basis functions and choosing appropriate weighting functions, the system equation is then reduced to a matrix equation which is solved numerically. Note, one needs to compute the Green's function of the waveguide for various arrangements of locations of source and receiver in the system equation. The hybrid ray-mode formulation is best suited for this purpose because it combines rays and modes self-consistently to optimize the advantages of each. The efficiency of the ray-mode Green's function makes use of the moment method practical in waveguides, and the flexibility of the moment method permits the application of the Green's function considered to arbitrarily shaped and placed apertures.

II. OVERVIEW OF HYBRID METHODS

A. Global Propagators

For propagation in a stratified medium, discrete or continuous transforms can be applied to the lateral coordinates. The remaining transverse coordinate can then be solved by numerical methods (such as finite elements and finite difference), analytic methods (such as characteristic Green's functions, invariant embedding, reflectivity, propagator, etc.), or asymptotic methods (such as WKBJ). Therefore, many alternative numerical schemes are available. According to the techniques employed for the three configurational coordinates, we have spectra - finite elements [1], spectra - finite difference [2], depth mode - lateral ray [3], etc. According to the ways of partitioning spectra and of performing inverse transform(s), we have spectral integral, mode, generalized ray, ray-mode [4,5], etc.

In the generalized ray representation the field is resolved into progression (traveling) waves along (ray) trajectories that chart the local progress of the motion via multiple reflections and refractions between source and observer. When observation is made at the source point, separation of the singular contribution, i.e., the direct ray term, is very convenient for this ray representation. This scheme becomes intractable for large source-observer separation where many rays must be included. When many layers are present, ray proliferation could be serious even for small source-observer separation.

The spectral integration approach highlights the features associated with the various plane waves (propagating and evanescent) that synthesize the source distribution. It represents all of the rays collectively and, therefore, can alleviate the difficulties of the ray proliferation. This method becomes inefficient at high frequencies and/or for large source-observer separation due to the strong oscillatory behavior of the integrand.

The modal expansion represents the field globally in terms of normal modes which propagate with different velocities in the direction parallel to the layers. It becomes inefficient at high frequencies and/or small source-observer separation where many modes are required.

These three conventional approaches for the Green's function of a layered structure have complementary properties. None of them are satisfactory over broad ranges of parameter regimes. Realizing that spectral intervals containing clusters of rays or modes are sparsely filled by modes or rays, respectively, the hybrid ray-mode method combines rays, modes and remainders self-consistently within a single framework, so as to optimize the advantages of each. Here the remainders are actually modified spectral representations accounting for truncation effects of ray or mode series, and are represented by partial sums of plane waves with complex spectra. This hybrid solution has advantages of the three conventional representations, i.e., separation of the singular term, remedy of the difficulties due to the ray proliferation or mode clustering, and efficiency for all possible arrangements of source and receiver locations. In some parameter regimes, the general hybrid ray-mode representation may reduce to one, or a combination of two, of the three conventional representations, when the reduced form is more convenient.

If the medium properties are not laterally invariant, the exact transform theory fails. Nevertheless, local modes [6], intrinsic modes [7], asymptotic rays [8], paraxial beams [8,9], and the adiabatic transform [10] are still useful in a weakly range-dependent medium. Some semi-numerical methods, such as parabolic equation - implicit finite difference [11], are also available.

When the medium does not have a preferential direction, the guided mode concept is no longer useful. However, the local plane wave spectra, such as asymptotic rays and paraxial beams [8,9], remain useful if, except across interfaces, the variations of the medium properties occur over a scale length much larger than the wavelength.

B. Scattering Operators

If the size of the scatterer (or aperture) is small, discrete coordinate methods such as finite difference and finite elements are suitable for modeling the interior responses. Coupling between the interior and exterior can be solved numerically by the boundary element or moment method.

C. Combination of Global Propagators and Local Scattering Operators

By imposing the boundary conditions on the the aperture or the surface of a scatterer, the global propagator and scattering operator will be integrated together [12]. One way of doing so is to formulate the problem in terms of integral equations by invoking Green's theorem. By expressing the unknown field distributions along the scattering surface in terms of appropriate basis functions, these integral equations are then discretized and reduced to algebraic equations that are solved numerically. In the integral equations, the exterior Green's function is evaluated in terms of global propagators, and the interior response is modeled by local scattering operators.

III. EXAMPLE

The analysis of wave scattering from an aperture coupled system is important in the areas of electromagnetic radiation, compatibility, scattering, interference, and shielding effects. We consider a two-dimensional problem where an incident plane wave is coupled through an aperture into a parallel plate waveguide (See Figure 1). The infinite conducting wall with an aperture of width $2w$, centered at the origin, separates the structure into two regions. The half space is denoted by region "a"; the interior of the waveguide is denoted by region "b". The time dependence $\exp(-i\omega t)$ is assumed and suppressed.

A. Formulation

Based on the equivalence principle, we divide the original structure into two problems. The aperture is replaced by an electric conducting surface with unknown magnetic currents M and $-M$ on the upper and lower sides, respectively, of this surface. Let E be the electric field on the aperture in Fig. 1, and U_z be a unit vector in the z-direction. The two new problems in are equivalent to the original problem in Figure 1 if we assign $M = U_z \times E$. In region "a" (i.e., the half space), the field is excited by the incident wave and M. In region "b" (i.e., the waveguide), the field is excited by $-M$. The magnetic current M is to be determined by imposing appropriate boundary conditions on the aperture. Across the aperture, the continuity of the tangential components of the electric field E_t is ensured by the arrangement of choosing M in region "a" and $-M$ in region "b". The continuity of the tangential components of magnetic field H_t on the aperture leads to the operator equation

$$\underline{H}_t^a(i) + \underline{H}_t^a(\underline{M}) = \underline{H}_t^b(-\underline{M}), \qquad |x| \leq w, \; z = 0 \tag{1}$$

where the superscript specifies the side of the aperture, and the argument in ( ) denotes the source which generates the field. Here, (i) denotes the incident source.

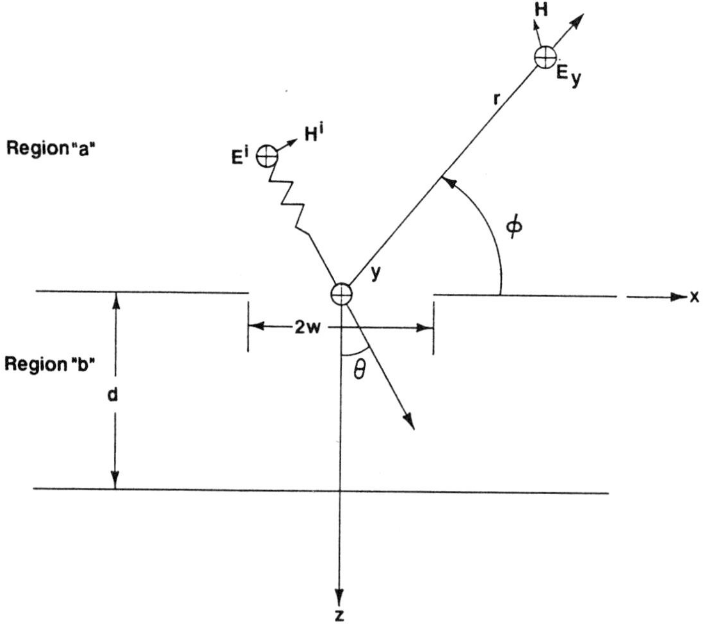

Fig. 1 A plane wave incident upon a slot aperture on one of the walls of a parallel-plate waveguide.

For an E-polarization plane wave (denoted by E^i and H^i in Fig. 1) incidence

$$\underline{H}_t^a(i) = 2H_0 \cos\theta \, e^{ikx\sin\theta} \, \underline{U}_x, \qquad \text{at} \quad z = 0 \tag{2}$$

where k, θ, and H_0 are the wavenumber, incident angle, and magnitude of the magnetic field of the incident wave (see Fig. 1). Here, U_x is a unit vector in the x-direction, and the factor two is due to the reflection at the conducting surface at $z = 0$. Since the magnetic current M has only an x-component, it is convenient to use the electric vector potential formulation for finding $\underline{H}_t^a$ and $\underline{H}_t^b$. Let $F^a \underline{U}_x$ and $F^b \underline{U}_x$ satisfy

$$\left(\nabla^2 + k^2\right)\begin{Bmatrix} F^a \\ F^b \end{Bmatrix}\underline{U}_x = \begin{Bmatrix} - \\ + \end{Bmatrix} M\underline{U}_x, \qquad \begin{Bmatrix} z \leq 0 \\ 0 \leq z \leq d \end{Bmatrix} \tag{3}$$

Then, the electric and magnetic fields are

$$\underline{E}^p = -\nabla \times F^p \underline{U}_x \ , \quad \underline{H}^p = i\omega\varepsilon F^p \underline{U}_x - \frac{1}{i\omega\mu}\nabla(\nabla\cdot F^p \underline{U}_x) \tag{4}$$

where the superscript p stands for either "a" or "b". To solve (3), let g^a and g^b be the Green's function in region "a" and "b", respectively, which satisfy the following wave equations

$$\left(\nabla^2 + k^2\right)\begin{Bmatrix} g^a \\ g^b \end{Bmatrix} = -\,\delta(x-x')\,\delta(z-z'), \qquad \begin{Bmatrix} z\leq 0 \\ d\geq z\geq 0 \end{Bmatrix} \tag{5}$$

and boundary conditions

$$\frac{\partial g^a}{\partial z}\Big|_{z=0} = 0, \quad \frac{\partial g^b}{\partial z}\Big|_{z=0,d} = 0 \tag{5a}$$

The tangential magnetic fields in (1) are then expressed

$$\begin{Bmatrix} \underline{H}_t^{\,a}(\underline{M}) \\ \underline{H}_t^{\,b}(-\underline{M}) \end{Bmatrix} = \underline{U}_x \left(i\omega\varepsilon - \frac{1}{i\omega\mu}\frac{\partial^2}{\partial x^2}\right)\int_{-w}^{w} \begin{Bmatrix} g^a(x,z,x',z') \\ -g^b(x,z,x',z') \end{Bmatrix} M(x')dx', \quad z'=0 \tag{6}$$

B. Global Propagators: Green's Functions

In region "a" (half space), the Green's function solution of (5) is two times the free space Green's function:

$$g^a = 2\left(\frac{i}{4}H_0^{(1)}\left(k\sqrt{(x-x')^2 + (z-z')^2}\right)\right) \tag{7}$$

where the factor two in (7) is from the image theory.

In region "b" (waveguide), the Green's function is represented in terms of a combination of rays, modes and a remainder

$$g^b = \left[\sum_{l=1}^{4}\sum_{n=0}^{N-1} g_{nl}\right]_{rays} + \left[\sum_{m}^{M} g_m\right]_{modes} + \left[R_N\right]_{remainder} \tag{8}$$

Here, a typical ray of the l*-th* species and with n reverberations is given by

$$g_{nl} = \frac{i}{4}H_0^{(1)}(kr_{nl}), \quad r_{nl} = \sqrt{(x-x')^2 + (z-z_{nl})^2}, \quad z_{nl} = \begin{cases} |z-z'| + 2nd, & l=1 \\ z+z' + 2nd, & l=2 \\ -z-z' + 2(n+1)d, & l=3 \\ -|z-z'| + 2(n+1)d, & l=4 \end{cases} \tag{8a}$$

and the m*-th* mode is specified by

$$g_m = \frac{i\alpha_m}{\varsigma d}\cos(\kappa z)\cos(\kappa z')\exp\left[i\varsigma\,|x-x'|\right], \quad \varsigma = \varsigma_m, \quad \alpha_m = \begin{cases} 1, & m>0 \\ 0.5, & m=0 \end{cases}, \quad \mathrm{Im}(\varsigma_m)\geq 0 \tag{8b}$$

Playing a key role of determining the numbers of rays and modes, the remainder is defined as

$$R_N = \int_{C'} A(\varsigma)\exp\left[i\,P(\varsigma)\right]d\varsigma \tag{9}$$

where the phase term

$$P(\varsigma) = 2N\kappa d + \varsigma(x-x'), \quad \kappa = (k^2-\varsigma^2)^{\frac{1}{2}}, \quad \mathrm{Im}(\kappa)\geq 0 \tag{9a}$$

and the amplitude function

$$A(\varsigma) = \frac{\left[e^{i\kappa|z-z'|} + e^{i\kappa(z+z')} + e^{i\kappa(2d-(z+z'))} + e^{i(2d-\kappa|z-z'|)}\right]}{-4\pi i\kappa\,(1-e^{i2\kappa d})}, \tag{9b}$$

The integration contour C' is the steepest descent path specified by the condition

$$P(\varsigma) = P(\varsigma_s) + iu \ , \ u > 0 \tag{10}$$

where ς_s is the saddle point

$$\varsigma_s = k\,(x-x')\,[(2Nd)^2 + (x-x')^2]^{-\frac{1}{2}} \tag{10a}$$

The contour C' in (9) partitions the entire spectra into two parts which are to be accounted by rays and modes, respectively. The mode sum in (8) is contributed from those modes g_m intercepted during the path deformation from the real axis C to the steepest descent contour C', and the ray sum in (8) is from those rays g_{nl} with n less than N. The ability of explicit separation of the direct ray g_{01} in (8) is important for evaluating the singular contribution in the method of moments in Sec. III.C. when source and receiver are overlapping each other. One criterion to determine the number N in (8) is related to how easy one can compute the remainder in (9) where a first order asymptotic approximation, when possible, is preferred. A rule of thumb is to keep the saddle point ς_s away from pole ς_m's and to make sure that the amplitude is relatively a slowly varying function with respect to the phase. The other criterion of choosing N is to minimize the total number of rays and modes. If computing rays is much easier than computing modes or vice versa, one may try to minimize the number of rays or modes, respectively.

For sufficiently large N, the saddle point approaches zero and no modes will be intercepted (unless $\varsigma=0$ is a pole of (9)). The hybrid ray-mode solution in (8) is reduced to the ray plus remainder solution where the remainder can be interpreted as a collective ray accounting for rays g_{nl} with $n \geq N$. If $N = \infty$, (8) is reduced to the conventional ray solution. If $N=0$, no saddle point exists for the phase in (9a). When choosing C' to be the real axis C, the hybrid ray-mode solution is reduced to the conventional plane wave spectral representation. Closing the integration path C by a semicircle at infinity and employing the residue theorem, the ray-mode solution is reduced to the conventional modal solution. Since the separation of amplitude and phase in a spectral integral such as (9) is not unique, the steepest descent contour is also not unique. Therefore the number of modes included in (8) may depend on the definition of phase in (9a). No approximation is made so far, and all options are exact solutions. They can be derived from the general solution in (8) by simply choosing proper values of a single parameter N. This feature is very suitable for numerical implementation to optimize the computational efficiency. Since the mode sum converges slowly for small $x-x'$ and the ray sum converges slowly for large $x-x'$, it is suggested to predetermine three threshold values X_1, X_2 and X_3, $X_1 \leq X_2 \leq X_3$, where the mode, ray-mode, ray-remainder, and ray formats are employed for $X_3 < |x-x'|$, $X_3 \geq |x-x'| \geq X_2$, $X_2 \geq |x-x'| \geq X_1$, and $X_1 > |x-x'|$, respectively. The determination of X_i, $i=1,2,3$, depends on the computational efficiency of each individual term and the converging efficiency of these options. Fortunately small variation of these threshold values does not affect the accuracy or efficiency of the algorithm.

C. Method of Moments (MOM)

Following the procedures of MOM, $\underline{M}$ is expressed as $\underline{M} \approx V_m \underline{M}_m$ where $\underline{M}_m$, $m=1,2,...,N$ is the expansion function for $\underline{M}$. We define a coefficient vector $\vec{V}$ as $\vec{V} = [V_m]_{N\times 1}$. Let $\underline{W}_m$, $m=1,2,3,...,N$ be the weighting functions and define an inner product of two functions as

$$< \underline{f}_1, \underline{f}_2 > = \int_{-w}^{w} \underline{f}_1 \cdot \underline{f}_2 \, dx \qquad \text{at } z = 0 \tag{11}$$

Henceforth, (1) is discretized and reduced to the following equation:

$$\left\{[Y^a] + [Y^b]\right\}\vec{V} = \vec{I}^i \tag{12}$$

where the admittance matrices are

$$[Y^a] = [Y_{mn}{}^a]_{N\times N} \ , \qquad Y_{mn}{}^a = < \underline{W}_m , \underline{H}_t{}^a(\underline{M}_n) > \tag{12a}$$

$$[Y^b] = [Y_{mn}{}^b]_{N\times N} \ , \qquad Y_{mn}{}^b = < \underline{W}_m , \underline{H}_t{}^b(\underline{M}_n) > \tag{12b}$$

for regions "a" and "b", respectively, and the incident source vector is

$$\vec{I}^i = [I_m{}^i]_{N\times 1} \qquad I_m{}^i = < -\underline{W}_m , \underline{H}_t{}^a(i) > \tag{12c}$$

The solutions of (12) are

$$\vec{V} = \left\{[Y^a] + [Y^b]\right\}^{-1} \vec{I}^i \tag{13}$$

From (4), the fields at any arbitrary location can be evaluated.

D. Numerical Results

The configuration employed for calculation is shown in Fig. 1. At first we choose three incident angles, two aperture widths and two waveguide heights. The normalized amplitude of the tangential electric field $\bar{E}_y = E_y/E^i$ at the aperture are shown in Fig. 2. Note that the equivalent magnetic current M has only an x component for the present case. In all cases, E_y (or M_x) approaches zero at $x = \pm w$, which is consistent with the edge condition. The scattered electric field E_y at a semi-circle with radius $r = 100\lambda$ due to the equivalent source M_x is shown in Fig. 3. The angles corresponding to the maximums of the scattered field in the far zone for various incident angles are coincidence with the geometric reflection angles. In these examples, the case with smaller incident angle, waveguide height, or aperture size excites a stronger peak scattered field. E_y on various cross-sections of the waveguide for $w/\lambda=1$, $d/\lambda=5.2$ and $\theta=30^o$ is shown in Fig. 4. Near the aperture, the field in the illuminated region (shown by dashed lines) is stronger than that in the shadow region. The mode-like interference pattern is due to the multiple reflection of the incident and diffracted waves in the waveguide. However, the pattern is not stabilized as the wave propagates along the waveguide. This is due to the fact that more than one propagating mode is excited. Those modes with modal angles which line up with or close to the incident angle are strongly excited and determine the periodicity of the "standing wave" pattern. This phenomena is more profound when the size of the aperture is larger. In the next example we choose $w/\lambda=9.6$ and $d/\lambda=5.2$. The magnitude of $\bar{E}_y$ at the cross-section at $x = 24\lambda$ for two incident angles is shown in Fig. 5. The solid curve corresponds to the incident angle $\theta=\cos^{-1}(4\lambda/d)=39.71^o$ which is the 8-th modal angle The dashed curve is corresponding to the incident angle $\theta=35.18^o$ which is between the 8-th and 9-th modal angles. Note that the solid curve has four full periods and is more like a mode, and the dashed curve has four and half periods which is caused by the interference between modes eight and nine.

E. Discussion

It has been demonstrated that aperture coupled waveguide systems can be efficiently analyzed using a new method which combines the moment method with the hybrid ray-mode Green's function formulation. The efficiency is mainly gained by using the hybrid ray-mode method over the conventional ray or mode method. This new method has involved proper partitions in both configurational and spectral domain to ensure efficiency and flexibility. The moment method's advantages are its flexibility to model arbitrarily shaped, and placed, finite apertures. The ray-mode method has to date only been considered desirable for large waveguides which support many propagating modes. Here we have shown that the concept of combining resonant modes and traveling rays is still useful in a small waveguide. The numerical efficiency of the ray-mode Green's function makes use of the moment method practical in a waveguide system. The concept of combining an efficient global propagator (ray-mode) and a flexible local coupling operator (moment method) is potentially very powerful for many problems with planar symmetry.

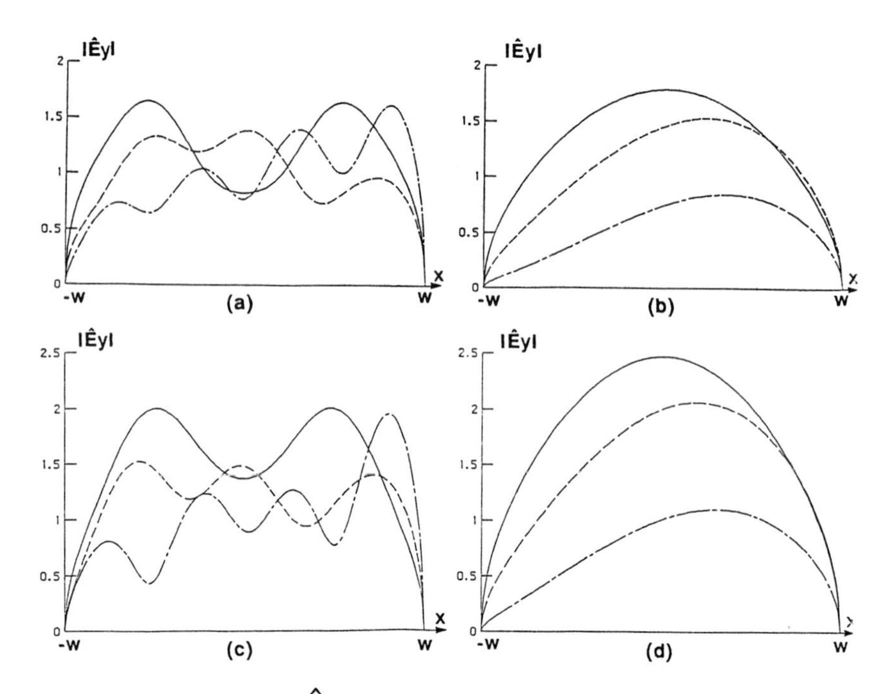

Fig. 2 The amplitude of $\hat{E}_y$ at the aperture of the configuration in Fig. 1 with
(a) $d/\lambda=5.2$ and $w/\lambda=1$;
(b) $d/\lambda=5.2$ and $w/\lambda=0.25$;
(c) $d/\lambda=2.4$ and $w/\lambda=1$;
(d) $d/\lambda=2.4$ and $w/\lambda=0.25$.
Legends: $\theta=0^o$ (________); $\theta=30^o$ (- - - - -); $\theta=60^o$ (_ . _ . _ .).

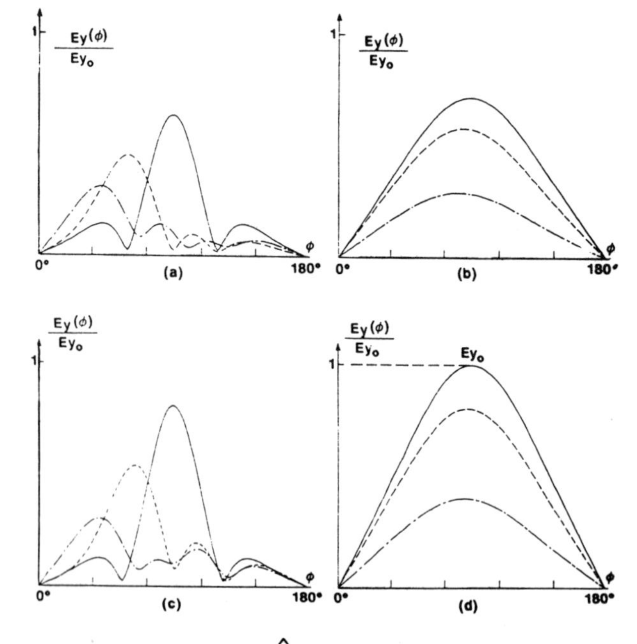

Fig. 3 The normalized magnitude of $\hat{E}_y$ at a semi-circle with radius $r=100\lambda$ versus ϕ (see Fig. 1). The parameters are the same as those in Fig. 2.

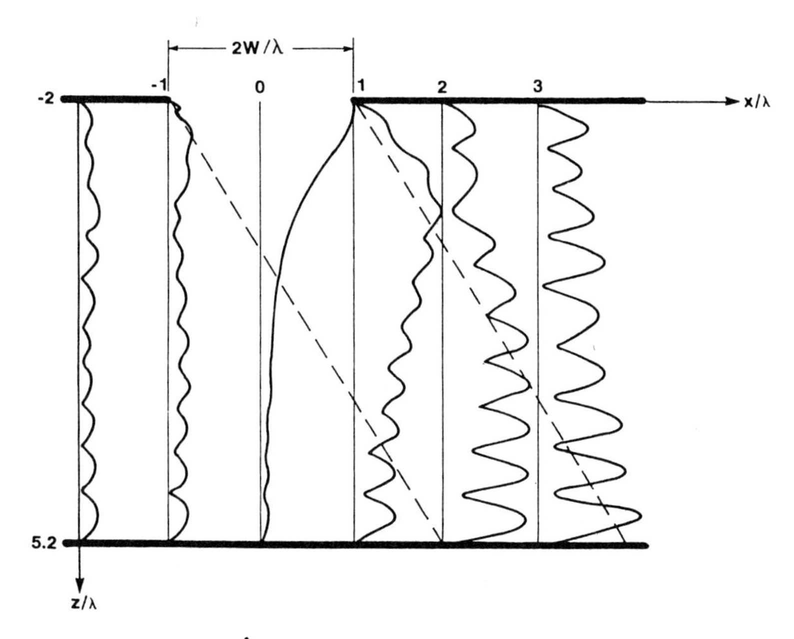

Fig. 4 The magnitude of $\hat{E}_y$ at various cross-sections of the waveguide in Fig. 1 with $\theta=30^o$, $d/\lambda=5.2$ and $w/\lambda=1$.

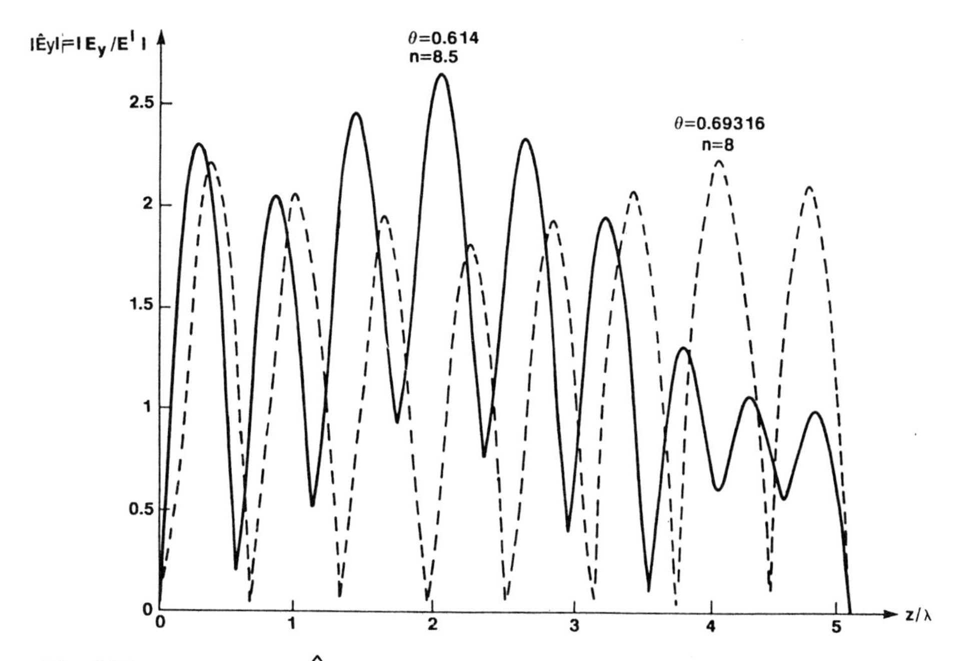

Fig. 5 The magnitude of $\hat{E}_y$ at the cross-section at $x=24\lambda$ of the waveguide in Fig. 1 with $d/\lambda=5.2$ and $w/\lambda=9.6$. The solid and dashed curves are corresponding to $\theta=39.71^o$ and $\theta=35.18^o$, respectively.

IV. CONCLUSION

Reliable and efficient algorithms for wave propagation and scattering in large and complicated structures are to be developed. Since no rigorous solution is available, a hybrid combination of various methods may be *the* way. To construct a hybrid scheme, the first step is to specify the applicable parameter regimes and to quantify the numerical efficiency and error of each individual method to be employed. Even these aspects are missing for many widely employed methods such as parabolic equation, beam shooting, local mode, etc. The second step is to determine a combining strategy which indicates *a priori* where to employ a certain method and how to combine various methods together. This decision making process has to be efficient, systematic, and physics oriented. (Otherwise, the hybrid method becomes a numbers game.) The last step is to have an overall assessment of various aspects (such as applicable ranges, errors, efficiency, compatibility with other methods, etc.) of the hybrid scheme. To make an algorithm robust and useful, this is a crucial step. In conclusion, to develop new hybrid schemes for unsolved problems and to make these new algorithms robust and efficient are the future directions.

ACKNOWLEDGEMENT

This work is supported in part by the National Science Foundation under grant No. ECS-8707615 and by the Joint Services Electronics Program under Contract No. F49620-88-C-0075.

REFERENCES

1. A.H. Olson, J.A. Orcutt and G.A. Frazier, "The discrete wave number - finite element method for synthetic seismograms," Geophys. J.R. Astr. Soc., 1983.
2. A.S. Alekseev and B.G. Mikhailenko, "The solution of dynamic problems of elastic wave propagation in inhomogeneous media by a combination of partial separation of variables and finite-difference methods," J. Geophys., 48, 161-172, 1980.
3. L. B. Felsen, I. T. Lu and K. Naishadham, "Ray formulation of waves guided by cylindrically stratified dielectrics", the proceeding of the URSI EM Theory Symposium 1989, CONGREX AB, Sweden.
4. I.T. Lu and L.B. Felsen, "Ray mode and hybrid options for source-excited propagation in an elastic plate," J. Acoust. Soc. Am., 78, 701-704 1985.
5. I.T. Lu and Felsen, "Matrix Green's for array-type sources and receivers in multiwave layered media," Geophys. J.R. Astron, Soc., 84, 31-42, 1986.
6. S.A. Chin-Bing, "Strategies for the solution of scattering from rough interfaces in a waveguide," Proceeding of the second IMACS Symposium on Computational Acoustics, Princeton, Mar. 1989.
7. I. T. Lu and L. B. Felsen, "Intrinsic modes in a wedge-shaped ocean with stratified bottom," Proceedings of the Conference on Ocean Seismo-Acoustics, edited by T. Akal and M. Berkson, Plenum Press, NY, 1986.
8. Cerveny, "Gaussian beam synthetic seismograms," J. Geophys. Res., 58, 44-72, 1985.
9. I. T. Lu, L. B. Felsen and Y. Z. Ruan, "Spectral aspects of the Gaussian method: reflection from an isovelocity half space," Geophys. J. R. Astron. Soc., 89, 915-932, 1987.
10. I.T. Lu and L.B. Felsen, "Adiabatic transforms for spectral analysis and synthesis of weakly range-dependent shallow ocean Green's functions," J.Acoust. Soc. Am., 81, 897-911, 1987.
11. D. Lee and J. Papadakis, "Numerical solutions of the parabolic wave equation: An ordinary-differential-equation approach," J. Acoust. Soc. Am., 68(5), 1482-1488.
12. I. T. Lu and H.K. Jung, "A Hybrid boundary elements - finite elements - ray - mode method for wave scattering by inhomogeneous scatterers in a waveguide," J. Acoust. Soc. Am., 87, 3, 988-996, 1990.

WAVE THEORY MODELING IN UNDERWATER ACOUSTICS

Henrik Schmidt

Department of Ocean Engineering
Massachusetts Institute of Technology
Cambridge, Massachusetts 02139

The ocean acoustics environment is a complicated waveguide bounded above by a rough sea surface or ice cover and below by an inhomogeneous, elastic sea bed. Further, the acoustic properties of the water column are dependent on temperature, pressure and salinity, giving rise to a significant spatial and temporal variation. Analytical treatment of the propagation is limited to a very few, canonical problems, with the only alternative being numerical approaches. However, due to the complexity of the environment, no single numerical approach is capable of treating the general propagation problem. Therefore, a suite of numerical techniques have been developed and implemented, each applicable to a limited class of problems. Thus, normal mode approaches have been applied to longer range propagation problems with little bottom interaction and weak range dependence of the environment; parabolic equation methods have been applied to strongly range-dependent problems; and wavenumber integration or integral transform methods have been applied for short range propagation involving significant interaction with the elastic bottom. However, in recent years, each of these approaches have been further developed to incorporate more and more environmental effects, allowing a significant overlap in the areas of applicability. Further, the traditional restriction to 2-dimensional, one-way propagation scenarios has been abandoned, with a more realistic treatment of 3-dimensional propagation and reverberation problems as a result. Here, the various methods are described, and the capabilities of the state-of-the art ocean seismic-acoustic models will be illustrated through a series of characteristic examples.

Directions in Electromagnetic Wave Modeling
Edited by H.L. Bertoni and L.B. Felsen, Plenum Press, New York, 1991

SEISMIC WAVEFORM MODELLING OF HIGH-FREQUENCY BODY WAVES

C.H. Chapman[1] and R.T. Coates[2]

Department of Earth Sciences
University of Cambridge
Cambridge CB3 0EZ, U.K.

INTRODUCTION

The Earth is a complicated inhomogeneous, anelastic and anisotropic medium. Even though the fundamental equations of elasticity are well known and understood, modelling the propagation of seismic waves is non-trivial, and the interpretation or inverse problem is difficult. Many aspects of seismic wave propagation have their counterparts in other fields, and similar techniques can be used to model the propagation. Nevertheless, for realistic Earth structures, certain problems unique to seismology arise.

Expensive numerical computations, e.g. finite difference methods, are feasible for realistic 2D and 3D models (although the latter stretch most existing computing facilities). If it were only necessary to study a few standard structures, such computations might be adequate, but given the great variety of scales and combinations of structures encountered in seismology, it is necessary to use more efficient approximate methods. Even using finite difference computations, it is unlikely that all features of seismic data can be modelled, and again approximate methods are more appropriate. Finite difference calculations have to be interpreted using the same approximate methods used to interpret real data.

Analytic studies of simple, canonical elastodynamic problems, e.g. reflection from a plane interface, diffraction by a sphere, wedge, etc., are instructive for seismic modelling, but similar analytic methods have limited use in realistic models. The details of these analytic methods usually depend on the exact analytic properties of the model. Earth structures can rarely be modelled using analytic functions, and numerically specified models have complicated, unstable analytic properties. The small variations in models required during interpretation cause large, unstable changes in the analytic properties of the model and intermediate results. The application of such methods to realistic models has been very limited.

Asymptotic ray theory (ART) and various extensions, e.g. Maslov asymptotic theory, Gaussian beam theory, geometrical theory of diffraction, etc., are widely used for seismic modelling and interpretation. Nevertheless, only the zeroth-order term, e.g. geometrical ray theory (GRT), is generally useful, and higher-order terms do not model

[1]On leave from: Department of Physics, University of Toronto, Toronto, Ontario M5S 1A7, Canada
[2]Now at: Earth Resources Laboratory, Department of Earth, Atmospheric and Planetary Science, Massachusetts Institute of Technology, Cambridge, MA 01242, U.S.A.

Directions in Electromagnetic Wave Modeling
Edited by H.L. Bertoni and L.B. Felsen, Plenum Press, New York, 1991

useful corrections. This is partly because models and results are only known numerically and not analytically, but more fundamentally is a general limitation of asymptotic methods. Signals reflected or scattered by smooth but rapid parameter changes are not modelled by ART as they are asymptotically small, but nevertheless they may be physically very significant. The errors of asymptotic methods are generally difficult to predict, and the accuracy of most seismic modelling methods are unknown for realistic models.

The basic theory for ART in anisotropic media is well known (e.g. Červený, 1972). The normal procedure for ART in anisotropic media is to introduce an ansatz for the displacement and eliminate the stress between the equations of motion and the constitutive equations. Second-order differential equations are then obtained connecting the amplitude coefficients and travel time. When we restrict our solution to the geometrical ray approximation, this leads to inconsistencies — the displacement only has one term but the stress contains two terms. In other words, the geometrical ray approximation for displacement will satisfy the constitutive equation exactly but not the equation of motion. This is physically illogical. The displacement and stress are equally significant, and neither set of equations has priority for an approximation. It is physically more logical to introduce ansätze for both displacement and stress, and approximate both series to the same order. This modification is described in the next section. The use of traction vectors $\mathbf{t}_j$ in ART is also new and simplifies the algebra. In the following 2 sections we briefly outline the solutions of the eikonal and transport equations, which in ray theory are known as *kinematic* and *dynamic ray tracing.*

GRT breaks down when the solution varies rapidly and this occurs in many situations in seismology. In general, higher-order terms in ART do not improve the situation. The break down may occur when geometrical spreading varies rapidly, e.g. at caustics, or due to inhomogeneities. The latter will cause non-geometrical, scattered signals. In addition, in anisotropic media, ray theory will break down near degeneracies in the ray types, e.g. when quasi-shear waves have similar velocities. In the final 3 sections of this paper, we describe 2 theories that solve these problems — Maslov asymptotic ray theory (MART) and a generalization of Born scattering theory. MART may be used at and near caustics, cusps, etc. (Chapman, 1985). A major simplification occurs in seismology as we require the impulse response — the transform integral cancels and need not be calculated in Maslov geometrical ray theory seismograms, described in the second section.

Born scattering theory has been widely used to describe scattering of elastic waves but normally requires the separation of the model into a background, reference model and a scattering perturbation. In the final section of this paper, a generalization is described where the scattered signal is generated by errors in the GRT Green's functions (Coates and Chapman, 1991). This allows scattered signals in inhomogeneous media to be studied. One application is to the coupling of quasi-shear rays that occurs near degeneracies (Coates and Chapman, 1990). Unlike conventional ray theory, the resultant coupling theory agrees with ray theory in isotropic media.

These 2 extensions of GRT are essential in seismology to describe the many non-geometrical signals observed. A combined theory generalizing Born scattering theory using MART is also needed and is currently under development.

ASYMPTOTIC RAY THEORY IN ANISOTROPIC MEDIA

The transformed equation of motion is

$$\frac{\partial \hat{\sigma}_{ij}}{\partial x_j} = -\rho\omega^2 \hat{u}_i - \delta_{ik}\delta(\mathbf{x} - \mathbf{x}_s) \tag{1}$$

where σ_{ij} is the stress tensor, ρ the density, ω the angular frequency, and u_i the infinitesimal displacement. For generality we have introduced a point force source at $\mathbf{x}_s$

in the k–th direction. The transformed constitutive equation in an anisotropic medium is

$$\hat{\sigma}_{ij} = c_{ijkl}\hat{e}_{kl} = c_{ijkl}\frac{\partial \hat{u}_k}{\partial x_l} \tag{2}$$

where c_{ijkl} is the 4–th order tensor of elastic parameters and e_{kl} is the strain tensor. In order to simplify the following equations, we introduce a notation close to that of Woodhouse (1974). We decompose the stress tensor into its cartesian traction vectors, i.e. $(\mathbf{t}_j)_i = \sigma_{ij}$, so $\mathbf{t}_j$ is the traction on a surface perpendicular to the j–th axis. The constitutive equation (2) can then be rewritten as

$$\hat{\mathbf{t}}_j = \mathbf{c}_{jk}\frac{\partial \hat{\mathbf{u}}}{\partial x_k} \tag{3}$$

where the $\mathbf{c}_{jk}$ are 3×3 matrices such that $(\mathbf{c}_{jk})_{il} = c_{ijlk}$.

From studies in homogeneous, non-attenuating media, it is well known that elastic waves propagate without dispersion with a velocity independent of frequency. At interfaces between homogeneous media, plane waves satisfy Snell's law. In inhomogeneous media, we expect similar behaviour at high frequencies when the wavelength is short compared with the scale of inhomogeneities in the medium. This physical picture is built into the ansatz of ART. We write the Green's functions as

$$\hat{\mathbf{u}}^{(\mathrm{ART})}(\mathbf{x},\omega;\mathbf{x}_s) = \sum_i \sum_{m=0}^{\infty} \frac{\mathbf{U}^{(m)}(\mathbf{x};\mathcal{L}_i^s)}{(-i\omega)^m} e^{i\omega T(\mathbf{x};\mathcal{L}_i^s)} \tag{4}$$

$$\hat{\mathbf{t}}_j^{(\mathrm{ART})}(\mathbf{x},\omega;\mathbf{x}_s) = \sum_i \sum_{m=0}^{\infty} \frac{\mathbf{T}_j^{(m)}(\mathbf{x};\mathcal{L}_i^s)}{(-i\omega)^{m-1}} e^{i\omega T(\mathbf{x};\mathcal{L}_i^s)}. \tag{5}$$

The *amplitude coefficients*, $\mathbf{U}^{(m)}(\mathbf{x};\mathcal{L}_i^s)$ and $\mathbf{T}_j^{(m)}(\mathbf{x};\mathcal{L}_i^s)$ (which are 3×3 matrices), and the *travel time*, $T(\mathbf{x};\mathcal{L}_i^s)$, depend on the initial conditions and observation point $\mathbf{x}$. The solutions may also be multi-valued. From a point source at $\mathbf{x}_s$, several ray paths may exist to the same point $\mathbf{x}$. This situation is quite common in seismology, e.g. the triplications for rays through the upper mantle. In addition anisotropy may generate multi-pathing directly, i.e. cusps on the qS wavefronts due to inflections in the slowness surfaces. We indicate a ray path from $\mathbf{x}_s$ to $\mathbf{x}$ by $\mathcal{L}_i(\mathbf{x},\mathbf{x}_s)$. The subscript i enumerates the possible ray paths. For brevity, we simplify the full notation by writing $(\mathbf{x};\mathcal{L}_i^s) = (\mathbf{x};\mathcal{L}_i(\mathbf{x},\mathbf{x}_s))$ when the first argument is the same as the end of the ray $\mathcal{L}_i$.

The frequency dependence of the amplitude functions is restricted to the sign. To obtain a real time series for the solution, we must have $\mathbf{U}^{(m)*}(\mathbf{x},\omega;\mathcal{L}_i^s) = \mathbf{U}^{(m)}(\mathbf{x},-\omega;\mathcal{L}_i^s)$. Since this is the only frequency dependence, we omit it from the amplitude coefficients and where necessary understand the value for positive frequencies. At high frequencies or short times, we can approximate the ART expansion by the first term. Thus we obtain

$$\hat{\mathbf{u}}^{(\mathrm{GRT})}(\mathbf{x},\omega;\mathbf{x}_s) = \sum_i \mathbf{U}^{(0)}(\mathbf{x};\mathcal{L}_i^s) e^{i\omega T(\mathbf{x};\mathcal{L}_i^s)}, \tag{6}$$

or in the time domain

$$\mathbf{u}^{(\mathrm{GRT})}(\mathbf{x},t;\mathbf{x}_s) = \sum_i \mathrm{Re}\left[\Delta\left(t - T(\mathbf{x};\mathcal{L}_i^s)\right) \mathbf{U}^{(0)}(\mathbf{x};\mathcal{L}_i^s)\right]. \tag{7}$$

where $\Delta(t) = \delta(t) - i\pi/t$ is the analytic delta function. We refer to this approximation as *geometrical ray theory* (GRT). Solutions for a specific source force can be obtained from the Green's solution by product and convolution.

Substituting equations (4) and (5) in the constitutive equation and the equation of motion, and equating the coefficient of each power of ω to zero, we obtain

$$\frac{\partial \mathbf{T}_j^{(m-1)}}{\partial x_j} = p_j \mathbf{T}_j^{(m)} + \rho \mathbf{U}^{(m)} \tag{8}$$

$$\mathbf{c}_{jk}\frac{\partial \mathbf{U}^{(m-1)}}{\partial x_k} = \mathbf{T}_j^{(m)} + p_k \mathbf{c}_{jk}\mathbf{U}^{(m)}, \qquad (9)$$

defining $\mathbf{U}^{(-1)} = \mathbf{T}_j^{(-1)} = \mathbf{0}$ and $\mathbf{p} = \nabla T$, the *phase slowness vector.* For brevity we omit the argument $(\mathbf{x}; \mathcal{L}_i^s)$.

THE EIKONAL EQUATION AND KINEMATIC RAY TRACING

When $m = 0$, the amplitude coefficients are related by

$$\mathbf{T}_j^{(0)} = -p_k \mathbf{c}_{jk}\mathbf{U}^{(0)} \qquad (10)$$

$$\rho \mathbf{U}^{(0)} = -p_j \mathbf{T}_j^{(0)}. \qquad (11)$$

Eliminating the traction between these equations, we obtain the eikonal equation for anisotropic media

$$(p_j p_k \mathbf{c}_{jk} - \rho \mathbf{I})\mathbf{U}^{(0)} = \mathbf{0}. \qquad (12)$$

This equation can be written as an eigenvector equation

$$(p_j p_k \mathbf{a}_{jk} - \mathbf{I})\mathbf{g} = 0 \qquad (13)$$

where $\mathbf{a}_{jk} = \mathbf{c}_{jk}/\rho$ are the anisotropic velocity parameters and $\mathbf{g}$ are the orthonormal eigenvectors (one corresponding to the actual polarization). Taking the product with $\frac{1}{2}\mathbf{g}^T$, the eikonal equation is equivalent to (Kendall and Thomson, 1989)

$$H(\mathbf{x}, \mathbf{p}) = \frac{1}{2}\left(p_j p_k \mathbf{g}^T \mathbf{a}_{jk}\mathbf{g} - 1\right) = 0. \qquad (14)$$

Substituting $\mathbf{p} = \nabla T$ in (14) we recognize the equation as equivalent to the Hamilton-Jacobi equation. The solution can be found solving the ordinary differential equations

$$\frac{dx_i}{dT} = \frac{\partial H}{\partial p_i} = p_k \mathbf{g}^T \mathbf{a}_{ik}\mathbf{g} = V_i \qquad (15)$$

$$\frac{dp_i}{dT} = -\frac{\partial H}{\partial x_i} = -\frac{1}{2} p_j p_k \mathbf{g}^T \frac{\partial \mathbf{a}_{jk}}{\partial x_i}\mathbf{g}, \qquad (16)$$

where the vector $\mathbf{V}$ is the *ray (group) velocity vector.* In mechanics these are known as the Hamilton equations, but in ray theory they are usually called the *kinematic ray equations.* Although they form a 6–th order differential system, not all the equations are independent. There are 2 constraints: the Hamiltonian (14) is zero, and

$$\mathbf{p} \cdot \mathbf{V} = 1 = \frac{V}{v}\cos\phi. \qquad (17)$$

Thus only a 4–th order system is necessary. Equation (17) can be obtained by taking the scalar product of equation (15) with $\mathbf{p}$ ($v = |\mathbf{p}|^{-1}$ is the *phase velocity* and $V = |\mathbf{V}|$ the *group velocity*). In anisotropic media, it is important to remember that the normal to the wavefront (in the direction of $\mathbf{p} = \nabla T$) is not parallel to the ray direction (in the direction of $\mathbf{V}$). The angle between these directions is ϕ. Given an initial position, $\mathbf{x}_s$, and an initial slowness vector, $\mathbf{p}_s$ (as $H = 0$ there are only 2 degrees of freedom specified by parameters q_1 and q_2, say), it is straightforward to solve the differential equations (15) and (16), either numerically or, for some special velocity structures, analytically.

The Lagrangian for the system can be obtained by the Legendre transform

$$L(\mathbf{x}, \dot{\mathbf{x}}) = \mathbf{p} \cdot \dot{\mathbf{x}} - H(\mathbf{x}, \mathbf{p}). \qquad (18)$$

By Hamilton's principle, the integral of the Lagrangian over the ray path, which is the travel time, is an extremum

$$T = \text{Ext} \int_{\mathcal{L}} L(\mathbf{x}, \dot{\mathbf{x}})\, dT = \int \mathbf{p} \cdot d\mathbf{x} = \text{Ext} \int_{\mathcal{L}} \frac{\cos\phi\, ds}{v} = \text{Ext} \int_{\mathcal{L}} \frac{ds}{V} \tag{19}$$

where ds is an element of ray length, using (17) and (18). In ray theory, Hamilton's principle is known as *Fermat's principle.*

THE TRANSPORT EQUATION AND DYNAMIC RAY TRACING

In order to obtain the amplitude coefficients, we consider equations (8) and (9) with $m = 1$. For simplicity we consider a single (vector) solution rather than the Green's function. Eliminating $\mathbf{T}_j^{(1)}$ between the equations, pre-multiplying by $\mathbf{U}^{(0)T}$ and using the transpose of the eikonal equation (12), we obtain

$$\nabla \cdot \mathbf{N} = 0 \tag{20}$$

where the vector $\mathbf{N}$ is the energy flux vector with $N_j = \mathbf{U}^{(0)T}\mathbf{T}_j^{(0)}$ (analogous to the Poynting vector in electromagnetism) and equation (20) states that energy is conserved. Substituting for the traction (10) in (20) and using (15), we obtain

$$\frac{d}{dT} \ln(\rho U^{(0)2}) = -\frac{\partial}{\partial x_j}\left(\frac{dx_j}{dT}\right) \tag{21}$$

where $\mathbf{U}^{(0)} = U^{(0)}\mathbf{g}$. This equation, or variants thereof, is known as the *transport equation.* For a differential equation like (15) with solutions $\mathbf{x} = \mathbf{x}(T, q_1, q_2)$ where q_1 and q_2 are *ray parameters* or *coordinates*, Smirnov's lemma (Smirnov 1942, p. 422) gives

$$\frac{d}{dT}\left(\ln \frac{\partial(x_1, x_2, x_3)}{\partial(T, q_1, q_2)}\right) = \frac{\partial V_i}{\partial x_i} = \frac{\partial}{\partial x_i}\left(\frac{dx_i}{dT}\right) \tag{22}$$

Combining (21) and (22), we obtain

$$\rho U^{(0)2} \frac{\partial(x_1, x_2, x_3)}{\partial(T, q_1, q_2)} = \text{constant along a ray.} \tag{23}$$

In order to find the Jacobian describing the ray spreading, we perturb the kinematic ray equations (15) and (16). The *dynamic ray equations* are

$$\frac{d}{dT}\begin{pmatrix} \delta\mathbf{x} \\ \delta\mathbf{p} \end{pmatrix} = \begin{pmatrix} \nabla_{\mathbf{p}}\nabla_{\mathbf{x}} H & \nabla_{\mathbf{p}}^2 H \\ -\nabla_{\mathbf{x}}^2 H & -\nabla_{\mathbf{x}}\nabla_{\mathbf{p}} H \end{pmatrix} \begin{pmatrix} \delta\mathbf{x} \\ \delta\mathbf{p} \end{pmatrix}. \tag{24}$$

Equation (24) is easily solved numerically, and for some special cases analytically. In general the solution at two points on a ray can be related by a propagator $\mathbf{P}(\mathbf{x}; \mathcal{L}_i^s)$ which we partition into 3×3 sub-matrices, i.e.

$$\begin{pmatrix} \delta\mathbf{x} \\ \delta\mathbf{p} \end{pmatrix} = \mathbf{P}(\mathbf{x}; \mathcal{L}_i^s) \begin{pmatrix} \delta\mathbf{x}_s \\ \delta\mathbf{p}_s \end{pmatrix} = \begin{pmatrix} \mathbf{X}_{\mathbf{x}} & \mathbf{X}_{\mathbf{p}} \\ \mathbf{P}_{\mathbf{x}} & \mathbf{P}_{\mathbf{p}} \end{pmatrix} \begin{pmatrix} \delta\mathbf{x}_s \\ \delta\mathbf{p}_s \end{pmatrix}. \tag{25}$$

At a point source $\delta\mathbf{x}_s = \mathbf{0}$, and the geometrical spreading is described by the matrix $\mathbf{X}_{\mathbf{p}}$. Matching into the solution for a point source, it can be shown (Kendall, Guest and Thomson, 1991) that the Green's solution is

$$\hat{\mathbf{u}}^{(\mathrm{GRT})}(\mathbf{x}, \omega; \mathbf{x}_s) = \sum_i \frac{\mathbf{g}(\mathbf{x}; \mathcal{L}_i^s)\mathbf{g}^T(\mathbf{x}_s; \mathcal{L}_i(\mathbf{x}, \mathbf{x}_s))}{4\pi(\rho\rho_s|\mathcal{R}(\mathbf{x}; \mathcal{L}_i^s)|)^{1/2}} e^{i\omega T(\mathbf{x}; \mathcal{L}_i^s) + i\,\mathrm{sgn}(\omega)\sigma(\mathbf{x}; \mathcal{L}_i^s)} \tag{26}$$

where $\mathcal{R}(\mathbf{x};\mathcal{L}_i^s) = \det(\mathbf{X}_\mathbf{p})\mathbf{V}_s^T\mathbf{X}_\mathbf{p}^{-1}\mathbf{V}$, and $\sigma(\mathbf{x};\mathcal{L}_i^s)$ is the KMAH index (Chapman and Drummond, 1982).

MASLOV ASYMPTOTIC RAY THEORY

If ART breaks down, we can hypothesise a more general ansatz. As always with an ansatz, "the proof of the pudding is in the eating". Let us write any field variable, e.g. displacement, traction, etc. as a Fourier transform. Thus we write

$$\mathbf{u}(\mathbf{x}) = \left(\frac{i\omega}{2\pi}\right)^{1/2} \int_{-\infty}^{\infty} \tilde{\mathbf{u}}(\tilde{\mathbf{x}}) e^{i\omega\tilde{x}_1 x_1} d\tilde{x}_1 \tag{27}$$

where the new mixed variable is $\tilde{\mathbf{x}} = (\tilde{x}_1, x_2, x_3)$, i.e. $\tilde{x}_2 = x_2$ and $\tilde{x}_3 = x_3$. Potentially we could transform with respect to any component, or with respect to any component in a rotated coordinate system (it is in fact best to rotate the coordinate system so x_1 is perpendicular to singular surfaces — caustics), or with respect to more than one coordinate. For notational simplicity we just transform with respect to the one coordinate x_1. The new transformed function $\tilde{\mathbf{u}}(\tilde{\mathbf{x}})$ can always be obtained from the function $\mathbf{u}(\mathbf{x})$ by

$$\tilde{\mathbf{u}}(\tilde{\mathbf{x}}) = \left(-\frac{i\omega}{2\pi}\right)^{1/2} \int_{-\infty}^{\infty} \mathbf{u}(\mathbf{x}) e^{-i\omega\tilde{x}_1 x_1} dx_1. \tag{28}$$

Normally in 1D models, the Fourier transform is used to remove partial derivatives (with respect to time and transverse coordinates) and reduce a partial differential equation to an ordinary differential equation. In inhomogeneous 3D models, such simplification does not occur as the wave equations contain products of functions depending on all coordinates. Nevertheless, we can always take the Fourier transform of any solution, e.g. (28).

We can now construct ansätze for the transformed functions as in ART, i.e.

$$\hat{\tilde{\mathbf{u}}}^{(\mathrm{MART})}(\tilde{\mathbf{x}},\omega;\mathbf{x}_s) = \sum_{\tilde{\imath}} \sum_{m=0}^{\infty} \frac{\tilde{\mathbf{U}}^{(m)}(\tilde{\mathbf{x}};\tilde{\mathcal{L}}_{\tilde{\imath}}^s)}{(-i\omega)^m} e^{i\omega\tilde{T}(\tilde{\mathbf{x}};\tilde{\mathcal{L}}_{\tilde{\imath}}^s)} \tag{29}$$

and similarly for the tractions, generalizing (4) and (5). This approximation is known as *Maslov asymptotic ray theory* (MART) after the asymptotic theory developed by Maslov (1965). As with ART, the possibility of multiple solutions exists. We denote a path from $\mathbf{x}_s$ to $\tilde{\mathbf{x}}$ by $\tilde{\mathcal{L}}_{\tilde{\imath}}(\tilde{\mathbf{x}},\mathbf{x}_s)$ which we often abbreviate as $\tilde{\mathcal{L}}_{\tilde{\imath}}^s$. The multiple solutions are enumerated by the index $\tilde{\imath}$, where we include the tilde to emphasize that the multipaths to $\tilde{\mathbf{x}}$ enumerated by $\tilde{\imath}$ are unrelated to those to $\mathbf{x}$ enumerated by i.

Physically, the justification for thinking that (29) may be a useful ansatz is that taking the Fourier transform of the solution (28) is equivalent to decomposing the solution into plane waves. Locally, plane wavefronts should behave just as any wavefronts, e.g. the ansatz of ART. If the wavefronts of ART behave in a singular fashion, then plane wavefronts may be better behaved. By Louiville's theorem, the rays cannot be singular in all dimensions in the $\mathbf{x} \times \mathbf{p}$ phase space. At high frequencies, we can approximate the solution by the first term — we call this *Maslov geometrical ray theory* (MGRT), i.e.

$$\hat{\mathbf{u}}^{(\mathrm{MGRT})}(\mathbf{x},\omega;\mathbf{x}_s) = \left(\frac{i\omega}{2\pi}\right)^{1/2} \int_{-\infty}^{\infty} \tilde{\mathbf{U}}^{(0)}(\tilde{\mathbf{x}};\tilde{\mathcal{L}}_{\tilde{\imath}}^s) e^{i\omega\left(\tilde{T}(\tilde{\mathbf{x}};\tilde{\mathcal{L}}_{\tilde{\imath}}^s)+\tilde{x}_1 x_1\right)} d\tilde{x}_1. \tag{30}$$

Modified eikonal and transport equations can be obtained for the transformed functions. It is also possible to transform the constitutive equation and equations of motion to obtain equations acting in the $\tilde{\mathbf{x}}$–domain, but this depends on the theory of pseudo-differential operators. For our purposes these complications are not necessary and we need not solve the modified eikonal or modified transport equations directly (Thomson and Chapman, 1985).

The approximation (30) is obviously different from GRT. To be useful it should be asymptotically equivalent in regions where GRT is valid (or at least in most regions) and remain valid in regions where GRT is invalid. If we evaluate the integral in (30) by the method of stationary phase, then we would expect the results to agree with GRT. Stationary phase points exist when

$$\frac{\partial \tilde{T}(\tilde{\mathbf{x}})}{\partial \tilde{x}_1} + x_1 = 0 \quad \text{i.e. when} \quad x_1 = -\frac{\partial \tilde{T}(\tilde{\mathbf{x}})}{\partial \tilde{x}_1} = x_1(\tilde{\mathbf{x}}), \text{ say,} \tag{31}$$

which occurs at $\tilde{x}_1 = \tilde{p}_1(\mathbf{x})$, say, which is a function of $\mathbf{x}$. Multiple solutions may occur but let us consider just one point and omit the argument $\tilde{\mathcal{L}}_i^s$ for brevity. Thus applying the method of stationary phase, we obtain

$$\hat{\mathbf{u}}^{(\mathrm{MGRT})}(\mathbf{x}, \omega; \mathbf{x}_s) \simeq \left[\tilde{\mathbf{U}}^{(0)}(\tilde{\mathbf{x}}) \left| \frac{\partial^2 \tilde{T}}{\partial \tilde{x}_1^2} \right|^{-1/2} e^{i\omega\left(\tilde{T}(\tilde{\mathbf{x}}) + \tilde{x}_1 x_1\right) + \frac{i\pi}{4}\mathrm{sgn}(\omega)\left[1 + \mathrm{sgn}\left(\partial^2 \tilde{T}/\partial \tilde{x}_1^2\right)\right]} \right] \Bigg|_{\tilde{x}_1 = \tilde{p}_1(\mathbf{x})} \tag{32}$$

For this to be equivalent to GRT, we must have

$$T(\mathbf{x}) = \tilde{T}\left(\tilde{p}_1(\mathbf{x}), x_2, x_3\right) + \tilde{p}_1(\mathbf{x}) x_1 \tag{33}$$

$$\mathbf{U}^{(0)}(\mathbf{x}) = \left[\tilde{\mathbf{U}}^{(0)}(\tilde{\mathbf{x}}) \left| \frac{\partial^2 \tilde{T}}{\partial \tilde{x}_1^2} \right|^{-1/2} e^{\frac{i\pi}{4}\mathrm{sgn}(\omega)\left[1 + \mathrm{sgn}\left(\partial^2 \tilde{T}/\partial \tilde{x}_1^2\right)\right]} \right] \Bigg|_{\tilde{x}_1 = \tilde{p}_1(\mathbf{x})} . \tag{34}$$

Differentiating equation (33) with respect to x_j, we obtain

$$p_1 = \frac{\partial T}{\partial x_1} = \tilde{p}_1(\mathbf{x}) + \left(\frac{\partial \tilde{T}}{\partial \tilde{x}_1} + x_1 \right) \frac{\partial \tilde{p}_1}{\partial x_1} = \tilde{p}_1(\mathbf{x}) \quad \text{and} \quad p_J = \frac{\partial T}{\partial x_J} = \frac{\partial \tilde{T}}{\partial x_J} = \tilde{p}_J \tag{35}$$

by equation (31), where $J = 2$ or 3. Thus $\tilde{p}_1(\mathbf{x})$, the stationary point of the integral (30), can be identified with $p_1(\mathbf{x})$, the ray slowness at $\mathbf{x}$ (as anticipated by the notation — sometimes we can drop the tilde without ambiguity). Equation (33) is thus a Legendre transform with respect to variables x_1 and p_1. To obtain $\tilde{T}(\tilde{\mathbf{x}})$, we can rewrite (33) as

$$\tilde{T}(\tilde{\mathbf{x}}) = T(x_1(\tilde{\mathbf{x}}), x_2, x_3) - p_1 x_1(\tilde{\mathbf{x}}) \tag{36}$$

where $x_1(\tilde{\mathbf{x}})$ is a ray coordinate expressed as a function of the slowness coordinate and the other coordinates, and the mixed coordinate is $\tilde{\mathbf{x}} = (p_1, x_2, x_3)$.

The transformed coefficient, $\tilde{\mathbf{U}}^{(0)}(\tilde{\mathbf{x}})$, is

$$\tilde{\mathbf{U}}^{(0)}(\tilde{\mathbf{x}}) = \mathbf{U}^{(0)}(x_1(\tilde{\mathbf{x}}), x_2, x_3) \left| \frac{\partial x_1(\tilde{\mathbf{x}})}{\partial p_1} \right|^{1/2} e^{-\frac{i\pi}{4}\mathrm{sgn}(\omega)\left[1 - \mathrm{sgn}(\partial x_1/\partial p_1)\right]} . \tag{37}$$

With these expressions substituted in (30), MGRT is asymptotically (i.e. by the method of stationary phase) equal to GRT. However, in general if the integral (30) is evaluated accurately, MGRT will differ from GRT. It is useful because it remains valid when GRT breaks down, e.g. at caustics. In the method of stationary phase, we have assumed that the transformed amplitude coefficient is constant, and that the transformed travel time is quadratic. If these functions have different forms we can use other asymptotic methods to evaluate (30), e.g. third-order saddle point method if $\tilde{T}(\tilde{\mathbf{x}})$ is cubic, incomplete saddle point method if $\tilde{\mathbf{U}}^{(0)}(\tilde{\mathbf{x}})$ is discontinuous, etc. Alternatively, and more generally, the integral can be evaluated numerically but this is difficult at high frequencies as the integrand is highly oscillatory. These difficulties can be avoided by evaluating the inverse Fourier transform with respect to frequency of (30) to obtain the impulse response directly, and this technique is discussed in the next section.

MASLOV GEOMETRICAL RAY THEORY SEISMOGRAMS

The MGRT approximation for the displacement is equation (30). The travel time in the transformed domain is obtained from the usual travel time by the Legendre transform (36). The transformed amplitude coefficient is obtained by the canonical transformation (37). Applying the inverse Fourier transform to (30) we obtain

$$\mathbf{u}^{(\mathrm{MGRT})}(\mathbf{x},t;\mathbf{x}_s)=\frac{1}{2\pi}\int_B\int_{-\infty}^{\infty}\left(\frac{i\omega}{2\pi}\right)^{1/2}\sum_i\tilde{\mathbf{U}}^{(0)}(\tilde{\mathbf{x}};\tilde{\mathcal{L}}_i^s)e^{i\omega\left(\tilde{T}(\tilde{\mathbf{x}};\tilde{\mathcal{L}}_i^s)+\tilde{x}_1x_1-t\right)}d\tilde{x}_1\,d\omega. \quad (38)$$

Note that we use $\tilde{x}_1$ for the variable of integration in (38) rather than p_1, as only at the saddle point is $\tilde{x}_1=p_1(\mathbf{x})$. We approximate the infinite $\tilde{x}_1$–integral by that part for which $\tilde{T}(\tilde{\mathbf{x}};\tilde{\mathcal{L}}_i^s)$ is real. Reversing the order of the integrals and evaluating the frequency transform first, we formally obtain

$$\mathbf{u}^{(\mathrm{MGRT})}(\mathbf{x},t;\mathbf{x}_s)=-\frac{1}{2^{1/2}\pi}\frac{d}{dt}\left[\bar{\lambda}(t)*\mathrm{Re}\sum_i\int_{\mathrm{Im}(\tilde{T})=0}\tilde{\mathbf{U}}^{(0)}(\tilde{\mathbf{x}};\tilde{\mathcal{L}}_i^s)\Delta\left(t-\tilde{T}(\tilde{\mathbf{x}};\tilde{\mathcal{L}}_i^s)-\tilde{x}_1x_1\right)d\tilde{x}_1\right] \quad (39)$$

where $\bar{\lambda}(t)=H(-t)(-t)^{-1/2}$. We describe this as a formal result as the operations of differentiation and Hilbert transformation may be ill-defined for certain singular functions but overall the expression is well-defined. Transferring the Hilbert transform outside the integral, we can evaluate the $\tilde{x}_1$–integral exactly and obtain

$$\mathbf{u}^{(\mathrm{MGRT})}(\mathbf{x},t;\mathbf{x}_s)=-\frac{1}{2^{1/2}\pi}\frac{d}{dt}\mathrm{Im}\left[\Lambda(t)*\sum_i\sum_{t=\tilde{T}(\tilde{\mathbf{x}};\tilde{\mathcal{L}}_i^s)+\tilde{x}_1x_1}\frac{\tilde{\mathbf{U}}^{(0)}(\tilde{\mathbf{x}};\tilde{\mathcal{L}}_i^s)}{\left|x_1-x_1(\tilde{\mathbf{x}};\tilde{\mathcal{L}}_i^s)\right|}\right] \quad (40)$$

where $\Lambda(t)=\lambda(t)+i\bar{\lambda}(t)$ is the analytic time series $t^{-1/2}$, and we have used (31). For any time t, we solve

$$t=\tilde{T}(\tilde{\mathbf{x}};\tilde{\mathcal{L}}_i^s)+\tilde{x}_1x_1 \quad (41)$$

for $\tilde{x}_1$. Each solution is substituted in the final term in (40), and all cases are summed. Thus the final term in (40) is a function of time through solving (41). It is important to note that although we have only used the MGRT approximation in the transformed domain, the $\tilde{x}_1$–integral has been evaluated exactly apart from restricting its range so $\tilde{T}(\tilde{\mathbf{x}};\tilde{\mathcal{L}}_i^s)$ is real. Between equations (39) and (40), no approximations are made and the evaluation is exact whatever the form of $\tilde{T}(\tilde{\mathbf{x}};\tilde{\mathcal{L}}_i^s)$ and $\tilde{\mathbf{U}}^{(0)}(\tilde{\mathbf{x}};\tilde{\mathcal{L}}_i^s)$. No prior knowledge of the behaviour of these functions is needed, and no special cases for different forms are necessary. Chapman (1985) has discussed various problems in seismology when MGRT is valid but GRT breaks down.

GENERALIZED BORN SCATTERING THEORY

The Green's functions (4) and (5) are only approximate and do not satisfy the constitutive equations (3) nor equations of motion (1) exactly. Rather they satisfy

$$\hat{\mathbf{t}}_j^{(\mathrm{GRT})}(\mathbf{x},\omega;\mathbf{x}_s)=\mathbf{c}_{jk}(\mathbf{x})\frac{\partial\hat{\mathbf{u}}^{(\mathrm{GRT})}(\mathbf{x},\omega;\mathbf{x}_s)}{\partial x_k}-\mathbf{c}_{jk}(\mathbf{x})\sum_i\frac{\partial\mathbf{U}^{(0)}(\mathbf{x};\mathcal{L}_i^s)}{\partial x_k}e^{i\omega T(\mathbf{x};\mathcal{L}_i^s)} \quad (42)$$

a modified constitutive equation, and

$$\frac{\partial\hat{\mathbf{t}}_j^{(\mathrm{GRT})}(\mathbf{x},\omega;\mathbf{x}_s)}{\partial x_j}=-\rho(\mathbf{x})\omega^2\hat{\mathbf{u}}^{(\mathrm{GRT})}(\mathbf{x},\omega;\mathbf{x}_s)-\mathbf{I}\delta(\mathbf{x}-\mathbf{x}_s)-i\omega\sum_i\frac{\partial\mathbf{T}_j^{(0)}(\mathbf{x};\mathcal{L}_i^s)}{\partial x_j}e^{i\omega T(\mathbf{x};\mathcal{L}_i^s)}. \quad (43)$$

a modified equation of motion. The final term in each equation can be regarded as due to errors in the Green's functions.

Changing the source in (1) to $\mathbf{x}_r$ and transposing the equation, we post-multiply it by $\hat{\mathbf{u}}^{(\mathrm{GRT})}(\mathbf{x},\omega;\mathbf{x}_s)$, pre-multiply the modified equation of motion (43) by $\hat{\mathbf{u}}^T(\mathbf{x},\omega;\mathbf{x}_r)$, subtract the two equations and integrate over the volume of the medium. A volume integral is integrated by parts and we use the reciprocal relation $\hat{\mathbf{u}}^T(\mathbf{x}_s,\omega;\mathbf{x}_r) = \hat{\mathbf{u}}(\mathbf{x}_r,\omega;\mathbf{x}_s)$, to obtain

$$\begin{aligned}\hat{\mathbf{u}}(\mathbf{x}_r,\omega;\mathbf{x}_s) &= \hat{\mathbf{u}}^{(\mathrm{GRT})}(\mathbf{x}_r,\omega;\mathbf{x}_s) \\ &+ \int_S \left(\hat{\mathbf{t}}_j^T(\mathbf{x},\omega;\mathbf{x}_r)\hat{\mathbf{u}}^{(\mathrm{GRT})}(\mathbf{x},\omega;\mathbf{x}_s) - \hat{\mathbf{u}}^T(\mathbf{x},\omega;\mathbf{x}_r)\hat{\mathbf{t}}_j^{(\mathrm{GRT})}(\mathbf{x},\omega;\mathbf{x}_s)\right) dS_j \\ &- \int_V \sum_i \left(i\omega\hat{\mathbf{u}}^T(\mathbf{x},\omega;\mathbf{x}_r)\frac{\partial \mathbf{T}_j^{(0)}(\mathbf{x};\mathcal{L}_i^s)}{\partial x_j} + \hat{\mathbf{t}}_j^T(\mathbf{x},\omega;\mathbf{x}_r)\frac{\partial \mathbf{U}^{(0)}(\mathbf{x};\mathcal{L}_i^s)}{\partial x_j}\right) e^{i\omega T(\mathbf{x};\mathcal{L}_i^s)} dV.\end{aligned} \tag{44}$$

This is an exact integral equation for the displacement at $\mathbf{x}_r$, but normally it can only be solved iteratively. We will assume that the surface S is sufficiently removed from the source and receiver that the surface integral can be neglected. Otherwise, it can be included as a Kirchhoff surface integral. In the volume integral, the first term arises from the error term in the modified equation of motion (43) and the second term from the modified constitutive equation (42).

The zeroth iteration to the equation comes from neglecting the volume integral and gives GRT (6). The first iteration is obtained by substituting this approximation into the integral to obtain

$$\begin{aligned}\hat{\mathbf{u}}^{(\mathrm{SRT})}(\mathbf{x}_r,\omega;\mathbf{x}_s) &= \hat{\mathbf{u}}^{(\mathrm{GRT})}(\mathbf{x}_r,\omega;\mathbf{x}_s) \\ &- \frac{i\omega}{2}\int_V \sum_{m,n} \left(\mathbf{U}^{(0)T}(\mathbf{x};\mathcal{L}_m^r)\frac{\partial \mathbf{T}_j^{(0)}(\mathbf{x};\mathcal{L}_n^s)}{\partial x_j} - \mathbf{T}_j^{(0)T}(\mathbf{x};\mathcal{L}_m^r)\frac{\partial \mathbf{U}^{(0)}(\mathbf{x};\mathcal{L}_n^s)}{\partial x_j}\right. \\ &\left. + \frac{\partial \mathbf{T}_j^{(0)T}(\mathbf{x};\mathcal{L}_m^r)}{\partial x_j}\mathbf{U}^{(0)}(\mathbf{x};\mathcal{L}_n^s) - \frac{\partial \mathbf{U}^{(0)T}(\mathbf{x};\mathcal{L}_m^r)}{\partial x_j}\mathbf{T}_j^{(0)}(\mathbf{x};\mathcal{L}_n^s)\right) \hat{F}_{mn}(\mathbf{x},\omega) dV,\end{aligned} \tag{45}$$

where we have manipulated the volume integral to obtain a result that is obviously reciprocal, and $\hat{F}_{mn}(\mathbf{x},\omega) = \exp\left(i\omega T(\mathbf{x};\mathcal{L}_n^s) + i\omega T(\mathbf{x};\mathcal{L}_m^r)\right)$. The first term in (45) contains a summation of ray types and paths, and the scattering integral contains a double summation over different ray types from the source to the scattering element, and from the scattering element to the receiver. In the simplest case of anisotropic media with 3 ray types and no multi-pathing, this means that there are 9 volume integrals for different types of scattering.

It is instructive to substitute the dyadic forms of the Green's functions (26)

$$\mathbf{U}^{(0)}(\mathbf{x};\mathcal{L}_n^s) = \mathbf{g}(\mathbf{x};\mathcal{L}_n^s)A(\mathbf{x};\mathcal{L}_n^s)\mathbf{g}^T(\mathbf{x}_s;\mathcal{L}_n(\mathbf{x},\mathbf{x}_s)) \tag{46}$$

$$\mathbf{T}_j^{(0)}(\mathbf{x};\mathcal{L}_n^s) = -\mathbf{Z}_j(\mathbf{x};\mathcal{L}_n^s)\mathbf{g}(\mathbf{x};\mathcal{L}_n^s)A(\mathbf{x};\mathcal{L}_n^s)\mathbf{g}^T(\mathbf{x}_s;\mathcal{L}_n(\mathbf{x},\mathbf{x}_s)) \tag{47}$$

for the displacement and tractions, where $\mathbf{Z}_j(\mathbf{x};\mathcal{L}_n^s) = p_k(\mathbf{x};\mathcal{L}_n^s)\mathbf{c}_{jk}(\mathbf{x})$ is a generalization of the scalar impedance and $A(\mathbf{x};\mathcal{L}_n^s)$ is defined in equation (26). Various terms arise in the scattering integral due to derivatives of different terms: derivatives of the source and receiver polarizations, derivatives of the amplitude functions, derivatives of the polarizations at the scattering point, and derivatives of the impedance at the scattering point. Although these are all connected as variations in the model affect the ray paths which influence all terms, they model different physically effects and can be considered separately. GRT breaks down on the nodes of the source radiation pattern and receiver directivity functions and the terms containing derivatives of the source and receiver polarizations correct this error. We shall not investigate these phenomena here and shall ignore the terms. If the amplitude function varies rapidly, as occurs at caustics and critical points, the geometrical ray approximation breaks down. The terms containing the derivative of the amplitude variation are significant in these circumstances. However, scattering between GRT solutions is not an efficient way to obtain the solution near caustics or critical points, and a scattering theory based on Maslov asymptotic theory is needed. Amplitude variations may also occur due to inhomogeneities and we retain this term to model signals scattered by inhomogeneities. Rapid changes in the ray direction which may occur at inhomogeneities cause the derivatives of the polarization vectors at the scattering point to be significant. In anisotropic media, the polarizations may also change rapidly near singularities in the slowness surfaces without needing strong inhomogeneities, and the term with the derivatives of the polarization vectors models the coupling that exists between quasi-shear waves (Coates and Chapman, 1990). Finally, the term with derivatives of the impedance models the reflections that occur from inhomogeneities. Substituting (46), (47) into (45), and neglecting the source and receiver terms

$$\begin{aligned}\hat{\mathbf{u}}^{(\mathrm{SRT})}(\mathbf{x}_r,\omega;\mathbf{x}_s) = \hat{\mathbf{u}}^{(\mathrm{GRT})}(\mathbf{x}_r,\omega;\mathbf{x}_s) - \frac{i\omega}{2}\int_V \sum_{m,n} A(\mathbf{x};\mathcal{L}^r_m)A(\mathbf{x};\mathcal{L}^s_n)\mathbf{g}(\mathbf{x}_r;\mathcal{L}_m(\mathbf{x},\mathbf{x}_r)) \quad (48)\\
\times\Bigg(\mathbf{g}^T(\mathbf{x};\mathcal{L}^r_m)\frac{\partial}{\partial x_j}\ln\left(\frac{A(\mathbf{x};\mathcal{L}^r_m)}{A(\mathbf{x};\mathcal{L}^s_n)}\right)\left(\mathbf{Z}^T_j(\mathbf{x};\mathcal{L}^r_m) - \mathbf{Z}^T_j(\mathbf{x};\mathcal{L}^s_n)\right)\mathbf{g}(\mathbf{x};\mathcal{L}^s_n)\\
+\frac{\partial \mathbf{g}^T(\mathbf{x};\mathcal{L}^r_m)}{\partial x_j}\left(\mathbf{Z}^T_j(\mathbf{x};\mathcal{L}^r_m) - \mathbf{Z}^T_j(\mathbf{x};\mathcal{L}^s_n)\right)\mathbf{g}(\mathbf{x};\mathcal{L}^s_n)\\
-\mathbf{g}^T(\mathbf{x};\mathcal{L}^r_m)\left(\mathbf{Z}^T_j(\mathbf{x};\mathcal{L}^r_m) - \mathbf{Z}^T_j(\mathbf{x};\mathcal{L}^s_n)\right)\frac{\partial \mathbf{g}(\mathbf{x};\mathcal{L}^s_n)}{\partial x_j}\\
+\mathbf{g}^T(\mathbf{x};\mathcal{L}^r_m)\frac{\partial}{\partial x_j}\left(\mathbf{Z}^T_j(\mathbf{x};\mathcal{L}^r_m) + \mathbf{Z}^T_j(\mathbf{x};\mathcal{L}^s_n)\right)\mathbf{g}(\mathbf{x};\mathcal{L}^s_n)\Bigg)\mathbf{g}^T(\mathbf{x}_s;\mathcal{L}_n(\mathbf{x},\mathbf{x}_s))\hat{F}_{mn}(\mathbf{x},\omega)dV.\end{aligned}$$

It is straightforward to interpret physically the significance of the various terms in the scattering integral. It is also straightforward to convert any of the above results into the time domain as the scattering terms are independent of frequency — for brevity we omit explicitly writing the results.

ACKNOWLEDGEMENTS

The research described in this paper was partly supported by a Natural Science and Engineering Research Council (Canada) Operating grant OGP0009130. Much of this work was done while one of us (CHC) was visiting the University of Cambridge. We gratefully acknowledge the facilities provided. Department of Earth Sciences publication number 1789.

REFERENCES

Červený, V., 1972, Seismic rays and ray intensities in inhomogeneous anisotropic media, *Geophys. J. R. astr. Soc.*, **29**, 1–13.

Chapman, C.H., 1985, Ray theory and its extensions: WKBJ and Maslov seismograms, *J. Geophys.*, **58**, 27–43.

Chapman, C.H. and Drummond, R., 1982, Body-wave seismograms in inhomogeneous media using Maslov asymptotic theory, *Bull. seism. Soc. Am.*, **72**, S277–S317.

Coates, R.T. and Chapman, C.H., 1990, Quasi-shear wave coupling in weakly anisotropic 3–D media, *Geophys. J. Int.*, **103**, 301–320.

Coates, R.T. and Chapman, C.H., 1991, Generalized Born scattering of seismic waves in 3–D media, *Geophys. J. Int.*, (submitted).

Kendall, J-M. and Thomson, C.J., 1989, A comment on the form of the geometrical spreading equations, with some examples of seismic ray tracing in inhomogeneous, anisotropic media, *Geophys. J. Int.*, **99**, 401–413.

Kendall, J-M., Guest, W.S. and Thomson, C.J., 1991, Ray theory Green's function reciprocity and ray-centered coordinates in anisotropic media, *preprint*.

Maslov, V.P., 1965, "Theory of Perturbations and Asymptotic Methods", (in Russian), Izd. MGU, Moscow.

Smirnov, V.I., 1964, "A Course in Higher Mathematics", Vol. 4, Trans. Brown, D.E. and Sneddon, I.N., Pergamon Press, Oxford.

Thomson, C.J. and Chapman, C.H., 1985, An introduction to Maslov's asymptotic method, *Geophys. J. R. astr. Soc.*, **83**, 143–168.

Woodhouse, J.H., 1974, Surface waves in a laterally varying layered structure, *Geophys. J. R. astr. Soc.*, **37**, 461–490.

CURRENT-BASED HYBRID ANALYSIS IN SURFACE-WAVE HYBRID FORMULATION FOR ELECTROMAGNETICS MODELING*

D.-S. Wang and L. N. Medgyesi-Mitschang

McDonnell Douglas Research Laboratories
P.O. Box 516
St. Louis, MO 63166

ABSTRACT

Current-based hybrid analyses have the potential to solve a broad class of complex electromagnetic scattering and radiation problems. In this paper, the discussion emphasizes the development of a new hybrid analysis that incorporates a surface-wave basis set. Representative examples illustrate the effectiveness of this new method for both scattering and radiation problems involving 2D and 3D objects with coatings.

INTRODUCTION

Modeling of electromagnetic (EM) waves interacting with large complex objects is difficult. Realistic models require accurate evaluation of a variety of edge diffractions and surface-wave effects, such as creeping and traveling waves. The electrical size of the scatterer severely limits the use of numerical methods such as the methods of moments (MM), and the geometric complexity circumscribes the applicability of optics-derived methods. Hybrid methods, combining both numerical and optics-derived methods, have the potential to substantially enlarge the class of scattering and antenna radiation problems that can be treated. Properly formulated, the hybrid methods can build on the strength of both the numerical and optics-derived methods and mitigate their limitations.

One of the principal hybrid methods currently in use is the current-based hybrid formulation that incorporates the Ansatz currents into the MM solution. The effectiveness of this formulation has been illustrated for both perfectly conducting and coated scatterers [1]-[3]. In this hybrid method, the accuracy of the Ansatz currents constrains the accuracy of the numerical results. Most of the Ansatz currents derived from physical optics (PO), the physical theory of diffraction (PTD), the geometrical theory of diffraction (GTD), and the Fock theory are inaccurate when multiple edge diffractions and surface-wave effects predominate. Available asymptotic techniques offer only limited improvement in this context.

In this paper, the foregoing hybrid formulation is briefly reviewed and then is used as a point of departure for a novel hybrid approach that overcomes the above shortcomings. In this approach, a surface-wave basis set is incorporated into the hybrid formulation to obtain current components arising from multiple interactions and surface-wave effects.

*This research was conducted under the McDonnell Douglas Independent Research and Development program.

Through these surface-wave bases, the hybrid formulation becomes highly efficient in treating complex diffraction effects. To demonstrate the application of this approach, we consider the scattering and antenna radiation from a variety of 2D and 3D complex objects.

CURRENT-BASED HYBRID FORMULATION

As originally formulated, a current-based hybrid analysis incorporates Ansatz currents derived from PO, GTD, PTD, and the Fock theory. These currents then serve as effective sources in the surface integral equation (SIE) formulation, solved by a method of moments (MM) procedure. We designate this formulation as the Ansatz hybrid formulation. We will briefly discuss the salient features here. Details can be found in [1]-[3]. Subsequently, the extension to the new hybrid formulation incorporating the surface-wave bases will be made.

Ansatz Hybrid Formulation

Surface integral equation representations of the electromagnetic problems are a convenient starting point for hybrid analyses. We restrict the present discussion to perfect electric conductors (pec). The formulation for dielectric and coated objects is relegated to the citations. The electric field integral equation (EFIE), giving the total electric field $\vec{E}(\vec{r})$ at a field point $\vec{r}$ in terms of the electric current density $\vec{J}(\vec{r}')$ on the surface of a scatterer and the incident field $\vec{E}^i(\vec{r})$, is

$$\vec{E}(\vec{r}) = \vec{E}^i(\vec{r}) - L\vec{J}(\vec{r}') \tag{1}$$

where the harmonic time dependence ($e^{j\omega t}$) has been suppressed and L is a integro-differential operator defined as

$$L\vec{J}(\vec{r}') = j\omega\mu \int_S (\vec{J}(\vec{r}') + \frac{1}{k^2} \nabla\nabla' \bullet \vec{J}(\vec{r}'))\, \Phi \, ds' \tag{2}$$

where k is the wave number in free space, S is the surface of the scatterer, and Φ is the free-space Green's function. Applying the boundary condition on a pec surface, $\vec{E}(\vec{r})|_{tan} = 0$, to (1) gives

$$\vec{E}^i(\vec{r})\Big|_{tan} = L\vec{J}(\vec{r}')\Big|_{tan} . \tag{3}$$

Conventional MM techniques employ subdomain, entire-, or mix-domain expansions to form a set of linear equations to solve for the unknown currents $\vec{J}$. Such techniques require large number of unknowns per wavelength and hence their application to large scatterers is precluded. One of the principal advantages of the hybrid methods is the reduction of unknowns by developing efficient representations of the unknown currents. For the Ansatz hybrid formulation, we begin by subdividing the scatterer (Fig. 1(a)) into subdomain (SD) and Ansatz (AN) regions. The former encompasses irregular surfaces including protrusions, concavities, apertures, and material discontinuities and the corresponding surface currents $\vec{J}_1$ are represented by subdomain bases. The AN region encompasses smooth surfaces for which Ansatz currents obtained from PO, PTD, GTD, and the Fock theory are applicable. Specific choices of Ansatz currents are determined by the details of the problem. According to Fig. 1(a), Eq. (3) reduces to the following integral equation:

$$L_{SD}\vec{J}_1(\vec{r}') = \vec{E}^i(\vec{r}) - L_{AN}\vec{J}_2(\vec{r}') \tag{4}$$

where L_{SD} and L_{AN} are operators over the SD and AN regions, respectively. The AN region may be further partitioned according to the choice of Ansatz currents in a specific area. For example, the AN region may be subdivided into an illuminated region with the PO based currents and a shadowed region

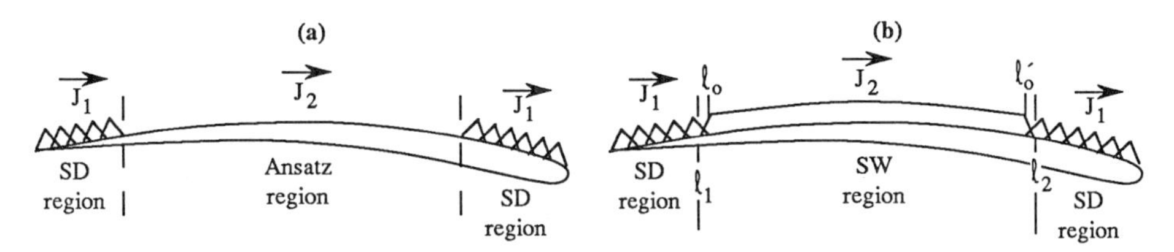

Fig. 1 Generic 2D scattering geometry. Partitioned regions for: (a) Ansatz hybrid analysis b) SW hybrid analysis

with the Fock theory based currents. Regardless of these choices, these currents are treated as known quantities and the currents $\vec{J}_1$ are expanded in terms of the SD bases consisting of triangle functions as shown in Fig. 1(a). Applying the Galerkin solution procedure to (4) yields

$$[Z][I] = [V] - [\bar{V}] \tag{5}$$

where [I] denotes the unknown expansion coefficients of $\vec{J}_1$; [V] and $[\bar{V}]$ represent the terms on the right side of (4), associated with the incident wave and the Ansatz currents, respectively. The inclusion of $[\bar{V}]$ is crucial to obtaining satisfactory results as will be illustrated later by contrasting the solutions with and without $[\bar{V}]$. Detailed expressions for [Z], [V], and $[\bar{V}]$ are given in [1]. We solve (5) for [I] to reconstruct the currents $\vec{J}_1$ in the SD region. Knowing both $\vec{J}_1$ and the Ansatz currents $\vec{J}_2$, the scattered far fields can be computed from the radiation integral [1]. The Ansatz hybrid approach has been generalized and applied to partially coated surfaces and penetrable ones satisfying the impedance boundary condition [2]-[4].

Surface-Wave (SW) Hybrid Formulation

The foregoing Ansatz hybrid solution depends on the validity of the Ansatz currents derived from the optics-derived methods. When the scattering effects from multiple interactions and surface waves predominate, the Ansatz currents are no longer valid leading to the deterioration of the Ansatz hybrid solution. The surface-wave (SW) hybrid formulation circumvents the foregoing deficiency by introducing a surface-wave expansion set for the currents arising from multiple interactions and surface-wave effects. A GTD-MM based approach described in [4]-[5] suggested a similar approach by introducing a diffraction current term based on GTD diffraction coefficients. This diffraction current outside the SD region is expressed in terms of a series of diffraction coefficients with unknown constants to be solved for by the MM technique. The use and limitation of this approach are detailed in [5]. Broadly speaking, the present method is a generalization of this approach. Fig.1 (b) illustrates the key concepts of the present formulation. The Ansatz region in Fig. 1(a) is now denoted as the surface wave (SW) region spanning S_{SW}. The surface currents, formerly derived from the Ansatz solutions, are decomposed into a PO term plus a surface wave term, i.e.

$$\vec{J}_2(\vec{r}') = \vec{J}_{PO}(\vec{r}') + \vec{J}_{SW}(\vec{r}'). \tag{6}$$

In contrast to the Ansatz hybrid formulation where the AN region is further partitioned according to the choices of Ansatz currents made, in the SW hybrid method the currents $\vec{J}_2$ are represented by (6) throughout the SW region, i.e. the representation is independent of geometry and illumination. This is advantageous in the development of a user-friendly computer software. Using (6), Equation (4) becomes

$$L_{SD}\vec{J}_1(\vec{r}') + L_{SW}\vec{J}_{SW}(\vec{r}') = \vec{E}^i(\vec{r}) - L_{SW}\vec{J}_{PO}(\vec{r}') \tag{7}$$

where $\vec{J}_1(\vec{r}')$ and $\vec{J}_{SW}(\vec{r}')$ are unknown currents to be determined using the MM technique to solve (7). As shown in [6], the subtraction of the PO term in (7) reduces the number of unknowns required to represent $\vec{J}_{SW}$ in the SW region. Instead of using the diffraction coefficients from the GTD solution to represent the diffraction current term as in [5], a more general, geometry independent representation is used here to represent the surface wave term. The $\vec{J}_{SW}$ is expanded in terms of a set of surface-wave bases as

$$\vec{J}_{SW}(\vec{r}') = \sum_{\alpha=1}^{2} \sum_{i=0}^{L} [a_{i+}^{\alpha} u_i^{+}(\ell) + a_{i-}^{\alpha} u_i^{-}(\ell)] \; \vec{v}^{\alpha} \tag{8}$$

where the superscripts α represent two cuvilinear coordinates tangent to S_{SW}, the unit vectors $\vec{v}^{\alpha}$ are the surface tangential vectors, and $a_{i\pm}^{\alpha}$ are unknown coefficients. The surface wave bases $u_i^{\pm}(\ell)$ are entire-domain (global) basis functions defined as

$$u_i^{\pm}(\ell) = T_i(\xi)e^{\pm jk\ell} \tag{9}$$

where the functions $T_i(\xi)$ are Chebyshev polynomials of the first kind, ℓ is the arc length measured from a reference point on S_{SW}, and ξ is a normalized parameter given as $\xi = (\ell - \ell_c)/\ \ell_s$ where ℓ_s is the arc length spanning S_{SW} in the $\vec{v}^{\alpha}$ direction and ℓ_c is the arc length at the center of S_{SW}. The two terms in (8) can be considered as currents arising from two sets of surface waves propagating in opposite directions. In fact, the $e^{-jk\ell}$ term in (8) closely resembles the expression for the diffraction current term in [4] and [5]. Adding the $e^{jk\ell}$ term in (8) generalizes the expression for $\vec{J}_{SW}$ for finite 3D arbitrary surfaces. If we follow the formulation in [4]-[5], the surface-wave bases $u_i^{\pm}$ would have been represented by a series of $(\sqrt{\ell})^{-n}$ terms for $-1 < n < N$. In the present formulation, we chose to express $u_i^{\pm}$ in terms of the Chebyshev polynomials because of their superior convergence properties [7]. Furthermore, the subtraction of the PO term in (7) and the separation of the oscillatory factor $e^{\pm jk\ell}$ lead one to conclude that only a few terms are needed in (8) to accurately represent $\vec{J}_{SW}$. This assertion will be demonstrated later for a variety of 2D and 3D objects.

Applying the Galerkin technique to (7) leads to a matrix equation as

$$\begin{bmatrix} Z_{DD} & Z_{WD} \\ Z_{DW} & Z_{WW} \end{bmatrix} \begin{bmatrix} I_D \\ I_W \end{bmatrix} = \begin{bmatrix} V_D \\ V_W \end{bmatrix} - \begin{bmatrix} \bar{V}_D \\ \bar{V}_W \end{bmatrix} \tag{10}$$

where the subscripts D and W stand for the SD and SW regions, respectively. As in (5), [V] and $[\bar{V}]$ represent the terms on the right side of (7), associated with the incident electric field and the PO term. Note that the Z elements in (10) represent all possible interactions between different parts of a scatterer. The inclusion of inter-coupling effects between the SD and SW regions represented by the off-diagonal blocks in Z are made possible by introducing $\vec{J}_{SW}$ in (6). Such inter-coupling effects were missing in the earlier Ansatz hybrid formulation. Specific expressions for [Z], [V], and $[\bar{V}]$ can be obtained following the same procedure as in [1]. Once the unknown current coefficients for both $\vec{J}_1$ and $\vec{J}_{SW}$ are obtained by solving (10), the scattered far field can be calculated as before using the radiation integral equation.

Antenna Radiation Problems

The extension of the foregoing hybrid formulation to antenna radiation problems is self-evident. The PO terms in (6) and (7) are eliminated and

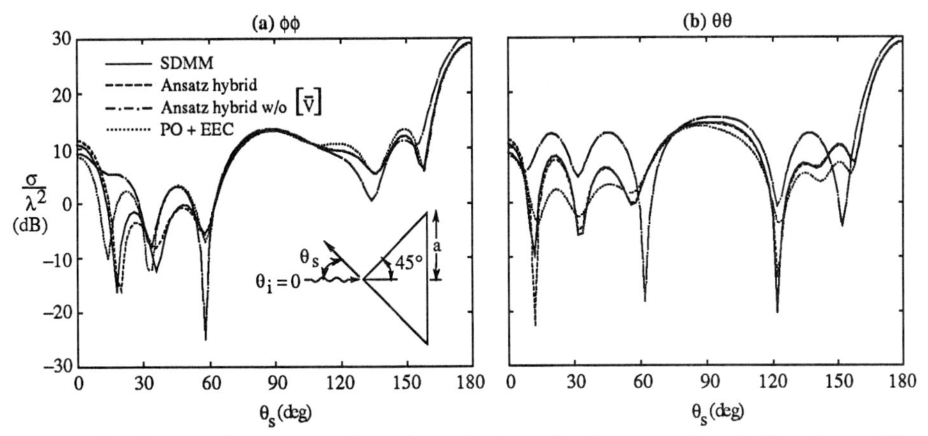

Fig. 2 Bistatic scattering cross sections of a perfectly conducting cone. Cone half angle $\alpha = 45°$, $a = 1.6\lambda$. Axial illumination. (a) $\phi\phi$ - polarization (b) $\theta\theta$ - polarization.

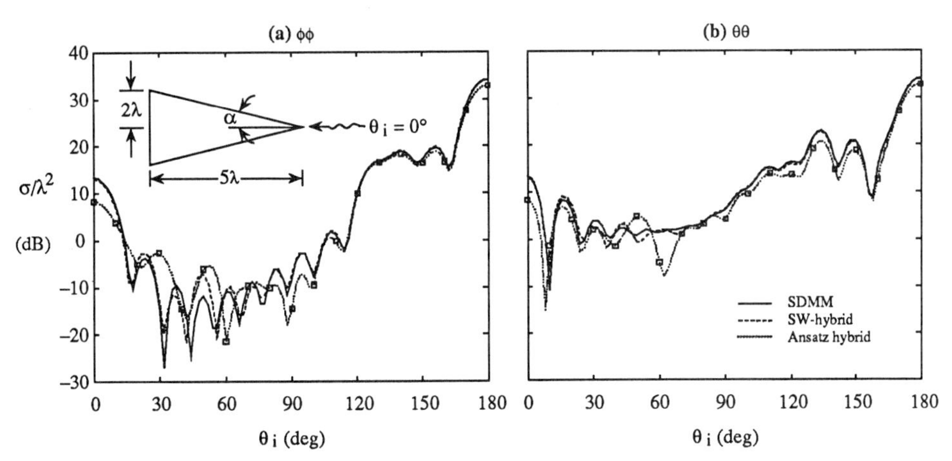

Fig. 3 Bistatic scattering cross sections of a small angle cone. $\alpha = 21.8°$, radius = 2λ, axia illumination. (a) $\phi\phi$ - polarization (b) $\theta\theta$ - Polarization

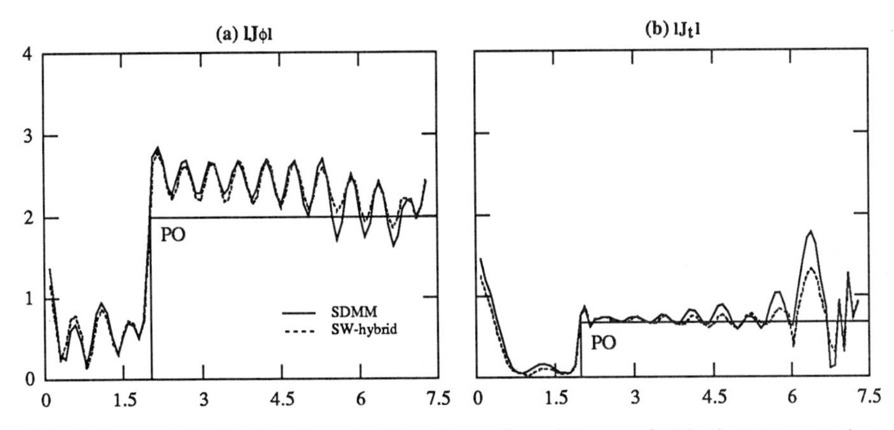

Fig. 4 Current distribution along $\phi = 0°$ on the surface of the cone in Fig. 3. (a) ϕ - polarized illumination (b) θ - polarized illumination.

[V] is expressed in terms of the excitation electric field $\vec{E}^a$ for embedded apertures or the excitation currents on wire antennas. Accordingly, Equation (10) becomes

$$\begin{bmatrix} Z_{DD} & Z_{WD} \\ Z_{DW} & Z_{WW} \end{bmatrix} \begin{bmatrix} I_D \\ I_W \end{bmatrix} = \begin{bmatrix} V_D \\ V_W \end{bmatrix} \tag{11}$$

The remaining formulation parallels that for the scattering problem.

RESULTS

Application of the Ansatz hybrid formulation is illustrated in Fig. 2 depicting the bistatic scattering cross sections of a pec circular cone illuminated axially. The cone base radius is 1.6 λ; the cone half-angle is 45°. The hybrid results correlate closely with the subdomain based MM (SDMM) results. For comparison, we also show two additional results from the Ansatz hybrid formulation without the term [V] and from the optics-derived method with the equivalent edge current (EEC) representation for the cone rim and using the PO approximation elsewhere. Only the Ansatz hybrid results correlate closely with the SDMM results. Apparently, using the SD bases near the tip and the base region and PO approximation elsewhere yields accurate results for this large angle cone in Fig. 2. This is no longer the case for small angle cones with significant surface wave effects. Fig. 3 depicts the bistatic scattering cross sections of a circular pec cone illuminated axially. The cone base radius is 2 λ; the cone half-angle is 21.8°. The correlation between the Ansatz hybrid and the SDMM results is not as good as in Fig. 2, especially for $0° < \theta_i < 90°$. By contrast, the SW hybrid results are in good agreement with the SDMM ones. The effectiveness of the SW hybrid formulation is further substantiated by Fig. 4 showing the surface current distribution along the $\phi = 0°$ plane of the cone for both θ- and ϕ- polarized illuminations. We note the marked difference between the SDMM and SW hybrid currents and the PO induced currents in Fig. 4. The magnitude of the former pair are in good agreement. Same assertion extends to the phase of the currents, not shown here for brevity. In the SW hybrid calculation, the SD bases are used near the tip and the base spanning a region 1 λ wide, and three Chebyshev functions are used to represent the surface wave currents in the SW region. Further applications of the present approach are shown in Figs. 5 and 6, namely, the backscatter cross sections of a dihedral and a curved

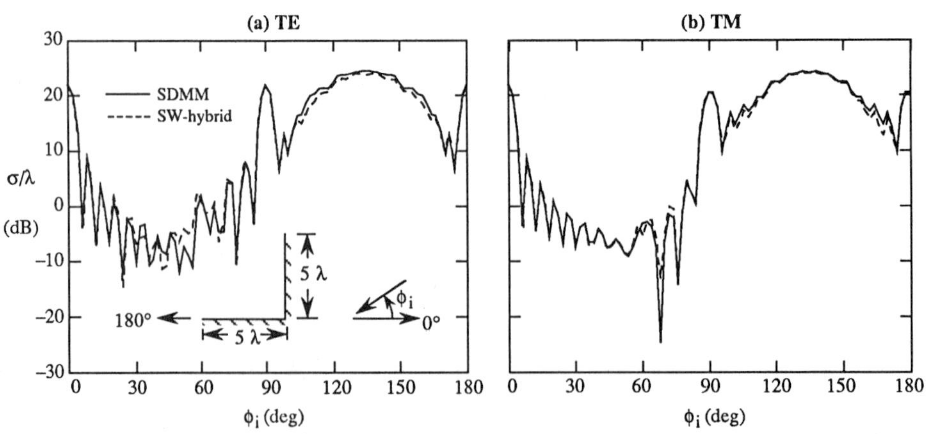

Fig. 5 Backsatter cross sections of a dihedral. (a) TE polarization (b) TM polarization

finite plate, respectively. High fidelity modeling of the scattering phenomena from these configurations require incorporation of the multiple bounce effects for the dihedral, and the traveling wave effects (arising from edge-edge interactions) on the edge-on illuminated curved plate. The SW hybrid results correlate well with the SDMM results.

Two examples illustrate application of the SW hybrid formulation to antenna radiation problems. Fig. 7 depicts the radiation patterns of two slot antennas on a curved strip, 10 λ wide. This configuration embodies the strong coupling between the edge and the slot antennas. The close correlation between the SW hybrid and SDMM results indicate the proper representation of this effect. Fig. 8 shows the radiation patterns in the vertical and azimuthal planes of a probe-fed annular slot antenna covered with a tapered dielectric (ϵ = 1.7) coating or window. For comparison, we also depict the results for the same slot antenna without the coating. In all

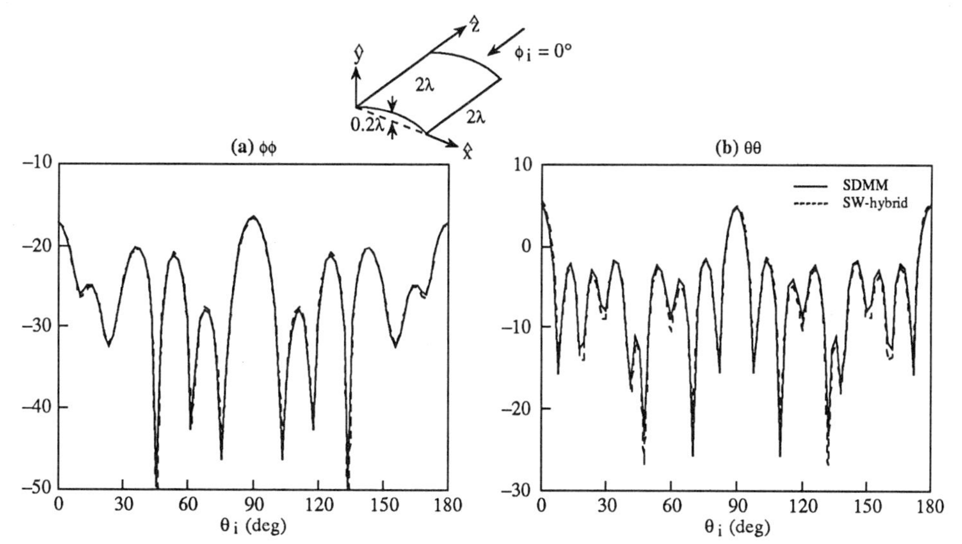

Fig. 6 Backscatter cross sections of curved plate. ϕ_i = 0°: (a) ϕϕ - polarization (b) θθ - polarization

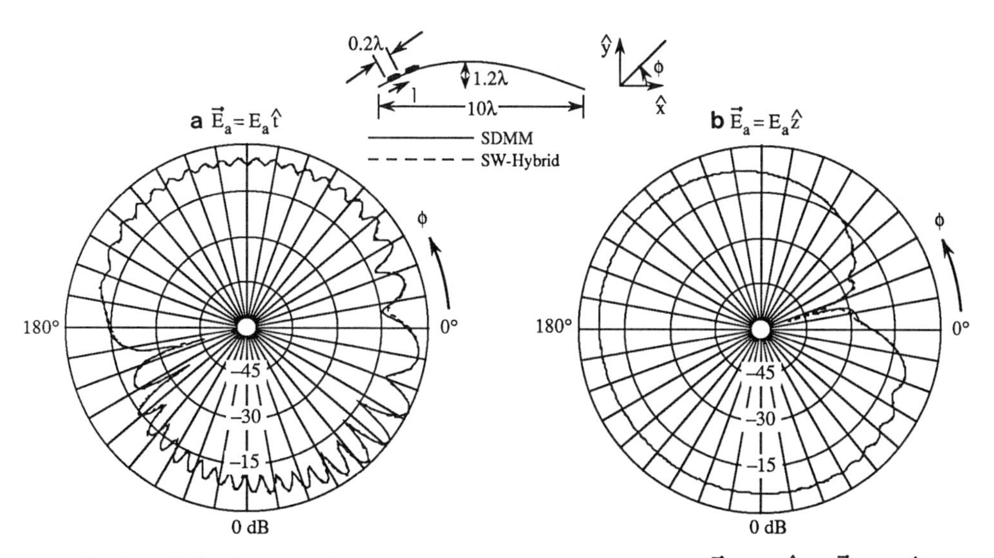

Fig. 7 Radiation Patterns of two slot antennas on a curved strip. (a) $\vec{E}_a = E_a\hat{t}$ (b) $\vec{E}_a = E_a\hat{z}$

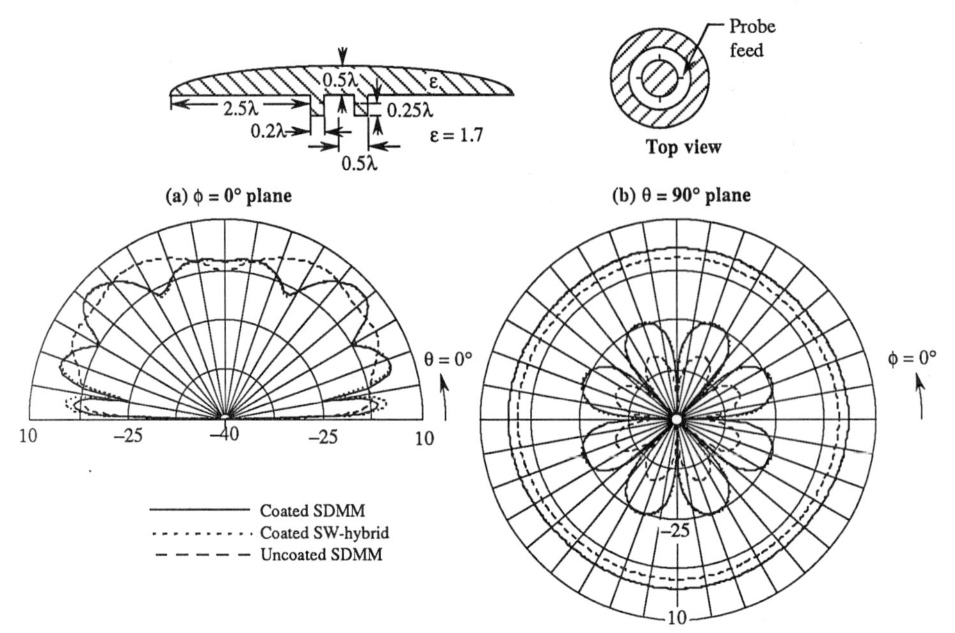

Fig. 8 Radiation patterns of a probe-fed annular slot antenna with tapered dielectric coating, ε =1.7. (a) vertical plane (b) azimuthal plane.

the foregoing cases, we observed substantial reductions in the number of unknowns. In most cases a reduction in excess of 70% was achieved over SDMM, leading to a concomitant reduction of required computer memory.

SUMMARY

This paper discussed the development of the new surface wave (SW) hybrid formulation. This method greatly expands the domain of validity of current-based hybrid formulations by incorporating a surface-wave basis set augmenting the optics-derived Ansatz currents. Specifically, large complex objects dominated by multiple interactions and surface wave effects can be analyzed with high accuracy. Numerical experiments confirm that this approach offers the same accuracy as MM solutions using SD bases, however, with substantial reduction of required computational resources. Representative examples were given demonstrating its potential to treat complex, electrically large scattering and radiation problems.

REFERENCES

1. L. N. Medgyesi-Mitschang and D. -S. Wang, "Hybrid solutions for scattering from perfectly conducting bodies of revolution," IEEE Trans. Antenna Propagat. AP-31, pp. 570-583, 1983.
2. L. N. Medgyesi-Mitschang and D. -S. Wang, "Hybrid solutions for scattering from large bodies of revolution with material discontinuities and coatings," IEEE Trans. Antenna Propagat. AP-32, pp. 717-723, 1984.
3. L. N. Medgyesi-Mitschang and D. -S. Wang, "Hybrid solutions for large impedance coated bodies of revolution," IEEE Trans. Antenna Propagat. AP-34, pp. 1319-1329, 1986.
4. W. D. Burnside, C. L. Yu, and R. J. Marhefka, "A technique to combine the geometric theory of diffraction and the moment method," IEEE Trans. Antenna Propagat. AP-23, pp. 551-558, 1975.

5. J. N. Sahalos and G. A. Thiele, "On the application of the GTD-MM technique and its limitation," IEEE Trans. Antenna Propagat. AP-29, pp. 780-786, 1981.
6. M. D. Tew and L. L. Tsai, "A method toward improved convergence of moment method solutions,' IEEE Proceedings, pp. 1436-1437, Nov. 1972.
7. C. A. J. Fletcher, Computational Galerkin Methods, New York: Springer-Verlag, 1984.

TRANSIENT PLANE WAVE SCATTERING FROM A CIRCULAR CYLINDER BACKED BY A SLIT-COUPLED COAXIAL CAVITY

G. Vecchi

Dipartimento di Electtronica
Politecnico
I - 10129 Torino
Italy

L. B. Felsen

Department of Electrical Engineering
Polytechnic University
Farmingdale, NY 11735

Abstract — Aperture coupling to enclosures with low interior losses can cause marked changes in the scattering characteristics of a target without perforation. The changes are attributable primarily to interior resonances. In a previous frequency domain study, a generally applicable self-consistent system format has been introduced for "observable-based" analytical modeling of the exterior-interior wave processes pertaining to cavity backed aperture geometries. The formulation has been applied there to plane-wave scattering from a slit-perforated thin perfectly conducting smooth convex cylindrical shell with an interior perfectly conducting smooth convex cylindrical load, modeled externally via the ray fields of the geometrical theory of diffraction, internally via guided modes in the annular waveguide between the inner and outer boundaries, and coupled self-consistently by the slit. This study is extended here to short-pulse plane wave excitation. The various wave constituents are now distinguished by their different arrival and turn-on times. Alternative hybrid wavefront-resonance formulations are explored and are shown to furnish well convergent parametrizations of the physical observables in the scattered signal.

I INTRODUCTION

The two-dimensional problem of time-harmonic plane wave scattering from a perfectly conducting cylindrical smooth convex thin outer shell with a narrow longitudinal slit that grants access to an interior with a perfectly conducting smooth convex cylindrical load has been studied in two previous publications [1], [2]. This study is extended here to determination of the scattered field when the incident plane wave is impulsive. Although applied to a coaxial circular cylinder geometry (Fig.1), the general method accommodates a broad class of scattering configurations with aperture-coupled waveguide-like interiors. The frequency domain results in [2] were found to be reconstructed efficiently, and interpreted physically, in terms of an observable-based parametrization (OBP) that treats the external wave phenomena by ray-type interactions and the internal wave phenomena by modes propagating in the annular waveguide between the two boundaries, both wave systems being coupled by the slit in a self-consistent system format.

Directions in Electromagnetic Wave Modeling
Edited by H.L. Bertoni and L.B. Felsen, Plenum Press, New York, 1991

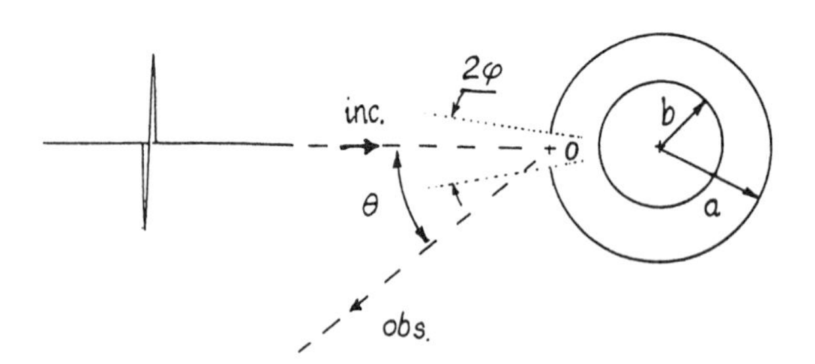

Fig. 1. Problem geometry. The plane wave is incident along the slit midplane (normal incidence, magnetic field along symmetry axis). The polar coordinate system is centered at the slit center O, and the observer is located along the direction θ at a distance r away from the slit center O. Problem parameters: inner radius b = 0.8, outer radius a = 1, and slit angular aperture $2\varphi = 1°$. The figure also contains the incident pulse.

To generalize the ray-mode-resonance hybrid model in [1] and [2], which has been described previously [3], from the frequency domain to the time domain, it is necessary to translate phase-dominated phenomena in the frequency domain into the time-resolved wave arrivals with their characteristic "turn-on" times related to causality. Discussion on how to embed resonant phenomena within the transient framework may be found in the literature (see [3]-[6] for example). For the present application, where the system parameters (slit scattering coefficients, utility of interior annular waveguide concepts, etc.) in [1] and [2] are restricted to a frequency window with low and high frequency cutoffs, *strict* causality and *true* turn-on of the *interior mode dominated* response is irrelevant because these phenomena occur well after the initial (but negligibly weak) causal arrivals that travel at the speed of light. Accordingly, we shall employ a phenomenological turn-on time for the interior pulsed modes based on the fastest modal group velocity. We shall also develop a self-consistent hybrid (traveling mode)-(resonant mode) expansion which incorporates these concepts. The resulting OBP, which does not appear to have been used before, is found to reconstruct with excellent accuracy the data generated by an independent numerical reference solution (for an independently performed related study dealing with an open-ended short-circuited plane parallel waveguide, see [6]).

To implement the above solution strategy, we return to [1] and decompose the frequency domain scattered field $u_s(\omega)$ into a *resonant* contribution $u_{res}(\omega)$ and a "background" contribution $u_b(\omega)$:

$$u_s(\omega) = u_{res}(\omega) + u_b(\omega) \tag{1}$$

The resonant term u_{res} contains all wave interactions resulting from multiple propagation and scattering events that result in high-Q global resonance poles dominated by the interior cavity effect. The background field u_b represents all *external* interactions which *do not involve* coupling to the interior, including geometrical optics (GO) reflection, creeping waves (CW) and direct slit diffraction. This separation allows the natural imposition of the above-noted OBP in the time domain. The exciting pulse will be chosen so that its frequency spectrum fits into the spectral window accommodated by [2], and the resulting fields are calculated accordingly. The time domain results, distinguished by a caret ($\check{}$), are generated from those in the frequency domain by Fourier inversion

$$\check{u}(t) = \frac{1}{2\pi}\int_{-\infty}^{\infty} d\omega e^{-i\omega t} u(\omega) \tag{2}$$

and analytic extension into the complex ω-plane. In what follows, we summarize the basic analytic buiding blocks and show a typical result. A much more detailed treatment, with additional results, has been submitted elsewhere for publication [4].

II EXTRACTION OF PROPAGATION DELAYS AND TURN-ON TIMES

II.A *The transient resonant field*

In the frequency domain, the resonant form of the scattered resonant field is given by (see (1.40) - (1.43))

$$u_{res} = \mathbf{Q}^T(\underline{\underline{\mathbf{1}}} - \underline{\underline{\mathbf{X}}}\,\underline{\underline{\mathbf{S}}})^{-1}\mathbf{I}\, u_0 \tag{3}$$

where u_0 is the incident field, $\underline{\underline{\mathbf{X}}}$ is the slit-to-slit modal propagator matrix along the interior waveguide, $\underline{\underline{\mathbf{S}}}$ is the mode coupling matrix representing the effect of the slit (including loading by the exterior CW), while $\mathbf{Q}^T$ (with ^{T} denoting the transpose) and $\mathbf{I}$ are projection vectors coupling the resonant field system to the observer and the incident field, respectively. Here and subsequently, the prefixes 1. and 2. identify equations, tables and figures in references [1] and [2], respectively. It is useful to extract the exterior propagator from the slit to the observer and the first round trip modal propagator in the interior waveguide from the projection vectors $\mathbf{Q}$ and $\mathbf{I}$, respectively, thereby rewriting the defining equations (1.42) and (1.43) (see also Tab. 1.BII) as follows

$$\mathbf{Q}(r,\theta) = \mathbf{Q}^{(s)}(\theta)\frac{e^{ikr}}{\sqrt{r/a}}, \quad k = \omega/c \tag{4}$$

$$\mathbf{I} = \underline{\underline{\mathbf{X}}}\mathbf{I}', \quad \underline{\underline{\mathbf{X}}} = \begin{bmatrix} \underline{\underline{0}} & \underline{\underline{\mathrm{p}}} \\ \underline{\underline{\mathrm{p}}} & \underline{\underline{0}} \end{bmatrix}, \quad \underline{\underline{\mathrm{p}}} = \mathbf{diag}\{p_\ell\}, \quad p_\ell = e^{i\nu_\ell(\omega)\Theta} \equiv \exp[ik\kappa_\ell(\omega)L] \tag{5}$$

with the form of the modified projection vectors $\mathbf{Q}'$ and $\mathbf{I}'$ inferred directly from (1.42) and (1.43).

The following definitions apply:

k: wavenumber in free space; c: speed of light in free space;

$\nu_\ell(\omega)$: azimuthal wavenumber for mode ℓ in the annular waveguide;

$\kappa_\ell(\omega)$: normalized propagation coefficient of mode ℓ;

Θ: total angular length ($\Theta \approx 2\pi$) of the annular waveguide; L: physical length ($L \approx 2\pi a$) of the the annular waveguide.

Restoring the omitted time dependence $e^{-i\omega t}$, u_{res} is composed of typical terms structured as follows:

$$f_\ell(\omega)\exp\{i\frac{\omega}{c}[r + \kappa_\ell(\omega)L - ct]\} \tag{6}$$

where the time reference $t = 0$ is chosen so as to correspond subsequently to the arrival of the incident pulse at the slit. The function $f_\ell(\omega)$ has no *explicit* exponential terms, but contains all scattering events described by $\underline{\underline{\mathbf{X}}}\,\underline{\underline{\mathbf{S}}}$ as well as all the resonance singularities of $[\underline{\underline{\mathbf{1}}} - \underline{\underline{\mathbf{X}}}\,\underline{\underline{\mathbf{S}}}]^{-1}$. Because in (6) the first round trip interior propagator has been extracted explicitly, all remaining interior propagation events in $f_\ell(\omega)$ sample the *global* internal environment. Staying within the frequency constraints imposed in [2], we shall need to consider only the lowest ($\ell = 0$) mode, the quasi-TEM (QTEM) mode, in the interior. All higher modes are nonpropagative in the chosen frequency window (see Sec.V). Nevertheless, we have retained the mode index ℓ because the formulation accomodates multimode operation.

From the above, u_{res} in (3) can be written via (4)-(6) as follows

$$u_{res}(\omega) = \frac{e^{ikr}}{\sqrt{r/a}}\{\mathbf{Q}^{(s)\mathrm{T}}(\underline{\underline{\mathbf{1}}} - \underline{\underline{\mathbf{X}}}\,\underline{\underline{\mathbf{S}}})^{-1}\underline{\underline{\mathbf{P}}}_0\mathbf{I}^{(s)}\}u_0 \tag{7}$$

with

$$\mathbf{I}^{(s)} \equiv \underline{\underline{\mathbf{X}}}^{(s)}\mathbf{I}' \tag{8}$$

$$\underline{\underline{\mathbf{X}}}^{(s)} \equiv \underline{\underline{\mathbf{P}}}_0^{-1}\underline{\underline{\mathbf{X}}}, \quad \underline{\underline{\mathbf{P}}}_0 = \begin{bmatrix} \mathrm{diag}\{p_0^{(\ell)}\} & 0 \\ 0 & \mathrm{diag}\{p_0^{(\ell)}\} \end{bmatrix}, \quad p_0^{(\ell)} = \exp[i\omega t_{0\ell}] \tag{9}$$

or, explicitly,

$$u_{res}(\omega) = \sum_\ell u_{res}^{(\ell)}(\omega) \equiv \sum_\ell \frac{e^{ikr}}{\sqrt{r/a}} e^{i\omega t_{0\ell}} \cdot w_{res}^{(\ell)}(\omega) \tag{10}$$

where

$$w_{res}^{(\ell)}(\omega) = h^{(\ell)}(\omega)u_0(\omega) \tag{11}$$

Here, $h^{(\ell)}(\omega)$ is the transfer function of the resonant field, which is regular at infinity in the half plane $Im(\omega) < 0$ and can therefore be expanded in a resonant mode series. In (10), the phase term $\exp(i\omega t_{0\ell})$ accounts for the minimum round trip group delay $t_{0\ell}$ of the ℓ–th mode relative to the reference phase $t = 0$ identified in (6); this group delay is taken as the turn-on time of the interior resonant mode expansion. For notational convenience, we introduce the following normalized parameters:

$$\xi \equiv ka, \quad \tau \equiv \frac{ct}{2\pi a}, \quad \tau_{0\ell} \equiv \frac{ct_{0\ell}}{2\pi a}, \quad \tau_{fs}(r) = \frac{r/c}{2\pi a/c}, \quad \mu_\ell(ka) = \frac{\nu_\ell(ka)}{ka}. \tag{12}$$

Here, τ_{fs} is the normalized free-space slit-to-observer time delay. Referring to (10), the transient scattered field $\breve{u}_{res}(\tau)$ at the far zone observation point location r from the slit center is thus related to the transient resonant mode field $\breve{w}_{res}(\tau)$ after turn-on through the propagation delays

$$\breve{u}_{res}^{(\ell)}(\tau) = \frac{1}{\sqrt{r/a}}\breve{w}_{res}^{(\ell)}(\tau - \tau_{fs}(r) - \tau_{0\ell}) \tag{13}$$

II.B The transient background field

Referring to (1.45), the background $u_b(ka)$ in (1) can be written as

$$u_b(ka) \approx u_{cyl}(ka) + d_f\, u_0(ka), \quad d_f \equiv D_f\frac{e^{ikr}}{\sqrt{r/a}} \tag{14}$$

where $u_{cyl}(ka)$ is the spectrum of the field scattered by the unslitted external cylinder and d_f is the diffraction coefficient for the directly illuminated slit (see also Table 1.B-II). For large ka, the term u_{cyl} tends to the frequency-independent asymptotic value representing the GO (specular) reflection. This term has to be extracted to guarantee convergence of the FT integral (2). Also extracting the free-space delay from slit to observer leads to

$$u_b(ka) = \frac{e^{ikr}}{\sqrt{r/a}} w_b(ka), \quad w_b(ka) = u_0(ka)\Gamma_{GO} e^{-i\,ka\,2\pi\tau_{GO}(\theta)} + D_f(ka)\,u_0(ka) + w_{cyl}^{(s)}(ka) \tag{15}$$

where $\Gamma_{GO}\exp[-i2\pi ka\tau_{GO}(\theta)]$ is the GO reflection coefficient (see [1], and references therein) with

$$\Gamma_{GO}(ka) = \sqrt{\frac{1}{2}\cos\frac{\theta}{2}}, \quad 2\pi\tau_{GO}(\theta) \equiv -[1 + \cos\theta - 2\cos\frac{\theta}{2}] \tag{16}$$

whereas

$$w_b^{(s)}(ka) = D_f(ka)\,u_0(ka) + w_{cyl}^{(s)}(ka) \tag{17}$$

incorporates direct slit diffraction and the non-GO (i.e., CW) cylinder scattering contribution. Insertion of (15), (16) into (2) yields

$$\breve{u}_b(\tau) = \breve{w}_b(\tau - \tau_{fs}) \tag{18}$$

$$\breve{w}_b(\tau) = \Gamma_{GO}\breve{u}_0(\tau + \tau_{GO}) + \breve{w}_b^{(s)}(\tau) \tag{19}$$

In the calculations that follow, the cylinder scattering as a whole will be represented by its GTD approximation as in [1]. The low frequencies, which are poorly accommodated by this high-frequency approximation, are rendered insignificant by choice of an input signal with low-frequency (LF) cutoff (see Sec.V).

II.C The total transient field

Combining the results in (13), (18) and (19), one obtains for the total scattered transient field

$$\breve{u}_s(\tau) = \frac{1}{\sqrt{r/a}}\left\{\Gamma_{GO}\breve{u}_0(\tau - \tau_{fs} + \tau_{GO}) + \breve{w}_b^{(s)}(\tau - \tau_{fs}) + \breve{w}_{res}(\tau - \tau_{fs}(r) - \tau_0)\right\} \tag{20}$$

Inspection of (20) reveals that the first arrival, reaching the observer in the slit-illuminated region at $\tau = \tau_{fs} - \tau_{GO}$ ($\tau_{GO} \geq 0$ for any $\theta < \pi$), is always due to specular reflection, which involves the shortest path length. Subsequent early time transients are due to direct slit diffraction and CW terms in $\breve{w}_b^{(s)}(\tau - \tau_{fs})$. The resonance-related oscillations arrive at the delayed time $\tau_{fs} + \tau_0$.

III THE RESONANT MODE EXPANSION

For the $\ell = 0$ mode regime, the transfer function of the resonant field, defined in (11), has the following spectral representation in terms of the eigenvectors of the system matrix $\underline{\underline{\mathbf{X}}}\,\underline{\underline{\mathbf{S}}}$ (see (2.6)),

$$h(ka) = \sum_m \frac{E_m(ka)}{1 - \lambda_m(ka)} \tag{21}$$

where

$$E_m(ka) = \frac{\mathbf{Q}^{(s)\mathrm{T}}(ka)\mathbf{\Psi}_m(ka)\mathbf{\Psi}_m^{\dagger}(ka)\mathbf{I}^{(s)}(ka)}{\|\ \mathbf{\Psi}_m(ka)\ \|^2} \tag{22}$$

We list below various relevant equations which have bee extracted from [2]. Eigenvalue problem (2.3):

$$(\underline{\underline{1}}\lambda_m - \underline{\underline{\mathbf{X}}}\,\underline{\underline{\mathbf{S}}})\mathbf{\Psi}_m = 0, \quad \mathbf{\Psi}_m^\dagger(\underline{\underline{1}}\lambda_m - \underline{\underline{\mathbf{X}}}\,\underline{\underline{\mathbf{S}}}) = 0 \tag{23}$$

λ_m, $\mathbf{\Psi}_m$ and $\mathbf{\Psi}_m^\dagger$ are the eigenvalues and right and left eigenvectors of $\underline{\underline{\mathbf{X}}}\,\underline{\underline{\mathbf{S}}}$, respectively.
Normalized resonance frequencies $k_{mn}a = \xi_{mn}$:

$$\lambda_m(\xi_\nu) - 1 = 0, \quad \nu \equiv (m,n), \quad Im(\xi_\nu) < 0 \tag{24}$$

Then by contour deformation, the real-ω continuum in (21) can be *contracted* around the lower half plane singularities to yield

$$h(ka) = \sum_\nu \left[\frac{E_m(ka)}{-\lambda_m'(ka)}\right]_{\xi_\nu} \frac{1}{ka - \xi_\nu} + B.P. \tag{25}$$

where the prime represents the derivative with respect to the argument, and $B.P.$ represents the contribution from the branch point at $ka = 0$ in this two-dimensional configuration. The $B.P.$ contribution can be neglected, as is confirmed by the comparison between the alternative calculations performed on (21) and (25) (see [4]). Moreover, in the present problem, $B.P.$ is de-emphasized further by the low frequency cutoff of the incident pulse spectrum (see (35) and Fig. 2). Invoking frequency domain symmetries for synthesizing real transient signals, and omitting $B.P.$, one may write

$$h(ka) = \sum_\nu \left(\frac{R_\nu}{ka - \xi_\nu} + \frac{-R_\nu^*}{ka + \xi_\nu^*}\right), \quad R_\nu = \left[E_m(ka)\frac{1}{-\lambda_m'(ka)}\right]_{ka=\xi_\nu} \tag{26}$$

where the asterisk denotes the complex conjugate. Inverting via (2) into the time domain yields

$$\check{h}(\tau) = \sum_\nu M_\nu e^{-A_\nu \tau} \sin(\Omega_\nu \tau - \Phi_\nu) \tag{27}$$

where

$$M_\nu \equiv -4\pi|R_\nu|, \qquad \Phi_\nu \equiv \mathrm{Arg}(R_\nu) \tag{28}$$

$$\Omega_\nu \equiv 2\pi Re(\xi_\nu), \quad A_\nu \equiv -2\pi Im(\xi_\nu) \tag{29}$$

IV Hybrid Traveling and Resonant Mode Representation

The resonant inverse $(\underline{\underline{1}} - \underline{\underline{\mathbf{X}}}\,\underline{\underline{\mathbf{S}}})^{-1}$ appearing in (3) may be expanded in a Neumann series when none of the eigenvalues of $\underline{\underline{\mathbf{X}}}\,\underline{\underline{\mathbf{S}}}$ equals unity. More generally one may employ the "hybrid" form resulting from expansion to a finite number of terms,

$$(1-x)^{-1} = \sum_{n=0}^{M} x^n + \frac{x^{M+1}}{1-x}$$

When used in (3) along with (9) and (8), such an expansion yields

$$w_{res}(ka) = u_0(ka)\left\{\sum_{n=0}^{M} \mathbf{Q}^{(s)\mathrm{T}}(\underline{\underline{\mathbf{X}}}\,\underline{\underline{\mathbf{S}}})^n \mathbf{I}^{(s)} + \mathbf{Q}^{(s)\mathrm{T}}(\underline{\underline{\mathbf{X}}}\,\underline{\underline{\mathbf{S}}})^{M+1}(\underline{\underline{1}} - \underline{\underline{\mathbf{X}}}\,\underline{\underline{\mathbf{S}}})^{-1}\mathbf{I}^{(s)}\right\} \tag{30}$$

Letting $M \to \infty$ in (30) makes the last term disappear, and one obtains the *traveling* transient mode expansion of the *resonant* transient mode field u_{res}, in which $(\underline{\underline{\mathbf{X}}}\,\underline{\underline{\mathbf{S}}})^n$

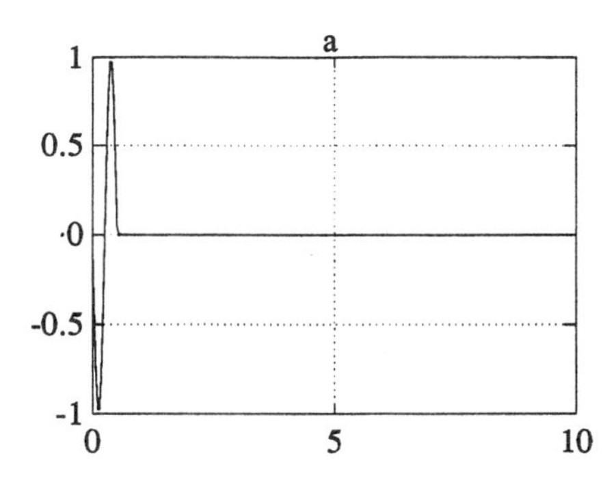

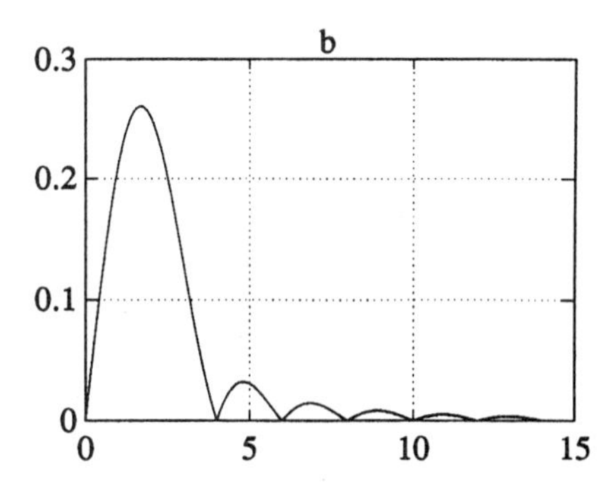

Fig. 2. Input waveform. a) Time domain: $\check{u}_0(\tau)$, b) frequency spectrum: $|u_0(ka)|$. The pulse parameters are $\tau_{on} = 0.5$, centerband frequency $ka = \xi_0 = 2$ (single-cycle wavelet).

represents n round-trip excursions ($\underline{\underline{\mathbf{X}}}$) around the body, including slit scattering (S) events. From (30) and (23), one obtains

$$w_{res}(ka) = p_0^{M+1} W_{res}^{(M)}(ka) + \sum_{j=0}^{M} p_0^j W_j(ka) \tag{31}$$

where

$$W_{res}^{(M)}(ka) = u_0(ka) \sum_m (\lambda_m^{(s)})^{M+1} \frac{E_m(ka)}{1 - \lambda_m(ka)}, \qquad \lambda_m^{(s)} \equiv \frac{\lambda_m}{p_0} \tag{32}$$

$$W_j(ka) = u_0(ka) \sum_m (\lambda_m^{(s)})^j E_m(ka) \tag{33}$$

The inverse FT (2) of (31) yields

$$\check{w}_{res}(\tau) = \sum_{j=0}^{M} \check{W}_j(\tau - j\tau_0) + \check{W}_{res}^{(M)}(\tau - (M+1)\tau_0) \tag{34}$$

The formulation clearly places in evidence the delayed turn-on times of the individual M mode pulses $\check{W}_j$, and of the collective resonant remainder $\check{W}_{res}^{(M)}$.

V Numerical Results

V.A *Input Signal*

We have chosen the sinewave pulse with centerband frequency ξ_0 and duration τ_{on},

$$\check{u}_0(\tau) = p(\tau) = \begin{cases} \sin(2\pi\xi_0(\tau - \tau_{on})), & |\tau - \tau_{on}| \le \tau_{on} \\ 0, & |\tau - \tau_{on}| > \tau_{on} \end{cases} \tag{35}$$

for which the transient resonant field $\check{w}_{res}(\tau)$ can be evaluated in closed form (see [4]). The pulse and its spectrum are shown in Figs.2a and 2b.

a

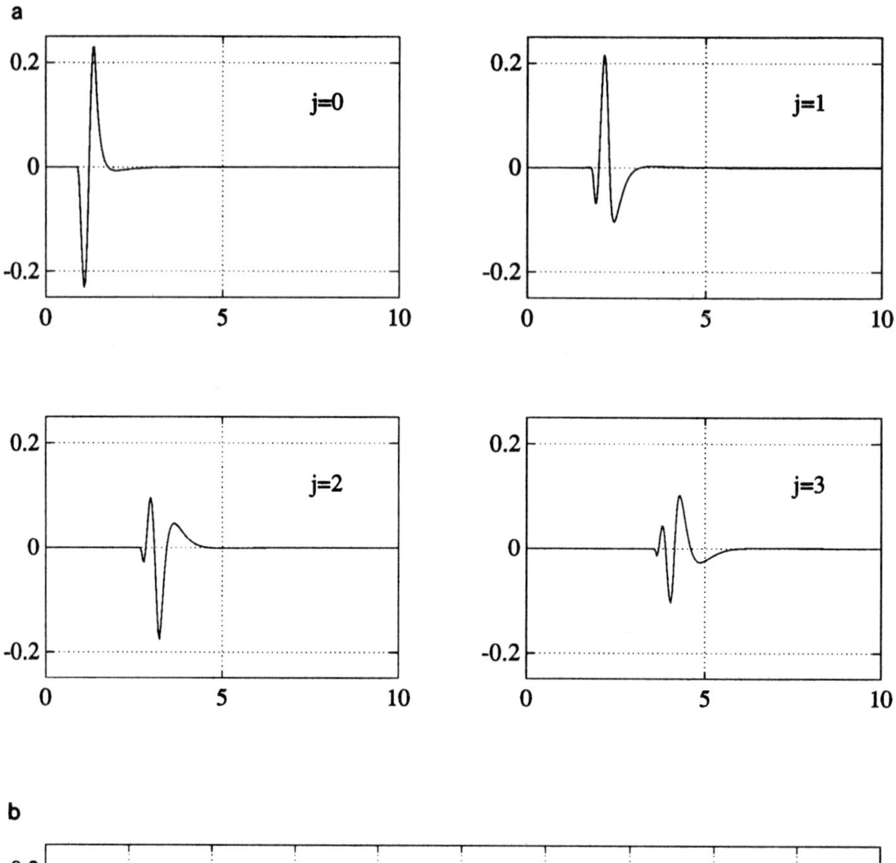

b

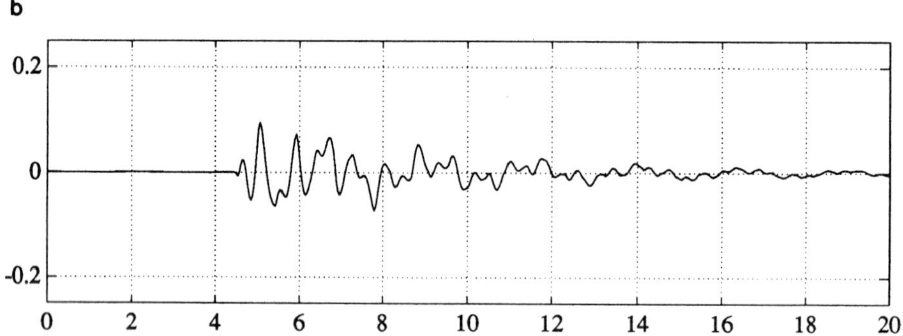

Fig. 3. Transient resonant field: hybrid traveling and resonant mode representation. Problem parameters as in Figs. 1 and 2. a) First four traveling mode contributions $\check{W}_j(\tau - (j+1)\tau_0)$ *in (31) and (33). Note the increasing dispersive distortion of the input pulse for multiple round trips. b) Resonant remainder* $\check{W}^{(4)}_{res}(\tau - \tau_0)$ *in (31) and (32), which accounts collectively for all traveling mode fields* $j \geq 4$.

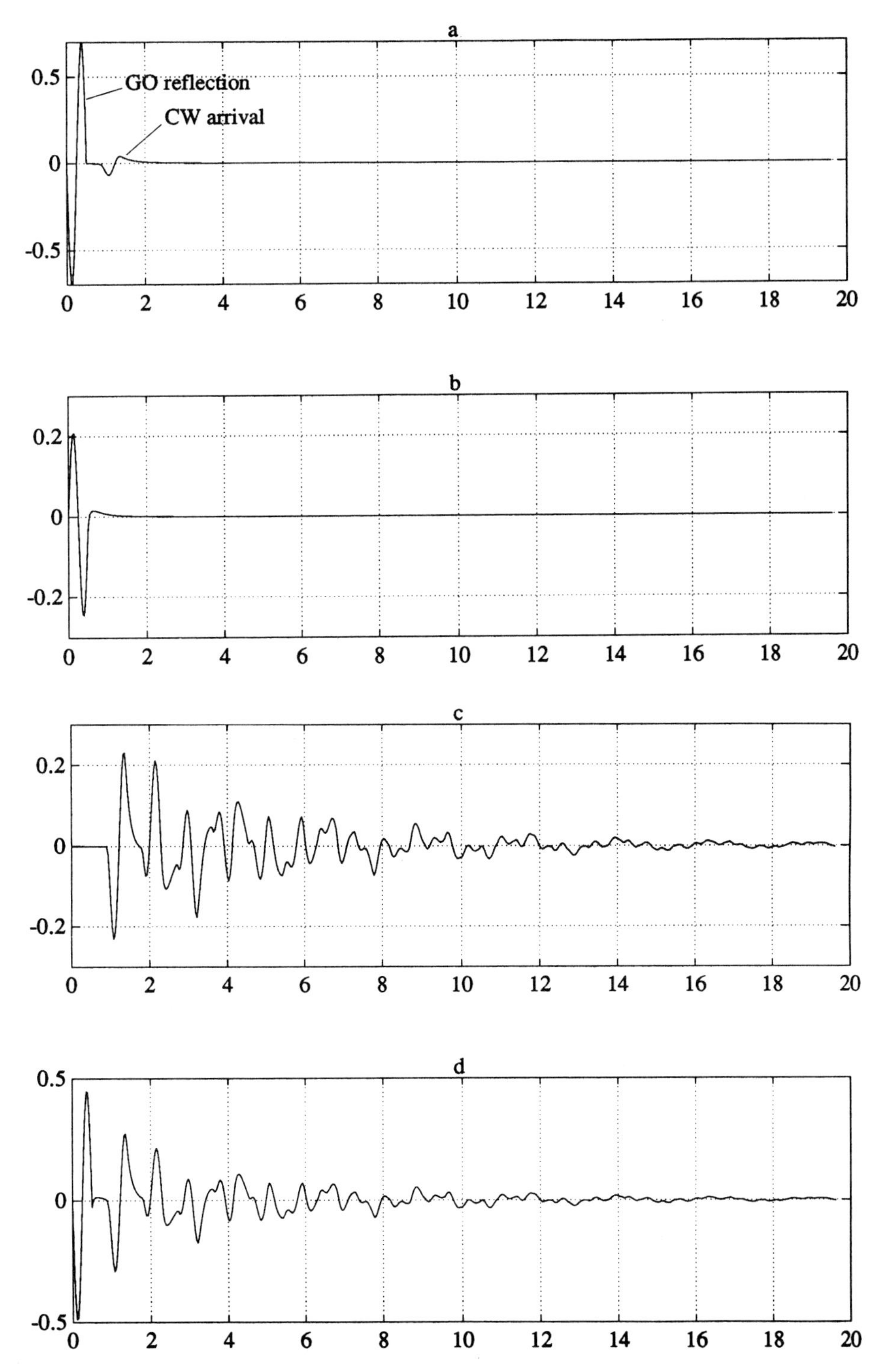

Fig. 4. Total backscattered transient field ($\theta = 0°$). a)$\breve{w}_{cyl}(\tau)$; b) direct slit diffraction; c) $\breve{w}_{res}(\tau - \tau_0)$; d) $\breve{w}_s(\tau)$ (note the various vertical scales). Geometrical parameters and incidence as in Figs. 1 and 2

V.B Backscattered field $\theta = 0$ (see Fig.1)

We show a typical sample of the more comprehensive calculations in [4]: the far field scattered back along the direction $\theta = 0$ of the incident plane wave pulse. Figures 3 and 4 show how the field is synthesized by the hybrid traveling and resonant mode expansion in (30). Figure 3a shows the contribution from the first four traveling wave pulses whereas Fig. 3b shows the contribution from the collective remainder. Note the different turn-on times of these constituents, and the greater pulse distortion due to dispersion, after multiple round trips. Figure 4 collects all contributions to synthesize the total scattered field $\check{w}_s(\tau)$ (note the different amplitude scales).The resulting synthesis shown in the bottom graph *agrees very well* with an independent reference solution obtained by numerical integration [4], via (2), of the real-ω continuum form of the transfer function in (21). This establishes the validity of O.B.P., and *interprets* the relevant wave phenomena in detail. One observes that the first arrivals are due to the exterior background field in (19), with (17), whereas the interior hybrid portion of Fig. 3 accounts for the later time response. Enlarged versions of these figures (not shown) reveal that *all* features of the total reference field $\check{w}_s$ are sequentially accounted for. In the synthesis, the time domain traveling mode pulses and the exterior background terms have been constructed by direct numerical inversion, via (2), whereas the resonant mode remainder has been computed from the resonant expansion corresponding to (27) but implemented for $\check{W}_{res}^{(M)}$.

VI CONCLUSIONS

Extending a previous investigation in the frequency domain [1],[2], we have shown in this study how transient scattering from targets with slit-coupled high-Q interior cavities can be well parametrized in terms of self-consistently combined exterior GTD and interior modes, which track the relevant observables (OBP). When the interior is waveguide-like, as in the structures under consideration, the OBP synthesis implemented in [2] for the frequency domain is validated even more convincingly by the sequential time-resolved buildup manifested in the transient case. Extension to multimode operation (i.e. higher frequency components) and to nonconcentric interiors is under consideration.

ACKNOWLEDGMENT

L.B.Felsen acknowledges partial support from the Joint Services Electronics Program under Contract No. F 49620-88-C-0075.

REFERENCES

[1] L.B. Felsen and G. Vecchi, "Wave scattering from slit-coupled cylindrical cavities with interior loading: I. Formulation by ray-mode parametrization", to be published in *IEEE Trans. Antennas Propagat.*

[2] L.B. Felsen and G. Vecchi, "Wave scattering from slit-coupled cylindrical cavities with interior loading: II. Resonant mode expansion", to be published in *IEEE Trans. Antennas Propagat.*

[3] E.Heyman and L.B.Felsen, "A wavefront interpretation of the singularity expansion method", *IEEE Trans. Antennas Propagat.*, vol. AP-33, no. 7, pp. 706-718, July 1985.

[4] G. Vecchi and L.B. Felsen, "Short-Pulse Wave Scattering from a Cylindrical Shell with Slit-Coupled Coaxial Interior", submitted to *IEEE Trans. Antennas Propagat.*

[5] M.A.Morgan, " Singularity expansion representation of fields and currents in transient scattering", *IEEE Trans. Antennas Propagat.*, vol. AP-32, no. 5, pp. 466-473, May 1984.

[6] E.Heyman and G.Friedlander: private communication

TRANSIENT ACOUSTIC RADIATION IN A CONTINUOUSLY LAYERED FLUID - AN ANALYSIS BASED ON THE CAGNIARD METHOD

A. T. de Hoop and M. D. Verweij

Laboratory of Electromagnetic Research
Department of Electrical Engineering
Delft University of Technology
P.O. Box 5031, 2600 GA Delft, The Netherlands

I. INTRODUCTION

In the present paper, the problem of the transient wave propagation in a continuously layered fluid is addressed directly with the aid of the Cagniard method. The standard integral transformations that are characteristic for this method (Cagniard, 1939, 1962; De Hoop, 1960, 1961, 1988; Aki and Richards, 1980) are applied to the first-order acoustic wave equations of a fluid. The resulting system of differential equations in the depth coordinate is next transformed into a system of integral equations. These integral equations admit a solution by a Neumann iteration. Each higher-order iterate can be physically interpreted as to be generated, through continuous reflection, by the previous one. To show the generality of the method, anisotropy of the fluid in its volume density of mass is included. This type of anisotropy is encountered in the equivalent medium theory of finely discretely layered media (Schoenberg, 1984). The compressibility is a scalar. Next, the transformation back to the space-time domain is performed using the Cagniard method, in which a number of steps can be carried out analytically even for the anisotropic case.

The iterative method is shown to be convergent for any continuous and piecewise continuously differentiable depth profile in the inertia and compressibility properties of the fluid. This is contrary to the frequency-domain analysis of the problem, where the corresponding Neumann series, which is also known as the Bremmer series (Bremmer, 1939, 1949a, 1949b, 1951), can only be shown to be convergent for profiles that vary within certain, frequency dependent, bounds.

The difficulties that are met with an inversion method based on a time Fourier transformation with real frequency variable can be ascribed to the fact that with this transformation causality is lost, while with a time Laplace transformation with a real transform parameter as used by Cagniard (1939, 1962), as a crucial point, causality is automatically taken care of by restricting the transform-domain counterparts of the physical quantities to being bounded functions of the remaining space variables. Also, in the modified Cagniard method the time variable is kept real all the way through, in accordance with its physical meaning. Further, no asymptotics is needed, and only convergent expansions occur. Another aspect of the propagation of transient waves in continuously layered media is covered by the Spectral Theory of

Directions in Electromagnetic Wave Modeling
Edited by H.L. Bertoni and L.B. Felsen, Plenum Press, New York, 1991

Transients that has been introduced by Heyman and Felsen (1984, 1987). This theory aims at a complete asymptotic expression for the total wave field in the neighborhood of the wave fronts, rather than an exact expression in terms of successively reflected wave constituents.

II. DESCRIPTION OF THE CONFIGURATION AND FORMULATION OF THE ACOUSTIC WAVE PROBLEM

Small-amplitude acoustic wave motion is considered in an unbounded inhomogeneous fluid, the properties of which vary in a single rectilinear direction in space only. This direction is taken as the vertical one. To specify position in the configuration the coordinates $\{x_1, x_2, x_3\}$ with respect to a fixed, orthogonal, Cartesian reference frame with the origin O and the three mutually perpendicular base vectors $\{\mathbf{i}_1, \mathbf{i}_2, \mathbf{i}_3\}$, of unit length each, are used; $\mathbf{i}_3$ points vertically downward. The subscript notation for vectors and tensors is used and the summation convention applies. Lowercase Latin subscripts are used for this purpose; they are to be assigned the values $\{1, 2, 3\}$. The time coordinate is denoted by t. Partial differentiation is denoted by ∂; ∂_m denotes differentiation with respect to x_m, ∂_t is a reserved symbol denoting differentiation with respect to t.

The acoustic properties of the (anisotropic) fluid are characterized by the tensorial volume density of mass ρ_{kr} and the scalar compressibility κ. Both are functions of x_3 only; these functions are assumed to be continuous and piecewise continuously differentiable. For $x_3 \leq x_{3;min}$ and $x_3 \geq x_{3;max}$ the medium is homogeneous, so both functions are constant in these intervals. At any x_3, the tensor ρ_{kr} is assumed to be symmetrical and positive definite and κ is positive. It is advantageous to distinguish, in the vectorial and tensorial quantities, between their horizontal and their vertical components. For the former, lowercase Greek subscripts will be used; for the latter, the subscript 3 will be written explicitly. The position of the point source is indicated by $x_{m;S}$, the receiver position by x'_m.

The acoustic wave motion in the configuration is characterized by its acoustic pressure p and its particle velocity v_r. These quantities satisfy the first-order acoustic wave equations

$$\partial_k p + \rho_{kr}\,\partial_t v_r = f_r, \tag{1}$$

$$\partial_r v_r + \kappa\,\partial_t p = q, \tag{2}$$

where f_r is the volume source densitiy of force and q is the volume source density of injection rate. Without loss of generality it is in the further analysis assumed that

$$\{f_r, q\} = \{F_{r;S}(t), Q_S(t)\}\,\delta(x_1, x_2, x_3 - x_{3;S}), \tag{3}$$

i.e., the point source is located at $x_1 = 0, x_2 = 0, x_3 = x_{3;S}$.

III. THE TRANSFORM-DOMAIN ACOUSTIC WAVE EQUATIONS AND THE WAVE-MATRIX FORMALISM

The acoustic wave equations (1) and (2) are subjected to a one-sided Laplace transformation with respect to time with real, positive transform parameter s, and a Fourier transformation with respect to the horizontal space coordinates with real transform parameters $s\alpha_1$ and $s\alpha_2$. For the acoustic pressure the two transformations are

$$\hat{p}(x_m, s) = \int_0^\infty \exp(-st)\,p(x_m, t)\,dt, \tag{4}$$

$$\tilde{p}(i\alpha_\mu, x_3, s) = \int_{x_\mu \in \mathbb{R}^2} \exp(is\alpha_\mu x_\mu)\,\hat{p}(x_m, s)\,dx_1 dx_2, \tag{5}$$

respectively. The extra factor of s in the spatial Fourier-transform parameters has been included for later convenience. In view of this, the transformation inverse to Eq. (5) is given by

$$\hat{p}(x_m, s) = \left(\frac{s}{2\pi}\right)^2 \int_{\alpha_\mu \in \mathbb{R}^2} \exp(-is\alpha_\mu x_\mu)\, \tilde{p}(i\alpha_\mu, x_3, s)\, d\alpha_1 d\alpha_2. \tag{6}$$

Under these transformations Eqs. (1) and (2) transform into

$$-is\alpha_\nu \tilde{p} + s\rho_{\nu r}\tilde{v}_r = \tilde{f}_\nu, \tag{7}$$

$$\partial_3 \tilde{p} + s\rho_{3r}\tilde{v}_r = \tilde{f}_3, \tag{8}$$

$$-is\alpha_\eta \tilde{v}_\eta + \partial_3 \tilde{v}_3 + s\kappa\tilde{p} = \tilde{q}. \tag{9}$$

Upon eliminating the horizontal components $\tilde{v}_\eta$ of the particle velocity from these equations, a system of two ordinary differential equations results with x_3 as independent variable and $\tilde{p}$ and $\tilde{v}_3$ as dependent variables. Let $[\tilde{F}]$ denote the acoustic field matrix, $[\tilde{A}]$ the acoustic system's matrix, and $[\tilde{N}]$ the notional source matrix, then the transform-domain matrix differential equation is

$$\partial_3 \tilde{F}_I + s\tilde{A}_{IJ}\tilde{F}_J = \tilde{N}_I, \tag{10}$$

in which the summation convention again applies, and the elements of the acoustic field matrix are given by

$$\tilde{F}_1 = \tilde{p}, \qquad \tilde{F}_2 = \tilde{v}_3, \tag{11}$$

the elements of the acoustic system's matrix by

$$\begin{aligned} \tilde{A}_{11} &= \rho_{3\eta}\,[\rho_{\parallel}^{-1}]_{\eta\nu}\, i\alpha_\nu, & \tilde{A}_{12} &= \rho_{33} - \rho_{3\eta}\,[\rho_{\parallel}^{-1}]_{\eta\nu}\,\rho_{\nu 3}, \\ \tilde{A}_{21} &= \kappa - i\alpha_\eta\,[\rho_{\parallel}^{-1}]_{\eta\nu}\, i\alpha_\nu, & \tilde{A}_{22} &= i\alpha_\eta\,[\rho_{\parallel}^{-1}]_{\eta\nu}\,\rho_{\nu 3}, \end{aligned} \tag{12}$$

and the elements of the notional source matrix by

$$\tilde{N}_1 = -\rho_{3\eta}\,[\rho_{\parallel}^{-1}]_{\eta\nu}\,\tilde{f}_\nu + \tilde{f}_3, \qquad \tilde{N}_2 = i\alpha_\eta\,[\rho_{\parallel}^{-1}]_{\eta\nu}\,\tilde{f}_\nu + \tilde{q}. \tag{13}$$

Here, $[\rho_{\parallel}^{-1}]_{\eta\nu}$ is the inverse of $\rho_{\nu\eta}$. Note that $[\tilde{A}]$ is independent of s.

Via an appropriate linear transformation to be carried out on the acoustic field matrix, a wave-matrix formalism will be arrived at from which the interaction between up- and downgoing waves in a region of inhomogeneity will be manifest. The relevant linear transformation is written as (cf. Chapman (1974) for the isotropic case)

$$\tilde{F}_I = \tilde{L}_{IJ}\tilde{W}_J, \tag{14}$$

where $[\tilde{W}]$ is the wave matrix and the matrix $[\tilde{L}]$ is to be chosen appropriately. On the assumption that the inverse $[\tilde{L}^{-1}]$ of $[\tilde{L}]$ exists, substitution of Eq. (14) into Eq. (10) yields

$$\partial_3 \tilde{W}_I + s\tilde{\Lambda}_{IJ}\tilde{W}_J = [\tilde{L}^{-1}]_{IJ}\tilde{N}_J - [\tilde{L}^{-1}]_{IJ}\,[\partial_3 \tilde{L}]_{JK}\,\tilde{W}_K, \tag{15}$$

where $[\tilde{\Lambda}] = [\tilde{L}^{-1}][\tilde{A}][\tilde{L}]$. Equation (15) indeed expresses the traveling-wave structure of the up- and downgoing waves provided that $[\tilde{\Lambda}]$ is a diagonal matrix. From the observation that $[\tilde{A}][\tilde{L}] = [\tilde{L}][\tilde{\Lambda}]$ it follows that $[\tilde{\Lambda}]$ is diagonal if $[\tilde{L}]$ consists of the eigencolumns of $[\tilde{A}]$; $[\tilde{\Lambda}]$ then has the eigenvalues of $[\tilde{A}]$ as its (diagonal) elements.

The elements of $[\tilde{L}]$ can be expressed in terms of the vertical acoustic wave admittance $Y = (\tilde{A}_{21}/\tilde{A}_{12})^{1/2}$, a quantity that, for $\alpha_\mu \in \mathbb{R}^2$, is real and positive. Note that this wave admittance is the same for up- and downgoing waves even in the case of an anisotropic fluid. The relevant expressions are

$$\tilde{L}_{11} = (2Y)^{-1/2}, \quad \tilde{L}_{12} = (2Y)^{-1/2}, \quad \tilde{L}_{21} = -(Y/2)^{1/2}, \quad \tilde{L}_{22} = (Y/2)^{1/2}. \tag{16}$$

With this, the coupling matrix becomes $[\tilde{L}^{-1}][\partial_3\tilde{L}] = R\,[C]$, where $C_{11} = C_{22} = 0$, $C_{12} = C_{21} = -1$ and $R = \partial_3 Y/2Y$ is the local reflection coefficient. Using these results and writing the elements of the wave matrix $[\tilde{W}]$ as $\tilde{W}_1 = \tilde{W}^-$ and $\tilde{W}_2 = \tilde{W}^+$, where $\tilde{W}^-$ is the local amplitude of the upgoing wave and $\tilde{W}^+$ is the local amplitude of the downgoing wave, the system of differential equations Eq. (15) leads to

$$\partial_3\tilde{W}^- + s\gamma^-\tilde{W}^- = \tilde{X}^- + R\tilde{W}^+, \tag{17}$$

$$\partial_3\tilde{W}^+ + s\gamma^+\tilde{W}^+ = \tilde{X}^+ + R\tilde{W}^-, \tag{18}$$

where

$$\gamma^- = \tfrac{1}{2}\,(\tilde{A}_{11} + \tilde{A}_{22}) - (\tilde{A}_{12}\tilde{A}_{21})^{1/2}, \qquad \gamma^+ = \tfrac{1}{2}\,(\tilde{A}_{11} + \tilde{A}_{22}) + (\tilde{A}_{12}\tilde{A}_{21})^{1/2}, \tag{19}$$

are the vertical slownesses of the up- and downgoing waves, respectively, while

$$\tilde{X}^- = [\tilde{L}^{-1}\tilde{N}]_1, \qquad \tilde{X}^+ = [\tilde{L}^{-1}\tilde{N}]_2. \tag{20}$$

In Section IV, the coupled wave propagation problem is recast in an integral-equation formulation that is equivalent to Eqs. (17) and (18). Next, these integral equations are solved iteratively and for the transform-domain acoustic pressure and particle velocity series expansions are obtained. The zero-order term in this expansion is representative for the direct wave generated by the source; the subsequent terms are representative for the waves that are successively reflected at the inhomogeneity levels.

IV. INTEGRAL-EQUATION FORMULATION AND ITERATIVE SOLUTION OF THE TRANSFORM- DOMAIN COUPLED WAVE PROBLEM

The integral-equation formulation of the transform-domain coupled wave problem follows from Eqs. (17) and (18) upon introducing appropriate one- sided Green's functions for the differential operators occurring at the left-hand side of these equations. The result can be written as

$$\tilde{W}^- = \tilde{W}_0^- + K^-\tilde{W}^+, \qquad \tilde{W}^+ = \tilde{W}_0^+ + K^+\tilde{W}^-, \tag{21}$$

where

$$\tilde{W}_0^-(x_3') = \int_{x_3'}^{\infty} \tilde{G}^-(x_3', x_3)\,\tilde{X}^-(x_3)\,dx_3, \tag{22}$$

$$\tilde{W}_0^+(x_3') = \int_{-\infty}^{x_3'} \tilde{G}^+(x_3', x_3)\,\tilde{X}^+(x_3)\,dx_3, \tag{23}$$

$$[K^-\tilde{W}^+](x_3') = \int_{x_3'}^{\infty} \tilde{G}^-(x_3', x_3)\,R(x_3)\,\tilde{W}^+(x_3)\,dx_3, \tag{24}$$

$$[K^+\tilde{W}^-](x_3') = \int_{-\infty}^{x_3'} \tilde{G}^+(x_3', x_3)\,R(x_3)\,\tilde{W}^-(x_3)\,dx_3, \tag{25}$$

$$\tilde{G}^-(x_3', x_3) = -H(x_3 - x_3')\,\exp\left[-s\int_{x_3}^{x_3'} \gamma^-(\varsigma)\,d\varsigma\right], \tag{26}$$

$$\tilde{G}^+(x_3', x_3) = H(x_3' - x_3)\,\exp\left[-s\int_{x_3}^{x_3'} \gamma^+(\varsigma)\,d\varsigma\right]. \tag{27}$$

The equations in (21) suggest the possibility of an iterative solution of the Neumann type by repeated substitution of the first equation in the second and vice versa. Since s is real and positive,

$$|K^-\tilde{W}| \le A(s)\,M\int_{x_3'}^{x_{3;max}} \exp[-s\gamma(x_3 - x_3')]\,dx_3 \le \frac{A(s)\,M}{s\gamma}, \tag{28}$$

$$|K^+\tilde{W}| \le A(s)\,M\int_{x_{3;min}}^{x_3'} \exp[-s\gamma(x_3 - x_3')]\,dx_3 \le \frac{A(s)\,M}{s\gamma}, \tag{29}$$

where

$$\left.\begin{array}{lcl} A(s) & = & \max\{|\tilde{W}^-|, |\tilde{W}^+|\}, \\ M & = & \max\{|R|\}, \\ \gamma & = & \max\{\mathrm{Re}(-\gamma^-), \mathrm{Re}(\gamma^+)\} > 0, \end{array}\right\} x_3 \in [x_{3;min}, x_{3;max}]. \tag{30}$$

The procedure is convergent if the real, positive time Laplace-transform parameter is taken in the semi-infinite interval $s > M/\gamma$, which can always be done since M is independent of s, positive and bounded, while γ is independent of s, positive and bounded away from zero. Upon summing the remaining convergent infinite series

$$\tilde{W}^-(x_3') = \sum_{n=0}^{\infty} \tilde{W}_n^-(x_3'), \qquad \tilde{W}^+(x_3') = \sum_{n=0}^{\infty} \tilde{W}_n^+(x_3'), \tag{31}$$

with

$$\begin{array}{ll} \tilde{W}_0^- = \tilde{G}^-(x_3', x_{3;S})\,\tilde{X}^-, & \tilde{W}_0^+ = \tilde{G}^+(x_3', x_{3;S})\,\tilde{X}^+, \\ \tilde{W}_{n+1}^- = [K^-\tilde{W}_n^+](x_3'), & \tilde{W}_{n+1}^+ = [K^+\tilde{W}_n^-](x_3'), \qquad (n = 0, 1, 2, \ldots), \end{array} \tag{32}$$

the inequality

$$A(s) \le \frac{A_0(s)}{1 - M/s\gamma} \tag{33}$$

results. In the latter, $A_0(s) = \max\{|\tilde{W}_0^-|, |\tilde{W}_0^+|\}$ on $[x_{3;min}, x_{3;max}]$. Hence, $A(s)$ is a bounded function in the interval $s > M/\gamma$ if $A(s)$ is so. The latter condition is satisfied if the source is taken to have at most a delta function (Dirac distribution) time dependence as Eqs. (13), (20), (32), and the property $\{|\tilde{G}^-|, |\tilde{G}^+|\} \le 1$ (cf. Eqs. (26) and (27)), show.

Note that the kind of reasoning here employed cannot be used when a Fourier transformation with respect to time, with real angular frequency transform parameter, is carried out and the inversion back to the time domain is based on the Fourier inversion integral. The latter employs, in fact, imaginary values of the time Laplace-transform parameter s, to which values Lerch's theorem does not apply, and for which, most importantly, the estimates of Eqs. (28) and (29) are lost.

The equations under (31) entail, via Eq. (14), the following transform-domain representations for the acoustic pressure and the vertical component of the particle velocity:

$$\tilde{p}(x_3') = [2Y(x_3')]^{-1/2} \left[\sum_{n=0}^{\infty} \tilde{W}_n^-(x_3') + \sum_{n=0}^{\infty} \tilde{W}_n^+(x_3'),\right] \tag{34}$$

$$\tilde{v}_3(x_3') = [Y(x_3')/2]^{1/2} \left[-\sum_{n=0}^{\infty} \tilde{W}_n^-(x_3') + \sum_{n=0}^{\infty} \tilde{W}_n^+(x_3'),\right] \tag{35}$$

A typical term of order n in the right-hand sides of Eqs. (34) and (35) now consists of an n-fold repeated integration in the vertical direction, the limits in which are the successive interaction levels of multiple reflection. In them, the exponential functions, that contain in their arguments additional integrations from the source level to the receiver level, are gathered to a single one. The factor that remains in the n-fold integral is the product of one of the source signatures that only depend on s, an s-independent coupling coefficient that describes the coupling of the source to the wave, the s-independent reflection coefficients R at the successive interaction levels of multiple reflection, and an s-independent coupling coefficient that describes the coupling of the wave to the receiver. For our further analysis, such a typical term is written as

$$\tilde{U} = \hat{S}(s)\,\tilde{\Pi}(i\alpha_\mu) \exp\left[-s\int_{Z^-} \gamma^-(i\alpha_\mu, \varsigma)\,d\varsigma - s\int_{Z^+} \gamma^+(i\alpha_\mu, \varsigma)\,d\varsigma\right]. \tag{36}$$

Here, $\hat{S}$ stands for the source signature, $\tilde{\Pi}$ for the product of coupling coefficients and reflection coefficients, Z^- for the accumulated vertical travel path traversed by the upgoing waves and

Z^+ for the accumulated vertical travel path traversed by the downgoing waves. Both Z^- and Z^+ may, in part or entirely, be multiply covered. In accordance with the property $\mathrm{Re}\{\gamma^-\} < 0$ and $\mathrm{Re}\{\gamma^+\} > 0$, the vertical travel paths can be written as $Z^- = \{\varsigma \in \mathbb{R} | x_3 < \varsigma < x_3^+\}$ for some x_3^+ and $Z^+ = \{\varsigma \in \mathbb{R} | x_3^- < \varsigma < x_3\}$ for some x_3^-, and consequently the signed vertical path lengths satisfy the inequalities $\int_{Z^-} d\varsigma \leq 0$ and $\int_{Z^+} d\varsigma \geq 0$, respectively. Expressions of the type (36) will, just as in the case of discretely layered media, be denoted as generalized-ray constituents (Wiggins and Helmberger, 1974). Their transformation back to the space-time domain with the aid of the modified Cagniard method will be described in section V.

It is to be noted that for obtaining the correct early time asymptotic expressions for the total wave amplitude at and immediately behind the wave front, all contributons that travel in a particular direction (up or down) must be added, since they all arrive at the same instant as the direct wave, be it with decreasing initial amplitudes. This kind of asymptotics is investigated in Singh and Chapman (1988) and Heyman and Felsen (1984). Our decomposition yields expressions for the successively reflected waves *at all times*; we have not been able yet to sum these contributions analytically, for example, at the wave front.

V. THE MODIFIED CAGNIARD METHOD APPLIED TO A GENERALIZED-RAY CONSTITUENT

The transformation back to the space-time domain of the generalized-ray constituent in Eq. (36) will now be discussed. Using Eq. (6), the s-domain expression corresponding to Eq. (36) is given by

$$\hat{U} = \left(\frac{s}{2\pi}\right)^2 \hat{S}(s) \int_{\alpha_\mu \in \mathbb{R}^2} \tilde{\Pi}(i\alpha_\mu) \exp\left[-s\left(i\alpha_\mu x_\mu + \int_{Z^-} \gamma^- d\varsigma + \int_{Z^+} \gamma^+ d\varsigma\right)\right] d\alpha_1 d\alpha_2. \quad (37)$$

For the present case of an anisotropic medium the most appropriate version of the Cagniard method seems to be the one where the variables of integration $i\alpha_\mu$ are replaced by

$$i\alpha_1 = p\cos(\theta + \psi), \qquad i\alpha_2 = p\sin(\theta + \psi), \quad (38)$$

where $\theta = \arctan(x_2'/x_1')$, p is positive imaginary and $0 \leq \psi \leq 2\pi$. Using the relevant symmetry properties, these transformations lead to

$$\begin{aligned}\hat{U} &= -\left(\frac{s}{2\pi}\right)^2 \hat{S}(s) \int_{-\pi/2}^{\pi/2} d\psi \,\mathrm{Re}\left\{\int_0^{i\infty} \tilde{\Pi}(p,\psi)\right.\\ &\quad \left.\exp\left[-s\left(pd\cos\psi + \int_{Z^-} \gamma^- d\varsigma + \int_{Z^+} \gamma^+ d\varsigma\right)\right] p\,dp\right\}. \end{aligned} \quad (39)$$

The essential feature of the modified Cagniard method consists of replacing the integration with respect to p along the positive imaginary axis, through continuous deformation, by one along a modified Cagniard contour that follows from

$$\tau = pd\cos\psi + \int_{Z^-} \gamma^-(p,\psi,\varsigma)\,d\varsigma + \int_{Z^+} \gamma^+(p,\psi,\varsigma)\,d\varsigma = \text{Real}. \quad (40)$$

The admissibility of the contour deformation rests on the applicability of Cauchy's theorem and Jordan's lemma. The latter only allows for a deformation into the right half of the complex p-plane. The only singularities of the integrand are the branch points due to the occurrence of $\tilde{A}_{21}^{1/2}$ in the expressions for $\tilde{\Pi}$, γ^+ and γ^-, i.e., the zeros of $\tilde{A}_{21}$. These zeros can easily be proved to reside on the real p-axis. From Eq. (40) it follows that the part of the real p-axis from the origin to the branch point nearest to the origin, as well as the complex path that satisfies the equation for $\tau \to \infty$, are candidates for modified Cagniard contours. As to the complex part of the modified Cagniard contours two possibilities exist: (a) it intersects the real p-axis at a regular point of the left-hand side of Eq. (40), in which case $\partial\tau/\partial p = 0$ at that point; (b) the modified Cagniard contour touches the real p-axis at the branch point nearest to the origin. These two cases are shown in Fig. 1.

Which of the two cases applies depends on the vertical profiles of the constitutive parameters and the mutual positions of the source, the reflection levels and the receiver. In the latter case the modified Cagniard contour must be supplemented by a circular arc around the relevant branch point since this branch point can be a simple pole of R (in this case the corresponding generalized ray is the time-domain conterpart of a turning ray in the asymptotic frequency-domain approach). Along the modified Cagniard contour τ is introduced as the variable of integration and the integrations with respect to τ and ψ are interchanged. The final result can be written as

$$\hat{U} = s^2\,\hat{S}(s)\,\hat{G}(Z^-, Z^+, s), \tag{41}$$

in which

$$\hat{G}(Z^-, Z^+, s) = \frac{-1}{2\pi^2}\int_{T_0}^{\infty} d\tau\, \exp(-s\tau)\int_{\Psi_1(Z^-,Z^+,\tau)}^{\Psi_2(Z^-,Z^+,\tau)} \mathrm{Re}\{\tilde{\Pi}(p,\psi)\,p\,\frac{\partial p}{\partial\tau}\}\,d\psi. \tag{42}$$

For details of the procedure we refer to Van der Hijden (1987), Sections 4.6, 6.4 and 7.4. The time-domain counterpart of Eq. (41) is

$$U = \partial_t^2\int_{T_0}^{t} S(t-\tau)\,G(Z^-, Z^+, \tau)\,d\tau, \tag{43}$$

where $G(Z^-, Z^+, t)$ can be recognized form Eq. (42) as

$$G(Z^-, Z^+, t) = \frac{-1}{2\pi^2}\,H(t-T_0)\int_{\Psi_1(Z^-,Z^+,t)}^{\Psi_2(Z^-,Z^+,t)} \mathrm{Re}\{\tilde{\Pi}(p,\psi)\,p\,\frac{\partial p}{\partial\tau}\}\,d\psi. \tag{44}$$

Obviously, T_0 is the arrival time of the generalized ray constituent under consideration.

VI. THE CASE OF AN ISOTROPIC FLUID

For an isotropic fluid, $\rho_{kr} = \rho\delta_{kr}$, where ρ is the scalar volume density of mass. In this case $\gamma^- = -\gamma$ and $\gamma^+ = \gamma$, where $\gamma = (c^{-2} + \alpha_\eta\alpha_\eta)^{1/2}$. In the latter expression $c = (\rho\kappa)^{-1/2}$ is the acoustic wave speed. The vertical acoustic wave admittance becomes $Y = \gamma/\rho$.

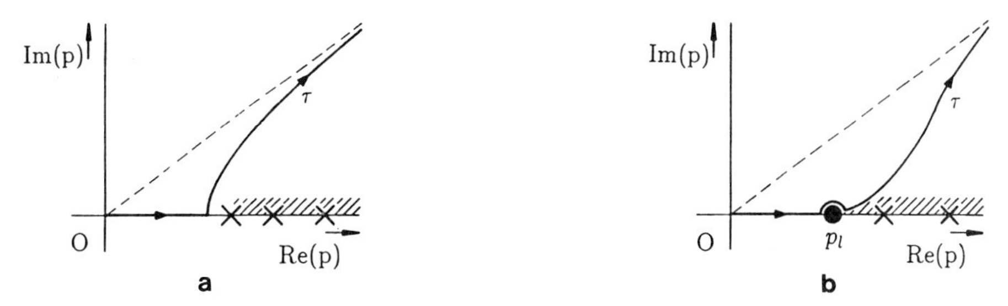

Fig.1. (a) Modified Cagniard countour with complex part intersecting the real p-axis in a point where $\partial\tau/\partial p = 0$; (b) Modified Cagniard contour with complex part touching the real p-axis in the leftmost branch point.

REFERENCES

Aki, K., and Richards, P. G., 1980, "Quantitative seismology", W. H. Freeman & Co., San Francisco, p. 224.

Bremmer, H., 1939, Geometrisch-optische benadering van de golfvergelijking, *in*: "Handelingen Natuur- en Geneeskundig Congres, Nijmegen," Ruygrok, Haarlem, The Netherlands (in Dutch; title in English: Geometric-optical approximation of the wave equation).

Bremmer, H., 1949a, The propagation of electromagnetic waves through a stratified medium and its WKB approximation for oblique incidence, *Physica*, 15:593-608.

Bremmer, H., 1949b, Some remarks on the ionospheric double refraction. Part II: Reduction of Maxwell's equations; WKB approximation, *Philips Research Reports*, 4:189-205.

Bremmer, H., 1951, The WKB approximation as the first term of a geometric-optical series, *Commun. Pure Appl. Math.*, 4:105-115 (also *in*: "The theory of electromagnetic waves", M. Kline, ed., Interscience, New York).

Cagniard, L., 1939, "Relection et refraction des ondes seismiques progressives", Gauthier-Villars, Paris (in French).

Cagniard, L., 1962, "Reflection and refraction of progressive seismic waves", McGraw-Hill, New York (translation and revision of Cagniard (1939) by E. A. Flinn and C. H. Dix).

Chapman, C. H., 1974, Generalized ray theory for an inhomogeneous medium, *Geophys. J. R. Astr. Soc.*, 36:673-704.

de Hoop, A. T., 1960, A modifiction of Cagniard's method for solving seismic pulse problems, *App. Sci. Res.*, B8:349-356.

de Hoop, A. T., 1961, Theoretical determination of the surface motion of a uniform elastic half-space produced by a dilatational, impulsive point sorce, *in*: "Proceedings Colloque International du C.N.R.S.", no. 111, Marseille, pp. 21-31.

de Hoop, A. T., 1988, Acoustic radiation from impulsive sources in a layered fluid, *Nieuw Archief voor Wiskunde*, 6:111-129.

Heyman, E., and Felsen, L. B., 1984, Non-dispersive approximations for transient ray fields in an inhomogeneous medium, *in*: "Hybrid formulation of wave propagation and scattering," L. B. Felsen, ed., Martinus Nijhoff Publishers, Dordrecht, (The Netherlands).

Heyman, E., and Felsen, L. B., 1987, Real and complex spectra – a generalization of WKB seismograms, *Geophys. J. R. Astr. Soc.*, 91:1087-1126.

van der Hijden, J. H. M. T., 1987, "Propagation of transient elastic waves in stratified anisotropic media", North-Holland, Amsterdam.

Schoenberg, M., 1984, Wave propagation in alternating solid and fluid layers, *Wave Motion*, 6:303-320.

Singh, S. C., and Chapman, C. H., 1988, WKBJ seismogram theory in anisotropic media, *J. Acoust. Soc. Am.*, 84:732-741.

Wiggins, R. A., and Helmberger, D. V., 1974, Synthetic seismogram computation by expansion in generalized rays, *Geophys. J. R. Astr. Soc.*, 37:73-90.

DIFFRACTED MICROWAVE FIELDS NEAR DIELECTRIC SHELLS: COMPUTATION, MEASUREMENT, AND DECOMPOSITION

R. A. Hayward, E. L. Rope, and G. Tricoles

General Dynamics Electronics
P. O. Box 81127
San Diego, CA 92186

INTRODUCTION

This paper describes the fields produced near finite dielectric slabs and cones by incident, plane electromagnetic waves. It presents calculated and measured fields and their decomposition into constituent waves. The purpose was to understand approximations in some methods for analyzing performance of pointed, dielectric radomes. These methods compute patterns of an enclosed aperture antenna by evaluating diffraction integrals that have T(P), the complex-valued transmittance for externally incident plane waves as a factor in the integrand, where P is a point in the radome-bounded region. T(P) depends on polarization because of curvature and large incidence angles. Overall, radome analysis accuracy also depends on antenna and radome dimensions and on their proximity. For radomes with diameters exceeding 10 wavelengths (λ), accuracy was good if T(P) had the value for an infinite flat sheet tangent to the surface at the intersection of an incident ray directly to P.[1] This direct ray method failed for radomes with diameters approximately 5λ.[2] Accuracy was high when T(P) was evaluated by integrating over a portion of the incident wave surface. Neither method accurately predicted fields very near thin radomes.

To study approximations we have compared measured and computed values of T(P) for finite dielectric slabs and shells that are idealized radomes. For slabs and hollow wedges, measurements suggest guided waves; T(P) calculated from a volume integral assumed internal plane and slab-guided waves.[3] A hollow cone also showed guided waves.[4] Slab-guided waves were measured in a thin shell with axial and longitudinal curvature.[5] These results showed that the fields consist of constituents that are plane, slab-guided, or edge-scattered waves.

This paper presents new results for finite slabs, and finite, hollow cones. Calculations were done with two versions of the moment method.[6] The calculations were experimentally verified. Constituent fields were identified. For slabs, Richmond's theory[7] was applied to calculate fields for grazing incidence. Results were verified by measurment. Free space and guided waves were identified. These waves are contained implicitly[7] and explicitly[8] in Richmond's theories. For normal incidence, calculations and measurements have been compared.[9] A new version of the moment method for axially symmetric shells is described. Calculations and measurement are presented for hollow cones. Guided waves are inferred, and polarization dependence is shown.

FINITE SLAB

Fields were computed with Richmond's formulation, Reference 7. The method solves an integral equation for the total field E, which is $E^S + E^I$, where E^I is the known incident field and the scattered field is

$$4\pi E^S = k_o^2 \int (\kappa - 1) g E \, dV + \nabla' \int g \cdot [(\kappa - 1) E] dV \tag{1}$$

where κ is the dielectric constant; g is the free space Green's function; and k_o is the free space propagation constant $2\pi/\lambda$. The gradient is evaluated at the field point. Equation 1 is reduced to algebraic equations by subdiving the scatterer into right circular cyliners of infinite length. The unknowns are the total fields at the center of the cylinders. Figure 1 shows the coordinate system.

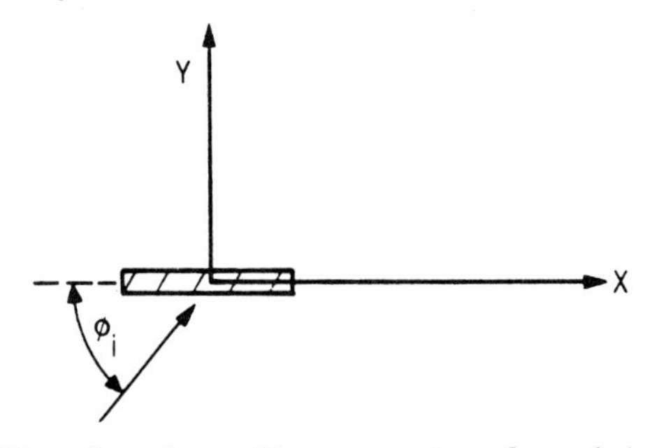

Fig. 1. Co-ordinate system for slab.

For perpendicular polarization at grazing incidence on a 10 inch long, 0.25 inch thick slab, Figure 2 shows computed values of E . The slab was divided into 40 cyliners. Figure 2 also shows values measured with an interferometer that had a half wave dipole connected through rotary joints to a network analyzer; the incident wave closely approximated a plane wave. The spacing of the deep minima approximates $\delta = 2\pi/(k_g - k_o)$ where k_o is the free space propagation constant and k_g is the theretical propagation constant[10] of a wave guided by an infinitely low slab with properties identical to those of our slab. This result suggests the external total field can be decomposed into the incident wave and a forward traveling guided wave. The small ripples in Figure 2 have period that corresponds to interference between the forward and backward guided waves.

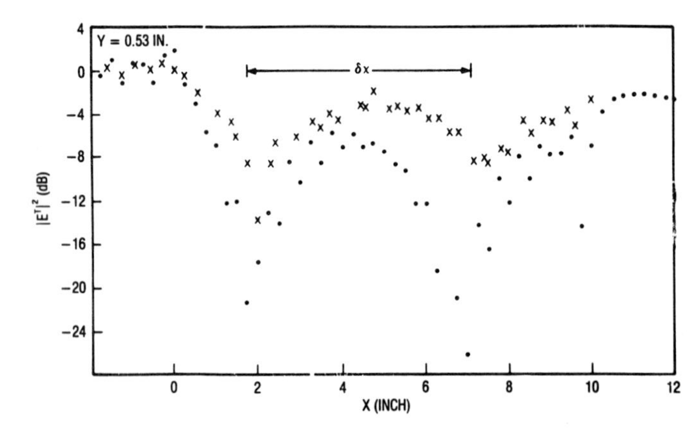

Fig. 2 Total field at grazing incidence and perpendicular polarization for 0.25 inch thick slab with dielectric constant 2.6. Frequency 9.375GHz. Calculated (.) for 40 cells, radius 0.143 inch. Measured (x).

Figure 3 shows computed $|E^S|$ that corresponds to Figure 2. The maxima are spaced by the same δ value as in Figure 2. This result suggests that the scattered wave can be decomposed into a forward guided wave and a free space wave, in agreement with Reference 8. The small ripples suggest interference of forward and backward guided waves.

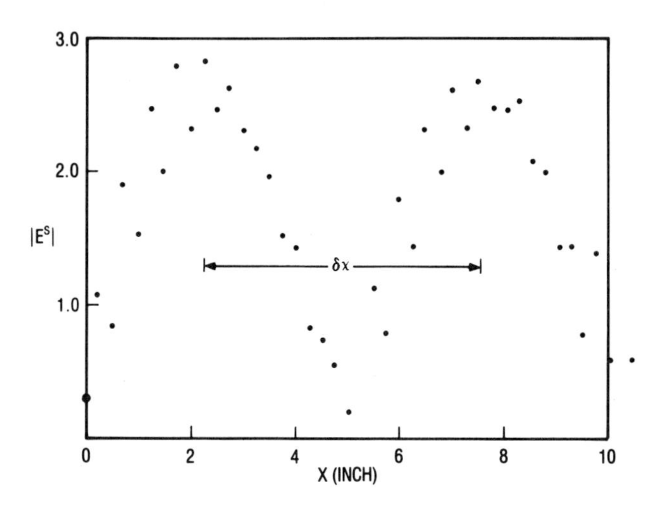

Fig. 3 Scattered field corresponding to Figure 2, for y equal 0.53 inch.

Figure 4 shows transverse variation of the forward guided wave, for x equals 5 inches. From Figure 2, the peak amplitude is 1.2 where E^S varies from 2.3 to 0.1 if the ripples are smoothed. We assume 1.2 for the amplitude of the forward guided wave E_G. On the basis of the slab boundary value problem [10] we assume the forward guided wave in the slab is

$$E_{Gi} = 1.2 \cos \pi (u-y) \tag{2}$$

Outside, the forward guided wave is assumed

$$E_{Go} = 0.96\ e^{-v\ (y-b)} \tag{3}$$

where 2b is slab thickness, u is 1.65/inch and v is 1.12/inch. Figure 4 shows values of E_{Go} and R_{Gi} from Equations (2) and (3). The guided wave amplitude 1.2 was deduced from Figure 3 by assuming the incident and guided waves were out of phase at x equals 5 inches. This result supports the decomposition of the scattered field into a forward guided wave and the incident wave within the slab.

The calculated results were plotted as contours, which suggest constitutent waves. Figure 5 shows equiphase contours of E^S; The figure suggests a cylindrical wave scattered from the lit edge. Figure 6 shows contours of $|E|$. The two parabolic curves are loci of constructive interference of the incident waves and a cylindrical wave.

HOLLOW CONE

Equation 1 was reduced to a set of algebraic equations by dividing a cone into rings like those in Figures 7 and 8. The rings were further

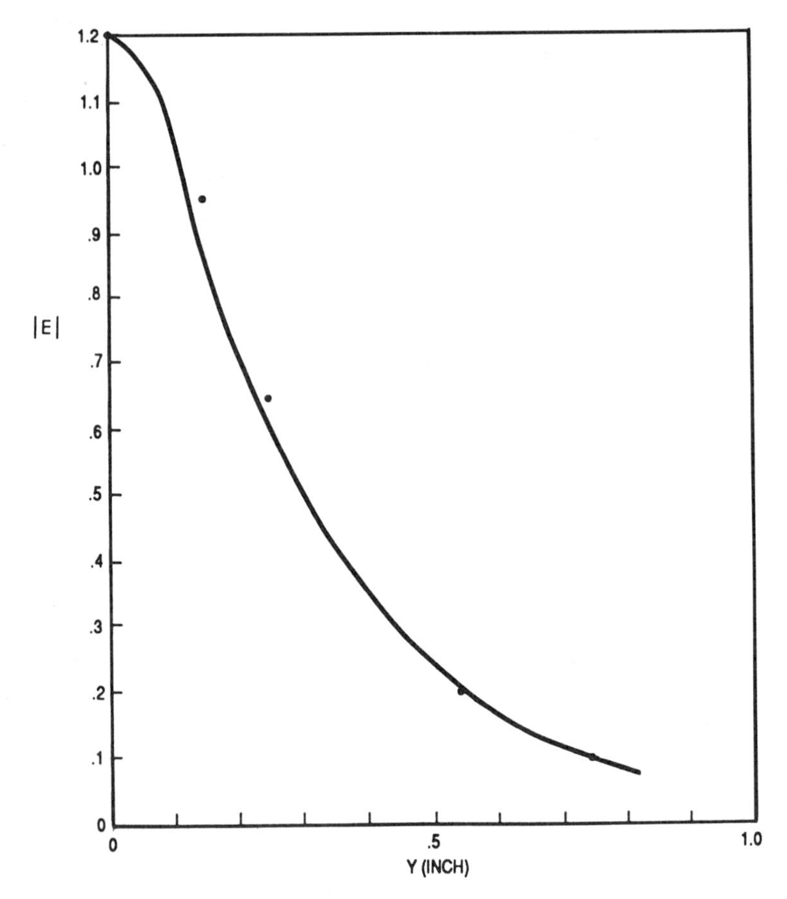

Fig. 4 Computed transverse variation of guided wave on slab. Moment method (.); Equations 2 and 3(-).

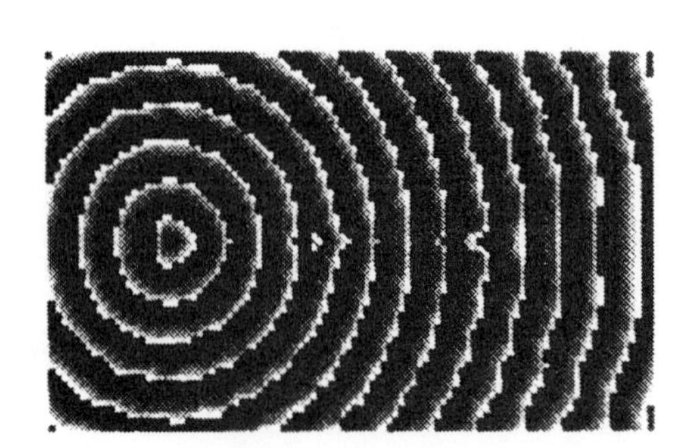

Fig. 5 Computed equiphase contours of E_S for slab of Figure 2.

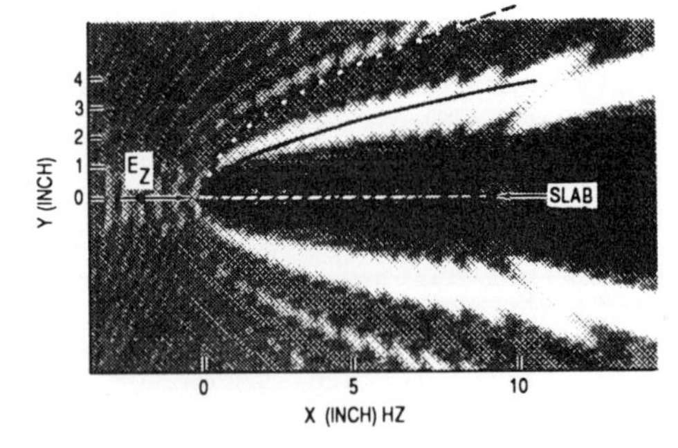

Fig. 6 Computed field magnitude for slab of Figure 2.

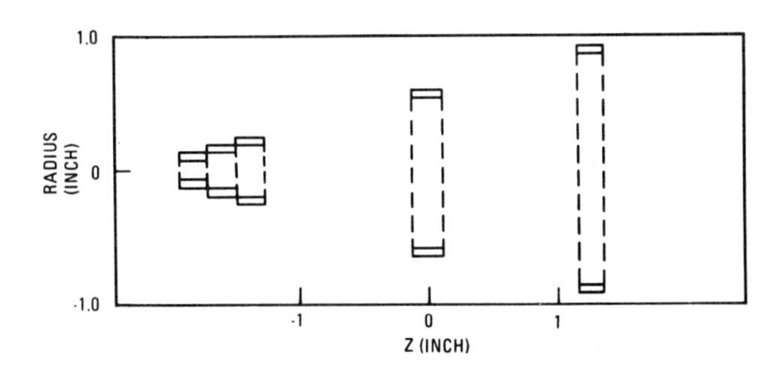

Fig. 7 Dimensions of experimental cone.

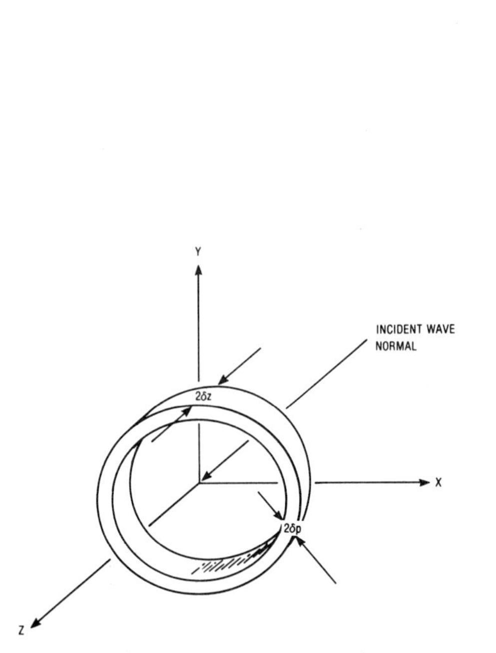

Fig. 8 A ring and co-ordinate system.

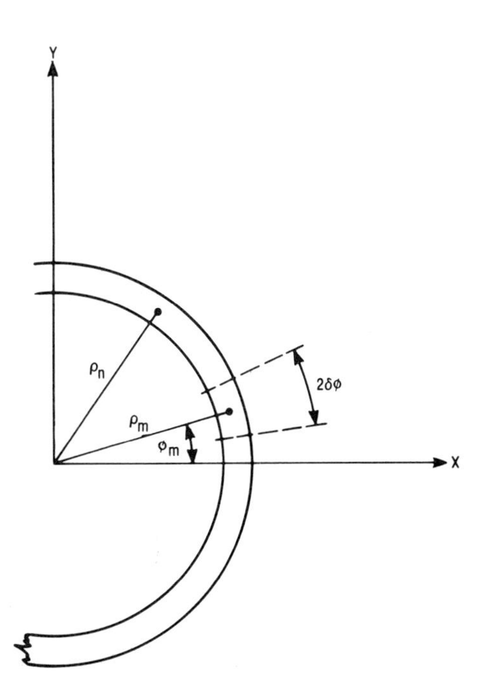

Fig. 9 Description of cells for the experimental cone. Cell lengths $2\delta z$ were 0.215 inch. Thickness $2\delta\rho$ was 0.065 inch. Angular extent was $\pi/4$.

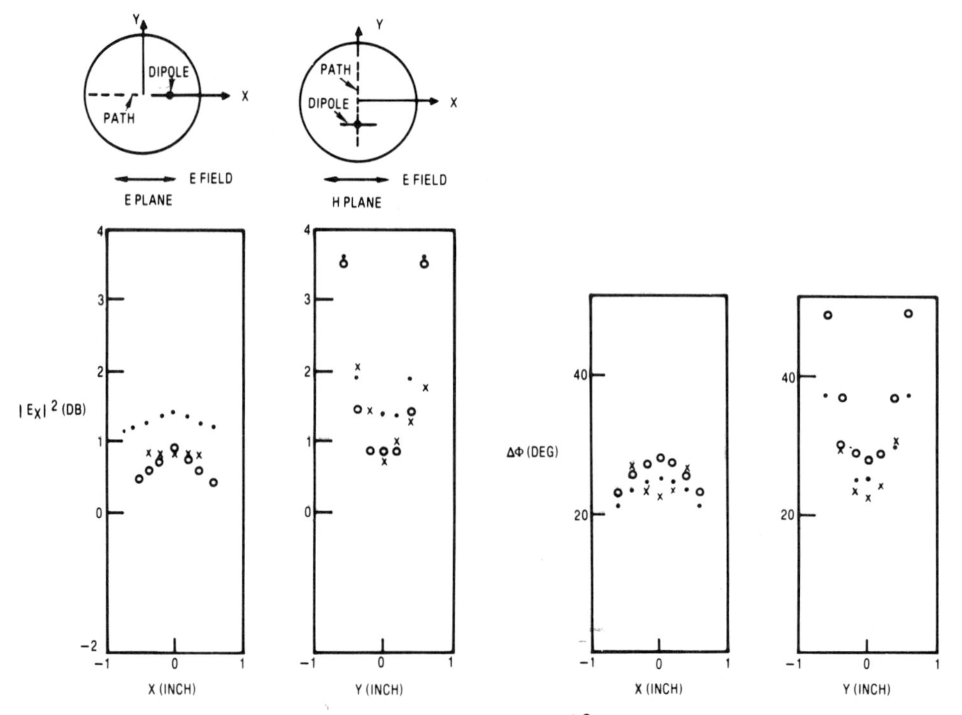

Fig 10. Phase change $\Delta\Phi$ and intensity $|E|^2$ measured (x) and computed in plane of twelfth ring (z equal 0.86 in) for cone at axial incidence. Dielectric constant 2.6. Frequency was 9.4GHz. Computed for $E_Z = 0$ (0); for $E_Z \pm 0$ (0).

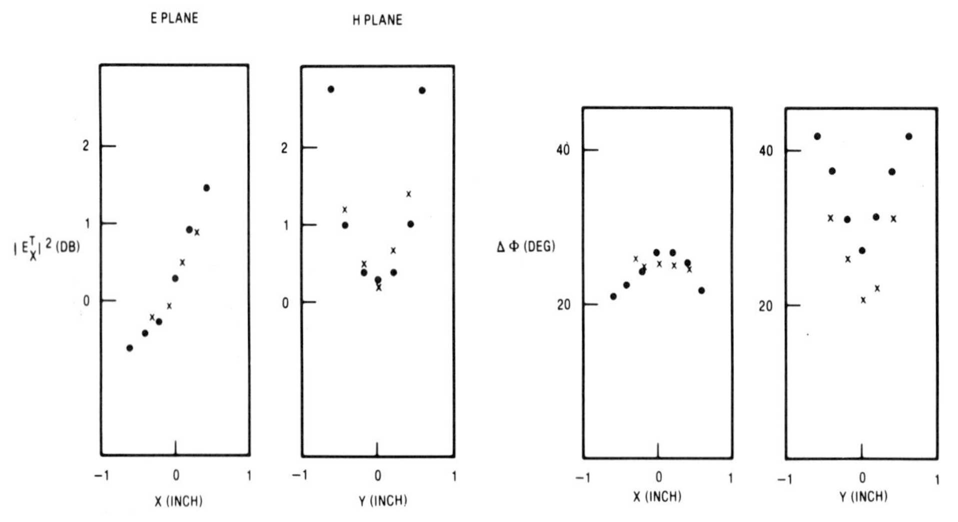

Fig. 11 As in Figure 10 but for the incident wave normal inclined 14.9 degrees to the cone axis.

divided into angular sectors, or cells, as in Figure 9. These cells were approximated by equal volume circular cylinders of length $2\delta z$ to express matrix elements by closed form integrals. The field in a cylinder was assumed constant. The unknowns in the algebraic equations were the fields at the centers of the cells.

Wave polarization was restricted by assuming the incident wave normal in the x-z plane and the electric field horizontally polarized; thus, E_y^I was zero and E_x^I and E_z^I non zero. We assumed that the y component of the total field E_y was zero. Two cases were considered. In one E_z the longitudinal component of the total field also was assumed to vanish; in the other both E_z and E_x were assumed non-zero.

Interferometer measurements were made. The cone was suspended by threads. Incidence angle was changed by moving the transmitting antenna.

For a 13 ring cone with the dimensions shown in Figure 7, Figure 10 shows measured and computed values in the plane of the next to the largest ring for axial incidence. Figure 11 shows values for the incident wave normal inclined 14.9^0 to z axis. The measured transmittance $|E_x|^2$ exceeds unit value in Figure 10 and for both planes in Figure 11; this result may arise from guided waves.

REFERENCES

1. G. Tricoles, J. Opt. Soc. Am., Vol. 54, p. 1094 (1964).
2. R. A. Hayward, E. L. Rope, G. Tricoles, Digest Antennas and Propagation Symposium, p. 598 (1979).
3. G. Tricoles and E. L. Rope, J. Opt. Soc. Am., Vol. 55, p. 1479 1965).
4. G. Tricoles and E. L. Rope, J. Opt. Soc. Am., Vol. 55, p. 328 (1965).
5. G. Tricoles, E. L. Rope, R. A. Hayward, Ch. V. 11 in _Inverse Methods in Electromagnetic Imaging_, ed. W-M, Boerner, Reidel 1983.
6. R. F. Harrington, _Field Computation by Moment Methods_, Ch. 1 (1968).
7. J. H. Richmond, IEEE Trans., Vol. AP-13, p. 334 (1965).
8. J. H. Richmond, IEEE Trans., Vol. AP-33, p. 64 (1985).
9. G. Tricoles, E. L. Rope, R. A. Hayward, _Optica Hoy y Manana_, Proceedings Eleventh Congress of ICO, J. Bescos et al. eds, Institute of Optics, Madrid, p. 731 (1978).
10. R. F. Harrington, _Time Harmonic Electromagnetic Fields_, McGraw-Hall, p. 163 (1961).

GREEN'S FUNCTION OF WAVE EQUATION FOR INHOMOGENEOUS MEDIUM

Evgueni G. Saltykov

Department of Computational Math. and Cybernetics
Moscow State University
Moscow USSR

INTRODUCTION

A method is proposed for constructing the Green's function for the wave equation describing wave propagation in the ionosphere with electronic density $N(z,x)$ depending on two independent variables. By means of the Fourier transform with respect to t, we obtain a reduced wave equation with an inhomogeneous term of the δ-function type. The last problem solution has the form of integral containing the eigenfunctions of one-dimensional problem. The eigenfunctions depend on the parameter x. The expansion coefficients satisfy the second kind integral equation. In the weak dependence case of electronic density on a coordinate x, the integral equation can be solved by the successive approximations method.

We imagine that the proposed method can have great application both for mathematical physics problems and in electromagnetic wave modeling. This method can be considered as the generalization of the separation of variables method in mathematical physics for the case when variables are not separable. By means of this method, wave propagation in different media and in the presence of different objects can be described. This method can be applied to problems of scattering by arbitrary bodies, situated in inhomogeneous media. By means of this method, wave propagation in irregular waveguides and the determination of eigen oscillations for arbitrary volumes can be considered.

PROBLEM POSITING AND PROBLEM SOLUTION METHOD

We construct the function v which satisfies the equation

$$\frac{1}{c^2}\frac{\partial^2 v}{\partial t^2} = \frac{\partial^2 v}{\partial z^2} + \frac{\partial^2 v}{\partial x^2} - N(z,x)v + 2\pi\delta(t-t')\delta(z-z')\delta(x-x') \quad (1)$$

and represents the Green's function for the wave equation. The function v is sought in the form

$$v = \int_{-\infty}^{\infty} e^{i\omega t}\ u(\omega,\ z,\ x)\ d\omega.$$

Upon inserting the function v into the equation (1) and taking the inverse transform, we obtain

$$\frac{\partial^2 U}{\partial z^2} + \frac{\partial^2 U}{\partial x^2} + (K_0^2 - N(z,x))U = -e^{-i\omega t'}\delta(z-z')\delta(x-x') \qquad (2)$$

where $K_0 = \omega/c$ is the wave number of a free space, $N(z,x)$ is the normalized electronic density. $N \geq 0$.

$$N(z,x) = \begin{cases} N(z,x), & 0 \leqslant z \leqslant z_0,\ |x| \leqslant x_0, \\ N(z), & |x| > x_0, \\ N_0, & z<0,\ z>z_0. \end{cases}$$

Here z_0 and x_0 are constants. The function $N(z,x)$ is twice differentiable with respect to z and x.

The equation solution is sought in the form

$$u(x,z) = \int_{\Lambda} \Phi(x,\lambda)\ \varphi(z,x,\lambda)\ d\lambda \qquad (3)$$

The function $\varphi(z,x,\lambda)$ is the eigenfunction of Sturm-Liouville's operator

$$-\frac{\partial^2 \varphi}{\partial z^2} + N(z,x)\varphi = \lambda^2\varphi \qquad (-\infty < z < \infty) \qquad (4)$$

Λ is the spectrum of the operator.

The integral sign in (3) means the integral with respect to the continuous spectrum and the summation with respect to the discrete spectrum. The variable x in (4) is considered as a parameter.

The operator spectrum Λ consists of the continuous spectrum values belonging to the real axis $|\lambda| \in (\sqrt{N_0},\ \infty)$. Besides of that, the spectrum consists of the discrete spectrum values belonging to a real axis:

$$\lambda_j \in (\sqrt{\bar{N}}, \sqrt{N_0}),\ \bar{N} = \min_{z\in(-\infty,\infty)} N(z,x).$$

The possibility justification of the function representation u in the form (3) is described by Levitan[1].

EIGENFUNCTIONS DESCRIPTION

The eigenfunctions $\varphi\ (z,x,\lambda)$ has the following asymptotic behavior at infinity.

For the functions φ corresponding to the waves coming from $-\infty$ and propagating in the positive direction of axis z $(\lambda>0)$

$$\varphi = a\, c(\lambda,x)\, e^{-i\lambda z} + O(1), \quad z\to +\infty ,$$
$$\varphi = a\, (e^{-i\lambda z} + b(\lambda,x)\, e^{i\lambda z}) + O(1), \quad z\to -\infty , \tag{5}$$

For the functions φ corresponding to the waves coming from $+\infty$ and propagating in the negative direction of axis $z(\lambda < 0)$

$$\varphi = a\, c(\lambda,x)\, e^{-i\lambda z} + O(1), \quad z\to -\infty ,$$
$$\varphi = a\, (e^{-i\lambda z} + b(\lambda,x)\, e^{i\lambda z}) + O(1), \quad z\to +\infty , \tag{6}$$

$b(\lambda,x), c(\lambda,x)$ are the constants not depending on z, a is the constant not depending on x and z $(a = 1/\sqrt{2\pi})$.

In the discrete spectrum points, the eigenfunctions $\varphi(z,x,\lambda_j(x)) = O(1)$ when $|z|\to \infty$.

PROBLEM REDUCING TO INTEGRAL-DIFFERENTIAL EQUATION

Substituting (3) in (2), we derive

$$\int_\Lambda (\Phi_{xx}\varphi + 2\Phi_x\varphi_x + \Phi\varphi_{xx})\, d\lambda + \int_\Lambda \Phi(\varphi_{zz} + K^2\varphi)\, d\lambda = -\delta(x-x')\delta(z-z')e^{-i\omega t'},$$

$K^2 = K_0^2 - N(z,x)$.

Using (4), we have

$$\int_\Lambda (\Phi_{xx}\varphi + 2\Phi_x\varphi_x + \Phi\varphi_{xx})\, d\lambda + \int_\Lambda \Phi(K_0^2 - \lambda^2)\varphi\, d\lambda = -\delta(x-x')\delta(z-z')e^{-i\omega t'} \tag{7}$$

We multiply (7) by $\varphi^*(z,x,\lambda)$ and integrate over z. Changing the integration order and using the equality

$$\int_{-\infty}^{\infty} \varphi(z,x,\lambda')\varphi^*(z,x,\lambda)\, dz = \delta(\lambda-\lambda') \tag{8}$$

where $\delta(\lambda-\lambda')$ - δ-function for the continuous spectrum and Kronecker's symbol for the discrete spectrum points, we derive

$$\Phi_{xx}(x,\lambda) + (K_0^2 - \lambda^2)\Phi(x,\lambda) = -\delta(x-x')\varphi^*(z',x,\lambda)e^{-i\omega t'}$$
$$- \int_\Lambda d\lambda' [K_1(x,\lambda',\lambda)\, \Phi_x(x,\lambda') + K_2(x,\lambda',\lambda)\Phi(x,\lambda')] \tag{9}$$

where

$$K_1(x,\lambda',\lambda) = 2\int_{-\infty}^{\infty} \varphi_x(z,x,\lambda')\varphi^*(z,x,\lambda)\, dz,$$

$$K_2(x,\lambda',\lambda) = \int_{-\infty}^{\infty} \varphi_{xx}(z,x,\lambda')\varphi^*(z,x,\lambda)dz,$$

EQUATION FOR FUNCTION Φ

We apply the computation method of the integrals K_1 and K_2 described by Sommerfeld[2] with the condition that the function φ has the asymptotic behavior when $|z|\to\infty$ determined by the formulae (5-6). We use the representation

$$\int_0^{\infty} e^{-i\alpha z}dz = \pi\delta(\alpha) + 1/i\alpha,$$

$$\int_{-\infty}^{0} e^{-i\alpha z}dz = \pi\delta(\alpha) - 1/i\alpha, \tag{10}$$

The equality must be understood in the generalized functions sense as shown by Fedorük[3].

We derive

$$K_1(x,\lambda',\lambda) = [b_x(\lambda,x)b^*(\lambda,x) + c_x(\lambda,x)c^*(\lambda,x)]\ \delta(\lambda-\lambda')-$$

$$-\frac{2}{(\lambda^2-\lambda'^2)}\int_{-\infty}^{\infty} N_x(z,x)\ \varphi(z,x,\lambda')\varphi^*(z,x,\lambda)dz, \tag{11}$$

$$K_2(x,\lambda',\lambda) = \frac{1}{2}\ [b_{xx}(\lambda,x)b^*(\lambda,x) + c_{xx}(\lambda,x)c^*(\lambda,x)]\ \delta(\lambda-\lambda')-$$

$$-\frac{1}{(\lambda^2-\lambda'^2)}\int_{-\infty}^{\infty}[2\ N_x(z,x)\ \varphi_x(z,x,\lambda') + \tag{12}$$

$$+ N_{xx}(z,x)\varphi(z,x,\lambda')]\varphi^*(z,x,\lambda)dz,\ \lambda > 0,\ \lambda' > 0.$$

By means of the representation (11-12), the formula (9) acquire the form

$$\Phi_{xx}(x,\lambda) + \pi A(x,\lambda,\lambda)\Phi_x(x,\lambda)+[\pi B(x,\lambda,\lambda)+K_0^2 - \lambda^2]\Phi(x,\lambda) =$$

$$= -e^{-i\omega t'}\delta(x-x')\ \varphi^*(z',x,\lambda) + \int_{\Lambda_1}\frac{d\lambda'}{(\lambda^2-\lambda'^2)}\ [\bar{A}(x,\lambda',\lambda)\Phi_x(x,\lambda') + \tag{13}$$

$$+ \bar{B}(x,\lambda',\lambda)\Phi(x,\lambda')]d\lambda' -$$

$$- \int_{\Lambda_2} d\lambda' [K_1 (x,\lambda',\lambda)\Phi_x (x,\lambda') + K_2 (x,\lambda',\lambda)\Phi(x,\lambda')]$$

where

$$\pi A(x,\lambda,\lambda) = b_x(\lambda,x)b^*(\lambda,x) + c_x(\lambda,x)c^*(\lambda,x) \ , \tag{14}$$

$$2\pi B(x,\lambda,\lambda) = b_{xx}(\lambda,x)b^*(\lambda,x) + c_{xx}(\lambda,x)c^*(\lambda,x), \tag{15}$$

$$\bar{A}(x,\lambda',\lambda) = 2 \int_{-\infty}^{\infty} N_x (z,x)\varphi(z,x,\lambda')\varphi^* (z,x,\lambda) \ dz,$$

$$\bar{B}(x,\lambda',\lambda) = \int_{-\infty}^{\infty} [2N_x (z,x)\varphi_x (z,x,\lambda') +$$

$$+ N_{xx}(z,x)\varphi(z,x,\lambda')]\varphi^*(z,x,\lambda)dz.$$

Here Λ_1 signifies the continuous spectrum and Λ_2 signifies the discrete spectrum if λ belongs to the continuous spectrum. If λ belongs to the discrete spectrum, the integral over Λ_1 must be considered equal to zero and the integral over Λ_2 must be understood as integration over the continuous spectrum and the summation over the discrete spectrum.

MAIN APPROXIMATION FOR FUNCTION $\Phi(x,\lambda)$

We use the solution $G(x,\lambda)$ for the equation

$$G_{xx}(x,\lambda) + \pi A(x,\lambda,\lambda)G_x(x,\lambda) + [\pi B(x,\lambda,\lambda) + K_o^2 - \lambda^2]G(x,\lambda) = \\ = - \delta(x-x') \tag{16}$$

(In the discrete spectrum case $\lambda = \lambda_j(x)$, the function $G(x,x',\lambda_j(x))$ is the Green's function satisfying the equation (16)). When the values $K_o^2 - \lambda^2$ are great enough, the equation (16) transforms in the equation

$$\bar{G}_{xx}(x,\lambda) + (K_o^2 - \lambda^2)\bar{G}(x,\lambda) = - \delta(x-x') \tag{17}$$

The solution of (17) in the continuous spectrum case has the form

$$\bar{G}(x,x',\lambda) = \frac{e^{-i\sqrt{K_o^2-\lambda^2}|x-x'|}}{2i\sqrt{K_o^2-\lambda^2}} \ .$$

By means of the function G, for the function Φ, we derive

$$\Phi(x,\lambda)= G(x,x',\lambda)\varphi^*(z',x',\lambda)e^{-i\omega t'} - \int_{-\infty}^{\infty} dx' \; G(x,x',\lambda) \times$$

$$\times \int_{\Lambda_1} \frac{d\lambda'}{(\lambda^2-\lambda'^2)} \; [\bar{A}(x',\lambda',\lambda)\Phi_x(x',\lambda') + \bar{B}(x',\lambda',\lambda)\Phi(x',\lambda')] +$$

$$+ \int_{-\infty}^{\infty} dx' G(x,x',\lambda) \int_{\Lambda_2} d\lambda' [K_1(x',\lambda',\lambda)\Phi_x(x',\lambda') + \tag{18}$$

$$+ K_2(x',\lambda',\lambda)\Phi(x',\lambda')].$$

The equation (18) represents the integral-differential equality for the determination of the function $\Phi(x,\lambda)$.

By means of the function $\Phi(x,\lambda)$, using the formula (3), we derive the Green's function of wave equation for inhomogeneous medium - the solution (1).

The functions $K_1, K_2, \bar{A}, \bar{B}$ contain the derivatives with respect to x of the eigenfunctions, reflection and refraction coefficients, the coefficient N. In the weak dependence case of the function N(z,x) on x, when the derivatives satisfy the condition

$$|N_x| \ll I, \qquad |N_{xx}| \ll I, \qquad -\infty < x < \infty,$$

the last functions are small. Thus, naturally, the function

$$\bar{\Phi}(x,\lambda) = G(x,x',\lambda)\varphi(z',x',\lambda)e^{-i\omega t'}$$

can be considered as the main approximation in the small dependence case of the coefficient N(z,x) of the variable x in the equation (1).

REDUCING TO SECOND-KIND INTEGRAL EQUATION

We consider the case when the spectrum of the equation (4) is continuous. In this case, the second integral term in (18) must be set equal to zero. We transform the term with the function $\bar{A}$ in the integral (18).

$$I(x,\lambda) = - \int_{-\infty}^{\infty} dx' \; G(x,x',\lambda) \int_{\Lambda} \frac{d\lambda'}{(\lambda^2-\lambda'^2)} \; \bar{A}(x'\lambda',\lambda)\Phi_x(x',\lambda') =$$

$$= - \int_{\Lambda} d\lambda' \int_{-\infty}^{\infty} \frac{dx'}{(\lambda^2-\lambda'^2)} \; \bar{A}(x'\lambda',\lambda)G(x,x',\lambda)\Phi_x(x',\lambda').$$

Since it is supposed that $N(z,x) = N(z)$ when $|x| > x_0$, $\bar{A}(x,\lambda',\lambda) \equiv 0$, when $|x|>x_0$. Integrating by parts in inner integral, we derive

$$I(x,\lambda) = \int_{\Lambda} d\lambda' \int_{-\infty}^{\infty} dx' \frac{\Phi(x',\lambda')}{(\lambda'^2-\lambda^2)} \frac{d}{dx'} [\bar{A}(x',\lambda',\lambda)G(x,x',\lambda)].$$

Finally, we derive the integral equation

$$\Phi(x,\lambda) = F(x,x',z',\lambda) + \int_{\Lambda} d\lambda' \int_{-\infty}^{\infty} dx' K(x,\lambda;x',\lambda')\Phi(x',\lambda') \quad (19)$$

where

$$F(x,x',z',\lambda) = G(x,x',\lambda)\varphi^*(z',x',\lambda)e^{-i\omega t'},$$

$$K(x,\lambda;x',\lambda') = -\frac{1}{(\lambda^2-\lambda'^2)} \left\{- \frac{\partial}{\partial x'} [\bar{A}(x',\lambda',\lambda)G(x,x',\lambda)] + \right.$$

$$\left. + \bar{B}(x',\lambda',\lambda)G(x,x',\lambda')\right\}.$$

REPRESENTATION FOR FUNCTION G

It can be shown that the function $G(x,x',\lambda)$ has the form

$$G(x,x',\lambda) = V \exp\left\{- \frac{1}{2} \int_{x'}^{x} [b_x(\lambda,x)b^*(\lambda,x)+c_x(\lambda,x)c^*(\lambda,x)]dx\right\}.$$

where the function V satisfies the equation

$$V_{xx} + [K_o^2 - \lambda^2 - \delta v] V = -\delta(x-x').$$

Here

$$\delta v(\lambda,x) = (b_x b^* + c_x c^*)^2/4 + [|b_x|^2 + |c_x|^2]/2.$$

Neumann's series for the equation (19) converges to the function Φ if N_x, N_{xx} are uniformly small with $-\infty < x < \infty$ according to Riesz's theorem on a limitation of singular operator with Cauchy's kernel[4].

REFERENCES

1. B. M. Levitan, "Inverse Sturm-Liouville's Problems," Nauka, Moscow (1984).
2. A. Sommerfeld, "Partielle Differentialgleichungen der Physik," Akademische Verlagsgesellschaft Geest & Portig K.-G., Leipzig (1966).
3. M. V. Fedorük, "Asymptotic Expansion Integrals and Series," Nauka, Moscow (1987).
4. N. Danford and J. T. Shwartz, "Linear Operators," Part 2, Interscience Publishers, New York, London (1963).

NUMERICAL SOLUTION OF MAXWELL'S EQUATIONS

DIRECT MAXWELL'S EQUATION SOLVERS

IN TIME AND FREQUENCY DOMAINS -- A REVIEW

Raj Mittra and Jin-Fa Lee

Electromagnetic Communication Laboratory
University of Illinois at Urbana-Champaign
Urbana, IL 61801-2991

1. INTRODUCTION

In this paper, we present a brief review of partial differential equation (PDE) techniques for solving the problems of electromagnetic scattering from complex bodies and circuit modeling of microwave components. Radar targets as well as microwave circuits in use today often possess intricate geometries and comprise of materials with highly inhomogeneous properties. One consequence of this is that the analytical and asymptotical techniques can seldom be used to solve the field problems in these geometries with sufficient accuracy, and even the application of the integral equation approach, e.g., the Method of Moments, becomes rather involved. However, the direct solution of Maxwell's equations using PDEs, such as the finite element (FEM) or the finite difference time domain (FDTD) methods, offers an attractive alternative for analyzing these complex problems conveniently, reliably and accurately.

In the frequency domain, FEM is probably the most versatile and general technique for constructing numerical solutions of Maxwell's equations. However, in the time domain, the FDTD approach has recently gained great popularity, and has proven to be highly successful for resolving many complex problems. In light of this, our focus in this paper would be to review some of the recent advancements of these two methods and to illustrate their applications using a few representative examples.

Recently, there have been three major advances in the use of FEM for the solution of Maxwell's equations and we enumerate them below. The first of these involves the use of the coupled azimuthal potentials (CAPs) for solving the problem of radar scattering from bodies of revolution (BORs). The CAP formulation was first introduced by Morgan, Chang and Mei[1] to analyze the problem of electromagnetic scattering from inhomogeneous axis-symmetric objects. The method utilizes two continuous potentials to generate the Fourier azimuthal modes of the time-harmonic vector fields, and generates two coupled, partial differential equations for these two potentials.

The second breakthrough in the area of FEM is the introduction of the edge elements (as opposed to node-based elements) and the tangential vector finite element methods (TVFEMs) to solve the vector wave equation. The TVFEM, first developed by Nedelec,[2] is the vector finite element formulation which conforms to the function space *H(curl)*, the linear space of vector fields whose curl is also square-integrable. This kind of vector FEM has a very important property that the unknown vector field (but not necessarily its normal component) is tangentially continuous across the element boundaries. The lowest order of the TVFEM turns out to be identical to the edge-elements, where all the vector basis functions are defined on the element edges. It has been proven that by choosing the TVFEMs as the vector finite element bases, the null space of the *curl* operator can be modeled exactly.[3,4] An important consequence of this that the spurious modes, which typically plague the conventional nodal type of FEMs, are eliminated altogether.[3,5]

The third significant advance in the FEM method is the introduction of the absorbing boundary conditions (ABCs) for truncating the infinite problem domain into a region which is of finite and

Directions in Electromagnetic Wave Modeling
Edited by H.L. Bertoni and L.B. Felsen, Plenum Press, New York, 1991

manageable size. There are two types of ABCs, viz., local and global or nonlocal. A local ABC possesses two distinct advantages over the global versions: (i) It only calls for local type of operations that are no more involved than those needed in the process of the normal FEM matrix assembly; and, (ii) It preserves the sparsity of the finite element matrix. A number of authors have investigated local ABCs, notably Bayliss, Gunzburger and Turkel,[6] Engquist and Majda,[7] and Mittra and Ramahi.[8] The material found in the last reference is helpful in elucidating the underlying physics of the ABCs. The global boundary conditions give rise to partly full matrices and are usually limited in application to small or moderately-sized scatterers, and/or to those possessing certain symmetries.

In the past, a majority of electromagnetic wave scattering computations have been carried out in the frequency domain. However, for a variety of reasons, there has been an increasing interest recently in the solution of transient EM scattering problems. For instance, it is sometimes more economical to obtain frequency domain results via the time domain approach, particularly when these results are needed over a band of many frequencies. There are other situations, e.g., in EMP calculations, where the final result of interest is the transient response. Finally, the case of nonlinear and/or time-varying scatterers is most conveniently addressed in the framework of the time domain.

Among all of the available numerical methods for analyzing the transient electromagnetic wave phenomena, the FDTD algorithm is probably the one that is used most widely. The FDTD algorithm was first introduced by Yee[9] and was later developed, among others, by Kunz and Lee,[10] and Taflove and Brodwin,[11] who successfully demonstrated its application to a number complicated three-dimensional problems. However, in its original form, the FDTD algorithm suffers from two serious limitations: (i) The geometry of the scatterer is modeled by square or cubic cells which may introduce significant discretization errors unless a very high mesh density, that severely tasks the computer memory and significantly increases the solution time, is employed; and, (ii) The computation domain is discretized in a uniform manner and, as a result, a large portion of this domain is overly meshed, especially if one attempts to model some fine features of the object, and one must again pay the price in terms of increased computer time and storage which can become prohibitive. A typical microwave circuit, which can often be modeled in the frequency domain by using only 15 to 30 thousand unknowns, may require upward of a million unknowns in the conventional FDTD algorithm. Since the FDTD approach is a recursive procedure which requires no matrix inversion, it has the capability of handling a much larger number of unknowns than does the FEM. Even so, this does not always offset the difficulties with the prohibitively large number of unknowns needed for accurate discretization with high resolution in the conventional, uniform-sample FDTD.

A number of variants of the FDTD algorithm have been developed with a view to circumventing the above difficulties. These are: the point-matching time domain finite element methods (PTDFEMs)[12]; the curvilinear FDTD algorithm[13]; the time-domain finite-volume (TDFV) algorithm[14]; and the generalized FDTD (GFDTD) on Delaunay-Voronoi complementary grids.[15] The PTDFEM is similar to the FEM in the frequency domain and it uses piecewise polynomial basis functions to approximate both the electric and magnetic fields in a conforming finite element mesh. Enforcing the two *curl* equations on the grid points, the PTDFEM results in an explicit time-marching scheme for obtaining the transient response of the electromagnetic scattering. In Ref. 13, Holland has modified the FDTD formulation to a general non-orthogonal curvilinear coordinate system. Later, Fusco[16] has applied the curvilinear FDTD to two-dimensional scattering problems. However, the application of this algorithm to three-dimensional waveguide discontinuity problems has not appeared in the literature and will shortly be presented in this paper. The TDFV algorithm has been derived from the proven computational fluid dynamics methods by Shankar, Mohammadian, and Hall.[14] This algorithm employs an explicit Lax-Wendroff upwind scheme to integrate Maxwell's equations in time, and results in second order accuracy in both space and time discretizations. All of the above approaches employ body-fitted conforming grids with non-rectangular bricks for meshing up the computation domain. Although these methods are more flexible than the conventional FDTD scheme, they still have some difficulties in modeling many realistic practical problems. On the other hand, the GFDTD method uses the Delaunay tessellation to discretize the problem region into a group of tetrahedra. Because of this, the GFDTD approach appears to offer the most flexibility in modeling the problem geometries. The counterpart of the Delaunay tetrahedra is the Voronoi polyhedra. Approximating the electric and magnetic fields on the Delaunay tetrahedra and Voronoi polyhedra, respectively, Butler[15] has generalized the FDTD algorithm to this unstructured mesh and has applied the GFDTD to solve some simple microwave circuits.

Having presented a brief survey of the past developments in the area of direct PDE solution of Maxwell's equations, we now outline the organization of the rest of this paper. Section 2 provides some details of the edge-elements and TVFEM for solving the Maxwell's equations in the frequency domain. In Sec. 3, we briefly describe the formulations of the FDTD on a non-uniform rectangular grid, the curvilinear FDTD, and the GFDTD, and their applications to the modeling of microwave circuits. A brief discussion of the ABCs is presented in Sec. 4. Finally, some suggestions for future directions of research are given in Sec. 5.

2. FREQUENCY DOMAIN SOLUTION USING FEM

As mentioned earlier, one of the most serious difficulties in applying the FEM to solve the vector wave equation is that the numerical solution is often corrupted by the presence of spurious modes,[17] and an approach to eliminating these is to use the edge-elements or the TVFEM. With these finite elements, only the tangential continuity of the vector field is enforced across the element boundaries. The advantages of this approach are threefold: (i) In conformity with the dictates of the physics of the problem, it imposes the continuity of only the tangential components of the electric and magnetic fields across the element boundaries; (ii) The interfacial boundary conditions are automatically satisfied through the natural boundary condition built into the variational principle; and (iii) Dirichlet boundary condition can be easily imposed along the element edges and on the element faces.

In this section, we describe some of the applications of the edge-elements and the TVFEM, for modeling the electromagnetic scattering problems in the frequency domain. Additional details on edge elements may be found in Ref. 4.

2a. Edge-Elements for Scattering Problem

In Fig. 1, we show the edge element on a tetrahedral finite element. There is a vector basis function $\vec{w}_e$ associated with each edge. For example, for edge {1,2}, the vector basis function $\vec{w}_e$ is

$$\vec{w}_e = \varsigma_1 \nabla \varsigma_2 - \varsigma_2 \nabla \varsigma_1 \tag{1}$$

Here ζ_i is the barycentric coordinate of node i, which is a piecewise linear function, equal to 1 at node i, and 0 at all other nodes. The direction of the vector along an edge is arbitrary and is defined by the global node numbering system in (1).

According to Fig. 1, we can approximate the vector field, for example the electric field E, within the tetrahedral element as

$$E = \sum_{i=0}^{3} \sum_{j>i} e_i^j \left(\varsigma_i \nabla \varsigma_j - \varsigma_j \nabla \varsigma_i \right) \tag{2}$$

From (2), the curl of the unknown vector E is given by

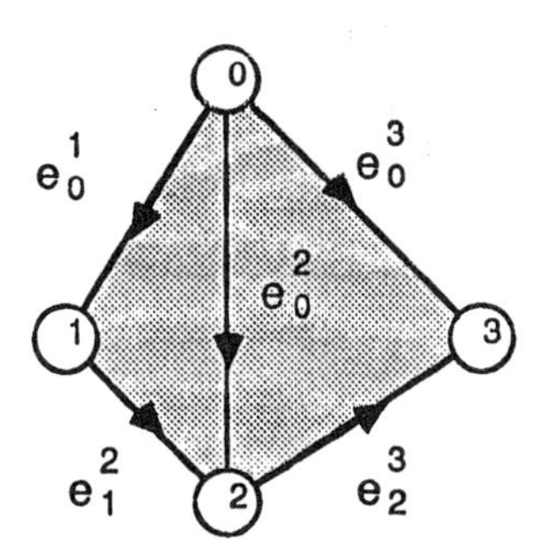

Fig. 1. Tetrahedral edge-element.

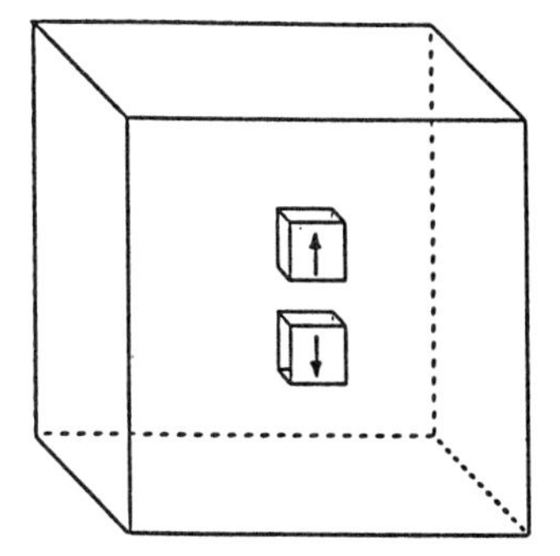
Fig. 2. A linear quadrupole radiator with the computation domain.

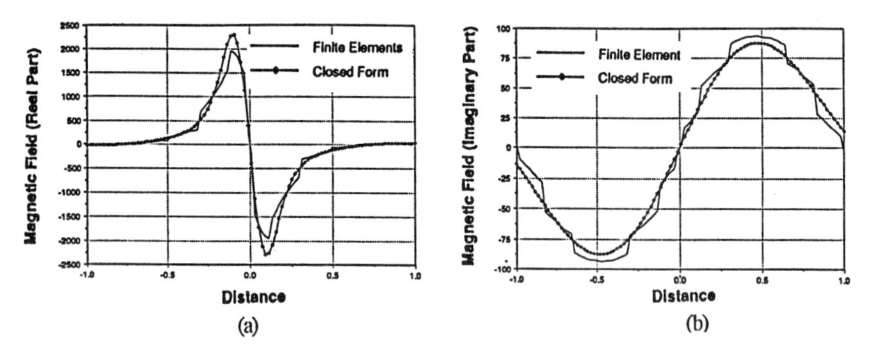

Fig. 3. The magnetic field for the linear quadrupole problem; (a) real part, and (b) imaginary part.

$$\nabla \times E = 2\sum_{i=0}^{3}\sum_{j>i} e_i^j \left(\nabla \varsigma_i \times \nabla \varsigma_j\right) \tag{3}$$

D'Angelo and Mayergoyz[18] have employed the edge-elements, in conjunction with the Engquist-Majda ABC on a rectangular outer boundary, to analyze a linear quadrupole radiator (see Fig. 2). The two cubic magnetized volumes are shown in Fig. 2. The side length of the cubes is 0.1 meters, the cubes are displaced 0.1 meters in the z-direction and are oppositely polarized in the same direction. The linear quadrupole is centered at the origin and the outer boundary of the finite element mesh is at 1.0 meter. The frequency of operation is 250 MHz. The near fields of the finite element solution are compared against the closed form solution in Fig. 3. Figure 3(a) shows the real part of the magnetic field along a z-directed line that starts at the point (0.1, 0.1, –1.0) and ends at (0.1, 0.1, 1.0) whereas Fig. 3(b) shows the imaginary part of the magnetic field along the same line.

2b. H_1(curl) TVFEM for Modeling Microwave Circuits

As mentioned in Section I, the lowest order of the TVFEM turns out to be identical to the edge-elements. However, in practical applications, we often need to use a higher-order version of the same in order to realize a satisfactory level of accuracy.

In Ref. 19, Lee has presented a first-order TVFEM, based upon the *curl* operator, and has applied it to analyze several three-dimensional passive microwave components. This TVFEM, which is complete to first-order in the range of the *curl* operator, was denoted by $H_1(curl)$ elements in Ref. 19 and was defined mathematically as

$$H_1(curl) = \left\{ \vec{u} \middle| \vec{u} \in \left[L_2(\Omega)\right]^3, \nabla \times \vec{u} \in \left[P_1(\Omega)\right]^3 \right\} \tag{4}$$

where $L_2(\Omega)$ is a set of square integrable functions, and $P_1(\Omega)$ is a set of piecewise linear functions in the discretized problem domain Ω. The unknowns in the three-dimensional $H_1(curl)$ TVFEM are designated as shown in Fig. 4. In each tetrahedron, there are two unknowns associated with each edge and with each face. Therefore, a tetrahedral element has twenty degrees of freedom available to describe the vector field. The two vector basis functions of edge $\{i,j\}$ are $\zeta_i \nabla \zeta_j$ and $\zeta_j \nabla \zeta_i$ for node i and node j, respectively. The two facial vector basis functions can be chosen for any two of the three edges. For example, in Fig. 4, we have chosen them to be $\zeta_i \zeta_j \nabla \zeta_k$ on the face $\{i,j,k\}$ and $\zeta_i \zeta_k \nabla \zeta_j$ for the edges $\{i,j\}$ and $\{i,k\}$, respectively. According to Fig. 4, the unknown field E within the tetrahedron can be written as

$$\begin{aligned} E = &\sum_{i=0}^{3}\sum_{j=0}^{3} e_i^j \varsigma_i \nabla \varsigma_j \qquad\qquad i \neq j \\ &+ \sum_{l=0}^{3} \left(f_1^l \varsigma_i \varsigma_j \nabla \varsigma_k + f_2^l \varsigma_i \varsigma_k \nabla \varsigma_j \right) \end{aligned} \tag{5}$$

where l,i,j,k form cyclic indices.

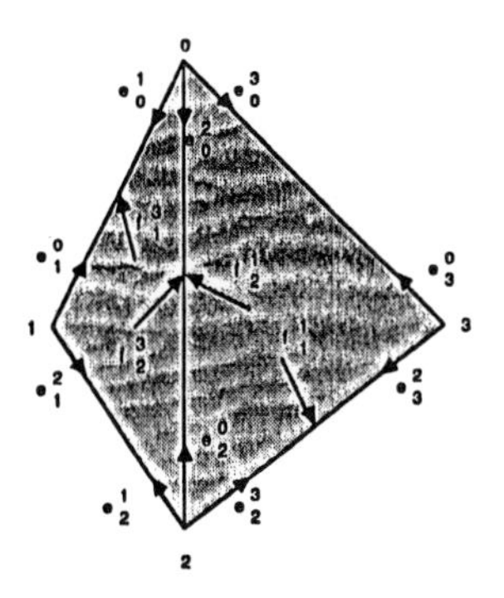

Fig. 4. Three-dimensional H_1 *(curl)* tangential finite element.

To verify that the E field defined in (5) is tangentially continuous across the element boundaries, we first take the projection of (5) along edge $\{i,j\}$, the result is

$$E \bullet \vec{r}_{ij}\Big|_{edge\{i,j\}} = e_i^j \varsigma_i \left(\nabla\varsigma_j \bullet \vec{r}_{ij}\right) - e_j^i \varsigma_j \left(\nabla\varsigma_i \bullet \vec{r}_{ij}\right)$$
$$= e_j^i \varsigma_j - e_i^j \varsigma_i \tag{6}$$

where $\vec{r}_{ij}$ is the position vector from node i to node j. From (6), it can be shown that the vector field E is tangentially continuous across the edge if the two edge variables, e_j^i and e_i^j, are set to be common unknowns for all the tetrahedral elements that share edge $\{i,j\}$. Secondly, the field value on the face $\{1,2,3\}$, which corresponds to $\zeta_0 = 0$, is obtained according to

$$E\Big|_{\varsigma_0=0} = \sum_{i=1}^{3}\sum_{j=1}^{3} e_i^j \varsigma_i \nabla\varsigma_j \qquad i \neq j$$
$$+ f_1^0 \varsigma_1\varsigma_2 \nabla\varsigma_3 + f_2^0 \varsigma_1\varsigma_3 \nabla\varsigma_2$$
$$+ \left[\sum_{j=1}^{3} e_j^0 \varsigma_j + f_1^1 \varsigma_2\varsigma_3 + f_2^2 \varsigma_1\varsigma_3\right] \nabla\varsigma_0 \tag{7}$$

Since $\nabla\zeta_0$ is normal to the face $\{1,2,3\}$, the tangential component of the field on this face is uniquely defined by the six edge variables, e_i^j, and the two facial variables, f_1^0 and f_2^0, on this face. Therefore, in addition to the edge variables, these facial variables are also required to be common unknowns, for the two tetrahedral elements that contain this triangular face, to assure the tangential continuity of the vector field.

When modeling passive microwave circuits, we also need to calculate $\nabla \times E$ within the elements. This can be accomplished by taking the *curl* of both sides of (5), which results in

$$\nabla \times E = \sum_{i=0}^{3}\sum_{j=0}^{3} e_i^j \left(\nabla\varsigma_i \times \nabla\varsigma_j\right) \qquad i \neq j$$
$$+ \sum_{l=0}^{3}\left[f_1^l\left(\varsigma_i \nabla\varsigma_j \times \nabla\varsigma_k + \varsigma_j \nabla\varsigma_i \times \nabla\varsigma_k\right) + f_2^l\left(\varsigma_i \nabla\varsigma_k \times \nabla\varsigma_j + \varsigma_k \nabla\varsigma_i \times \nabla\varsigma_j\right)\right] \tag{8}$$

Various microwave structures have been studied by using the H_1*(curl)* TVFEM. Figure 5 shows a coax to waveguide connector. Note that in view of the symmetry of the connector, it is necessary to model only half of the geometry. Also, shown in Fig. 5 is the Delaunay tessellation of the geometry that is used in the current analysis. The computed electric field on the symmetry plane is shown in Fig. 6 at the frequency 10 GHz. It is of interest to mention that this problem was originally analyzed by using the edge-elements. At a frequency of 10 GHz, the best result obtained was about 20 dB for the return loss, obtained by using a 32 Mbytes RAM workstation. Using the same hardware, the application of the H_1*(curl)*

TVFEM yielded a result of 39 dB for the return loss which compares very closely with the measured result. This leads us to conclude that the accuracy of the $H_1(curl)$ TVFEM approach is far superior to that of the edge-element method. In Fig. 7, we show a planar microstrip low-pass filter on a substrate with a dielectric constant of 2.2. This circuit was discretized as shown in Fig. 8, with approximately 17,000 unknowns in the finite element analysis. This discretization is employed for the calculation throughout the entire frequency domain from 0 to 20 GHz. The computed return loss and the insertion loss, together with the measure data, are shown in Figs. 4(a), and 4(b), respectively. We observe that, in general, good agreements are obtained between the two.

3. TIME DOMAIN ANALYSIS

The issue of how to discretize the computational space when using the FDTD algorithm to solve for wave interaction with a scatterer, or wave propagation in a guided-

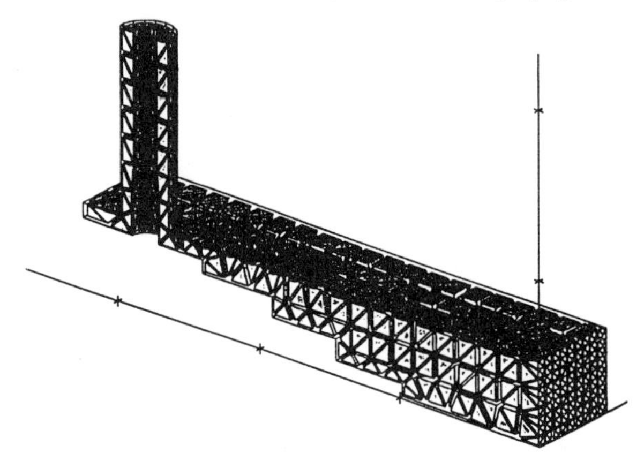

Fig. 5. Finite element mesh of a coax to waveguide connector.

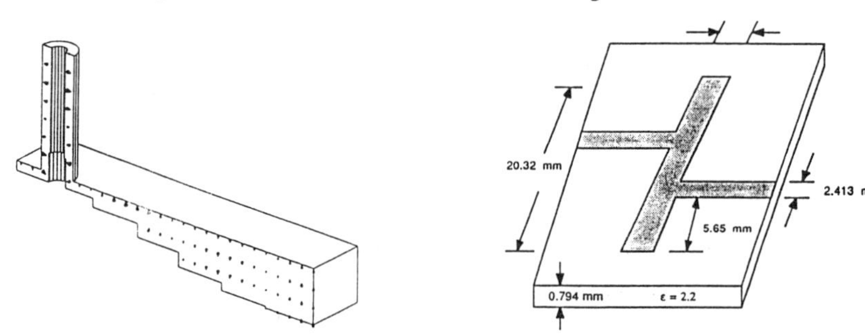

Fig. 6. Electric Field plot on the symmetry plane at 10 GHz.

Fig. 7. The dimensions of a non-symmetric microstrip low-pass filter.

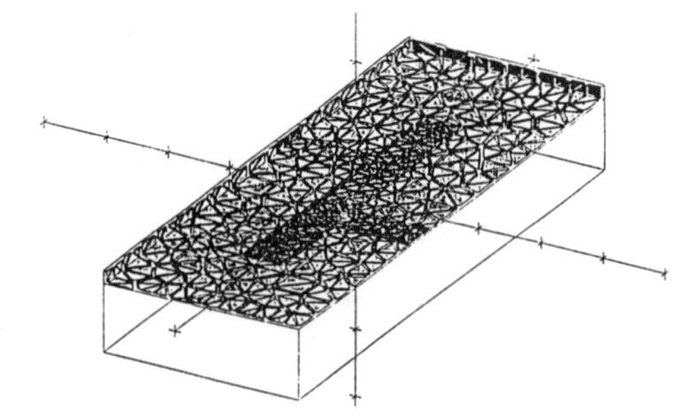

Fig. 8. The three-dimensional Delaunay tessellation of the low-pass filter in Fig. 7.

—— FDTD
- - - measurement
°$H_1(curl)$ TVFEM

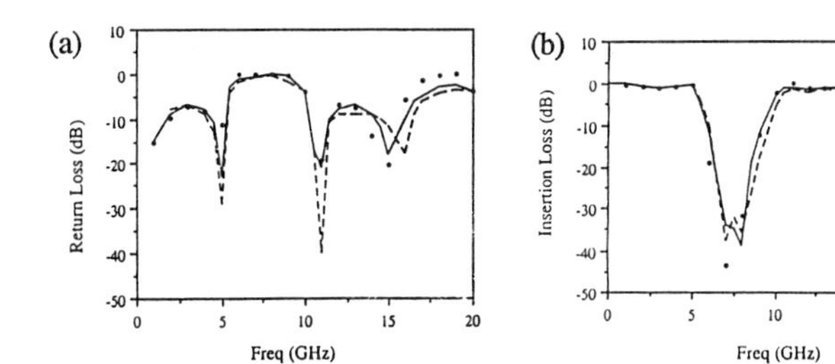

Fig. 9. (a) The return loss, and (b) the insertion loss of the low-pass filter.

wave structure, has been a concern in efforts to improve the efficiency of this algorithm. While the basic Yee algorithm[9] can handle general electrical properties of media filling the region of computation, the discretization of problem geometry into a uniform lattice of unknowns may, as pointed out earlier, lead to some difficulties for a class of problems. In this section, we discuss three generalizations of the FDTD algorithm that can obviate some of these problems. These are, in order of increasing flexibility, the variable size mesh FDTD; the curvilinear FDTD; and the GFDTD algorithm on the Delaunay-Voronoi complementary grids.

3a. Variable-Size Mesh FDTD

In the modeling of geometric features which are much smaller than the wavelength of interest, it is often beneficial to employ different mesh sizes for different regions. By varying the grid size of the cells, one can model the small features with a fine mesh, while use larger cells in regions where the fields vary relatively slowly. However, the use of the non-uniform grid necessarily introduces the problem that the four points, whose field values are used to iterate the curl equations at a particular node, are no longer equidistant to this node. Previous incorporations of the non-uniform grid in the FDTD algorithm have appealed to the integral form of Maxwell's equations for field updating. However, the integral approach is accurate only to the first order while the variable mesh FDTD algorithm, described herein, preserves the second-order accuracy.

Shown in Fig. 10 is an example where a region has been discretized in the x-direction with different lengths for each cell. From Taylor's expansion, the field values E_1, E_2, E_{-1}, and E_{-2} can be approximated by

$$
\begin{aligned}
E_1 &= E_0 + \partial_x E|_0 \Delta_1 + \partial_x^2 E|_0 \frac{\Delta_1^2}{2} + O\left(\Delta_1^3\right) \\
E_2 &= E_0 + \partial_x E|_0 (\Delta_1 + \delta_1) + \partial_x^2 E|_0 \frac{(\Delta_1 + \delta_1)^2}{2} + O\left([\Delta_1 + \delta_1]^3\right) \\
E_{-1} &= E_0 - \partial_x E|_0 \Delta_2 + \partial_x^2 E|_0 \frac{\Delta_2^2}{2} + O\left(\Delta_2^3\right) \\
E_{-2} &= E_0 - \partial_x E|_0 (\Delta_2 + \delta_2) + \partial_x^2 E|_0 \frac{(\Delta_2 + \delta_2)^2}{2} + O\left([\Delta_2 + \delta_2]^3\right)
\end{aligned}
\tag{9}
$$

From (9), we can obtain a second-order accurate expression for $\partial_x E|_o^R$ from E_2, E_1, and E_{-1} by using

$$
\partial_x E\Big|_o^R = \frac{-(2\Delta_1 + \delta_1)}{(\Delta_1 + \Delta_2)(\Delta_1 + \Delta_2 + \delta_1)} E_{-1} + \frac{\Delta_1 - \Delta_2 + \delta_1}{\delta_1(\Delta_1 + \Delta_2)} E_1 + \frac{\Delta_2 - \Delta_1}{\delta_1(\Delta_1 + \Delta_2 + \delta_1)} E_2 \tag{10}
$$

A similar expression for $\partial_x E|_o^L$, derived from E_1, E_{-1}, and E_{-2}, reads

$$
\partial_x E\Big|_o^L = \frac{(2\Delta_2 + \delta_2)}{(\Delta_1 + \Delta_2)(\Delta_1 + \Delta_2 + \delta_2)} E_1 + \frac{\Delta_1 - \Delta_2 - \delta_2}{\delta_2(\Delta_1 + \Delta_2)} E_{-1} + \frac{\Delta_2 - \Delta_1}{\delta_2(\Delta_1 + \Delta_2 + \delta_2)} E_{-2} \tag{11}
$$

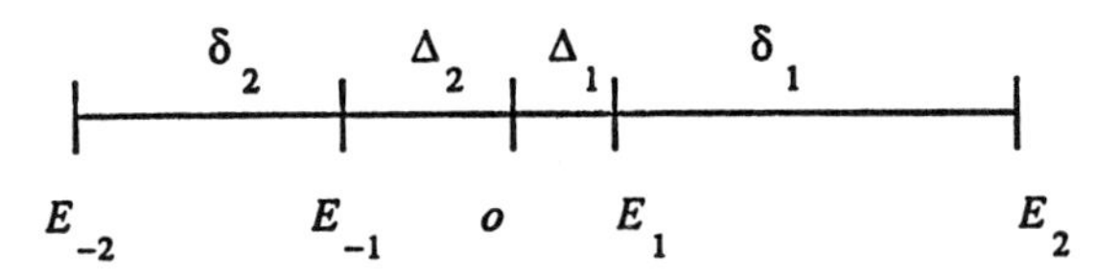

Fig. 10. A general variable size mesh in the x direction.

Finally, to render the algorithm stable, we need to take the average of equations (10) and (11), and we let

$$\partial_x E|_o = \frac{1}{2}\left(\partial_x E\Big|_o^R + \partial_x E\Big|_o^L \right) \tag{12}$$

Similar expressions can be obtained for the other derivatives following the same approach presented here.

To study the differences between the present formulation and the contour integral approach, we apply both methods to simulate the wave propagation within an air-filled waveguide. The transverse dimensions of the guide are 0.33 λ by 1 λ at the frequency of operation. We note from the mesh employed, which is shown in Fig. 11, that the spatial discretization is staggered in each dimension, resulting in cells of dimension axaxa, axax4a, ax4ax4a, and 4ax4ax4a. Figure 12 compares the exact sinusoidal distribution of the E field along the longitudinal direction to the results, obtained from the integral-form of the FDTD algorithm and the present formulation. At a distance of two and a half free-space wavelengths, the phase advance of the field obtained by using the formulation based upon the contour integration scheme, which does not preserve the second-order accuracy is seen to be off by at least 30 degrees. However, those derived by using the formulas presented here are seen to compare well with the true solution.

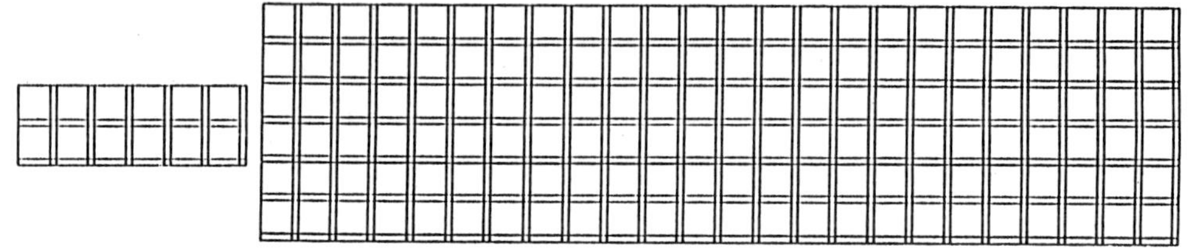

Fig. 11. Non-uniform mesh for a rectangular waveguide.

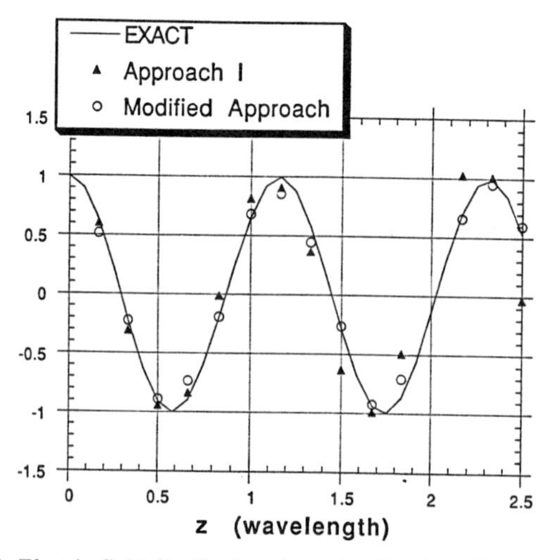

Fig. 12. Electric field distribution along the direction of propagation.

3b. FDTD in the Curvilinear Coordinates

A formulation of the FDTD algorithm in the generalized curvilinear coordinate system has been given by Holland[13] and, in Ref. 16, Fusco has successfully used this approach to solve some two-dimensional scattering problems. Our formulation in this paper is similar to that of Holland, except that we have introduced the concept of normalized covariant fields and have rewritten the formulation in a manner which is computationally more efficient.

In the curvilinear FDTD algorithm, the first components of the contravariant magnetic field H^1, and electric field E^1 are updated from the normalized covariant electric field e, and magnetic field h, respectively, according to

$$
\begin{aligned}
H^1(i,j,k)^{n+1} &= H^1(i,j,k)^n - w(i,j,k) \\
&\quad [e_3(i,j+1,k) - e_3(i,j-1,k) - e_2(i,j,k+1) + e_2(i,j,k-1)] \\
E^1(i,j,k)^{n+1/2} &= E^1(i,j,k)^{n-1/2} + w(i,j,k) \\
&\quad [h_3(i,j+1,k) - h_3(i,j-1,k) - h_2(i,j,k+1) + h_2(i,j,k-1)]
\end{aligned}
\tag{13}
$$

The other component equations can be obtained through straightforward index permutation.

The generation of the normalized covariant field components requires both an interpolation as well as formation of a linear combination of the contravariant field components. For example, the normalized covariant components e_1 and h_1 are given by

$$
\begin{aligned}
e_1(i,j,k) &= \bar{g}_{11}(i,j,k)\cdot E^1(i,j,k) \\
&+ \frac{\bar{g}_{12}(i,j,k)}{4}\left[E^2(i+1,j+1,k) + E^2(i-1,j+1,k) + E^2(i+1,j-1,k) + E^2(i-1,j-1,k)\right] \\
&+ \frac{\bar{g}_{13}(i,j,k)}{4}\left[E^3(i+1,j,k+1) + E^2(i-1,j,k+1) + E^2(i+1,j,k-1) + E^2(i-1,j,k-1)\right] \\
h_1(i,j,k) &= \bar{g}_{11}(i,j,k)\cdot H^1(i,j,k) \\
&+ \frac{\bar{g}_{12}(i,j,k)}{4}\left[H^2(i+1,j+1,k) + H^2(i-1,j+1,k) + H^2(i+1,j-1,k) + H^2(i-1,j-1,k)\right] \\
&+ \frac{\bar{g}_{13}(i,j,k)}{4}\left[H^3(i+1,j,k+1) + H^2(i-1,j,k+1) + H^2(i+1,j,k-1) + H^2(i-1,j,k-1)\right]
\end{aligned}
\tag{14}
$$

The other covariant components are obtained by index permutation of (14). The weighting factor w as well as the geometric parameters $\bar{g}$ are determined from

$$
w(i,j,k) = \begin{cases} \dfrac{\Delta t}{\varepsilon(i,j,k)}\dfrac{1}{area(i,j,k)} & E\text{ - grid} \\[2ex] \dfrac{\Delta t}{\mu(i,j,k)}\dfrac{1}{area(i,j,k)} & H\text{ - grid} \end{cases}
$$

$$
\bar{g}_{lm} = \frac{g_{lm}}{\sqrt{g_{mm}}}
\tag{15}
$$

where g_{lm} is defined as in Ref. 13.

Note that, in this formulation, all the parameters specific to the curvilinear grid algorithm, viz., w and $\bar{g}$, need be computed only once and stored along with the locationof each grid point for future use. Consequently, the only additional steps over and above the conventional FDTD algorithm, required in the leap-frog time-marching process based upon the present formulation, are the transformations given in (14).

In Fig. 13, we show a circular waveguide with a circular iris discontinuity. The waveguide is filled with an isotropic dielectric material of dielectric constant 9.7, except for the iris region which is filled with air. This problem has been analyzed by using two techniques, viz., the mode matching method and the curvilinear FDTD approach. The comparisons between the computed reflection and transmission coefficients derived from these two approaches are shown in Figs. 14 and 15 and the agreement is seen to be quite favorable.

The above approach has also been applied to the problems of RCS computation and excellent results have been achieved.[20]

3c. FDTD in Delaunay-Voronoi Tessellation

In his recent Ph. D. dissertation,[15] Butler has generalized the FDTD algorithm to the Delaunay-Voronoi complementary grids. The construction of the Delaunay tessellation, for any given point distribution in three dimensions, is currently a major topic in the area of computer graphics and the interested reader is referred to Refs. 21 and 22 for further details on this subject. The generalization of the FDTD algorithm to the Delaunay-Voronoi meshing scheme offers the following three advantages: (i) The tetrahedral mesh is both simple and geometrically flexible. From analytic geometry,[23] we know that any polygon in 2D, or any polyhedron in 3D, can be discretized into triangles or tetrahedra, respectively. (ii) The flexibility of the tetrahedral mesh allows an efficient use of the resources and permits the modeling of geometrically complex problems. In contrast, the use of rectangular grids presents difficulties in modeling curved boundaries, or even wedge-shaped objects. (iii) Several authors[21,22] have succeeded in meshing up arbitrary geometries in a fully automatic manner by using the Delaunay tessellation scheme. Thus, a combination of such a mesh generating scheme with the integral form FDTD algorithm offers an excellent potential for the development of general purpose, user-friendly, electromagnetic field analysis softwares in the time domain.

A simple example of a Delaunay tetrahedron and a Voronoi polyhedron is shown in Fig. 16. As seen from this figure, one of the important properties of the Delaunay-Voronoi complementary meshes is that the edges of the Delaunay tetrahedra are orthogonal to the faces of the Voronoi polyhedra and vice versa. In view of this, the generalization of the FDTD algorithm in the Delaunay-Voronoi system becomes straightforward.

In Fig. 16, let $\vec{e}_i$, designated by black arrows, be the tangential component of the electric field on the edge i of the tetrahedron. Also let $\vec{h}_j$, (white arrows) be the magnetic field tangential to the edge j of the Voronoi polyhedron. Due to the Delaunay-Voronoi duality, $\vec{h}_j$ is also normal to the face j of the tetrahedron. Using this property, we can obtain the following discretized form for the integral form of Faraday's law:

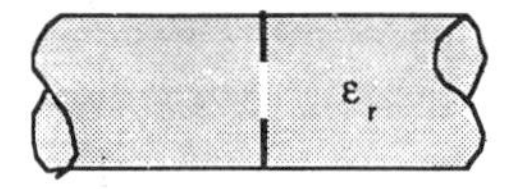

Fig. 13. A circular waveguide with a circular iris.

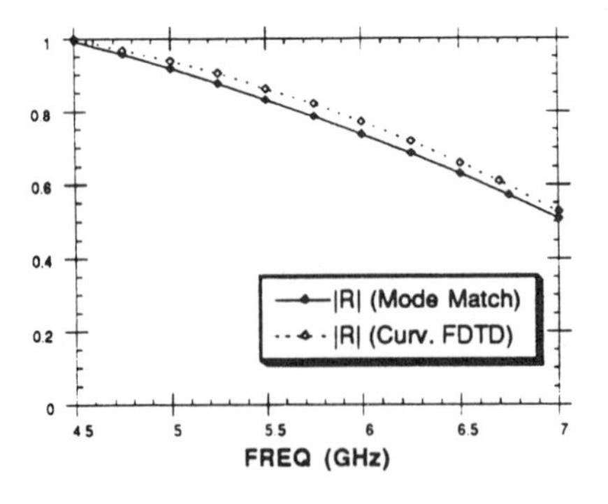

Fig. 14. Numerical results of the reflection coefficient.

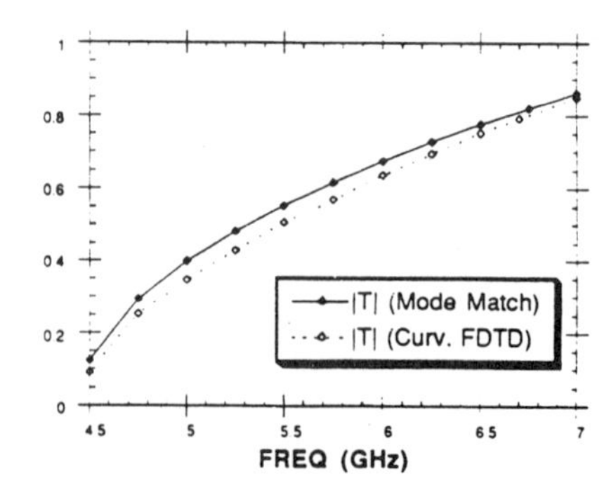

Fig. 15. Numerical results of the transmission coefficient.

$$\sum_{i=0}^{2} \vec{e}_i^{\,n} \bullet \vec{\ell}_{di} = -\mu \vec{\Delta}_{fd} \bullet \frac{\vec{h}_j^{\,n+\frac{1}{2}} - \vec{h}_j^{\,n-\frac{1}{2}}}{\Delta t} \tag{16}$$

where Δ_{fd} is the area of the face of the Delaunay tetrahedron and l_{di} is the directed length vector of the edge i. The subscripts d and v refer to the Delaunay and Voronoi structures, respectively. Likewise, for the Voronoi polyhedron, we have

$$\sum_{i=0}^{m-1} \vec{h}_i^{\,n+\frac{1}{2}} \bullet \vec{\ell}_{vj} = \varepsilon \vec{\Delta}_{fv} \bullet \frac{\vec{e}_i^{\,n+1} - \vec{e}_i^{\,n-1}}{\Delta t} \tag{17}$$

where Δ_{fv} is the area of the face of the Voronoi polyhedron and m is the number of edges on the face in question.

The same approach can also be applied in the frequency domain. A two-dimensional version of this algorithm in the frequency domain has been developed by McCartin, Bahrmasel and Meltz.[24] They have applied this technique to investigate the coupling between the scattered fields in a cavity-backed aperture (Fig. 17). Shown in Fig. 18 is the computed normalized backscatter cross section and its comparison with the generalized dual series solution obtained by Ziolkowski and Grant.[25]

4. ABSORBING BOUNDARY CONDITIONS

Various absorbing boundary conditions, i.e., ABCs, have been developed for the truncation of FEM and FDTD meshes and a number of these were mentioned (see Refs. 6–8) in Section 1. The interested reader may refer to these papers for further details. In this section, we present a new approach for mesh truncation which has been inspired by a recent paper by Deveze et al.,[26] although the derivation presented is markedly different from that given in the above reference.

Let us assume, without loss of generality, that the mesh is located in the region $0 \le z$, and we are seeking to derive an ABC for the plane $z = 0$.

We begin by expressing the transverse components E_x and E_y in terms of the longitudinal components E_z and H_z (see for instance the TE/TM representation of fields in Harrington[27]) as follows:

$$\begin{aligned} \left(\frac{1}{c^2}\partial_t\partial_t - \partial_z\partial_z\right)E_x &= -\partial_z\partial_x E_z + \mu\partial_y\partial_t H_z \\ \left(\frac{1}{c^2}\partial_t\partial_t - \partial_z\partial_z\right)E_y &= -\partial_z\partial_y E_z - \mu\partial_x\partial_t H_z \end{aligned} \tag{18}$$

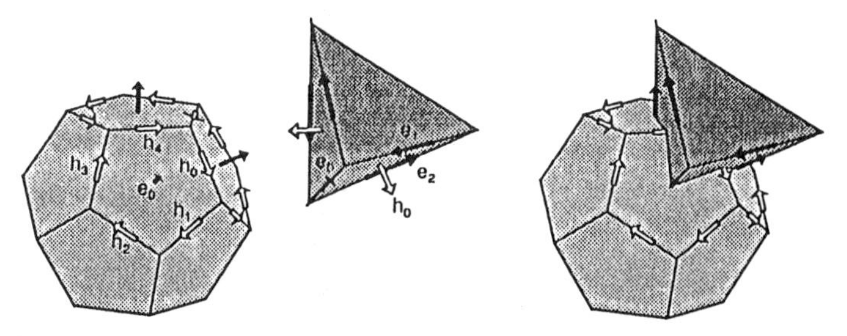

Fig. 16. A Delaunay tetrahedra and Voronoi polyhedra with corresponding field vectors.

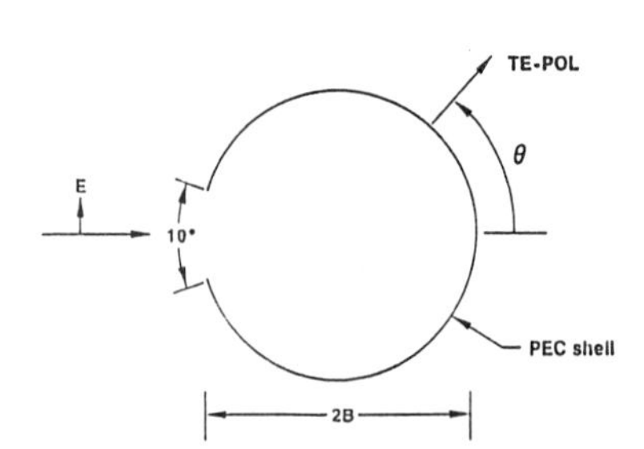

Fig. 17. TE scattering from an axially slotted circular cylinder.

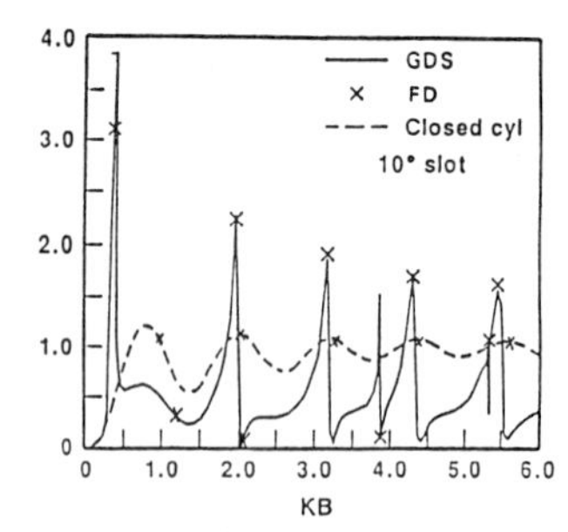

Fig. 18. Normalized backscatter cross section of a cavity-backed axial slot in a cylinder.

Next, we split the operator, $\frac{1}{c^2}\partial_t\partial_t - \partial_z\partial_z$, into two parts, corresponding to the incoming and outgoing waves, respectively. We then rewrite (18) as

$$\left(\frac{1}{c}\partial_t + \partial_z\right)E_x = \frac{-\partial_z\partial_x E_z}{\frac{1}{c}\partial_t - \partial_z} + \frac{\mu\partial_y\partial_t H_z}{\frac{1}{c}\partial_t - \partial_z}$$
$$\left(\frac{1}{c}\partial_t + \partial_z\right)E_y = \frac{-\partial_z\partial_y E_z}{\frac{1}{c}\partial_t - \partial_z} - \frac{\mu\partial_x\partial_t H_z}{\frac{1}{c}\partial_t - \partial_z} \tag{19}$$

Using (19), we now derive an ABC operator that completely suppresses the reflection of plane waves incident upon the boundary plane at angles $\theta = 0°$, and $\theta = \theta_1$, where θ_1 is prescribed by the user. The incident angle θ is measured from the normal of the boundary plane. To this end, we first observe that for a plane wave incident at an angle θ_1, the following relationship holds

$$\left(\partial_z + \frac{\cos\theta_1}{c}\partial_t\right)W = 0 \tag{20}$$

where W represents any component of the electric or the magnetic field. Substituting (20) into the r.h.s. of (19), the desired ABC is then obtained as

$$\left(\frac{1}{c}\partial_t + \partial_z\right)E_x = \frac{\cos\theta_1\partial_x E_z}{1+\cos\theta_1} + \mu c\frac{\partial_y H_z}{1+\cos\theta_1}$$
$$\left(\frac{1}{c}\partial_t + \partial_z\right)E_y = \frac{\cos\theta_1\partial_y E_z}{1+\cos\theta_1} - \mu c\frac{\partial_x H_z}{1+\cos\theta_1} \tag{21}$$

Following the procedure presented in this paper, we have generalized the above ABC to the case where the two angles for reflectionless condition can be prescribed arbitrarily. Numerical tests with this boundary condition have shown good potential for mesh truncation, without sacrificing too much accuracy, even when the truncation boundary is in the Fresnel field of the scatterer. Further investigations of this boundary condition are being continued.

5. FUTURE RESEARCH AREAS

We conclude this paper by offering the following suggestions for possible areas of future research and development in the area of PDE techniques, that would improve their practicality as useful tools for the design engineer:

(1) A user-friendly automatic mesh generation scheme for both the frequency and time domain analyses.

(2) A fast and memory-efficient solver for sparse matrix equations that result from the application of either the edge-elements or the TVFEM in three-dimensional FEM formulation.

(3) New basis functions that model the wave nature of the fields in an efficient manner, using only about one element per wavelength, allowing us to deal with large objects.

(4) Extensions of the ABCs to non-orthogonal meshes which can be used together with the curvilinear FDTD, TDFV, and GFDTD algorithms.

(5) Accurate and convenient ABCs for three-dimensional FEM calculations useful for mesh truncation at distances not too far from the surface of the scatterer.

(6) Graphical visualization of numerical solutions in three dimensions via post-processing.

REFERENCES

1. M. A. Morgan, S. K. Chang, and K. K. Mei, Coupled azimuthal potentials for EM field problems in inhomogeneous penetrable bodies of revolution, IEEE Trans. Antennas Propagat. 27:202 (1979).
2. J. C. Nedelec, Mixed finite element in R^3, Numer. Math. 35:315 (1980).
3. S. H. Wong and Z. J. Cendes, Combined finite element – modal solution of three-dimensional eddy current problems, IEEE Trans. Magn. 24:2685 (1988).
4. A. Bossavit and I. Mayergoyz, Edge-elements for scattering problems, IEEE Trans. Magn. 25:2816 (1989).
5. J. F. Lee, D. K. Sun, and Z. J. Cendes, Full-wave analysis of dielectric waveguides using tangential vector finite elements, to appear in IEEE Trans. Microwave Theory and Tech.
6. A. Bayliss, M. Gunzburger, and E. Turkel, Boundary conditions for the numerical solution of elliptic equations in exterior regions, SIAM J. Appl. Math. 42:430 (1982).
7. B. Engquist and A. Majda, Radiation boundary conditions for the numerical simulation of waves, Math. Comp. 31:629 (1977).
8. R. Mittra and O. Ramahi, Absorbing boundary conditions for the direct solution of partial differential equations arising in electromagnetic scattering problems, in: "PIER2: Finite Element and Finite Difference Methods in Electromagnetic Scattering," M. A. Morgan, ed., Elsevier, New York (1990).
9. K. S. Yee, Numerical solution of initial boundary value problems involving Maxwell's equations in isotropic media, IEEE Trans. Antennas Propagat. 14:302 (1966).
10. K. S. Kunz and K. M. Lee, A three-dimensional finite-difference solution of the external response of an aircraft to a complex transient EM environment: Part I – The method and its implementation, IEEE Trans. Electromagn. Compat. 20:328 (1978).
11. A. Taflove and M. E. Brodwin, Numerical solution of steady-state electromagnetic scattering problems using the time-dependent Maxwell's equations, IEEE Trans. Microwave Theory and Tech. 23:623 (1975).
12. A. C. Cangellaris, C. C. Lin, and K. K. Mei, Point-matching time domain finite element methods for electromagnetic radiation and scattering, IEEE Trans. Antennas Propagat. 35:1160 (1987).
13. R. Holland, Finite difference solutions of Maxwell's equations in generalized nonorthogonal coordinates, IEEE Trans. Nuc. Sci. 30:4689 (1983).
14. V. Shankar, A. H. Mohammadian, and W. F. Hall, A time-domain finite-volume treatment for the Maxwell equations, Electromagnetics 10:127 (1990).
15. A. J. Butler, "Time domain solutions of two- and three-dimensional transient electromagnetic fields," Ph.D. thesis, Carnegie-Mellon University (1990).
16. M. Fusco, FDTD algorithm in curvilinear coordinates, IEEE Trans. Antennas Propagat. 38:76 (1990).
17. A. Konrad, Vector variational formulation of electromagnetic fields in anisotropic media, IEEE Trans. Microwave Theory and Tech. 24:553 (1976).
18. J. D'Angelo and I. Mayergoyz, Finite element methods for the solution of RF radiation and scattering problems, Electromagnetics 10:177 (1990).
19. J. F. Lee, Analysis of passive microwave devices by using three-dimensional tangential vector finite elements, to appear in Int. J. Numer. Model. (1990).
20. J. F. Lee, R. Mittra, and J. Joseph, Finite difference time domain algorithm in curvilinear coordinates for solving three-dimensional open-region scattering problems, to appear in IEEE AP-S Int. Symposium (1991).
21. N. Phai, Automatic mesh generation with tetrahedral elements, Int. J. for Numer. Meth. in Eng. 18:273 (1982).
22. D. F. Watson, Computing the n-dimensional Delaunay tessellation with application to Voronoi polytopes, The Comp. J. 24:167 (1981).
23. S. S. Carins, "Introductory Topology," The Ronald Press Company, New York (1961).

24. B. J. McCartin, L. J. Bahrmasel, and G. Meltz, Application of the control region approximation to two-dimensional electromagnetic scattering, in: "PIER2: Finite Element and Finite Difference Methods in Electromagnetic Scattering," M. A. Morgan, ed., Elsevier, New York (1990).
25. R. W. Ziolkowski and J. B. Grant, Scattering from cavity-backed apertures: The generalized dual series solution of the concentrically loaded E-pol slit cylinder problem, IEEE Trans. Antennas Propagat. 35:504 (1987).
26. T. Deveze, F. Clerc, and W. Tabbara, Second order pseudo-transparent boundary equations for FDTD method, AP-S Symposium Digest 1627 (1990).
27. R. F. Harrington, "Time-harmonic Electromagnetic Fields," McGraw-Hill, New York (1961).

ASSESSING THE IMPACT OF LARGE-SCALE COMPUTING ON THE SIZE AND COMPLEXITY OF FIRST-PRINCIPLES ELECTROMAGNETIC MODELS

E. K. Miller

Los Alamos National Laboratory
Group MEE-3, MS J580, PO Box 1663, Los Alamos, NM 87545

INTRODUCTION

There is a growing need to determine the electromagnetic performance of increasingly complex systems at ever higher frequencies. The ideal approach would be some appropriate combination of measurement, analysis, and computation so that system design and assessment can be achieved to a needed degree of accuracy at some acceptable cost. Both measurement and computation benefit from the continuing growth in computer power that, since the early 1950s, has increased by a factor of more than a million in speed and storage. For example, a CRAY2 has an effective throughput (not the clock rate) of about 10^{11} floating-point operations (FLOPs) per hour compared with the approximate 10^{5} provided by the UNIVAC-1. The purpose of this discussion is to illustrate the computational complexity of modeling large (in wavelengths) electromagnetic problems.

In particular, we make the point that simply relying on faster computers for increasing the size and complexity of problems that can be modeled is less effective than might be anticipated from this raw increase in computer throughput. This is because the required FLOP count for integral-equation (IE) and differential-equation (DE) models grows with frequency, f, as f^{x}, where $3 \leq x \leq 9$. The effect of this exponential dependence on frequency is demonstrated by examining the computation time that is needed to model a generic problem, a perfectly conducting, space-shuttle-size object, using various models and approximations. We choose as a reference the CRAY2 computer operating at 10^{11} FLOPs/hour, for comparison with the massively parallel computers becoming available, for which computation bandwidths of 10^{16}/hour and 10^{18}/hour are being projected by the year 1995 and 2,000 respectively. The impact of anticipated computer performance projected to the year 2050, where a simple extrapolation would suggest a throughput of about 10^{26} FLOPS/hour, is also considered.

We suggest that rather than depending on faster computers alone, various analytical and numerical alternatives need development for reducing the overall FLOP count required to acquire the information desired. One approach is to decrease the operation count of the basic model computation itself, by reducing the order of the frequency dependence of the various numerical operations or their multiplying coefficients. Another is to decrease the number of model evaluations that are needed, an example being the number of frequency samples required to define a wideband response, by using an auxiliary model of the expected behavior.

COMPUTATIONAL ELECTROMAGNETICS

In CEM, as in other kinds of real-world phenomena, a continuous physical reality is represented by a discretized, sampled, numerical approximation intended to replicate behavior of the actual problem to some appropriate accuracy and resolution. The *fidelity (F)* with which this is accomplished

Directions in Electromagnetic Wave Modeling
Edited by H.L. Bertoni and L.B. Felsen, Plenum Press, New York, 1991

involves a solution effort, or *model complexity (C)* (defined below), which is driven by:

*The number of *spatial samples* (X_S), needed to achieve acceptable fidelity;
*The number of *numerical operations* (*N*) or FLOPs required per sample in the initial solution process;
*The number of *bits (b)* required per operation to maintain roundoff error at or below the intrinsic modeling error;
*The number of *operations needed per additional right-hand side (S)*; and
*The number of *Right-Hand Sides (H)* needing solution to acquire the total information sought.

The effect of each of these operations on the total FLOP count required when using several different CEM modeling approaches is now discussed, beginning with the computer time involved, and a proposed definition of model complexity.

Computer Time, Model Complexity and a Figure of Merit

The elapsed *time* (*T*) needed to perform the total number of *bit operations (BitFLOPs or BLOPs)* involved in the model computation, is determined by the effective throughput or *FLOP rate R* of the computer used in the modeling process and its *word length W* relative to *b*, whose product we call the *computation bandwidth* ($B = RW$). Although the specific dependence of *T* on these factors is a function of formulation, algorithm, and computer, it will increase with increasing X_S, *N*, *b*, *S* and *H* and decrease with increasing *R* and possibly *W*, for which a generic functional form might be

$$T \sim (X_S Nb + SH)/RW \equiv C/B, \tag{1a}$$

where $X_S Nb + SH$ denotes the model complexity. The various terms in *C* are all possibly functions of *frequency (f)*. Because there is a practical acceptable upper limit on *T* dependent on the application and value of the numerical result, an ongoing challenge in CEM is to find formulations and approximations by which X_S, *N*, *b*, *S* and/or *H* can be reduced without sacrificing acceptable fidelity and by exploiting computers with larger values of *R* and *W*. We also propose a *Figure of merit (M)* to characterize application of a particular model to a given problem as

$$M \equiv F/C, \tag{1b}$$

where the model which maximizes the ratio of *fidelity* to *complexity*, everything else being equal, would be the best choice for that application.

Spatial-Sampling Requirements

Although any model which yields numerical results must involve a sampling process of some sort, even classical, separation-of-variables techniques, the kinds to be considered here are specifically integral- (IE) and differential-equation (DE) models. For problems having homogeneous subregions, IE models in the frequency domain require sampling only over bounding surfaces, so that the *spatial* sample count varies as f^{D-1} where *D* is the number of problem dimensions (1, 2, or 3). For IE models of inhomogeneous media and for all DE models of penetrable media in the frequency domain, the sample count varies as f^D. This is because the sources for an IE model of an inhomogeneous region are distributed throughout it as induced polarization currents, while a DE model is formulated as a spatial sampling of the space in which the fields occur.

Temporal-Sampling Requirements

When time-domain versions of these same models are considered, the spatial sampling is repeatedly updated for a total of X_t *time samples*. Generally speaking, the response is needed for a total *observation time* at least as long as a field traveling at *light speed (c)* takes to propagate across the object being modeled. Further, since explicit models require the interval between time samples (the time step) to be less than the distance between space samples divided by *c*, we can deduce that $X_t \sim f$. Consequently, the total number of spatial samples needed for a time-domain (TD) model is a factor approximately *f* times its frequency-domain (FD) counterpart, becoming f^D for IE models of homogeneous problems, and f^{D+1} for DE and inhomogeneous-media IE models. Since a TD model is capable of providing information across of band of frequencies up to some maximum frequency, unless otherwise stated we use *f* to mean the maximum frequency for which the model provides useful results, when referring to TD models. For any of these models, it should be clear that the solution effort as a function of f must grow at least as fast as the total number of source/field samples being computed.

FLOP-COUNT DEPENDENCE ON MODEL TYPE AND FREQUENCY

As a means of comparing the expected performance of some of the various CEM model types available, it is useful to summarize how their FLOP counts depend on the frequency at which a given model is applied. We first consider FD models, followed by their TD counterparts.

Modeling in the Frequency Domain

There are three basic steps in FD modeling: 1) setting up the the matrix that is the discretized, sampled, approximate representation of the physical problem to be modeled, which we refer to as the *system* matrix; 2) solving that system of equations for the desired observables; and 3) repeating whatever part of the solution process needs re-doing for *H RHSs*. Each of these steps is considered in turn in the following.

Computing the frequency-domain system matrix. The FLOP count for "filling" the system matrix for both DE and IE models is always exceeded by that for the solution process for large enough X_S. For DE models, the matrix-fill FLOP count is proportional to X_S, because the local-sampling nature of PDEs means that only some small number of the spatial samples are numerically related at any field point. Because an IE model uses a Green's function, on the other hand, all unknowns are related at each field point and the matrix-fill FLOP count is proportional to $(X_S)^2$. We note furthermore, that a DE matrix is very sparse, having mostly nonzero coefficients, whereas an IE matrix is dense, i.e., all of its coefficients are nonzero. The IE matrix for an electric-field integral equation (EFIE) is commonly called the *impedance* matrix, but we use the term system matrix as a generic name for the linear system of equations, whatever might be the physical units carried by its coefficients.

Solving the frequency-domain system matrix. The process of solving the system matrix can be implemented using a *direct* technique such as factorization or inversion, or using an *iterative* technique. A direct technique produces a RHS-independent solution matrix, whereas an iterative technique yields a solution for only a single RHS, i.e. only a solution vector, rather than a solution matrix. Direct techniques for IE matrices require a FLOP count proportional to $(X_S)^3$, whereas an iterative solution has an associated FLOP count that varies as $I(X_S)^2 \approx (X_S)^{2+x}$, where $0 \leq x \leq 1$, depending on the number of iterations, I, that are needed to achieve acceptable convergence. This dependency results because the number of FLOPs per iteration is proportional to $(X_S)^2$, the number of coefficients in the matrix. Iteration would normally not be a good choice except when $I \ll X_S$ and/or only a few RHSs are of interest, since otherwise the FLOP count can approach that required of a direct solution.

The FLOP-count requirements of direct-solution techniques for DE matrices can be reduced substantially when the *bandedness* of these matrices can be exploited by using banded-matrix solvers. For a one-dimensional problem, for example, where the matrix could have as few as three nonzero elements in a row, centered on the main diagonal, the matrix is tridiagonal with the solution FLOP count being proportional to X_S. For a two-dimensional (2D) problem, a bandwidth, B, results which is determined by the problem geometry and mesh-numbering scheme, but which generally would be on the order of $\sqrt{X_S}$, assuming approximately the same number of unknowns are used in the two spatial directions. Since the number of FLOPs required for solution of a banded matrix is proportional to $X_S(B)^2$, the 2D DE-model FLOP count grows as $(X_S)^2$. A three-dimensional problem similarly leads to a bandwidth proportional to $(X_S)^{2/3}$, resulting in a FLOP count that varies as $(X_S)^{7/3}$.

Solution of a DE system matrix using an iterative technique offers even further possibilities for reducing the FLOP count, assuming as before that the number of iterations is much less than the number of unknowns. Since there are only some number of nonzero coefficients, say, K, per row of the DE matrix, the FLOP count per iteration grows as KX_S, and the total FLOP count for a single RHS is proportional to IX_S.

Obtaining frequency-domain solutions for H right-hand sides. All iterative solutions provide a

RHS-dependent result, so that the total for H different RHSs will be the sum of their separate FLOP counts. For direct solutions, on the other hand, the FLOP count is dominated by factorization or inversion of the system matrix with subsequent computation of the unknown solution vector requiring fewer operations. In the case of IE models, each new solution requires of order $(X_S)^2$ operations in general, although there could be as few as X_S operations when using variable excitation points for antenna models for which there would be only one nonzero entry in the RHS vector. For the banded matrices of DE models, the corresponding FLOP count is proportional to $(X_S)^{(2D-1)/D}$. Therefore, unless a very large number of RHSs were to be used in the model, approaching X_S for IE models and $(X_S)^{(D-1)/D}$ for DE models, solution of the system matrix would always dominate the overall FLOP count.

Frequency-domain model complexity and summary. For iterative solutions of DE models when $I \ll X_S$, we thus obtain a model complexity given by $C \sim f^D + (H-1)S \sim Hf^D$. The corresponding result for a banded-matrix algorithm used in a DE model is $C \sim f^{3D-2}$. For an IE model solved using iteration when $I \ll X_S$ is true, we have $C \sim Hf^{2(D-1)}$ and $\sim Hf^{2D}$ respectively for homogeneous and inhomogeneous problems. When solving an IE matrix using factorization, this changes to being of order $C \sim f^{3(D-1)} + Hf^{2(D-1)}$ and $f^{3D} + Hf^{2D}$, respectively. The frequency dependence of FD DE and IE model complexities, when no approximations are used, thus spans f^3 (iterative, DE) to f^9 (LU decomposition, inhomogeneous, IE) for 3D problems, a tremendously wide range. For convenience, we summarize the frequency dependence of the various operations required of the FD models being considered in Table I below. Note that only the highest-order term is included everywhere except in the bottom line, where the largest FLOP count may depend on how many RHSs are computed.

TABLE I Frequency Dependence of X_S, N, and S for IE and DE Frequency-Domain Models

	IE Homogeneous	IE Inhomogeneous	DE
Sample Count X_S	f^{D-1}	f^D	f^D
Operation Count per Unknown N			
Iteration	f^{D-1}	f^D	*Constant*
LU Decomposition	$f^{2(D-1)}$	f^{2D}	$f^{2(D-1)}$
Total Operations for Single Solution			
Iteration	$f^{2(D-1)}$	f^{2D}	f^D
LU Decomposition	$f^{3(D-1)}$	f^{3D}	f^{3D-2}
Operations per Additional RHS S			
Iteration	$f^{2(D-1)}$	f^{2D}	f^D
LU Decomposition	$f^{2(D-1)}$	f^{2D}	f^{2D-1}
Total Operations for H RHSs, or Model Complexity			
Iteration	$Hf^{2(D-1)}$	Hf^{2D}	Hf^D
LU Decomposition	$f^{3(D-1)} + Hf^{2(D-1)}$	$f^{3D} + Hf^{2D}$	$f^{3D-2} + Hf^{2D-1}$

Modeling in the Time Domain

Time-domain models include time as an explicit variable in addition to space variables. As briefly mentioned above, this increases the overall spatial sample count by a factor of f, because a sequence of solutions are needed to develop a transient response over $X_t \sim f$ time steps. The DE and IE models can utilize *explicit* or *implicit* solutions. In explicit models, samples of the spatial solution at the *present* time step are wholly determined by the *present* value of the excitation and *past* sampled values of the solution, or response. In implicit models, conversely, some of the spatial samples at the present time step are permitted to interact. Whether a model is explicit or implicit depends on the size of the time step, δt, compared with the space-sample spacing, Δ, divided by the field-propagation speed, c. An explicit model requires that $\delta t \leq \Delta/c$, and is otherwise implicit. Explicit models are intrinsically easier to solve because the present response depends only on past, already-known values of the response, a direct consequence of causality.

Computing the time-domain system matrix. Because of causality, the system matrices of both the DE and IE models are diagonal for explicit models in the TD. This means that an algebraic solution is possible, i.e., one that does not require matrix solution via factorization or iteration. For the implicit model, not only does the DE approach produce a sparse matrix as in the FD, but the TD IE is also sparse. Furthermore, the coefficients of these system matrices in the TD are all real numbers whereas they are complex for FD models. The sparsity of the IE system matrix diminishes as δt increases, as this allows more space samples to interact within the same time step. However, an implicit DE model in the TD retains its sparsity as δt increases, being comparable with the FD counterpart whose time step, in effect, is infinite.

Because response values at all X_s space-sample locations contribute to the "scattered" field over the entire object being modeled, filling an IE TD system matrix requires computing $(X_s)^2$ interaction coefficients. The DE TD model on the other hand, requires computing KX_s matrix coefficients, where K is the average number of nonzero coefficients per field point. Not only does the IE model require X_s/K more coefficients to be computed, but they are generally more complicated to evaluate.

Solving the time-domain system matrix. For TD explicit models, solution of the diagonal system matrix is trivial, involving a FLOP count at each time step proportional to $(X_s)^2$ and X_s for IE and DE models, respectively, with each repeated X_t times. Implicit DE models yield system matrices similar to their FD versions, and so can be solved via factorization or iteration for the FLOP counts already indicated, i.e., $(X_s)^{(3D-2)/D}$ and X_s, respectively. For the former, there would be the additional cost of solving the factored matrix at each time step, which yields an additional FLOP count of $X_t(X_s)^{(2D-1)/D}$. The latter would entail a total FLOP count of X_tX_s, since the iterative solution must be repeated at each time step. Implicit IE models could also be solved using factorization or iteration, with the former involving X_t additional matrix-vector multiplications for a total FLOP count proportional to $(X_s)^3 + X_t(X_s)^2$. Since the system matrix is sparse for the implicit IE model, an iterative solution for the *interacting* space samples proportional to X_s is possible at each time step. However, adding in the effects of the time-retarded interactions remains of order $(X_s)^2$ at each time step, leading to an overall FLOP count proportional to $X_t(X_s)^2$.

Obtaining time-domain solutions for H right-hand sides. When solving TD models for additional RHSs, it is always necessary to repeat the time-stepping part of the solution, i.e., all terms which include X_t above would be multiplied by H. Only those operations needed for filling the system matrix and factoring it for non-iterative solutions are not redone.

Time-domain model complexity and summary. We summarize in Table II the frequency dependence of the various solution components involved in TD modeling.

TABLE II Frequency Dependence of X_s, N, and S for IE and DE Time-Domain Models

	IE Homogeneous	IE Inhomogeneous	DE
Spatial Sample Count X_s	f^{D-1}	f^D	f^D
Time Sample Count X_t	f	f	f
Operation Count per Unknown N			
Explicit	f^{D-1}	f^D	*Constant*
Implicit Iteration	f^{D-1}	f^D	*Constant*
Implicit LU Decomposition	f^{2D-3}	f^{2D}	f^{2D-3}
Total Operations for Single Solution			
Explicit	f^{2D-1}	f^{2D+1}	f^{D+1}
Implicit Iteration	f^{2D-1}	f^{2D+1}	f^{D+1}
Implicit LU Decomposition	$f^{3(D-1)}$	f^{3D}	f^{3D-2}
Operations per Additional RHS S			
Explicit	f^{2D-1}	f^{2D+1}	f^{D+1}
Implicit Iteration	f^{2D-1}	f^{2D+1}	f^{D+1}
Implicit LU Decomposition	f^{2D-1}	f^{2D+1}	f^{2D}
Total Operations for H RHSs, or Model *Complexity*			
Explicit	Hf^{2D-1}	Hf^{2D+1}	Hf^{D+1}
Implicit Iteration	Hf^{2D-1}	Hf^{2D+1}	Hf^{D+1}
Implicit LU Decomposition	$f^{3(D-1)} + Hf^{2D-1}$	$f^{3D} + Hf^{2D+1}$	$f^{3D-2} + Hf^{2D}$

COMPUTATIONAL IMPLICATIONS

The computational implications of the techniques discussed here in terms of the required computer resources to model a given problem, or the problem size that might be modeled for given computer resources, are worth examining. For this purpose, we have chosen as a generic target, a perfectly conducting, space-shuttle-size object. The numerical models to be compared include: 1) an IE solved using standard LU decomposition of the system matrix; 2) solution of an IE using iteration, assuming convergence is achieved after 100 and 10 iterations; 3) an IE solved using a near-neighbor approximation (NNA) (described further below) also assuming convergence after 100 and 10 iterations; 4) a finite-difference, time-domain DE model; and 5) a physical-optics approximation.

We present several different comparisons of these various models, as follows:

1) the frequency for which a solution can be obtained in one hour on a CRAY2 computer;

2) the solution time versus model used at a frequency of 4 GHz;

3) the solution time versus frequency;

4) projected one-hour solution frequency as a function of calendar year, based on anticipated growth in computer speed; and

5) a conceptual plot to illustrate the relationship between model complexity, computation bandwidth, and model formulation.

The following assumptions/conditions apply to the derivation of these results:

1) The effective CRAY2 FLOP rate is taken to be 10^{11} per hour based on its factoring a complex matrix of about 6,000 complex unknowns in an hour.

2) The reference, approximately "shuttle-size" target is assumed to have a total surface area of *540* m^2, which leads to 6,000 unknowns at a frequency of 100 MHz using 100 samples per square wavelength.

3) A sampling density of *100* unknowns per square wavelength is assumed for all models considered.

4) Computation time for the *100*- and *10*-step iteration models is based on achieving acceptable convergence in *100* and *10* iterations, respectively.

5) Computation time for the *100* by *100* and *10* by *10* NNA models is based on solutions involving *100* iterations and *100* interaction coefficients, and *10* iterations and *10* coefficients respectively, where the second number refers to the number of interaction coefficients retained in the impedance matrix.

6) Computation time for the physical-optics solution assuming *10* FLOPS are required per current sample.

7) No provision is made for the possible cost of I/O when using virtual memory for those problems that do not fit in core.

8) The possible effect of the system matrix becoming increasingly illconditioned as X_S increases has not been taken into account.

In Fig. 1, we present the frequency at which various approaches can model the backscatter radar cross section (RCS) for a single look angle of our generic target in one hour of CRAY2 time. The range of frequencies for which solutions are available in one hour spans more than three orders-of-magnitude. It should be observed that although LU decomposition is by far the "least productive" of the methods considered here, it would provide solutions for additional RHSs of about $1/X_S$ the FLOP count of the first solution, whereas all the other models would required the computation to be entirely repeated.

Presented in Fig. 2 is a plot of the CRAY2 computer time required to obtain the backscatter RCS at a single look angle for the same target at a frequency of 4 GHz, a problem solvable in one hour using the 100 by 100 NNA model. Also included in Fig. 2 is a comparison with one of the more widely used differential-equation models, Finite-Difference, Time-Domain (FDTD). Assuming that our generic target can be modeled in a cube *50* m on a side but excluding a center volume of *1000* m^3 occupied by the object, we have a space of *124,000* m^3 to be sampled. Assuming that the sample spacing $\Delta L = \lambda/10 = c\delta t$ where δt is the time step, then *6,000* samples per cubic wavelength are needed (for 3 electric fields and 3 magnetic fields). We further assume that the number of time steps needed for the FDTD model to reach steady state at a single frequency is $L/\Delta L = 25m/\Delta L = 250m/\lambda = L/c\delta t$, i.e., about one transit time of the object. Finally, we assume that the equivalent of a single multiply/divide is needed to update each of the fields per time step. The total number of operations then varies as ~

$2x10^{11}/\lambda^4$, which means that the one-hour CRAY2 run would be at about *120* MHz (Fig. 1) and that FDTD is comparable to an iterative, moment-method solution. It should be noted ,however, that being a time-domain model, FDTD is intrinsically capable of providing wideband information from a single calculation for impulsive excitation, thus potentially reducing the solution-time dependency to ~ f^3. Furthermore, if the closure boundary can be moved near the object being modeled, then the a further reduction in operation count might be achieved. On the other hand, there is some evidence [Cangellaris (1991)]that as the problem size increases, the spatial-sampling density may have to be increased to control field-propagation errors, with the result that $X_S \sim f^{D+x}$, rather than varying as f^D, where x is about unity for three-dimensional problems.

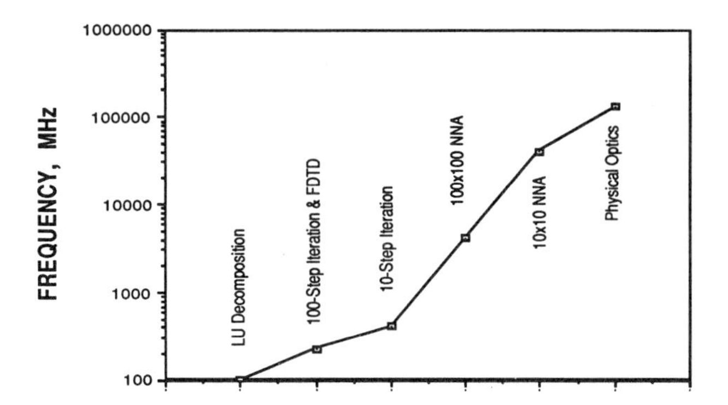

Fig. 1. One-hour solution frequency versus model used. Horizontal axis is used to separate various models whose accuracy could generally be expected to decrease with distance from the origin. Note that all of these models produce RHS-dependent results except for LU decomposition.

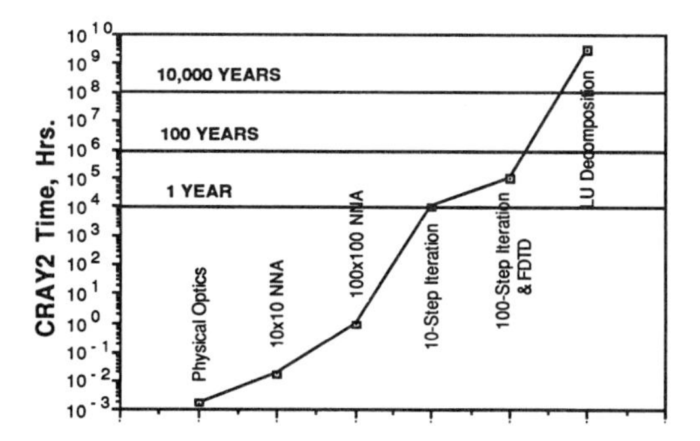

Fig. 2. Solution time versus model used for the perfectly conducting, space-shuttle-size object at 4 GHz. The time range spanned here is a factor of more than 10^{12}, emphatically demonstrating the influence of the frequency dependence of these various models.

In Fig. 3 we plot the CRAY2 time needed for each of these models to solve the generic problem as a function of frequency. The costs of increasing the model frequency are seen to be substantial, growing at a minimum rate of f^2 for the PO and NNA models, f^4 for the iteration solutions, and f^6 for the moment-method model solved using LU decomposition without any approximations.

Although the CRAY2 offers an effective computation bandwidth of about 10^{11}/hour, future supercomputers can be anticipated to continue the approximate two-order-of-magnitude improvement per decade that has prevailed since the introduction of Univac1 about 1950. However, whereas these speed increases have been largely due to use of faster components as tube technology gave way to transistors, which were in turn replaced by integrated circuits of greater and greater density, future speed growth will increasingly depend on improvements in computer architecture. Parallelism is the key to increasing computation bandwidth, ranging from the modest degree of parallelism of the Hypercube architecture that has a few hundreds of complex CPUs to the massive parallelism of the Connection Machine architecture that can have many thousands of simpler CPUs. Presently, both kinds of parallel machines are factoring or inverting complex matrices of *20,000* unknowns in a few hours, and projections are for computation bandwidths of 1 *TeraFLOP* by 1995 and *100 TeraFLOPS* by the year 2,000 [Ratner (1989)].

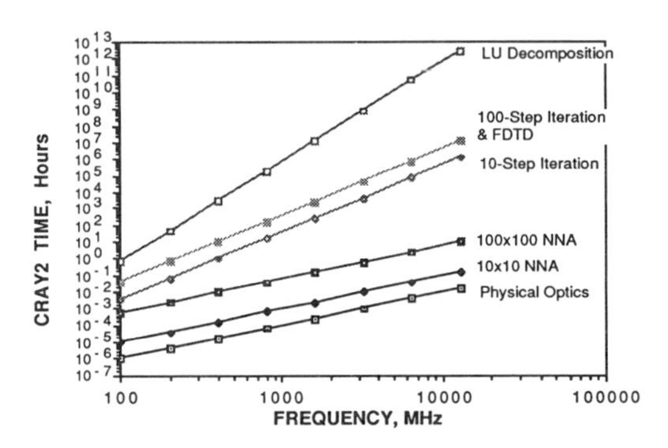

Fig. 3. Solution time versus frequency for various models. While it should be apparent that computer times exceeding 100 or 1000 hours are unrealistic, this graph is included to dramatize the time requirements of even modest problems as the frequency approaches the GHz range.

Based on these projections and assuming the historical growth rate continues to 2050, the centenary of the Univac1, we obtain the results plotted in Fig. 4. Although by 2050 the per hour throughput would be 3.6×10^{26} per hour (!), the frequency of our 1-hour Shuttle LU-decomposition solution would only be about 20 GHz or so, emphasizing the need for alternate formulations and models that are less operation-count intensive.

This point is made graphically in Fig. 5, where, borrowing from a similar plot developed by Jameson (1989) to illustrate the impact of computer speed and model formulation on aerodynamics modeling, we show the same idea in an electromagnetics context. Clearly, as model fidelity and problem complexity increase, the computing requirements can become formidable, so that even an unimaginable 10^{26} FLOPs per hour would be strained by computing costs that increase as f^6 (or even faster).

SOME POSSIBILITIES FOR REDUCING THE FLOP COUNT IN CEM

In the following, we briefly discuss some ways by which the problem complexity might be decreased through reducing X_S, N, S and H via formulation and algorithm development. As problem size in wavelengths increases, we have observed that the number of spatial samples or unknowns X_S can grow as fast as f^D where D is the dimension of the sampling space, with an additional factor of f occurring when modeling in the time domain. Clearly, the most fruitful approach for modeling with larger values of X_S is to reduce the impact of the terms which determine the overall computer time for IE and DE models, given approximately by

$$T_{IE} \sim A_{1IE}X_S + A_{2IE}(X_S)^2 + A_{3IE}(X_S)^3, \tag{2a}$$

$$T_{DE} \sim A_{1DE}X_S + A_{3DE}\,(X_S)^{(3D-2)/D} \tag{2b}$$

where A_1, A_2 and A_3 are algorithm- and computer-dependent coefficients and account for the various factors contributing to the operation count required per sample. The last terms in each equation account for matrix solution via factorization, where a banded procedure is assumed to be used for the DE model. Since the square (matrix fill) and factor terms must eventually dominate the IE solution time as X_S becomes arbitrarily large, any attempt to model larger problems must confront those dependencies and try to minimize or eliminate their impact. This is one reason why iterative methods, the conjugate-gradient technique (CGT) for example, have become widely studied for IE modeling. They offer, without approximation, the tantalizing possibility of reducing the highest-order term to $(X_S)^2$, although at the expense of making the solution RHS dependent, i.e., for a scattering application, a new angle of incidence requires an entirely new computation. The CGT has the advantage, however, that it offers a numerically-rigorous solution technique, subject only to roundoff during the course of the computation. The alternative, which affects all terms in Eq. (2), is to reduce the number of unknowns needing solution. We briefly examine the impact of some possible methods that might reduce the overall computing resources required, as summarized in Table III. Further, more detailed discussion of this topic can be found in Miller (1991).

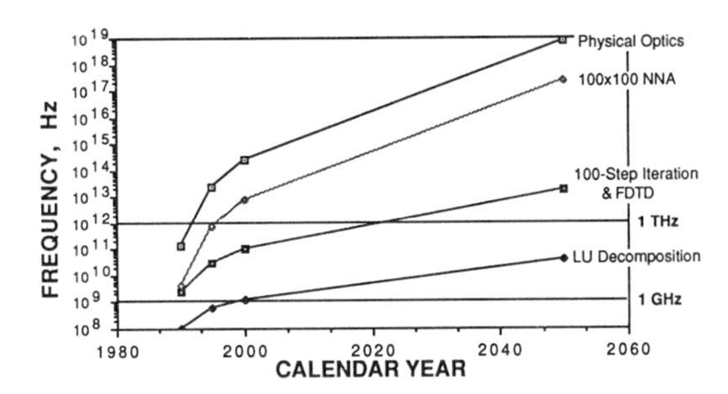

Fig. 4. Projected 1-hour solution frequency as a function of calendar year for models of 3D perfectly conducting bodies which vary as f^2 (PO and NNA), f^4 (iteration and FDTD), and f^6 (LU decomposition).

Reducing the Spatial-Sample Count

The spatial-sample count in DE and IE models in electromagnetics is driven by the wave-nature of the sources and fields described by Maxwell's equations. The sampling rate and, consequently, the total number of samples needed for a given problem are thus determined primarily by the size of the solution space measured in wavelengths. This means that the number of spatial samples must ultimately scale in proportion to f^D for models which require volumetric sampling in D dimensions. Therefore, unless a sampling procedure can be devised which decreases the frequency exponent of the spatial-sample count, merely reducing the sampling density can only decrease the total number of samples for a given problem but not the growth rate as frequency increases. Although there may be some possibilities for reducing the frequency exponent of the DE and IE model operation dependencies (one is described below), the opportunities for doing so appears rather limited as compared with an approach such as the GTD (Geometrical Theory of Diffraction) techniques whose sampling rate is driven by geometrical complexity rather than electrical size.

TABLE III Effect of Various Techniques for Reducing the IE Operation Count

IMPACT ON----> TECHNIQUE	A_1X_S Term	$A_2(X_S)^2$ Term	$A_3(X_S)^3$ Term
Reduced Spatial Sample Count (using $Y_S < X_S$ samples)	$x\ (Y_S/X_S)$	$x\ (Y_S/X_S)^2$	$x\ (Y_S/X_S)^3$
MBPE in Filling Impedance Matrix	–	Reduce A_2	–
Iterative solution	–	–	$x\ (I/X_S)$
Iterative solution with MBPE (where $I' < I$)	–	–	$x\ (I'/X_S)$
NNA alone	–	$x\ (1/X_S)$	Possible sparse-matrix solution
NNA with iterative solution	–	$x\ (1/X_S)$	$x\ (1/X_S)^2$

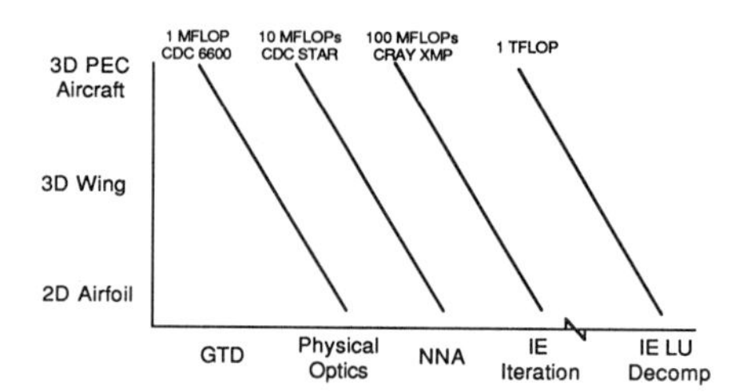

Fig. 5. Conceptual plot after Jameson (1989) to illustrate qualitative relationship between problem complexity, model formulation, and computational bandwidth.

Reducing the spatial sample count nevertheless can offer a significant reduction in overall operation count, one possibility being use of adaptive spatial sampling. This idea is routinely employed in DE models, where finite-element meshes of highly variable element size and shape are used in regions needing more detailed geometrical definition or having spatially rapid variations of electrical parameters. A similar technique would seem likely to offer computational advantages in IE modeling using subsectional bases. One approach would be to use variable patch or segment sizes in modeling surfaces and wires such the some measure of model quality, the integrated boundary-field error for example, would be minimized for a given number of unknowns. Little work has been reported on this kind of approach, one exception being due to Tabet et. al (1990). Development of IE models for surfaces and wires having the flexibility of DE finite-element meshing is an area that seems worth pursuing.

One other way of reducing the spatial sample count might be use of Model-Based Parameter Estimation [MBPE, Miller and Burke (1989)], in connection with increasing X_S as a means of seeking a converged solution. Convergence studies usually examine the behavior of an observable of interest, or some error measure associated with that observable, as X_S is systematically increased, usually by uniformly decreasing the spatial-sample size [Miller, et. al. (1971), Butler and Wilton (1975)], yielding a sequence of results that might be compared with the successive iterates produced during an iteration process. Another way of viewing the sequence of solution values with increasing X_S might be as a sort of generalized "time" series or "convergence" waveform. In this interpretation, an "acceleration" in convergence, to an estimate of a converged answer, is sought by applying some suitable signal-processing model, i.e. MBPE, to the convergence waveform, with the goal of estimating the converged answer without having to increase X_S to extent that might otherwise be necessary. One MBPE approach that has been found useful in computing Sommerfeld integrals [Lager and Lytle (1975)] is Shanks' Method [Shanks (1955)] which was developed specifically for application to problems that, in Shanks' words exhibit "physical" or "mathematical" transients. A key to successfully using such a technique is finding a suitable model, preferably one that is physically based, so that it can represent the mathematical transient with acceptable accuracy while using a minimum of parameters requiring estimation.

Reducing the Operation Count per Sample in IE Models

Reducing the operation count per sample might involve a variety of approaches. One approach using MBPE, is to reduce the operation count of filling the moment-method impedance matrix by simplifying computation of the interaction coefficients. MBPE might also be used for developing better starting values for use in an iterative solution with the goal of reducing the number of iterations required to achieve an acceptably converged solution. Another method is that of employing the Near-Neighbor Approximation (NNA) or its equivalent, which reduces both the operation count of filling the impedance matrix and its subsequent solution using sparse- or iterative-solution procedures. In order to place the potential computational advantages of the NNA in perspective, we discuss it below as one among a number of related approaches ranging from highly approximate, such as physical optics, to a complete moment-method model solved using standard matrix techniques.

Reducing the Operation Count per Sample in DE Models

Because DE models naturally produce matrices that are extremely sparse, there being usually just a few nonzero coefficients per row of the linear system into which the problem is discretized, they provide an opportunity for exploiting this sparseness by using special solution procedures. One of the more commonly used for this purpose is a banded-matrix algorithm, which takes advantage of the fact that there is a regularity to the pattern of nonzero coefficients that occur in the DE matrix. The nonzeros are confined to a "band" of width $B = X_S^{(D-1)/D}$ centered about the main diagonal making possible a factored solution of the linear system to be obtained in an operation count proportional to $X_S B^2$ (or f^4 and f^7 for 2D and 3D problems as shown in Table I) rather than the usual $(X_S)^3$ required of a full matrix. An iterative solution procedure can also be used for the banded DE matrix, offering a potential further reduction in the operation count, but at the expense of developing a right-hand-side dependent solution.

A recently reported alternate approach [Miller and Gilbert (1991)] also exploits the bandedness of the DE matrix but avoids the fill-in of coefficients that are originally zero within the band that otherwise occurs when using a standard banded algorithm. This approach further reduces the operation count of the DE-matrix solution to produce right-hand-side independent solutions having on the order f^{-1} fewer operations than the standard banded approach, to become of order f^3 and f^6 respectively in two and three dimensions. This is achieved by successively "propagating" linearly-independent starting values for the fields at one boundary of the spatial mesh to the other boundary (e.g., the radiation-closure boundary and the surface of a perfectly conducting object coated with inhomogeneous dielectric), and solving the resulting "reduced" matrix at the second boundary.

Reducing the Number of Model Evaluations

The approach described above for using MBPE to begin an iterative solution with a better initial "guess" or for reducing the impedance matrix fill time is one that can be used as well to decrease the number of model evaluations needed to obtain a transfer function or scattering pattern. This approach has been described by Burke et al. (1989) for obtaining EM transfer functions from at most two frequency samples per resonance by using a rational function as a model. The basic idea is to exploit knowledge of electromagnetic physics to reduce to a minimum the number of samples in frequency, angle, spatial coordinate, etc., for which the detailed moment-method model must be run. For the transfer-function application, the physics being exploited is knowledge of the fact that the frequency response is well approximated by a pole series, for which a more general form is a rational function. Not only frequency samples, but samples of frequency derivatives can be used for this purpose, the advantage being that computing a frequency-derivative sample requires a FLOP count proportional to $(X_S)^2$ rather than the $(X_S)^3$ cost of the frequency sample itself.

DISCUSSION AND CONCLUDING REMARKS

In the above discussion we have characterized computational electromagnetics from two viewpoints. First, we developed a measure of CEM model complexity from the viewpoint of the total number of computational operations needed to acquire the information sought when using that model, and then discussed a variety of approaches by which the operation count might be reduced. Second, we examined the present and anticipated future capabilities of various modeling approaches when applied to a reference problem in terms of the expected computational bandwidths that might be seen in the future. The point to be made was that the total operation count can grow so rapidly with increasing frequency that even orders of magnitude speedups in computer FLOP rate translate to only small increases in the frequency for which a given problem can be solved in a given clock time. To be more specific, integral-equation models of perfectly conducting 3D objects in the frequency domain when solved using LU matrix decomposition require an operation count proportional to f^6, so that a one-million times speed increase in computer throughput leads to only a times-ten increase in frequency. For an iterative solution of the same model and problem, or a finite-difference time-domain approach, the corresponding frequency dependence is on the order of f^4, so that the one-million speed increase produces a times-thirty frequency increase. While the set of problems for which numerical techniques are practically useful continues to grow, it is clear that these computational realities will continue to limit what can be modeled for the foreseeable future when using present IE- and DE-based numerically rigorous models.

It is also worth noting that the computer speed increases anticipated in the near future rely for the most part on using an increasing degree of parallelism in computer design. When mapping these new designs to the present numerical models, can probably expect that truly massive parallelism, where hundreds of thousands to multi-million simple microprocessors are employed in a single machine, will be better suited to DE models for which a single processor might handle one to several nodes in the model mesh. On the other hand, more moderate parallelism employing more sophisticated microprocessors will probably be better suited to the kinds of matrices which arise from IE models. Possibly new kinds of computer designs and strategies will also evolve, one of potential for wave-equation problems potentially being based on optical designs. But one of the ways by which the modeling of larger, more complex problems is likely to become more feasible will be through alternate formulations and computational approaches which offer reductions in operation count of the kinds discussed here.

REFERENCES

Burke, G. J., E. K. Miller, S. Chakrabarti, and K. Demarest (1989), "Using model-based parameter estimation to increase the efficiency of computing electromagnetic transfer functions", *IEEE Trans. Magnetics*, **25**(4), pp. 2807-2809.

Butler, C. M. and D. R. Wilton (1975), "Analysis of various numerical techniques applied to thin-wire scatterers, *IEEE Trans. Antennas and Propagat.*, **AP-22**, pp. 534-540.

Cangellaris, A. C. (1991), "Time-domain finite methods for electromagnetic wave propagation and scattering," to be published in *IEEE Trans. on Magnetics.*

Jameson, A. (1989), "Computational aerodynamics for aircraft design," *SCIENCE*, **245**, pp. 361-371.

Lager, D. and R. J. Lytle (1975), "Fortran subroutine for the numerical evaluation of Sommerfeld integrals unter andem," Lawrence Livermore Laboratory Report, UCRL-51821.

Miller, E. K. (1991), "Solving bigger problems--by decreasing the operation count and increasing the computation bandwidth," submitted to special issue of *IEEE Proc. on Electromagnetics.*

Miller, E. K., G. J. Burke and E. S. Selden (1971), "Accuracy-modeling guidelines for integral-equation evaluation of thin-wire scattering structures," *IEEE Trans. Antennas Propagat.*, **AP-19**, pp. 534-536.

Miller, E. K. and M. A. Gilbert (1991), "Solving the Helmholtz equation using multiply-propagated fields," to be published in *International Journal of Numerical Modeling.*

Ratner, J. (1989), "Towards the teraflop machine", presented at Workshop on Highspeed Simulation and Visualization, California Institute of Technology, June 19-21, Pasadena, CA.

Shanks, Daniel (1955), "Nonlinear transformations of divergent and slowly convergent sequences," *Journal of Mathematics and Physics*, The Technology Press, Massachusetts Institute of Technology, Cambridge, MA, pp. 1-42.

Tabet, S. N., J. P. Donohoe and C. D. Taylor (1990), "Using nonuniform segment lengths with NEC to analyze large wire antennas," 6th Annual Review of Progress in Applied Computational Electromagnetics, Naval Postgraduate School, Monterey, CA.

STATE OF THE ART AND FUTURE DIRECTIONS IN FINITE-DIFFERENCE AND RELATED TECHNIQUES IN SUPERCOMPUTING COMPUTATIONAL ELECTROMAGNETICS

Allen Taflove

Department of Electrical Engineering and Computer Science
McCormick School of Engineering
Northwestern University
Evanston, IL 60208-3118

INTRODUCTION

The numerical modeling of electromagnetic (EM) phenomena can be a computationally intensive task. To date, the design and engineering of aerospace vehicles has been the primary application driving the development of large-scale methods in computational electromagnetics (CEM). Efforts in the area have been aimed primarily at minimizing the radar cross section (RCS) of aerospace vehicles. RCS minimization enhances the survivability of vehicles that are subjected to precision-targeted ordnance. The physics of RCS is determined by Maxwell's equations and the constitutive properties of a vehicle's materials. As a result, an interesting situation arises in which the effectiveness and cost of state-of-the-art aerospace systems in part depends upon the ability to develop an efficient engineering understanding of 120-year old equations that describe the propagation and scattering of EM waves.

This paper summarizes the present state and future directions of applying finite-difference and finite-volume techniques for Maxwell's equations on supercomputers to model complex EM wave interactions with structures, including but not limited to RCS. A number of topics are selected as focus areas.

1. Background

   a. The state of the full-matrix method of moments (or, why we should consider alternative Maxwell's equations solvers)

   b. Development of the finite-difference time-domain method

2. Space-grid time-domain codes

   a. General ideas behind these approaches

   b. Algorithms and meshes. Which is best? Or, does the optimum algorithm/mesh depend upon the problem?

Directions in Electromagnetic Wave Modeling
Edited by H.L. Bertoni and L.B. Felsen, Plenum Press, New York, 1991

c. Dynamic range. We have solid evidence for 40-dB dynamic ranges of existing codes. Can we get to 60 dB or more?

d. Physics of dispersive materials. Can we handle physical dispersion for key classes of materials?

3. Computer resource and architecture issues

a. Vectorization and concurrency, processing speed and storage requirements for FD-TD;

b. Is there one best architecture for supercomputers implementing grid-based Maxwell's equations solvers?

c. Must we bend our EM simulation algorithms to fit the computer architecture? Does Fortran have any place in programming machines having many processors?

4. Horizons through the mid 1990's.

a. Computer processing speeds

b. New applications for space-grid time-domain Maxwell's solvers in high-speed integrated electronics and optics

c. Supercomputing personal workstations

BACKGROUND

The State of the Full-Matrix Method of Moments

The modeling of engineering systems involving EM wave interactions has been dominated by frequency-domain integral equation techniques and high-frequency asymptotics. This is evidenced by the almost universal use of the method of moments (MoM)[1,2] to provide a rigorous boundary value analysis of structures, and the geometrical theory of diffraction (GTD)[3,4] to provide an approximate analysis valid in the high-frequency limit.

However, a number of important contemporary problems in EM wave engineering are not adequately treated by such models. Complexities of structure shape and material composition confound the GTD analysts; while structures of even moderate electrical size (spanning five or more wavelengths in three dimensions) present very difficult computer resource scaling problems for MoM.

The latter problem is particularly serious since MoM has provided virtually the only means of dealing with the non-metallic materials now commonly used in aerospace design. To illustrate this problem, consider the present state-of-the-art of applying Cray Research supercomputing hardware and software to full-matrix MoM solutions. Cray analysts have developed an asynchronous, out-of-core, complex LU decomposition matrix solver (with partial pivoting) specifically for ultra-large-matrix MoM problems[5]. When automatically multitasked to run on all eight processors of the Cray Y-MP/8 system, this solver runs at an average computation rate exceeding 2.1 Gflops.

Only 1.99 hours are needed to fully process a 20K x 20K MoM matrix. During this run, 138 GBytes of I/O are discharged to and from seven DD-40 disk drives. Yet, only 228 seconds (3.8 minutes) represent I/O wait time. In fact, 90% of the actual I/O operations are performed concurrently with the floating point arithmetic by virtue of the asynchronous I/O scheme, and therefore do not contribute to the observed wait time. As matrix size increases, the relative efficiency of the asynchronous scheme *improves*, with the I/O concurrency factor rising to 95% for the 40K matrix. This shows that the massive I/O associated with solving huge, dense, complex-valued MoM matrices can be almost completely buried.

Consider the RCS modeling capabilities implied by this highly efficient Cray matrix solver. We first define d_{span} as the characteristic span of a flat-configured target, and λ_o as the wavelength of the impinging radar beam. The electrical surface area of the two sides of the target (in square wavelengths) is given by $2(d_{span}/\lambda_o)^2$. Assuming a standard triangular surface patching implementation of the electric field integral equation, with the surface area of the target discretized at R divisions per λ_o, the number of MoM triangular surface patches is given by $4R^2(d_{span}/\lambda_o)^2$; and N, the number of field unknowns, is given by $6R^2(d_{span}/\lambda_o)^2$. In fact, N is the order of the MoM system matrix subject to LU decomposition. With R usually taken as 10 or greater to properly sample the magnitude and phase of the induced electric current distribution on the target surface, we see that N rises above 10K for d_{span} greater than $4\lambda_o$. Further, we see that N increases quadratically as λ_o drops (radar frequency increases), and the computational burden for LU decomposition increases to the *sixth power* of frequency (order of N^3).

To see what this means, assume that d_{span} is in the order of 16 meters, perhaps typical of a modern jet fighter. N rises above 10K for λ_o in the order of 4 meters, a radar frequency of 75 MHz, with a measured Cray Y-MP/8 running time of 0.24 hour. This is quite tolerable. But, N rises above 40K at a radar frequency of 150 MHz, and the Y-MP/8 running time balloons to a measured value of 16.8 hours, a bit more than a 64-fold (2^6) increase. This is less tolerable. At 300 MHz, the Y-MP/8 running time leaps to an estimated 1100 hours (45 days), and at 600 MHz to an estimated *8 years*.

--- The latter assumes that: (a) we have enough disk drives to store the trillion-word MoM matrix; (b) we find acceptable the error accumulation resulting from the *million-trillion* floating-point operations working on individual MoM matrix elements having precisions of only perhaps 1 part per ten-thousand; and (c) that the computer system stays up *continuously* for 8 years.

It is quite clear that the traditional, full-matrix MoM computational modeling of an entire aerospace structure such as a fighter plane is presently impractical at radar frequencies much above 150 MHz, despite advances in supercomputer hardware and software. Unfortunately, radar frequencies of interest go very much higher than 150 MHz, in fact, up to 10 GHz or more. Much research effort has been expended in deriving alternative iterative frequency-domain approaches (conjugate gradient, spectral, etc.) that preserve the rigorous boundary-integral formulation of MoM while realizing dimensionally reduced computational burdens. These would permit, in principle, entire-aircraft modeling at radar frequencies well above 150 MHz. However, these alternatives may not be as robust as the full-matrix MoM, that is, providing results of engineering value for a wide class of structures without the user wondering if the algorithm is converged.

Development of the Finite-Difference Time-Domain Method

Since 1973, I have worked to establish the finite-difference time-domain (FD-TD) method as an alternative to MoM for rigorous, detailed modeling of EM wave interactions with structures. FD-TD is the first approach to obtain results for such problems by directly integrating Maxwell's time-dependent curl equations. It implements the spatial derivatives of the curl operators using finite differences in a regular Cartesian space mesh, as proposed in 1966 by Yee[6], modifying only the space cells adjacent to a structure to conform with the exact structure shape. It employs simple leapfrog time integration[6]. The rationale for investigating FD-TD (and subsequent alternative grid-based Maxwell's equations solvers) has been that it conceptually permits modeling much larger problems than MoM because of dimensionally reduced computational burdens. This will be explored later.

Over the years, significant developments in FD-TD theory and applications include the following:

1. The correct numerical stability criterion for Yee's algorithm, and the first grid-based time-integration of an EM scattering problem all the way to the sinusoidal steady state[7];

2. The first 3-D grid-based computational model of EM wave absorption in complex, inhomogeneous biological tissues[8];

3. Coining of the term "FD-TD," and the first validated FD-TD models of EM wave penetration into a 3-D metal cavity[9];

4. An accurate and numerically stable second-order radiation boundary condition for the Yee grid[10];

5. The first grid-based EM wave scattering models computing near fields, far fields, and RCS for 2-D structures[11] and 3-D structures[12];

6. Contour-path, subcell techniques to permit FD-TD modeling of EM wave coupling to thin wires and wire bundles[13]; EM wave penetration through cracks in conducting screens[14]; and conformal surface treatments of curved structures[15];

7. Introduction of modern radiation boundary condition theory to the engineering EM community[16,17];

8. Validated FD-TD models of antennas[18,19];

9. Application to a wide range of EM bioeffects and hyperthermia problems for realistic models of the entire human body[20–23];

10. Validated FD-TD models of media having first and second-order dispersions of permittivity with frequency[24,25]; and

11. Application to high-speed interconnects and microstrips[26–28], high-speed electronic and optoelectronic switches[29–31], and integrated optics[32].

In the section below, the characteristics of FD-TD and the major competing approach, finite-volume time-domain (FV-TD), will be discussed.

SPACE-GRID TIME-DOMAIN CODES

The problems involved in applying frequency-domain, full-matrix MoM technology to large scale RCS modeling have prompted a rapid recent expansion of interest in an alternative class of *non-matrix* approaches: direct space-grid, time-domain solvers for Maxwell's time-dependent curl equations. These approaches appear to be as robust and accurate as MoM, but have dimensionally-reduced computational burdens to the point where whole-aircraft modeling for RCS can be considered in the near future. Currently, the primary approaches in this class are the finite-difference time-domain (FD-TD) method (see summary above) and finite-volume time-domain (FV-TD) techniques[33,34]. These are analogous to existing mesh-based solutions of fluid-flow problems in that the numerical model is based upon a direct, time-domain solution of the governing partial differential equation. Yet, FD-TD and FV-TD are *very* nontraditional approaches to CEM for detailed engineering applications, where frequency-domain methods (primarily full-matrix MoM) have dominated.

Basis

FD-TD and FV-TD are direct solution methods for Maxwell's equations. They employ no potentials. Rather, they are based upon volumetric sampling of the unknown near-field distribution (E and H fields) within and surrounding the structure of interest, and over a period of time. The sampling in space is at sub-wavelength (sub-λ_0) resolution set by the user to properly sample (in the Nyquist sense) the highest near-field spatial frequencies thought to be important in the physics of the problem. Typically, 10 to 20 samples per illumination wavelength are needed. The sampling in time is selected to insure numerical stability of the algorithm.

Overall, FD-TD and FV-TD are marching-in-time procedures which simulate the continuous actual EM waves by sampled-data numerical analogs propagating in a computer data space. Time-stepping continues as the numerical wave analogs propagate in the space grid to causally connect the physics of all regions of the target. All outgoing scattered wave analogs ideally propagate through the lattice truncation planes with negligible reflection to exit the region. Phenomena such as induction of surface currents, scattering and multiple scattering, aperture penetration, and cavity excitation are modeled time-step by time-step by the action of the curl equations analog. Self-consistency of these modeled phenomena is generally assured if their spatial and temporal variations are well resolved by the space and time sampling process. In fact, the goal is to provide a self-consistent model of the mutual coupling of all of the electrically-small volume cells comprising the structure and its near field, even if the structure spans tens of λ_o in 3-D and there are tens of millions of space cells.

Time-stepping is continued until the desired late-time pulse response or steady-state behavior is observed. An important example of the latter is the sinusoidal steady state, wherein the incident wave is assumed to have a sinusoidal dependence, and time-stepping is continued until all fields in the sampling region exhibit sinusoidal repetition. This is a consequence of the limiting amplitude principle[35]. Extensive numerical experimentation has shown that the number of complete cycles of the incident wave required to be time-stepped to achieve the sinusoidal steady state is a function of two factors:

1) *Target electrical size.* For many targets, this requires a number of time steps sufficient to permit at least two front-to-back-to-front traverses of the target by a wave analog. For example, assuming a target spanning $10\lambda_o$, at least 40 cycles of the incident wave should be time-stepped to approach the sinusoidal steady state. For a grid resolution of $\lambda_o/10$, this corresponds to 800 time steps for FD-TD.

2) *Target Q factor.* Targets having well-defined low-loss cavities or low-loss dielectric compositions may require the number of complete cycles of the incident wave to be time-stepped to approach the Q factor of the resonance. Because Q can be large even for electrically moderate size cavities, this can dictate how many time steps the FD-TD or FV-TD code must be run.

In the RCS area, target electrical size may often be the dominant factor. Cavities for RCS problems (such as engine inlets) tend to be open, and therefore moderate Q; and the use of radar-absorbing material (RAM) serves further to reduce the Q factors of structures.

Algorithms and Meshes

The primary FD-TD and FV-TD algorithms used today are fully explicit second-order accurate grid-based solvers employing highly vectorizable and concurrent schemes for time-marching the six vector components of the EM near field at each of the volume cells. The explicit nature of the solvers is maintained by either leapfrog or predictor-corrector time integration schemes. Second-order accurate radiation boundary conditions are used to simulate the extension of the problem space to infinity, thereby minimizing error due to the artifactual reflection of outgoing numerical wave modes at the mesh outer boundary.

Present methods differ primarily in how the space grid is set up. In fact, gridding methods can be categorized according to the degree of structure or regularity in the mesh cells:

1) Almost-completely structured-- Space cells more than one or two cells from the structure of interest are organized in a completely regular manner, for example, using a uniform Yee cartesian mesh. Only the cells adjacent to the structure are modified in size and shape to conformally fit the structure surface. This approach is used by my group in our FD-TD codes, and has also been reported by Madsen and Ziolkowski[34] in their finite-volume research.

-- Advantage: Computationally efficient, because there are relatively few modified cells requiring special storage to locate the cells in the mesh and special computations to perform the field updates. In fact, the number of modified cells (proportional to the surface area of the structure) becomes arbitrarily small compared to the number of regular mesh cells as structure size increases. As a result, the computer memory and running time needed to implement a fully conformal model can be virtually indistinguishable from that required for a stepped-surface model. Further, an ultra-fast, minimal memory, Yee-like algorithm can be used for the regular space cells.

-- Advantage: Artifacts due to refraction and reflection of numerical wave modes propagating across global mesh distortions are not present. This is especially important for 3-D structures having substantial EM coupling between electrically disjoint sections, or reentrant regions.

-- Disadvantage: Geometry generation software is not available and must be constructed from the ground up. This is completely nontrivial.

-- Disadvantage: Thin target surface coatings do not conform to mesh boundaries. These are best modeled here using surface impedances.

2) Body-fitted-- The space grid is globally distorted to fit the shape of the structure of interest. Curvilinear space cell boundaries are pre-computed by solving a potential problem: with the structure assumed to be held at a dc potential relative to infinity, the cell surfaces represent intersecting equipotential and gradient loci. Currently, the most advanced work in this area is being reported by Shankar et al[33] at Rockwell International.

-- Advantage: Well-developed geometry generation software is immediately available from the computational fluid dynamics community. Aerodynamic shapes appropriate for the RCS problem are nicely handled.

-- Advantage: Thin target surface coatings naturally conform to mesh boundaries, and are easily modeled.

-- Disadvantage: Relative to the baseline Yee algorithm, extra computer storage must be allocated to account for the 3-D position and stretching factors of each space cell. Further, extra computer arithmetic operations must be performed to implement Maxwell's equations at each cell and/or to enforce EM field continuity at the interfaces of adjacent cells. As a result, the number of floating point operations needed to update the six field components at a space cell over one time step can exceed that of the Yee algorithm by as much as 20:1, thereby increasing running times by the same amount with respect to FD-TD.

-- Disadvantage: Artifacts due to refraction and reflection of numerical wave modes propagating across global mesh distortions will be present. These errors arise because the phase velocity of numerical wave modes propagating in the mesh is a function of position in the mesh, as well as angle of propagation. These artifacts are especially important for 3-D structures having substantial EM coupling between electrically disjoint sections, or reentrant regions.

3) Completely unstructured-- The space containing the structure of interest is completely filled with a collection of solid cells of varying sizes and shapes, but conforming to the structure surface.

-- Advantage: Geometry generation software is available. This software is appropriate for modeling complicated 3-D shapes possibly having volumetric inhomogeneities, for example, inhomogeneous radar absorbing material.

-- Disadvantages: The same as for the body-fitted meshes.

At present, the choice of computational algorithm and mesh is not at all straightforward. Clearly, there are important tradeoff decisions to be made. For the next several years, we can expect considerable progress in this area as various groups develop their favored approaches and perform validations.

Predictive Dynamic Range

For computational modeling of the RCS of aerospace vehicles (especially mitigated-RCS vehicles) using space-grid time-domain codes, it is useful to consider the concept of predictive dynamic range. Perhaps for lack of a previous published definition, I shall state my own:

--- Let the power density of a modeled incident plane wave in the space grid be P^{inc} W/m^2. Let the "minimum observable" local power density of a modeled scattered wave at any bistatic angle be $P^{scat, min}$ W/m^2, where "minimum observable" means that the accuracy of the scattered field computation degrades to poorer than 1 dB (or some other accuracy criterion) at lower levels than $P^{scat, min}$. Then, let us define the *predictive dynamic range* as $10 \log(P^{inc}/P^{scat, min})$ decibels.

This definition seems nicely suited for space-grid time-domain codes for two reasons: (1) It squares nicely with the concept of a "quiet zone" in an experimental anechoic chamber, which is intuitive to most EM engineers; and (2) It succinctly quantifies the reality that weak (but physical) numerical wave analogs propagating in the space grid exist in an additive noise environment due to nonphysical propagating wave analogs caused by the imperfect radiation boundary conditions. In addition to additive noise, the desired physical wave analogs undergo gradual progressive deterioration while propagating due to accumulating numerical dispersion artifacts, including phase velocity anisotropies and inhomogeneities within the mesh.

Over the past ten years, researchers have accumulated solid evidence for a predictive dynamic range in the order of 40 dB for existing space-grid time-domain codes. This value is reasonable if one considers the additive noise due to imperfect radiation boundaries to be the primary limiting factor, since existing second-order radiation boundary conditions yield effective reflection coefficients of about 1% (-40 dB), with an additional factor of perhaps -10 dB provided by the normal $r^{-1/2}$ rolloff (in 2-D) or r^{-1} rolloff (in 3-D) experienced by the outgoing scattered waves before reaching the radiation boundaries. Examples of 1 dB accuracy over 42-dB ranges of monostatic and bistatic RCS are given in Figs. 1 and 2 for the FD-TD method. Fig. 2 is especially noteworthy in that it shows that the range of FD-TD accuracy is not degraded by using partial mesh unstructuring to conform to curved target surfaces.

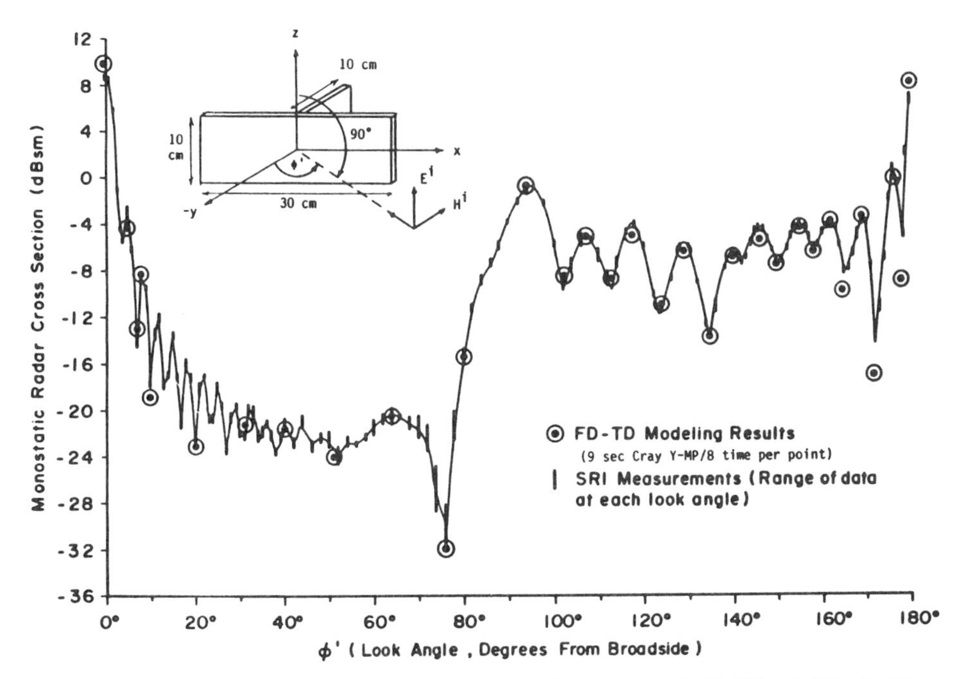

Fig. 1. Agreement of FD-TD and measured monostatic RCS within 1 dB over a 42-dB range for a $9\lambda_o$ x $3\lambda_o$ x $3\lambda_o$ conducting T-shaped target.

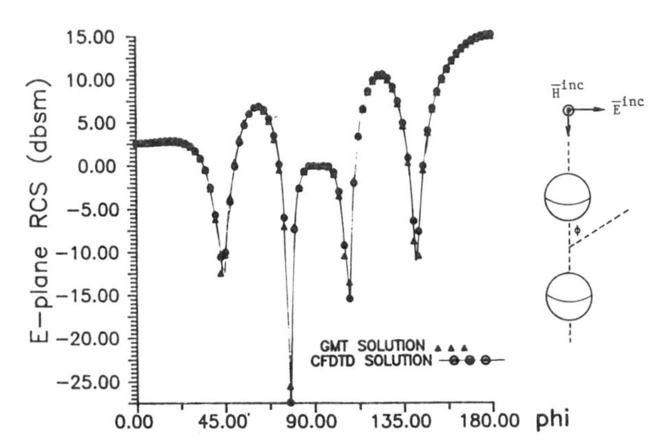

Fig. 2. Agreement of FD-TD and generalized multipole technique (GMT)[36] bistatic RCS within 1 dB over a 42-dB range for a pair of 1-λ_o diameter conducting spheres separated by a 1-λ_o air gap.

Is more dynamic range needed? Almost certainly yes, if a better job is to be done in modeling certain specially shaped targets. A good example of such a shape is the NASA almond[37], which has been demonstrated to have monostatic RCS variations of 60 or more dB occurring over broad angular ranges. However, going from 40 dB to 60 dB will not be simple:

--- We will require the development of advanced radiation boundary conditions having effective reflection coefficients of 0.1%, thereby reducing this contribution to the grid noise by 20 dB.

--- We probably will have to shift to space-sampling, time-integration algorithms having dimensionally better accuracy than the second-order procedures common today. Probably, a 20 dB reduction in the grid noise contribution due to dispersive effects accumulating on propagating numerical waves is needed to permit the use of grid resolutions no finer than those of today's algorithms.

Physics of Dispersive Materials

A key issue in scoping the advantages and limitations of space-grid time-domain Maxwell's solvers involves the modeling of materials having frequency varying (dispersive) electrical properties. This issue does not arise when considering the merits of frequency domain techniques such as MoM, since such approaches assume monochromatic illumination. But, time-domain approaches permit impulsive (broadband) illumination, so the ability to model material dispersion directly in the time domain would permit efficient broadband determination of RCS for practical engineering problems, a significant advantage over MoM.

Currently, evidence is emerging that this can be accomplished without sacrificing the simplicity and computational efficiency of the basic time-domain solver. For example, workers at Penn State[24] have published an accurate and efficient approach for first-order material dispersions (Debye relaxations) using FD-TD. This method recursively implements the classical time-domain convolutional relation between the displacement field, D(t), and the electric field, E(t), taking advantage of the simple decaying-exponential nature of the material susceptibility impulse response. The recursive process avoids the very troublesome backstorage in time that would otherwise be required to implement the convolution. Excellent results have been obtained relative to exact analyses (Fourier-transformed monochromatic data) for 1-D material slabs having canonical first-order dispersions, and extension to 2-D and 3-D material structures is simple with little computational penalty.

My group has developed an alternative FD-TD procedure suitable for material dispersions of arbitrary order, including resonant dispersions.[25] Our approach recasts the time-domain convolutional relation between D(t) and E(t) as an ordinary differential equation in time. (That this is valid can be seen upon Fourier transforming this differential equation, whereupon we obtain the desired functional behavior of D(f) vs. E(f), i.e., the permittivity vs. frequency.) This differential equation is simply integrated in time concurrently with Maxwell's curl equations. The backstorage in time needed to implement this process is minimal. Excellent results have been obtained for 1-D material slabs having canonical second-order dispersions, and, like the Penn State approach, extension to 2-D and 3-D material structures is simple with little computational penalty.

COMPUTER RESOURCE AND ARCHITECTURE ISSUES

Vectorization and Concurrency, Processing Speed for FD-TD

Both the FD-TD algorithm of my group and the FV-TD algorithm of the group at Rockwell International[33] have been highly vectorized, having been benchmarked at over 200 Mflops on a single Cray Y-MP processor for real models. However, the attainment of even higher Mflop rates may be hampered by the fact that the the space grids have an unavoidable number of non-standard cells requiring either scalar or odd-lot vector operations. These non-standard cells result from the need to program a near-field radiation condition at the outermost grid boundary, and the need to stitch varying types of meshes together to accommodate complex structure shapes.

Despite this, on the Cray Y-MP/8, it has been found possible to achieve nearly 100% concurrent utilization of all eight processors using autotasking for both my group's 3-D FD-TD code and Rockwell's 3-D FV-TD code. Relatively minor modifications to the original single-processor Fortran were required. Average processing rates exceeding 1.6 Gflops have been achieved for both 3-D FD-TD and FV-TD codes.

In one example[5], FD-TD was used to conformally model an electrically large ($25\lambda_0 \times 10\lambda_0 \times 10\lambda_0$) 3-D serpentine jet engine inlet. The solution, involving the time-marching of 23-million vector field components over 1,800 time steps, was completed in only 3 minutes, 40 seconds per illumination angle. A 7.97/8 utilization factor of the eight Cray Y-MP processors was achieved.

--- We note in comparison that if the full-matrix MoM were applied to model this engine inlet, a complex-valued linear system involving approximately *450K equations* would have to be set up and solved. This assumes a standard triangular surface patching implementation of the electric field integral equation, with the $1500\lambda_0^2$ area of the engine inlet discretized at 10 divisions per λ_0. Using the 2.1-Gflop multiprocessing out-of-memory subroutine for LU decomposition discussed earlier, the Cray Y-MP/8 running time for this matrix would be about 2.6 years for 5,000 illumination angles. This compares to about 12.7 days for FD-TD for the same number of illumination angles, a speedup of *75:1*.

Although the next-generation Cray Y-MP/16 will not be available until early 1992, it is possible to firmly estimate that its average processing speed will be 6 to 8 times that of the Y-MP/8 for the FD-TD, FV-TD, and MoM solvers (10 - 16 Gflops). Applying this factor, the FD-TD inlet model drops to only 28 - 37 sec per illumination angle, or 1.6 - 2.1 days for 5000 angles. The corresponding LU decomposition time for MoM would be 3.9 - 5.2 months.

Storage Requirements for FD-TD

Much modeling experience with FD-TD and FV-TD has shown that $\lambda_0/10$ to $\lambda_0/20$ is the range of grid space resolution suitable for engineering accuracy (± 1 dB) in predictions of the near and far fields. Thus, at least 1,000 volume cells per λ_0^3, equivalently 6,000 vector-field components per λ_0^3, must be maintained. For the simplest solver, FD-TD, applied to model non-dispersive media, the required storage is 18,000 real numbers per λ_0^3 (each real field component value having two associated real coefficients that must be stored).

Consider again modeling a jet fighter, but now in the context of FD-TD. Assuming dimensions of 20 x 20 x 5 meters for the fighter, and assuming a radar frequency of 1 GHz ($\lambda_0 = 0.3$ meter), the space grid would be in the order of 75,000 λ_0^3 (450-million vector field components), and the FD-TD memory requirement would be about 1.3 Gword. Although this requirement exceeds Cray Y-MP/8 possibilities for central memory and solid-state device (SSD), it likely will be feasible with the Cray Y-MP/16. Running time on the latter would be in the range of 20 - 25 minutes per illumination angle.

For modeling RCS at even higher frequencies, we can apply out-of-memory techniques similar to those used for the very large MoM matrices discussed earlier. In fact, an out-of-memory code has already been developed for the 3-D FD-TD algorithm using as a host a single-processor of the Cray X-MP. This code uses asynchronous I/O, variable buffering, and disk-striping to achieve I/O concurrency of better than 95% relative to the computer arithmetic operations when 12 drives are used. When adapted to the Cray Y-MP/16, the resulting software should permit modeling a complete jet fighter to 2 GHz or more, a factor of six higher in frequency than that conceivable using the full-matrix MoM on the same computer.

Is There One Best Architecture for Supercomputers Implementing Grid-Based Maxwell's Equations Solvers?

A number of groups have implemented grid-based Maxwell's equations solvers on supercomputers having widely varying architectures. As stated above, my group and the Rockwell group have acquired substantial Cray experience. Other notable efforts include those of a group at MRJ, Inc. in porting a 3-D FD-TD code to the CM-2 Connection Machine[38], and a group at Jet Propulsion Laboratory (JPL) in porting FD-TD to the JPL and Intel Hypercubes[39].

My impressions of the principal conclusions of this work are summarized below:

1. Computers having many processors working concurrently represent the best platforms for implementing grid-based Maxwell's equations solvers at speeds exceeding 1 Gflop.

2. Concurrencies very close to 100% (i.e., an algorithm speedup factor equal to 8 if the number of available processors equals 8) can be achieved with the Cray Y-MP/8 under autotasking for FD-TD and FV-TD. Performance scaling looks good at least through 16 high-performance (Cray Y-MP class) processors.

3. Good to excellent concurrencies for FD-TD can also be achieved using the JPL/Intel Hypercube. Performance scaling looks good into the hundreds of moderate-performance (Intel I-860) processors.

4. The coarser-grained processor concurrencies permitted by the Cray and JPL/Intel Hypercube architectures appear to be more robust means of attaining Gflop rates for grid-based Maxwell's solvers than the fine-grained concurrency of the Connection Machine (tens of thousands of low-performance processors).

In my opinion, the two leading supercomputer architectures for implementing grid-based Maxwell's solvers at the largest scales are presently: (a) the Cray Y-MP design, now achieving average rates of 1.6 - 2.1 Gflops with the Y-MP/8 and proceeding to 10 - 15 Gflops with the Y-MP/16 by early 1992; and (b) the Intel Hypercube design, in particular the Touchstone Delta, claimed to attain average rates of 5 - 15 Gflops by mid-1991[40]. With both companies working seriously on follow-up machines reaching toward average rates exceeding 100 Gflops (see later discussion), both types of architectures have to be taken seriously.

In addition to basic arithmetic speed (Gflops) in performing the Maxwell's solvers, an important point that is often overlooked in evaluating computer architectures is the need to achieve a balance between the Gflop rate and the aggregate rate of input/output (I/O). Simply, the fastest processors will be starved (and left waiting) for stored data regarding the vector field components if there is insufficient I/O bandwidth between the processors and the memory. As discussed earlier, the grid-based Maxwell's solvers require Gwords of memory to permit modeling of entire aircraft at GHz frequencies. With time-stepping mandating the flow of these Gwords between the central and/or mass memory and the multiprocessors at each time step, it is clear that aggregate I/O bandwidths measured in the tens of Gbytes/sec are mandatory. As supercomputer architectures evolve, it will be vital to perform benchmarks to determine whether or not they continue to achieve balanced performance. Ultimately, balanced performance (or lack thereof) may determine the superiority (inferiority) of any one of these approaches.

Must We Bend Our EM Algorithms to Fit the Computer Architecture?

Does Fortran Have Any Place in Programming Multiprocessor Machines?

Briefly, I would answer these two questions "no" and "yes," respectively. Given our experience to date, and given the emerging highly capable multiprocessing hardware and software, I believe that:

1. Space-grid time-domain Maxwell's solvers will be ported to either type of machine with little difficulty. Multiprocessor concurrencies of better than 90% will be achieved using ordinary compilers (such as Cray's current autotasking compiler) with minimal work. Moderate efforts (order 1 or 2 person-weeks) will result in >95% concurrencies.

2. Matrix fill algorithms for MoM will probably also be automatically multiprocessed with little difficulty. Library algorithms for large-scale, multiprocessing LU decomposition will be turnkey.

3. By implication of the above points, Fortran (latest version) will remain as useful as it is today.

HORIZONS THROUGH THE MID 1990'S

Computer Processing Speeds

The most capable machines will provide average computation rates of at least *0.1 teraflop*. Contending machines at this throughput level will include the Cray C-95, possibly having 64 advanced Y-MP-type processors; and the Intel Sigma, possibly having 4096 advanced I-860-type processors. *Full-teraflop* rates are likely by the end of this decade as Cray, Intel, and possibly others proceed to develop effective massively parallel architectures.

New Applications for Space-Grid Time-Domain Maxwell's Solvers

In addition to implementing whole-aircraft RCS models, we will probably be using these ultra-computers to model important EM wave problems that arise in non-defense commercial and industrial applications. For example, we will be using space-grid time-domain Maxwell's solvers to model from first principles:

a. Operation of high-speed interconnects in digital computers. See Refs. 26 - 28 for the first literature publications in this area.

b. Operation of electronic, electro-optic, and all-optical devices and circuits switching in less than 1 nanosecond. See Refs. 29 - 31 for the first literature in this area.

c. Design of antennas and UHF/microwave data links for worldwide personal wireless telephony, cellular communications, remote computing, and advanced automotive electronics (particularly the car location and navigation functions).

Supercomputing Personal Workstations

Finally, we will see inexpensive personal workstations having at least the speed of the original Cray-1. That this is feasible is clearly indicated by the price and performance of the current IBM reduced instruction set (RISC) series of workstations (in the order of 6 - 12 Mflops of average performance for real Fortran codes, perhaps 40% - 50% of the Cray-1 throughput).

As desktop workstation processing speeds reach Cray-1 rates by the mid-1990's, a serious user will actually have at his disposal computing capabilities *exceeding* those of his counterparts working at the best-equipped U.S. national laboratories only 10 years earlier. This is because his workstation probably will have much more user-accessible central memory (in the tens of millions of words), more user-accessible mass storage (in the hundreds of millions of words), and well-integrated, high-resolution, color graphics hardware/software. Using space-grid time-domain Maxwell's solvers, such a user could conduct detailed studies in all of the defense and non-defense areas mentioned above, albeit at electrical size regimes scaled well down from that possible with the 0.1-teraflop machines representing the state-of-the-art. I believe that this will be an unprecedented opportunity for students, faculty, and industrially-employed engineers anywhere in the country to bring to bear the fundamental physics of Maxwell's equations to their academic and professional studies. I would caution government agencies, however, that the widespread availability of this hardware/software capability will proliferate knowledge and capability in the EM observables area.

ACKNOWLEDGEMENTS

The author wishes to acknowledge the support of National Science Foundation Grant ASC-8811273; Office of Naval Research Contract N00014-88-K-0475; the NASA Ames University Consortium; and Cray Research Inc.

REFERENCES

1. R. F. Harrington, "Field Computation by Moment Methods," Macmillan, New York (1968).
2. K. R. Umashankar, Numerical analysis of electromagnetic wave scattering and interaction based on frequency-domain integral equation and method of moments techniques, Wave Motion 10:493 (1988).
3. J. B. Keller, Geometrical theory of diffraction, J. Opt. Soc. Amer. 52:116 (1962).
4. R. G. Kouyoumjian and P. H. Pathak, A uniform geometrical theory of diffraction for an edge in a perfectly conducting surface, Proc. IEEE 62:1448 (1974).
5. J. P. Brooks, K. K. Ghosh, E. Harrigan, D. S. Katz, and A. Taflove, Progress in Cray-based algorithms for computational electromagnetics, in: "Progress in Electromagnetics Research (PIER) 4," T. Cwik, ed., Elsevier, New York (1991).
6. K. S. Yee, Numerical solution of initial boundary value problems involving Maxwell's equations in isotropic media, IEEE Trans. Antennas Propagat. 14:302 (1966).
7. A. Taflove and M. E. Brodwin, Numerical solution of steady-state electromagnetic scattering problems using the time-dependent Maxwell's equations, IEEE Trans. Microwave Theory Tech. 23:623 (1975).
8. A. Taflove and M. E. Brodwin, Computation of the electromagnetic fields and induced temperatures within a model of the microwave-irradiated human eye, IEEE Trans. Microwave Theory Tech. 23:888 (1975).
9. A. Taflove, Application of the finite-difference time-domain method to sinusoidal steady-state electromagnetic penetration problems, IEEE Trans. Electromagn. Compat. 22:191 (1980).
10. G. Mur, Absorbing boundary conditions for the finite-difference approximation of the time-domain electromagnetic field equations, IEEE Trans. Electromagn. Compat. 23:377 (1981).
11. K. R. Umashankar and A. Taflove, A novel method to analyze electromagnetic scattering of complex objects, IEEE Trans. Electromagn. Compat. 24:397 (1982).
12. A. Taflove and K. R. Umashankar, Radar cross section of general three-dimensional scatterers, IEEE Trans. Electromagn. Compat. 25:433 (1983).
13. K. R. Umashankar, A. Taflove, and B. Beker, Calculation and experimental validation of induced currents on coupled wires in an arbitrary shaped cavity, IEEE Trans. Antennas Propagat. 35:1248 (1987).
14. A. Taflove, K. R. Umashankar, B. Beker, F. Harfoush, and K. S. Yee, Detailed FD-TD analysis of electromagnetic fields penetrating narrow slots and lapped joints in thick conducting screens, IEEE Trans. Antennas Propagat. 36:247 (1988).
15. T. G. Jurgens, A. Taflove, K. R. Umashankar, and T. G. Moore, Finite-difference time-domain modeling of curved surfaces, submitted to IEEE Trans. Antennas Propagat.

16. G. A. Kriegsmann, A. Taflove, and K. R. Umashankar, A new formulation of electromagnetic wave scattering using an on-surface radiation boundary condition approach, IEEE Trans. Antennas Propagat. 35:153 (1987).
17. T. G. Moore, J. G. Blaschak, A. Taflove, and G. A. Kriegsmann, Theory and application of radiation boundary operators, IEEE Trans. Antennas Propagat. 36:1797 (1988).
18. J. G. Maloney, G. S. Smith, and W. R. Scott, Jr., Accurate computation of the radiation from simple antennas using the finite-difference time-domain method, IEEE Trans. Antennas Propagat. 38:1059 (1990).
19. D. S. Katz, M. J. Piket, A. Taflove, and K. R. Umashankar, FD-TD analysis of electromagnetic wave radiation from systems containing horn antennas, submitted to IEEE Trans. Antennas Propagat.
20. D. M. Sullivan, O. P. Gandhi and A. Taflove, Use of the finite-difference time-domain method in calculating EM absorption in man models, IEEE Trans. Biomed. Eng. 35:179 (1988).
21. C.-Q. Wang and O. P. Gandhi, Numerical simulation of annular phased arrays for anatomically based models using the FD-TD method, IEEE Trans. Microwave Theory Tech. 37:118 (1989).
22. D. Sullivan, Three-dimensional computer simulation in deep regional hyperthermia using the finite-difference time-domain method, IEEE Trans. Microwave Theory Tech. 38:204 (1990).
23. M. J. Piket-May, A. Taflove, W.-C. Lin, D. S. Katz, V. Sathiaseelan, and B. Mittal, Computational modeling of electromagnetic hyperthermia: three-dimensional and patient specific, submitted to IEEE Trans. Biomed. Engrg.
24. R. Luebbers, F. P. Hunsberger, K. S. Kunz, R. B. Standler, and M. Schneider, A frequency-dependent finite-difference time-domain formulation for dispersive materials, IEEE Trans. Electromagn. Compat. 32:222 (1990).
25. R. Joseph, S. Hagness, and A. Taflove, FD-TD computational modeling of dispersive media having second-order and higher relaxations," presented at 1991 IEEE International AP-S Symposium, London, ON.
26. G.-C. Liang, Y.-W. Liu, and K. K. Mei, Full-wave analysis of coplanar waveguide and slotline using the time-domain finite-difference method, IEEE Trans. Microwave Theory Tech. 37:1949 (1989).
27. D. M. Sheen, S. M. Ali, M. D. Abouzahra, and J. A. Kong, Application of the three-dimensional finite-difference time-domain method to the analysis of planar microstrip circuits, IEEE Trans. Microwave Theory Tech. 38:849 (1990).
28. T. Shibata and E. Sano, Characterization of MIS structure coplanar transmission lines for investigation of signal propagation in integrated circuits, IEEE Trans. Microwave Theory Tech. 38:881 (1990).
29. R. H. Voelker and R. J. Lomax, A finite-difference transmission line matrix method incorporating a nonlinear device model, IEEE Trans. Microwave Theory Tech. 38:302 (1990).
30. E. Sano and T. Shibata, Fullwave analysis of picosecond photoconductive switches, IEEE J. Quantum Electron. 26:372 (1990).
31. S. M. El-Ghazaly, R. P. Joshi, and R. O. Grondin, Electromagnetic and transport considerations in subpicosecond photoconductive switch modeling, IEEE Trans. Microwave Theory Tech. 38:629 (1990).
32. S.-T. Chu and S. K. Chaudhuri, A finite-difference time-domain method for the design and analysis of guided-wave optical structures, IEEE J. Lightwave Tech. 7:2033 (1989).

33. V. Shankar, A. H. Mohammadian, and W. F. Hall, A time-domain, finite-volume treatment for the Maxwell's equations, Electromagnetics 10:127 (1990).
34. N. K. Madsen and R. W. Ziolkowski, A three-dimensional modified finite volume technique for Maxwell's equations, Electromagnetics 10:147 (1990).
35. G. A. Kriegsmann, Exploiting the limiting amplitude principle to numerically solve scattering problems, Wave Motion 4:371 (1982).
36. Generalized multipole technique (GMT) RCS data for the double-sphere problem was provided by Arthur Ludwig of General Research Corp., Santa Barabara, CA.
37. S. W. Lee, ed., "High-Frequency Scattering Data Book," Electromagnetics Laboratory, Univ. of Illinois at Urbana-Champaign (1989).
38. A. T. Perlik, T. Opsahl, and A. Taflove, Predicting scattering of electromagnetic fields using FD-TD on a Connection Machine, IEEE Trans. Magnetics 25:2910 (1989).
39. J. E. Patterson, T. Cwik, R. D. Ferraro, N. Jacobi, P. C. Liewer, T. G. Lockhart, G. A. Lyzenga, J. W. Parker, and D. A. Simoni, Parallel computation applied to electromagnetic scattering and radiation analysis, Electromagnetics 10:21 (1990).
40. D. P. Hamilton, World's fastest computer, Science 250:1203 (1990).

APPLICATION OF CONJUGATE GRADIENT METHOD FOR THE SOLUTION OF LARGE MATRIX PROBLEMS

Tapan K. Sarkar
Saila Ponnapalli
Peter Petre

Department of Electrical and Computer Engineering
Syracuse University
Syracuse, New York 13244-1240

ABSTRACT

The conjugate gradient method (CGM) has found a wide variety of applications in electromagnetics and in signal processing. In addition, CGM when used in conjunction with FFT (CGFFT) is extremely efficient for solving Hankel and Toeplitz or block Toeplitz matrix systems which frequently arise in both electromagnetics and signal processing applications. The FFT may be utilized because of the convolutional nature of the matrix. CGM has also been used in adaptive spectral estimation. In this paper, a novel application of the conjugation gradient method in the determination of far-field antenna patterns via a near-field to far-field transformation will be discussed.

I. INTRODUCTION

The conjugate gradient method was initially presented in electromagnetics as an efficient means of solving matrix equations [1]. In some special cases, the matrix is Toeplitz or block Toeplitz. In such cases, the FFT may be incorporated into CGM to perform the convolution as a multiplication in the frequency domain. CGFFT has been used to solve scattering problems for thin wires, and conducting strips or plates using a surface formulation [2], or arbitrary shaped dielectric complexity of the problem from N^2 to 0(NLOG(N)). CGM and CGFFT have also been used in several signal processing applications such as fitting an all-pole model of data [6], or adaptive spectral estimation [7].

The application of CGM and CGFFT to near-field to far-field transformations has very recently been explored [8-11]. The accurate determination of far-fields without the used of large outdoor test ranges has necessitated the use of near-field measurements, which are performed in a controlled environment. An efficient and accurate method of transforming from near-fields to far-fields is required. The

Directions in Electromagnetic Wave Modeling
Edited by H.L. Bertoni and L.B. Felsen, Plenum Press, New York, 1991

established technique for near-field to far-field transformations has been the modal expansion method [12-17]. However, there are several drawbacks in this approach, foremost which is that when a Fourier transform method is used, the fields outside the measurement surface are assumed to be zero in the formulation, particularly in the planar and cylindrical measurement case. Hence, far-fields accurate only in a particular angular sector, which is dependent on the measurement configuration, are obtained.

The novel idea presented here is to replace a radiating antenna by equivalent currents which reside on a fictitious surface which encompasses the antenna. The equivalent currents are assumed to radiate identical fields as the original antenna in the region of interest. The theoretical explanation is found in [9-11] and will not be elaborated in this paper. The electric field integral equation (EFIE) is written to relate the near-fields to the equivalent currents. A method of moments procedure is used in which the currents are discretized in terms of known basis functions with unknown coefficients, and a point-matching procedure is used [18]. If an Nth order approximation is made for the currents, then only N complex near-field points anywhere in the region of interest need to be known, and no assumption is made that the fields are zero in any region of space.

CGM can be used to efficiently solve the resulting matrix equation, and for a special case, CGFFT can be utilized to further enhance computational efficiency and storage requirements. The efficiency in applying CGM arises from the fact that the normal form of $Ax=y$, that is $A^H Ax=A^H y$ is solved without explicitly forming it. Therefore, the number of currents can be chosen to be far fewer than the number of measured near-field points, and a least-squares solution for the currents can be found.

The advantages of the equivalent current approach are that measurements on simple surfaces such as a plane can be used even for antennas which are not highly directive. This is because the fields outside the measurement surface are not assumed to be zero, and hence the far-fields can be extrapolated over large elevation and azimuthal ranges. All first-order and higher-order effects of the measurement probe can be compensated for regardless of the measurement configuration. The issue of probe-compensation in conjunction with the equivalent current approach is addressed in [9-11]. In addition, the near-field measurements can be made over an arbitrarily shaped surface. An indication of the actual current distribution is also obtained. This can be used for applications such as the determination of faculty elements in an antenna array. Finally, the far-fields are obtained with any desired resolution and interpolation is not required unlike the modal expansion method.

The formulation of the matrix, and the application of CGM and CGFFT are summarized in Section II. Numerical results for several antenna configurations are presented in Section III.

II. FORMULATION OF MATRIX EQUATION

Consider the case of a planar dipole antenna where the planar currents have been replaced by equivalent electric currents, and a probe has been placed in the right half plane in order to measure the near-fields radiated by the test antenna. Let the principle fields radiated by the test antenna be $\vec{E}_a$ and those radiated by the probe be $\vec{E}_b$. Let the current on the probe in the absence of the test antenna be $\vec{J}_b$ and the current induced on the probe when the test antenna is turned on be $\vec{J}_{bs}$. The near-field data is obtained by measuring the voltage across a load on the terminals of the probe. To within a constant of proportionality, the received signal when the probe is positioned at $\vec{r}_o$ is [19-20],

$$P_B(\vec{r}_o) \, \Delta \int_V (\vec{E}_a \cdot \vec{J}_b - \vec{E}_b \cdot \vec{J}_{bs}) \, dV \tag{2}$$

The second term on the right arises from the induced current on the test antenna due to the presence of the probe. If this term is neglected,

$$P_B(\vec{r}_o) \approx \int_V \vec{E}_b \cdot \vec{J}_a dV = \int_V \vec{E}_a \cdot \vec{J}_b \, dV \tag{3}$$

One can approximate the currents on the probe with known basis functions and determine $\vec{J}_b$. If the probe is assumed to be an ideal Hertzian dipole located at (x_f, y_f, z_f), the current $\vec{J}_b$ is,

$$\vec{J}_b = \delta(x'-x_f) \; \delta(y'-y_f) \; \delta(z'-z_f) \tag{4}$$

Then evaluating (4) is equivalent to finding the scattered fields due to $\vec{J}_a$ at a point (x_f, y_f, z_f) in the $\hat{a}_b$ direction from

$$P_B(\vec{r}_o) = E(x_f, y_f, z_f) \cdot \hat{a}_b \quad (5)$$

$$P_B(\vec{r}_o) = (-j\omega\mu\vec{A} - \nabla\phi) \cdot \hat{a}_b \quad (6)$$

If a method of moments procedure is used to discretize $\vec{J}_a$ in terms of known basis functions, then (6) is equivalent to using a point matching procedure to form a matrix equation.

Consider the equivalent source distribution to be a plate in the x-z plane composed of equally spaced rooftop function representations for the currents. The currents are given by,

$$\vec{J} = \hat{a}_x \sum_{I=1}^{M} \sum_{j=1}^{N} \alpha_{ij} P_x(x_i, z_j) + \hat{a}_z \sum_{i=1}^{M} \sum_{j=1}^{N} \beta_{ij} P_z(x_i, z_j) \quad (7)$$

where P_x and P_z are rooftop basis functions [9-11]. α_{ij} and β_{ij} are the unknown coefficients to be solved for. The field at any point (x_f, y_f, z_f) can be found from,

$$\vec{E}(x_f, y_f, z_f) = -j\omega\mu [\sum_{k=1}^{M} \sum_{l=1}^{N} \vec{J}(k,l) \iint g(\vec{r} - \vec{r}_{kl}) ds'_{kl} + \frac{1}{k_o^2} \iint - \nabla' \cdot \vec{J}(k,l) \nabla' g(\vec{r} - \vec{r}_{kl}) ds'_{kl}] \quad (10)$$

where, g is the three dimensional Green's function,

$$g(\vec{r} - \vec{r}_{kl}) = g(x, y, z, x', z') \frac{e^{-jkR}}{4\pi R} \quad (11)$$

$$R = [(x-x')^2 + (y)^2 + (z-z')^2]^{1/2} \quad (12)$$

The first term due to the vector potential may be simply evaluated by approximating the rooftop functions to be pulses where the integration is performed over the kl^{th} current patch. The second term due to the charge may be evaluated by finding the divergence of the current from (7) using,

$$\nabla \cdot \vec{J} = \hat{a}_x \sum_{i=1}^{M} \sum_{j=1}^{N} \alpha_{ij} \frac{\partial P_x(x_i, z_j)}{\partial x'} + \hat{a}_z \sum_{i=1}^{M} \sum_{j=1}^{N} \beta_{ij} \frac{\partial P_z(x_i, z_j)}{\partial z'} \tag{13}$$

Equation (10) represents the EFIE, and when a point matching procedure is used, a component of the electric field in any direction at the given point may be found. This is proportional to the signal on a probe measuring the electric field in that direction, when it is assumed that the probe is a small dipole. For the special case of planar scanning, the following matrix equation is obtained -

$$\begin{bmatrix} E_x(J_x) & E_x(J_z) \\ E_z(J_x) & E_z(J_z) \end{bmatrix} \begin{bmatrix} J_x \\ J_z \end{bmatrix} = \begin{bmatrix} E_x(x_f, y_f, z_f) \\ E_z(x_f, y_f, z_f) \end{bmatrix} \tag{14}$$

The matrix equation is solved using CGM (case A), the algorithm for which is outlined in [2].

In planar measurements in which the plane containing the current patches and that over which the field points are taken are parallel, and of the same size, the computation and storage requirements can be tremendously reduced by using CGFFT. Most of the computation at each iteration in CGM occurs in the calculation of A P and $A^H R$, where P is the search direction and R is the residual. When the discretization of the current patches and field points is chosen to be equal, the resulting matrix is block Toeplitz. In this case, A P and A^H R represent discrete convolutions, and can be evaluated by multiplying their discrete Fourier transforms and inverse Fourier transforming the result [21]. This can be performed efficiently using a two dimensional FFT.

Consider the matrix equation (14), which must be evaluated when computing AP or AJ. This is of the form,

$$\begin{bmatrix} E_x(J_x) & E_x(J_z) \\ E_z(J_x) & E_z(J_z) \end{bmatrix} \begin{bmatrix} P_x \\ P_z \end{bmatrix} = \begin{bmatrix} AP_x(x_f, y_f, z_f) \\ AP_z(x_f, y_f, z_f) \end{bmatrix} \tag{15}$$

Then $[AP_x]$ can be found from,

$$[AP_x] = F_\tau^{-1}\{ F\{[E_x(J_x)]_p\}F\{[P_x]_0\} + F\{[E_x(J_z)]_p\}F\{[P_z]_0\}\} \quad (16)$$

and $[AP_z]$ can be found from,

$$[AP_z] = F_\tau^{-1}\{ F\{[E_z(J_x)]_p\}F\{[P_x]_0\} + F\{[E_z(J_z)]_p\}F\{[P_z]_0\}\} \quad (17)$$

Here F denotes a two dimensional discrete Fourier transform and F^{-1} is a two dimensional inverse discrete Fourier transform. The subscript p denotes a periodic extension of the matrix, the subscript 0 denotes the matrix with zero padding, and the subscript τ denotes the truncation operator. A similar expression is also found for the adjoint operator acting on the residual. Using CGFFT, for N X N current patches and field points, only six matrix of size (2N-1) X (2N-1) need to be stored, and the order of computation is $(2N-1)^2 \mathrm{LOG}_2(2N-1)$ times the number of iterations required.

III. NUMERICAL INVESTIGATION

In the following examples, synthetic near-field data is used, and a comparison is made with analytic far-fields. In all examples shown here, CGFFT has been employed.

In the first example a uniform array consisting of 20 X 20 dipoles uniformly spaced on a 10λ X 10λ surface is considered. The array is replaced by 50 X 50 equivalent currents on a 10.5λ X 10.5λ fictitious planar surface. Near-field measurements (50 X 50) on a 10.5λ X 10.5λ planar surface at a distance of 20λ from the source plane are utilized to determine the equivalent currents. Figure 1 shows the vertical cut with the analytic solution and the horizontal cut with the analytic solution. Also shown are the magnitude of the computed electric field over all θ and ϕ and the analytic electric field over all ranges. There is reasonably good agreement between analytic and computed results over all elevation and azimuthal ranges.

In the second example, a current distribution which is sinusoidal in the z-direction and uniform in the x-direction is considered. The support for the current is 10½λ in the z-direction and 2λ in the x-direction and 2λin the x-direction. The currents are replaced by 189 X 21 equivalent currents on a 20λ X 3λ fictitious planar surface. Near-fields measurements (189 X 21) on a 20λ X 3λ planar surface at a distance of 3λ from the source plane are utilized to determine the equivalent currents. Figure 2 shows the vertical cut with the analytic solution and the horizontal cut with the analytic solution. Also shown are the magnitude of the computed electric field over all θ and ϕ and the analytic electric field over all ranges. There is excellent agreement between analytic and computed results over all elevation and azimuthal ranges.

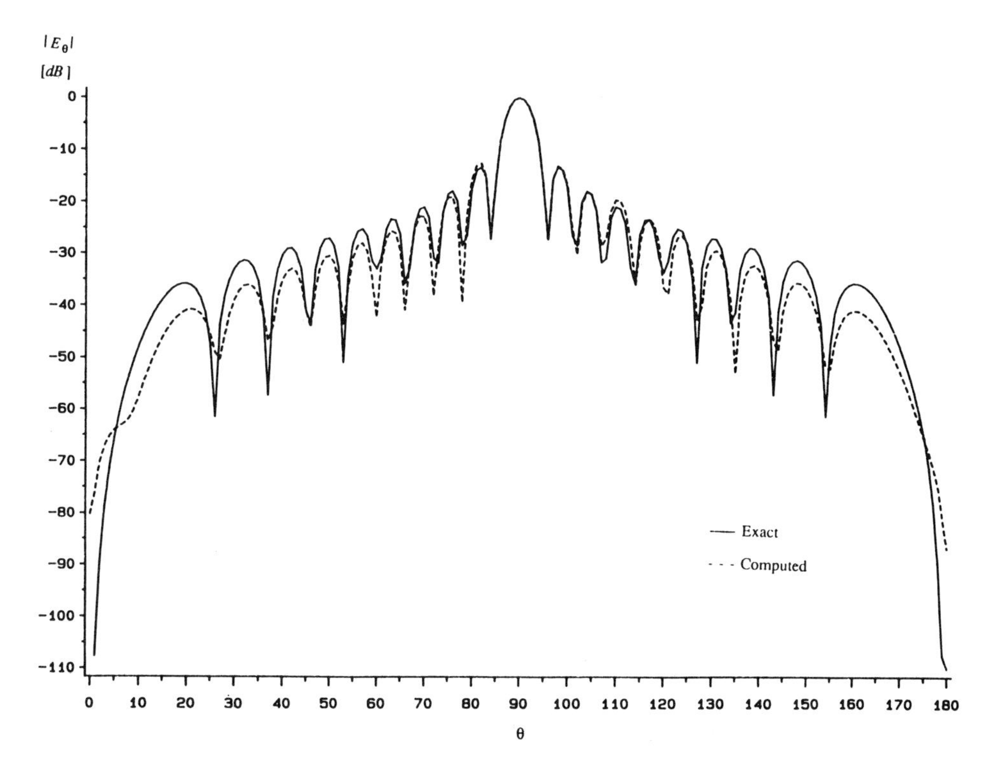

Figure 1a. Vertical (E-Plane) Cut at $\phi=90^o$ for 20X20 Dipoles on a 10λ X 10λ Surface

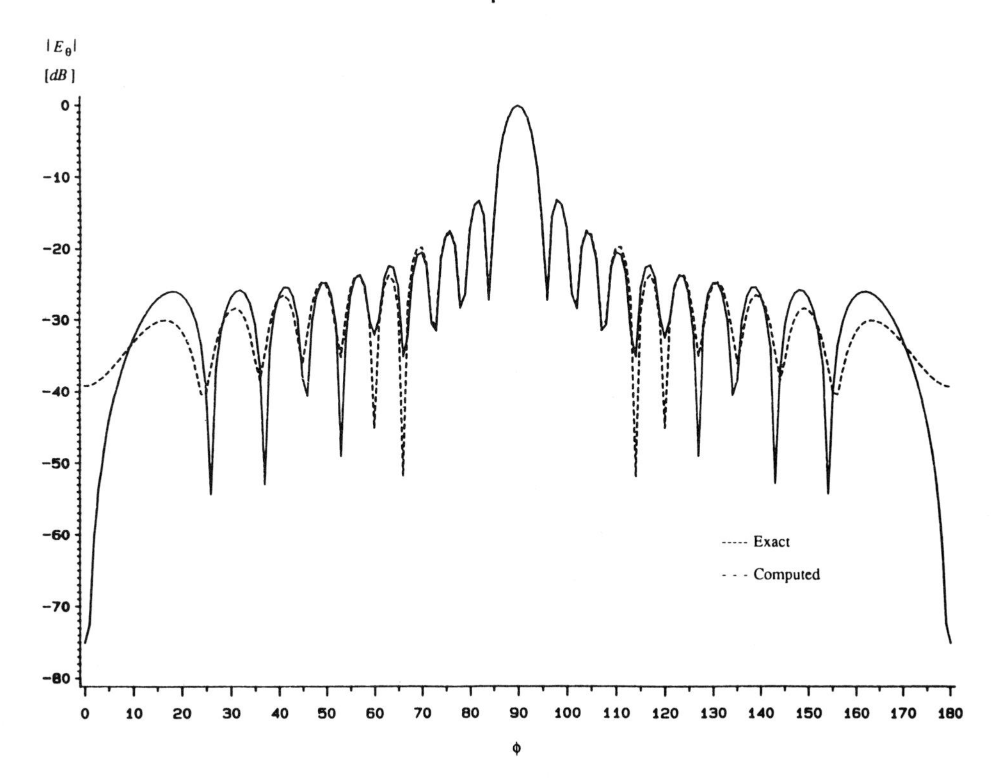

Figure 1b. Horizontal (H-Plane) Cut at for 20X20 Dipoles on a 10λ X 10λ Surface

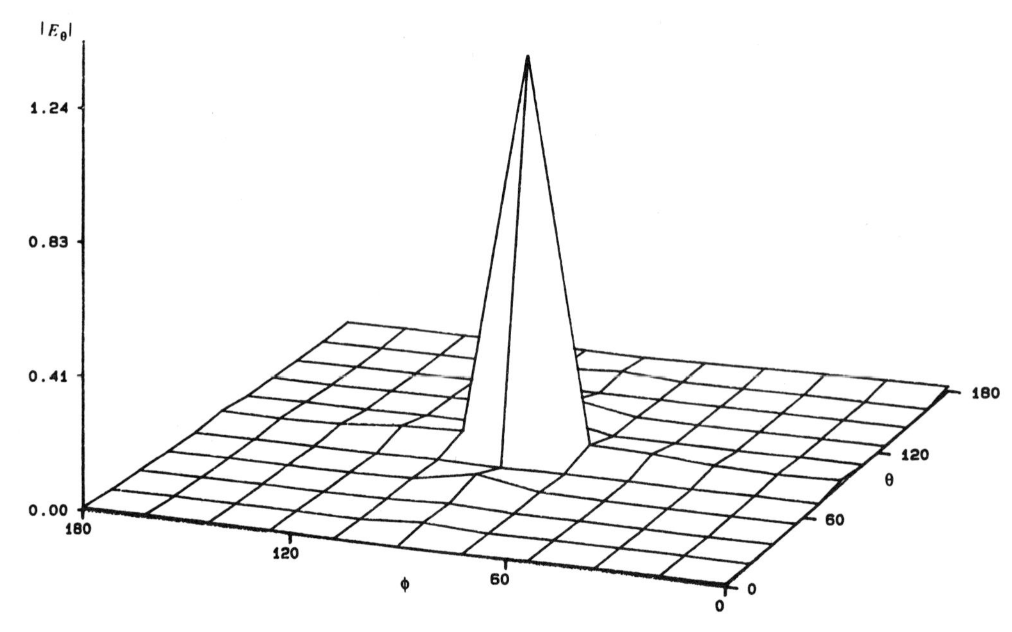

Figure 1c. Computed normalized $|E_\theta|$ vs θ,ϕ for 20X20 Dipoles on a 10λ X 10λ surface

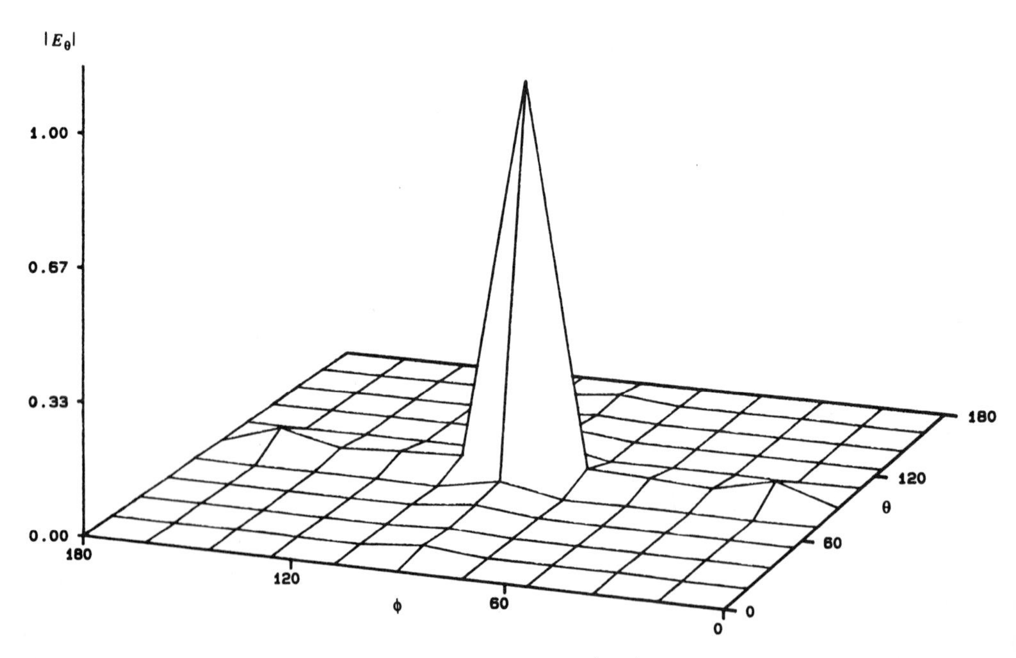

Figure 1d. Analytic Normalized $|E_\theta|$ vs θ,ϕ for 20X20 Dipoles on a 10λ X 10λ surface

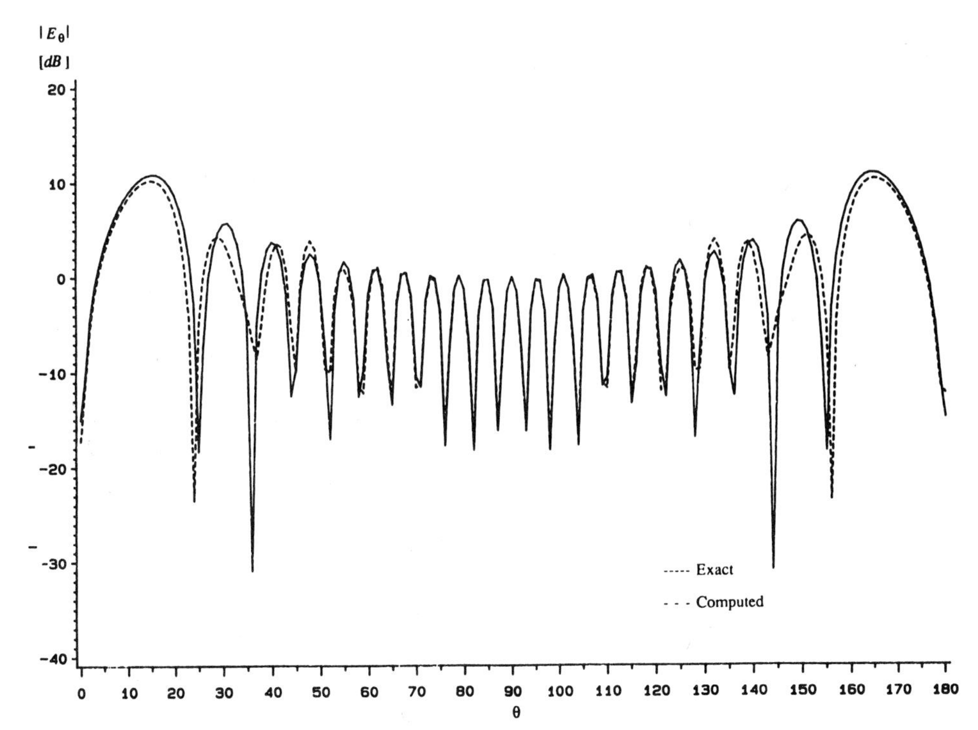

Figure 2a. Vertical Cut at $\phi=90^o$ for Sinusoidal Current of Length 10.5λ and Width 2λ

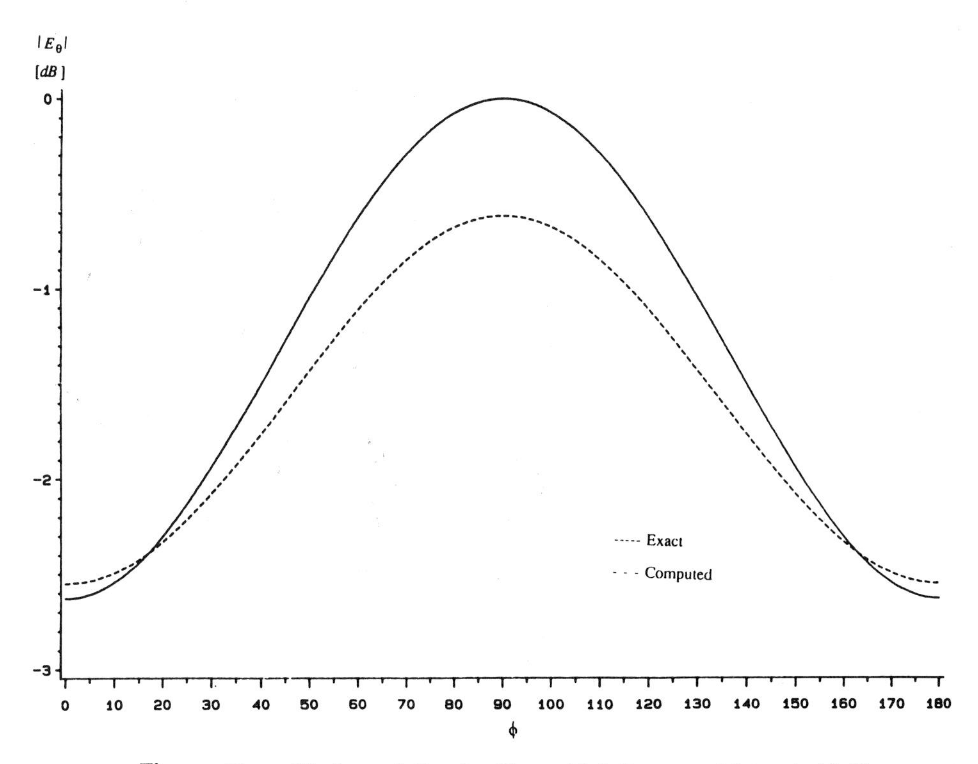

Figure 2b. Horizontal Cut for Sinusoidal Current of Length 10.5λ and Width 2λ

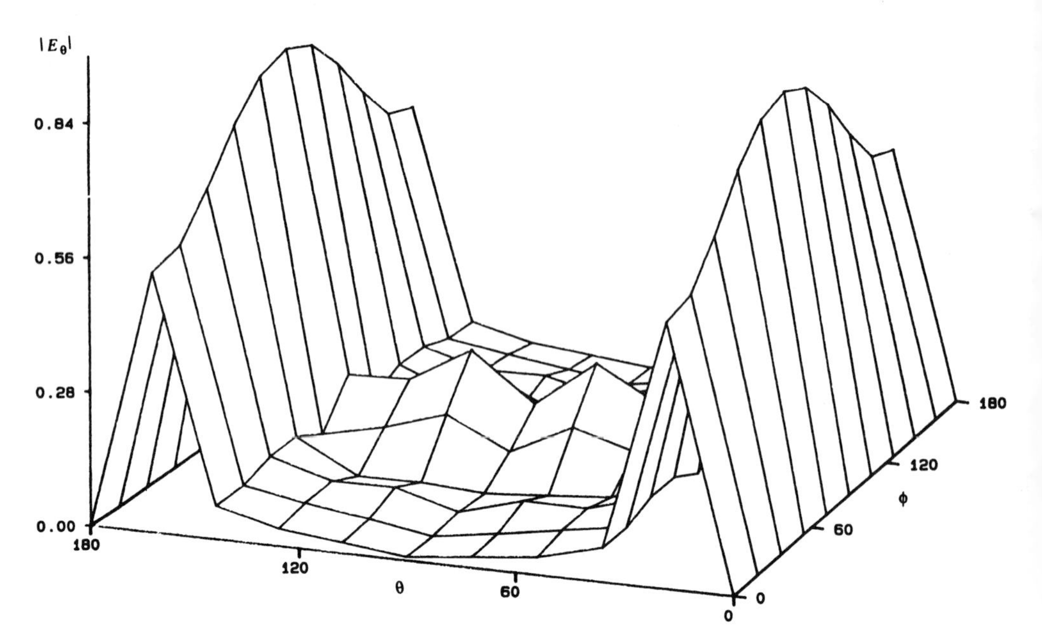

Figure 2c. Computed normalized $|E_\theta|$ vs θ,ϕ for Sinusoidal Current of Length 10.5λ and Width 2λ

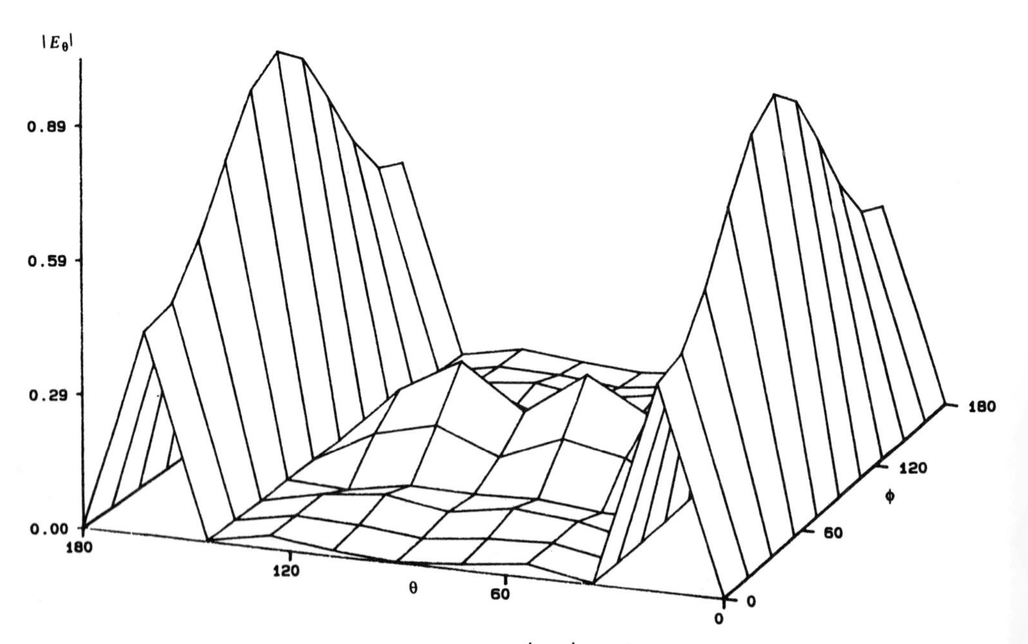

Figure 2d. Analytic normalized $|E_\theta|$ vs θ,ϕ for Sinusoidal Current of Length 10.5λ and Width 2λ

IV. CONCLUSION

A simple, accurate method is presented for computing far-fields from near-field data. The method can incorporate probe compensation which takes into account all induced currents and fields due to the presence of the probe. It is amenable to any measurement configuration, and the far-fields outside of the solid angle subtended by the measured region can be accurately approximated. Using this method it is possible to find the far-fields of antennas which are not highly directive over large elevation and azimuthal ranges without using spherical scanning. Furthermore, the far-field may be found with any desired resolution, and interpolation is not required. The drawback of the method is computation - when CGFFT is not used, the method is computationally intense. For planar scanning using CGFFT, the equivalent current approach is computationally efficient. Since planar measurements are the easiest to perform, and accurate results are obtained over large elevation and azimuthal ranges with the technique presented in this paper, planar near-field measurements used in conjunction with the equivalent current approach and CGFFT provide a viable and efficient technique for the determination of far-field antenna patterns.

REFERENCES

[1] T. K. Sarkar, K. R. Siarkiewicz, and R. F. Stratton, "Survey of numerical methods for solution of large systems of linear equations for electromagnetic field problems", IEEE Trans on Antennas and Propagat., vol. AP-29, pp. 847-856, Nov. 1981.

[2] T. K. Sarkar, E. Arvas, and S. M. Rao, "Application of FFT and the conjugate gradient method for the solution of electromagetic radiation for electrically large and small conducting bodies", IEEE Trans on Antennas and Propagat., vol. AP-34, pp. 635-640, May 1986.

[3] C. C. Su, "Calculation of electromagnetic scattering from a dielectric cylinder using the Conjugate Gradient Method and FFT", IEEE Trans. on Antennas and Propagat., vol. AP-35, pp. 1418-1425, Dec. 1987.

[4] Su, C. C., "Electromagnetic scattering by a dielectric body with arbitrary inhomogeneity and anistropy", IEEE Trans. on Antennas and Propagat., vol. AP-37, 384-389, Mar. 1989.

[5] M. F. Catedra, E. Gago, and L. Nuno, "A numerical scheme to obtain the RCS of three-dimensional bodies of resonant size using the Conjugate Gradient Method and the Fast Fourier Transform", IEEE Trans on Antennas and Propagat., vol. AP-37, pp. 528-537, May, 1989.

[6] T. K. Sarkar and X. Yang, "Efficient solution of Hankel systems utilizing FFTs and the Conjugate Gradient Method", Proc. of International Conf. on Acoustics, Speech and Signal Processing (ICASSP 86), Dallas, TX, pp. 1835-1838, May 1987.

[7] H. Chen, T. K. Sarkar, S. A. Dianat and J. D. Brule, "Adaptive spectral estimation by the Conjugate Gradient Method", IEEE Trans. on ASSP, vol. ASSP-34, pp. 272-283, Apr. 1986.

[8] T. K. Sarkar, S. Ponnapalli and E. Arvas, "An accurate, efficient method of computing far-field antenna patterns from near-field measurements", Proc. of International Conf. on Antennas and Propagation (AP-S 90), Dallas, TX, May 1990.

[9] S. Ponnapalli, "The Computation of Far-field Antenna Patterns from Near-field Measurements Using an Equivalent Current Approach", Ph.D. dissertation, Syracuse University, December 1990.

[10] S. Ponnapalli, "Near-field to far-field transformation utilizing the conjugate gradient method", in Application of Conjugate Gradient Method in Electromagnetics and Signal Processing, vol. 5 in PIER, T. K. Sarkar, ED. New York: VNU Science Press, Ch. 11, December 1990.

[11] S. Ponnapalli and T. K. Sarkar, "Near-field to far-field transformation using an equivalent current approach", submitted to IEEE Trans. on MTT, September 1990.

[12] J. Brown and E. V. Jull, "The prediction of aerial radiation patterns from near-field measurements", Proc. Inst. Elec. Eng., vol. 108B, pp. 635-644, Nov. 1961.

[13] D. M. Kerns, "Plane-wave scattering-matrix theory of antennas and antennas-antenna interaction", NBS Monograph 162, U.S. Govt. Printing Office, Washington, DC, June 1981.

[14] F. Jensen, "Electromagnetic near-field far-field correlations", Ph.D. dissertation, Techn. Univ. Denmark, July 1970.

[15] P. F. Wacker, "Near-field antenna measurements using a spherical scan: efficient dat reduction with probe correction", in Inst. Elec. Eng. Conf. Publi. 113, Conf. Precision Electromagn. Measurements, London, July 1974, pp. 286-288.

[16] __________, "Non-planar near-field measurements: spherical scanning", NBSIR 75-809, June 1975.

[17] W. M. Leach and D. T. Paris, "Probe-compensated near-field measurements on a cylinder", IEEE Trans. on Antennas and Propagat., vol. AP-21, pp. 435-445, July 1973.

[18] R. F. Harrington, "Field Computation by Moment Methods", Malabar: Robert E. Kreiger Publishing, 1968.

[19] D. T. Paris, W. M. Leach, and E. B. Joy, "Basic theory of probe-compensated near-field measurements", IEEE Trans. Antennas and Propagat., vol. Ap-26, pp. 373-379, May 1978.

[20] R. E. Collin and F. J. Zucker, Antenna Theory, Part I., New York: McGraw-Hill, 4, 1969.

[21] A. V. Oppenheim and R. W. Shafer, "Digital Signal Processing", Englewood Cliffs, NJ: Prentice-Hall, 1975.

ELECTROMAGNETIC MODELING IN ACCELERATOR DESIGNS*

Richard K. Cooper and K. C. Dominic Chan

Los Alamos National Laboratory
Los Alamos, NM 87545

INTRODUCTION

Through the years, electromagnetic modeling using computers has proved to be a cost-effective tool for accelerator designs. Traditionally, electromagnetic modeling of accelerators has been limited to resonator and magnet designs in two dimensions. In recent years, with the availability of powerful computers, electromagnetic modeling of accelerators has advanced significantly.

In the last few years, two conferences were organized to review the state-of-the-art of electromagnetic modeling of accelerators. These conferences were the first in a series which has as its goal an exchange of information about codes between those writing and those using these codes for the design and analysis of accelerators and their components. The first conference[1] was held in San Diego in January 1988, and concentrated on beam-dynamics codes and Maxwell's-equation solvers. The second conference[2] was held in Los Alamos in January 1990, and concentrated on three-dimensional codes and techniques to handle the large amounts of data required for three-dimensional problems.

Through the above conferences, it is apparent that breakthroughs have been made during the last decade in two important areas: three-dimensional modeling and time-domain simulation. Success in both these areas have been made possible by the increasing size and speed of computers. In this paper, the advances in these two areas will be described.

Three-dimensional modeling has been making steady progress in the last five years. The codes are now capable of producing credible results and three-dimensional calculations are routinely performed by accelerator designers. Available codes

---

*Work supported by the US Department of Energy, Office of High Energy and Nuclear Physics.

have been developed principally by workers in the plasma and accelerator physics communities. Among them, most notably, are the ARGUS[3] and SOS[4] codes from the plasma community and the MAFIA[5] codes from the accelerator community. Their capabilities are similar and will be illustrated in this paper using modeling experiences of actual devices in Los Alamos.

Time-domain simulations have grown in popularity in the last ten years. They, beside complementing the traditional frequency-domain computations, allow accelerator designers to achieve a better understanding of the physics by letting them visualize the development of electromagnetic fields in accelerator components. In this paper, we will show, in particular, time-domain simulations for computing beam-induced effects. These simulations are essential for designing high-brightness accelerators.

THREE-DIMENSIONAL (3-D) MODELING

Various groups of people have been involved in writing computer codes for modeling accelerator components in three dimensions since 1985. Their efforts have been successful because of the advances in computer hardwares both in the memory sizes and computing speed. At this time, the accelerator-physics designers are reaping the benefits of the foresight of these pioneers.

The capabilities of the 3-D codes will be illustrated in this section using analyses of accelerator devices in Los Alamos. These analyses will be presented roughly in the order in which a particle beam would encounter the devices, i.e. from the low-energy radio-frequency quadrupole (RFQ) focusing and accelerating device, to the drift-tube linear accelerator (DTL), to the coupled-cavity linear accelerator (CCL), to the traveling wave "jungle gym" structure. All of these devices involve asymmetric geometries, so that the use of 3-D codes for their analyses becomes necessary. The analyses were performed using the MAFIA codes. The theory of these codes is well documented in ref. 6 and will not be repeated here.

RFQ. The radio-frequency quadrupole provides both focusing and acceleration of low-energy ion beams (~ 100 keV for protons). Figure. 1 shows one quadrant of the end region of a four-vane

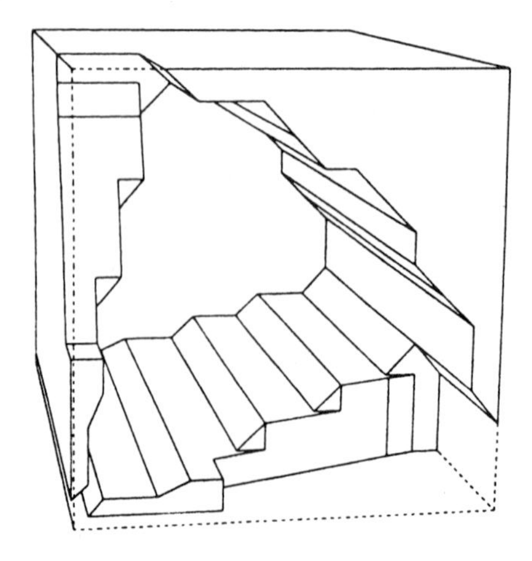

Fig. 1. One quadrant of a four-vane RFQ.

RFQ as approximated by the MAFIA mesh generator. The codes were used to design the undercutting of the vane for tuning the end region so that it matches properly the extended portion of the RFQ structure.[7] Figure 2 shows the joining of two RFQs to form a compensated structure. The coupling between the RFQs has improved the separation between the quadrupole modes, and has improved the longitudinal rf stability. A long RFQ made by joining short RFQs can offer more robust operation than a single long RFQ.[8]

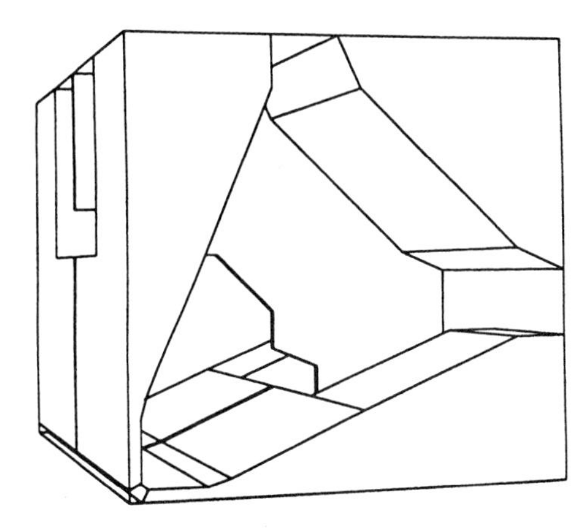

Fig. 2. The coupling of two RFQs.

DTL. A drift-tube linear accelerator is efficient for the acceleration of low-energy ions. Figure. 3 shows a four-gap DTL with a slug tuner, the drift-tube supports, and post couplers used for achieving a flat accelerating-field profile.

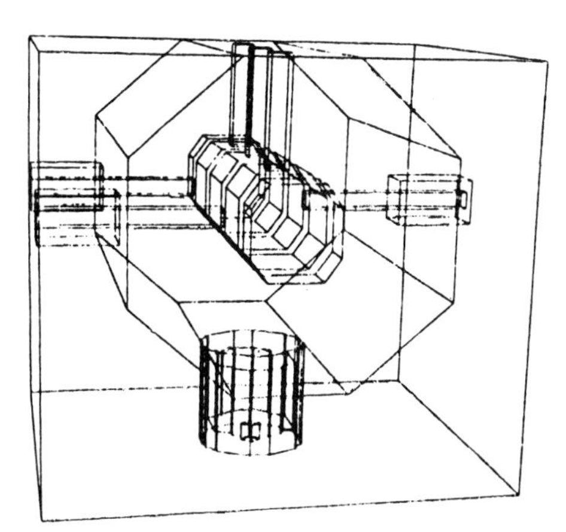

Fig. 3. A three-dimensional approximation to a four-gap DTL.

A study by Shmoys and Li[9] has represented a DTL as a coaxial line with microwave circuit loading representing the gaps and post couplers. One application of the MAFIA codes is the determination of microwave circuit parameters of such a structure. Details of such an analysis can be found in ref. 9.

The MAFIA codes have also allowed the designers to investigate the mode structures around the operating mode of a DTL. The analysis has shown that no harmful modes (Figs. 4 and 5) have been introduced by the supports of the drift tubes and

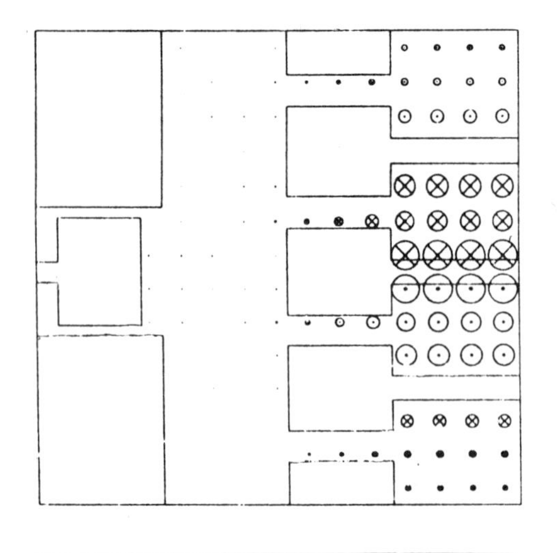

Fig. 4. The magnetic field of a stem mode.

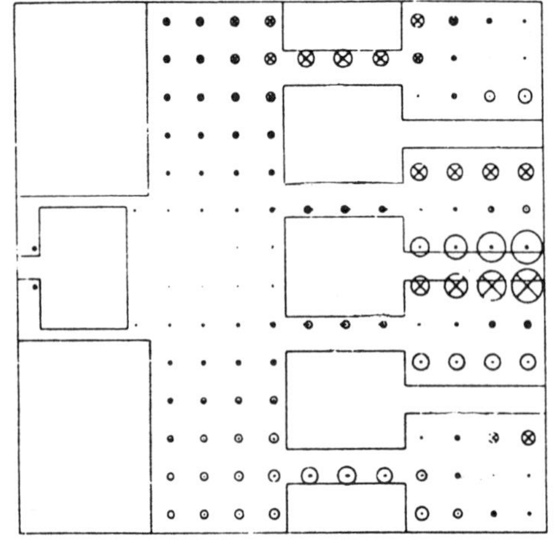

Fig. 5. The magnetic field of a higher-order mode.

post couplers in the vicinity of the operating mode. The analysis has also provided considerable insight without the need for expensive fabrication and rf testing.

CCL. Coupled-cavity linear accelerators are efficient for accelerating high-energy ion beams. The popular side-coupled and on-axis-coupled structures are shown in Fig. 6 and 7 respectively. Recently, there have been experimental reports of the asymmetry effects introduced by the coupling slots.[10] The asymmetry introduces quadrupole focusing fields near the axis. Figure 8 shows these additional quadrupole fields as calculated using the MAFIA codes.[11] These field patterns were obtained by taking the difference of two field solutions generated for structures with and without the coupling slots.

Jungle-gym Structure. The Jungle-gym structure is an advanced bar-loaded traveling-wave accelerating structure for high-energy ($v \approx c$) particles. The structure has been studied experimentally by Tigner at Cornell.[12] A unit cell of this structure consists of a pair of vertical bars followed by a pair of horizontal bars. An approximation of a three-cell structure

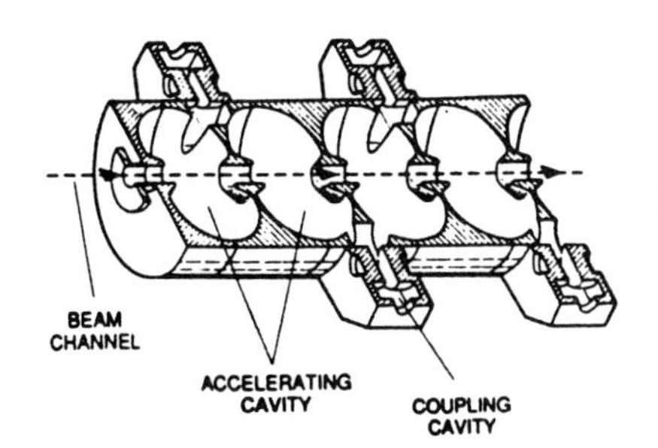

Fig. 6. A cutaway drawing of a side-coupled linac structure.

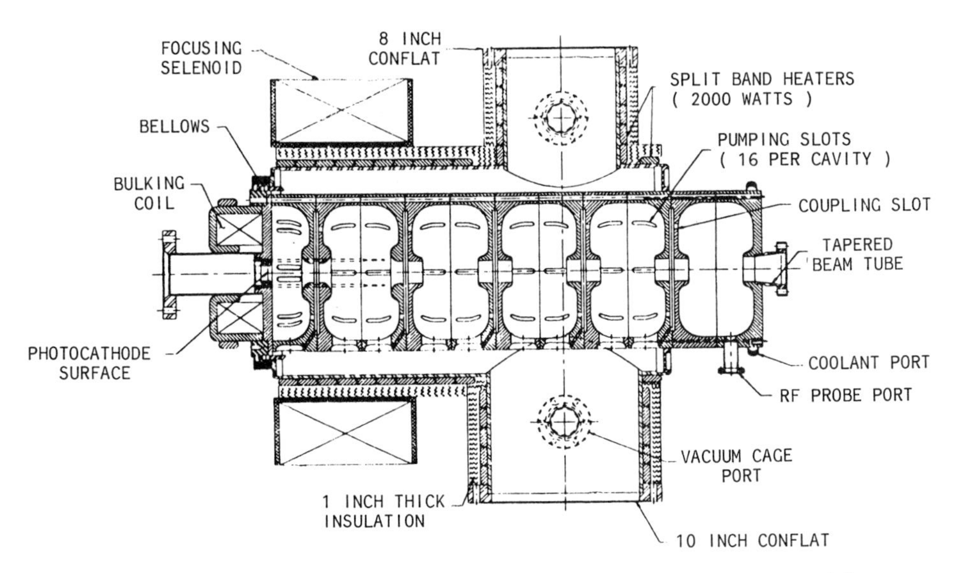

Fig. 7. A cutaway drawing of an on-axis-coupled linac structure.

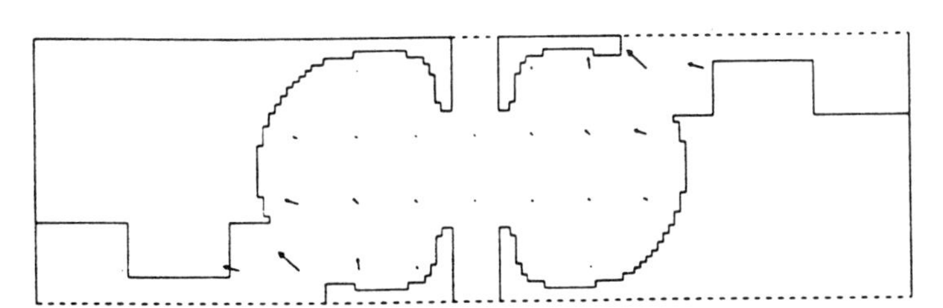

Fig. 8. Difference fields between a symmetrical cavity and a side-coupled cavity, showing the effect of the coupling slots and cells.

by the MAFIA codes is shown in Fig. 9. Only one-fourth of the structure is shown; the x and y planes are planes of mirror symmetry as far as the structure is concerned. Figure 10 shows plots of the accelerating field on the axis for several of the modes calculated. The dispersion curve calculated by the MAFIA codes was in good agreement with that measured by Tigner. Details of the analysis are given in ref. 13.

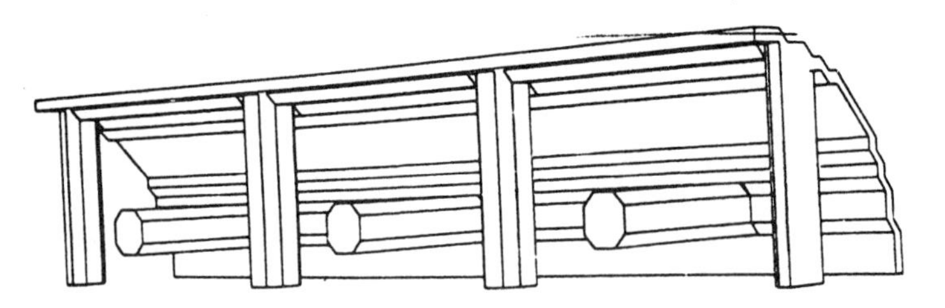

Fig. 9. One quarter of the jungle gym structure with three cells.

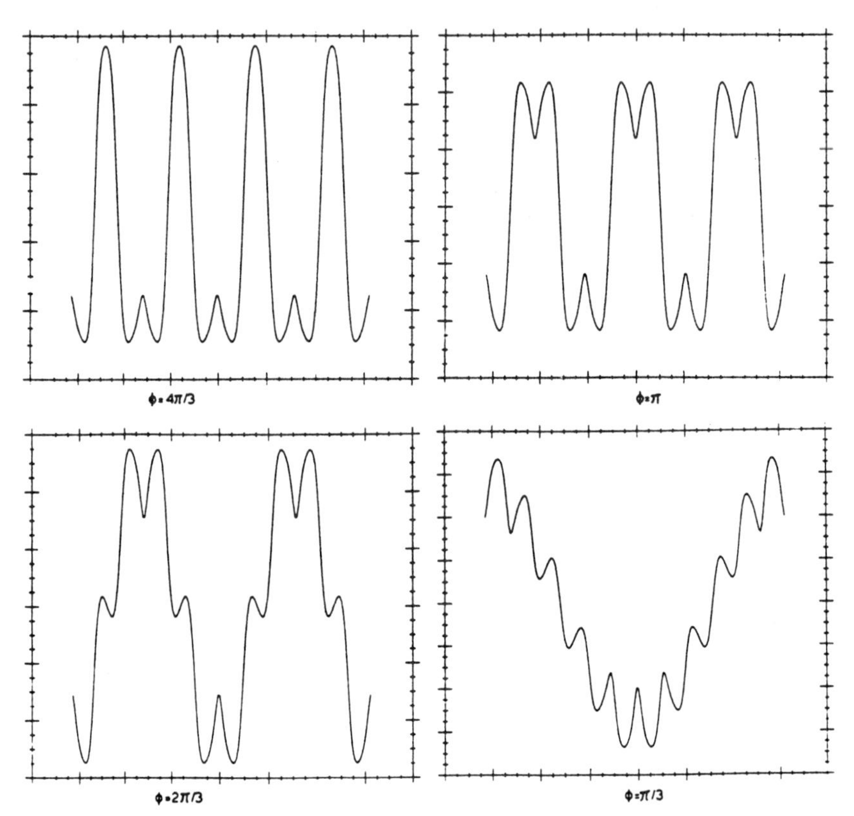

Fig. 10. Plots of the longitudinal field on axis for several modes of the jungle gym structure. The various values of φ denote the phase shifts per cells.

TIME-DOMAIN SIMULATION

In accelerator design, electromagnetic modeling can be performed in either the frequency domain or the time domain. In the past, accelerator modeling was mainly performed in the frequency-domain by calculating eigenfunctions, similar to analyses as described in the last section except they were for two dimensions. Time-domain modeling has become popular only in the last ten year. It has become popular because of the availability of computer codes and because of the needs for calculating beam-induced effects of short beam bunches with high charge densities in accelerators with high beam brightness. Frequently, the time-development of the electromagnetic fields as simulated in time-domain modeling is captured in a time-sequence of pictures or movies, giving a vivid representation of the physics interplay of the problem.

In this paper, we will show two analyses in the time domain: the radiations of a sub-nanosecond pulse of protons exiting a beam pipe; and the excitation of a relativistic klystron with a field-transformer accelerator structure. We will also comment briefly about the computer codes and results of direct comparisons of calculations using these computer codes to experiments.

<u>First Analysis</u>: Figure 11 is a time sequence of pictures showing the development of radiations when a sub-nanosecond proton bunch moves from a beam pipe into open space. This calculation models an experiment performed at the Los Alamos Meson Physics Facility. The comparison of the measured radiations to calculations has been described in detail in ref. 14 and will be mentioned briefly later in this section.

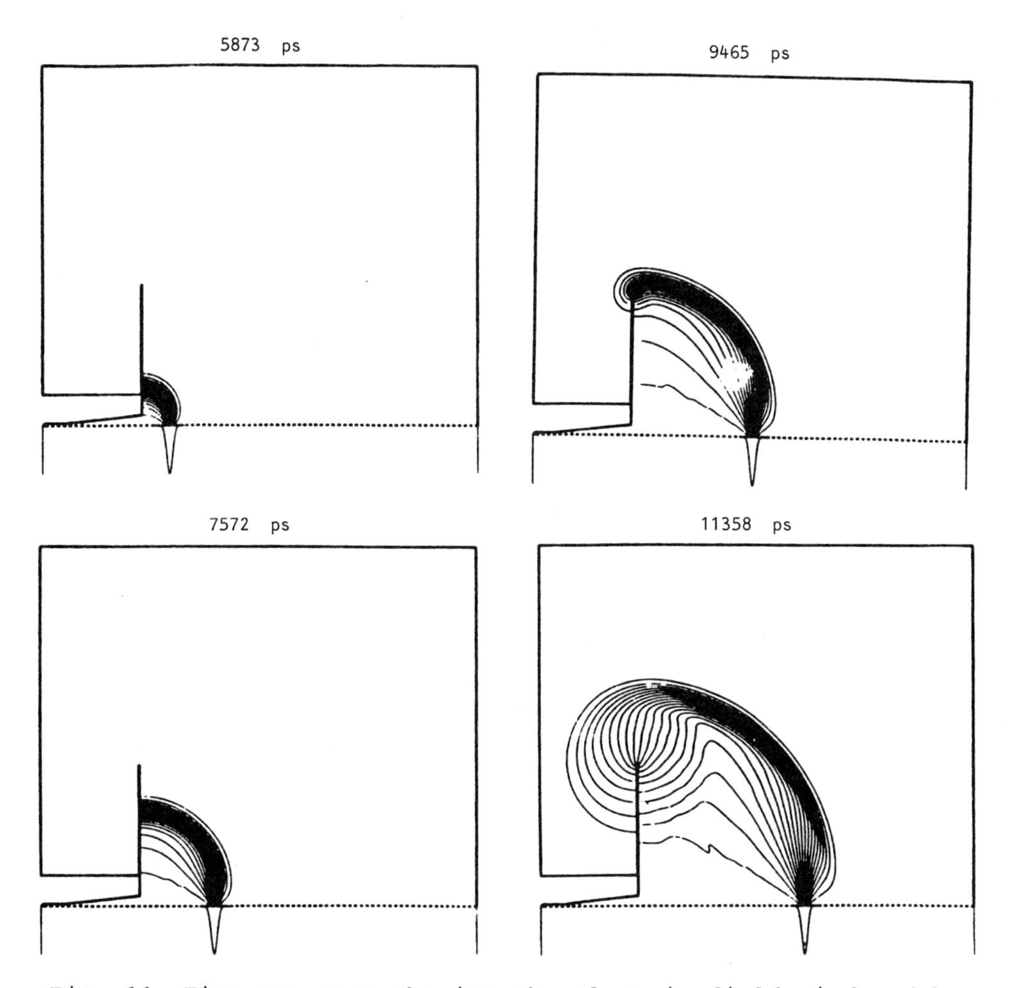

Fig. 11. Time sequence showing the electric fields induced by a proton beam.

Second Analysis: Figure 12 shows a sequence of pictures which simulates an experiment being conducted at the Naval Research Laboratory.[15] A high-intensity bunch of relativistic electrons was used to excite a coaxial cavity via an accelerating gap. The electromagnetic energy in the cavity built up and was coupled to an accelerating structure for the acceleration of a high-energy low-intensity beam. The excitation of the coaxial cavity and then the structure has been clearly depicted in Fig. 12.

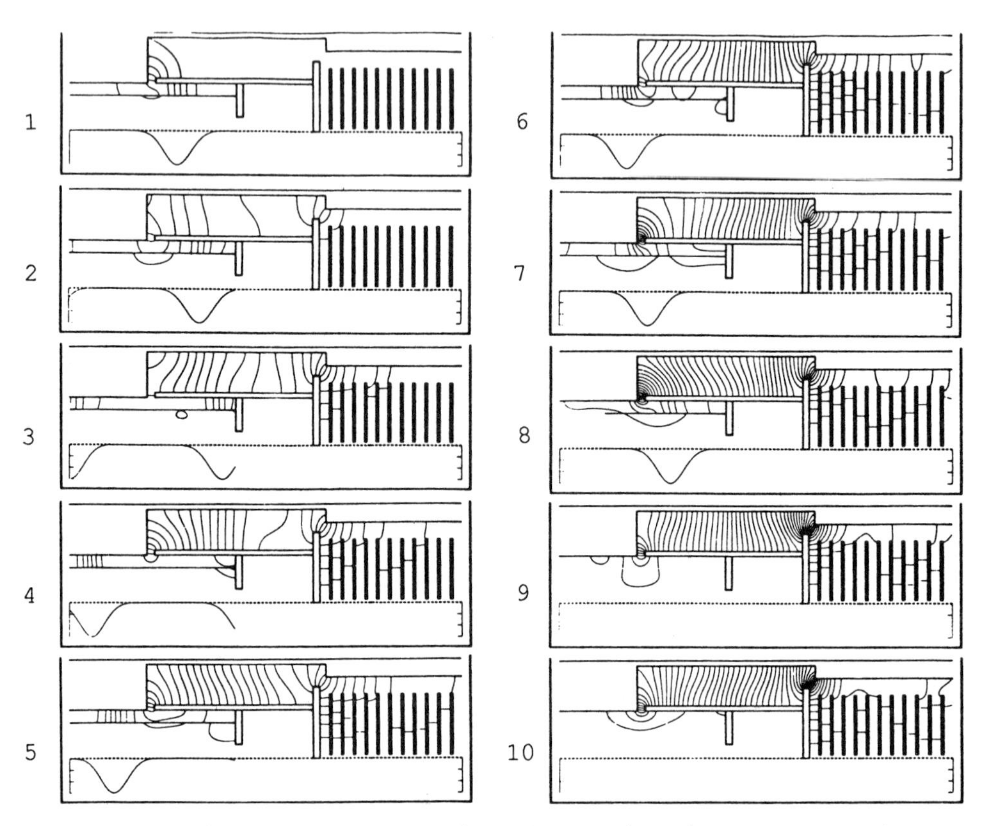

Fig. 12. Time sequence showing the excitation of a coaxial cavity and an accelerating structure.

Computer Codes. Although time-domain particle-in-cell (PIC) codes have been used by plasma physicists for a long time, the time-domain code most used in the accelerator physics community, TBCI[16], written by Thomas Weiland, appeared only in 1981. This code uses the finite-difference method and is designed for the analysis of axially symmetric structures. IBCI incorporates features for saving CPU time, such as the open-boundary condition for modeling infinite beam pipes, and the window option, exploiting causality of particles traveling at the speed of light. This code has been extended to three dimensions and is a part of the MAFIA codes.

A time-domain code usually uses the finite-difference forms of the Faraday's and Ampere's laws, and uses a leapfrog integration method to advance the electric field and magnetic field, alternatively, in time. Such an algorithm is a natural candidate for use in a massively parallel computer. Each node on the mesh corresponds to a single processor in such a machine. Recently, such programs have already been successfully implemented in the Connection Machine at the Argonne National Laboratory, with impressive performance.[17]

The results of TBCI have been compared directly to experiments. Figure 13 shows a typical comparison for radiations emitted from a sub-nanosecond proton bunch as in an experiment described earlier in this section. The comparison shows excellent agreement, considering the 15% error bars of the experimental data. Recently, experimenters of the Advanced Accelerator Test Facility at Argonne National Laboratory measured the longitudinal and transverse fields induced by a beam bunch in a dielectric-lined waveguide. The experimental data are compared to simulations using the MAFIA and ARCHON[17] codes in Fig. 14 showing good agreement.

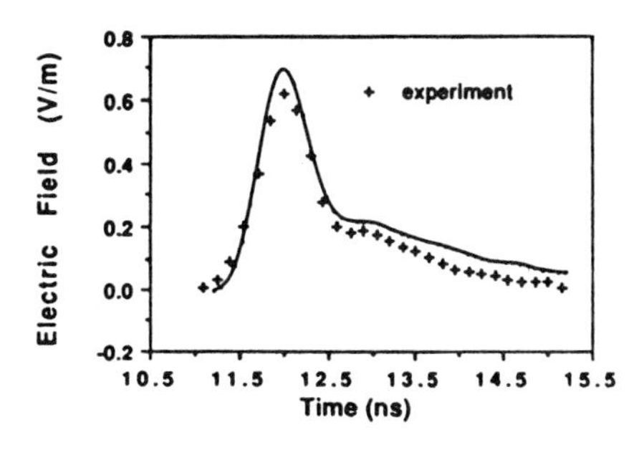

Fig. 13. Electric field induced by a 800 MeV proton beam exiting a beam pipe. The calculated results are shown as a solid curve.

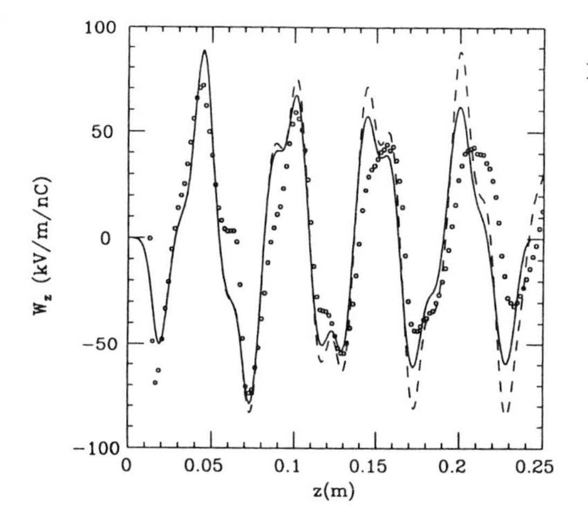

Fig. 14. Comparison of the longitudinal W_z and transverse field W_t induced by a beam bunch in a dielectric-lined waveguide. The open circles are the measurement. The solid and dashed curves are, respectively, calculations using MAFIA and ARCHON.

SUMMARY

Electromagnetic modeling in accelerators has been an expanding area of research in the last ten years with great progress in three-dimensional modeling and time domain simulation.

REFERENCES

1. C. R. Eminhizer, Ed., "AIP Conference Proceedings 177: Linear Accelerator and Beam Optics Codes, La Jolla Institute, 1988," American Institute of Physics, New York, (1988).
2. R. K. Cooper, Compiler, "Proceedings of the Conference on Computer Codes and the Linear Accelerator Community, Los Alamos National Laboratory, January 22-25, 1990," Los Alamos National Laboratory report LA-11857-C, (1990).
3. C. L. Chang, D. Chernin, A. Drobot, K. Ko, M. Kress, A. Mankofsky, A. Mondelli, and J. Petillo, "Three-Dimensional Modeling of Accelerators," pp 27-55 in ref. 2.
4. B. Goplen, K. Heaney, J. McDonald, and G. Warren, "SOS Reference Manual Version October 1988," Mission Research Corporation final report MRC/WDC-R-190 (March 1989).
5. T. Weiland, "Solving Maxwell's Equations in 3D and 2D by means of MAFIA," pp 3-25 in ref.2.
6. T. Weiland, "On the Numerical Solution of Maxwell's Equations and Applications in the Field of Accelerator Physics," Particle Accelerators **15**, (1984), pp 245-292 and references therein.
7. M. J. Browman, G. Spalek and T. C. Barts, "Studying the End Regions of RFQs Using the MAFIA Codes," Proceedings of the 1988 Linear Accelerator Conference, Continuous Electron Beam Accelerator Facility, Newport News, Virginia, Oct. 3-7, 1988, CEBAF-Report-89-001, p 64, (1989).
8. M. J. Browman and L. M. Young, "Coupled Radio-Frequency Quadrupoles as Compensated Structures," Presented at the 1990 Linear Accelerator Conference, Sept. 10-14, 1990, Albuquerque, NM.
9. J. Shmoys and R. Li, "Transmission Line Modeling of Drift Tube Linear Accelerators," Los Alamos National Laboratory internal report AT6:ATN-88-16.
10. H. Euteneuer, "Notiz zur Drehung der MAMI Beschleunigungssektionen um Ihre Langsachse," University Mainz Internal Note MAMI 1/89, 1989.
11. M. J. Browman, private communication.
12. M. Tigner, "Bar Loaded Waveguide for Accelerator Service," IEEE Trans. Nucl. Sci. NS-18, No. 3, p 249, (1971).
13. J. Loo, M. J. Browman, K. C. D. Chan and R. K. Cooper, "Numerical Studies of the Jungle Gym Slow-wave Accelerator Structure," Particle Accelerator **23**, p 279, (1988).
14. K. C. D. Chan and R. K. Cooper, "Time-Domain Calculation of Sub-nanosecond Pulse Launched by a Proton Beam", presented at the First Los Alamos Symposium on Ultra-Wideband Radar, Los Alamos, NM, March 5-8, 1990. Also in Los Alamos National Laboratory report LA-CP-89-263.
15. M. Friedman, V. Serlin, A. Drobot, and L. Seftor, "Self-Modulation of an Intense Relativistic Electron Beam," J. Appl. Phys. 56 (9), p 2459, (1984).
16. T. Weiland, "Transverse Beam Cavity Interaction, Part I: Short Range Forces," Nucl. Inst. Meth. 212, pp 13-24, (1983).
17. P. Schoessow, "Wakefield Calculations on Parallel Computer," pp 377-386, in ref. 2.

APPLICATION OF ADAPTIVE REMESHING TECHNIQUES TO THE FINITE ELEMENT ANALYSIS OF NONLINEAR OPTICAL WAVEGUIDES

R. D. Ettinger, F. A. Fernandez and J. B. Davies

Department of Electronic and Electrical Engineering
University College London
Torrington Place, London, WC1E 7JE, England

INTRODUCTION

There is rapidly growing interest in the finite element analysis of optical waveguide modes. The generality of the variational formulation of the method allows the treatment of an enormously wide range of problems, ranging from closed metallic microwave guides to integrated optical structures with extremely intricate cross-sectional form.

It is obviously crucial to avoid repetitious or unnecessary steps for both finite element programmers and users. This means that software should not be highly problem-specific, but must at the same time not make burdensome demands upon the user (who is likely to be much more interested in the results than the method). In our workstation-based environment, a user-friendly graphical interface allows great freedom in entry of geometry and material inputs, but the mesh generation which follows on from this is made largely automatic. Assembly and matrix solution of the discretised problem is then followed by interactive examination of the results.

The main theoretical complications in electromagnetic modal analysis (described by a 2D Helmholtz equation) are due to having an eigenvalue problem, which is furthermore vectorial in **E** or **H** and may also be nonlinear. Computer analysis is vital in this field and will hopefully be as "user-friendly" as packages already available for solving the Laplace and Poisson equations in 2D[1].

Adaptive mesh generation consists of an algorithm that automatically selects a mesh distribution that is (in some sense) best suited to a particular mode or field distribution. One feature of adaptive techniques is of interest even in the linear case, namely that non-uniform and hopefully near-optimum layout of the mesh does not require any intervention on the user's part. Adaptive mesh generation methods are specially attractive when the modal solution depends on a parameter, such as power in a nonlinear optical guide. The optimum mesh then differs greatly from one power to another, where a non-adaptive program would simply use the same mesh in every case, hence sacrificing either accuracy or speed.

In the area of optical waveguide analysis, nonlinearity is of special interest because integrated and fibre optic nonlinear components exhibit novel physical behaviour with

Directions in Electromagnetic Wave Modeling
Edited by H.L. Bertoni and L.B. Felsen, Plenum Press, New York, 1991

applications to all-optical signal processing. We believe that finite elements combined with adaptive remeshing techniques, whose importance is already appreciated in nonlinear stress analysis, will be a natural tool for this nonlinear modeling. The task is usually to calculate not just a particular nonlinear result, but the solution space representing the varying behaviour over a wide range of operating powers. We present below efficient techniques for this purpose, which we have implemented in our computer package, POMME (Program for Optical and Microwave Mode Evaluation) running on a SUN 4/110 computer.

Two general features of nonlinear optics are particularly important to remember. The first is that comparisons between reality and approximate models which retain one instead of two transverse dimensions, can show fundamental differences. In particular, the well-observed threshold for in-bulk self-focusing does not occur under this 1D approximation. The preferable treatment for waveguides must therefore use a full 2D cross section, which is indeed the case in the present work.

The second, and even more fundamental feature, is a consequence of non-superposition. It is that the nonlinear "modes" cannot be used to describe the behaviour of a waveguide under arbitrary excitation, although they are of fundamental importance because they exist as solutions, either stable or unstable. A parallel use of nonlinear modal analysis and initial value (propagation) treatment[2,3] is the best way to understand unstable situations such as the emission of optical spatial solitons. Finite element methods have been applied to the "z-transient"[2-4] problem and we believe that adaptive meshing techniques will prove useful here as well.

THE METHOD OF ADAPTIVE REMESHING

Given an approximate finite element solution obtained with some ad hoc discretization, adaptive remeshing[5] allows one to obtain a more accurate solution without increasing the size of the matrix calculation. The way is to reduce the global error with a completely new mesh derived from the earlier approximate solution by placing more elements where accuracy was locally low and fewer elements where accuracy was higher than relevant. This 'equipartition' of a suitable error norm amongst the elements leads to an optimized mesh. A series of such meshes will converge to the true solution much more rapidly than a series of uniform meshes, for given computing resources.

A less general approach than adaptive remeshing is adaptive refinement where refinement (with no coarsening) of only selected elements takes place at each step. The only advantage of this approach may be that less calculation would be necessary in reassembling the matrices for the new calculation. However, assembly time is not the major expense in our calculations. Therefore we have considered the more powerful adaptive remeshing approach. Certainly this approach is more relevant when a solution is sought with some parameter (power in the case of nonlinear optics) being varied.

In the finite element analysis of waveguides, absolute discretization error arises mainly in the regions of maximum guided power. Conversely, if anywhere the field tends to zero, even a locally coarse mesh detracts little from excellent results. Hence we find that a convenient and effective error indicator is the Poynting vector integral across the area of an element.

If we divide the total integral from a previous calculation by a given minimum number of elements in the new calculation, we fix the elemental power flow. A new mesh is now generated, using a procedure of forcing elements to subdivide until each elemental power flow is smaller than or equal to the fixed reference. Local equipartition of power flow is not exact in this method, but globally the new mesh is certainly close to an optimum.

It is well known that adaptive methods are essential for the satisfactory treatment of nonlinear problems, because the optimal mesh changes[6]. Likewise, the examples from nonlinear optics given below demonstrate the case of the optimal mesh changing with power level.

A mesh generator to support adaptive remeshing may be organised by analogy to an existing non-adaptive algorithm. In our implementation, there are two main steps in mesh generation. Firstly, all nodes required on the various different material boundaries must be pre-selected. Secondly, automatic generation of an adequate mesh from the "outline" mesh defined on these nodes proceeds by the Delauney algorithm with a stopping criterion. Adaptivity comes in differently in the pre-selection and in the infilling.

Under interactive operation, the pre-selection step on each boundary involves choosing the number of subdivisions for either uniform or logarithmic placement. In the adaptive case, which is not interactive but automatic, the previous field solution should select the number of subdivisions and the algorithm place nodes individually to satisfy a pre-defined power criterion. This selection should also take into account the eventual requirement for a fixed total number of elements. The stopping criterion (in step 2) is also different for the adaptive case. Instead of employing a maximum permissible area, a maximum permissible Poynting vector integral controls when to divide further a particular element.

NONLINEAR WAVEGUIDE MODAL ANALYSIS

The nonlinear modes are power-dependent solutions whose fields are given in terms of time t and propagation direction z,

$$E = e(x,y,p)e^{j(\omega t-\beta z)} \tag{1}$$

$$H = h(x,y,p)e^{j(\omega t-\beta z)} \tag{2}$$

where $p = \frac{1}{2}\int e \times h^* \, dxdy$. The full solution of the mode defines a surface in 3-dimensional space $\{\omega, \beta, p\}$ consistent with (1), (2) and with Maxwell's equations, the boundary conditions and the nonlinear constitutive relationship. Assuming the permeability to be constant, H and h will be continuous even at dielectric interfaces, making it appropriate to solve for h rather than e.

If the nonlinearity is non-diffusive, it can still be very complicated due to the tensor character of the linear and higher-order terms. The method used here can deal with a wide variety of laws, but will be illustrated for the following simple saturable nonlinearity (which becomes Kerr-like for low fields):

$$D = \varepsilon_0 \varepsilon E$$

$$\varepsilon = \varepsilon_{lin} + \Delta\varepsilon_{sat}\left[1 - e^{-\alpha|E|^2 / \Delta\varepsilon_{sat}}\right] \tag{3}$$

as well as for a pure Kerr law, where

$$\varepsilon = \varepsilon_{lin} + \alpha|E|^2 \tag{4}$$

The nonlinear eigenvalue equation for H in all cases is as follows:

$$\nabla \times \left(\varepsilon^{-1} \nabla \times H\right) = k_0^2 H \tag{5}$$

where $k_0 = \omega/c$. We may consider β constant and ω as the eigenvalue, or vice versa, whichever is more convenient.

To solve eqn. (5) we have used an iterative approach, consisting of the successive modal solution of easier linear eigenvalue problems, to generate a series of solutions as follows:

$$\begin{cases} \nabla \times \left(\varepsilon_{i-1}^{-1} \nabla \times H_i \right) = k_0^2 H_i & \text{with eigenvalue } \beta_i \\ \varepsilon_i = \varepsilon(H_i) & \end{cases} \tag{6}$$

This series will either converge to a limit, which would be an exact solution of the nonlinear equation (5), or it will fail to settle down. If a stable nonlinear solution exists and if the series of solutions starts with a "guess" close to that solution, then convergence of the series into that particular solution is assured. If the "guess" is inadequate, the limit can be some other stable solution of eqn. (5). The series does not converge onto any physically unstable solutions - instead regular oscillations[7] or chaotic behaviour[8] can occur.

A continuation approach ensures that a suitable "guess" is provided for the iterative solution method outlined above. We select a trajectory on the solution surface, by fixing ω, and look for the modal solution with gradually changing power. As the power is stepped, the final solution of the iterations for the previous power is used as the "guess" for starting the new iterative solution of eqn. (5). At certain special powers, gradual continuation is not possible, such as where the stable mode bifurcates, disappears, or suddenly changes its character (analogous to snap-through in nonlinear mechanical deformation[6,9]). Then the new "guess" is no longer a good first approximation and one must expect the series to take a larger number of terms to settle down.

In the possible case of coexisting bistable nonlinear modes in a certain power range, the "guess" obtained by the continuation approach influences which of the two possible limits the iterations settle onto for given power, causing hysteresis with respect to power[10]. Physically, the possibility of hysteresis depends not on having "slow-response" material but on possessing independent global solutions (for the equilibrium between field and material), their occurrence being dependent on the system's history and not only on power.

The necessary linear eigenvalue finite element solution has been described elsewhere[11]. Recalling that the sparse finite element matrices are solved by subspace iteration (an efficient type of inverse iteration), it is obvious that the closeness of the initial "guess" affects the computing time (number of subspace iterations) for an individual linear solution in the series needed for the nonlinear analysis at given power. To obtain H_i the normalized linear solution is scaled to the required total power.

The finite element mesh could remain the same during the entire calculation, but this would be grossly inefficient, as the later results show. Whenever the modal profile has changed noticeably, an adaptive remeshing should be carried out. In order not to disturb the continuation approach, the "guess" for the new mesh is provided from the old mesh by interpolation, thus helping fast solution in the matrix solver.

TRANSVERSE SWITCHING BEHAVIOUR

Analytic studies of planar (one-transverse-dimensional) nonlinear waveguides provided the earliest examples of multiple field solutions for the same power[12].

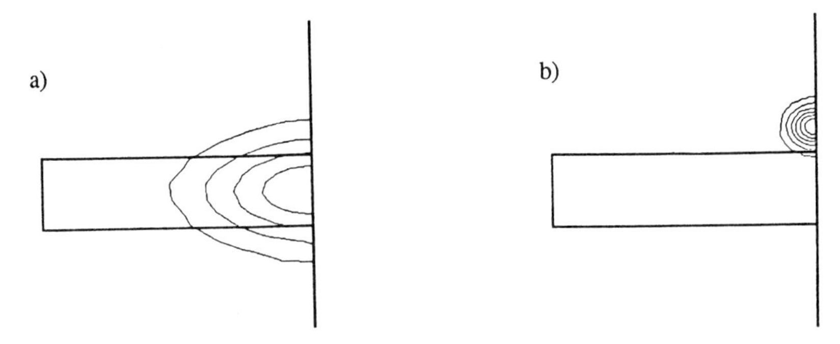

Fig 1. Power-dependent switching. Dominant quasi-TE mode of a rectangular guide $5.31 \times 0.67\ \mu m$, at wavelength $0.515\ \mu m$. Poynting vector contours on the half cross-section at $p/2 = 0.24\ \mu W$ (a) and $p/2 = 0.34\ \mu W$ (b). Outside $n = 1.55$, $\alpha = 1.29 \times 10^{-9}\ m^2/V^2$. Inside $n = 1.57$ (linear material).

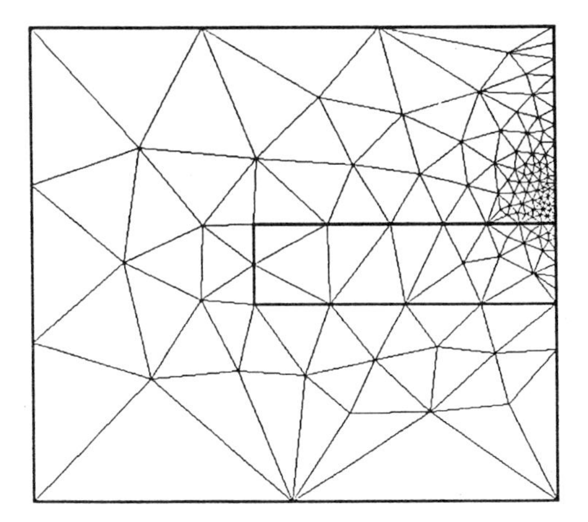

Fig 2. Adaptive mesh generated in calculation of figure 1b.

In the 1D case of a thin film with nonlinear cladding, the analytic β versus power characteristic for the fundamental mode folds back on itself to give an S shaped curve, which includes both stable and unstable regions. The stable portions are those with positive slope, as shown both by the Beam Propagation Method[13] and from first principles[14].

When the same nonlinear structure is solved using our continuation approach with the finite element method[10,15], only stable solutions are generated. Thus one observes a realistic sudden "switching" to a new transverse pattern with different β, when the power reaches a critical value. This effect is of obvious interest for optical signal processing.

The first two-transverse-dimensional results published for transverse switching were by Hayata and Koshiba[16] with a finite element model requiring significant time on a supercomputer. Because of our use of adaptive remeshing, and our efficient sparse matrix software, we are now able to study problems of the same type routinely on a SUN4 workstation (equipped with 8Mb of RAM).

Figure 1 shows a typical two-transverse-dimensions mode at powers before and after the sudden switching effect for a linear guide surrounded by nonlinear cladding. The final mesh produced by adaptive remeshing is shown in figure 2.

STABLE SELF-FOCUSING EFFECT

It is well known that, allowing for two-transverse-dimensional field variation, the onset of self-focusing in bulk material is virtually catastrophic at a critical power. This arises from a basic singularity in the 2D scalar nonlinear Schrödinger equation, which approximates propagation in homogeneous Kerr material as in eqn. (4). Self-trapped guided modes can be calculated for this bulk case, but are unstable and have never been observed experimentally.

The nonlinear waveguide consisting of inhomogeneous Kerr and linear material distribution has a very much more interesting behaviour. In transverse-1D, the mode can focus indefinitely (approaching a spatial soliton[18]) without any instability, as was indeed the case in 1D bulk mentioned before. In transverse-2D nonlinear waveguides, however, our finite element results illustrate a very different effect to that in bulk. Up to a critical power, there is a very pronounced reduction in the area of the mode, but the mode retains stability. This 2D effect is not catastrophic at all even though almost all the power is in the Kerr material. This stabilized self-focusing effect does not seem to have been stressed before, and may be useful in situations which require a power-controlled spot-size.

Figure 3 gives the spot-size versus total power for a two-transverse-dimensions nonlinear guide with linear cladding. Spot-size is defined as the half-power area of the beam. Note that the stable characteristic permits a 30-fold variation of spot-size to be achieved.

This calculation would have been impossible without adaptive remeshing, because mesh-dependence would then be bound ultimately to affect the results. It should also be noted that this particular problem is not soluble using separable variables in cylindrical coordinates, although the mode does become gradually more circular.

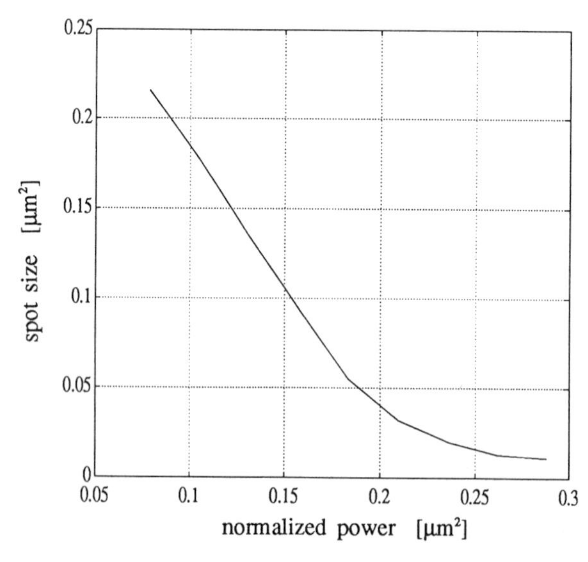

Fig 3. Stable self-focusing effect in saturation-free (Kerr) regime. Spot-size as function of normalized power for the dominant mode of a 2.0 μm square guide, at wavelength 0.85 μm. Inside $n = 1.55+\alpha|E|^2$. Outside $n = 1.5$ (linear material). Actual power is given by normalized quantity times $(Z_0\alpha)^{-1}$.

The example in figure 4 illustrates our use of remeshing to follow the stable self-focusing effect in a rectangular guide.

An alternative method for stabilization of self-focusing is to use a 1D film geometry, as demonstrated experimentally recently by Aitchison et al.[17] Our results suggest that it will, after all, be possible also to work in 2D. Furthermore, high power concentration, needed for large nonlinear effects, is easier to obtain under 2D confinement conditions. Only after reaching a critical power does the stable 2D self-focusing effect cease, and no stable numerical solution was found (this is connected with our choice of a pure Kerr material). The curve in figure 3 therefore ends at this point.

CONCLUSIONS

Theoretical analysis of novel phenomena and, similarly, applied analysis for design, demand ever more accurate results. However, even with modern computers, rapid turnaround is only possible in complex situations by using efficient algorithms. Our work has shown that the use of adaptive remeshing in finite element analysis allows a very efficient solution for the case of nonlinear optical waveguide modeling. By means of this innovative approach, fully vectorial two-transverse-dimensional modal analysis can be carried out on a small workstation. The new results shown here illustrate the usefulness of this approach.

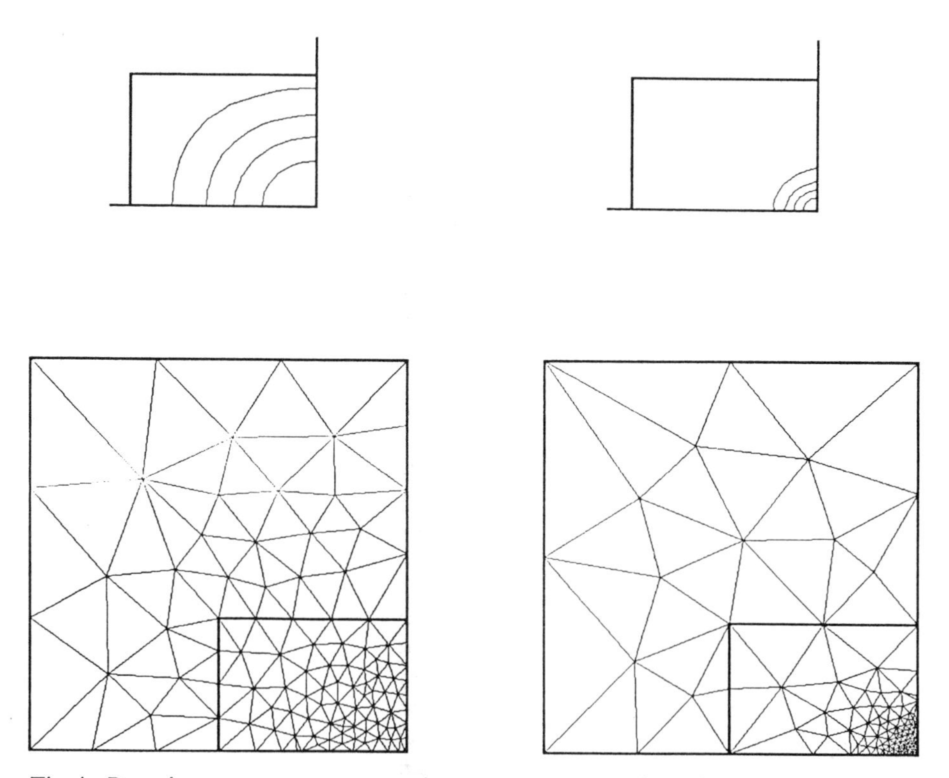

Fig 4. Poynting vector contours on the quarter cross-section of a rectangular guide 3.0×2.0 μm at two different powers. The meshes shown are the results of adaptive remeshing using these distributions.

ACKNOWLEDGEMENTS

We would like to acknowledge Vector Fields Ltd. for provision of software used in the development of our programs. We are also grateful to Dr. B.M.A. Rahman for helpful comments. This work was supported financially by the UK Science and Engineering Research Council.

REFERENCES

1. B. Colyer and C. W. Trowbridge, Finite element analysis using a single user computer, Computer-Aided Design 17:142 (1985).
2. T. B. Koch, J. B. Davies, and D. Wickramasinghe, Finite element/finite difference propagation algorithm for integrated optical device, Electron. Lett. 25:514 (1989).
3. T. B. Koch, R. März, and J. B. Davies, Beam propagation method using z-transient variational principle, 16th European Conf. on Opt. Comm. (ECOC), Amsterdam, (16-20 Sept. 1990).
4. T. B. Koch, Computation of wave propagation in integrated optical devices, PhD thesis, University of London (1989).
5. H. Jin and N.-E. Wiberg, Two-dimensional mesh generation, adaptive remeshing and refinement, Int. J. Numer. Methods Eng. 29:1501 (1990).
6. I. Babuska and W. C. Rheinboldt, Computational error estimates and adaptive processes for some nonlinear structural problems, Comput. Methods Appl. Mech. Eng. 34:895 (1982).
7. A. C. P. Zabeu and J. R. Souza, The stability of nonlinear TE_1 guided waves revisited, Microwave Opt. Tech. Lett. 3:298 (1990).
8. K. Hayata and M. Koshiba, Self-focusing instability and chaotic behavior of nonlinear optical waves guided by dielectric slab structures, Opt. Lett. 13:1041 (1988).
9. W. C. Rheinboldt and E. Riks, Solution techniques for nonlinear finite element equations, in: "State-of-the-art surveys on finite element technology," A. K. Noor and W. D. Pilkey, eds., The American Society of Mechanical Engineers, New York (1983).
10. B. M. A. Rahman and J. B. Davies, Finite element solution of nonlinear bistable optical waveguides, Int. J. Optoelectron. 4:153 (1989).
11. B. M. A. Rahman and J. B. Davies, Finite-element solution of integrated optical waveguides, J. Lightwave Technol. 2:682 (1984).
12. C. T. Seaton, J. D. Valera, R. L. Shoemaker, G. I. Stegeman, J. T. Chilwell, and S. D. Smith, Calculations of nonlinear TE waves guided by thin dielectric films bounded by nonlinear media, IEEE J. Quantum Electron. 21:774 (1985).
13. J. V. Moloney, J. Ariyasu, C. T. Seaton, and G. I. Stegeman, Stability of nonlinear stationary waves guided by a thin film bounded by nonlinear media, Appl. Phys. Lett. 48:826 (1986).
14. C. K. R. T. Jones and J. V. Moloney, Instability of standing waves in nonlinear optical waveguides, Phys. Lett. A 117:175 (1986).
15. B. M. A. Rahman, J. R. Souza, and J. B. Davies, Numerical analysis of nonlinear bistable optical waveguides, IEEE Photonics Tech. Lett. 2:265 (1990).
16. K. Hayata and M. Koshiba, Full vectorial analysis of nonlinear-optical waveguides, J. Opt. Soc. Am. B 5:2494 (1988).
17. J. S. Aitchison, A. M. Weiner, Y. Silberberg, M. K. Oliver, J. L. Jackel, D. E. Leaird, E. M. Vogel, and P. W. E. Smith, Observation of spatial optical solitons in a nonlinear glass waveguide, Opt. Lett. 15:471 (1990).
18. G. I. Stegeman and E. M. Wright, All-optical waveguide switching, Opt. and Quantum Electron., 22:95 (1990).

MODELING OF PRINTED ELEMENT STRUCTURES

NUMERICAL MODELING OF PASSIVE NETWORKS AND COMPONENTS IN MONOLITHIC MICROWAVE INTEGRATED CIRCUITS (MMICs)

David C. Chang, Doris I. Wu and Jian X. Zheng

MIMICAD Center
Dept. of Electrical and Computer Engineering
University of Colorado at Boulder
Boulder, CO 80309-0425

ABSTRACT

A review of the vital issues affecting the accurate and efficient simulation of the electrical behavior of passive MMIC networks and circuit components is given. With emphasis on the microstrip structures, we discuss what constitutes an efficient algorithm in developing a numerical solver using a mixed-potential, integral equation approach.

INTRODUCTION

In the design of monolithically integrated microwave and millimeter-wave circuits, multiple design iterations often are due more to the unknown nature of mismatch and parasitic coupling between circuit elements than to statistical variations in the fabrication process or limitations in the active device models (although they do exist). Numerical modeling of microstrip structures using full-wave methods has been limited to canonical problems involving simple junctions and discontinuities[1-4]. Existing solvers developed based on these techniques may not adequately provide good physical insight into large, complex layouts where both parasitic couplings and discontinuities exist and interact. The tools needed to execute layout compaction and sensitivity analysis simply do not exist.

In this paper, we will first review factors that affect the computational efficiency of layout simulation. Several algorithms which are capable of providing highly accurate numerical modeling of complex circuit configurations will be introduced, highlighted by a newly developed algorithm called P(seudo)-Mesh. This algorithm is based on a mixed-potential integral equation approach with piece-wise linear basis functions for rectangular and triangular cells. The approximation is carried out in such a manner that it is equivalent to a pseudo-wire mesh structure formed by the cells. In P-Mesh, the scattering matrix of a complex structure can be extracted from the full-wave solution using "de-embedding arms", which are extended feed strips attached automatically to the microstrip structure in the solution process. Other issues concerning the simulation of large layouts in a workstation environment will also be examined briefly in this paper. Numerical results and experimental verification for several typical circuit configurations will be presented as well.

Directions in Electromagnetic Wave Modeling
Edited by H.L. Bertoni and L.B. Felsen, Plenum Press, New York, 1991

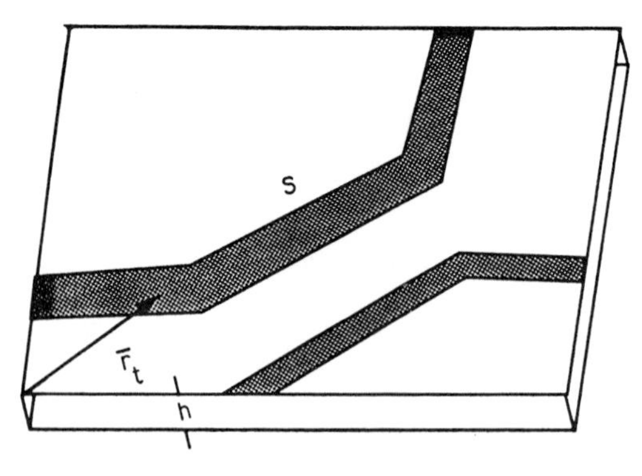

Figure 1

SOLUTION PROCESS

In this section, we shall outline the basic approach of an integral equation formulation. Consider the current distribution $\vec{J}(\vec{r})$ on a microstrip structure as the physical quantity of interest. The induced tangential electric field, $\vec{E}_t^s(\vec{r})$, on the microstrip can be expressed in terms of this current and its associated charge distribution, $\rho(\vec{r}) = \frac{j}{\omega}\nabla t \cdot \vec{J}(\vec{r})$, according to[3-5]

$$\vec{E}_t^s(\vec{r}) = \left(\frac{-2\pi j}{k_0\eta_0}\right)^{-1} \left\{ \int_s G_m(\vec{r}, \vec{r}')\vec{J}(\vec{r}') + \frac{j\omega\mu}{k_0^2}\nabla_t \int_s G_e(\vec{r}, \vec{r}')\rho(\vec{r}')dS' \right\}. \tag{1}$$

As shown in Figure 1, S is the surface area of the microstrip structure, $(\vec{r}, \vec{r}')$ are positions vectors denoting the observation and the source points, respectively, k_0 is the free-space wave number, and η_0 is the free-space characteristic impedance. A time convention of $\exp(j\omega t)$ has also been adopted, and the metallic surfaces are assumed to be perfect. The functions G_m and G_e are the Green's functions of the magnetic and electric type, respectively. They are the fields produced by a unit point current source and charge source, respectively. The expressions for G_m and G_e are rather complicated and will be deferred to a later section.

A mixed-potential integral equation for the current distribution $\vec{J}(\vec{r}')$ can be obtained by equating the negative of $\vec{E}_t^s(\vec{r})$ to the known impressed electric field, $\vec{E}_t^i(\vec{r})$, on the surface of the microstrip:

$$\int_s \left[G_m(\vec{r}, \vec{r}') - \frac{1}{k_0^2}\nabla_t \nabla_t' \cdot G_e(\vec{r}, \vec{r}') \right] \vec{J}(\vec{r}')dS' = \frac{2\pi j}{k_0\eta_0}\vec{E}_t^i(\vec{r});\ \vec{r}\varepsilon S. \tag{2}$$

To find the solution, we assume that $\vec{J}(x,\ y)$ can be approximated by a set of basis functions, $\vec{H}_m(\vec{r});\ m = 1,\ 2\ldots\ M$,

$$\vec{J}(\vec{r}) = \sum_{m=1}^{M} I_m \vec{H}_m(\vec{r}). \tag{3}$$

Substituting (3) into (1) and using the Galerkin method, the integral equation can be discretized into the following matrix equation[2]:

$$\sum_{n}^{M} Z_{mn} I_n = V_m;\ m = 1,\ 2, \ldots\ M, \tag{4}$$

where Z_{mn} is referred to as the moment integral defined as

$$Z_{mn} = \int_s \vec{H}_m(\vec{r}) \cdot \left\{ \int_s \left[G_m(\vec{r},\ \vec{r}') - \frac{1}{k_0^2}\nabla_t \nabla_t' \cdot G_e(\vec{r},\ \vec{r}') \right] \vec{H}_n(\vec{r}')dS' \right\} dS. \tag{5}$$

Table 1. Spatial-Domain Integral Equation Solver.

Choices of Basis Functions	Cell Shape: Rectangular	Cell Shape: Triangular
Constant Current/Constant Charge	OK	---
Linear Current/Constant Charge (Nodal Current)	---	OK
Linear Current/Constant Charge (Mesh Current)	OK	OK
Bilinear Current/Linear Charge (Nodal Current)	OK	---

Dimensionally, Z_{mn} can be identified as the mutual impedance between the mth and n^{th} elements; V_m, on the other hand, is the equivalent "voltage" source on the structure defined as:

$$V_m = \frac{2\pi j}{k_0 \eta_0} \int_s \vec{H}_m(\vec{r}) \cdot \vec{E}_t^i(\vec{r}) dS. \tag{6}$$

In principle, Equation (4) can now be used to find the expansion coefficients, I_m; $m = 1, 2, \ldots M$. However, the questions which we still must address are: (i) what are the appropriate choices for the basis functions; (ii) how can the evaluation of the moment integrals and the inversion of the resultant matrix be done efficiently; and (iii) how can the structural geometry be specified conveniently for simulation purposes. We shall discuss these issues in the sections that follow.

SELECTION OF BASIS FUNCTIONS

In developing a general solver, the basis functions selected must be flexible enough to adapt to microstrip geometries of arbitrary shape. Basis functions which contain special functions to describe the current distribution near the edges of a microstrip, for instance, are not convenient to use because they give rise to a different set of moment integrals for the edge portion and the interior portion of the microstrip. On the other hand, basis functions which describe the global behavior of the current over the entire structure, such as a Fourier series expansion, also would not be appropriate for a general solver because they are ill-suited for complex geometries. These considerations lead us to conclude that a more optimal set of basis functions would be one which expands the current distribution locally using low-order polynomials.

Accordingly, the most logical choice of discretization is to first subdivide the microstrip structure into small rectangular or triangular cells and then approximate the two-dimensional current and charge distributions locally in each cell by subsectional functions such as a constant, a linear, or even a bilinear, function. A summary of what may or may not be feasible is given in Table 1. In what follows, we shall briefly discuss some of the more convenient choices:

Nodal Current Representation for Triangular Cells

As in the case of the finite-element method, one of the most common ways of modeling is to discretize the entire geometry into triangular cells and express the unknown distribution in each cell by a linear function of x and y. The distribution in each cell is then expressed in terms of the nodal values at the three vertices of the triangle. For instance, consider a linear function $\lambda_i(\vec{r})$ given by

$$\lambda_i(\vec{r}) = 1 - \frac{\hat{z}_i \cdot (\vec{r} - \vec{r}_i)}{h_i}; \qquad \vec{r} = x\hat{x} + y\hat{y}, \tag{7}$$

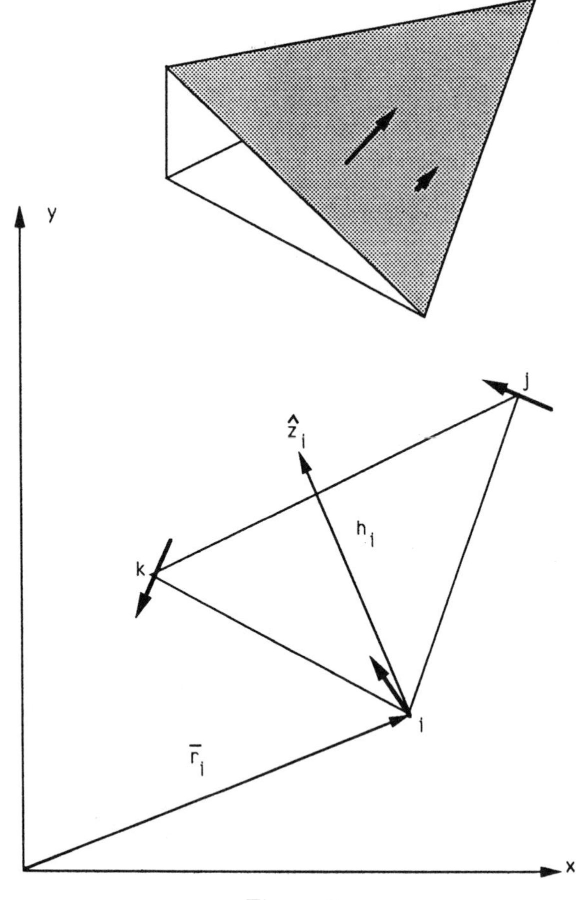

Figure 2

where $\vec{r}_i$ is the position vector for the ith vertex, $\hat{z}_i$ is the unit normal vector pointing from the ith vertex to the opposite edge, and h_i is the height measured from the same vertex to the opposite edge as shown in Figure 2. It is not difficult to show that $\lambda_i(\vec{r})$ describes a linear function which has an amplitude of unity at $\vec{r} = \vec{r}_i$ and zero at the opposite edge. The nodal representation of the current distribution inside a triangular cell can be written as

$$\vec{J}(\vec{r}) = \sum_{i=1}^{3} \vec{J}_i \lambda_i(\vec{r}); \qquad \vec{r} \in S. \tag{8}$$

To obtain the charge distribution associated with this current, we invoke the charge continuity equation to yield

$$\rho(\vec{r}) = \frac{-j}{w} \sum_{i=1}^{3} \vec{J}_i \bullet \left(\frac{\vec{h}_i}{h_i^2} \right); \qquad \vec{h}_i = h_i \hat{z}_i = \left(1 - \hat{t}_{jk}\hat{t}_{jk}\bullet\right)(\vec{r}_j - \vec{r}_i). \tag{9}$$

Here, we assume $i \neq j \neq k$, and $\hat{t}_{jk}$ is a unit vector tangent to the edge connecting the jth and k^{th} nodes. To find the equivalent global basis function, $\vec{H}_m$, referred to earlier in (3), we only need to define a unit-step function, $U_p(\vec{r})$; $p = 1,\ 2, \dots P$, for each cell and recognize that there are a finite number of cells sharing a common node m so that

$$I_m \vec{H}_m(\vec{r}) = \sum_{p \in p_m} \lambda_m^{(p)}(\vec{r}) U_p(\vec{r}) \vec{J}_m; \qquad m = 1,\ 2, \dots\ M, \tag{10}$$

where $\vec{J}_m$ is the current at the mth node shared by all the cells in p_m.

Mesh Current Representation for Triangular Cells

Another convenient way to describe the current distribution in a cell is to characterize it by the three edge currents, each of which is normal to a cell side and is assumed to be constant along the normal cell side. Mathematically, it is not difficult to show that the normal component of the vector function, $\vec{\Omega}_{i,\ i+1}$, as defined below for the cell shown in Figure 3, is unity along the boundary edge connecting the vertices i and $i+1$, and vanishes at the other two edges.

$$\vec{\Omega}_{i,\ i+1}(\vec{r}) = \hat{z} \times \{\lambda_i(\vec{r})\nabla_t\ \lambda_{i+1}(\vec{r}) - \lambda_{i+1}(\vec{r})\nabla_t\ \lambda_i(\vec{r})\}. \tag{11}$$

$\hat{z}$ is defined as the unit vector normal to the cell plane, i.e., $\hat{z} = \hat{x} \times \hat{y}$, and $i = 1,\ 2,\ 3$ are the vertices arranged in a counter-clockwise manner so that $i = i-3$ for $i > 3$, and $i = i+3$ for $i < 0$. We should also note that such a representation is closely associated with the so-called edge-elements in the finite-element method[6]. The current distribution inside a cell can now be characterized by these normal side currents, $J_{i,\ i+1}$; $i = 1 \ldots 3$:

$$\vec{J}(\vec{r}) = \sum_{i=1}^{3} J_{i,\ i+1}\ \vec{\Omega}_{i,\ i+1}(\vec{r}); \qquad \vec{r} \in S. \tag{12}$$

Since the normal component of the current at a cell's boundary edge is constant, the representation is topographically the same as replacing the current distribution in the cell by three mesh currents. However, unlike the nodal current method, this case has only two adjacent cells for each "mesh" current. Denoting the cell with an incoming current and the cell with an outgoing current as "+" and "–", respectively, we can again define an equivalent global basis functions for this mesh representation as

$$I_m\vec{H}_m(\vec{r}) = I_m \sum_p \vec{\Omega}_m^{(p)}(\vec{r})U_p(\vec{r}); \qquad p = \text{"+" or "-"}, \tag{13}$$

where $m = 1,\ 2, \ldots\ M$, I_m is the mesh current at the mth mesh shared by a pair of "+" and "–" cells, i.e. $m \rightarrow (i,\ i+1)$ for the "+" cell and $m \rightarrow (i,\ i-1)$ for the "–" cell.

The boundary condition at the edges of a microstrip structure can be easily implemented by setting the "mesh" currents to zero along the true boundary edges. The mesh representation, in this sense, offers a distinct advantage over the nodal representation since the value of the tangential component of the current at a true boundary node is, in principle, infinitely large. Furthermore, because the current distribution is not conformal to the corners of a microstrip structure in the nodal method, numerically unstable solutions can happen unless a modification is made to the method.

To further illustrate this point, let us consider two cells, α and α', as shown in Figure 4. Both cells border a true edge defined by $\angle 412$, and both share a common boundary l_{31}. Since the normal component of the current has to vanish along sides l_{41} and l_{12}, we have $\vec{J_1} = J_0\hat{t}_{12}$ for $\vec{r} = \vec{r}_1 + d\hat{t}_{12},\ d \rightarrow 0_+$, and $\vec{J_1} = J_0\hat{t}_{41}\ \vec{r} = \vec{r}_1 + d\hat{t}_{12}, d \rightarrow 0_-$. The normal component of the current flowing across the cell boundary l_{31} is therefore discontinuous by the amount of $(1-u)J_0\hat{n}_{31} \bullet [\hat{t}_{12} + \hat{t}_{41}]$ where $\vec{r}_{13} = \vec{r}_1 + u\ (\vec{r}_3 - \vec{r}_1)$, and $0 < u < 1$.

Pseudo-Mesh Representation

The use of triangular cells offers the advantage of being more adaptable to microstrip structures of arbitrary shape. However, most MMIC circuit layouts consist of microstrip segments which are rectangular in nature. This raises the question of whether it would be possible to develop a representation which uses both rectangular and triangular cells in a consistent manner. In the case of a nodal method, it is clear that the same linear approximation cannot be used to characterize the current distribution uniquely in a rectangular cell because we have eight nodal current components, or two at each vertex, but only six unknown coefficients. In the case of a mesh

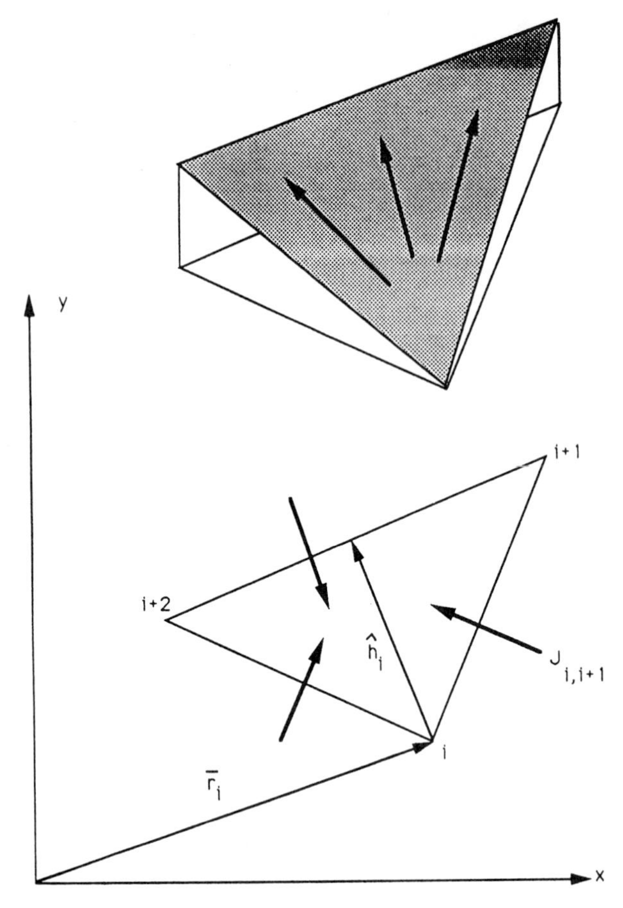

Figure 3

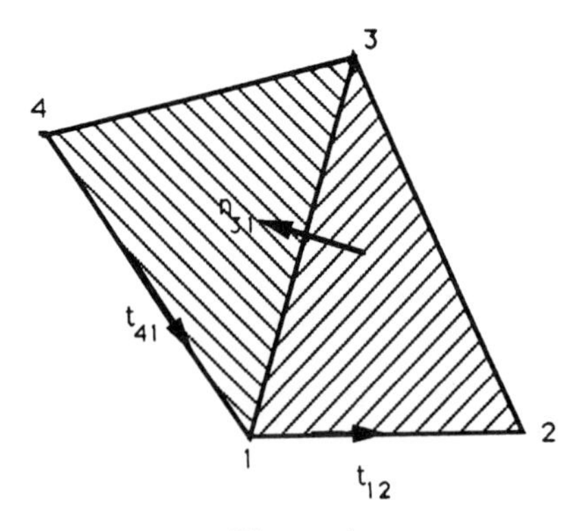

Figure 4

method, the constraint that the normal component of the current must be constant along a boundary edge gives rise to two, rather than four, constraint equations for a rectangular cell, leaving only six equations for the six unknowns. More specifically, consider the function

$$\vec{\Psi}_{i,\ i+1}(\vec{r}) = \left\{ \frac{1 - \hat{n}_{i,\ i+1} \bullet (\vec{r} - \vec{r_i})}{h_{i,\ i+1}} \right\} \hat{n}_{i,\ i+1}; \qquad i = 1, \dots 4, \tag{14}$$

where, as shown in Figure 5, $\hat{n}_{i,\ i+1}$ is a unit vector normal to the edge connecting

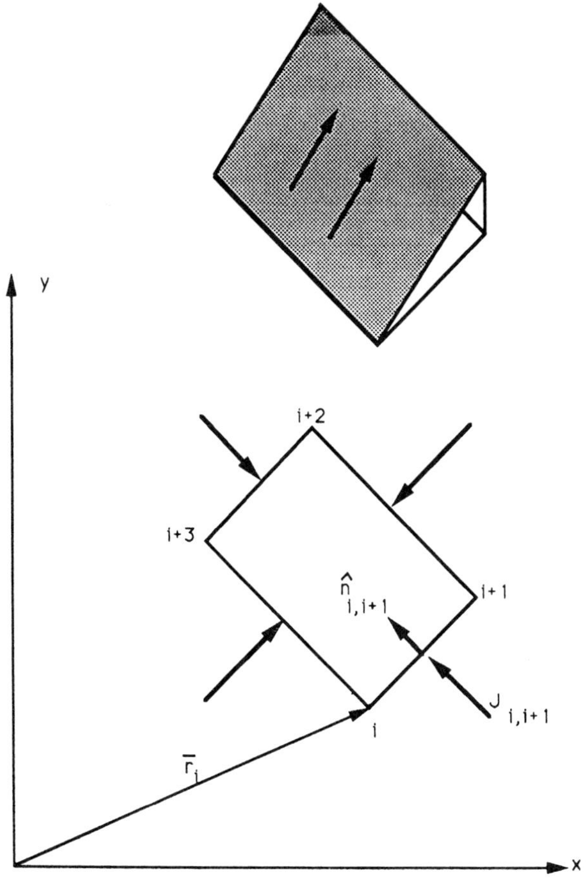

Figure 5

the ith and $(i+1)^{th}$ vertices, and $h_{i,\ i+1}$ is the length of that edge. $\vec{\Psi}_{i,\ i+1}$ is a vector pointing inward from the boundary edge to the opposite edge. It is not difficult to show that $\vec{\Psi}_{i,\ i+1}$ has a magnitude of one on side $(\mathrm{i}, \mathrm{i}+1)$ and vanishes on the opposite side $(\mathrm{i}+2, \mathrm{i}+3)$. The current distribution inside a rectangular cell can be characterized by the four normal currents, $J_{i,\ i+1};\ i = 1,\ 2, \ldots\ 4$ as

$$\vec{J}(\vec{r}) = \sum_{i=1}^{4} J_{i,\ i+1} \vec{\Psi}_{i,\ i+1}(\vec{r}); \qquad \vec{r} \in S. \tag{15}$$

Here the index i is again chosen in a counter-clockwise direction so that $i = i - 4$ for $i > 4$, and $i = i + 4$ for $i < 0$. As in the case of a triangular cell, we can designate the cell with an incoming current and the cell with an outgoing current as "+" and "−", respectively. Consequently, the expression given in (13) is equally applicable as the global representation of the basis functions for both triangular and rectangular cells.

The mixed representation outlined above gives us the flexibility of using triangular cells to better conform to the boundary of a complex microstrip in the discretization process, while still retaining the advantage of the regularity provided by rectangular cells for the interior region. For the chamfered bend example shown in Figure 6, a user can, in essence, follow his/her instinct in discretizing the geometry according to the boundary shape without the precise knowledge of the overall current distribution. The current flux in this approach follows the contour of the microstrip structure naturally. This method, as previously reported[7], is known as the P(seudo)-Mesh method.

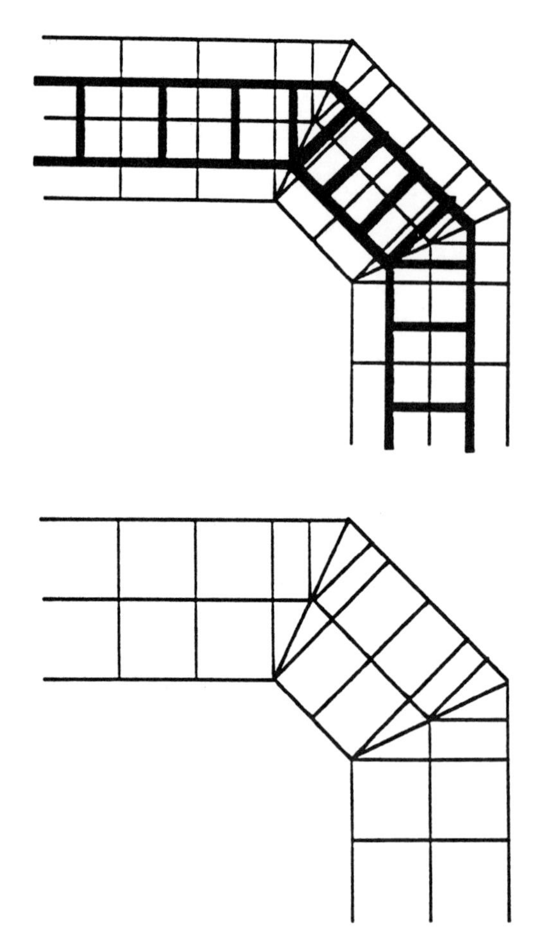

Figure 6

COMPUTATION OF THE GREEN'S FUNCTIONS

As we stated earlier, the Green's functions G_m and G_e represent the tangential electric fields produced by a current point dipole source and a charge point source, respectively, in the presence of the substrate. Because both the tangential electric and magnetic fields have to be continuous at the dielectric interface, the expressions for G_m and G_e can be obtained only in terms of the Fourier-Bessel transforms:[7]

$$G_m(\vec{r}) = \int_0^{\infty} \left\{ \frac{\mu_t}{\mu_r U_0 + U_n \coth U_n k_0 h} \right\} J_0(k_0 r)\tau d\tau \tag{16}$$

$$G_e(\vec{r}) = \int_0^{\infty} \left\{ \frac{U_0 + \mu_r \ U_n \tanh \ U_n k_0 h}{(\varepsilon_r U_0 + U_n \tanh \ U_n k_0 h)(\mu_r U_0 + U_n \coth U_n k_0 h)} \right\} J_0(k_0 r)\tau d\tau, \tag{17}$$

where $U_0 = (\tau^2 - 1)^{1/2}$, $U_n = (\tau^2 - \mu_r \varepsilon_r)^{1/2}$ and $Re\,\{U_0,\, U_n\} \geq 0$; r is the radial distance between the source and the observation points, and h is the thickness of the substrate. As shown in Figure 7, various deformed contours can be used to compute these functions. For example, by making the following transformation,

$$\int_0^{\infty} \ldots J_0(k_0 r\tau)\tau d\tau \Rightarrow \int_{-\infty}^{\infty} \ldots H_0^{(2)}(k_0 r\tau)\tau d\tau,$$

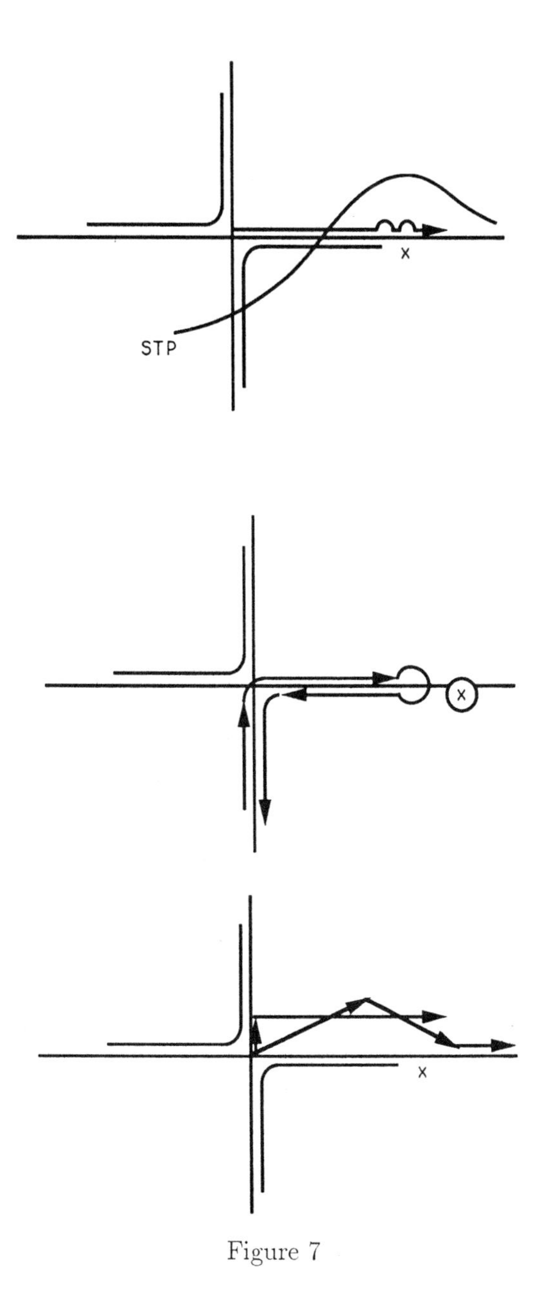

Figure 7

the Green's functions can be computed by deforming the contour onto both sides of the branch cut in the lower half of the τ-plane and adding to the integration the residue contribution at the poles, if any, of the integrand[8]. However, in choosing a numerical scheme to evaluate these Green's functions, we must again be sensitive to the issue regarding the tradeoffs between computational efficiency and versatility of the solver. As it will become clear later in the section, these Green's functions need to be computed only once at each frequency in the mixed-potential integral equation approach. As a result, the up-over-and-down contour indicated in Figure 7 is often preferred because it is readily extendable to multilayer substrates and other more complicated guiding structures[9].

Let us further assume that the Green's functions, $G_{e,m}(\vec{r}, \vec{r}')$, can be approximated *locally* by a polynomial of $|\vec{r}-\vec{r}'|$ in the α and α' cells associated with $\vec{H}_m(\vec{r})$ and $\vec{H}_n(\vec{r}')$, respectively,

$$G_{e,m}(\vec{r},\ \vec{r}') = \sum_{p=-1}^{P} C_p^{e,\ m}(\omega;\ \alpha,\ \alpha')|\vec{r}-\vec{r}'|^P. \tag{18}$$

The moment integrals Z_{mn} can then be expressed as

$$Z_{mn}(\omega) = \sum_{\alpha,\ \alpha'} \sum_{p=-1}^{P} \left\{ C_p^m(\omega;\ \alpha,\ \alpha')\xi_{m,\ n}^p(\alpha,\ \alpha') - \frac{1}{k_0^2} C_p^e(\omega;\ \alpha,\ \alpha')\sigma_{m,\ n}^p(\alpha,\ \alpha') \right\}; \tag{19}$$

where

$$\begin{aligned} \xi_{m,\ n}^p(\alpha,\ \alpha') &= \sum_{p=-1}^{P} \int_{S_\alpha} \int_{S'_\alpha} \vec{H}_m(\vec{r}) \bullet \vec{H}_n(\vec{r}')|\vec{r}-\vec{r}'|^p dS dS' \\ \sigma_{m,\ n}^p(\alpha,\ \alpha') &= \int_{S_\alpha} \int_{S'_\alpha} \nabla \bullet \vec{H}_m(\vec{r}) \nabla \bullet \vec{H}_n(\vec{r}')|\vec{r}-\vec{r}'|^p dS dS', \end{aligned} \tag{20}$$

are two geometric factors independent of the operating frequency and material properties. As shown in reference 7, both $\xi_{m,\ n}^p$ and $\sigma_{m,\ n}^p$ can be evaluated analytically in closed form using the pseudo-mesh algorithm discussed previously. Thus, we can pre-compute the values of G_m and G_e for a range of radial distances r, and curve-fit them locally using low-order polynomials for any pair of cells. For most practical cases of interest, the expansion coefficients $C_p^{m,\ e}(\omega;\ \alpha,\ \alpha')$, often are insensitive to the change in ω so that the computational scheme outlined above becomes particularly efficient for multi-frequency simulation.

SIMPLE MICROSTRIP DISCONTINUITIES

Numerical results for simple microstrip junctions are presented in this section along with experimental data. Since the physical parameter of interest for most circuit simulations is the scattering matrix rather than a detailed description the current distribution over the entire structure, some post-processing is needed to determine the forward and reflected wave amplitudes. This de-embedding process involves essentially the same procedures as those practiced in the laboratory. For instance, we can either use the physical distance between two minimums to determine the guided wavelength and the conventional VSWR method to determine the reflection coefficient[1], or we can simply use the computed current values at three adjacent locations to extract the same information[7]. The electric length of a U-bend with multiple chamfered corners is computed using the P-Mesh algorithm as a function of frequency. The computed results are shown in solid line in Figure 8 along with the measured results and the results obtained by the conventional transmission-line analysis (dashed line). The difference between the solid and dashed lines can be as high as $10° - 15°$ at high frequencies. Figure 9, on the other hand, shows the magnitude of S_{12} for a double-stub filter, again as a function of frequency. Because the two stubs are identical in length, conventional transmission-line analysis predicts only a sharp resonance at 10 GHz. Both measured and simulated results, however, show two coupled resonances with substantially broader bandwidths. Again, the agreement between the two results is excellent[10]. A more comprehensive investigation of the high-frequency behavior of simple microstrip junctions is given in the literature[7]. These junctions typically can be computed using 50 to 100 rectangular or triangular cells, with a computation time of 5 to 15 minutes per frequency point on an HP-300 series workstation.

LAYOUT SIMULATION

The number of cells required for the simulation of passive microstrip circuit components and layouts typically varies from a few hundred to a few thousand over a wide range of frequencies. Because the resultant matrix is typically very full, both the

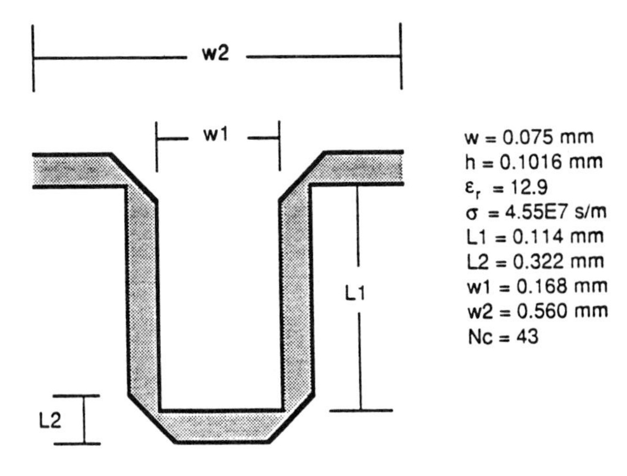

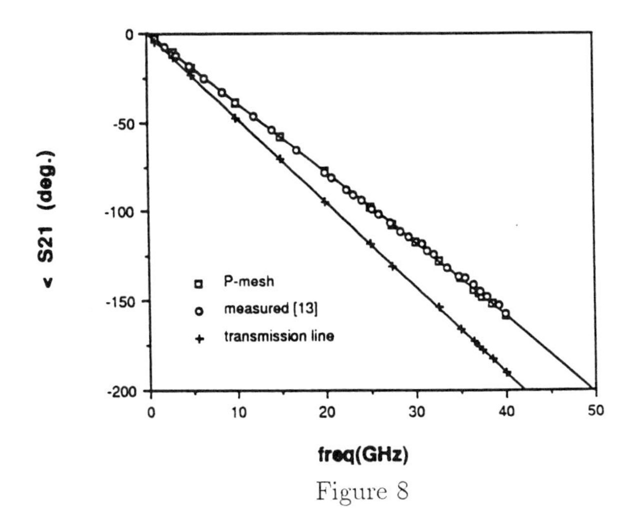

Figure 8

computation time and communication overhead for filling and inverting the matrix can become prohibitively large in a workstation environment. Therefore, the question of whether one can efficiently simulate a layout without resorting to a supercomputer by utilizing some of the underlying physics arises. In particular, one can ask the question of whether parasitic or mutual couplings among circuit segments (which is, after all, the essence of a layout simulation) can in fact be accounted for without inverting a full matrix.

One such method is to define a circle of influence for each cell so that the mutual couplings between mesh currents located inside and outside of the circle are ignored[11]. This then results in a sparse matrix where the inversion can be accomplished using efficient sparse algorithms. Although the solution thus obtained is necessarily approximate in nature, it may be sufficiently accurate for practical applications if the radius is large enough. Alternatively, the sparse matrix solution can also be used as a good trial function for other iterative methods to further improve the accuracy.

The microstrip meandered line shown in Figure 10 is a good example for demonstrating the effectiveness of the sparse method. The radius for the circle of influence, r, is chosen to be $r = h$, $3h$ and $7h$ where h is the thickness of the substrate. The

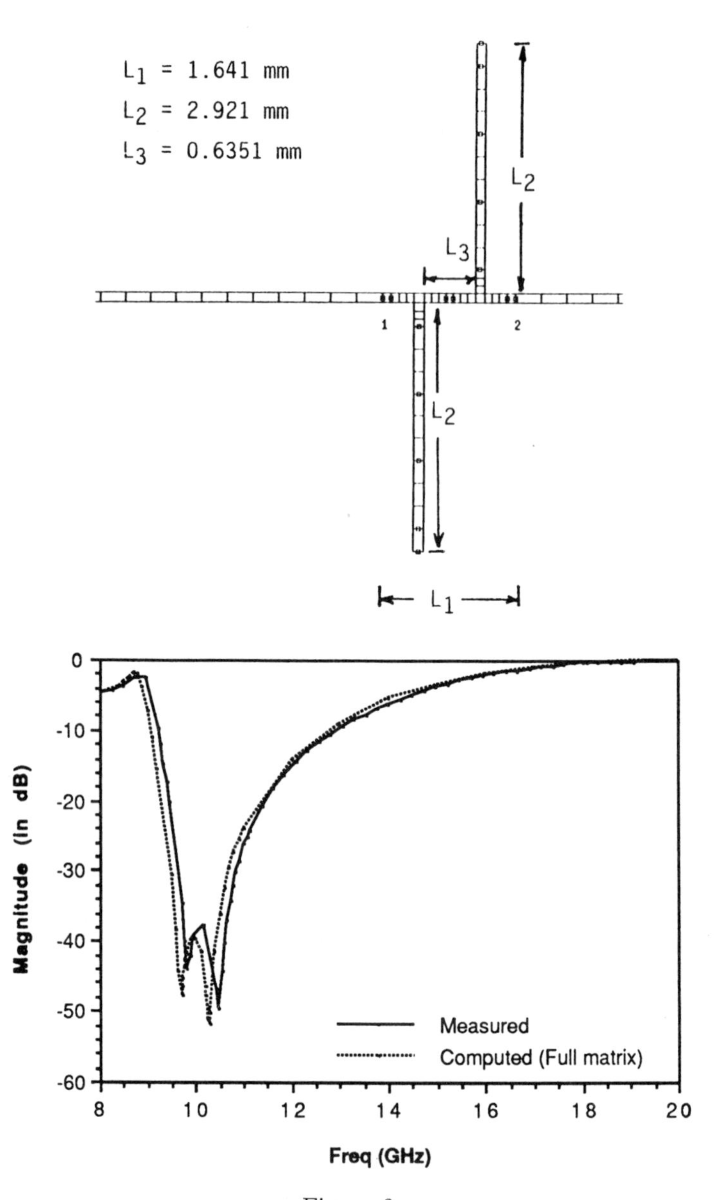

Figure 9

results are compared with the solution of the full matrix approach. It is of interest to note that for the given width of the microstrip, the mutual coupling between two adjacent parallel lines has been ignored for r less than h so that the case of $r = h$ is what can be expected from a transmission line analysis which takes into account the junction discontinuities but not mutual coupling. We should further note that the computation time for the case of $r = 7h$ is only 18 minutes for each frequency point on an HP 375 workstation versus 29 minutes for the full matrix. The number of cells in this case is about 200 which is still relatively small and the circuit layout is still rather rudimentary. The savings in the computation time, of course, will be substantially greater whenever the size of the matrix exceeds the internal memory of the workstation. To further improve the computational accuracy, we can use either the results obtained from the $r = 3h$ case, or the $r = 7h$ case as the starting point of a conjugate gradient method for the current distribution over the entire structure, or an adjoint operator method for the current or the scattering matrix parameters over specific portions of the circuit.

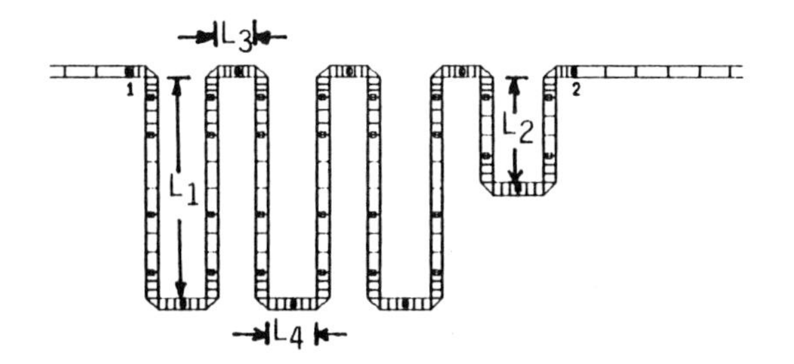

L_1 = 1.258 mm
L_2 = 0.592 mm
L_3 = 0.222 mm
L_4 = 0.296 mm

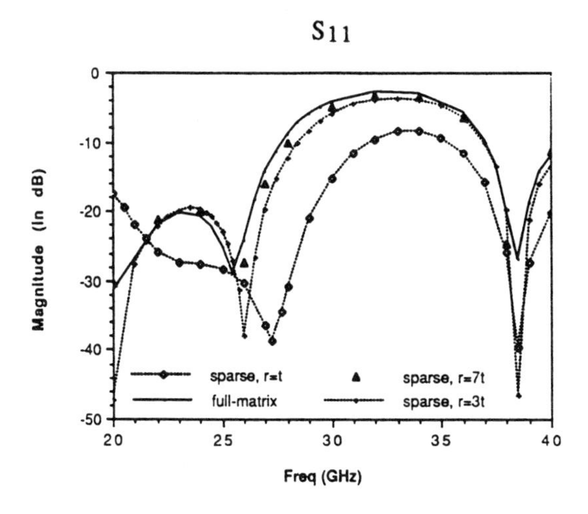

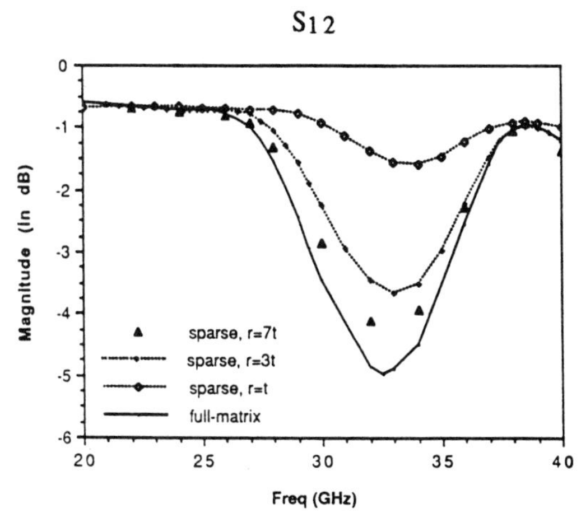

Figure 10

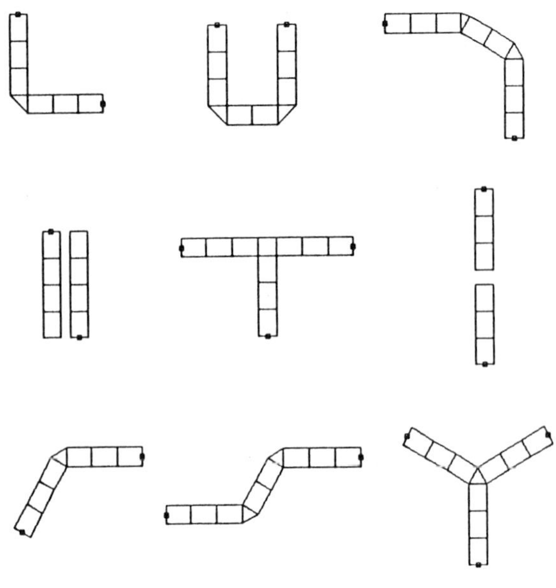

Figure 11

In addition to matrix inversion, large layout simulation also requires efficient graphical construction of the layout in terms of rectangular and triangular cells. To simplify the geometry specification process, one graphical tool that is currently being developed utilizes a set of standard, pre-defined, pre-gridded circuit blocks commonly encountered in a layout. Using this kind of tool, circuit construction becomes a simple task, particularly since the gridding process is eliminated. Moreover, by implementing this tool in a menu-driven, X-window environment, the connection of blocks into a layout can be executed easily[12]. Sample blocks and screen format of such a construction tool are shown in Figure 11.

CONCLUSION

We hope we have demonstrated some of the vital issues related to the electromagnetic simulation of passive MMIC circuit components and networks. The specific method we illustrated in this paper falls into the general category of a moment-method, mixed-potential integral-equation formulation. The accurate and efficient simulation of MMIC layouts are strongly influenced by numerical techniques, graphical construction and physical insight. While an accurate numerical simulation no doubt could provide us with a better understanding of the underlying electromagnetic phenomena, the utilization of the knowledge gained can in turn facilitate a more efficient numerical simulation scheme.

REFERENCES

1. P. B. Katehi and N. G. Alexopoulos, Frequency-Dependent Characteristics of Microstrip Discontinuities in Millimeter-Wave Integrated Circuits, IEEE Transactions on Microwave Theory Tech., MTT-33(10) (October 1985) 1029-1035.

2. R. W. Jackson, Full-Wave, Finite Element Analysis of Irregular Microstrip Discontinuities, IEEE Transactions on Microwave Theory Tech., MTT-37(1) (January, 1989) 81-89.

3. J. R. Mosig, Arbitrary Shaped Microstrip Structures and their Analysis with a Mixed Potential Integral Equation, IEEE Transactions on Microwave Theory Tech., MTT-36(2) (February 1988) 314-323.

4. D. I. Wu, D. C. Chang and B. L. Brim, Accurate Numerical Modeling of Microstrip Junctions and Discontinuities, International Journal of Microwave and Millimeter-Wave Computer-Aided Engineering (January 1991) Volume 1 Number 1, 48-58.

5. D. C. Chang and J. X. Zheng, Electromagnetic Modeling of Passive Circuit Elements in MMIC, to appear in IEEE Transactions on Microwave Theory Tech., MTT-41 (1991).

6. A. Bossavit, Solving Maxwell Equations in a Closed Cavity, and the Question of Spurious Modes, IEEE Trans. on Magnetics, 26(2) (March 1990) 702-705.

7. J. X. Zheng, "Electromagnetic Modeling of Microstrip Circuit Discontinuities and Antennas of Arbitrary Shape", Ph.D. Thesis, University of Colorado at Boulder (December 1990).

8. B. L. Brim and D. C. Chang, Accelerated Numerical Computation of the Spatial Domain Dyadic Green's Functions of a Grounded Dielectric Slab, National Radio Science Meeting Digest, (January, 1987) 164.

9. A. Hoorfar, J. X. Zheng and D. C. Chang, Accurate Characterization of Cross-Over and other Junction Discontinuities in Two-Layer Microstrip Circuits, Proceedings International Microwave Symposium, Boston, MA (June 1991).

10. M. Herman, Hughes Microwave Products Division, private communication.

11 D. I. Wu and D. C. Chang, Accurate and Efficient Simulation of MMIC Layouts, to appear in Proceedings of SPIE Symposium on Aerospace Sensing, Volume 1475 (April, 1991).

12. G. Wagner, "UBUILD User's Guide," Version 1.0, MIMICAD Center, Dept. of Electrical and Computer Engineering, University of Colorado at Boulder (November, 1990).

ON THE MODELING OF PRINTED CIRCUITS AND ANTENNAS ON CURVED SUBSTRATES

Akifumi Nakatani

Phraxos Research & Development Inc.
Santa Monica, California 90405, USA
(213) 450-4459

N. G. Alexopoulos

University of California, Los Angeles
Los Angeles, California 90024, USA
(213) 825-1027

1 INTRODUCTION

In many practical applications, non-planar circuit and antenna structures are conformal to cylindrical shapes. These coordinate systems are to be used to analyze applications of microstrip to antennas on curved surface, transition adapters, and baluns. A brief historical development for microstrip on cylindrical geometries is given in this section.

The transmission line on cylindrical or elliptic bodies has been analyzed by many authors. At low frequencies (Quasi-TEM Mode assumption), the potential field is found by solving Laplace's equation. Cylindrically curved striplines have been solved by Y. Wang [1]. The elliptic Laplace's equation has been treated by K.K.Joshi *et al* by using a change of variables [2]. L. Zeng *et al* introduced conformal mapping techniques to characterize the microstrip line [3]. The multiple layered circular cylindrical structure has been presented by C. J. Rebby *et al* [4]. At the same time, the multiple conductor on multilayered cylindrical substrates was analyzed by C. H. Chan *et al* [5]. Shortly after, C. J. Rebby *et al* published a coupled line analysis on a multiple layered structure [6]. A more general analysis for multiple layers and conductors has been introduced by using the variational technique by F. Medina [7]. The fullwave analyses based on the vector wave equations have been introduced by A. Nakatani and N.G. Alexopoulos for single and coupled microstrip lines [8-9] where attention has been focused on the development of an algorithm which computes the dyadic Green's functions with extremely high accuracy.

The analysis of conformal arrays has been presented by R. E. Munson where a paper-thin substrate board was used in the design of wraparound antennas [10]. A more rigorous analysis on cylindrical-rectangular microstrip antennas has been given by C. M. Krowne [11], who used an arc cavity model. A far field study for the wraparound antenna has been presented by S. B. Fonseca *et al* [12]. The dyadic

Directions in Electromagnetic Wave Modeling
Edited by H.L. Bertoni and L.B. Felsen, Plenum Press, New York, 1991

formulation of electric field/electric current has been investigated by N. G. Alexopoulos *et al* [13], where the Fourier Transform is used to formulate the Green's function for the dipole case. A similar electric surface current model has been adopted by J. Ashkenazy *et al* to compute the far-field pattern for the wraparound radiator [14]. The near-field computation to find the accurate input impedances for microstrip antennas has been presented by A. Nakatani *et al* [15] for the axial orientation. The same algorithm has been used to characterize microstrip lines with extremely high accuracy [8-9]. The radiators on the metallic cylinder were extensively investigated by E. V. Sohtell [16]. The solution for azimuthally oriented antennas on a cylindrical substrate has been presented by A. Nakatani *et al* who developed the Transverse Field Transmission Matrix (TFTM) formulation for application to microstrip circuits and antennas on cylindrical substrates [17-18]. This matrix formulation method was presented by C. M. Krowne for the solution of multiple layered anisotropic materials [24]. The radiation pattern, quality factor, and input resistance for circular patch antennas have been calculated [19]. The fullwave analysis for the solution of a rectangular patch and wraparound antenna was also presented by T. M. Habashy [20]. The same technique was adapted to characterize the resonance properties by S. M. Ali [21].

In this paper, the TFTM formulation is described as it has been adopted to solve multiple layered structures for mixed boundary value problems (including slot / microstrip configurations). The paper includes an overview of the numerical approach which is central to accurate and efficient computation of the dyadic Green's function in the Fourier transform domain.

2 THEORY

The fullwave dyadic Green's function formulation is presented in this section for a multi-layer cylindrical geometry. The Green's function is given in the form of a Transverse Field Transmission Matrix (TFTM) for each layer. The boundary value problem is formulated in such a way that the dyadic Green's function is generated numerically. This TFTM method provides the current and field relations for any interface. The geometry of interest is shown in Figure 1-(a). The cylindrical body consists of N layers. The layer interfaces are indexed from 1 to $N+1$. Each layer is characterized by ϵ_{ri} and μ_{ri}, where the permeability and permittivity are in general complex. The radii from the center of the cylinder to the ith and $i+1$ interfaces are defined as ρ_i and ρ_{i+1} respectively. Figure 1-(b) is the drawing of the cylindrical geometry to the rectangular coordinate system showing the interface and layer indices. The transverse fields at the ith interface can be related to the transverse fields at the $i+1^{th}$ interface as:

$$\begin{bmatrix} \vec{e}^{(i+1)} \\ \vec{h}^{(i+1)} \end{bmatrix} = \mathbf{T}(\epsilon_{ri}, \mu_{ri}, \rho_{i+1}, \rho_i) \begin{bmatrix} \vec{e}^{(i)} \\ \vec{h}^{(i)} \end{bmatrix} . \tag{1}$$

With the TFTM relation, the outer and inner boundary conditions are formulated in characteristic equation form at the jth source interface as

$$\prod_{k=j}^{N} \mathbf{T}^{-1}(\epsilon_{rk}, \mu_{rk}, \rho_{k+1}, \rho_k) \begin{bmatrix} \vec{e}^{out} \\ \vec{h}^{out} \end{bmatrix} - \prod_{k=j-1}^{1} \mathbf{T}(\epsilon_{rk}, \mu_{rk}, \rho_{k+1}, \rho_k) \begin{bmatrix} \vec{e}^{in} \\ \vec{h}^{in} \end{bmatrix} = \begin{bmatrix} \hat{\rho} \times \vec{M}_{sj} \\ \vec{J}_{sj} \times \hat{\rho} \end{bmatrix} . \tag{2}$$

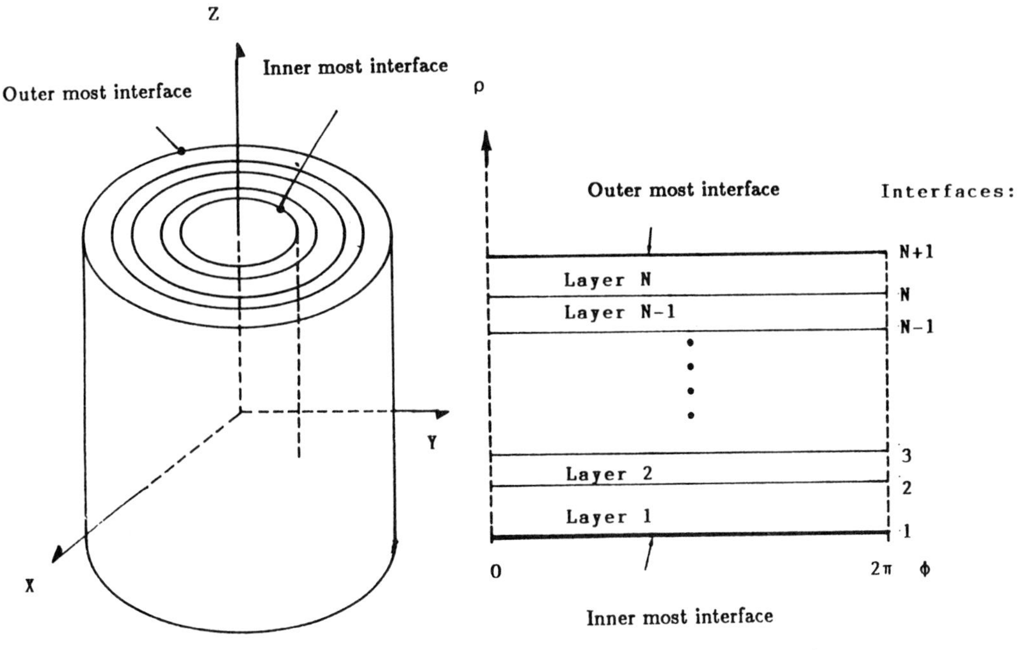

(a) Multi-layer Cylindrical Geometry. (b) Mapped Cylindrical Geometry.

Figure 1. Multi-Layer Cylindrical Substrates Geometry

The above equation relates 4 unknowns, two TE and TM unknowns at the inner and outermost boundaries, to the 4 sources, two electric and two magnetic sources. Therefore, the above equation can be written as

$$c_{TM}^{out}\mathbf{X}_{TM}^{out}(N,j)+c_{TE}^{out}\mathbf{X}_{TE}^{out}(N,j)+c_{TM}^{in}\mathbf{X}_{TM}^{in}(1,j-1)+c_{TE}^{in}\mathbf{X}_{TE}^{in}(1,j-1)=\begin{bmatrix}\hat{\rho}\times\vec{\tilde{M}}_{sj}\\ \vec{\tilde{J}}_{sj}\times\hat{\rho}\end{bmatrix} \quad (3)$$

where

$$\mathbf{X}_{TM}^{out}(N,j)=\prod_{k=j}^{N}\mathbf{T}^{-1}(\epsilon_{rk},\mu_{rk},\rho_{k+1},\rho_k)\begin{bmatrix}\bar{e}_{TM}^{out}\\ \bar{e}_{TM}^{out}\end{bmatrix}, \quad (4)$$

$$\mathbf{X}_{TE}^{out}(N,j)=\prod_{k=j}^{N}\mathbf{T}^{-1}(\epsilon_{rk},\mu_{rk},\rho_{k+1},\rho_k)\begin{bmatrix}\bar{e}_{TE}^{out}\\ \bar{e}_{TE}^{out}\end{bmatrix}, \quad (5)$$

$$\mathbf{X}_{TM}^{in}(1,j-1)=-\prod_{k=j-1}^{1}\mathbf{T}(\epsilon_{rk},\mu_{rk},\rho_{k+1},\rho_k)\begin{bmatrix}\bar{e}_{TM}^{in}\\ \bar{e}_{TM}^{in}\end{bmatrix}, \quad (6)$$

and

$$\mathbf{X}_{TE}^{in}(1,j-1)=-\prod_{k=j-1}^{1}\mathbf{T}(\epsilon_{rk},\mu_{rk},\rho_{k+1},\rho_k)\begin{bmatrix}\bar{e}_{TE}^{in}\\ \bar{e}_{TE}^{in}\end{bmatrix}. \quad (7)$$

The unknown coefficients are found by solving the 4×4 equation:

$$\vec{\mathbf{c}} = \left[\mathbf{X}_{TM}^{out}(N,j), \mathbf{X}_{TE}^{out}(N,j), \mathbf{X}_{TM}^{in}(1,j-1), \mathbf{X}_{TE}^{in}(1,j-1)\right]^{-1} \begin{bmatrix} \hat{\rho} \times \vec{M}_{sj} \\ \vec{J}_{sj} \times \hat{\rho} \end{bmatrix} \tag{8}$$

where the unknown coefficients are defined as $\vec{\mathbf{c}} = \left[c_{TM}^{out}, c_{TE}^{out}, c_{TM}^{in}, c_{TE}^{in}\right]^T$.

Once the coefficients are represented by current density, the tangential fields at the ith interface can be found as a function of current density.

- For $i \leq j$ case

$$\begin{bmatrix} \vec{e}_i \\ \vec{h}_i \end{bmatrix} = c_{TM}^{in} \mathbf{X}_{TM}^{in}(1, i-1) + c_{TE}^{in} \mathbf{X}_{TE}^{in}(1, i-1) \tag{9}$$

- For $i \geq j$ case

$$\begin{bmatrix} \vec{e}_i \\ \vec{h}_i \end{bmatrix} = c_{TM}^{out} \mathbf{X}_{TM}^{out}(N, i) + c_{TE}^{out} \mathbf{X}_{TE}^{out}(N, i) \tag{10}$$

Therefore, the final form of the 4×4 (electric field and magnetic field) dyadic Green's function can be written as

$$\begin{bmatrix} \vec{e}_i \\ \vec{h}_i \end{bmatrix} = \mathbf{G}^{4\times4}(i,j) \begin{bmatrix} \hat{\rho} \times \vec{M}_{sj} \\ \vec{J}_{sj} \times \hat{\rho} \end{bmatrix}, \tag{11}$$

where

$$\mathbf{G}^{4\times4}(i,j) = \begin{cases} \mathbf{X}^{-}(i) \cdot \mathbf{X}^{-1}(j) & i \leq j \\ \mathbf{X}^{+}(i) \cdot \mathbf{X}^{-1}(j) & i \geq j \end{cases}, \tag{12}$$

$$\mathbf{X}^{-}(i) = \left[\vec{0}, \vec{0}, \mathbf{X}_{TM}^{in}(1, i-1), \mathbf{X}_{TE}^{in}(1, i-1)\right], \tag{13}$$

$$\mathbf{X}^{+}(i) = \left[\mathbf{X}_{TM}^{out}(N, i), \mathbf{X}_{TE}^{out}(N, i), \vec{0}, \vec{0}\right], \tag{14}$$

and

$$\mathbf{X}^{-1}(j) = \left[\mathbf{X}_{TM}^{out}(N,j), \mathbf{X}_{TE}^{out}(N,j), \mathbf{X}_{TM}^{in}(1,j-1), \mathbf{X}_{TE}^{in}(1,j-1)\right]^{-1}. \tag{15}$$

It is noted that the dyadic Green's function can be evaluated from above or below the field interface, and the distinction is very important in the mixed boundary value problem.

Key Algorithms

In the above evaluation of 4×4 matrices, the common exponential factor, $J_m(\gamma_i \rho_i)$ $N_m(\gamma_i \rho_{i+1}) - J_m(\gamma_i \rho_{i+1}) N_m(\gamma_i \rho_i)$, is extracted so that the resulting TFTM matrix exhibits only bounded behavior. This avoids the numeric overflow or underflow in the computations.

The TFTM elements are organized in such a way that only ratios of the Bessel functions are computed. These ratio relations are computed by using either the continued fraction method (for Bessel function of the first kind) or a recurrence relation (for Bessel function of the second kind) [25-27].

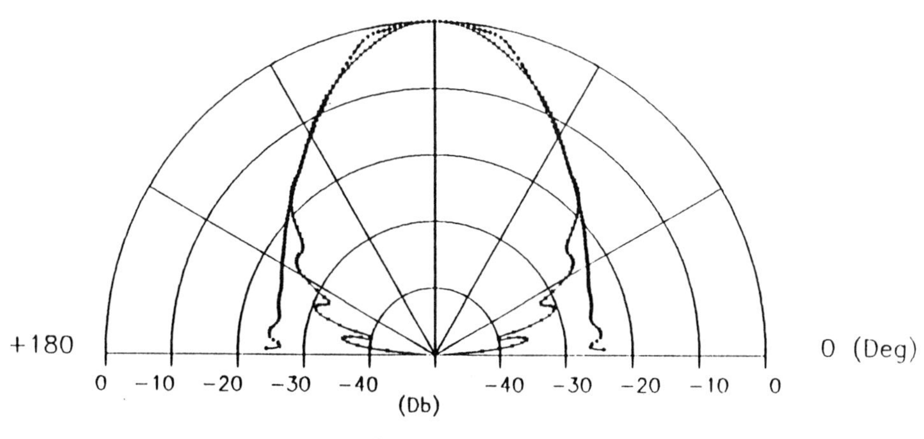

Figure 2. High Gain Condition for Printed Dipole Excitation: Solid Line is for E_ϕ by J_ϕ and Dashed Line is for E_θ by J_z. Radii of the inner ground and each interfaces are $\rho_1 = 1.0$, $\rho_2 = 1.110840$, $\rho_3 = 1.21679$, and $\rho_4 = 1.23250$. Materials are characterized by $\epsilon_{r1,2} = 2.1$ $\epsilon_{r3} = 100$ at $k_o = 10.0$.

For the case of three dimensional circuits, the integration contour is conveniently deformed into complex k_z to avoid algorithm and cylindrical surface wave singularities.

3 NUMERICAL RESULT

Far-field radiation can be computed easily once the transverse fields are characterized in the cylindrical Fourier Transform domain [22]. The material high gain condition may be achieved for an axial pattern by properly choosing the permittivity and the thickness of the material [23]. In Figure 2, the high gain axial and azimuthal patterns are shown. In this example, $\epsilon_{r1,2} = 2.1$, $\epsilon_{r3} = 100$, $\rho_1 = 1.0$, $\rho_2 = 1.110840$, $\rho_3 = 1.21679$, $\rho_4 = 1.23250$, and $k_o = 10.0$ are used to duplicate the high gain conditions. The high gain condition for the cylindrical substrates geometry is very similar to that of the planar geometry. However the azimuthal pattern is completely different from that of the planar geometry. The narrow beam resonance gain phenomenon may be explained by the leaky waves on the cylindrical structure. However, for the simple explanation, narrow beam condition can be satisfied by effectively widen the radiating aperture in axial direction. This case does not hold for the azimuthal direction since the cylindrical waves exist as a standing wave in azimuthal direction.

4 CONCLUSION

A highly accurate numerical approach has been established for the computation of the dyadic Green's function of the multiple layered circular cylindrical geometry. The method can be easily applied to mixed boundary value problems, such as the microstrip-slot-strip excitation mechanism on a cylindrical substrate geometry.

5 REFERENCES

Quasi-static Models:Microstrip Lines

1. Y. Wang, "Cylindrical and Cylindrically Warped Strip and Microstriplines," *IEEE Trans, Microwave Theory Tech.*, VOL. MTT-26, NO. 1, pp. 20-23, January 1978.
2. K. K. Joshi and B. N. Das, "Analysis of Elliptic and Cylindrical Striplines Using Laplace's Equation," *IEEE Trans, Microwave Theory Tech.*, VOL. MTT-28, NO. 4, pp. 381-386, April 1980.
3. L. Zeng and Y. Wang, "Accurate Solutions of Elliptical and Cylindrical Striplines and Microstrip Lines," *IEEE Trans, Microwave Theory Tech.*, VOL. MTT-34, NO. 2, pp. 259-265, February 1986.
4. C. J. Reddy and M. D. Deshpande, "Analysis of Cylindrical Stripline with Multilayer Dielectrics," *IEEE Trans, Microwave Theory Tech.*, VOL. MTT-34, NO. 6, pp. 701-706, June 1986.
5. C. H. Chan and R. Mittra, "Analysis of a Class of Cylindrical Multiconductor Transmission Lines Using an Iterative Approach," *IEEE Trans, Microwave Theory Tech.*, VOL. MTT-35, NO. 4, pp. 415-424, April 1987.
6. C. J. Reddy and M. D. Deshpande, "Analysis of Coupled Cylindrical Striplines Filled with Multilayered Dielectrics," *IEEE Trans, Microwave Theory Tech.*, VOL. MTT-36, NO. 9, pp. 1301-1310, September 1988.
7. F. Medina and M. Horno, "Spectral and Variational Analysis of Generalized Cylindrical and Elliptical Strip and Microstrip Lines," *IEEE Trans, Microwave Theory Tech.*, VOL. MTT-38, NO. 9, pp. 1287-1293, September 1990.

Frequency Dependent Models:Microstrip Lines

8. N. G. Alexopoulos and A. Nakatani, "Cylindrical Substerate Microstrip Line Characterization," *IEEE Trans, Microwave Theory Tech.*, VOL. MTT-35, NO. 9, pp. 843-849, September 1987.
9. A. Nakatani and N. G. Alexopoulos, "Coupled Microstrip Lines on a Cylindrical Substrate," *IEEE Trans, Microwave Theory Tech.*, VOL. MTT-35, NO. 12, pp. 1392-1398, December 1987.

Microstrip Antennas on a Cylindrical Structure

10. R. E. Munson, "Conformal Microstrip Antennas and Microstrip Phased Arrays," *IEEE Trans, Antennas Propagat.*, VOL. AP-22, NO. 1, pp. 74-77, January 1974.
11. C. M. Krowne, "Cylindrical-Rectangular Microstrip Antenna," *IEEE Trans, Antennas Propagat.*, VOL. AP-31, NO. 1, pp. 194-199, January 1983.
12. S. B. Fonseca and A. J Ciarola, "Analysis of Microstrip Wraparound Antennas Using Dyadic Green's Functions," *IEEE Trans, Antennas Propagat.*, VOL. AP-31, NO. 2, pp. 248-253, March 1983.

13. N. G. Alexopoulos, P. L. E. Uslenghi, and N. K. Uzunoglu, "Microstrip Dipoles on Cylindrical Structures," *Electromagnetics*, VOL. 3, pp. 311-326, 1983.

14. J. Ashkenazy, S. Shtrikman, and D. Treves, "Electric Surface Model for the Analysis of Microstrip Antennas on Cylindrical Bodies," *IEEE Trans, Antennas Propagat.*, VOL. AP-33, NO. 3, pp. 295-300, March 1985.

15. A. Nakatani, N. G. Alexopoulos, N. K. Uzunoglu, and P. L. E. Uslenghi, "Accurate Green's Function Computation for Printed Circuit Antennas on Cylindrical Substrates," *Electromagnetics*, VOL. 6, pp. 243-254, 1986.

16. E. V. Sohtell, "Microwave Antennas on Cylindrical Structures," *Ph.D Dissertation*, Chalmers University of Technology, Göteborg, Sweden 1988.

17. A. Nakatani, "Microstrip Circuits and Antennas on Cylindrical Substrates," *Ph.D Dissertation*, University of California, Los Angeles, USA 1988.

18. A. Nakatani and N. G. Alexopoulos "Microstrip Elements on Cylindrical Substrates - General Algorithm and Numerical Results - *Invited Paper*," *Electromagnetics*, VOL. 9, pp. 405-426, 1989.

19. K. Luk and K. Lee, "Characteristics of the Cylindrical-Circular Patch Antennas," *IEEE Trans, Antennas Propagat.*, VOL. AP-38, NO. 7, pp. 1119-1123, July 1990.

20. T. M. Habashy, S. M. Ali, and J. A. Kong, "Input Impedance and Radiation Pattern of Cylindrical-Rectangular and Wraparound Microstrip Antennas," *IEEE Trans, Antennas Propagat.*, VOL. AP-38, NO. 5, pp. 722-731, May 1990.

Microstrip Resonator on Cylindrical Structure

21. S. M. Ali, T. M. Habashy, J. Kiang, J. A. Kong, "Resonance in Cylindrical-Rectangular and Wraparound Microstrip Structures," *IEEE Trans, Microwave Theory Tech.*, VOL. MTT-37, NO. 11, pp. 1773-1783, Novenber 1989.

Other References

22. R. F. Harrington, Time-Harmonic Electromagnetic Fields, *McGraw-Hill Book Company*, 1961.

23. R. S. Elliott, Antenna Theory and Design, Englewood Cliffs, N.J.: *Prentice-Hall*, 1981.

24. P. B. Katehi "A Generalized Solution to a Class of Printed Circuit Antennas," *Ph.D Dissertation*, University of California, Los Angeles, USA 1984.

25. D. R. Jackson and N. G. Alexopoulos, "Gain Enhancement Methods for Printed Circuit Antennas," *IEEE Trans, Antennas Propagat.*, VOL. AP-33, NO. 2, pp. 976-987, September 1985.

26. C. M. Krowne, "Determination of the Green's Function in the Spectral Domain Using a Matrix Method: Application to Radiators or Resonators Immersed in a Complex Anisotropic Layered Medium," *IEEE Trans, Antennas Propagat.*, VOL. AP-34, NO. 2, pp. 247-253, February 1986.

27. M. Abramowitz and I. A. Stegun, Handbook of Mathematical Functions - with Formula, Graphs, and Mathematical Tables, New York, NY: *Dover Publications, Inc.* 1970.

28. I. S. Gradshteyn and I. M. Ryzhik, Table of Integrals, Series, and Products, Corrected and Enlarged Edittion by A. Jeffrey, New York, NY: *Academic Press, Inc.*, 1980.

29. G. N. Watson, A Treatice on the Theory of Bessel Functions, Second Edition, *Cambridge University Press*, 1941.

A GENERAL, FULL-WAVE APPROACH FOR MODELING SIGNAL LINES AND DISCONTINUITIES IN COMPUTER PACKAGES

Barry J. Rubin

IBM Research
T. J. Watson Research Center
Yorktown Heights, NY 10598

ABSTRACT

A full-wave moment-method approach is presented for obtaining the electrical characteristics of the 3D signal line and discontinuity structures typically found in computer packages. Rooftop current elements are used to represent both the surface current on conductors and the polarization current in any dielectric regions that may be present. The approach is applied to the computer module used in the IBM 3090 processor unit. The propagation characteristics of signal lines and vias, and coupled noise between signal lines are calculated and compared to results obtained from scale models and specifically designed test vehicles. The approach is then applied to representative, non-TEM waveguide structures that may appear in other package-related environments.

INTRODUCTION

The typical high-performance computer involves several levels of packaging. For instance, combinations of single-chip modules, cards, multi-chip modules, cables, and boards appear in some hierarchy to form the central processor unit (CPU); Fig. 1 gives the package configuration [1] for the IBM 3090 system. Independent of the particular design, the propagation characteristics of the signal line and discontinuity structures must be known so that the machine can be designed before test hardware is available. For instance, it must be known in advance whether the discontinuities associated with stubs or the connectors will cause severe reflections. In high-performance computers, all the receivers on an electrical net must correctly switch at the *first incidence* of the signal waveform, and not after multiple reflections have occurred. [2] Further, the crosstalk between the nets, due to parasitic coupling, must not cause circuits to falsely switch. [2]

Signal lines typically run between reference planes that are not solid, but perforated so that vias, which interconnect signal lines located on different layers, can readily pass (Fig. 1b). Other signal lines and vias lie close to, but not in direct contact with, the signal line of interest. These other line and vias (OLVs), as well as the imperfect ground planes and any attached stubs, increase the inhomogeneity of the environment and significantly modify the signal propagation characteristics. Thus, the electrical design of a computer package generally requires information regarding the propagation delays and characteristic impedances associated with the signal lines, equivalent circuits for the stubs and other discontinuities that may be present, propagation delays and impedances associated with the vias, crosstalk between the signal lines, and changes in the above resulting from the presence of OLVs.

Directions in Electromagnetic Wave Modeling
Edited by H.L. Bertoni and L.B. Felsen, Plenum Press, New York, 1991

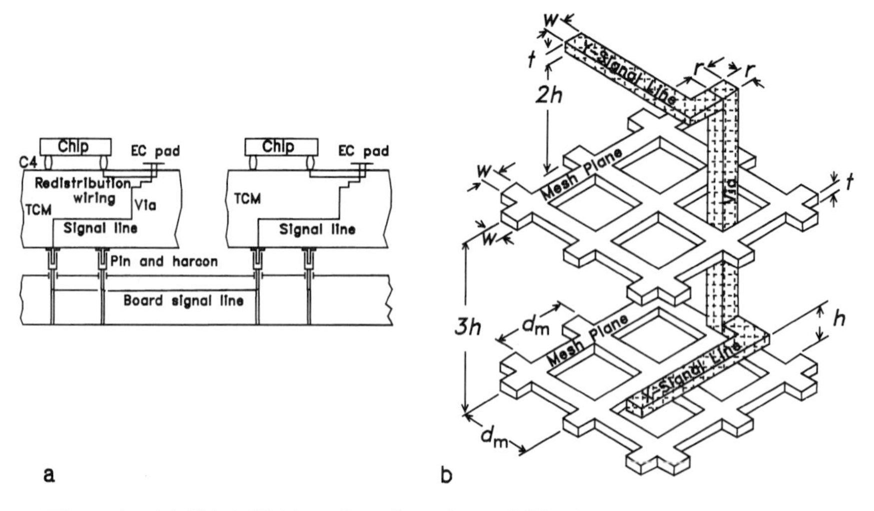

Figure 1. (a) Chip-TCM-board configuration and (b) TCM signal line environment.

Such information is generally provided in the form of equivalent circuits that can later be incorporated with circuit representations of the various drivers and receivers for subsequent circuit simulations.

Because such equivalent circuits are desired, frequency-domain analysis has been the preferred course of action for modeling the package. Time-domain approaches do not directly yield such equivalent circuits, and at present the computational power does not exist for a combined analysis of package structures and the nonlinear circuit models for the drivers and receivers (which may involve many thousands of circuit elements). Furthermore, the package analysis tools must be able to handle 3D, arbitrarily shaped geometries, because typical structures do not fall into the category of simple canonical structures. The tools must also handle a large enough section of the particular package structure, to allow for the tight coupling that generally exists in packages. These factors rule out many of the analysis approaches that have been developed to analyze microwave structures.

Circuit approaches and finite element solution are among the remaining possibilities. The latter, to the authors knowledge, has not been demonstrated for the complexity of structures that are later dealt with. The former has been the approach of choice for computer systems. In the circuit approach, programs calculate the capacitance, inductance and resistance of two-dimensional signal lines [3,4] and their three-dimensional discontinuities. [5,6] Virtually any geometry of interest can be analyzed, provided that it falls within the groundrules of the programs; for instance, many programs require that the structures must be defined through steps along the Cartesian axes. Because accuracy is limited only by the numerical grid size specified by the user, which in turn is limited by the computer resources available, highly accurate circuit models can be obtained. This circuit approach, however, has some fundamental limitations.

In a circuit model, all coupling effects appear instantaneously; they do not display the true physical time retardation associated with the finite velocity of light. Further, a discontinuity modeled through lumped elements contains various electrical nodes at which the circuit elements are connected. This artificial constraint of forcing current to flow through nodes is often not suitable for structures where current flow and coupling are distributed over large areas. A serious limitation also exists for predicting far-end noise in transmission line structures; because this noise is directly related to the difference between capacitive and inductive coupling coefficients, which often differ typically by only a few percent, poor results will be obtained unless both the capacitance and inductance are calculated with extreme accuracy. The package geometries and frequencies associated with today's, and even projected, systems in most cases, however, can be accommodated by the LC approach. Nevertheless, to provide solutions at the higher frequencies, to handle unconventional package structures not especially suited to circuit approaches, or to handle non-TEM structures such as waveguides and their discontinuities, alternative techniques must be developed. An alternative approach, based on the boundary-element method, or method-of-moments, that works within the constraints previously defined, is chosen.

APPROACH

A moment-method technique that employs a rooftop current [7] expansion is used. The current is represented as a linear combination of rooftop functions. The electric field is then expressed as a function of the current density and appropriately tested to generate a matrix equation, which is then solved.

A rooftop function (Fig. 2) is defined in a plane containing a surface current and has triangle dependency along the direction of current flow and pulse dependency in the transverse direction. The charge produced by a rooftop function, through the continuity equation, is constant over each rectangular patch it covers; by properly overlapping the rooftop functions on a surface, variations in the current flow can be represented smoothly and without producing the fictitious charge that would be generated if pulse functions, for instance, were used instead. [8] These rooftop functions are perhaps the simplest functions that adequately model the physical current in typical packages. And, as to be described later, these same rooftop functions can be used to represent the dielectric regions through their three-dimensional polarization currents. The tetrahedral [9] and 3D rooftop [10] functions, which do not produce fictitious charge and therefore could also be used to represent the polarization current, are not well-suited for representing surface current; to model composite conductor-dielectric structures, a second type of basis function would therefore be needed. In the following, the time dependence is $e^{j\omega t}$, where ω is the angular frequency, and t is time.

Modeling the dielectric

To represent the volume polarization with surface currents, the dielectric region is first replaced by a 3D version of the thin-wall mechanism employed by Harrington and Mautz to model dielectric shells.[11] As shown in Fig. 2a,b the entire structure including dielectric regions is subdivided along the Cartesian coordinates into subsections having dimensions τ_x, τ_y, and τ_z. The dielectric region is then replaced by an interlocking array of thin-wall sections (Fig. 2b), each of which are attributed appropriate sheet impedances. [12] Where a conductor surface and thin-wall overlap, the the thin-wall region is effectively removed (at a perfect conductor the tangential component of electric field is zero, so that the polarization current adjacent and tangential to the conductor should vanish). Provided the grid is sufficiently fine with respect to wavelength and to feature size, and provided that an appropriate sheet impedance is used to describe the cell walls, this new structure is electrically equivalent to the solid dielectric. Composite structures, having their dielectric regions so replaced, will yield equivalent electrical results. Because the cell walls have zero thickness, the currents that flow are precisely 2D surface currents; they may be represented by rooftop functions.

A sheet impedance is chosen such that the impedance (or capacitance) is the same between two planes that sandwich a cell of either the solid dielectric or the cellular replacement; only that contribution related to the volume polarization, or in other words, is included; the free space contribution [13] is omitted. From Fig. 2, the total impedance R_x along the x direction for a single cell of solid dielectric is

$$R_x = \frac{\tau_x}{j\omega\varepsilon_0(\varepsilon_r - 1)\,\tau_y\,\tau_z} \tag{1}$$

where ε_0 is the permittivity of free space and ε_r is a relative dielectric constant. For the cellular structure, the surface impedance must be such that when multiplied by length τ_x and divided by perimeter $2(\tau_y + \tau_z)$, the result is again R_x. Thus, the surface impedance along x, R_{sx}, is given by

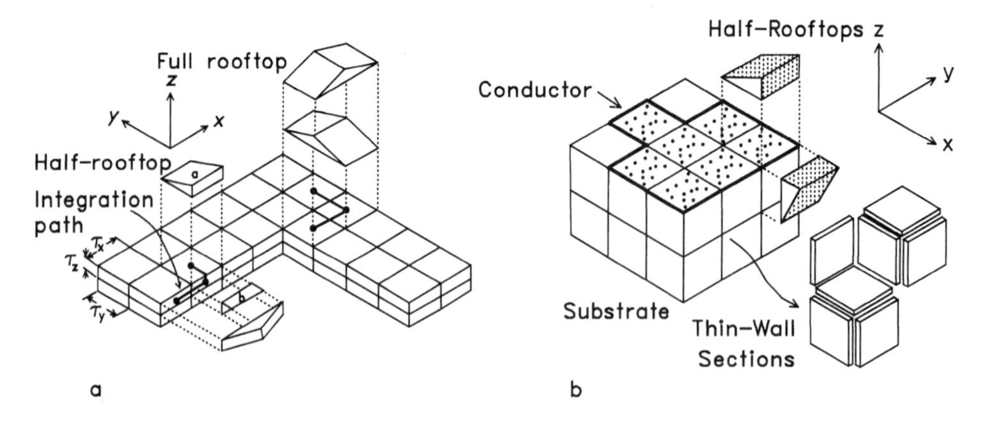

Figure 2. Use of (a) rooftop functions and (b) thin-wall sections.

$$R_{sx} = \frac{2\left(\frac{1}{\tau_y} + \frac{1}{\tau_z}\right)}{j\omega\varepsilon_0(\varepsilon_r - 1)} \tag{2}$$

For walls common to two cells, the impedance would be an appropriate parallel combination. Through permutation of x,y, and z, (2) also gives the surface impedances along y and z, namely R_{sy} and R_{sz}. Lossy dielectrics (or for that matter, lossy conductor volumes) could be handled through appropriate choice of a complex permittivity.

The rooftop functions are placed over each conductive surface. At corners, where only half rooftop functions exist, corresponding half rooftop functions are combined (by making their coefficients dependent) to form corner functions, as are half-rooftop functions a and b in Fig. 2a, and as shown in Fig. 3a. This prevents the development of line charges that would yield an electric field more discontinuous than desired and would slow numerical convergence. On a dielectric or conductive surface, both full and half-rooftops (Fig. 3) may appear. Internal to dielectric volumes are edges, or junctions, where three or four cell walls may intersect. At an external edge, only one corner function is needed (Fig. 3a). At three- and four-junctions (Fig. 3b and Fig. 3c), respectively, two and three corner functions are used. Because current flows continuously around each corner function, the total current into a junction must be zero; using more than two and three corner functions, respectively, would ultimately lead to a singular matrix. As a demonstration, the normalized current flow in the cross-section of a dielectric filled coaxial structure is shown in Fig. 3d; as discussed, either one, two, or three corner functions appear at a corner or junction. The details of the associated structure, analysis, and explanation of the slight asymmetry in the displayed current flow have been given. [12] Because of the roughly eight to sixteen grid sections per wavelength usually required for moment method solutions, the dielectric is limited to perhaps several cubic wavelengths for 3D problems and perhaps several tens of of square wavelengths for 2D problems. (This rules out analysis of structures involving dielectric half spaces). A further choice must be made regarding the boundary conditions of the structure that is to be modeled.

The propagation characteristics of a signal-carrying structure should not be obscured by the presence of source or termination regions. Such end regions, being electromagnetically coupled to the rest of the structure, modify its propagation characteristics. The author chooses to enforce periodic boundary conditions so that end regions need not be included. A region of the original structure is selected and periodically repeated along the x and y directions; the repeated region is designated as a unit cell. The direction of propagation is forced to be in the x direction by representing the current as a linear combination of of rooftop functions that is multiplied by $e^{-jk_x x}$. The parameter k_x is referred to as the propagation constant and essentially gives the propagation velocity of the wave. This current, because of the exponential factor, automatically satisfies Floquet's (Bloch's) theorem is thus a legiti-

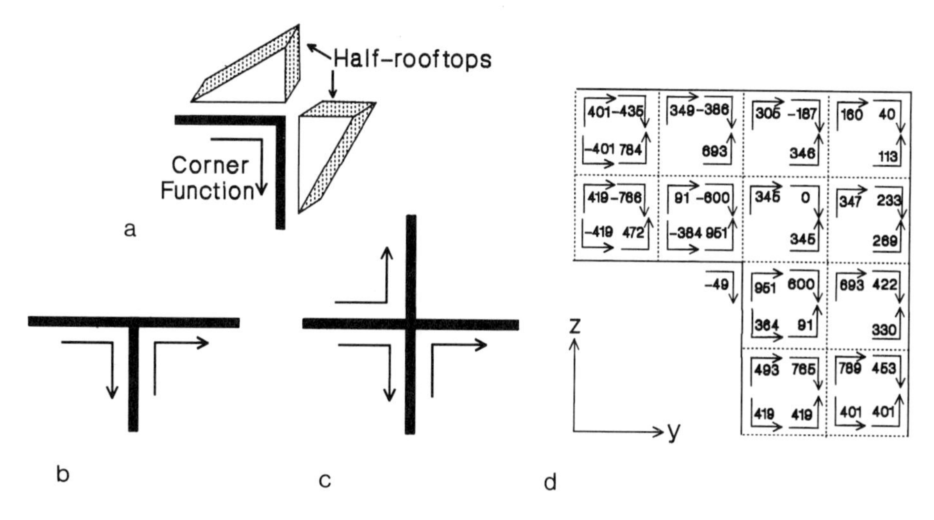

Figure 3. Representation of junction current. (a) At external edge. (b) At three-junction. (c) At four-junction. (d) Normalized current in coaxial structure.

mate choice. Further, because the above exponential dependency is precisely that expected in a TEM structure and roughly that expected in many quasi-TEM structures (which include most packages), it makes sense to include this factor explicitly. The unit cell must include at least two paths continuous along the x direction so that signal and return current can flow along the x direction.

The periodicity along the x direction is necessary for waves to propagate. The periodicity along the y direction, however, gives rise to coupling between unit cells that is generally undesirable but can be minimized. For instance, to analyze an isolated signal line above a mesh plane, a unit cell is defined having a y-periodicity much greater than the height of the signal line above the mesh plane; the coupling between signal lines in neighboring unit cells is small. To model two adjacent signal lines that run parallel along the x direction, a unit cell is defined that has three adjacent signal line channels but one of these channels is kept vacant. Thus, coupling is essentially limited to those adjacent lines that reside within the same unit cell.

Generating and Solving the Matrix Equation

The electric field is calculated using Maxwell's equations and facilitated, because the structure is periodic, by a Fourier analysis. [8,14–17] Once expressed, the electric field is then tested by integrating the electric field over line intervals that overlap the rooftop functions (Fig. 2a). This involves satisfying the electric field boundary condition over each dielectric cell wall and conductor surface,

$$\mathbf{E}_t - \mathbf{J}_s R_s = 0 \tag{3}$$

where $\mathbf{E}_t$ is the tangential electric field, $\mathbf{J}_s$ is the surface current density, and R_s is the appropriate surface impedance. For dielectric volumes, R_s is either R_{sx}, R_{sy}, or R_{sz}; for perfect conductors, R_s is zero, but for imperfect conductors it may be determined through skin-effect considerations. Line intervals were chosen because at the same time they provide, through Maxwell's equations, a test of the normal component of magnetic field wherever four line intervals form a closed rectangle. This consequence is believed by the author to yield stability in the numerical solution at low frequencies. By forcing the electric field to vanish over each rooftop function, a matrix equation is obtained having the form

$$\mathbf{Z}(k_x)\,\mathbf{I} = 0, \tag{4}$$

where **I** is a column vector of current coefficients and **Z** is matrix of impedances. The derivation of the Z matrix elements have appeared. [8]

The matrix equation (4) is an eigenvalue problem, with k_x as the eigenvalue and **I** as the eigenvector. The values of k_x that satisfy Equation (4), from elementary linear algebra, are those for which the determinant of **Z** vanishes. A Newton search [15] may be used to find k_x. Substitution of k_x back into Equation (4) gives **I**. To obtain the starting point, or guess, for the Newton search, a coarse plot of det(**Z**) against k_x may be needed; the zero crossings, which are the solutions, may be visually picked off and used as guesses. More recently, an Arnoldi-based algorthim has been effectively used to obtain the guesses much more efficiently. [18]

For structures involving only a single signal line and one reference conductor, only one quasi-TEM solution exists. In a crosstalk analysis, where multiple signal lines are present in the unit cell, multiple solutions, or modes, generally exist [15] ; the overall solution is given by an appropriate linear combination of these modes. If multiple reference conductors exist, as is the case when two mesh planes surround a signal line, these mesh planes must be shorted together through conductive straps. Otherwise, an undesired mode will exist (effectively, one of the mesh planes serves as the reference and the other acts like another signal line, introducing an extra mode). In practice, such modes are precluded because the mesh planes of a TCM are tied together either directly through vias or indirectly through paths that may include vias, pins, circuits, and capacitors.

For most package structures, which are small compared to a wavelength, the dispersion curve is linear; only one frequency needs to be considered. Thus, the propagation velocity v is equal to the phase velocity of the propagating wave, ω/k_x; the propagation delay t_0 relative to that of light may be expressed as $t_0 = c/v$ where $c = 1/\sqrt{\mu_0 \varepsilon_0 \varepsilon_r}$ is the speed of light in the dielectric medium and μ_0 is the permeability of free space. Voltage, which for non-TEM structures are path dependent, may be found by integrating the electric field and used to find the characteristic impedance, Z_0. Effective values of per-unit-length capacitance C and per-unit-length inductance L may be calculated, for lossless structures, through $t_0=\sqrt{LC}$ and $Z_0 = \sqrt{L/C}$. For structures involving multiple signal lines, the propagation constants and associated current and electric field distributions may be used to find the capacitance and inductance matrices, and the near- and far-end coupled noises. [15] A model for a finite size discontinuity can be inferred from computations performed on two signal-line models, one having the discontinuity and the other not; for instance, the differences in capacitance and in inductance between the models is attributed to the discontinuity. This procedure of tying a discontinuity to a signal line is physically justified since the surrounding environment of the discontinuity must be taken into account. Almost any package structure can be represented through a unit cell for subsequent analysis.

DESCRIPTION OF PACKAGE USED IN THE ANALYSIS

The TCM used in the IBM 3090 is a 10-cm-square substrate containing 36 molybdenum conductive layers in an alumina dielectric, comprising the signal wiring and power distribution to support the 100 chips that are mounted on its surface. Illustrative portions of a chip-TCM-board configuration is shown in Fig. 1a, where chips on different TCMs are interconnected through a path that includes C4 solder balls, engineering change (EC) pads, vias, signal lines, harcon connectors, and pins. More detailed accounts of this package have appeared. [1] Fig. 1b focuses on the signal wiring, showing a signal line, situated between mesh reference planes, that is connected to a second signal line through a via. The various structures are shown as rectangular because the analysis technique requires that the structure be defined by steps along the Cartesian axes. For the TCM used in the 3090, the signal pitch d_m is 0.5 mm. Signal lines have width $w = 0.1$ mm and thickness $t = 0.025$ mm. The mesh plane segments, which are aligned with the signal lines so that their projections on the $x-y$ plane coincide, are assumed to have the same dimensions. The vias are centered in the mesh plane openings, and for the sake of analysis, are approximated as squares with side $r = 0.15$ mm. Vertical spacing h is 0.2 mm, and the relative dielectric constant is 9.5.

The basic signal line structure, as shown in Fig. 1b, involves a signal line between two mesh reference planes. To minimize unwanted coupling to nearby signal lines, the unit cell includes two vacant signal line positions. For modeling purposes and if present, the adjacent vias and crossing signal lines that constitute the OLVs appear periodically at, unless otherwise noted, every available unfilled position within the unit cell. To calculate the propagation parameters of the via, a via that runs through an infinite array of mesh planes is considered; the unit cell is oriented with x along the length of the via, and y and z as shown in Fig. 4a. The unit cell must contain two vias, one of which carries the signal current and the other the return current.

For coupled noise calculation, unit cells must include the signal lines of interest and surrounding mesh planes (Mesh planes only imperfectly shield signal lines located on opposite sides.) The unit cell for coupling between two signal lines vertically separated by a mesh plane is shown in Fig. 4b. Here, three mesh planes are modeled, and some signal-line (shown dashed) positions are intentionally left vacant so that coupling is reduced between unit cells adjacent along the y direction. Signal-line positions 1 and 2 only are occupied. Similarly, unit cells for horizontally or diagonally coupled signal lines are defined. The effect of OLVs is shown in Fig. 4c,d; though only the capacitive effects are indicated in the figures, inductive effects, through eddy currents, also occur. A computer program has been written to perform the electromagnetic algorithm for any structure that can be defined through steps along the Cartesian coordinates. The set of structures now described has not been previously analyzed, either through circuit-based or other techniques.

NUMERICAL RESULTS AND COMPARISON WITH MEASURED RESULTS

Table 1 gives the propagation characteristics of the signal lines and vias in the TCM used in the IBM 3090. The propagation delay, t_0, is normalized to the delay of light in the dielectric. The coupled noise is given for the three cases of horizontally, vertically, and diagonally adjacent signal lines. A pulse having an amplitude of 1V and risetime of 1 ns is assumed, and all signal lines (as conventionally assumed) are terminated in their characteristic impedance. As indicated by the column labels, the signal environment may include y-lines and/or vias. The far-end coupled noise V_{FE} and saturated near-end coupled noise V_{NE}(sat) are maximum when OLVs are respectively present and not present.

The vias have lower impedance but greater propagation delay than the signal lines. The delay is not as great as one might expect, however. The additional capacitance added by running through mesh planes is offset by a reduction in inductance caused by the eddy currents that flow around the segments of the mesh plane. When adjacent y-lines are introduced, which corresponds to signal lines running close to the via, delay is increased by 5% and impedance is reduced by about 4%. The signal lines support waves that are nearly TEM. Even with no OLVs, though, the mere presence of mesh, as opposed to solid, reference planes causes the propagation delay to exceed the TEM value by 2%. A full array of crossing y-lines increases the delay to 1.06, while the full complement of vias increase delay to 1.10. A saturation effect is noted, since the presence of both vias and y-lines increase delay to only 1.11. The corresponding values of Z_0, C, and L are also given in the table. For the coupled noise, OLVs are considered only for horizontally adjacent lines. The near-end noise is greatest for horizontally coupled lines, and least for diagonally coupled lines. When OLVs are present, the near-end noise decreases by 41%, but the far-end noise increases by 315%.

A comparison to results obtained through capacitance calculation and measurements on scale models and test vehicles is presented in Table 2, where the parameters calculated using the method described here are listed in the column labeled "EM". Aside from the layout of their signal wiring (which was specially designed for parameter extraction) these test vehicles are actual TCMs, having

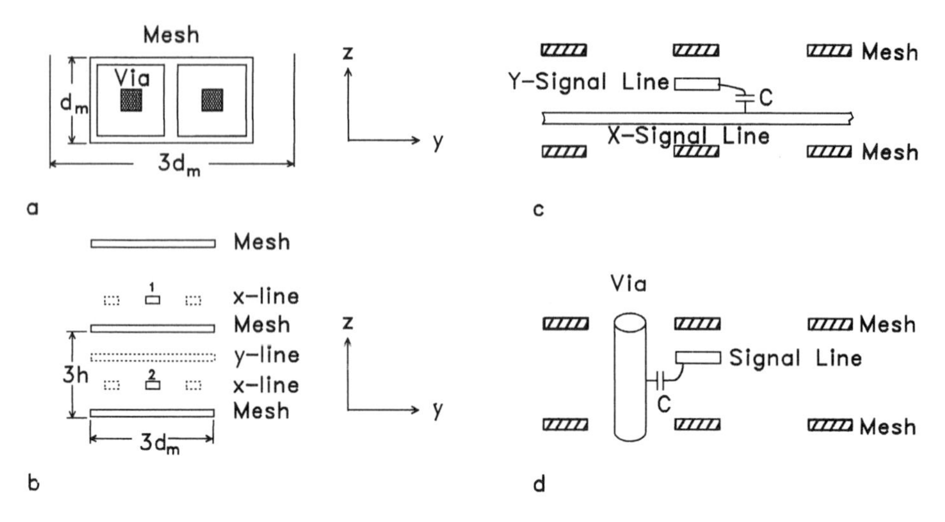

Figure 4. Unit cells for (a) via and (b) vertical signal line coupling. OLV effects of (c) crossing lines and (d) adjacent vias.

Table 1. Propagation parameters for TCM. Copyright 1990 by International Business Machines Corporation; reprinted with permission.

Parameter	No OLVs	y-lines	Vias	All
via t_0 (norm)	1.12	1.17	----	----
via Z_0 (Ω)	28.0	26.9	----	----
via C (pF/cm)	4.12	4.48	----	----
via L (nH/cm)	3.23	3.25	----	----
sig t_0 (norm)	1.02	1.06	1.10	1.11
sig Z_0 (Ω)	45.4	43.4	40.5	39.8
sig C (pF/cm)	2.31	2.51	2.79	2.87
sig L (nH/cm)	4.76	4.73	4.57	4.55
Hor V_{NE}(sat) %	3.64	----	----	2.16
Hor V_{FE} (mV/cm)	-1.01	----	----	-4.19
Vert V_{NE}(sat) %	0.98	----	----	----
Vert V_{FE} (mV/cm)	-0.82	----	----	----
Diag V_{NE}(sat) %	0.55	----	----	----
Diag V_{FE} (mV/cm)	-0.49	----	----	----

electrical characteristics representative of TCMs in general. For the 2D capacitance analysis, the transverse conducting elements cannot be included; capacitances are smaller than expected, impedances are greater than expected, and coupling parameters, because of reduced shielding, are also greater than expected. These capacitances may be viewed as bounds and do indeed appear that way when compared with corresponding data in the other columns. Regarding the scale models, because of practical considerations in their fabrication, they are only reasonable facsimiles of those TCM regions electromagnetically modeled. Nevertheless, the differences are relatively minor and this column should be given the heaviest weight. Because of process variations, the test vehicle measurements should be given less weight. Details of the measurements and layout of these test vehicles have been given. [8] The next structures considered are inhomogeneous, involving finite-size dielectric regions.

Table 2. Comparison of calculated and measured propagation parameters.

Parameter	OLVs	C(2D)	EM	Scale models	Test Vehicle (1979)	Test Vehicle (1985)
via C (pF/cm)	none	-----	4.12 *	3.50	3.15 - 3.29	3.48 - 3.56
	y-lines	-----	4.48 *	-----	-----------	3.77 - 3.81
	x,y-lines	-----	-----	4.23	-----------	-----------
sig C (pF/cm)	none	2.21	2.309	2.404	2.25 - 2.38	2.20 - 2.36
	y-lines	-----	2.514	2.654	-----------	-----------
	vias	-----	2.791	2.807	-----------	-----------
	all	-----	2.869	2.942	2.41 - 2.68 +	2.51 - 2.65 +
sig Z_0 (Ω)	none	46.5	45.4	-----	49.0 - 50.5	47.3 - 51.9
	all	-----	39.8	-----	44.5 - 45.5	46.2 - 50.1
Hor C_{12} (pF/cm)	none	0.188	0.145	0.143	0.096 - 0.131	-----------
Vert C_{12} (pF/cm)	none	0.070	0.027	0.037	0.024 - 0.028	-----------
Diag C_{12} (pF/cm)	none	0.038	0.014	-----	0.011 - 0.015	-----------
Hor C_{12} (pF/cm)	all	-----	0.014	0.031	-----------	-----------
Hor V_{NE}(sat) %	none	4.23	3.64	-----	2.9 - 3.0	2.18 - 2.57
Vert V_{NE}(sat) %	none	1.58	0.98	-----	0.9 - 1.0	0.93 - 0.98
Diag V_{NE}(sat) %	none	0.86	0.55	-----	0.5	0.44 - 0.64

* adjacent return via   + via density of 70 %.

Non-TEM structures

Though the signal structures dealt with so far lie deep within the module, and therefore involve a homogeneous dielectric, non-TEM, structures may also be handled. Signal line segments appearing on the top or bottom surface of the module, where there is an air-ceramic interface, are microstrips. In other packaging applications, as for fiber optic communication, there may be dielectric waveguides. The same computer program can be used to analyze these structures as well.

The first structure, shown in Fig. 5a, is a microstrip (w_1=0.01 cm, t=0.002 cm, d_2=0.1 cm) which displays an interesting peak in effective relative dielectric constant ε_{eff} that is not predicted from static capacitance calculation. The effective dielectric constant at frequency f = 1 GHz is plotted in Fig. 5a, and compared with results obtained using the approach of Gurel and Chew [19] gives agreement to within four percent. This noticeable difference is attributed to the edge singularity. Those basis functions [19] , which were carefully selected to model this particular structure, represent the edge current

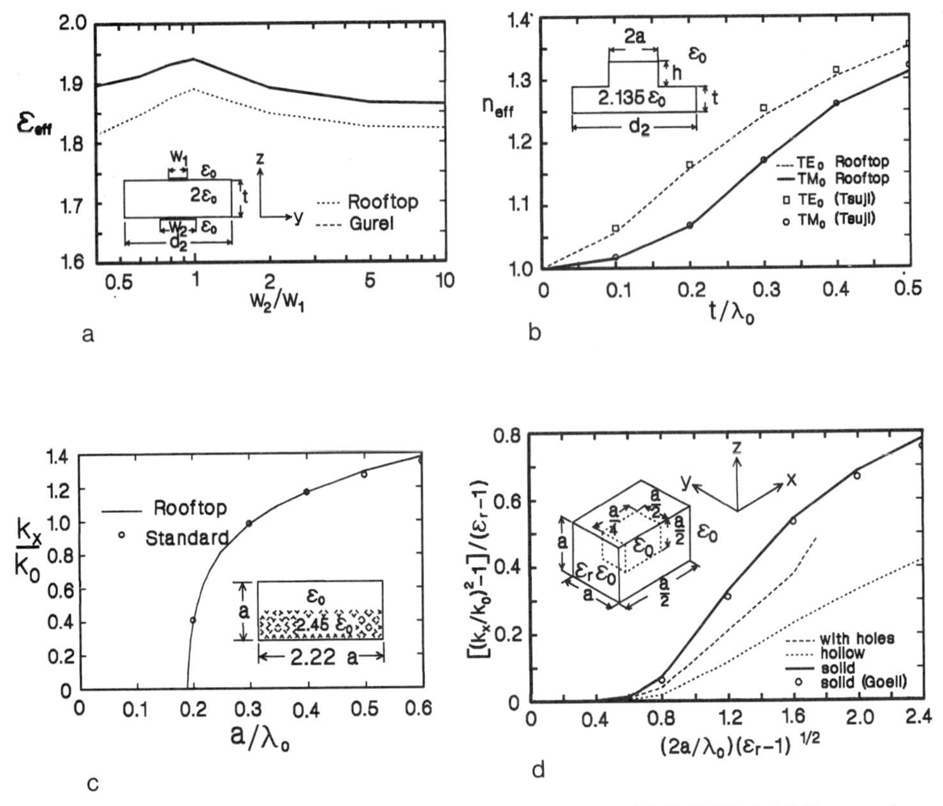

Figure 5. Waveguides. (a) Microstrip. (b) Ribbed dielectric. (c) Half-filled. (d) Rectangular dielectric.

better than the rooftop. The above accuracies, however, are in line with moment method solutions that employ a single type of subsectional basis function. Fig. 5b shows a ribbed dielectric waveguide (2a/t =1, h/t =1, d_2/t =16) having a refractive index $n_\varepsilon = \sqrt{\varepsilon_r} = 1.461$ and investigated by Tsuji et al..[20] The effective refractive index, n_{eff}, is closely predicted for the lowest order modes (denoted by a 0 subscript), which are described [20] as being essentially TE or TM to the vertical direction. The higher order TE and TM modes can be calculated, though their more rapid spatial variation requires that the grid be finer. In this and the following figure, the axes are normalized to be consistent with those appearing in the references. The next structure (Fig. 5c) is a half-filled rectangular waveguide; the normalized propagation constant is seen to agree closely with that calculated using a standard modal approach. The last structure (Fig. 5d) is a square dielectric waveguide, that is modeled as solid and then as a hollow. When a periodic array of rectangular holes are introduced (d_1/a = 0.5, d_2/a =10, ε_r=2.25), stopband-related behavior results; as expected, this last curve falls between that for the solid and hollow structures. Further details of work on dielectric waveguides have appeared. [12]

Computer Run Parameters and Accuracy Considerations

Table 3 gives the **Z** matrix size and run times for some of the representative structures discussed. The run times are for a highly vectorized computer code performed on an IBM 3090 Mod 600 system, and assumes three Newton iterations are required for each case. For the coupled noise case, the time is for two runs, one for each mode. All of these cases use structure symmetry to reduce the size of the **Z** matrix by factors of two or four. As discussed [16, 17] , the computer algorithm has been specially developed to dramatically reduce matrix fill-in time, so that the run time for complex structures is usually

Table 3. Computer run parameters for various cases.

Case	Unit Cell Grid	Symmetry	Matrix Order	Run time (minutes)
Via (no OLVs)	24-by-60-by-20	y, z	969	10.6
Signal Line (no OLVs)	10-by-60-by-25	y	1287	10.6
Signal Line (all OLVs)	10-by-60-by-25	y	2618	71.5
Vertical Coupling (no OLVs)	5-by-30-by-49	y	617	4.8
Sq. Diel. Waveguide (with holes)	4-by-80-by- 8	y, z	554	3.2

defined by the Gauss elimination step. Since a Gauss elimination requires $N^3/3$ complex multiplications, run times generally increase by factors of eight for each doubling in the number of independent rooftop functions employed.

The accuracy of this approach, as for other moment methods that involve subsectional representations, cannot be stated in an absolute sense; there is too wide a variety of structures for an error bound to be established. The confidence in such an approach can only be built up slowly, based on applying it to numerous problems and comparing with closed form solutions, other approaches, and measured results. The rooftop approach described here has been used on literally hundreds of different structures. Numerical convergence (related to the number of subsections used) and agreement with results from other sources has been observed, without exception, for every case tested. Based on this experience, we are very confident that the approach is valid and feel that the results obtained, even with coarse grids, fall well within engineering accuracy for the purposes of package analysis. Because of the variations (Table 2) in package parameters due to process and other variations, engineering accuracy for characteristic impedance and propagation delay (after subtracting the TEM delay) is probably about five or ten percent. Higher degrees of accuracy can be obtained by using finer grids. For the purposes of comparison of similar structures, as in sensitivity analyses where only relative difference matter, the accuracy is far better. As for any approach, there are certain structures or geometries that are more difficult to handle, and yield larger errors. Subsectional approaches, as ours, generally yield greater errors in structures where currents vary dramatically near edges, where conductors appear in very close proximity, or where segments with extreme aspect ratios are present.

CONCLUDING REMARKS

The rooftop approach described here has been successfully applied to a number of canonical structures found in a TCM, as well as to several non-TEM waveguide structures. As described, agreement well within engineering accuracy is observed when results are compared against measurements on scale models, actual hardware, and to results obtained through other approaches. Though not considered here, such structures as stubs and right angle bends of signal lines could be represented by a suitable unit cell and thus analyzed using this approach. Signal line structures that involve skin-effect can be handled either through appropriate conductor surface impedances or through modeling entire conductor volume as a dielectric region with complex dielectric constant. Many other structures could be handled, including those composed of spatially varying or even anisotropic dielectric constant, provided that the the structures can be defined by steps along the Cartesian axes. Recently, this rooftop approach has been applied to modeling various connectors in computer packages. It was first successfully applied by Spring and Smith [21] for complex connector structures, and later by Cases and Rubin.[22] The volume polarization aspect has also been successfully applied to source-fed and scattering problems. [23] The value of this approach lies in its full-wave nature and in its versatility; this single approach can be used to solve, within engineering accuracy, a large class of package-related problems.

ACKNOWLEDGEMENTS

The author expresses deep appreciation to Sam Diehl, IBM East Fishkill, Alina Deutsch, Al Ruehli, and Gerard Kopcsay, of the T. J. Watson Research Center, and Professor Henry Bertoni of the Polytechnic University of New York, for help in various apsects of the work presented here.

REFERENCES

1. A. J. Blodgett and D. R. Barbour, "Thermal Conduction Module: A High-Performance Multilayer Ceramic Package, *IBM J. Res. Develop.,* 26:30 (1982).

2. E. E. Davidson, "Electrical Design of a High Speed Computer Package," *IBM J. Res. Develop.,* 26:349 (1982).

3. W. T. Weeks, "Calculation of Coefficients of Capacitance of Multiconductor Transmission Lines in the Presence of a Dielectric Interface," *IEEE Trans. Microwave Theory Tech.,* 18:35 (1970).

4. W. T. Weeks, L. L. Wu, M. F. McAllister, and A. Singh, "Resistive and Inductive Skin Effect in Rectangular Conductors," *IBM J. Res. Develop.,* 23:652 (1979).

5. A. E. Ruehli and P. A. Brennan, "Efficient Capacitance Calculations for Three Dimensional Multiconductor Systems," *IEEE Trans. Microwave Theory Tech.,* 21:76 (1973).

6. A. E. Ruehli, "Inductance Calculations in a Complex Integrated Circuit Environment," *IBM J. Res. Develop.,* 16:470 (1972).

7. A. W. Glisson and D. R. Wilton, "Simple and Efficient Numerical Methods for Problems of Electromagnetic Radiation and Scattering from Surfaces," *IEEE Trans. Antennas Propagat.,* 25:593 (1980).

8. B. J. Rubin, "An Electromagnetic Approach for Modeling High-Performance Computer Packages," *IBM J. Res. Develop.,* 34:585 (1990).

9. D. H. Schaubert, D. R. Wilton, and A. W. Glisson, "A Tetrahedral Modeling Method for Electromagnetic Scattering by Arbitrarily Shaped Inhomogeneous Dielectric Bodies," *IEEE Trans. Antennas Propagat.,* 32:77 (1984).

10. M. F. Câtedra, E. Gago, and L. Nuño, "A Numerical Scheme to Obtain the RCS of Three-Dimensional Bodies of Resonant Size Using the Conjugate Gradient Method and the Fast Fourier Transform," *IEEE Trans. Antennas Propagat.,* 37:528 (1989).

11. R. F. Harrington and J. R. Mautz, "An Impedance Sheet Approximation for Thin Dielectric Shells," *IEEE Trans. Antennas Propagat.,* 23:531 (1975).

12. B. J. Rubin, "Electromagnetic Modeling of Waveguides Involving Finite-Size Dielectric Regions," *IEEE Trans. Microwave Theory Tech.,* 38:807 (1990).

13. T. K. Sarkar, E. Arvas, and S. Ponnapalli, "Electromagnetic Scattering from Dielectric Bodies," *IEEE Trans. Antennas Propagat.,* 37:673 (1989).

14. B. J. Rubin and H. L. Bertoni, "Waves Guided by Conductive Strips Above a Periodically Perforated Ground Plane," *IEEE Trans. Microwave Theory Tech.,* 31:541 (1983).

15. B. J. Rubin, "The Propagation Characteristics of Signal Lines in a Mesh-Plane Environment," *IEEE Trans. Microwave Theory Tech.,* 32:522 (1984).

16. B. J. Rubin, "Scattering from a Periodic Array of Apertures or Plates Where the Conductors Have Arbitrary Shape, Thickness, and Resistivity," *IEEE Trans. Antennas Propagat.,* 34:1356 (1986).

17. B. J. Rubin, "Modeling of Arbitrarily Shaped Signal Lines and Discontinuities," *IEEE Trans. Microwave Theory Tech.,* 37:1057 (1989).

18. B. Rubin, "Electrical Characterization of the Interconnects Inside a Computer," *SPIE International Symposium on Advances in Interconnects and Packaging,* 1389: Boston MA, Nov. 4-9, 1990; to be published.

19. L. Gurel and W. C. Chew, "Guidance or Resonance Conditions for Strips or Disks Embedded in Homogeneous and Layered Media," *IEEE Trans. Microwave Theory Tech.,* 36:1498 (1988).

20. M. Tsuji, S. Suhara, H. Shigesawa, and K. Takiyama, "Submillimeter Guided-Wave Experiments with Dielectric Rib Waveguides," *IEEE Trans. Microwave Theory Tech.,* 29:547 (1981).

21. C. Spring and H. H. Smith, "Coupled Noise Analysis of a Complex Connector Structure Using a Full Wave Modeling Approach", *Conference Proceedings for the Progress in Electromagnetics Research Symposium,* Cambridge MA, July 1-5, 1991; to be published.

22. M. Cases and B. J. Rubin, "Full-Wave Analysis of Connectors in High-Speed Computer Packages" *15th International Conference of Infrared and Millimeter Waves,* Orlando FL., Dec. 10-14, 1990; to be published.

23. B. J. Rubin and S. Daijavad, "Radiation and Scattering from Structures Involving Finite-Size Dielectric Regions," *IEEE Trans. Antennas Propagat.;* to be published.

NUMERICAL MODELING OF FREQUENCY DISPERSIVE BOUNDARIES IN THE TIME DOMAIN USING JOHNS MATRIX TECHNIQUES

Wolfgang J.R. Hoefer

Laboratory for Electromagnetics and Microwaves
Department of Electrical Engineering
University of Ottawa, Ottawa, Canada K1N 6N5

1 INTRODUCTION

In this paper, the time domain representation of frequency dispersive boundaries in TLM field modeling will be described. It is based on the technique of time domain diakoptics introduced by P.B. Johns [1] in 1981, has been generalized in terms of an impulse scattering matrix formulation, and called "Johns Matrix Technique" in memory of this pioneer of the TLM method [2].

The TLM algorithm generates the digital response of a dense mesh of transmission lines, interconnected at so-called nodes, to an impulsive excitation. This excitation can be either a single impulse or a sequence of impulses which are scattered at the nodes, propagate into the network, and are reflected at the boundaries. The impulsive node voltages at the instants of scattering are samples of a continuous function in space and time. The corresponding frequency domain characteristics can be obtained by discrete or fast Fourier transform of the TLM response.

The ability to model dispersive boundaries is crucial for the efficient use of such a time domain method. A sinusoidal excitation in the time domain is not economical because it fails to exploit the inherent broadband capability of time domain processes which allow impulsive excitation. Furthermore, the response of a system to transients and waveforms other than sinusoidal can only be computed if the dispersive characteristics of the system are properly modeled in the time domain. The use of complex quantities is obviously excluded.

The natural way to model a frequency-dispersive boundary in the time domain is to simulate the penetration of the field into the space behind that boundary. For instance, a wideband absorbing boundary for all modes in a rectangular waveguide is modeled naturally by the input plane of a long section of that same waveguide. Obviously, such a boundary representation considerably increases the numerical effort, and more economical ways must be found to model such an absorbing boundary. Other examples of dispersive boundaries are lossy metallic walls, or simply any arbitrary interface between subregions of a computational domain.

Directions in Electromagnetic Wave Modeling
Edited by H.L. Bertoni and L.B. Felsen, Plenum Press, New York, 1991

If the behaviour of a dispersive wall can be described by a differential equation, the latter may be discretized in time and solved by stepwise integration during the field simulation. However, a more general solution which does not require an analytical description of the boundary, resides in the technique of "Time Domain Diakoptics". This technique also opens new possibilities for the large-scale numerical preprocessing of substructures, the partitioning of large time domain problems at the field level, and for combining other numerical techniques with the TLM method.

2 TIME DOMAIN DIAKOPTICS

An important development in electromagnetic field modeling has been the introduction of diakoptics. Employed first in steady-state network theory by Kron [3], "diakoptics" means breaking down, or partitioning, a large network into sub-networks for individual analysis and subsequent reconnection [4],[5]. This technique has been used by Brewitt-Taylor and Johns [6] for steady-state network modelling, and was then extended to time-discrete field models by Johns and Akhtarzad [1],[7].

One way of looking at the diakoptics procedure is to consider its similarity with the Green's function technique. A Green's function $G(r,r',t-t')$ is defined as the field response to an impulsive (Dirac) excitation. It depends on both the field point and the source point coordinates and is an analytical function defined everywhere in the structure. The purpose of a Green's function is to permit the computation of the field in a structure for an arbitrary excitation in space and time. In classical terms, the field function $\Phi(r,t)$ due to a source function $K(r',t')$ is obtained as:

$$\Phi(r,t) = \int_{V'}\int_{t'} G(r,r',t-t')\; K(r',t')\; dV'\; dt' \qquad (1)$$

where the integration is taken over the source volume and the time history of the excitation.

Since the TLM algorithm is also based on impulsive excitation, the Green's function concept appears to be well suited for implementation in TLM. However, there are two differences between TLM and traditional field analysis:

(a) In TLM, field propagation is described by incident and reflected voltage impulses on the mesh lines, rather than in terms of total node voltages. Thus, it is appropriate to formulate the Green's function concept in these terms as well.

(b) The Greens function in TLM will be a discrete rather than a continuous function in space and time. Furthermore, since our goal is to characterize the behaviour of fields at boundaries and interfaces rather than in overall space, the Green's function needs to be known only for points lying on these boundaries (which are the points at which the boundaries intersect the TLM mesh). The branches cut by the boundaries are called "removed branches" in the language of diakoptics.

In a number of recent publications [2], [8] – [13] it has been shown how the discrete Green's function or Johns matrix can be obtained by TLM analysis of the structure it describes. The result is a three-dimensional matrix $[G(m,n,k)]$ (see eq. (5)) where m and n designate the output and input removed branches (field and source points), and k is the discrete time variable.

In the discrete TLM formulation, the convolution integral in eq. (1) becomes a numerical convolution of the form

$$[V^r(m,k)] = [G(m,n,k-k')] * [V^i(n,k')] \tag{2}$$

where $[V^i(n,k')]$ is the column vector of impulses incident upon the boundary, $[V^r(m,k)]$ the column vector of the resulting "reflected" output impulse functions, and $[G(m,n,k-k')]$ is the discrete Green's function or Johns matrix. The symbol "$*$" describes a restricted numerical convolution. The m-th term of the output column vector is computed as follows:

$$v^r(m,k) = \sum_{n=1}^{N}\sum_{k'=0}^{k} g(m,n,k-k')\ v^i(n,k') = \sum_{n=1}^{N}\sum_{k'=0}^{k} g(m,n,k')\ v^i(n,k-k') \tag{3}$$

It accounts for the inputs at all N removed branches. From the computational point of view, this convolution process is a simple multiply–shift–add procedure which is easily programmed.

3 BASIC PROPERTIES OF THE JOHNS MATRIX

In the case of a structure with a single boundary, the Johns Matrix has the characteristics of a reflection coefficient matrix, with the difference that the impulses reflected in the reference plane are not only functions of the impulses incident at a given instant, but also of all previously incident impulses.

In the case of a structure with two separate interfaces, such as a microwave two-port, the Johns matrix can be partitioned into two reflection and two transmission sub-matrices as follows.

Let the removed branches 1 to L form the interface 1, and branches $L+1$ to M form the interface 2 of a two-port, as shown in Fig. 1.

$[V_1^i(n,k')]$ and $[V_2^i(n,k']$ are the vectors of voltage impulse functions incident on the removed branches in interface 1 and interface 2, respectively, while $[V_1^r(n,k)]$ and $[V_2^r(n,k)]$ are the impulse streams emerging from these interfaces. We can then write:

$$\begin{pmatrix}[V_1^r(m,k)]\\ [V_2^r(m,k)]\end{pmatrix} = \begin{pmatrix}G_{11}(m,n,k-k') & G_{12}(m,n,k-k')\\ G_{21}(m,n,k-k') & G_{22}(m,n,k-k')\end{pmatrix} * \begin{pmatrix}[V_1^i(n,k')]\\ [V_2^i(n,k')]\end{pmatrix} \tag{4}$$

This partitioned Johns matrix resembles the scattering matrix of a two-port in traditional frequency domain network theory. For structures with "i" separate interfaces, the Johns matrix can be partitioned into i times i submatrices. The submatrices on the main diagonal always represent reflection coefficient matrices of the respective interfaces, while the others are transmission matrices.

As indicated previously, the concept of the Johns Matrix allows the partitioning of a large TLM mesh into smaller submeshes. The reconnection process can be executed in two ways. Either the large structure can be analysed in a sequential fashion, starting with a substructure at one extremity, finding the Johns matrix of its input plane, then terminating the next substructure with that Johns matrix as a boundary, finding its input Johns matrix,

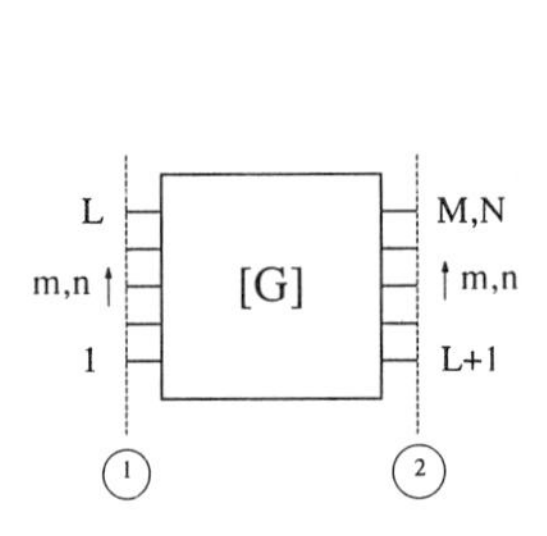

Fig. 1 Removed branches of a structure with two separate interfaces, characterized by its Johns Matrix [G].

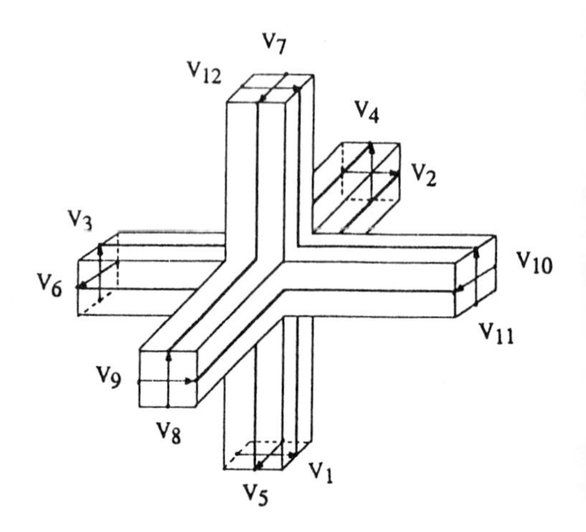

Fig. 2a Johns's 3D condensed TLM node

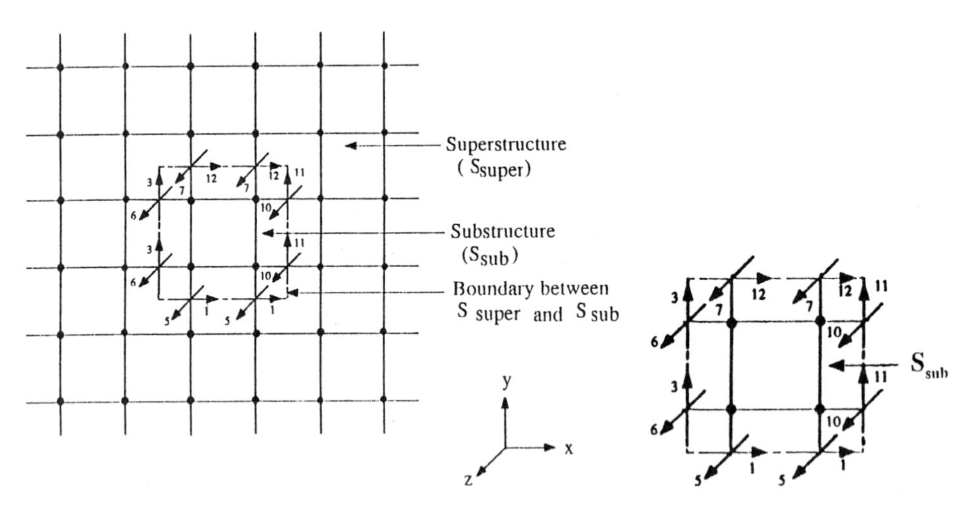

Fig. 2b Partitioning of a TLM structure

Fig. 2c Substructure bounded by Johns matrix boundary

and so on, until the total structure has been analyzed. Or, the Johns Matrices of the substructures can be computed separately and then reconnected [2].

The first approach is simpler from a computational point of view, but requires a sequence of TLM analyses. It proceeds like a traditional Smith Chart analysis where the load is transformed toward the generator into a complex impedance which then becomes the load impedance of the preceding subsection, and so forth. At the boundary which is modelled by the Johns matrix, all incident impulses are convolved with that Johns matrix as described by eqs. (2) and (3). TLM is then operating like a time domain boundary integral method.

The second approach allows to compute the Johns Matrices of the subsections separately, and to cascade them in arbitrary sequence. Note that the cascading of Johns Matrices is essentially a pre-processing tool and could thus be performed on a much larger processor than the final TLM analysis which uses the pre-computed Johns matrix as a boundary description. For more details the reader is referred to [2].

The storage of the Johns matrix in a computer is analogous to the storage of a microwave circuit component on a shelf: whenever it is needed, it can be connected with other components in a new circuit.

4 APPLICATIONS OF JOHNS MATRIX BOUNDARIES

In the following, the power of the Johns matrix concept will be demonstrated by means of two applications: (a) The partitioning of a large structure into smaller substructures, and (b) The time domain modelling of wideband absorbing loads in non-TEM waveguides.

4.1 Partitioning of TLM Networks into Substructures

This procedure has been discussed for the two-dimensional case by Johns and Akhtarzad [1] in their paper on time domain diakoptics. Fig. 2a shows the symmetrical condensed 3D-TLM node proposed by Johns [14],[15] for 3-D field modeling.

Fig. 2b shows a large TLM network made up of such nodes, which has been divided into two substructures designated S_{super} and S_{sub}. A cross-section in the $x-y$ plane is shown. A total of six such cross-sections form the interface between S_{super} and S_{sub}. Assume that the substructure S_{sub} represents a small portion of the network which needs to be modified repeatedly (for example during an optimization process), and the structure S_{super} is the major portion of the network which remains unchanged. First, the time domain response (impulse response) of S_{super} with respect to the N interconnection ports is computed and stored (where N is the total number of interconnection ports across the interface; in the example shown in Fig. 2b, there are 16 interconnection ports in the $x-y$ plane). Then, only the substructure S_{sub} (whenever a change is made in the geometry) is discretized and excited. The impulses emerging from the " removed branches " across the interface are convolved with the impulse response of S_{super} (Fig. 2c).

To compute the impulse response of S_{super}, the transmission lines across the interface are terminated with matched loads (zero local reflection coefficient). Note that each connection shown in Fig. 2b represents two transmission lines carrying two orthogonal polarizations as indicated by the arrows. A single impulse injected at any of these branches across the interface will cause impulses separated by twice the iteration time interval to flow out of this structure. The impulse function $g(m,n,k)$, represents the output impulse stream emerging at the m-th port (originating from the node $(x = i\Delta l, y = j\Delta l, z = l\Delta l)$) at $t = k\ \Delta t$ due to a unit excitation of the n-th port (originating from the node $(x = i'\Delta l, y = j'\Delta l, z = l'\Delta l)$) at $t = 0$. The combination of all possible unit excitations yields the Johns matrix of S_{super}; it is a three-dimensional array of dimension $(M \times N \times K)$,

g(1,1,K) - - - g(1,n,K) - - - g(1,N,K)
g(1,1,k) - - - g(1,n,k) - - - g(1,N,k)
g(1,1,0) - - - g(1,n,0) - - - g(1,N,0)
g(m,N,K)
g(m,N,k)
m ↓ g(m,1,0) - - - g(m,n,0) - - - g(m,N,0) (5)
g(M,N,K)
g(M,N,k)
g(M,1,0) - - - g(M,n,0) - - - g(M,N,0) ↗ k
→ n

K is the total number of iterations, and $M = N$ is the total number of ports across the interface. Note that each port corresponds to one polarization of a branch.

When impulses are injected into the substructure S_{sub} to excite it (Fig. 2c), they are scattered at nodes and reach the branches at the Johns matrix boundary after some time. Any impulse incident upon the boundary will give rise to streams of impulses, separated by $2\Delta t$, which flow back into the structure through all the branches. These reflected impulse voltages are computed, using equation (3), by convolving the incident impulses with the Johns matrix of S_{super}. Note that the Johns matrix boundary can be of arbitrary geometrical shape, provided that it is conformal with the TLM mesh.

The TLM algorithms with and without diakoptics approach are shown in Fig. 3. Note the extra module to be implemented for convolution purposes with the diakoptics approach. The computer run time and memory required with the conventional TLM algorithm is proportional to

$$(NX^{super} \times NY^{super} \times K) \tag{6}$$

while that with diakoptics technique is

$$(NX^{sub} \times NY^{sub} \times K) + (K \times (K+1) \times N^2)/2 \tag{7}$$

where NX is the number of grids along the x-axis and NY is the number of grids along the y-axis and K is the total number of iterations. In equation (7), the first term corresponds to the discretization of the structure S_{sub} and the second part corresponds to the convolution with the Johns matrix. For very big structures and small number of removed branches, the quantity given by (6) becomes larger than that given by (7) and hence the diakoptics procedure is more economical.

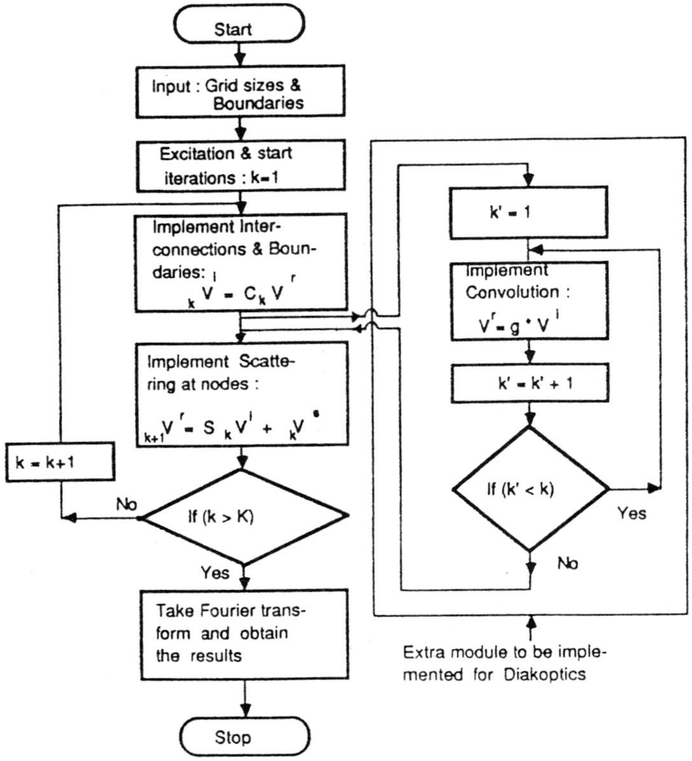

Start
Input : Grid sizes & Boundaries
Excitation & start iterations : k=1
Implement Inter-connections & Boun-daries: ${}_kV^i = C_k\,{}_kV^r$
Implement Scatte-ring at nodes : ${}_{k+1}V^r = S\,{}_kV^i + {}_kV^s$
If (k > K)
No
Yes
k = k+1
Take Fourier trans-form and obtain the results
Stop
k' = 1
Implement Convolution : $V^r = g * V^i$
k' = k' + 1
If (k' < k)
Yes
No
Extra module to be imple-mented for Diakoptics

Fig. 3 TLM algorithm without and with diakoptics (From Eswarappa [11].

Circuit or Discontinuity
Input Reference Plane
Output Reference Plane

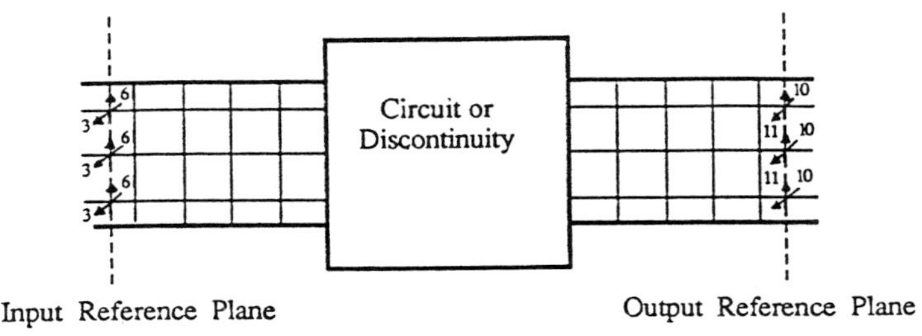

Fig. 4 3-D waveguide discontinuity region with wideband absorbing walls modeled by a Johns matrix (From Eswarappa [11].

Regular Johns Matrix
$\alpha_1 = 0.0001,\ \alpha_2 = 0.05$
$\alpha_1 = 0.0001,\ \alpha_2 = 1.0$
$\alpha_1 = 0.001,\ \alpha_2 = 0.5$
$\alpha_1 = 0.0002,\ \alpha_2 = 0.2$
$\alpha_1 = 0.001,\ \alpha_2 = 0.2$
Magnitude of Reflections
Frequency (GHz)

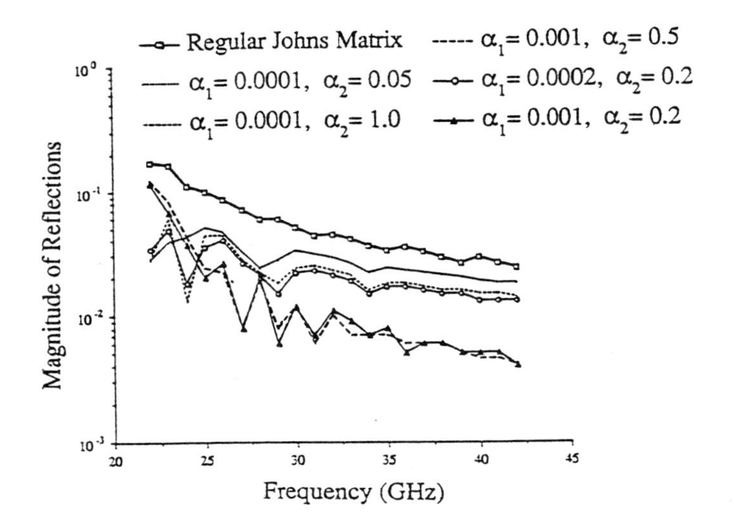

Fig. 5 Reflection from dispersive absorbing boundary modeled by a tapered Johns matrix (From Eswarappa [11].

However, for wideband S-parameter extraction of microwave circuits in waveguide systems, this algorithm becomes very efficient; this application will be discussed next.

4.2 Wideband Absorbing Boundaries for Non-TEM Waves

The impedance of absorbing boundaries for non-TEM waves depends on frequency. (Examples are a matched load in a TE_{10} waveguide, or an absorbing boundary for the near field of an antenna or a scatterer). Such wideband absorbing boundaries are desirable in order to cover a wide frequency range with a single TLM simulation followed by a Fourier analysis. (S-Parameters of a waveguide component, surface current distribution on a scatterer).

Indeed, a Johns matrix can represent such a boundary, keeping in mind that its frequency dependent character is really a consequence of the time history of field penetration into the space behind the boundary. We can thus model this space by a TLM network and obtain its Johns matrix for the branches cut by the boundary. Note that the full Johns matrix description is mode independent, and the representation of the boundary is thus accurate for any incident field configuration, including hybrid fields.

However, if the mode of propagation across the interface is well defined (a requirement for the definition of S-parameters), and if the transverse field distribution is frequency independent, (such as in TE or TM - modes of homogeneous waveguides), the Johns matrix can be reduced to a single impulse stream. This is so because for a specific mode excitation profile, the reflection coefficient is the same in all removed branches: it is characteristic of one specific mode of propagation across the interface. In such a case, the single mode Johns matrix is computed by simultaneously exciting each removed branch with an impulse having an amplitude proportional to the transverse amplitude distribution of the propagating mode. Under these specific excitation conditions, the impulse response is extracted from one removed branch only, which is closest to the field maximum in the interface. The impulses emerging from all the other removed branches are then computed from that response by space interpolation according to the transverse mode function.

The limitation to a single mode leads to considerable reductions in memory and computer time. If N is the number of removed branches in the interface, then the generation of the Johns matrix is N times faster, and the convolution process is accelerated by a factor N^2. The memory required to store the Johns Matrix is also reduced by N^2.

A typical example is shown in Fig. 4. It shows a three-dimensional discontinuity region with rectangular waveguide ports carrying the TE_{10} mode. In the reference planes which are placed out of reach of evanescent higher order modes, the structure is terminated by modal Johns Matrix boundaries which emulate a very long uniform waveguide section.

The representation of dispersive mode impedances by a single term Johns matrix is extremely efficient. It allows us to realize wideband matched impulsive sources and loads with a minimum of computational effort, and thus to extract S-parameters from a single impulsive TLM analysis of a structure. This procedure has been described in [9]. It has also been shown [11] that the return loss of such a wideband absorbing boundary can be considerably improved by "tapering" the Johns matrix of the long waveguide section. This amounts to gradually increasing the attenuation in the waveguide in time, thus progressively absorbing stray reflections at the mesh nodes due to finite time and space discretization. The tapered Johns matrix $J'(k)$ is obtained from the original Johns matrix $J(k)$ as follows:

$$J'(k) = J(k)\, e^{-\sum_{k'=1}^{k} \alpha(k')} \tag{8}$$

where $\alpha(k')$ is

$$\alpha(k') = \alpha_1\, e^{\frac{k'-1}{NI} ln(\alpha_2/\alpha_1)} \tag{9}$$

where α_1 is the attenuation constant for $k = 1$ (i.e. first iteration), and α_2 the attenuation constant for $k = NI$, the total number of terms in the Johns matrix. The values of α_1 and α_2 can be optimized for minimum return loss over the operating bandwidth. Fig. 5 shows the reflections from two opposing absorbing Johns matrix boundaries (WR-28) for different combinations of α_1 and α_2. It can be seen that reflections less than one percent can be achieved over the entire operating bandwidth.

To demonstrate the improvement due to tapering of the Johns matrix, Fig. 6 compares the magnitude and phase of the S-parameters of an inductive waveguide iris obtained with an untapered (a) and a tapered (b) Johns matrix boundary. Note that the phase in Fig. 6a shows strong periodic variations, indicating insufficient return loss of the termination. The tapered Johns matrix, however, provides a practically perfect match over the entire operating band of the waveguide. Note that each figure has been extracted by discrete Fourier transform from a single 3-D TLM simulation. Results for the same discontinuity from Marcuvitz's Waveguide Handbook have been included for comparison.

Obviously, the tapering of the Johns matrix not only improves the quality of the absorbing boundaries, but it also reduces the number of terms required in the Johns matrix, and thus the time needed for convolution.

5 DISCUSSION AND CONCLUSION

The modeling of arbitrary electromagnetic boundary conditions by a numerical Green's function or Johns matrix, in particular wideband absorbing boundary conditions for non-TEM fields, is an essential step towards efficient time domain modelling of microwave structures. It becomes particularly effective when applied to homogeneous structures carrying only one mode, because the Johns matrix is then reduced to a single characteristic impulse function representing the mode reflection coefficient of a boundary in the time dimension. This, in turn, opens the way for S-parameter extraction over a wide frequency range from a *single* impulsive TLM analysis. Waveguide discontinuity parameters obtained with tapered Johns matrix boundaries are in excellent agreement with published and measured results [11].

The partitioning of large structures at the field level using time domain diakoptics also represents a major progress towards CAD applications of time domain methods. It allows for large-scale preprocessing of substructures, the storage of their characteristics in digital form, and their subsequent recombination. This procedure is equivalent to the combination of subnetworks in the frequency domain using generalized complex scattering parameters. Further important applications of this technique include the representation of imperfectly conducting walls in the time domain, the limitation of the computational domain in open TLM problems, and the combination of the TLM method with other numerical techniques.

Future challenge in this field include the time domain modelling of frequency dispersive passive and active materials and devices, nonlinear and time-dependent media and

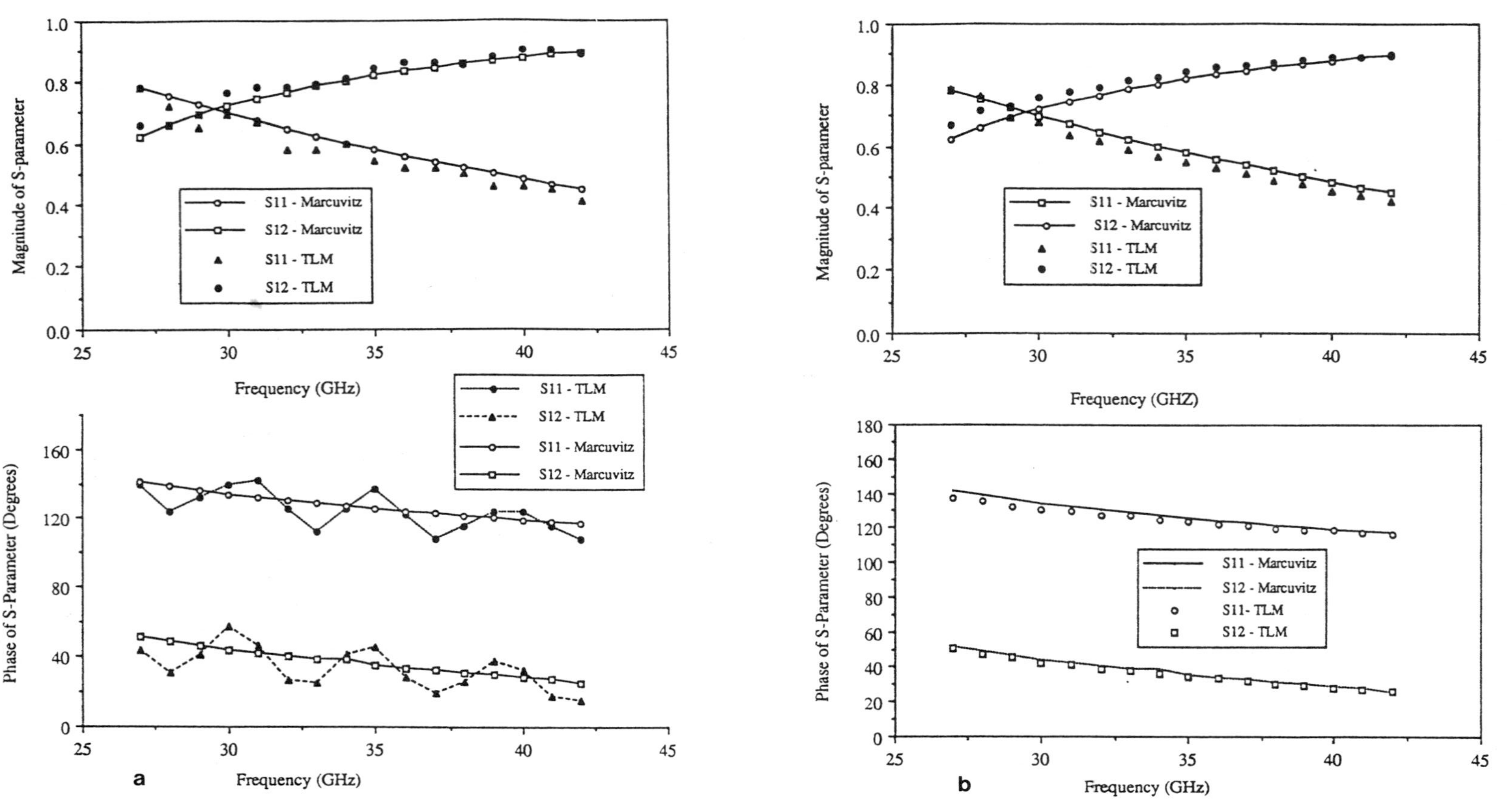

Fig. 6 S-Parameters of an inductive waveguide iris extracted from a single impulsive 3-D TLM analysis with wideband absorbing boundaries modeled by a Johns matrix .
(a) Results obtained with regular Johns matrix. (b) Results obtained with tapered Johns matrix.
(From Eswarappa [11])

boundaries, and optimization and synthesis techniques in the time domain. The development of new algorithms and procedures must go hand-in-hand with the creation of powerful, user-friendly graphics interfaces for efficient data input and output. From the computational point of view, the local properties of time domain scattering algorithms make them highly suitable for parallel processing. Furthermore, the control over time, in concert with Johns matrix techniques, will allow us to optimize bandwidth, memory requirements and computer time so as to approach the efficiency of frequency domain methods which are far more developed at the present time than the existing time domain methods.

6 ACKNOWLEDGEMENTS

The author acknowledges the many contributions of, and discussions with Z. Chen, Eswarappa, I. Kim, S. Liang, P. Russer, P. So, R. Sorrentino, and J. Uher. Financial support by the Natural Science and Engineering Research Council of Canada (NSERCC) and the Telecommunications Research Institute of Ontario (TRIO) is gratefully acknowledged.

7 REFERENCES

1. Johns, P.B., Akhtarzad, 'S., The Use of Time Domain Diakoptics in Time Discrete Models of Fields, Int.J.Numer. Methods Eng., vol. 17, pp. 1-14, (1981).

2. Hoefer, W.J.R., The Discrete Time Domain Green's Function or Johns Matrix - A New Powerful Concept in TLM, Intl. Journal of Numerical Modelling, vol 2, no. 4, pp. 215-225, (1990).

3. Kron, G. Diakoptics, MacDonald, (1963).

4. Braemellar,A., M.N. John, M.R. Scott, Practical Diakoptics for Electrical Networks, Chapman and Hall, London, (1969).

5. Happ,H.H., Diakoptics and Networks, Academic Press, New York and London, (1971).

6. Brewitt-Taylor, C.R., and P.B. Johns, On the Construction and Numerical Solution of Transmission-Line and Lumped Network Models of Maxwell's Equations, Int. J. Num. Meth. in Eng., vol. 15, pp. 13-30, (1980).

7. Johns,P.B., S. Akhtarzad, Time Domain Approximations in the Solution of Fields by Time Domain Diakoptics, Int. J. Num. Meth. in Eng., vol. 18, pp. 1361-1373, (1982).

8. So, P., Eswarappa, Hoefer, W.J.R., A Two-dimensional TLM Microwave Field Simulator using New Concepts and Procedures, IEEE Trans. Microwave Theory Techniques, vol. MTT-37, no. 12, pp. 1877-1884, (1989).

9. Eswarappa, Costache, G., Hoefer, W.J.R., TLM Modeling of Dispersive Wideband Absorbing Boundaries with Time Domain Diakoptics for S-Parameter Extraction, IEEE Trans. Microwave Theory Techniques, IEEE Trans. Microwave Theory Techniques, Vol. MTT-38, no. 4, (1990).

10. Eswarappa, So, P., Hoefer, W.J.R., New Procedures for 2D and 3D Microwave Circuit Modeling with the TLM Method, in 1990 IEEE Intl. Microwave Symp. Dig., Dallas, Texas, May 8-10, (1990).

11. Eswarappa, New Developments in the Transmission Line Matrix and the Finite Element Methods for Numerical Modeling of Microwave and Millimeter-Wave Structures, PhD Thesis, University of Ottawa, Canada, (1990).

12. Eswarappa, Hoefer, W.J.R., Application of Time Domain Diakoptics to 3-D TLM Method with Symmetrical Condensed Nodes, 1990 IEEE-AP-S Symposium and URSI Meeting, Dallas, Texas, May 7-11, (1990).

13. Eswarappa, Hoefer, W.J.R., Novel Field Partitioning Techniques in the Time Domain for Efficient Analysis of Microwave Structures, 1990 Asia Pacific Microwave Conference - APMC '90, Tokyo, Japan, Sept. 18-21, (1990).

14. Johns, P.B., Use of condensed and symmetrical TLM nodes in computer-aided electromagnetic design, IEE Proc., vol. 133, Pt.H, no. 5, pp. 368-374, (1986).

15. Johns, P.B., A Symmetrical Condensed Node for the TLM Method, IEEE Trans. Microwave Theory Techniques, vol. MTT-35, no. 4, pp. 370-377 (1987).

WAVE INTERACTIONS IN PLANAR ACTIVE CIRCUIT STRUCTURES

J. Birkeland, S. El-Ghazaly * and T. Itoh

Electrical Engineering Research Laboratory
The University of Texas at Austin
Austin, TX 78712

*Center for Solid State Electronics Research
Arizona State University
Tempe, AZ 85287

ABSTRACT

As the operating frequency of the integrated circuits is increased toward millimeter-waves, it becomes necessary to characterize the wave interactions with active devices. This is due to the following reasons. As the frequency is increased, the device size can be a substantial fraction of the wavelength. Hence, the distributed effect in the device can no longer be negligible. An example of modeling for such a situation is presented. On the other hand, wave interactions with solid state devices in an integrated circuit environment can be used for development of new device structures and components. In addition, understanding of the wave interaction can lead to finding a favorable environment to maximize the capability of the active devices. Several examples for modeling and design of such configurations are presented. Due to the infancy of this thrust, there remain a number of problems and limitations which will have to be addressed.

INTRODUCTION

In the microwave techniques, the passive configurations and the active devices have been treated separately. For instance, the guided wave phenomena such as the propagation characteristics are analyzed as the solution of the Maxwell's equations under specific boundary conditions. Based on the solution, various passive devices such as stubs, filters and directional couplers are designed and their terminal characteristics such as the S parameters are found. On the other hand, the active devices has been characterized separately based on the electron dynamics of the device structures and presented to the design engineers in the form of S parameters, for example. The circuit designers then combine the active and passive devices based on the individual S parameters.

Completely different situations can arise in which the procedures described above are no longer valid or adequate. This paper presents such situations and provide some attempts to develop new procedures. These approaches can lead to better understanding of the wave phenomena in the active guided wave structures and to development of new types of components and devices. In what follows, three examples of planar active circuit structures are presented.

Directions in Electromagnetic Wave Modeling
Edited by H.L. Bertoni and L.B. Felsen, Plenum Press, New York, 1991

DISTRIBUTED THREE-TERMINAL DEVICES

As the frequency of operation is increased toward millimeter-waves, the gate width may have a substantial fraction of the wavelength. Therefore, under such a situation, the wave phenomena need to be included in the characterization of the FET. In addition, it is conceivable to use such wave phenomena intentionally to develop a traveling wave transistor. In this chapter, an approach developed by the authors based on the small signal model is presented for the analysis of a traveling wave transistor. Although the method is valid for a general class of the traveling wave transistor, we proceed here by means of an example with a traveling wave inverted-gate FET (INGFET).[1] This is a traveling wave transistor with its cross section shown in Fig.1. By taking advantage of the structural symmetry of the FET, this device has a gate conductor on the underside of the GaAs active layer. This arrangement makes possible to ground the gate so that the source and the drain are used as the input and output lines.

The most comprehensive analysis might require a solution of the electrodynamic systems of equations for the device mechanism in conjunction with Maxwell's equation describing a traveling wave. Such a method, however, requires very extensive numerical calculations with very slow convergence. In addition, the physical mechanism of traveling wave phenomena is not clearly seen. It is found impractical to take this approach.

An alternative approach has been conceived. The analysis consists of four steps. The first step the two-dimensional FET is analyzed accurately. The second step computes the parameters of a cylindrical transmission line. The third step uses the coupled mode theory to formulate the wave-equation along the transmission line using the results of the first two steps. In the fourth step the resultant eigenvalue problem is solved and the propagation characteristics in this traveling-wave INGFET are obtained.

The first step is carried out by means of a two-dimensional modeling FET. In this model, the continuity equation, the energy conservation equation and Poisson's equation are solved simultaneously. The electron mobility and temperature are expressed as functions of the average electron energy. The model is found to be capable of accurately simulating the sub-micron gate GaAs MESFETs including the nonstationary conditions such as velocity overshoot. By means of this analysis, the potential distribution, carrier densities and energy distributions in the device can be found out. Subsequently, the main components of the small signal equivalent circuit can be derived.

In the next step the three electrodes of the traveling wave transistor are considered to constitute a cylindrical coupled transmission line. Since the cross sectional dimensions are much smaller than the operating wavelength, the use of a quasi-TEM analysis is justified. In this analysis, Laplace equation for the potential distribution in the transverse cross section is solved numerically.

To improve numerical efficiency, magnetic walls are placed on the top and side of the structure. (See Fig.2.) This is justifiable due to the fact that most of the energy is concentrated in the gap between the conductors and the field becomes negligible as one moves away from the central region. The above equation with appropriate boundary conditions is solved by a finite-difference scheme. From the potential distribution, the charge distribution and the capacitance are readily computed. The problem is solved for the even and odd excitations so that the self and mutual capacitances can be identified. The self and mutual inductances can be identified by solving the same problem with even and odd excitations in the structure where all the dielectric materials are replaced with air. The capacitances in such a hypothetical structure are found from the finite-difference scheme. By virtue of the fact that the phase velocity of the TEM transmission line is independent of the geometric configuration, the inductances can be derived.

The conductor loss can be estimated by the incremental inductance rule if the conductor is thick enough or by the phenomenological loss equivalence method.[2] The channel loss is accounted for by the source and drain resistances, R_s and R_d, The former is

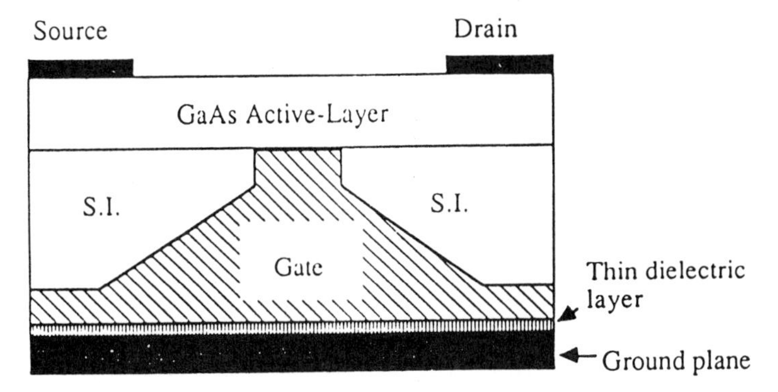

Fig. 1. The INGFET structure.

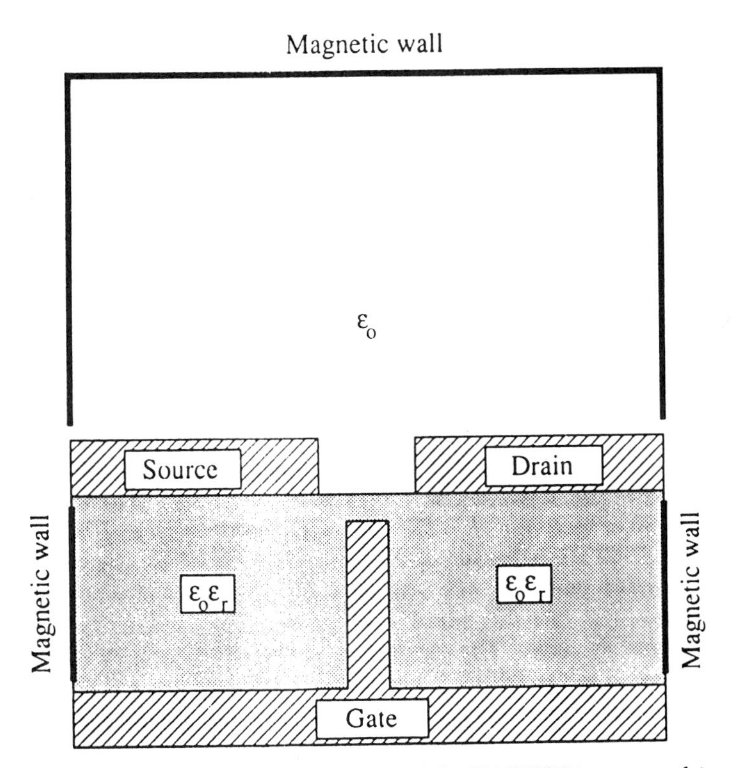

Fig. 2. The transmission line model of the INGFET (not to scale).

estimated by considering the source-gate region as ohmic resistance while R_d can be estimated from the I-V characteristics of the FET in the saturation region.

The coupled mode equation of the traveling wave FET can be obtained by adding the small signal FET parameters distibuted along the coupled transmission line. (See Fig.3.) The voltage and current relationship in this active coupled transmission line can be written as

$$\partial V_s/\partial z + I_s Z_s + I_d Z_m = 0 \quad (1a)$$

$$\partial I_s/\partial z + V_s(Y_s + Y_{as}) + V_d Y_m = 0 \quad (1b)$$

$$\partial V_d/\partial z + I_d Z_d + I_s Z_m = 0 \quad (1c)$$

$$\partial I_d/\partial z + V_d Y_d + V_s(Y_m + Y_{ad}) = 0 \quad (1d)$$

where V_s and V_d are the source and drain voltage waves whereas I_s and I_d are the source and drain current waves, respectively. When the above set of equations is transformed to that of the forward and backward waves, the equations to be solved become

$$\partial \mathbf{x}/\partial z + A\mathbf{x} = 0 \quad (2)$$

where $\mathbf{x}^T$, or $[S^+ \ S^- \ D^+ \ D^-]$, is a vector consisting of the coefficients of the incident and reflected waves on the source and drain lines and A is a 4 x 4 complex matrix contining the active and passive network parameters per unit length.

The matrix equation (2) above is solved as an eigenvalue problem. The eigenvalues are the possible propagation constants and the eigenfunctions are the corresponding modal configurations. Once the modal solution is found, the description of the traveling wave FET of a length of W can be determined for a given terminations and excitations. Hence, the total gain from the transistor unit and the optimal input and output impedances on the source and drain lines can be evaluated.

An example of the computed results is provided here for a device with an active layer of 0.1 m thickness. The source-to-drain spacing is 1.5 μm and the gate length is 0.5 μm. The source and drain lengths are 10 μm. Fig.4 shows that the device supports two modes. One of the mode is a growing mode with a negative value of attenuation constant while another is a decaying mode. The phase velocity of the growing mode is slightly smaller than the decaying mode.

The present approach is based on the small signal analysis. In the future, it is necessary that a large signal model is developed so that a more practical situation including the saturation effect is evaluated.

WAVE INTERACTIONS IN PLANAR CIRCUITS

Two examples are presented in this chapter. The first is a distributed transmission line which is periodically loaded with active devices. This structure provides a injection locking to all devices by virtue of stop band phenomenon associated with a periodic structure. In addition, the power output from the devices can be combined for generation of a larger output. The second example is an interaction of a passive periodic structure with an active device. The role of the passive structure is to provide a frequency selective feedback to the active device.

Transmission line loaded periodically with active devices

Fig.5 shows a schematic of the periodic power combining oscillator. [3] The structure consists of a microstrip line periodically loaded with two-terminal devices and associated networks. In this particular example, the active devices used are Gunn diodes.

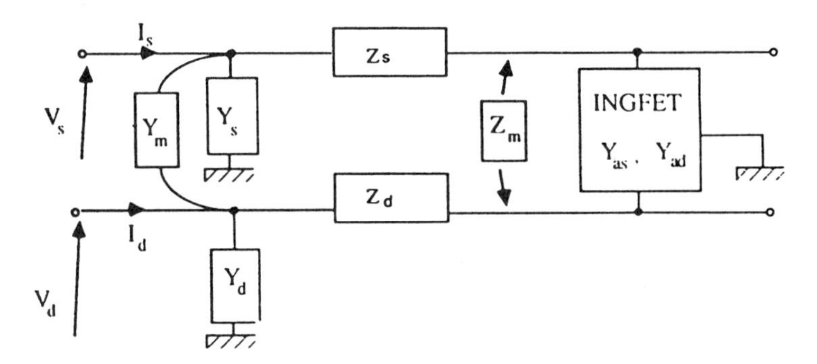

Fig. 3. The coupled-active transmission line circuit.

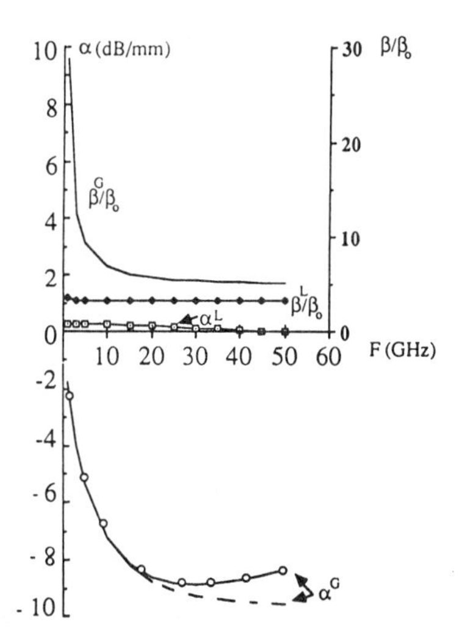

Fig. 4. The propagation constants of the possible modes on the traveling-wave INGFET. Superscript G denotes the gainful mode and superscript L denotes the lossy mode. For α^G, the line (—o—o—) is obtained considering both conductor and channel losses; the other curve (– – –) is obtained considering conductor losses only.

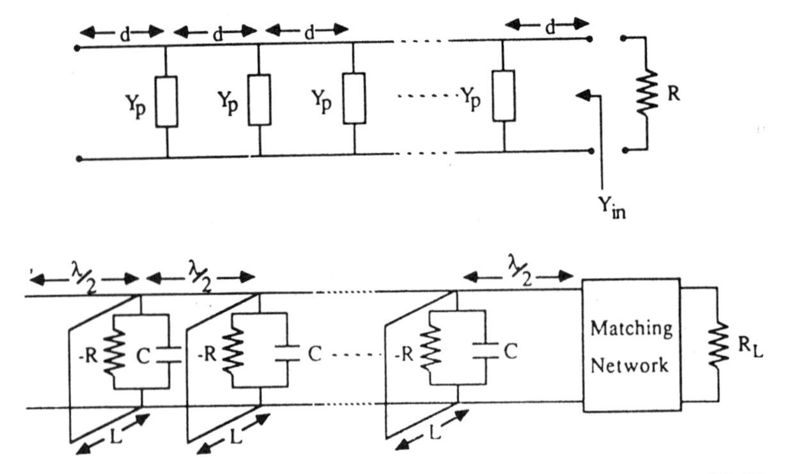

Fig. 5. (a) Transmission line loaded periodically with admittance Y_p. (b) The periodic power combining oscillator. The negative resistance diodes are modeled as a parallel-R/C network. The shorted stubs are to compensate the capacitive part of the device.

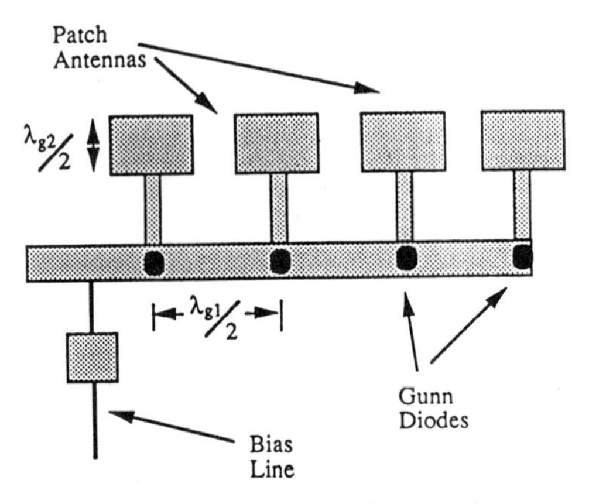

Fig. 6. Diagram of a four diode spatial second harmonic power combiner. λ_{g1} is the guided wavelength at fundamental frequency. λ_{g2} is the guided wavelength at fundamental frequency.

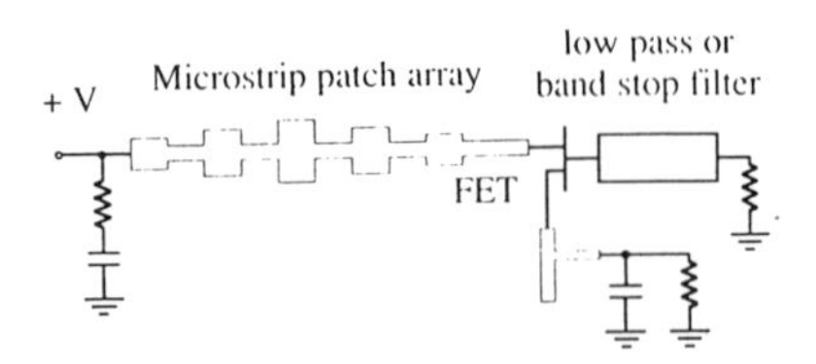

Fig. 7. Schematic view of oscillator circuit.

The spacing of the Gunn diodes is chosen to be one-half of the guide wavelength in the microstrip line at the desired frequency of oscillation. The inductive stubs connected to each diode cancel the reactive part of the diode which is usually capacitive. Hence, at the desired frequency of oscillation, the microstrip line is loaded periodically with pure negative resistances. At this frequency, all the pure negative resistances are combined in parallel and the outputs from all the devices are combined and delivered to the load. The oscillation frequency is determined by the two factors. The first factor is the combination of the inductive stub and the capacitance of the device itself. The second factor is the spacing of the devices along the periodically loaded microstrip line. The sharpness of the resonance is determined by these two factors. Outside of the resonance region, the devices do not constructively interfere and hence no oscillation takes place.

Many active devices exhibit nonlinear characteristics even at frequencies at which the devices no longer provide negative resistances. This characteristics can be used for generation and power-combining of the harmonics. Fig.6 shows one example. [4] The microstrip line is loaded periodically with negative resistance devices, Gunn diodes in this example. The circuit is designed in such a way that the structure oscillates at the frequency corresponding to the surface wave stop band. The second harmonic generated in each diode falls into the leaky wave stop band. The outputs at the second harmonic from each device are in phase and are led to the microstrip patch antennas which are resonant at the second harmonic. Therefore, the second outputs from each device are power-combined in free space.

The future problem to be addressed is the development of the network theory in which the nonlinear large signal model can be incorporated.

Oscillator with a leaky wave antenna

One of the problems in a planar circuit is the lack of a high Q element. An attempt to alleviate this situation was demonstrated by means of a leaky wave antenna.[5] The leaky wave antenna made of a periodic structure exhibits a leaky wave stop band at which the period of the structure is one guide wavelength. At the leaky wave stop band, the input VSWR of the antenna becomes very high due to the increase of the reactance of the antenna. Since the radiation efficiency becomes poor, the use of a leaky wave antenna at frequencies in the leaky wave stop band is ordinarily avoided. In the present application, this undesirable feature as an antenna is intentionally used for realizing a frequency selective feedback to an active device connected to the input terminal of the antenna as shown in Fig.7. In this example, an FET is brought into an unstable condition by a stub connected to the source terminal. The antenna is connected to the drain terminal. A resistive load is connected to the gate terminal. The purpose of the low pass filter is to load the device at the frequency corresponding to the surface wave stop band of the antenna which is located at the frequency approximately one half of the one corresponding to the leaky wave stop band.

Since the FET receives a positive feedback only at the leaky wave stop band at which it is unstable, a stable oscillation takes place. The oscillation output is extracted in the form of radiation from the leaky wave antenna in the broadside direction.

CONCLUSIONS

The wave interactions with active media and active devices have been presented by way of several examples. The wave propagation in a traveling wave transistor has been analyzed by means of the device equation and the transmission line equation followed by solution of the eigenvalue equation. Periodic structures have been shown to play an important role in developing new active components based on the two-terminal or three-terminal devices. It is believed that understanding the wave interaction with the device is essential for analysis and design of components for many millimeter-wave applications.

REFERENCES

[1] S. M. El-Ghazaly and T. Itoh, "Traveling-wave inverted-gate field-effect transistors: Concept, analysis and potential," IEEE Trans. Microwave Theory and Techniques, Vol.37, No.6, pp. 1027-1032, June 1989.

[2] H.-Y. Lee and T. Itoh, "Phenomenological loss equivalence method for planar quasi-TEM transmission lines with a thin normal conductor or superconductor," IEEE Trans. Microwave Theory and Techniques, Vol37, No.12, pp.1904-1909, December 1989.

[3] A. Mortazawi and T. Itoh, "A periodic planar Gunn diode power combining oscillator," IEEE Trans. Microwave Theory and Techniques, Vol.38, No.1, pp.86-87, January 1990.

[4] A. Mortazawi and T. Itoh, "A periodic second harmonic spatial power combining oscillator," 1990 IEEE MTT-S International Microwave Symposium, May 8-10, 1990, Dallas, Texas, pp.1213-1216.

[5] J. Birkeland and T. Itoh, "Planar FET oscillators using periodic microstrip patch antennas," IEEE Trans. Microwave Theory and Techniques, Vol.37, No.8, pp.1232-1236, August 1989.

HIGH FREQUENCY NUMERICAL MODELING OF PASSIVE MONOLITHIC CIRCUITS AND ANTENNA FEED NETWORKS

Linda P.B. Katehi

Radiation Laboratory
The University of Michigan
Ann Arbor, MI 48109

INTRODUCTION

During the past ten years, the requirements for low cost systems with small volume and low weight introduced the need for high operating frequencies and led to the development of monolithic technology. An important characteristic of this technology is that all antenna elements, active devices and passive circuits including components and interconnections are formed on the surface of a semi-insulating substrate by some deposition method.

The ability of the monolithic technology to deliver circuits and systems with the above characteristics depends heavily on the availability of appropriate design tools. The need for efficient algorithms capable to design complex circuits dominated the '80's and largely influenced the research in the area of numerical modeling. As a result, during the past decade, many techniques with various degrees of complexity and accuracy were developed and applied to a wide variety of problems. Due to the driving need for accurate design tools, the emphasis of the research was placed on the development of the techniques rather than the understanding of the physical phenomena involved in the performance of the monolithic structures.

In the 1990's the scopes of the research in circuit modeling needs to be reoriented. Circuit optimization through new innovative designs will attract more interest than circuit design through new approaches. The realization of this goal is influenced by the availability of techniques which can effectively take into account important physical phenomena such as electromagnetic coupling, free space radiation, surface and leaky waves, and ohmic losses.

A technique which can provide high frequency characterization of three dimensional planar structures including the phenomena mentioned above is the integral equation method [1], [2], [3], [4]. The space domain integral equation method ($SDIE$) has been employed to analyze circuit and antenna problems and has provided superb accu-

Directions in Electromagnetic Wave Modeling
Edited by H.L. Bertoni and L.B. Felsen, Plenum Press, New York, 1991

racy and efficiency when the considered monolithic structures are of small or medium electric size.

This paper describes the above technique as it applies to specific planar structures and discusses its effectiveness both in terms of accuracy and computer efficiency.

THEORY

The development of the space domain integral equation method is based on mathematical treatment of Maxwell's equations without incorporating any simplifying assumptions. As a result, the method itself is not limited by frequency or geometry as it happens with the so called quasi-static techniques or dispersion models. In this approach an integral equation for an unknown electric parameter of the circuit under study is formulated and solved numerically with the method of moments. The integral equation is derived from Maxwell's equations with the use of Green's second identity, the reciprocity theorem and/or other electromagnetic theorems. In its final form, the integral equation employs the dyadic Green's function for the pertinent boundary value problem and/or other scalar potentials. The use of these potential functions varies according to the geometry of the specific problem.

A widely used form of the integral equation is obtained from the so called Pocklington's equation for the unknown electric or magnetic fields:

$$\begin{bmatrix} \vec{E}(\vec{r}) \\ \vec{H}(\vec{r}) \end{bmatrix} = \int\int_{S'} [k^2 \bar{\bar{I}} + \bar{\nabla}\bar{\nabla}] \cdot \begin{bmatrix} \bar{\bar{G}}_{ej}(\vec{r}/\vec{r}\,') & \bar{\bar{G}}_{em}(\vec{r}/\vec{r}\,') \\ \bar{\bar{G}}_{hj}(\vec{r}/\vec{r}\,') & \bar{\bar{G}}_{hm}(\vec{r}/\vec{r}\,') \end{bmatrix} \cdot \begin{bmatrix} \vec{J}(\vec{r}\,') \\ \vec{M}(\vec{r}\,') \end{bmatrix} ds' \qquad (1)$$

where k is the wavenumber, $\bar{\bar{G}}_{\mu,\nu}$ with $\mu = e, h$ and $\nu = j, m$ are the electric and magnetic dyadic Green's function for the problem, and $\vec{J}$, $\vec{M}$ are the unknown electric and magnetic current densities.

The form of the Green's function depends on the problem under study. If the considered structure is open (microstrip antennas, slot apertures or antenna feed networks), then this function is expressed into the form of slowly converging Sommerfeld integrals with discrete singularities in their integrand. The treatment of the integrals has been studied extensively and has been presented in many publications [5], [6], [7]. When the circuit is shielded by a cavity or a section of a waveguide, these integrals are replaced by slowly converging summations that need special care in order to speed up the calculations.

The dyadic Green's functions of equation (1) and the unknown electric or magnetic fields satisfy all the boundary conditions except the ones on the surface of the strip conductors or slot apertures. In order to transform (1) into an appropriate integral equation and assure uniqueness of the solution, the remaining boundary conditions have to be imposed on the unknown fields. These conditions are given by the following relations:

$$\hat{n} \times \hat{n} \times \vec{E}(\vec{r}_{strip}) = \bar{\bar{Z}} \cdot \vec{J}(\vec{r}_{strip}) \qquad (2)$$

on the surface of the strip conductors, and

$$\hat{n} \times \begin{bmatrix} \vec{E}(\vec{r}_{slot}) \\ \vec{H}(\vec{r}_{slot}) \end{bmatrix}^{above\ the\ slot} = \hat{n} \times \begin{bmatrix} \vec{E}(\vec{r}_{slot}) \\ \vec{H}(\vec{r}_{slot}) \end{bmatrix}^{below\ the\ slot} \qquad (3)$$

on the slot apertures. In equation (2), $\hat{n}$ is a unit vector normal to the surface of the strip or slot and $\bar{\bar{Z}}$ is a dyad representing the overall effect of the field penetrating inside the strip conductors on the electromagnetic fields excited outside the strips. This dyad can be evaluated accurately by solving an appropriate integral equation for the electric current distributed within the conductor [8].

To obtain the unknown electric and magnetic current densities, the method of moments is applied. The area which contains nonzero currents is subdivided into smaller surface elements and the currents are expressed as a superposition of known basis functions multiplied by unknown coefficients:

$$\begin{bmatrix} \vec{J}(\vec{r'}) \\ \vec{M}(\vec{r'}) \end{bmatrix} = \sum_{n=1}^{N} \sum_{m=1}^{M} [I_{nm} \, , \, M_{nm}] \begin{bmatrix} \vec{f}_{nm}(\vec{r'}) \\ \vec{g}_{nm}(\vec{r'}) \end{bmatrix} \tag{4}$$

where the pairs (n,m) indicate the nodes in the mesh. The vector functions $\vec{f}_{nm}$ and $\vec{g}_{nm}$ are separable with respect to the planar coordinates and are chosen so that the corresponding current distribution satisfies the edge conditions on the strips or slot apertures.

In view of equations (2)-(4) and with the application of Galerkins method for error minimization, the original integral equation (1) reduces to a simple matrix equation of the form:

$$\begin{bmatrix} [Z_i] & [Z_m] \\ [Y_i] & [Y_m] \end{bmatrix} \cdot \begin{bmatrix} [I] \\ [M] \end{bmatrix} = \begin{bmatrix} [V_{exc}] \\ [I_{exc}] \end{bmatrix} \; . \tag{5}$$

In this equation, $[V_{exc}]$ and $[I_{exc}]$ are vectors providing the interaction between the excitation and the electric and magnetic currents, respectively. Furthermore, $[Z_{i,m}]$ and $[Y_{i,m}]$ are square submatrices with elements given in the form of multiple sums or integrals or combinations of both, and are evaluated by a combination of numerical and analytical techniques in order to improve accuracy and assure efficiency in the computations. The convergence of the elements in the above submatrices is very critical and greatly affects the accuracy of the solution.

With the field or current distribution available through the method of moments solution, an appropriate transmission line model is implemented to find a network representative of the planar discontinuity [1], [9]. In this network, each port represents one mode propagating under a feed line and, thus, the number of electric ports equals the number of propagating modes multiplied by the number of physical ports. In practice, however, the operating frequency permits only one propagating mode under each feeding line thus reducing the complexity of the network.

In the case of open circuits, the network parameters cannot provide a complete characterization of the discontinuity. The derived field or current distribution has to be substituted back into the integral equation to compute the space and surface wave far-field radiation.

CONVERGENCE CONSIDERATIONS

As with any other numerical approach, the accuracy that the space domain integral equation provides is subject to specific convergence criteria. Depending on the electrical parameters to be evaluated, the criteria for convergence define different ranges of values for the parameters introduced through the solution. The overlap of these ranges provides values for the various parameters which can assure overall convergent results.

The convergence criteria vary with the geometry. Specifically, in circuit elements where the scattering or other network parameters are of interest, the solution is more sensitive to the size of the mesh introduced for each planar component of the electric or magnetic current, and the relative position between these meshes. However, the same solution is rather insensitive to the form of the basis functions. On the contrary, in radiating elements where resonant properties are of interest, the accuracy of the solution depends on how successfully the chosen basis functions satisfy the edge condition.

It is worth noticing that convergence does not assure accuracy, while the lack of convergence indicates the presence of a mathematical, numerical or programing error. As a conclusion, the validity of the method and the accuracy of the results should be verified through extensive convergence tests and comparisons to experimental data or theoretical results derived through other independent techniques.

NUMERICAL RESULTS

As mentioned above, the importance of the *SDIE* method lies in the capability to provide accurate results in combination with the flexibility and computational efficiency of the associated computer codes. Several planar structures have been characterized via this approach some of which are discussed below

Open Microstrip Discontinuities

An important consideration in open microstrip circuits is the radiation loss which occurs in the form of space or surface waves. If we classify the open planar circuits into more or less radiating structures we will find that stubs radiate more as they approach their resonant frequency, while junctions radiate less with the cross junction providing the lowest amount of radiation. Since stubs are indisposable circuit elements, their geometries have to be optimized for best performance and minimum radiation loss.

The space domain integral equation method has been employed to evaluate single rectangular, radial and triangular stubs and, single and cross rectangular stubs in terms of their bandwidth, resonant characteristics and radiation properties. This comparative study indicates that the radial stub radiates less then the rectangular and triangular, and provides the largest bandwidth (Figure 1) [10]. Furthermore, the same study has shown that the cross rectangular stub radiates much less than the single rectangular due to phase cancellation of space and surface waves excited at the two open ends (Figure 2) [10], [11].

High Frequency Conductor Losses

Other undesired mechanisms associated with planar circuits are conductor and dielectric losses. Such losses vary with frequency and depend on the combination of the dielectric materials and the size and geometry and composition of the metallization. For high operating frequencies and pure dielectric or intrinsic semiconducting substrates conductor losses dominate and, in fact, impose serious frequency limitations.

In high frequency and high speed applications, where the conductors are very narrow and have thicknesses comparable to their skin depth, the pertubation method or other simplified low frequency techniques fail to provide accurate results. For this

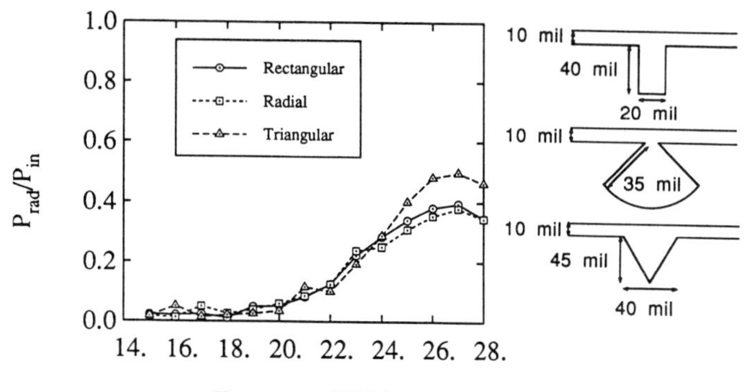

Fig.1. Radiation Loss of Microstrip Stubs Printed on a Grounded Single-layered Substrate ($\epsilon_r = 12, h = 25mils$)

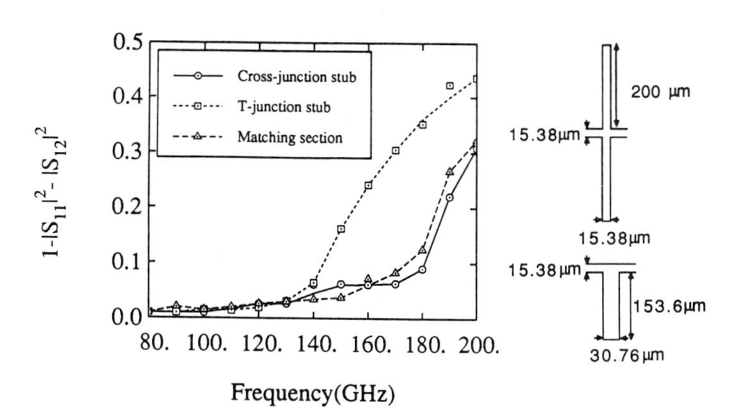

Fig.2. Radiation Loss of Single and Cross Rectangular Stubs Printed on a Grounded Single-layered Substrate ($\epsilon_r = 12, h = 25mils$)

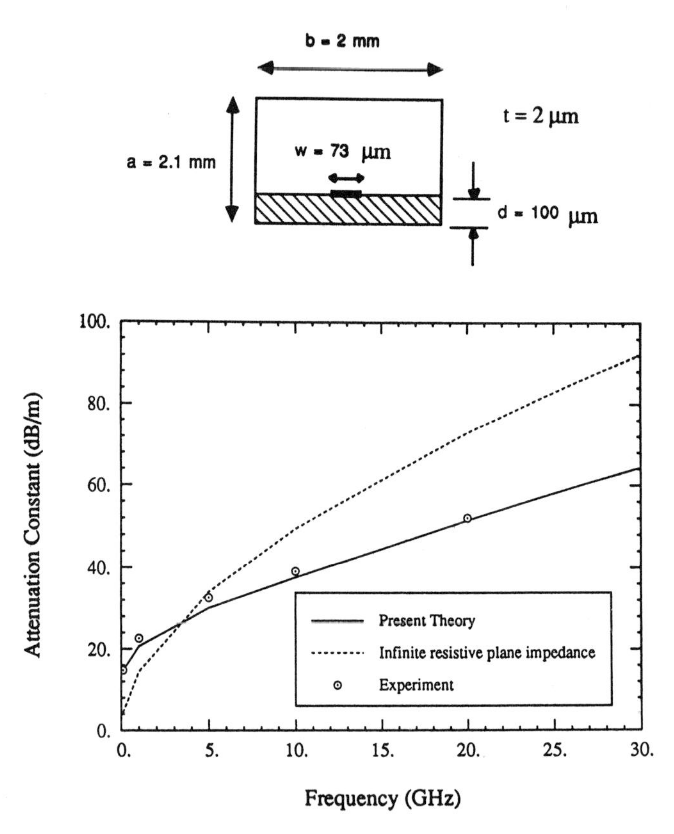

Fig.3. Conductor Losses as a Function of the Frequency

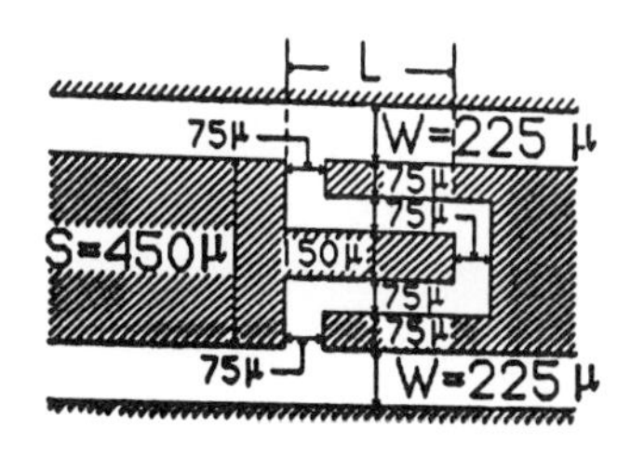

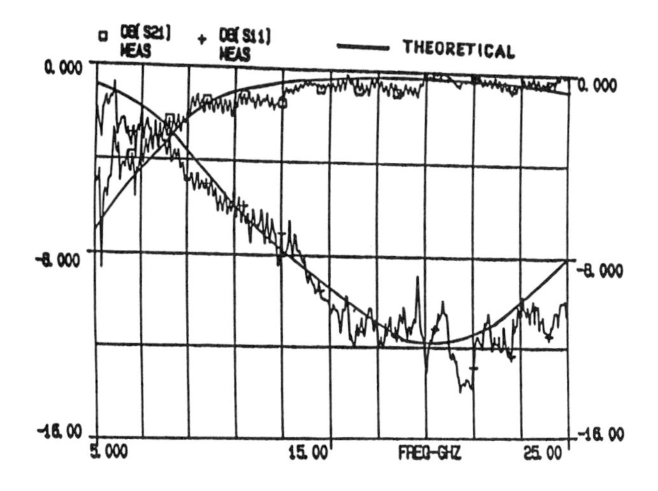

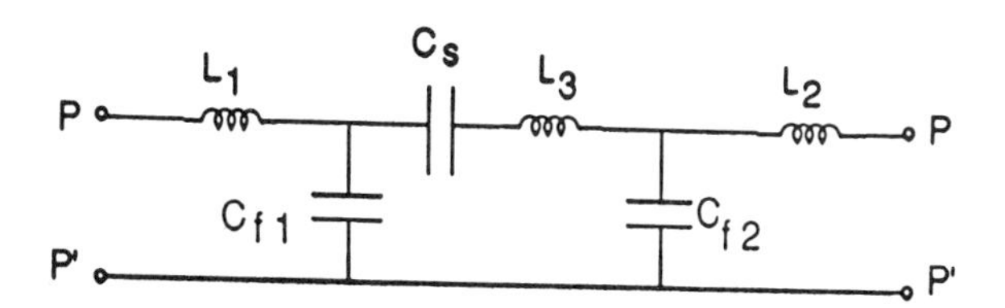

$$
\begin{aligned}
C_s &= 1.01 \times 10^{-4}\, L + 1.642 \times 10^{-2} \\
C_{f1} &= 0.39 \times 10^{-4}\, L + 1.765 \times 10^{-2} \\
C_{f2} &= 0.883 \times 10^{-4}\, L + 1.765 \times 10^{-2} \\
L_1 &= 1.22 \times 10^{-4}\, L \\
L_2 &= 1.43 \times 10^{-4}\, L \\
L_3 &= 3.26 \times 10^{-4}\, L
\end{aligned}
$$

Fig.4. Scattering Parameters and Equivalent Circuit for an Open-end CPW Stub ($D_1 = 25mils, D_2 = 125mils, \epsilon_{r1} = 9.9, \epsilon_{r2} = 2.2$)

purpose, a combined integral equation method has been developed to solve for the fields inside and outside the conductors permitting the evaluation of ohmic losses for any frequency and conductor thickness [8].

The validity of this method has been verified thoroughly through extensive comparisons to other theoretical and experimental data. Figure 3 shows the dependence of conductor losses on frequency for the case of very narrow and thin strips, compared to the pertubation method and other experimentally derived data [12].

CPW Discontinuities

Recently, the coplanar waveguide (CPW) technology has attracted a great deal of interest for RF circuit design due to several advantages over the conventional microstrip lines. Such an advantage is the capability to create shunt connections without the presence of via-holes, and to wafer probe at millimeter wave frequencies.

Stubs are important circuit elements in coplanar waveguide technology due to their capability to create optimum impedance match when used in appropriate configurations. Using the space domain integral equation approach, the scattering parameters for open-end and short-end stub configurations have been evaluated as a function of the stub length L and frequency [13]. Figure 4 shows the derived results for an open-end stub as they compare to experiments performed in the 5-25 GHz frequency range. The same figure shows an appropriate equivalent circuit for this discontinuity with its elements given in terms of the stub length. This equivalent circuit is valid up to the first resonance of the stub.

Dielectric Lines for Submillimeter and Terahertz Applications

Technology based on the frequency range of 0.1-2.0 THz offers narrow-beam, high resolution antennas which are essential for intelligent computer control guidance, command systems for space applications, and sensors which operate in optically opaque media. Since these systems require that the generated THz-power be guided to the antenna front-end through complex feeding networks, the design and construction of low-loss monolithic transmission lines is critical.

A very good candidate for such applications is the monolithic layered dielectric waveguide. This waveguide is constructed from dielectric materials which are available in monolithic technology so that its size is a fraction of the guided wavelength and integration with active devices or other passive circuits or radiating elements is possible [14]. An example of such a waveguide is given in Figure 5.

CONCLUSIONS

A generalized method for analyzing planar open and shielded circuits operating at high frequencies has been presented. The space domain integral equation method has demonstrated excellent performance in terms of accuracy and computational efficiency when applied to structures of small or medium electric size. A powerful extension of this method allows for the evaluation of the effects conductor loss has on circuit performance. The validity of the technique has been demonstrated through extensive convergence tests and comparisons to other theoretical or experimental data.

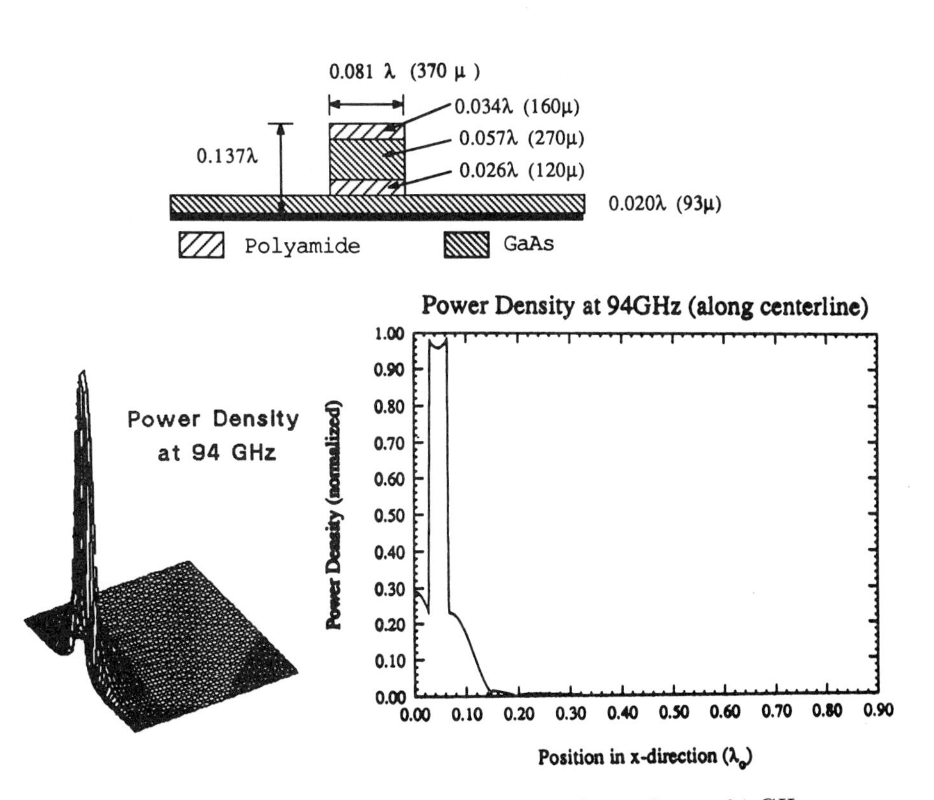

Fig.5. Low-loss Layered Dielectric Waveguide Operating at 94 GHz.

ACKNOWLEDGEMENTS

The presented work has been performed by the author and her collaborators: W. Harokopus, N. VandenBerg, T. Vandeventer, A. Engel, G. Eleftheriades, N. Dib and G. Ponchak who, at the present time, are studying towards their Ph.D. degree in the Electrical Engineering Department of the University of Michigan, Ann Arbor, MI.

References

[1] P.B. Katehi and N.G. Alexopoulos, "Frequency-Dependent Characteristics of Microstrip Discontinuities in Millimeter-Wave Integrated Circuits," *IEEE Trans. Microwave Theory and Techniques*, Vol.MTT-33, Oct.1985, pp. 1029-1035.

[2] R.J. Jackson and D.M. Pozar, "Full-Wave Analysis of Microstrip Open-End and Gap Discontinuities," *IEEE Trans. Microwave Theory and Techniques*, Vol.MTT-33, Oct.85, pp. 1036-1042.

[3] H. Yang, N.G. Alexopoulos and D.R. Jackson, "Analysis of Microstrip Open-end and Gap Discontinuities in a Substrate-Superstrate Configuration," *IEEE MTT-S Microwave Symposium Digest*, June 1988, pp. 705-708.

[4] N.I. Dib and P.B. Katehi, "Modeling of Shielded CPW Discontinuities Using the Space Domain Integral Equation Method (SDIE)," accepted for publication in *the Journal of Electromagnetic Waves and Applications, Special Issue on Electromagnetism and Semiconductors.*

[5] P.B. Katehi and N. G. Alexopoulos, "Real Axis Integration of Sommerfeld Integrals With Applications to Printed Circuit Antennas," *J. Math. Phys.*, vol. 24(3), Mar. 1983.

[6] D.M. Pozar, "Input Impedance and Mutual Coupling of Rectangular Microstrip Antennas," *IEEE Trans. Antennas and Propagation*, Vol. AP-30, No.6, Nov. 1982, pp.1191-1196.

[7] J. R. Mosig, "Arbitrarily Shaped Microstrip Structures and Their Analysis with a Mixed Potential Integral Equation," *IEEE Trans. Microwave Theory Tech.*, Vol. MTT-36, pp. 314-323, Feb. 1988.

[8] T.E. Vandeventer, P.B. Katehi and A.C. Cangellaris, "An Integral Equation Method for the Evaluation of Conductor and Dielectric Losses in High Frequency Interconnects," *IEEE Trans. Microwave Theory and Techniques*, Vol. MTT-37, pp. , Dec. 1989.

[9] W.P. Harokopus and P.B. Katehi, "Characterization of Microstrip Discontinuities on Multilayer Dielectric Substrates Including Radiation Losses," *IEEE Trans. on Microwave Theory and Techniques*, Vol. MTT-37, pp. 2058-2065, Dec. 1989.

[10] W.P. Harokopus, Jr. and P.B. Katehi, "Electromagnetic Coupling and Radiation Loss Considerations in Microstrip (M)MIC Design," accepted for publication in the *IEEE Trans. Microwave Theory and Techniques.*

[11] W. P. Harokopus, Jr. and P. B. Katehi, "Radiation Losses in Microstrip Antenna Networks Printed on Multi-Layer Substrates," accepted for publication in *International Journal of Numerical Modelling.*

[12] T. Meyers, Ball Aerospace: Private Communication.

[13] N.I. Dib, P.B. Katehi, G.E. Ponchak, R.N. Simons, "Theoretical and Experimental Characterization of Coplanar Waveguide Discontinuities for Filter Applications," submitted for publication in the *IEEE Trans. Microwave Theory and Techniques.*

[14] A.G. Engel, Jr. and P.B. Katehi, "Low-Loss Monolithic Transmission Lines for Sub-Mm and Terahertz Frequency Applications," submitted for publication in the *IEEE Trans. Microwave Theory and Techniques.*

Leaky-Waves on Multiconductor Microstrip Transmission Lines

Lawrence Carin

Weber Research Institute
Polytechnic University
Farmingdale, New York 11735

Abstract- The spectral domain technique with a Galerkin moment method solution is used to study leakage from modes on multilayered microstrip transmission lines. An asymptotic technique is used to improve the convergence of oscillatory spectral integrals involving distant expansion and testing functions. Several multiconductor structures are considered, and it is shown that in some cases leakage can occur at low frequencies in realistic structures for which leakage was previously unrecognized.

I. INTRODUCTION

Leakage from printed interconnects has been investigated on a limited class of transmission lines. Rutledge *et al.* [1] used an approximate reciprocity based analysis for leakage from coplanar transmission lines on substrates of infinite thickness. Shigesawa *et al.* [2] used a mode matching technique to investigate leakage from conductor backed coplanar waveguide and coplanar strips. Tsuji *et al.* [3] have also recently shown that microstrip can be leaky on a properly oriented uniaxial anisotropic substrate. This early work on leakage from printed transmission lines was very important because it brought to the forefront a new mechanism for energy loss and possible cross-talk in high-speed integrated circuits. There remains, however, a need to investigate more complicated structures in an effort to evaluate the importance of leakage. This is especially true since over the last two decades quasi-TEM and full-wave approaches have been used to study a wide range of complicated printed transmission lines without directing any attention to possible problems caused by leakage.

In this work realistic multiconductor, multilevel microstrip transmission lines are studied to address the importance of leakage from practical interconnects. Concentration is placed on analyzing relatively complicated structures that heretofore have been assumed to be nonleaky. The numerical analysis is implemented in the spectral domain with a Galerkin moment method solution [4]. Although this method is well known, special care must be taken when studying leaky waves. One must carefully account for leaky wave poles encountered along the path of integration in the complex plane [5]. Additionally, the reaction integrals which must be evaluated in the moment method solution are often highly oscillatory [6]-[9]. This is especially true for the problem at hand: since a multiconductor system is considered, expansion and testing functions will often be relatively far apart in space. This, coupled with the fact that leakage usually occurs at relatively high frequencies, leads to spectral integrals which are often highly oscillatory and hence CPU intensive. To alleviate this problem, an asymptotic technique is used to efficiently evaluate integrals involving expansion and testing functions with centers spaced greater than $0.01\lambda_0$ apart.

The analysis in Section II is divided into three subsections. In part A, the spectral domain procedure with a moment method solution is reviewed for transversely open layered structures. Part B describes the treatment of leaky-waves in this formalism, and finally an efficient asymptotic procedure is discussed in part C. In Section III results are given for leaky-waves on broad-side coupled microstrip transmission lines. Conclusions from this work are given in Section IV.

II. ANALYSIS

A. Formulation

The spectral domain technique with a Galerkin moment method solution will be used to calculate the complex propagation constants for several multiconductor printed transmission lines, such as that depicted in Fig. 1.

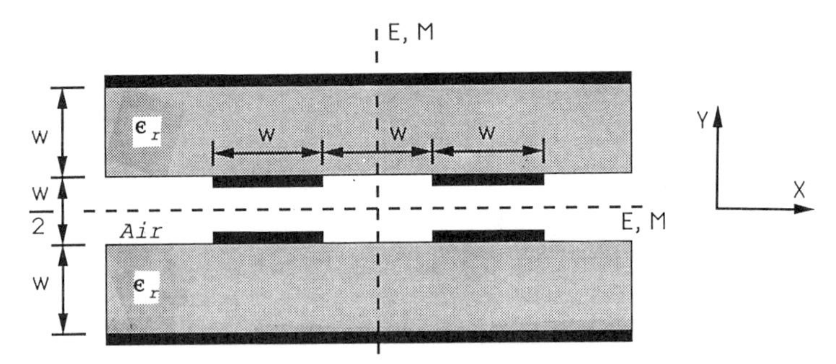

Figure 1. Example multiconductor microstrip geometry. The dashed lines represent symmetry planes which can be referred to as electric (E) or magnetic (M) walls. These symmetry planes are useful for mode designation.

Since this technique is well known [4], only special properties important to the present analysis will be stressed. The problem is formulated by taking a Fourier transform in the transverse (x) direction while assuming an $e^{-j\gamma z}$ z - (longitudinal) dependence $(\gamma=\beta-j\alpha)$. By working in the spectral domain, a convolution integral equation in the space domain is transformed to an algebraic equation in the spectral domain:

$$\sum_n \tilde{\mathbf{G}}(\gamma,k_x,y;y'_n)\cdot\tilde{\mathbf{J}}(\gamma,k_x,y'_n)=\tilde{\mathbf{E}}(\gamma,k_x,y), \tag{1}$$

where $\tilde{\mathbf{G}}(\gamma,k_x,y;y'_n)$ is the spectral domain dyadic Green's function for the electric field at y due to a surface current at y'_n, $\tilde{\mathbf{J}}(\gamma,k_x,y'_n)$ is the surface current at y'_n in the spectral domain, and $\tilde{\mathbf{E}}(\gamma,k_x,y)$ is the electric field tangent to the x-z plane and located at y . The summation is over the number of layers with conducting strips. The dyadic Green's function can be efficiently determined in the spectral domain by employing transmission line theory [4]. In the moment method solution, the unknown surface current on the strips is expanded in terms of a known set of basis functions with unknown coefficients. The vanishing of tangential electric field on the strips (assuming perfect electric conductors) is enforced by using a Galerkin procedure. This results in a matrix equation of the form

$$\mathbf{Z}(\gamma)\mathbf{c}=0 \tag{2}$$

where $\mathbf{c}$ is a vector comprised of the unknown basis function coefficients and $\mathbf{Z}(\gamma)$ is a matrix which is a function of the unknown propagation constants γ for the modes of the structure under study. To obtain a nontrivial solution for $\mathbf{c}$, the determinant of $\mathbf{Z}(\gamma)$ must vanish, thus allowing a means of determining γ and subsequently $\mathbf{c}$. Each component of $\mathbf{Z}(\gamma)$ is of the form

$$\int_C \tilde{J}_{tk}^*(k_x,y)\tilde{G}_{kl}(\gamma,k_x,y;y')\tilde{J}_{el}(\gamma,k_x,y')e^{jk_x\Delta}dk_x \tag{3}$$

where $\tilde{G}_{kl}(\gamma,k_x,y;y')$ is the spectral dyadic Green's function component for a k directed electric field component located at y due to an l directed surface current located at y', and $\tilde{J}_{el}(\gamma,k_x,y')$ is an l directed expansion current in the spectral domain located at y', $\tilde{J}_{tk}(k_x,y)$ is a k directed testing function in the spectral domain located at y, and Δ is the distance in x between the centers of the testing and expansion functions in the space domain. In (3), l and k can both be either x or z and the superscript $*$ denotes complex conjugate. The spectral integral exists on the real axis unless poles are encountered and/or integration around branch cuts is required.

In the evaluation of the integral in (3), one must properly account for branch cuts and poles encountered along the path of integration. As mentioned above, this paper will concentrate on the N-1 zero-cutoff-frequency modes in an N conductor system. Since these modes are slow-waves, no radiation into space-waves will exist. Hence an open physical structure (no top and/or bottom conducting surfaces) can be accurately modeled by placing conducting covers far from the transmission lines where all evanescent fields have decayed sufficiently. An adequate distance from the transmission lines can be determined numerically by successively increasing the distance of the conducting surfaces from the transmission lines until the solution stops changing. Therefore, in the analysis to follow, only geometries with top and bottom conducting surfaces will be considered and leakage will be in the form of parallel plate modes. Since only shielded (on top and bottom) geometries are to be studied, there will be no branch cuts in the complex k_x plane [10]. This simplification, in addition to being appropriate for the problem under study, significantly reduces numerical complexity. This will become evident in Section II.C, where the asymptotic evaluation of (3) is considered.

It can be shown that each component of the dyadic spectral Green's function for planar strips in a layered dielectric medium can be written in the form

$$\tilde{G}_{kl}(\gamma,k_x,y;y')=\frac{f(\gamma,k_x,y;y')}{D_{TE}(\gamma^2+k_x^2)D_{TM}(\gamma^2+k_x^2)} \tag{4}$$

where $D_{TE}(\Gamma^2)$ and $D_{TM}(\Gamma^2)$ are transcendental equations obtained for TE and TM modes, respectively, in the layered dielectric medium under study (with no conducting strips). If there are top and bottom conducting plates (as in the case considered here), these modes correspond to parallel plate modes; otherwise, the modes correspond to surface waves. In either case, the propagation constant (imaginary if below cutoff) of the TE or TM mode is Γ.

It can be shown that the poles of $\tilde{G}_{kl}$ come from the zero's of D_{TM} and D_{TE} and therefore one must concentrate on D_{TM} and D_{TE} when considering possible poles which may be encountered when evaluating (3) [10]. Assume Γ_r is a real root of $D_{TM}(\Gamma^2)$ or $D_{TE}(\Gamma^2)$ and (initially) that γ is real ($\alpha=0$). The Green's function $\tilde{G}_{kl}$

will therefore have a pole at $k_x=\pm\sqrt{\Gamma_r^2-\beta^2}$. If $\beta>\Gamma_r$, the transmission line mode is slower than the parallel plate mode and the pole exists on the imaginary axis of the complex k_x plane. Hence, if $\beta>\Gamma_r$ for all real roots of $D_{TE}(\Gamma^2)$ and $D_{TM}(\Gamma^2)$, the integration in (3) can be carried out along the real k_x axis since no poles will be encountered (the poles corresponding to cutoff parallel plate modes, or imaginary Γ_r, will also exist exclusively on the imaginary k_x axis). As discussed by Oliner [11], $\beta>\Gamma_r$ for all real Γ_r is the condition under which no leakage occurs ($\alpha=0$). It has recently been demonstrated, however, that for some geometries there can be a sufficiently high frequency such that $\Gamma_r=\beta$ (introducing a pole at $k_x=0$) and at still higher frequencies Γ_r can exceed β (in fact, for conductor-backed slotline, the propagation constant of the TEM parallel plate mode will exceed that of the slotline mode at all frequencies [2]). The condition $\beta<\Gamma_r$ has been shown to correspond to leakage; this leakage is in the form of the parallel plate mode with propagation constant Γ_r and occurs at an angle approximately $\cos^{-1}(\beta/\Gamma_r)$ from the strip (or slot) [11]. Note that the longitudinal propagation constant γ for the transmission line mode is now complex, signifying a nonspectral leaky-wave solution. One must now determine how to evaluate the integral in (3) to properly account for this leakage.

B. Leaky-Wave Contribution

To determine how to handle leaky-waves in the spectral domain formulation, initially the discussion will be confined to the case of leakage to a single parallel plate mode. Let Γ_{ri} represent the roots of $D_{TE}(\Gamma^2)$ and $D_{TM}(\Gamma^2)$ with $i=0,1,2,\ldots,\infty$. There will therefore be poles at $k_x=\pm\sqrt{\Gamma_{ri}^2-\gamma^2}$ in the complex k_x plane. Assume that at some frequency $Real(\Gamma_{r0})>\beta$ and $Real(\Gamma_{ri})<\beta$ for $i=1,2,3,\ldots,\infty$. Under such circumstances there is leakage to the mode represented by Γ_{r0} and therefore γ is now complex. The phase match condition dictates that the field of such a leaky-wave grows transversely as $exp(\,|\,xIm(\sqrt{\Gamma_{r0}^2-\gamma^2})\,|\,)$. Consider the inverse spectral integral for the electric field of such a leaky-wave:

$$\mathbf{E}(\gamma,x,y)=\sum_n\int_C \tilde{\mathbf{G}}(\gamma,k_x,y;y'_n)\cdot\tilde{\mathbf{J}}(\gamma,k_x,y'_n)e^{jk_xx}\,dk_x. \tag{5}$$

A real axis evaluation of this integral is inappropriate for an unbounded transverse solution, such as exists for a leaky-wave. Therefore, under leakage conditions, the integral must be evaluated by a proper deformation of the integral in the complex k_x plane. By deforming the integral around the poles at $\pm\sqrt{\Gamma_{r0}^2-\gamma^2}$ and evaluating the subsequent residues, the resultant integral in (5) satisfies the proper boundary conditions at $x=\pm\infty$. To ensure the proper behavior at $x=\pm\infty$, no additional poles can be picked up in the deformation, and the final integral path is therefore as shown in Fig. 2.

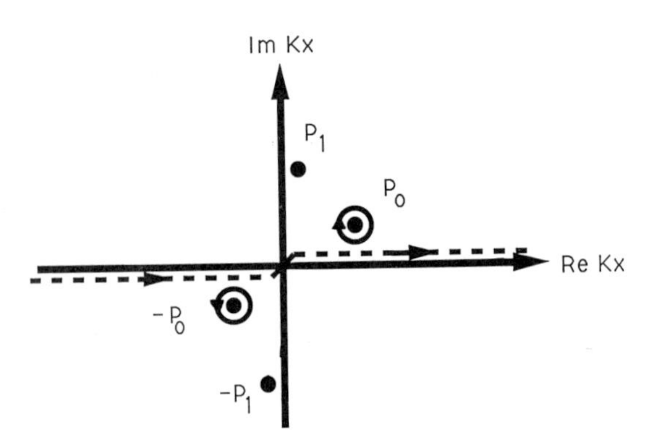

Figure 2. Integral path and two leaky-wave poles captured for spectral representation of leaky-waves on printed transmission lines.

Now consider the numerical solution of (5) using basis functions $\mathbf{j}_{in}(\gamma,x,y'_n)$ with unknown coefficients c_{in} for the i th basis function representing the current on layer y'_n:

$$\mathbf{E}(\gamma,x,y)\approx\sum_n\sum_i c_{in}\int_C \bar{\mathbf{G}}(\gamma,k_x,y;y'_n)\cdot\tilde{\mathbf{j}}_{in}(\gamma,k_x,y'_n)e^{jk_x x}\,dk_x, \tag{6}$$

where $\tilde{\mathbf{j}}_{in}(\gamma,k_x,y'_n)$ is the Fourier transform of $\mathbf{j}_{in}(\gamma,x,y'_n)$. In the Galerkin procedure (6) is multiplied by the complex conjugate of a testing function $\mathbf{j}_k(x,y)$ to obtain

$$\int_S \mathbf{E}(\gamma,x,y)\cdot\mathbf{j}_k^*(x,y)\,dx\approx\sum_n\sum_i c_{in}\int_C \tilde{\mathbf{j}}_k^*(k_x,y)\cdot\bar{\mathbf{G}}(\gamma,k_x,y;y'_n)\cdot\tilde{\mathbf{j}}_{in}(\gamma,k_x,y'_n)e^{jk_x\Delta}\,dk_x \tag{7}$$

where Δ is the difference in x between the centers of the testing and expansion functions. The left hand integral in (7) is over the width S of the testing function and the right side is obtained via Parseval's theorem. For (7) to be valid for all testing functions, independent of their position in x , the exponential growth of $\mathbf{E}(\gamma,x,y)$ must be accounted for. Therefore all integrals in (7), which are of the form in (3), are evaluated under the leakage condition by using the contour discussed above and shown in Fig. 2.

The above discussion was restricted to the special (but practically important) case of leakage to a single parallel plate mode. At a sufficiently high frequency, another parallel plate mode with real propagation constant Γ_{r1} may satisfy the condition $\Gamma_{r1}>\beta$. It is not clear how to treat the onset of leakage to a second parallel plate mode since now γ is complex rather than real (as was the case before the onset of leakage to the first parallel plate mode). It has been suggested by Phatack *et al.* [5] (using a perturbation technique in which a small amount of dielectric loss is added) that two criterions must be met upon deciding how to treat the integration in (3) under such a situation. In addition to requiring $\beta<\Gamma_{r1}$, the poles $\pm\sqrt{\Gamma_{r1}^2-\gamma^2}$ must also exist in the first and third quadrants of the complex k_x plane. If both conditions are met, the integral in (3) is deformed around $\pm\sqrt{\Gamma_{r0}^2-\gamma^2}$ and $\pm\sqrt{\Gamma_{r1}^2-\gamma^2}$. Such an approach can be extended to an arbitrary number of parallel plate modes. For most geometries and frequencies of practical interest, however, leakage will usually not occur to more than one parallel plate mode.

C. Asymptotic Technique

Consider the real axis integration in (3) for $\Delta>0$. This integral is evaluated by enclosing the top half ($\mathrm{Im}(k_x)>0$) of the k_x plane, as shown in Fig. 3, and hence capturing all poles with $\mathrm{Im}(k_x)>0$.

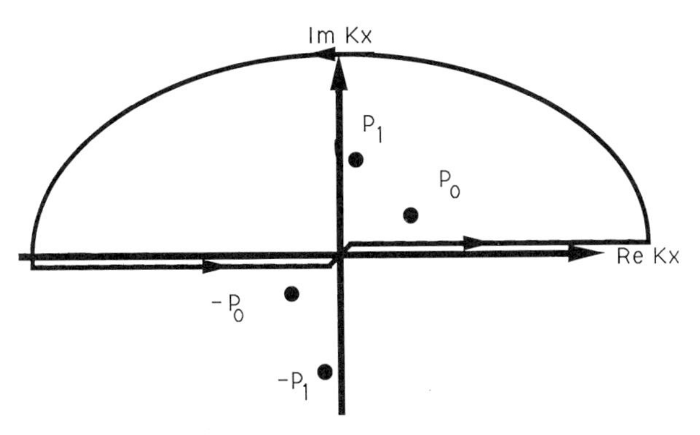

Figure 3. Integral path for asymptotic evaluation of the real axis contribution in Fig. 2.

If $\Delta<0$, the bottom half of the complex k_x plane is enclosed. The real axis contribution to (3) therefore becomes a summation of an infinite number of residues (each of which can be expressed in close form):

$$I=2\pi j\sum_{i=0}^{\infty}Res(\sqrt{\Gamma_{ri}^2-\gamma^2}) \tag{8}$$

where $\mathrm{Im}(\sqrt{\Gamma_{ri}^2-\gamma^2})>0$. Implicit in (12) is that all poles in (3) are from $D_{TE}(\Gamma^2)$ and $D_{TM}(\Gamma^2)$. It is well known that the poles of a Green's function for a layered dielectric correspond to the modes (source-free solutions) of that structure [10]. Hence, as long as the basis functions have no poles (as is usually the case), the representation in (12) is valid. One also now sees the advantage (when appropriate, as in this study) of adding top and bottom shields to the numerical solution. By doing so, no branch cuts need be evaluated. This technique can, however, be extended to cases in which radiation to space waves is important. Again using standard asymptotic techniques, the integral is deformed to the path of steepest descent, while picking up appropriate residues [10]. The steepest descent integral is evaluated around the saddle point and represents the space wave contribution to the integral. The finite number of residues represent the surface waves (proper and sometimes improper modes).

In general, an infinite number of terms are required in (12). For $\Delta/\lambda_o>0.01$, however, it has been found that only a small number of terms are required (as will be shown below). The advantage of this approach is that as Δ/λ_o increases a real axis integration of (3) becomes increasingly inefficient due to the oscillatory nature of the integral while for such cases fewer and fewer residues are required to obtain convergence in (12). It is important to make the separation between the top and bottom conducting shields as small as possible such that only a few parallel plate modes are above cutoff, accelerating the convergence of (12).

III. RESULTS

A. Numerical Considerations

Before presenting computed results, it is important to discuss the convergence properties of the method outlined above. Finding γ (in general complex) requires the computation of the components of $\mathbf{Z}(\gamma)$ in (2), each component of which is of the form in (3). Trigonometric basis functions modified by the edge condition are used in this work [12]. It was found that two even and two odd basis function for both the longitudinal and transverse current components on each strip was sufficient to yield convergence in all cases studied. The computation in (3) involves a real axis integration plus the addition of a finite number of closed form leaky wave poles when appropriate (for complex γ). When Δ in (3) satisfies $\Delta>0.01\lambda_o$, the asymptotic procedure outlined above is used, significantly increasing the efficiency of the computations. In the results to be presented, 30 TE and 30 TM parallel plate modes were used in (12) for all $\Delta>0.01\lambda_o$. The roots of the TE and TM parallel plate modes, Γ_{ri}, need only be computed once for each frequency and all residues in (12) can be expressed in closed form. The use of 30 TE and TM modes was done to assure convergence for all $\Delta>0.01\lambda_o$ encountered but as Δ/λ_o increases, far fewer parallel plate modes need actually be considered. For the cases for which $\Delta<0.01\lambda_o$, a standard real axis integration was applied. Taking advantage of the odd or even nature of the kernel in (3), only a semi-infinite integration from 0 to infinity need be evaluated along the real axis. The real axis integration was truncated at a sufficiently high value of

k_z (determined numerically) and the integration interval was broken up into small subsections where Gauss-Legendre integration was used. It was found during the course of this work that the computed γ was very sensitive to where the real axis integration was truncated and to how it was subsequently discretized. This was especially true for complex (leaky) γ. One must therefore take care to assure that the truncation and discretization of the real axis integration is sufficient to obtain convergence at all frequencies of interest. The roots of the determinant of $\mathbf{Z}(\gamma)$ were computed using Muller's method. Typically, less than 5 minutes of CPU time was required for the computation of γ at each frequency on an IBM RISC 6000 workstation.

B. Broad-side Coupled Microstrip

Example results are given here for a broad-side coupled microstrip transmission line. This structure has been analyzed from both the quasi-TEM and full-wave points of view. It has not been recognized, however, that this structure can be leaky. In Fig. 1 the dashed lines represent fictitious electric (E) and magnetic (M) walls, which due to the symmetry of this geometry are useful for mode designation. In Fig. 4 are shown the real and imaginary parts of the propagation constants of the five (zero-cutoff-frequency) fundamental modes of this structure.

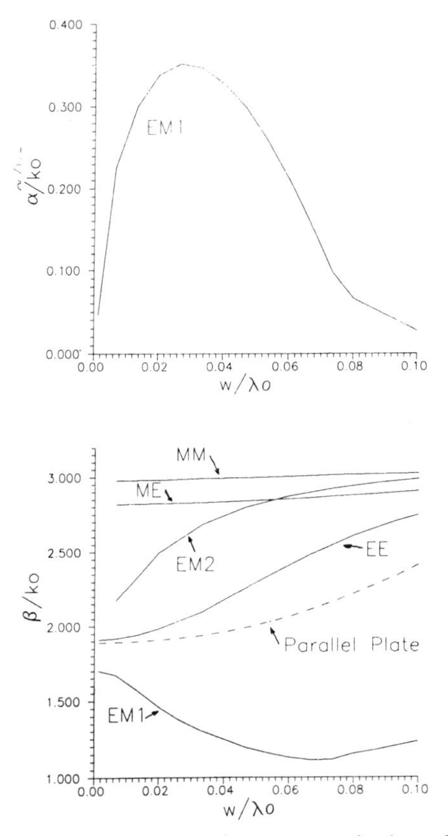

Figure 4. Propagation and attenuation constant for broadside-coupled microstrip in Fig. 1. The dielectric constant of the shaded region in Fig. 1 is 10.

The dispersion curve of the TMo parallel plate mode is also given. In the mode identifications, the first letter represents the symmetry of the horizontal plane (horizontal dashed line) and the second letter represents the symmetry of the vertical plane (vertical dashed line). Notice that the EM mode leaks at all frequencies. The leakage is manifested in the form of a TMo parallel plate mode, and this occurs when the phase velocity of a broad-side coupled microstrip mode exceeds that of the TMo parallel plate mode.

IV. CONCLUSIONS

An efficient numerical procedure has been described for the analysis of leaky-waves on multilayered printed transmission lines. The problem was formulated in the spectral domain, which resulted in a matrix equation with components represented as spectral integrals. Making use of the fact that the fundamental modes are slow waves, only leakage in the form of surface waves or parallel plate modes is possible. In the analysis, this allowed the placement of conducting shields above and below the transmission line system. This simplification, in addition to being appropriate for the problem under study, leads to a simple asymptotic expression in terms of a sum of closed form residues (no branch cut integrals). Upon comparison with results from rigorous numerical integration along the real axis, it was determined that for the geometry studied in this work, the asymptotic procedure was accurate for separations as small as .01 λ_o. Using this procedure, leaky-waves on broad-side coupled microstrip were investigated. It was shown that for such structures, leakage may occur at all frequencies.

REFERENCES

[1] D. B. Rutledge, D. P. Neikirk, and D. P. Kasilingam, "Integrated circuit antennas," in Infrared and Millimeter Waves, vol. 10, pp. 1-87, K. J. Button, Ed., Academic Press, Inc., 1983.

[2] H. Shigesawa, M. Tsuji, and A. A. Oliner, "Conductor-backed slot line and coplanar waveguide: Dangers and full-wave analyses," *1987 IEEE MTT Int. Symp. Dig.*, pp. 199-202.

[3] M. Tsuji, H. Shigesawa, and A. A. Oliner, "Printed-circuit waveguides with anisotropic substrates: a new leakage effect," *1989 IEEE MTT Int. Symp. Dig.*, pp. 783-786.

[4] T. Itoh, "Spectral domain immitance approach for dispersion characteristics of generalized printed transmission lines," *IEEE Trans. Microwave Theory Tech.*, vol. MTT-28, pp. 733-736, July 1980.

[5] D. S. Phatak, N. K. Das, and A. P. Defonzo, "Dispersion characteristics of optically excited coplanar striplines: Comprehensive full wave analysis," *IEEE Trans. Microwave Theory*, vol. MTT-38, Nov. 1990, to be published.

[6] D. M. Pozar, "Input impedance and mutual coupling of rectangular microstrip antennas," *IEEE Trans. Antennas Prop.*, vol. AP-30, pp. 1191-1196, Nov. 1982.

[7] K. A. Michalski and D. Zheng, "Electromagnetic scattering and radiation by surfaces of arbitrary shape in layered media, Part 1: Theory," *IEEE Trans. Antennas Prop.*, vol. AP-38, pp. 335-344, Mar. 1990.

[8] M. Marin, S. Barkeshli, and P. H. Pathak, "Efficient analysis of planar microstrip geometries using a closed-form asymptotic representation of the grounded dielectric slab Green's function," *IEEE Trans. Microwave Theory Tech.*, vol. MTT-37, pp. 669-679, Apr. 1989.

[9] K. A. Michalski, "On the efficient evaluation of integrals arising in the Sommerfeld halfspace problem," *Inst. Elec. Eng. Proc.*, vol. 132, pt. H, pp. 312-318, Aug. 1985.

[10] L. B. Felsen and N. Marcuvitz, Radiation and Scattering of Waves, Prentice-Hall, Inc., Englewood Cliffs, N. J., 1973.

[11] A. A. Oliner and K. S. Lee, "The nature of leakage from higher order modes on microstrip line," *1986 IEEE MTT Int. Symp. Dig.*, pp. 119-122.
[12] L. Carin and K. J. Webb, "Isolation effects in single- and dual-plane VLSI interconnects, *IEEE Trans. Microwave Theory Tech.*, vol. MTT-38, pp. 396-404, April 1990.

CLOSED-FORM ASYMPTOTIC REPRESENTATIONS FOR THE GROUNDED PLANAR SINGLE AND DOUBLE LAYER MATERIAL SLAB GREEN'S FUNCTIONS AND THEIR APPLICATIONS IN THE EFFICIENT ANALYSIS OF ARBITRARY MICROSTRIP GEOMETRIES

Sina Barkeshli

Sabbagh Associates, Inc
4639 Morningside Drive
Bloomington, IN 47401

P. H. Pathak

ElectroScience Laboratory
Department of Electrical Engineering
The Ohio State University
Columbus, OH 43212

I. INTRODUCTION

Closed form asymptotic representations are developed for the grounded planar single and double layer material slab Green's functions. These asymptotic Green's function are useful for the efficient determination of the currents on arbitrarily shaped microstrip configurations when employing a Moment Method (MM) solution of the integral equation for these currents. Previous work has in most cases employed either the Sommerfeld type integral, or the plane wave spectral (PWS) representation for these Green's functions.

In the present work, a transformation of the standard Sommerfeld type integral representation is appropriately introduced in the complex plane so that a radial-propagation integral representation of the surface dyadic Green's function is obtained. Then two different asymptotic closed form representations are obtained for the surface Green's functions such that they are valid for small and large lateral separations of the source and observation points, respectively.

It is interesting that the asymptotic closed-form representation of the Green's function for the case of large lateral separations remains accurate even for very small lateral separations between source and field points (a few tenths of the free space wavelengths). For observation points in the immediate vicinity of the source, the other asymptotic representation of the Green's function is accurate (up to a few hundredths of the free space wavelength). For the electrically thin grounded (single) dielectric/magnetic slabs, it is possible to blend the two asymptotic representations of the surface Green's functions, which are accurate in the two different ranges of

Directions in Electromagnetic Wave Modeling
Edited by H.L. Bertoni and L.B. Felsen, Plenum Press, New York, 1991

the lateral separations, to obtain a closed form representation of the Green's function which remains valid over the entire range of lateral separations of the source and observation points. This is of course due to presence of the overlapping range of the validity of these two asymptotic representations.

In this paper we outline the analytical techniques that have been utilized to develop such a closed form asymptotic solution of the grounded material slab Green's function. We also compare the efficiency and accuracy of this newly developed asymptotic closed form Green's function to its (PWS) and Sommerfeld integral representation counterparts when they are applied to analyze various microstrip geometries.

II. THE RADIAL PROPAGATION INTEGRAL REPRESENTATION OF THE GROUNDED PLANAR SINGLE AND DOUBLE LAYER MATERIAL SLAB GREEN'S FUNCTION

Consider an electric current point source $\mathbf{J}(\mathbf{r}')$ of strength $\mathbf{p}_e$ at $\mathbf{r} = \mathbf{r}'$; i.e., $\mathbf{J}(\mathbf{r}') = \mathbf{p}_e\delta(\mathbf{r} - \mathbf{r}')$, within the double layered grounded dielectric slab, (i.e., region 2), as shown in Figure 1. In this configuration, the medium (0) refers to free space with constitutive parameters μ_0, ϵ_0, whereas medium (1), and medium (2) correspond to the isotropic, homogenuous material layers with constitutive parameters and thicknesses of μ_1, ϵ_1, d_1 and μ_2, ϵ_2, d_2 respectively. The electric field $\mathbf{E}_m$ produced by $\mathbf{p}_e$ is given by

$$\mathbf{E}_m = -j\omega\mu_1 \mathcal{G}^m(\mathbf{r}, \mathbf{r}') \cdot \mathbf{p}_e \quad , \tag{1}$$

where $\mathcal{G}^m$ is the dyadic Green's function for the region m, (i.e., $m = 0, 1$, and 2). The transverse components of the dyadic Green's function due to the $\hat{\mathbf{x}}$-directed point source can be expressed as, [1]

$$\mathcal{G}^m_{xx}(\mathbf{r}, \mathbf{r}') = -\frac{1}{j\omega\mu_1}\left[\nabla_t^2 G''_m - \frac{\partial^2}{\partial x^2}\left(G''_m - \frac{1}{\omega^2\epsilon_m\epsilon_1}\frac{\partial^2}{\partial z \partial z'}G'_m\right)\right], \tag{2}$$

$$\mathcal{G}^m_{yx}(\mathbf{r}, \mathbf{r}') = -\frac{1}{j\omega\mu_1}\left[-\frac{\partial}{\partial x}\frac{\partial}{\partial y}\left(G''_m - \frac{1}{\omega^2\epsilon_m\epsilon_1}\frac{\partial^2}{\partial z \partial z'}G'_m\right)\right], \tag{3}$$

where (∇_t^2) operator is defined as a transverse Del square operator (i.e. $\nabla_t^2 \equiv \frac{\partial^2}{\partial x^2} + \frac{\partial^2}{\partial y^2}$), and

$$G_m = \frac{1}{4\pi}\left[\int_{-\infty}^{\infty} g_m(\zeta, z, z') H_0^{(2)}\left(\rho\sqrt{k_0^2 - \zeta^2}\right)\zeta \frac{d\zeta}{k_0^2 - \zeta^2}\right.$$
$$\left. -2\pi j\left(\sum_p Res\,(g_m(\xi_p, z, z)\frac{H_0^{(2)}(\xi_p\rho)}{\xi_p}\right)\right] \quad , \tag{4}$$

where the prime ($'$) and double prime ($''$) symbols have been omitted for convenience. The above result in (4) represents the ρ-propagation representation of G_m which is derived directly from its Sommerfeld (z-propagating) integral representation upon enclosing all the relevant singularities including surface wave poles and branch cuts in the lower half-plane of integration in the Sommerfeld representation [1,2].
The $\frac{\partial^2}{\partial z \partial z'} g'_m$, which results from interchanging the integration and the differentiation of prime (') terms in (2) and (3), and g''_m for $m = 1, (z = z' = 0)$ in (4) are given by

$$-\frac{1}{\omega^2\epsilon_1^2}\frac{\partial^2}{\partial z \partial z'}g_1' = -\frac{1}{\omega\epsilon_0}\frac{\kappa_1}{2\epsilon_{1,r}}\left[\frac{(1 + R_1{}')(1 - e^{-j2\kappa_1 d_1})}{1 + R_1{}' e^{-j2\kappa_1 d_1}}\right] \tag{5}$$

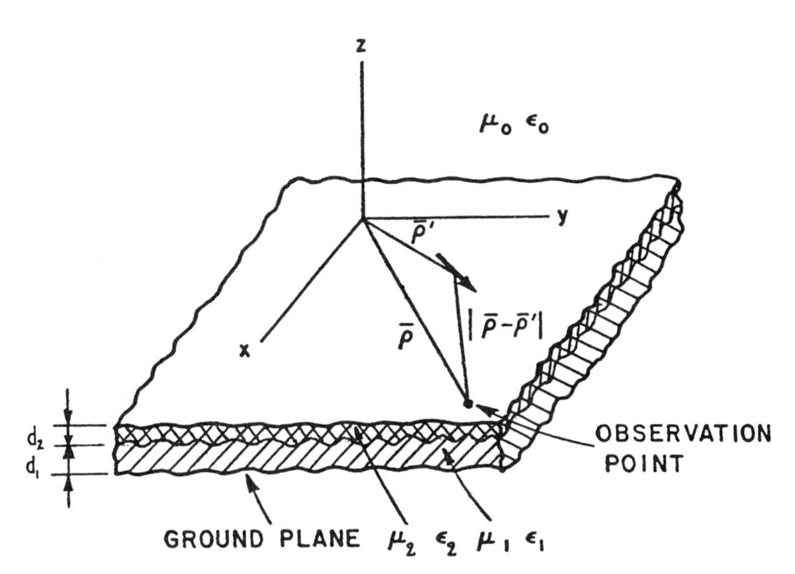

Figure 1. Geometrical configuration of grounded planar double layer material Green's function; electric point source embedded in a grounded double layer material slab.

$$g_1'' = -\frac{k_0^2}{\omega\epsilon_0}\frac{\mu_{1,r}}{2\kappa_1}\left[\frac{(1+R_1'')(1-e^{-j2\kappa_1 d_1})}{1+R_1''e^{-j2\kappa_1 d_1}}\right] , \quad (6)$$

where the modal reflection coefficients are defined as $\Gamma_1 = \frac{\eta_2-\eta_1}{\eta_2+\eta_1}$, $\Gamma_2 = \frac{\eta_0-\eta_2}{\eta_0+\eta_2}$ and effective reflection coefficient R_1 is given by

$$R_1 = \frac{\Gamma_1 + \Gamma_2 e^{-j2\kappa_2 d_2}}{1+\Gamma_1\Gamma_2 e^{-j2\kappa_2 d_2}} , \quad (7)$$

where

$$\eta_m' = \frac{\kappa_m}{\omega\epsilon_m} \quad ; \quad \eta_m'' = \frac{\omega\mu_m}{\kappa_m} \quad ; \quad \kappa_m = \sqrt{k_0^2(\epsilon_{m,r}\mu_{m,r}-1)+\zeta^2} \quad (8)$$

The radial propagation type dyadic Green's function for $z = z' = 0$, can be found explicitly after incorporating $(\nabla_t^2 G_1'')$ and $(G_1'' - \frac{1}{(\omega\epsilon_0)^2}\frac{\partial^2}{\partial z\partial z'}G_1')$ into (2) and (3).

The radial propagation representation of the grounded material slab dyadic Green's function exhibits fast convergence for laterally separated source and field points. This is due to the fact that for the $\zeta > k_0$, the Hankel functions associated with the integrand in (2) and (3) decay exponentially, [1].

III. UNIFORM ASYMPTOTIC CLOSED-FORM REPRESENTATION OF THE GROUNDED DOUBLE-LAYER MATERIAL SLAB GREEN'S FUNCTION

The radially propagating integral representation given in the previous section can be used as a starting point to find two different approximate asymptotic closed form representations for the Green's function based on the small and large lateral separations of the source and field points. The asymptotic representation of the surface dyadic Green's function for the case of large lateral separation remains accurate even for very small lateral separation of the source and observation points (a few tenths of the free space wavelengths). For the observation points in the immediate vicinity of the source, the other asymptotic representation of the Green's function is accurate (up to a few hundredths of the free space wavelength). As was mentioned previously, for an electrically thin grounded (single) dielectric/magnetic slab, it is possible to blend the two asymptotic representations of the surface Green's functions, which are accurate in the two different ranges of the lateral separations, to obtain a closed form asymptotic representation of the Green's function which remains valid for the entire range of lateral separations of the source and observation points. The details of these asymptotic developments are shown in the following.

Consider the integral I, which has the following functional form as shown:

$$I = \int_{-\infty}^{\infty} f(\zeta) H_0^{(2)}(k_0\rho\sqrt{1-\zeta^2})e^{-jk_0\zeta z}d\zeta , \quad (9)$$

in which $(k_0\rho)$ is assumed to be a large parameter. Although a conventional procedure can be used in general to construct a uniform asymptotic expansion of I in (9) from a formal saddle point technique, the following procedure has been found to provide more accurate results for small lateral separation of source and observation points. One may assume in general that $f(\zeta)$ has a finite number of poles, (surface and leaky wave poles), at ζ_p's close to $\zeta = 0$, then one can expand $f(\zeta)$ around zero as:

$$f(\zeta) \approx \sum_n a_n\zeta^n + \sum_p \frac{Res_f(\zeta_p)}{\zeta - \zeta_p} \quad (10)$$

where Res_f is the residue of $f(\zeta)$ at $\zeta = \zeta_p$; a_n are the coefficients of the power series, and the radius of convergence of the power series in uninfluenced by the presence of these poles. Then I can be approximated asymptotically as,

$$I \sim \sum_n a_n \frac{1}{(-jk_0)^n} \frac{\partial^n}{\partial \zeta^n} \int_{-\infty}^{\infty} H_0^{(2)}(k_0\rho\sqrt{1-\zeta^2}) e^{-jk_0\zeta z} d\zeta$$
$$+ \sum_p Res_f(\zeta_p) \int_{-\infty}^{\infty} \frac{H_0^2(k_0\rho\sqrt{1-\zeta^2})}{\zeta - \zeta_p} e^{-jk_0\zeta z} d\zeta \ . \tag{11}$$

The integral in the first summation can be evaluated in closed form as

$$\int_{-\infty}^{\infty} H_0^{(2)}\left(k_0\rho\sqrt{1-\zeta^2}\right) e^{-jk_0 z\zeta} d\zeta = 2j \frac{e^{-jk_0\sqrt{\rho^2+z^2}}}{k_0\sqrt{\rho^2+z^2}} \ , \tag{12}$$

hence, keeping only the first two terms and equating z to zero, we will get

$$I \sim a_0 \left[2j \frac{e^{-jk_0\rho}}{k_0\rho}\right] + a_2 \left[\frac{2}{k_0\rho}\left(1 - \frac{j}{k_0\rho}\right) \frac{e^{-jk_0\rho}}{k_0\rho}\right]$$
$$+ \sum_p Res_f(\zeta_p) \mathcal{M}(k_0\rho, \zeta_p) \ , \tag{13}$$

where $\mathcal{M}$ is the uniform asymptotic expansion of the integral in the second sum of (11), and for $\zeta_p = \mp j|\zeta_p|$, it is given by [2],

$$\mathcal{M}(k_0\rho, \zeta_p) = 2\pi j N(\zeta_p) H_0^{(2)}(\rho\sqrt{1-\zeta_p^2}) e^{-jk_0\zeta_p z} - \tag{14}$$
$$2j \frac{e^{-jk_0\rho}}{k_0\rho\zeta_p}\left(1 + \frac{j}{8k_0\rho}\right) \cdot \left(1 - \frac{|\zeta_p|}{\sqrt{2\alpha_p(\alpha_p - 1)}}\left[1 - F\left(\mp\sqrt{k_0\rho\,(1-\alpha_p)}\right)\right]\right) \ ,$$

where, $\alpha_p = \sqrt{1 + |\zeta_p|^2}$, and the transition function F is defined as

$$F(\mp\sqrt{x}) = \mp j\sqrt{x} e^{jx} \int_{\pm\sqrt{x}}^{\infty} e^{-ju^2} du \ . \tag{15}$$

The numerical value of $N(\zeta_p)$ in (14) depends on the location of ζ_p in the complex ζ-plane and takes the value of 1, -1 or 0, [2].

The technique discussed above for evaluating I is utilized to find the uniform asymptotic approximation of the planar grounded single and double layer material slab Green's functions whose components are given in (2) and (3), [2-4].

For the derivation of the other asymptotic expansion of the surface dyadic Green's function for the small parameter range $k_0\rho \ll 1$, one can use the geometric series representation of the effective reflection coefficient R_1 of (7) as

$$R_1 = \frac{\Gamma_1 + \Gamma_2 e^{-j2\kappa_2 d_2}}{1 + \Gamma_1\Gamma_2 e^{-j2\kappa_2 d_2}} = (1 - \Gamma_1^2) \sum_{n=1}^{\infty} (-\Gamma_1)^{n-1} \Gamma_2^n e^{-j2\kappa_2 n d_2} \ . \tag{16}$$

Thus, the substitution of (16) into the denominators of (5) and (6) , yields

$$\frac{1}{1 + R_1 e^{-j2\kappa_1 d_1}} = 1 + \tag{17}$$
$$\sum_{m=1}^{\infty} (-1)^m \left[\Gamma_1 + (1-\Gamma_1^2) \sum_{n=1}^{\infty} (-\Gamma_1)^{n-1} \Gamma_2^n e^{-j2\kappa_2 n d_2}\right]^m e^{-j2\kappa_1 m d_1} \ .$$

Next, expanding κ_m as a power series around zero as a function of k_0, one obtains:

$$\kappa_m \approx \zeta + \frac{1}{2}(\epsilon_{0,r}\mu_{0,r} - 1)\frac{k_0^2}{\zeta} + \cdots \quad . \tag{18}$$

Then, after employing the first term in the expansion of (18) for κ_m, and incorporating (16) into (5) and (6), the closed form series of the surface dyadic Green's functions can be obtained. For a single layer dielectric/magnetic slab with $\Gamma_2 = 0$

$$\begin{aligned}\nabla_t^2 G_1'' \approx\ & \frac{-1}{4\pi}\frac{k_0^2}{\omega\epsilon_0}2j\frac{k_0\mu_{1,r}}{\mu_{1,r}+1}\left[\frac{e^{-jk_0\rho}}{k_0\rho} - \frac{e^{-jk_0\sqrt{\rho^2+(2d_1)^2}}}{k_0\sqrt{\rho^2+(2d_1)^2}}\right. \\ & \left. + \sum_{n=1}^{\infty}\left(\frac{\mu_{1,r}-1}{\mu_{1,r}+1}\right)^n \left(\frac{e^{-jk_0\sqrt{\rho^2+(2nd_1)^2}}}{k_0\sqrt{\rho^2+(2nd_1)^2}} - \frac{e^{-jk_0\sqrt{\rho^2+(2(n+1)d_1)^2}}}{k_0\sqrt{\rho^2+(2(n+1)d_1)^2}}\right)\right]\end{aligned} \tag{19}$$

$$\begin{aligned}G_1'' - \frac{1}{\omega^2\epsilon_1^2}\frac{\partial^2}{\partial z\partial z'}G_1' \approx\ & \frac{-1}{4\pi}\frac{1}{\omega\epsilon_0}2j\frac{k_0}{\epsilon_{1,r}+1}\left[\frac{e^{-jk_0\rho}}{k_0\rho} - \frac{e^{-jk_0\sqrt{\rho^2+(2d_1)^2}}}{k_0\sqrt{\rho^2+(2d_1)^2}}\right. \\ & \left. + \sum_{n=1}^{\infty}\left(-\frac{\epsilon_{1,r}-1}{\epsilon_{1,r}+1}\right)^n \left(\frac{e^{-jk_0\sqrt{\rho^2+(2nd_1)^2}}}{k_0\sqrt{\rho^2+(2nd_1)^2}} - \frac{e^{-jk_0\sqrt{\rho^2+(2(n+1)d_1)^2}}}{k_0\sqrt{\rho^2+(2(n+1)d_1)^2}}\right)\right] \quad ,\end{aligned} \tag{20}$$

and for a double layer dielectric/magnetic slab,

$$\nabla_t^2 G_1'' \approx \frac{-1}{4\pi}\frac{k_0^2}{\omega\epsilon_0}2j\frac{k_0\mu_{1,r}}{\mu_{1,r}+1}\left[\frac{e^{-jk_0\rho}}{k_0\rho} - \frac{e^{-jk_0\sqrt{\rho^2+(2d_1)^2}}}{k_0\sqrt{\rho^2+(2d_1)^2}} + \mathcal{P}''\right] \tag{21}$$

$$G_1'' - \frac{1}{\omega^2\epsilon_1^2}\frac{\partial^2}{\partial z\partial z'}G_1' \approx \frac{-1}{4\pi}\frac{1}{\omega\epsilon_0}2j\frac{k_0}{\epsilon_{1,r}+1}\left[\frac{e^{-jk_0\rho}}{k_0\rho} - \frac{e^{-jk_0\sqrt{\rho^2+(2d_1)^2}}}{k_0\sqrt{\rho^2+(2d_1)^2}} + \mathcal{P}'\right] \quad . \tag{22}$$

In deriving (19), (20), (21) and (22) use has been made of the closed form integration of (12). The parameter $\mathcal{P}$ in (21) and (22) is defined as

$$\begin{aligned}\mathcal{P} =\ & -\Gamma_1 Q(d_1) + \Gamma_1^2 Q(2d_1) - \Gamma_1^3 Q(3d_1) + \Gamma_1^4 Q(4d_1) - \frac{\Gamma_1^5}{1+\Gamma_1}Q(5d_1) + \\ & (1-\Gamma_1^2)\Gamma_2\left[Q(d_1) - \Gamma_1\Gamma_2 Q(2d_1) - (1+2\Gamma_1)Q(d_1+d_2) + \right. \\ & \left. \Gamma_1(2+3\Gamma_1)Q(2d_1+d_2) + \Gamma_1\Gamma_2 Q(d_1+2d_2) - 3\Gamma_1^2 Q(3d_1+d_2)\right]\end{aligned} \tag{23}$$

and Γ_n and $Q(d)$ are defined as

$$\Gamma_1' = \frac{\epsilon_{1,r}-\epsilon_{2,r}}{\epsilon_{1,r}+\epsilon_{2,r}} \quad ; \quad \Gamma_2' = \frac{\epsilon_{2,r}-1}{\epsilon_{2,r}+1} \tag{24}$$

$$\Gamma_1'' = \frac{\mu_{2,r}-\mu_{1,r}}{\mu_{2,r}+\mu_{1,r}} \quad ; \quad \Gamma_2'' = \frac{1-\mu_{2,r}}{1+\mu_{2,r}} \tag{25}$$

and

$$Q(d) = \frac{e^{-jk_0\sqrt{\rho^2+4(d)^2}}}{k_0\sqrt{\rho^2+4(d)^2}} - \frac{e^{-jk_0\sqrt{\rho^2+4(d_1+d)^2}}}{k_0\sqrt{\rho^2+4(d_1+d)^2}} \quad . \tag{26}$$

It is noted that the asymptotic result for the Green's dyadic given above for $k_0\rho \ll 1$ is somewhat similar to the quasi-static result obtained by Mosig and Sarkar [5].

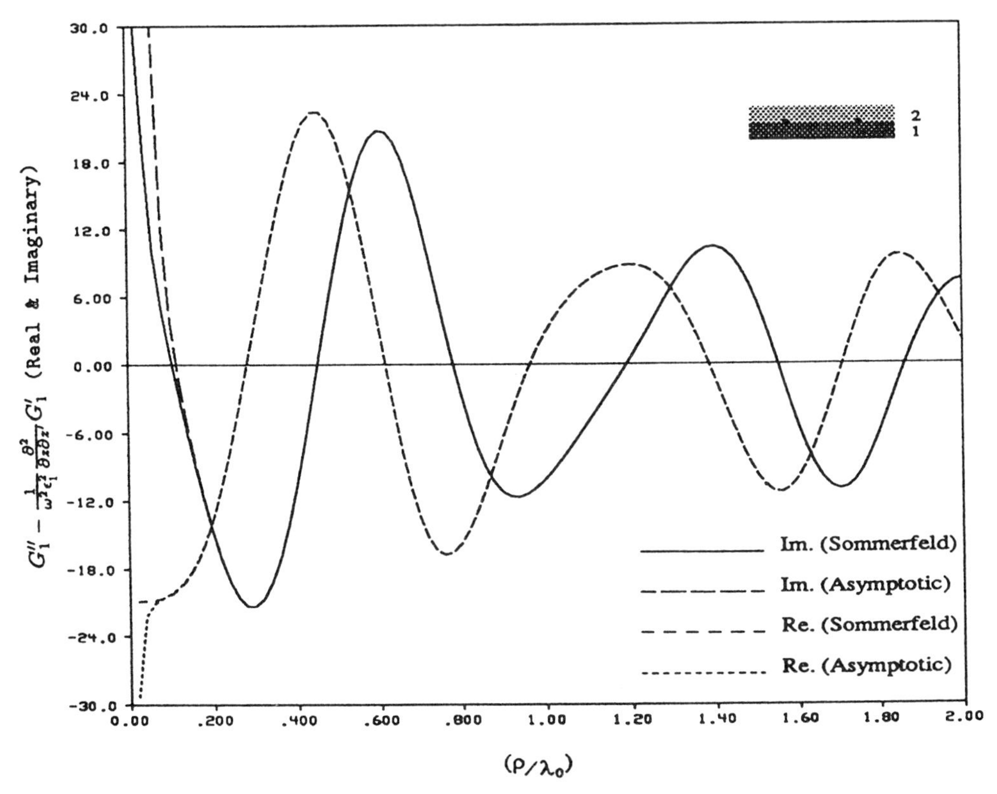

Figure 2. Comparison between the numerical integration and the asymptotic value of $G_1'' - \frac{1}{\omega^2\epsilon_1^2}\frac{\partial^2}{\partial z\partial z'}G_1'$ versus ρ/λ_0 in (2) and (13) for $\epsilon_{1,r} = 9.6, \mu_{1,r} = 1.0, d_1 = 0.09\lambda_0$, and $\epsilon_{2,r} = 3.25, \mu_{2,r} = 1.0, d_2 = 0.1\lambda_0$.

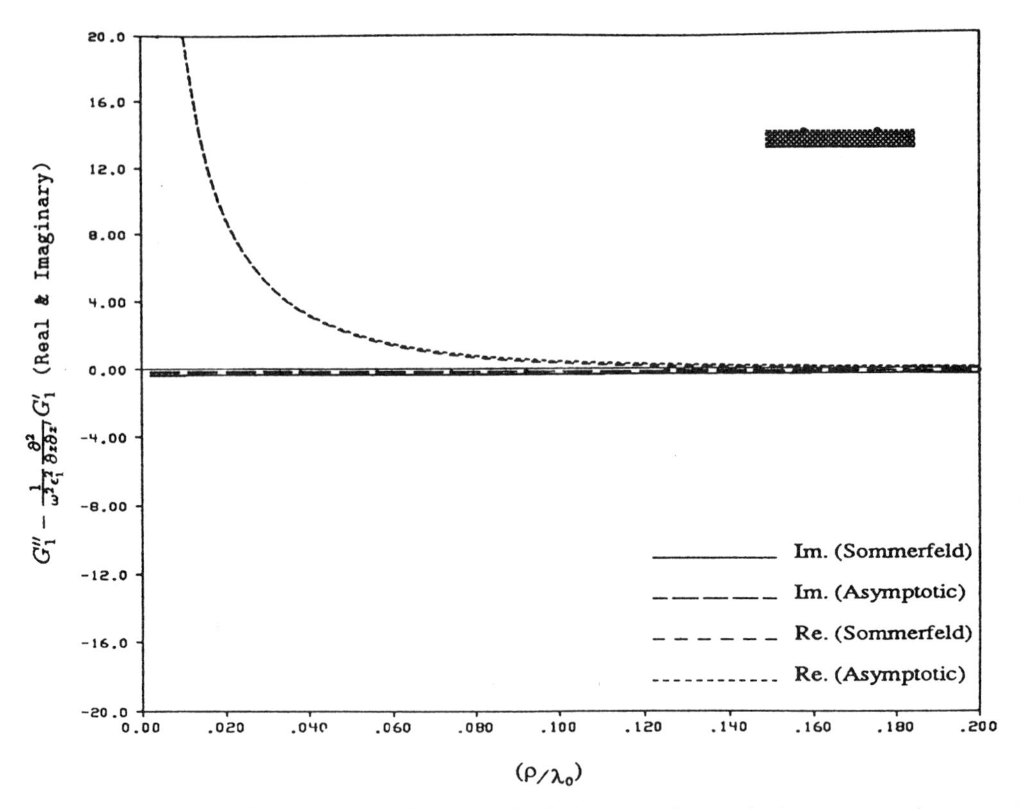

Figure 3. Comparison between the numerical integration and the asymptotic value of $G_1'' - \frac{1}{\omega^2 \epsilon_1^2} \frac{\partial^2}{\partial z \partial z'} G_1'$ versus ρ/λ_0 in (2) and (20) for $\epsilon_{1,r} = 3.25, \mu_{1,r} = 1.0, d_1 = 0.05\lambda_0$.

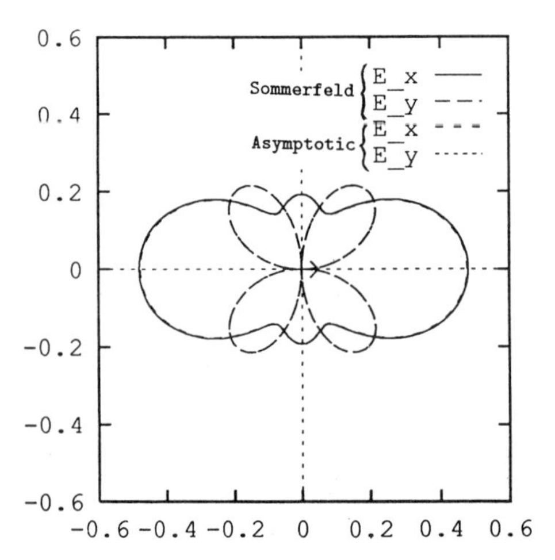

Figure 4. Comparison between the numerical integration and the asymptotic evaluation of surface electric fields, E_x, and E_y, over the planar interface of a grounded single dielectric at observation point (ρ, ϕ) versus ϕ at $\rho = \lambda_0$ in (2), (3), and (13), with $\epsilon_{1,r} = 3.25, \mu_{1,r} = 1.0$, and $d_1 = 0.1\lambda_0$.

IV. NUMERICAL RESULTS

Figures 2 and 3 show $G_1'' - \frac{1}{\omega^2\epsilon_1^2}\frac{\partial^2}{\partial z \partial z'}G_1'$ versus ρ/λ_0 based on the numerical integration of (2) and the asymptotic evaluation of (13), for $k_0\rho$ large, and of (20), for $k_0\rho$ small, for grounded double and single dielectric layer slabs, respectively. A comparison between the numerical and the asymptotic evaluations of surface electric fields, E_x, and E_y, over the planar interface of a grounded single dielectric slab, for large $k_0\rho$, versus ϕ at $\rho = \lambda_0$ are shown in Figure 4. Parameters of geometries of grounded dielectric slabs are given in captions of the Figures. It can be seen from the Figures, that the accuracy of the new representation is quite impressive. This new asymptotic solution (for large $k_0\rho$) remains valid even for field points very close to the source.

Two different closed-form asymptotic representations of the grounded double and single layer material slab Green's functions based on the small and large lateral separations of the source and observation points are presented. This work is particularly useful in the moment method analysis of large microstrip antenna arrays [2,3], amd monolithic millimeter and microwave integrated circuits (MMIC) [4], where the mutual coupling effects are important.

Acknowledgment

This work was supported by a grant from the Joint Services Electronic Program under Contract N00014-85-C-0049 with The Ohio State University Research Foundation.

V. REFERENCES

[1] S. Barkeshli, P. H. Pathak, ''Radial propagation and steepest descent path integral representations of the planar microstrip dyadic green's function,'' J. Radio Sci., vol. 25, no 2, pp 161-174, March-April 1990.

[2] S. Barkeshli, P. H. Pathak and M. Marin, ''An asymptotic closed form microstrip surface green's function for the efficient moment method analysis of mutual coupling in microstrip antennas'' IEEE Antennas Propagat., vol. 38, pp. 570-573, Sept. 1990.

[3] S. Barkeshli, ''An efficient moment method analysis of finite phased arrays of strip dipoles in a substrate/superstrate configuration using an asymptotic closed form approximation for the planar double layered microstrip green's function,'' presented at Int. IEEE Antennas Propagat. and Nat. Radio Sci. Meet. at San Jose, CA, June 26-30, 1989.

[4] S. Barkeshli, ''Efficient analysis of planar single and double layered microstrip geometries using asymptotic closed form microstrip surface green's functions,'' presented at National Radio Science Meeting at Boulder, CO, January 3-5, 1990.

[5] J. R. Mosig and T. K. Sarkar, ''Comparison of quasi-static and exact electromagnetic fields from a horizontal electric dipole above a lossy dielectric backed by an imperfect ground plane,'' IEEE Trans. Microwave Theory Tech., vol. MTT-34, pp. 379-387, April 1986.

IMPROVEMENTS OF SPECTRAL DOMAIN ANALYSIS TECHNIQUES FOR ARBITRARY PLANAR CIRCUITS

T. Becks and I. Wolff

Department of Electrical Engineering and
Sonderforschungsbereich 254, University of Duisburg
Bismarckstr. 69, D-4100 Duisburg 1, FRG

Abstract

Spectral domain analysis techniques using roof-top functions as expansion functions for the surface current density have proofed to lead to a flexible tool for the calculation of arbitrarily shaped planar microwave structures. Several improvements of this method e.g. the introduction of new integration paths and analytic integration of a separated part of the dyadic function which reduce the computation time and which for the first time introduce losses (without using perturbation techniques) into the spectral domain analysis will be described. Furthermore the influence of surface waves and radiation is considered so that the transmission properties of planar microwave components can be described more realistically.

Introduction

Spectral domain analysis techniques using roof-top functions [1] as expansion functions for the surface current density have proofed to lead to a flexible tool for the calculation of arbitrarily shaped planar microwave structures [2, 3, 4, 6], they are very helpful especially in the cases of electromagnetically closely coupled planar lines and discontinuities. On the other hand these methods still have high expense in computer time and storage. The requirement for storage can be reduced by using iterative methods like the conjugate gradient method [5, 7, 8]. To reduce the computing time, three approaches will be described in this paper:

- New integration paths using linear curve sections are introduced to reduce the numerical effort while integrating the dyadic Green's function near the surface wave poles.
- An asymptotic part of the dyadic function characterized by a simple function is separated and integrated analytically.
- It will be discussed whether the elemens of the Z-matrix can be calculated by using an efficient FFT-algorithm if, as will done below, losses are considered in the spectral domain calculation and thereby the surface wave poles are shifted away from the real axis.

Directions in Electromagnetic Wave Modeling
Edited by H.L. Bertoni and L.B. Felsen, Plenum Press, New York, 1991

As has already been mentioned above, for the first time losses will be considered in the spectral domain analysis of arbitrary planar structures. This is done by formulating the dyadic Green's function considering the dielectric losses and the current losses in the backside metallization (if it is available). The current losses of the metallic strip structure in the layered dielectric medium are formulated through the boundary conditions. Additionally surface waves and radiation losses can be considered by using a structure with an infinite shielding top plane or even an open space above the circuit. Their influence compared to the current losses will be discussed. By introducing the loss-calculation into the spectral domain techniques, the numerical results are more realistic, compared to experiments.

Method of Analysis

Impressed current source distributions generate a respective field $\vec{E}_{S,inc}$ equivalent to the incoming field of an scattering problem. A reaction of the system $\vec{E}_{S,d}$ has to compensate the portion of the source field tangential to the strip metallization of a multilayer structure in the sense of

$$\vec{E}_{S,inc} + \vec{E}_{S,d} = Z_S \, \vec{J}_S. \tag{1}$$

Z_S is the surface impedance and $\vec{J}_S$ is the surface current density. The wellknown surface impedance is extended with an additional term in order to get a correct behavior of the surface impedance down to lower frequencies.

$$Z_S = \frac{(1+j)\sqrt{\pi f \mu_0 \rho}}{1.0 - e^{-t\sqrt{\frac{\pi f \mu_0}{\rho}}}} \tag{2}$$

On the assumption of having a planar structure, eq.(1) leads -reduced by the well known spectral domain formalism- to a *Fredholm* integral equation of the second kind which behaves numerically as one of the first kind. This integral equation is solved by the *Galerkin* procedure leading to a system of algebraic equations

$$\begin{pmatrix} \overleftrightarrow{Z}_{xx} + \overleftrightarrow{Z}_{\Omega,xx} & \overleftrightarrow{Z}_{xy} \\ \overleftrightarrow{Z}_{yx} & \overleftrightarrow{Z}_{yy} + \overleftrightarrow{Z}_{\Omega,yy} \end{pmatrix} \begin{pmatrix} \vec{J}_{Sx} \\ \vec{J}_{Sy} \end{pmatrix} = \begin{pmatrix} \vec{V}_{inc,x} + \vec{V}_{\Omega,x} \\ \vec{V}_{inc,y} + \vec{V}_{\Omega,y} \end{pmatrix}. \tag{3}$$

As already mentioned above, the unknown surface current density is expanded in a set of roof-top functions. The choice of rooftop functions as expansion function leads to a discretization procedure. **Fig.1** shows a rooftop-element and an example for this procedure, the staircase approximation of a curved contour.

Important for an effective solution of the matrix-eq.(3) -discussed later on- is a property of the $\overleftrightarrow{Z}$-matrix. The $\overleftrightarrow{Z}_{uv}$-matrices $(u, v = x, y)$ are of block-*Toeplitz* type, *Toeplitz* submatrices

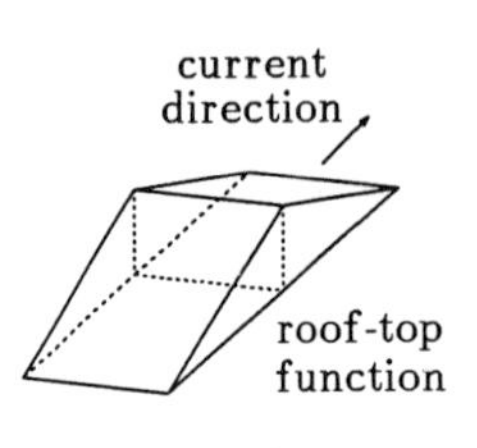

Fig.1. Rooftop as expansion function for the surface current

⇒

staircase approximation of the real contour.

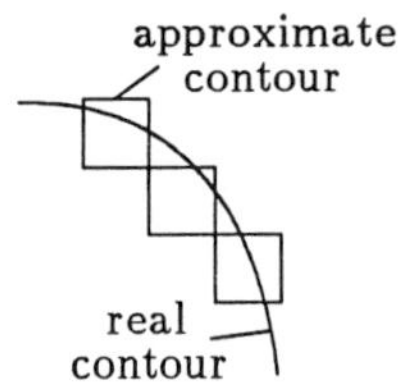

within due to the choice of subsectional current basis functions. A typical element of the Z-matrix looks like

$$Z \sim \iint_{-\infty}^{\infty} \tilde{G}_{E,J}\, si^4\left(\frac{k_x \Delta x}{2}\right) si^2\left(\frac{k_y \Delta y}{2}\right) e^{jk_x \acute{m} \Delta x}\, e^{jk_y \acute{n} \Delta y} dk_x dk_y \tag{4}$$

with Δx and Δy the length of the subsectional discretization scheme.

Solving this integral by changing to polar coordinates β, α

$$\text{Substitution:} \begin{cases} k_x &= k_0 \beta \cos\alpha \\ k_y &= k_0 \beta \sin\alpha \end{cases} \tag{5}$$

and numerical integration leads to the difficulty of integrating a singular function. The singularities are determined by the roots of the characteristic functions of the transverse electric (TE) and the tranverse magnetic (TM) surface waves or parallel-plate waves assuming a triplate structure. Assuming propagating modes, these poles occur for real values of β in the range of $1.0 \leq \beta \leq \sqrt{\epsilon_{r,max}}$ (lossless case). Ohmic losses in the upper or the backside metallization (both, if available) are included in the formulation of the dyadic *Green's* function. In case of nonideal metallization a relation between tangential E- and H-field can be derived in the following form.

$$\vec{\tilde{E}}_S = Z_S \vec{\tilde{J}}_S = Z_S(\vec{e}_{S,normal} \times \vec{\tilde{H}}_S) \tag{6}$$

If losses are present, the poles move into the negative complex halfplane. Dispersion curves of these poles are plotted in **Fig.2**. The solid line curves represent the solution for backside metallization with a specific surface resistance of $\varrho = 2.4 \cdot 10^{-8}\Omega m$. The broken line curves represent the solution for ideal backside metallization. Geometrie and material parameter of the structure are shown within the figure.

To avoid numerical difficulties when numerically integrating on β, the integration path is expanded into the complex $\underline{\beta}$-plane shown in **Fig.3**.

Further aspect in the choice of the integration range of β is the *Nyquist*-theorem. In order to fulfil this theorem the infinite integration in β has to be terminated at a relatively high upper bound shown in eq.(7).

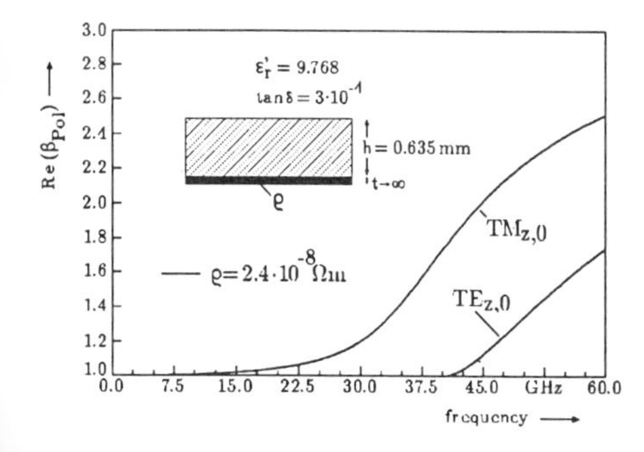

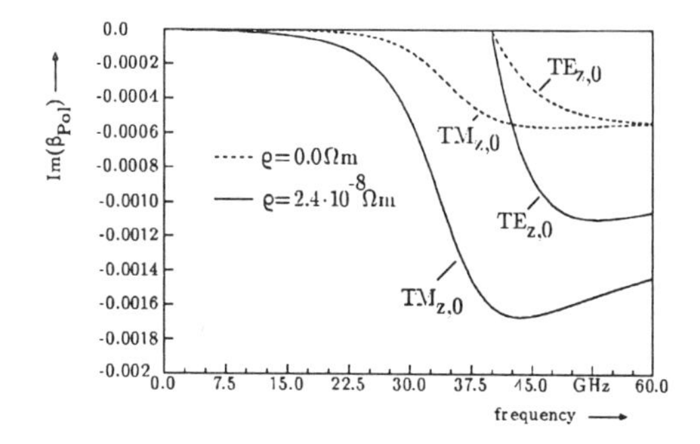

Fig.2. Dispersion curves for the $TM_{z,0}$- and $TE_{z,0}$- modes of a dielectric coated conductor. Geometrie and material parameters of the structure are shown within the figure. (- - -) ideal backside metallization, (—) lossy backside metallization.

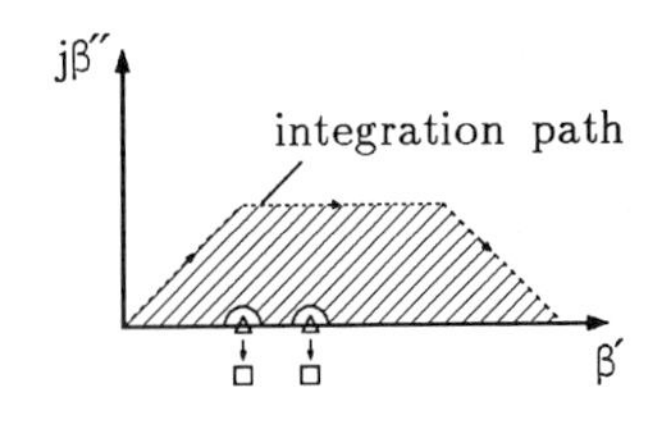

Fig.3. Integration path in the complex $\underline{\beta}$-plane.
Location of poles:
△ (losless case),
□ (lossy case).
The hatched area encloses no surface wave poles.

$$\beta_{max} \geq \frac{\pi}{k_0 min(\Delta u)} \qquad u = x, y \tag{7}$$

An asymptotic part of the dyadic function characterized by a simple function proportional to β is separated and integrated analytically in order to reduce this effort.

$$\int_0^{\infty}\int_0^{\frac{\pi}{2}} \tilde{G}_{E,J_S} \cdots d\alpha\, d\beta = \underbrace{\int_0^{\beta_{asym.}}\int_0^{\frac{\pi}{2}} \left[\tilde{G}_{E,J_S} - \tilde{G}^{asym.}_{E,J_S}\right] \cdots d\alpha\, d\beta}_{\text{numerical integration}} + \underbrace{\int_0^{\infty}\int_0^{\frac{\pi}{2}} \tilde{G}^{asym.}_{E,J_S} \cdots d\alpha\, d\beta}_{\text{analytical integration}} \tag{8}$$

The resulting upper numerical integration bound for β is reduced to $\beta_{asym.}$ and the numerical integration becomes much more faster.

Application of FFT-algorithm to the numerical solution of eq.(3) can be divided into two parts:

1. The calculation of the $\overleftrightarrow{Z}$-matrix-elements. From eq.(4) it is evident that all Z-matrix elements can be described by a twodimensional backward *Fourier* transform. If assuming a uniform discretization scheme any submatrix in eq.(3) can be performed efficiently by application of FFT techniques. In order to get correct results by using FFT to do this, three facts have to be taken into account:

   - *Nyquist*-theorem (eq.(7)) must be fulfilled.
   - Choice of a fine sampling rate around the poles is important in order to get the influences of surface waves into the calculation.
   - Choice of a fine sampling rate in the range of $0.0 \leq \beta \leq 1.0$ is important in order to get the influence of radiation into free space into the calculation.

   Because of this facts FFT-dimensions would become large. To overcome these disadvantages two different ideas have been used:

   - No location of sampling points near to the poles $\Rightarrow$ use of interpolation or translation theorem.
   - Including surface wave effects by using the residue theory near the poles to get an additional part to the 'FFT' integral.

2. An efficient solution of the linear matrix system from eq.(3). The conjugate gradient (cg-)method is proofed to be an efficient method to reduce:

   - requirement for storage
   - computation time

   during solving large linear systems. Most time consuming part of the cg-algorithm is the matrix-vector-multiplication needed once or twice each iteration step. As already mentioned above, the $\overleftrightarrow{Z}$-matrix has some special properties. Because of these properties matrix equation can be seen as a set of convolution equations and therefor an efficient use of FFT-algorithm becomes possible [11].

Results

To compare the results of the improved method to those spectral domain techniques described in the literature [2]-[5], two special structures which have been presented by Wertgen [2] have been produced and measured [9] (using the time-domain option of a network analyzer), the numerical results of this method for a radial stub and a meander line structure are compared to those of [2, 5] in **Figs.5** and **7**.

Fig.4 shows a radial stub structure and the calculated current distribution within this structure at a frequency of 21 GHz.

Fig.5 shows the calculated and measured scattering parameters of a radial stub structure shown in the insert compared to the theoretical results of [2]. As can be seen, several differences can be recognized in the theoretical results:

1. A package resonance (e.g. at about 15.5 GHz) occurs in the results of [2] because the structure is totally enclosed by an electric shielding. This resonance does not occur in the theoretical solution presented here and in the measurements.
2. The magnitude of the scattering parameters especially at resonant frequencies of the structure cannot exactly be calculated if the losses, especially through stimulation of surface waves and radiation are not taken into account (e.g. 13.5 GHz and at frequencies higher than 18 GHz).

Fig.6 shows an example how the dc-block using the radial stub can be made more broadband by using two 90°-radial stubs with different radii in a shifted butterfly-structure. As the figure shows, a very broadband dc-block with a bandwith of more than 7 GHz can be realized. The agreement of the measured and calculated results is very good about the whole frequency range shown in the figure. At the same time this special structure is an example for the possibility of handling diagonal edges and curved contours with subsectional basis functions which cover small rectangular areas. The analysis of this kind of circuits containing two very closely coupled resonant structures can only be done with high accuracy using a full wave solution like the spectral domain approach with roof-top functions.

Finally in **Fig.7** a comparison of different methods [2, 10] for analyzing a meander line structure as well as measured results [9] are shown. Again it can be recognized that the method discussed here, considering the losses, leads to very good results for the magnitudes and the phases. Differences between the different methods are especially recognizable in the magnitude of the reflection coefficient S_{11} in the stop band (12-22 GHz), again because in the method

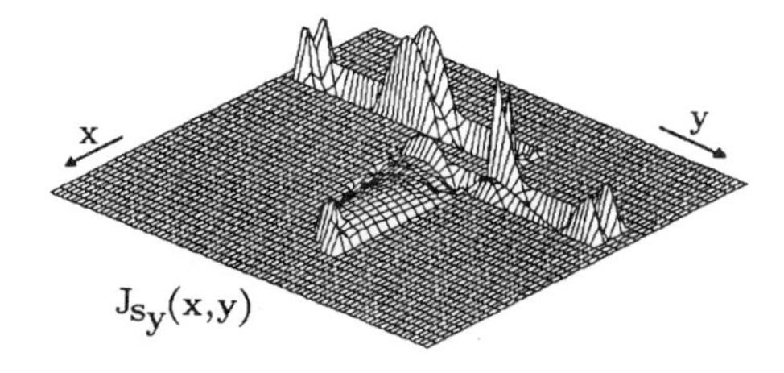

Fig.4. $J_{Sy}(x,y)$ component (real part) of surface current distribution.
Structure: radial stub (shown in insert of Fig.3).

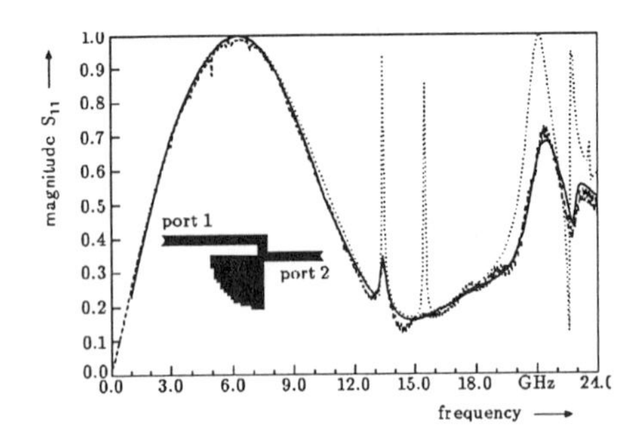

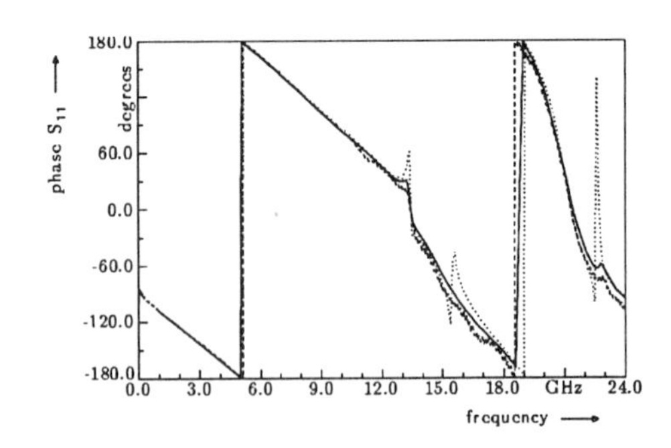

Fig.5. Calculated [this method (—), by Wertgen [2] (···)] and measured (- - -) S-parameter of the radial stub.

Geometry parameter:

$$w = 610\mu m$$
$$R = 3.253mm$$

Material parameter:

$$\text{height} = 635\mu m$$
$$\acute{\epsilon}_r = 9.768 \quad \text{(measured)}$$
$$\tan\delta = 3 \cdot 10^{-4}$$
$$\rho_{\text{strip}} = \rho_{\text{back}} = 2.4 \cdot 10^{-8}\Omega m$$
$$t_{\text{strip}} = 5\mu m \quad , \quad t_{\text{back}} \mapsto \infty$$

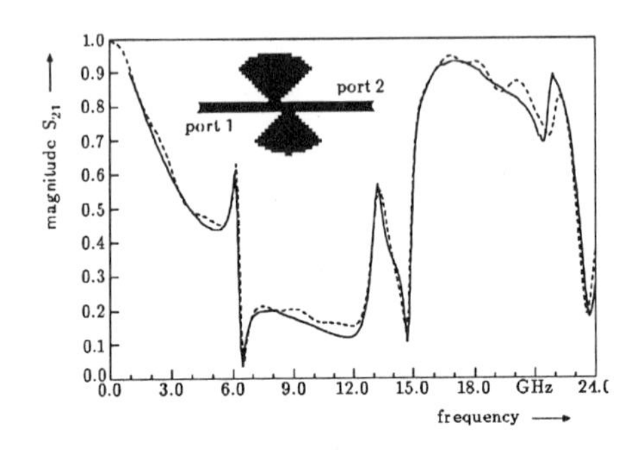

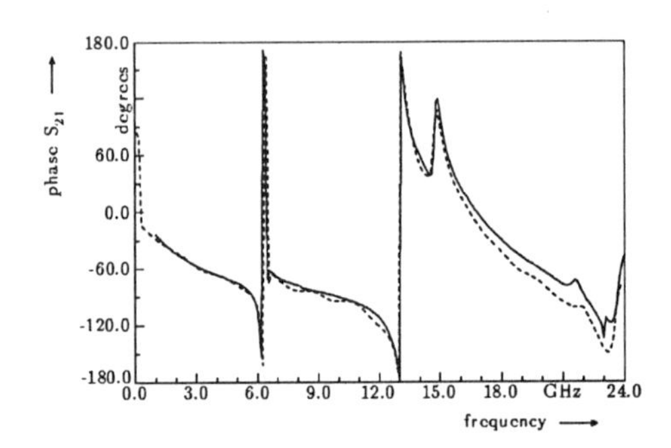

Fig.6. Calculated (—) and measured (- - -) S-parameter of the shifted butterfly-structure.

Geometry parameter:

$$w = 610\mu m \quad , \quad s = 813\mu m$$
$$R_1 = 3.355mm$$
$$R_2 = 2.949mm$$

Material parameter:

$$\text{height} = 635\mu m$$
$$\acute{\epsilon}_r = 9.779 \quad \text{(measured)}$$
$$\tan\delta = 3 \cdot 10^{-4}$$
$$\rho_{\text{strip}} = \rho_{\text{back}} = 2.4 \cdot 10^{-8}\Omega m$$
$$t_{\text{strip}} = 5\mu m \quad , \quad t_{\text{back}} \mapsto \infty$$

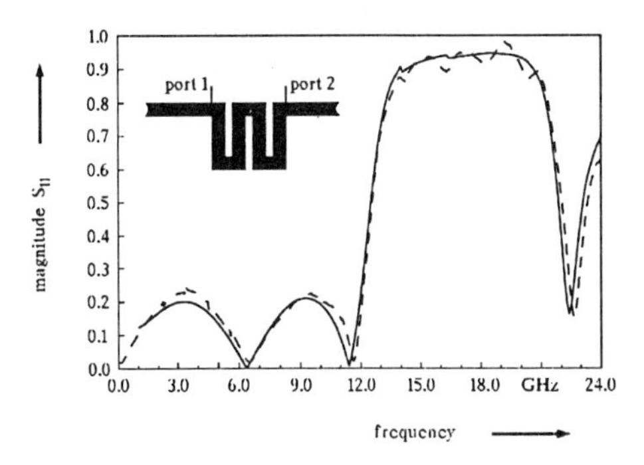

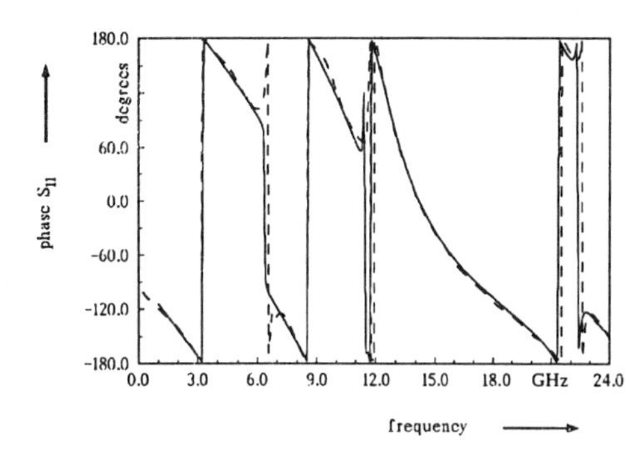

Fig.7a. Calculated (—) and measured (- - -) S-parameter of the meander line.

Geometry parameter:	Material parameter:
w = $610\mu m$	height = $635\mu m$
	$\acute{\varepsilon}_r$ = 9.978 (measured)
	$\tan\delta$ = $3 \cdot 10^{-4}$

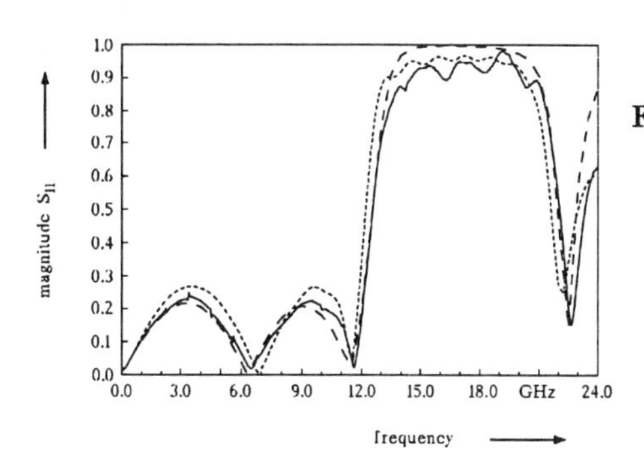

Fig.7b. Measured by Gronau [9] (—) and calculated [by Wertgen [2] (- - -) and by Rittweger [10] (···)] S-parameter of the meander line.
Substrate and geometry: shown in Fig.5a.

presented here, the losses are considered. It can also be recognized that all full wave analyses show a small frequency set-off to the measured results. This leads to the assumption that the produced hardware structure was not 100% identical with the calculated structure.

Conclusion

It has been shown that essential improvements can be made to the state of the art of spectral domain analysis of arbitrary planar structures. The computer time which is needed can be reduced, and the accuracy can be improved taking into account losses in the substrate and the metallization layers. Especially the consideration of surface wave and radiation losses is important for an accurate analysis.

References

[1] A. W. Glisson and D. R. Wilton, "Simple and efficient numerical methods for problems of electromagnetic radiation and scattering from surface," *IEEE Trans. Antennas Propagat.*, vol. AP-28, pp. 593-603, September 1980.

[2] W. Wertgen, *Elektrodynamische Analyse geometrisch komplexer (M)MIC-Strukturen mit effizienten numerischen Strategien.* Ph.D. Thesis, Duisburg University, FRG, 1989.

[3] W. Wertgen and R. H. Jansen, "A 3d fieldtheoretical simulation tool for the CAD of mm-wave MMICs," *Alta Frequenza*, LVII-N.5, pp. 203-216, 1988.

[4] R. W. Jackson, "Full-wave, finite element analysis of irregular microstrip discontinuities," *IEEE Trans. Microwave Theory Tech.*, vol. MTT-37, pp. 81-89, January 1989.

[5] W. Wertgen and R. H. Jansen, "Novel Green's function database technique for the efficient full-wave solution of complex irregular (M)MIC-structures," in *19th European Microwave Conf. Proc.*, (London, England), 1989, pp. 199-294.

[6] A. Skriverik and J. R. Mosig, "Equivalent circuits of microstrip discontinuities including radiation effects," in *1989 IEEE MTT-S Digest*, (Long Beach, USA), 1989, pp. 1147-1150.

[7] R. Chandra, *Conjugate gradient methods for partial differential equations.* Ph.D. Thesis, Yale University, USA, 1978.

[8] A. F. Peterson, *On the implementation and performance of iterative methods for computational electromagnetics.* Ph.D. Thesis, University of Illinois at Urbana-Champaign, USA, 1986.

[9] G. Gronau and I. Wolff, "A simple broad-band device de-embedding method using an automatic network analyzer with time-domain option," *IEEE Trans. Microwave Theory Tech.*, vol. MTT-37, pp. 479-483, March 1989.

[10] M. Rittweger and I. Wolff, "Analysis of complex passive (M)MIC components using the finite difference time-domain approach," in *GAAS' 90, Gallium Arsenide Application Symposium, Conference Proceedings*, (Rom, Italy), 1990, pp. 162-167.

[11] H. L. Nyo and R. F. Harrington, *The discrete convolution method for solving some large moment matrix equations.* Technical Report No.21, Department of Electrical and Computer Engineering, Syracuse University, Syracuse, New York, USA, July 1983.

INTEGRAL EQUATION ANALYSIS OF MICROWAVE INTEGRATED CIRCUITS

Krzysztof A. Michalski

Department of Electrical Engineering
Texas A&M University
College Station, Texas

INTRODUCTION

Most of the analysis methods presently available for the analysis of microstrip structures are either limited to or optimized for planar geometries with simple, regular shapes.[1–4] The modern microwave and millimeter–wave integrated circuits, however, cannot always be considered planar and they are usually far from being simple.[5,6] Therefore, it is increasingly important to have at one's disposal techniques, which would make it possible to accurately and rigorously analyze more complex circuit geometries. In this paper, we describe an integral equation approach, which we find especially suitable for this task.

FORMULATION

Consider an arbitrarily shaped object embedded in a layered dielectric medium, as illustrated in Fig. 1. The dielectric layers, which are assumed to be of infinite extent along the x and y coordinates, may be uniaxially anisotropic with the optic axis parallel to the z axis,[7] and are characterized by relative permittivity and permeability tensors

$$\underline{\underline{\epsilon}} = \underline{\underline{I}}\,\epsilon_t + \hat{\boldsymbol{z}}\hat{\boldsymbol{z}}\,(\epsilon_z - \epsilon_t)\ , \quad \underline{\underline{\mu}} = \underline{\underline{I}}\,\mu_t + \hat{\boldsymbol{z}}\hat{\boldsymbol{z}}\,(\mu_z - \mu_t) \tag{1}$$

where $\underline{\underline{I}}$ is the idemfactor and where the subscript t denotes quantities transverse to $\hat{\boldsymbol{z}}$ (we denote unit vectors by carets). We also introduce electric and magnetic anisotropy ratios and the wavenumber for the ordinary waves, defined respectively as

$$\nu^e = \frac{\epsilon_t}{\epsilon_z}\,, \quad \nu^h = \frac{\mu_t}{\mu_z}\,, \quad k_t = k_0\sqrt{\mu_t\epsilon_t} \tag{2}$$

The wavenumber and intrinsic impedance of free space are denoted by k_0 and η_0, respectively. The integral equation for the current density $\boldsymbol{J}$ on the surface S of the object (which is assumed to be perfectly conducting) can be expressed in the convenient mixed–potential form

$$\hat{\boldsymbol{n}}\times\left\{\int_S \underline{\underline{G}}^A(\boldsymbol{r}|\boldsymbol{r}')\cdot\boldsymbol{J}(\boldsymbol{r}')\,dS' + \nabla\int_S G^\phi(\boldsymbol{r}|\boldsymbol{r}')\,\nabla'\cdot\boldsymbol{J}(\boldsymbol{r}')\,dS'\right\} = \hat{\boldsymbol{n}}\times\boldsymbol{E}^i(\boldsymbol{r}), \quad \boldsymbol{r}\in S \tag{3}$$

where $\hat{\boldsymbol{n}}$ is normal to S at a point specified by the position vector $\boldsymbol{r}$, $\boldsymbol{E}^i$ is the "incident" electric field (i.e., the field that exists in the layered medium in the absence of the

Directions in Electromagnetic Wave Modeling
Edited by H.L. Bertoni and L.B. Felsen, Plenum Press, New York, 1991

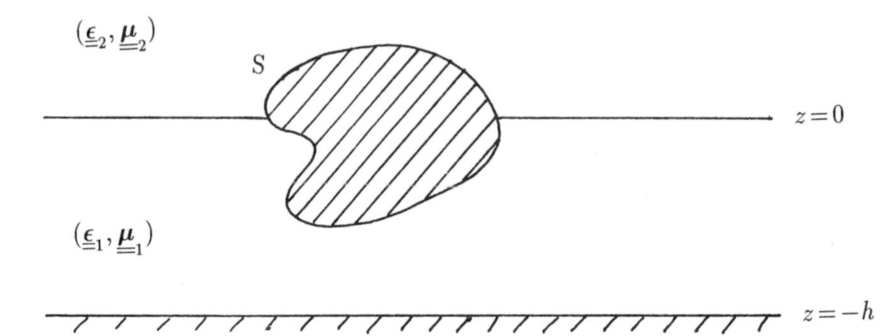

Figure 1. Arbitrarily shaped object embedded in a layered medium.

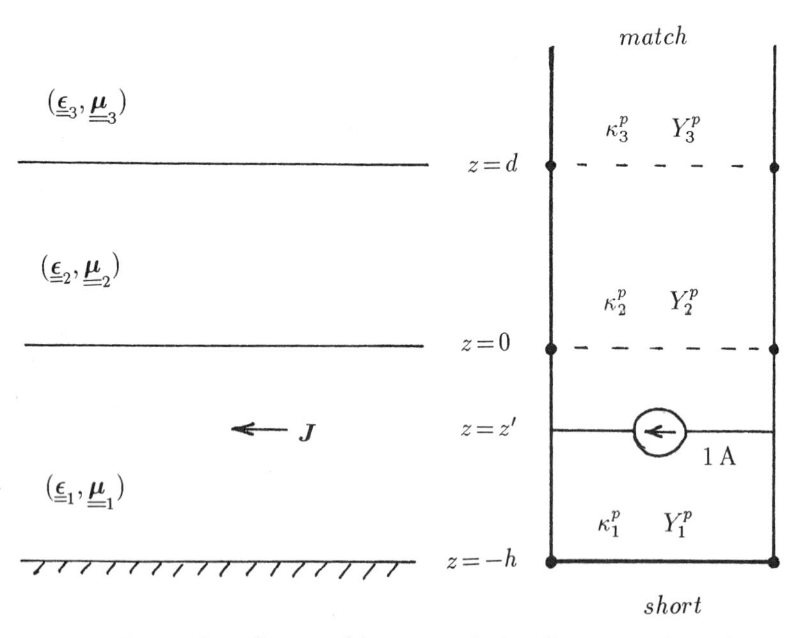

Figure 2. Layered medium and its transmission–line network analogue.

object), and where the dyadic kernel (which is a suitably modified vector potential Green's function of the layered medium[8,9]) has the form

$$\underline{\underline{G}}^A(\boldsymbol{r}|\boldsymbol{r}') = (\hat{\boldsymbol{x}}\hat{\boldsymbol{x}} + \hat{\boldsymbol{y}}\hat{\boldsymbol{y}})G^A_{xx} + \hat{\boldsymbol{z}}\hat{\boldsymbol{x}}G^A_{zx} + \hat{\boldsymbol{z}}\hat{\boldsymbol{y}}G^A_{zy} + \hat{\boldsymbol{x}}\hat{\boldsymbol{z}}G^A_{xz} + \hat{\boldsymbol{y}}\hat{\boldsymbol{z}}G^A_{yz} + \hat{\boldsymbol{z}}\hat{\boldsymbol{z}}G^A_{zz} \tag{4}$$

Using the notation

$$\mathcal{S}_n\{f(k_\rho)\} = \frac{1}{2\pi}\int_0^\infty f(k_\rho)J_n(k_\rho\xi)\,k_\rho^{n+1}\,dk_\rho\,,\quad n = 0,1 \tag{5}$$

$$\xi = \sqrt{(x-x')^2+(y-y')^2}\,,\quad \zeta = \tan^{-1}\left(\frac{y-y'}{x-x'}\right) \tag{6}$$

where J_n is the Bessel function of order n, the elements of the dyadics (4) and the scalar potential kernel G^ϕ in (3) can be expressed as (an $e^{j\omega t}$ time convention is implied throughout)

$$G^A_{xx}(\boldsymbol{r}|\boldsymbol{r}') = \mathcal{S}_0\left\{V_i^h(z|z')\right\} \tag{7}$$

$$G^A_{zx}(\boldsymbol{r}|\boldsymbol{r}') = -jk_0\eta_0\mu_t\cos\zeta\,\mathcal{S}_1\left\{\frac{I_i^h(z|z')}{k_\rho^2} - \frac{I_i^e(z|z')}{k_\rho^2}\right\} \tag{8}$$

$$G^A_{xz}(\boldsymbol{r}|\boldsymbol{r}') = -jk_0\eta_0\mu_t'\cos\zeta\,\mathcal{S}_1\left\{\left[1+\left(1-\nu^h\right)\left(\frac{k_\rho}{k_t}\right)^2\right]\frac{V_v^h(z|z')}{k_\rho^2} - \frac{V_v^e(z|z')}{k_\rho^2}\right\} \tag{9}$$

$$G^A_{zz}(\boldsymbol{r}|\boldsymbol{r}') = \frac{\eta_0^2\mu_t}{\epsilon_z'}\mathcal{S}_0\left\{\left[1-\frac{\mu_t'\epsilon_z'}{\mu_t\epsilon_t}\left(\frac{\kappa^e}{k_\rho}\right)^2\right]I_v^e(z|z') + \eta_0^2\frac{\mu_t'}{\epsilon_t}\left(\frac{k_t}{k_\rho}\right)^2\left[1+\left(1-\nu^h\right)\left(\frac{k_\rho}{k_t}\right)^2\right]I_v^h(z|z')\right\} \tag{10}$$

$$G^\phi(\boldsymbol{r}|\boldsymbol{r}') = \mathcal{S}_0\left\{\left[1+\left(1-\nu^h\right)\left(\frac{k_\rho}{k_t}\right)^2\right]\frac{V_i^h(z|z')}{k_\rho^2} - \frac{V_i^e(z|z')}{k_\rho^2}\right\} \tag{11}$$

with the remaining kernels, G^A_{zy} and G^A_{yz}, given by (8) and (9), respectively, with $\cos\zeta$ replaced by $\sin\zeta$. Here, we have adopted the convention that primed quantities are evaluated in the layer where the source is located, i.e., $\epsilon_t' \equiv \epsilon_t(z')$, etc., while the unprimed parameters pertain to the layer in which the field is being observed. Also, we have employed in the above the transmission line analogue of the layered medium,[7] as illustrated in Fig. 2. This network actually represents two networks that arise from the decomposition of the electromagnetic field into E–waves and H–waves,[10–13] which are, respectively, transverse–magnetic (TM) and transverse–electric (TE) to $\hat{\boldsymbol{z}}$. The quantities associated with the TM and TE networks are distinguished by superscripts e or h, respectively. The propagation constants κ^p and characteristic admittances Y^p, where $p = e$ or h, are given as

$$\kappa^e = \sqrt{k_t^2-\nu^e k_\rho^2}\,,\quad Y^e = \frac{k_0\epsilon_t}{\eta_0\kappa^e} \tag{12}$$

$$\kappa^h = \sqrt{k_t^2-\nu^h k_\rho^2}\,,\quad Y^h = \frac{\kappa^h}{\eta_0 k_0\mu_t} \tag{13}$$

It should be noted that there is a two-sheeted Riemann surface associated with the square root function κ^p. We identify the "proper" sheet by the requirement that Im $\kappa^p < 0$. The opposite holds on the other, "improper" sheet. In (7)–(11), we have also introduced the transmission line Green's functions V_i^p, V_v^p, I_v^p, and I_i^p, where $V_i^p(z|z')$ and $V_v^p(z|z')$ denote the voltage at a point z due, respectively, to a unit-strength current source i and a unit-strength voltage source v, located at a point z' on the corresponding transmission line network, and similarly for $I_v^p(z|z')$ and $I_i^p(z|z')$.[12] If, for simplicity, we limit attention to the three-layer configuration of Fig. 1 and assume that the object is confined to the first two layers, the pertinent transmission line Green's functions are easily found as[7]

$$V_i^p(z|z') = \frac{\overleftarrow{V_\alpha^p}(z_<)\overrightarrow{V_\alpha^p}(z_>)}{\overleftarrow{Y_1^p}(0)+\overrightarrow{Y_2^p}(0)}, \quad -h < z, z' < d \tag{14}$$

where $z_<$ and $z_>$ denote, respectively, the lesser and the greater of z and z', and where the subscript α stands for 1 or 2 if the corresponding variable z or z' lies within the first or the second layer, respectfully. The other symbols in (14) are defined as

$$\overleftarrow{V_1^p}(z) = \cos\kappa_1^p z + j\,\frac{\overleftarrow{Y_1^p}(0)}{Y_1^p}\,\sin\kappa_1^p z\,, \quad \overrightarrow{V_1^p}(z) = \cos\kappa_1^p z - j\,\frac{\overrightarrow{Y_2^p}(0)}{Y_1^p}\,\sin\kappa_1^p z \tag{15}$$

$$\overleftarrow{V_2^p}(z) = \cos\kappa_2^p z + j\,\frac{\overleftarrow{Y_1^p}(0)}{Y_2^p}\,\sin\kappa_2^p z\,, \quad \overrightarrow{V_2^p}(z) = \cos\kappa_2^p z - j\,\frac{\overrightarrow{Y_2^p}(0)}{Y_2^p}\,\sin\kappa_2^p z \tag{16}$$

$$\overleftarrow{Y_1^p}(0) = -jY_1^p\cot\kappa_1^p h\,, \quad \overrightarrow{Y_2^p}(0) = Y_2^p\,\frac{Y_2^p - jY_3^p\cot\kappa_2^p d}{Y_3^p - jY_2^p\cot\kappa_2^p d} \tag{17}$$

To obtain $I_v^p(z|z')$, we invoke the principle of duality and simply replace in the above all voltages and admittances by the corresponding currents and impedances. The remaining Green's functions then follow from the telegraphers' equations

$$I_i^p(z|z') = -\frac{Y_n^p}{j\kappa_n^p}\frac{d}{dz}V_i^p(z|z')\,, \quad V_v^p(z|z') = -\frac{1}{j\kappa_n^p Y_n^p}\frac{d}{dz}I_v^p(z|z') \tag{18}$$

Observe that, for notational simplicity, we do not explicitly indicate the dependence of the transmission line Green's functions on k_ρ.

The geometry of Fig. 1 may also represent a microstrip transmission line that is uniform along, say, the y axis, but whose upper conductor S has an arbitrary cross-section profile C. In that case, we are interested in the fields (modes) that can be supported by the structure in the absence of external excitation. We postulate that all field components and the current density on S vary with y as $e^{-jk_y y}$, where k_y is the sought-after propagation constant, which is real for purely bound modes, and complex for attenuated (leaky) modes.[14] Hence, in general $k_y/k_0 = \beta - j\alpha$, where β and α are, respectively, the phase and attenuation constants normalized to k_0. The transmission line modes can be found as the homogeneous solutions of the integral equation (3), in which S is replaced by the contour C and $\partial/\partial y$ by $-jk_y$. Also, the following substitutions must be made in (7)–(11):

$$\mathcal{S}_0\{f(k_\rho)\} \longrightarrow \frac{1}{\pi}\int_0^\infty f(k_\rho)\cos k_x(x-x')\,dk_x\,, \quad k_\rho = \sqrt{k_x^2+k_y^2} \tag{19}$$

$$\sin\zeta\,\mathcal{S}_1\{f(k_\rho)\} \longrightarrow \frac{jk_y}{\pi}\int_0^\infty f(k_\rho)\cos k_x(x-x')\,dk_x \tag{20}$$

$$\cos\zeta\, \mathcal{S}_1\{f(k_\rho)\} \longrightarrow \frac{1}{\pi}\int_0^\infty k_x f(k_\rho) \sin k_x(x-x')\, dk_x \tag{21}$$

Observe that by eliminating the variables y and y' from (3), we have reduced the latter to a two-dimensional problem.

SOLUTION PROCEDURES

Being in a mixed-potential form, the integral equation (3) is amenable to the well-established numerical solution procedures,[15,16] originally developed for conducting surfaces of arbitrary shape in homogeneous media. We note that except for the presence of the spectral integrals (5) or (19)–(21), the only major difference between (3) and its homogeneous-space counterpart is the dyadic character of the vector potential kernel, which must be properly accounted for in the solution procedure. In the three-dimensional case, the surface S of the arbitrarily shaped object is approximated by triangular finite elements and the unknown current density on each element is expanded in terms of suitable vector basis functions.[16,17] In the transmission line case, the arbitrary cross-section profile C is modeled by straight line segments and the current density on each segment is represented by linear basis functions.[15,18] In either case, a suitable testing procedure reduces the resulting integral equation to a matrix equation, which is then solved by standard techniques. To determine the propagation constants of the microstrip modes, a numerical search is performed for the zeros of the matrix determinant in the complex k_y-plane.

The matrix elements comprise spectral integrals (5) or (19)–(21), which are evaluated by a numerical quadrature. The integration paths must be chosen carefully to avoid the pole and branch-point singularities of the integrands.[17] The situation is especially delicate in the transmission line problem, where the location of these singular points may change drastically for different values of k_y. For the leaky modes, a part of the integration path ends up on the improper sheet of the k_x-plane.[18]

SAMPLE RESULTS

In Fig. 3 are shown sample computed results for a coax-fed annular sector microstrip antenna[19] on an isotropic substrate with thickness 1.57 mm, relative dielectric constant 2.484, and loss tangent 6×10^{-4}. As illustrated in Fig. 3*a*, the conducting patch was approximated by 320 nonuniformly distributed triangular elements. Observe the expected radial current flow near the feed point. In Fig. 3*b*, the computed input impedance data for this antenna are compared with measured results.

In Fig. 4 are shown sample computed results for a microstrip transmission line with a conductor of trapezoidal cross-section (e.g., to simulate the effect of underetching or epitaxial growth) on an electrically uniaxial substrate. These data illustrate the dependence of the characteristic impedance (Fig. 4*a*) and the effective dielectric constant (Fig. 4*b*) on the frequency and the angles of the trapezoid. Results are also given for an infinitesimally thick microstrip and are seen to agree well with data computed by another technique.[20] In Figs. 5(*a*) and 5(*b*) we show, respectively, the transverse and longitudinal components of the surface current on the upper conductor of the transmission line of Fig. 4 at $f = 30$ GHz and for $\varphi = 90°$. The arclength ℓ is measured counterclockwise from the lower left corner of the conducting strip.

CONCLUSIONS

A technique based on the mixed-potential integral equation (MPIE) formulation has been applied to analyze three-dimensional and two-dimensional microstrip

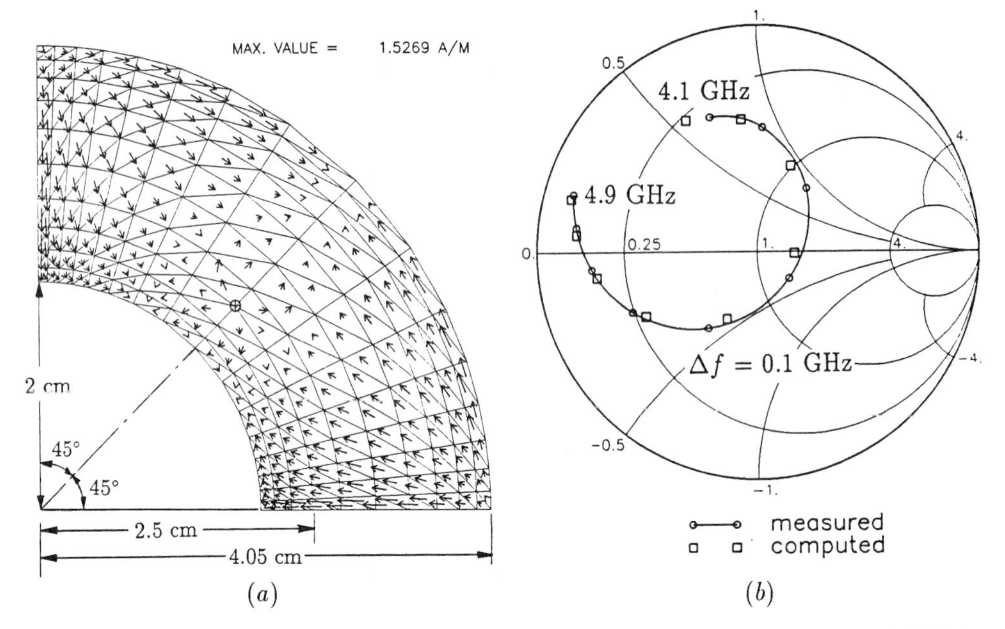

Figure 3. Sample results for a coax–fed annular sector microstrip antenna. (a) Real part of the surface current distribution at the first resonance frequency 4.3 GHz. (b) Smith chart plot of the measured and computed input impedance normalized to 50 Ω.

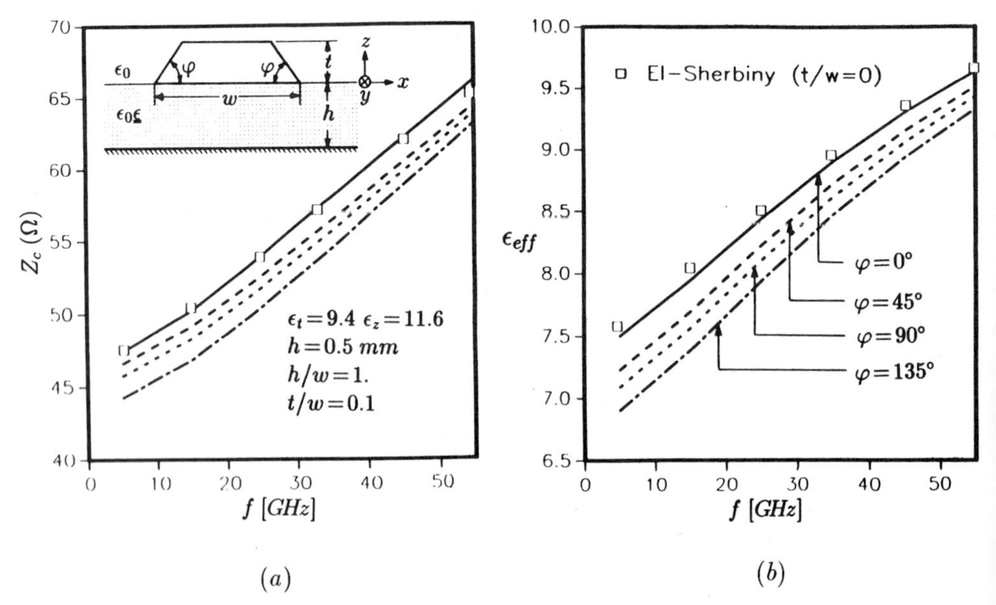

Figure 4. Sample results for a microstrip transmission line with a trapezoidal cross-section profile. (a) Characteristic impedance. (b) Effective dielectric constant $\epsilon_{\text{eff}}=\beta^2$.

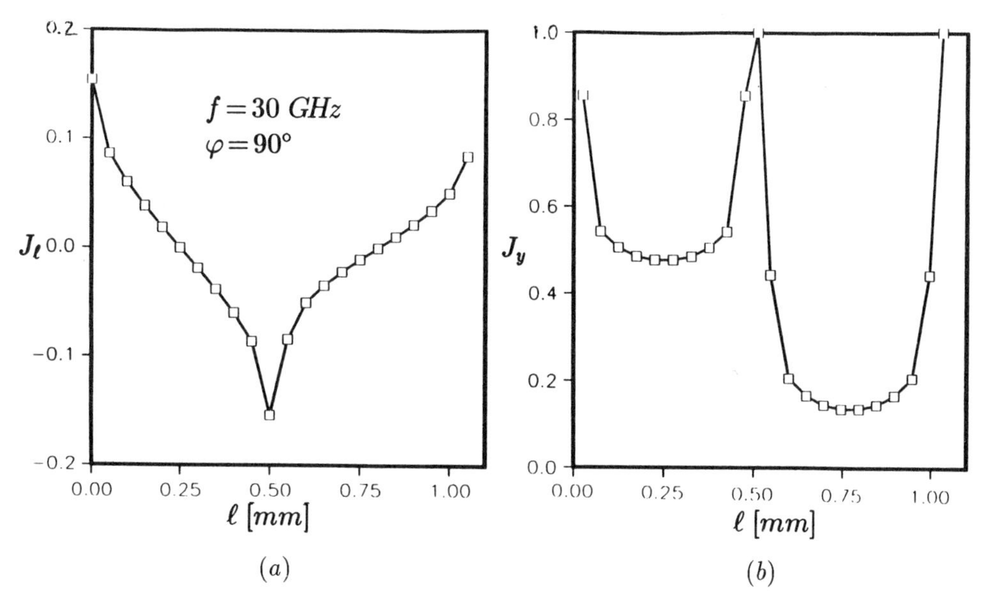

Figure 5. Current distribution on the microstrip transmission line of Fig. 4 with $\varphi=90°$ at $f=30$ GHz. (*a*) Transverse current. (*b*) Longitudinal current.

structures of arbitrary shape. The MPIE approach has the advantage that it involves kernels which are less singular and converge faster than the kernels encountered in other integral equation formulations. Also, the MPIE is amenable to well–established numerical solution procedures originally developed for scatterers of arbitrary shape in free space. Although the MPIE approach may not be the most efficient one to use in the case of planar structures with simple shapes, it offers the flexibility and accuracy which make it well suited for the analysis of nonplanar and/or irregularly shaped microstrip transmission lines and antennas frequently encountered in modern microwave and millimeter–wave circuits. The MPIE is also a promising candidate for use in hybrid formulations, which combine differential and integral equation methods.[21,22]

ACKNOWLEDGMENTS

This work was supported in part by the Office of Naval Research under Contract No. N00014-90-J-1197 and by the Texas Advanced Research Program. The author acknowledges with gratitude the aid of Dr. Dalian Zheng, who prepared computer programs implementing the procedures presented in this paper.

REFERENCES

1. T. Itoh, ed., *Planar Transmission Line Structures.* New York: IEEE Press, 1987.

2. K. C. Gupta and A. Benalla, eds., *Microstrip Antenna Design.* Norwood, MA: Artech House, 1988.

3. T. Itoh, ed., *Numerical Techniques for Microwave and Millimeter–Wave Passive Structures.* New York: Wiley, 1989.

4. R. Sorrentino, ed., *Numerical Methods for Passive Microwave and Millimeter Wave Structures.* New York: IEEE Press, 1989.

5. R. K. Hoffmann, *Handbook of Microwave Integrated Circuits.* Norwood, MA: Artech House, 1987.

6. I. Bahl and P. Bhartia, eds., *Microwave Solid State Circuit Design.* New York: Wiley, 1988.

7. L. B. Felsen and N. Marcuvitz, *Radiation and Scattering of Waves.* Englewood Cliffs, N.J.: Prentice Hall, 1973.

8. K. A. Michalski, "The mixed-potential electric field integral equation for objects in layered media," *Arch. Elek. Übertragung.*, vol. 39, pp. 317–322, Sept.-Oct. 1985.

9. K. A. Michalski and D. Zheng, "Electromagnetic scattering and radiation by surfaces of arbitrary shape in layered media, Part I: Theory," *IEEE Trans. Antennas Propagat.*, vol. 38, pp. 335–344, Mar. 1990.

10. N. K. Das and D. M. Pozar, "A generalized spectral-domain Green's function for multilayer dielectric substrates with application to multilayer transmission lines," *IEEE Trans. Microwave Theory Techn.*, vol. MTT-35, pp. 326–335, Mar. 1987.

11. L. Vegni, R. Cicchetti, and P. Capece, "Spectral dyadic Green's function formulation for planar integrated structures," *IEEE Trans. Antennas Propagat.*, vol. 36, pp. 1057–1065, Aug. 1988.

12. R. Kastner, E. Heyman, and A. Sabban, "Spectral domain iterative analysis of single– and double–layered microstrip antennas using the conjugate gradient algorithm," *IEEE Trans. Antennas Propagat.*, vol. 36, pp. 1204–1212, Sept. 1988.

13. Y. T. Lo, S. M. Wright, and M. Davidovitz, "Microstrip antennas," in *Handbook of Microwave and Optical Components* (K. Chang, ed.), vol. 1, pp. 764–888, New York: Wiley, 1989.

14. A. A. Oliner, "Leakage from higher modes on microstrip line with application to antennas," *Radio Sci.*, vol. 22, pp. 907–912, Nov. 1987.

15. A. W. Glisson and D. R. Wilton, "Simple and efficient numerical methods for problems of electromagnetic radiation and scattering from surfaces," *IEEE Trans. Antennas Propagat.*, vol. AP–28, pp. 593–603, Sept. 1980.

16. S. M. Rao, D. R. Wilton, and A. W. Glisson, "Electromagnetic scattering by surfaces of arbitrary shape," *IEEE Trans. Antennas Propagat.*, vol. AP–30, pp. 409–418, May 1982.

17. K. A. Michalski and D. Zheng, "Electromagnetic scattering and radiation by surfaces of arbitrary shape in layered media, Part II: Implementation and results for contiguous half-spaces," *IEEE Trans. Antennas Propagat.*, vol. 38, pp. 345–352, Mar. 1990.

18. K. A. Michalski and D. Zheng, "Rigorous analysis of open microstrip lines of arbitrary cross-section in bound and leaky regimes," *IEEE Trans. Microwave Theory Techn.*, vol. 37, pp. 2005–2010, Dec. 1989.

19. D. Zheng and K. A. Michalski, "Analysis of arbitrarily shaped coax–fed microstrip antennas—A hybrid mixed-potential integral equation approach," *Microwave & Opt. Techn. Lett.*, vol. 3, pp. 200–203, June 1990.

20. A. M. A. El-Sherbiny, "Hybrid mode analysis of microstrip lines on anisotropic substrates," *IEEE Trans. Microwave Theory Techn.*, vol. MTT–29, pp. 1261–1266, Dec. 1981.

21. C. C. Su, "A combined method for dielectric waveguides using the finite–element technique and the surface integral equation method," *IEEE Trans. Microwave Theory Techn.*, vol. MTT–34, pp. 1140–1146, Nov. 1986.

22. X. Yuan, D. R. Lynch, and J. W. Strohbehn, "Coupling of finite element and moment methods for electromagnetic scattering from inhomogeneous objects," *IEEE Trans. Antennas Propagat.*, vol. 38, pp. 386–393, Mar. 1990.

SEMI-DISCRETE FINITE ELEMENT METHOD

ANALYSIS OF MICROSTRIP STRUCTURES

Marat Davidovitz and Zhiqiang Wu

Department of Electrical Engineering
University of Minnesota
Minneapolis, MN

INTRODUCTION

Numerical modeling and characterization of passive components for Microwave and Millimeter-wave Integrated Circuit (MMIC) applications is an active area of research at the present time. The requirements of versatility, accuracy and computational efficiency have been met only partially by the existing numerical solutions. Therefore these issues continue to motivate the development of new numerical techniques for MMIC component characterization.

Among the most versatile demonstrated solutions are those based on the finite element analysis of the governing differential equations. The Finite Element Method (FEM) is applicable to a wide range of geometries and material parameter distributions. However, the generality of FEM solutions is obtained at a significant cost to computational efficiency. FEM discretization of three-dimensional problems results in large numbers of unknowns, and hence implementation of FEM solutions requires considerable memory and CPU resources.

In this paper a semi-discrete variant of the finite element method is described and applied to several classes of microstrip problems, including the calculation of capacitances for multi-strip transmission lines on anisotropic substrates, and, in three-dimensions, capacitance of arbitrary microstrip elements. In the Semi-Discrete Finite Element Method (SD-FEM), the solution domain is discretized only along two of the Cartesian coordinates, with the third coordinate variable acting as a parameter. Thus, the original partial differential equation is reduced to a set of coupled ordinary differential equations in the undiscretized coordinate, which can be solved analytically. Compared to the conventional FEM solutions, this approach results in significantly fewer unknowns and, therefore greater computational efficiency.

FORMULATION OF THE SOLUTION

In the initial phase of the investigation a set of microstrip circuit problems governed by the Poisson's equation was chosen in order to establish the viability of the SD-FEM, and study its numerical characteristics. A typical planar circuit element, formed by single conductive sheet embedded in a stratified dielectric medium, is shown in Figure 1. The conductive strip is assumed to be infinitesimally thin, but otherwise no restriction is placed on its shape. The lateral extent of the structure is limited by a cylindrical metallic enclosure, which may have an arbitrary cross-section. Metallic plates, parallel to the dielectric interfaces, are introduced to terminate the structure on the top and the

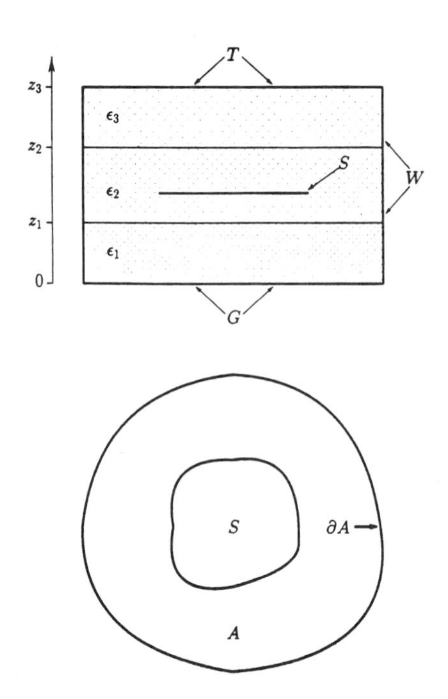

Figure 1. Shielded microstrip patch in a stratified dielectric medium.

bottom. It can be shown that although numerical considerations dictate that the model have finite lateral dimensions, vertically it may extend to infinity in either direction. Moreover, the analysis can be easily extended to circuits composed of any number of dielectric layers and parallel metallic strips.

In order to evaluate the capacitance of the circuit element shown in Fig. 1 it is necessary to solve a boundary value problem governed by the Poisson's equation

$$\nabla^2 \phi(x, y, z) = -\frac{1}{\epsilon_n} \rho(x, y, z) \tag{1}$$

In the preceding equation $\phi(x, y, z)$ denotes the potential, ϵ_n is the permittivity of the n-th layer, and $\rho(x, y, z)$ represents the charge density distribution, defined as $\rho(x, y, z) = \sigma(x, y)\delta(z - z')$ if $n = 2$, and $\rho(x, y, z) = 0$ if $n = 1, 3$. The boundary conditions for the potential can be stated as follows

$$\phi(x, y, z) = 0 \text{ if } (x, y, z) \in \text{W,T,G} \tag{2}$$

$$\phi(x, y, z) = \phi_0 \text{ if } (x, y, z) \in \text{S} \tag{3}$$

Additional transition conditions must be enforced at the dielectric interfaces ($z = z_k, k = 1, 2$). Solution of the posed problem yields $\sigma(x, y)$ - the charge density distribution on the strip. The capacitance can be then calculated using the standard definition.

The development of the SD-FEM solutions is based on a weak statement of the governing equation (1). An appropriate weak formulation of (1) requires $\phi(x, y, z)$ to satisfy

$$\int_A [\, -\nabla_t \phi \cdot \nabla_t \psi + \frac{\partial^2 \phi}{\partial z^2} \psi + \frac{1}{\epsilon_n} \rho \psi \,] \, dx dy + \oint_{\partial A} \frac{\partial \phi}{\partial n} \psi \, dc = 0 \tag{4}$$

for all values of z and all weighting functions $\psi(x, y)$. The latter are selected from a space of functions satisfying certain regularity conditions [1]. The symbol ∇_t denotes the transverse-to-z component of the "del" operator.

Equation (4) can be solved by a combination of numerical and analytical methods. In the first stage of analysis, the cross-section A of the structure is divided into a finite number of triangular sub-domains or elements, collectively known as the finite element mesh. Distribution of the nodes - vertices of the elements - is governed by several considerations, most important of which is the local behavior of the solution. A priori information about the solution behavior can be used to guide the process of discretization. A set of low-order polynomial basis functions, each spanning a small number of adjacent elements, is generated. This basis set is used in the application of the Galerkin method to the formulation in (4). The sought solution is expanded (interpolated) in terms of the constructed basis set $\{\phi_j(x,y)\}_{j=1}^{N}$ as follows

$$\phi(x,y,z) = \sum_j v_j(z)\phi_j(x,y) = \tilde{\phi}(x,y)\mathbf{v}(z) \tag{5}$$

where $v_j(z)$ is the value of the approximate solution at the j-th mesh node (x_j, y_j), and $\phi_j(x,y)$ denotes the j-th basis function. The boldface letters represent matrix quantities and tilde denotes transposition. As required in the Galerkin approach, representation (5) is substituted into the weak statement (4) and the weight functions $\psi = \phi_i$, $i = 1, N$ are used to test the resulting equation over the transverse domain A.

The outlined procedure yields the following system of N coupled, ordinary differential equations in the N unknowns $v_j(z)$

$$\mathbf{B}\frac{d^2\mathbf{v}(z)}{dz^2} - \mathbf{A}\mathbf{v}(z) = -\frac{1}{\epsilon_n}\mathbf{s}\delta(z-z') \tag{6}$$

where $\mathbf{v}, \mathbf{s}$ are $N \times 1$ vectors and $\mathbf{A}, \mathbf{B}$ are sparse, symmetric, positive-definite $N \times N$ matrices, whose elements are defined by the following equations

$$a_{ij} = \int_A \nabla_t\phi_i \cdot \nabla_t\phi_j \, dxdy\,, \quad b_{ij} = \int_A \phi_i\phi_j \, dxdy\,, \quad s_i = \int_A \sigma\phi_i \, dxdy \tag{7}$$

Note that the boundary integral term in equation (4) does not contribute to (7). This has been achieved by requiring that the weight functions have zero values on nodes located on the boundary ∂A of the cross-section. Thus, the boundary condition (2) on the lateral wall W (Fig. 1) is satisfied.

The differential equation set (6) can be solved analytically after the unknowns are decoupled by a linear transformation of the solution $\mathbf{v}$ to the principal axis, as proposed in [2]. A very expedient approach for solving the decoupled differential equations uses an analogy between the potential and its derivative with respect to z on the one hand, and voltage and current on a distributed transmission line on the other.

The boundary conditions at $z = 0, z_3$, as well as the transition conditions at dielectric interfaces, are applied in the process of solving for $\mathbf{v}(z)$. The remaining boundary condition on the strip S is enforced last. This is accomplished by matching the elements of $\mathbf{v}(z)$ corresponding to nodes located on the strip to the prescribed potential, i.e. $v_j(z') = \phi_0$ if $(x_j, y_j) \in S$. The result of this procedure is a matrix equation for $\mathbf{s}$ - the elements of which are moments of the charge density distribution $\sigma(x,y)$ with the weighting functions. After solving for $\mathbf{s}$, the charge Q on the strip is determined simply by summing the elements of $\mathbf{s}$. A more detailed derivation of the outlined solution can be found in [3].

NUMERICAL RESULTS

Multi-Strip Transmission Lines

The formulation described in the preceding section can be easily modified to apply to two-dimensional problems involving multi-strip transmission lines on isotropic and anisotropic substrates [4],[5]. The geometry of such problems is illustrated in Figure 2. Among the problems which were analyzed using the two-dimensional version of SD-FEM

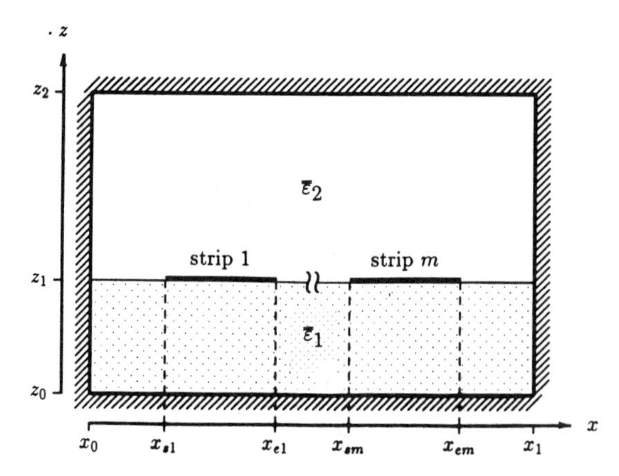

Figure 2. Multi-strip transmission line with anisotropic substrate.

is that of edge-coupled, shielded microstrip lines on Sapphire substrate. The dimensions of the structure for which the capacitance matrix was calculated are defined as follows (see Fig. 2)

$$m = 2 \quad x_{s1} = -W - \frac{S}{2} \quad x_{e1} = -\frac{S}{2} \quad x_{s2} = \frac{S}{2} \quad x_{e2} = W + \frac{S}{2}$$

$$x_0 = -11.5W \quad x_1 = 11.5W$$

$$z_0 = 0 \quad z_1 = H \quad z_2 = H + A$$

and the relative permittivity matrices are given by

$$\bar{\bar{\varepsilon}}_2 = \begin{pmatrix} 1 & 0 \\ 0 & 1 \end{pmatrix}$$

$$\bar{\bar{\varepsilon}}_1 = \begin{pmatrix} 9.4 & 0 \\ 0 & 11.6 \end{pmatrix}$$

Comparison of the capacitance values computed using SD-FEM with data in reference [6] is presented in Table I. C_e^+, C_e^- are the upper and lower bounds of even-mode capacitances reported in Table I of reference [6] and C_e, C_o are the even- and odd-mode capacitances computed using the method of this paper.

Table I. Comparison of even- and odd-mode capacitances for edge-coupled strip lines.

A/H	S/H	W/H	SAPPHIRE SUBSTRATE $\epsilon_{11} = 9.40, \epsilon_{22} = 11.60$			
			$\frac{C_e^+}{\epsilon_0}$	$\frac{C_e^-}{\epsilon_0}$	$\frac{C_e}{\epsilon_0}$	$\frac{C_o}{\epsilon_0}$
0.5	0.5	1.0	21.54	21.02	21.34	27.44
		2.0	35.69	34.37	34.98	41.20
	1.0	1.0	23.05	22.32	22.69	25.01
		2.0	37.17	35.65	36.52	38.94
1.0	0.5	1.0	20.34	19.85	20.14	26.62
		2.0	33.37	32.22	32.78	39.39
	1.0	1.0	21.83	21.16	21.49	24.08
		2.0	34.86	33.50	34.31	36.99
2.0	0.5	1.0	19.82	19.34	19.62	26.40
		2.0	32.29	31.23	31.77	38.75
	1.0	1.0	21.28	20.63	20.96	23.79
		2.0	33.77	32.51	33.28	36.27

The even-mode capacitance data were found to be consistent with [6]. However, the odd-mode capacitance results reported in [6] appear to be incorrect, and therefore only the results obtained with SD-FEM are presented in Table I. In all cases the number of elements in the mesh implemented along the x-axis was between 150 and 180.

Capacitance of Microstrip Patches

Although the applicability of SD-FEM to microstrip problems was illustrated in the preceding section, its versatility becomes fully apparent only in the solution of three-dimensional problems. Analysis of the Poisson's equation for elements with complex metallization geometries is used to illustrate the technique. Validation of the method was carried out by comparing the results of capacitance calculations using the SD-FEM formulation with the integral equation solutions for two canonical geometries, namely the circular and rectangular patches. It was found that the capacitance data resulting from the two methods of solution were in very close agreement - the differences in all cases being less than 1%. The solution was subsequently used to calculate the capacitance of more complex geometries. The results for the case of an annular strip bounded by concentric circles with a 2:1 radius ratio are presented here. A magnified central portion of the finite element mesh for this geometry, showing the discretization in the proximity to the patch, is plotted in Figure 3.

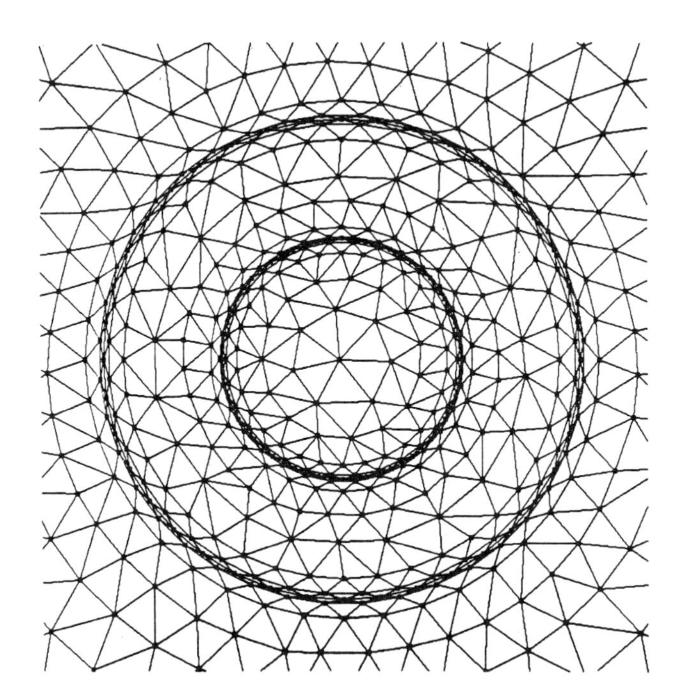

Figure 3. Magnified central portion of the mesh for the annular patch.

The geometry and the material parameters of the analyzed structure are given in the legend to Figure 4, in which the normalized capacitance values are plotted as functions of the normalized substrate thickness for several values of substrate permittivity. The expression for C_N - the factor used to normalize the calculated data - is given by

$$C_N = \frac{\epsilon_{r1} + \epsilon_{r2}}{2} C_\infty + \epsilon_{r1} \frac{A}{t_1}$$

where $C_\infty = 37.2a$ and $A = 0.433b^2 - \pi a^2$, and the remaining quantities are as defined in Figure 4. The number of elements used in the calculation of the results was 870.

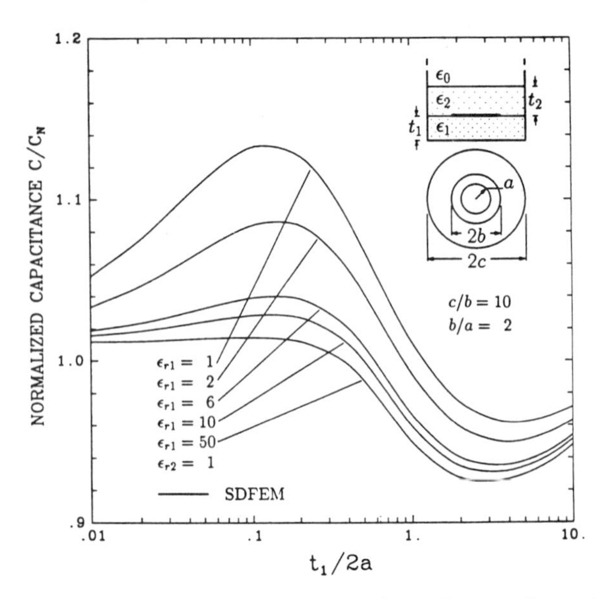

Figure 4. Capacitance of a circular annulus with 2:1 radius ratio.

The effect of the lateral walls on the patch capacitance is minimal, provided that the ratio $t_1/2a$ is smaller than a certain number, which depends on the patch-to-wall distance and the permittivity of the dielectric. It has been found that for $t_1/2a \leq 1.$ - a number which exceeds most encountered in microstrip circuit design - the capacitance is insensitive to the enclosure's presence if $c/b \geq 10$. Moreover, it should be noted that this conclusion holds for all values of substrate permittivity. As indicated in the legend to Fig. 4, the capacitance calculations were performed with $c/b = 10$.

CONCLUSIONS

The application of the Semi-Discrete Finite Element Method to microstrip circuit characterization was demonstrated. For problems involving planar structures, this method, while retaining the flexibility of the conventional FEM solutions, is much more economical and simpler to implement. Efforts are currently under way to develop a full-wave modeling package for microstrip circuit elements based on this approach. Full exploitation of the sparsity of the matrices appearing in the formulation will lead to a highly efficient and flexible modeling tool for circuit designers. A pre-processing interface is also being developed to assist in the initial stages of geometry definition and finite element mesh generation.

REFERENCES

1. E.B. Becker, G.F. Carey, J.T. Oden,*Finite Elements: An Introduction*, New Jersey: Prentice-Hall, 1981.

2. R. Pregla and W. Pascher, "The Method of Lines", Ch. 6, *Numerical Techniques for Microwave and Millimeterwave Passive Structures*, T. Itoh,ed., New York: Wiley Interscience, 1989.

3. M. Davidovitz and Z. Wu, "Semi-Discrete Finite Element Method Analysis of Arbitrary Microstrip Elements - Static Solution", submitted to IEEE Trans. on Microwave Theory Tech., Aug. 1990.

4. Z. Wu and M. Davidovitz, "Capacitance of Microstrip Lines on Anisotropic Substrate using the Semi-Discrete FEM Analysis", submitted for review in International Journal on MIMICAE, Sept. 1990.

5. M. Davidovitz, "Calculations of Multi-Conductor Microstrip Line Capacitances Using the Semi-Discrete Finite-Element Method", to be published in IEEE Microwave and Guided-Wave Letters, Jan. 1991.

6. M. Horno, "Quasi-static Characteristics of Covered Coupled Microstrips on Anisotropic Substrates: Spectral and Variational Analysis", IEEE Trans. Microwave Theor. Techniques , MTT-30, pp. 1888-1892, Nov. 1982.

SPECTRAL-DOMAIN MODELING OF RADIATION AND GUIDED WAVE LEAKAGE IN A PRINTED TRANSMISSION LINE

Nirod K. Das

Weber Research Institute
Polytechnic University
Route 110, Farmingdale, NY 11735

INTRODUCTION

The dominant mode of an infinite transmission line printed on a layered substrate and/or ground plane configuration can be lossy due to coupling of power to radiation or to source free guided modes of the layered structure [1-3]. Such 'non-conventional' lossy modes of a printed transmission line are undesirable for an integrated circuit application. However, these modes can sometimes be inevitably excited for certain specific layered configurations at all frequencies, or only at higher frequencies for certain other geometries. Similar lossy modes can also occur for higher order modes of printed transmission lines which are otherwise dominantly non-leaky (bound) [4]. Unlike the dominant mode of a conventional transmission line (a regular microstrip line, for example) where the field is bound to the guiding structure, for these leaky modes the field profile of the infinite transmission line has been found to behave in a non-standard way by growing exponentially in transverse directions [3]. For such growing behavior, the Fourier transforms for various field components in the transverse dimension do not exist for real values of the spectral argument. Hence, the standard spectral-domain formulations [5-7], where existence of Fourier transform on the real axis is assumed, can not be directly used to characterize this class of leaky modes of printed transmission lines.

A modified spectral-domain formulation for such leaky transmission line modes, as applied to specific example geometries, has been recently presented in [3], with suitable qualitative explanation of different analytical steps involved. Generally, the analysis is based on an analytic extension of the Fourier domain method to complex spectral plane. This involves suitably including the 'leaky' poles and the corresponding residues on the complex spectral plane, and accounting for the proper branch cuts. The present paper describes the detailed analytical considerations for such a complex Fourier domain analysis in a more general environment, discusses the practical implications of such modes with growing field components (and hence the meaning of impedance for such leaky transmission lines), and proposes possible ways to avoid/suppress excitation of such modes for certain geometries for multilayer integrated circuit applications.

ANALYSIS

In performing a spectral-domain moment method solution for the complex propagation constant, k_e, of a printed transmission line geometry, one needs to expand the transverse (y) variation of the unknown electric currents or the equivalent magnetic currents (electric field across a slot on a ground plane) of the system using a known basis set with unknown amplitudes. An $\exp(-jk_e x)$ variation along the direction of propagation (x) is assumed. By enforcing proper boundary condition via a suitable testing procedure one obtains a set of linear equations:

$$\sum_{i}^{N} I_i Z_{ij}(k_e) = 0; \quad j=1,N \tag{1}$$

As a result, it is required to solve for the root, k_e, of a determinant:

$$\mathrm{Det}[Z_{ij}(k_e)] = 0 \tag{2}$$

where Z_{ij} is the reaction of suitable field component(s) due to the ith expansion function on the jth testing function. In spectral domain the Z_{ij}'s are obtained by evaluating Fourier integrals involving the transforms of the expansion and testing functions chosen, and the spectral domain Green's functions for the (layered) substrate geometry [8] that account for the field components for proper boundary condition testing. The above Fourier integrals are usually defined and evaluated along the real spectral axis, k_y (corresponding to the transverse dimension, y), [5-7]. But as discussed in [3] the integration path needs to be deformed off the real k_y axis to account for the radiation or source free guided mode leakage. Fig. 1 shows the contour of integration, C, for a general case of a multilayer transmission line with multiple poles and branch cuts on the compex k_y plane. The following points should be carefully noted in determining the correct contour of integration:

(1) The integration contour, C, should enclose (between C and the real axis) a particular pole if and only if a) the pole is in the first or the third quadrant, and b) the pole corresponds to a guided substrate mode propagation constant, β, such that $\mathrm{Re}(\beta) > \mathrm{Re}(k_e)$. Referring to the Fig. 1 as an example case of a transmission line coupling power to two characteristic substrate modes, the pole pair $\pm P_4$ are not enclosed by C because it is in the second and the fourth quadrants. Among the poles in the first and the third quadrants, $\pm P_3$ is a possible pole pair that does not satisfy the above leakage condition, (b), and hence is not enclosed by C. On the other hand, the pole pairs $\pm P_1$ and $\pm P_2$ correspond to characteristic source free modes of the layered substrate geometry that satisfy the above leakage condition and need to be enclosed by C to properly account for the associated leakage.

(2) For a general multilayer substrate and/or ground plane configuration with different dielectric constants and loss tangents, these different poles in Fig. 1 represent surface waves, parallel plate waves, or various forms of trapped guided modes.

(3) A branch cut corresponds to possible radiation into a semi-infinite medium (if any) on top or bottom of the layered structure. Thus a layered structure can have maximum two branch cuts. For example, a regular microstrip line geometry has only one branch cut corresponding

to the semi-infinite air medium above, a coplanar stripline at the interface between two semi-infinite mediums (air, and another semi-infinite dielectric medium, ε_r, for example) [2,3] has two branch cuts, and a stripline with its top and bottom covered by conducting ground planes do not have any branch cut. These branch cuts can be chosen in many different ways. In any case, from analyticity considerations the integration contour, C, can not cross the branch cuts (as always). Depending on which branch cut one chooses across the real axis, it determines which branch points the deformed integration contour, C, need to go around. A branch cut should be selected across the real axis (thus forcing deformation of C) if and only if a) the corresponding branch points are in the first and the third quadrants, and b) the propagation constant of the corresponding semi-infinite medium, β, is such that $Re(\beta) > Re(k_e)$.

Thus, any branch cut pair that has the branch points in the second and the fourth quadrants should not be chosen across the real axis. In Fig. 1 there are two branch cut pairs (B_1, B_1', and B_2, B_2') with the branch point pairs ($\pm b_1$ and $\pm b_2$) in the third and the first quadrants. $\pm b_2$ represent a possible branch point pair that do not satisfy the above radiation leakage condition, b), and so the corresponding branch cuts are selected away from the real axis. On the other hand $\pm b_1$ correspond to the semi-infinite medium of the structure that satisfy the above radiation leakage condition, b), and so the corresponding branch cuts are chosen to intersect the real axis forcing C to be deformed around $\pm b_1$. For example, in the case of a coplanar stripline on the interface of air and a semi-infinite dielectric medium, $\pm b_1$ correspond to the propagation constant of the dielectric medium into which the coplanar strip line radiates, and $\pm b_2$ correspond to the propagation constant of the air medium.

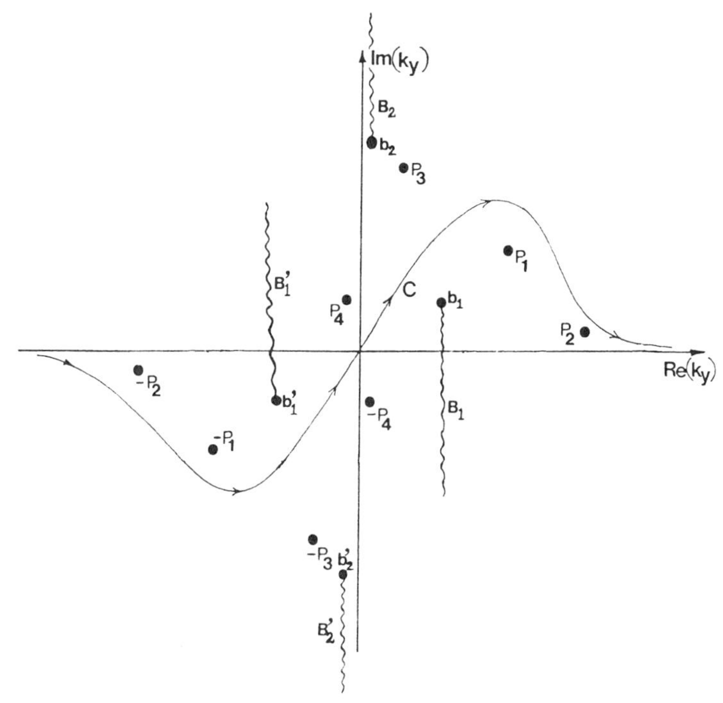

Fig. 1. Deformation of the integration contour on the complex spectral plane to account for radiation and/or characteristic mode leakage.

4) It should be noted that for the above considerations on the proper selection of the branch cuts it is assumed that only the top branch plane is used.

5) Due to the above choice of branch cuts, it can be seen that a part of the real k_y axis does not satisfy the radiation condition (i.e. all field components need to be zero at large distances) in the semi-infinite medium corresponding to the branch point pair $\pm b_1$. Physically, this is because in this medium the field components do grow exponentially in the outward direction [3]. In contrast, the entire real k_y axis satisfy the radiation condition for the semi-infinite medium corresponding to the branch point pair $\pm b_2$. This implies an outward decaying field in the corresponding semi-infinite medium.

DISCUSSION

Such spectral-domain analyses always assume an infinite length transmission line along the direction of propagation. Also, the substrates are assumed to be of infinite lateral extent. Thus, the idealistic mode solution the spectral-domain analysis provides always correspond to a non-practical structure. In practice, the transmission lines used in an integrated circuit application are always of finite length and are printed on substrates of finite lateral extent. However, if the transmission line is dominantly non-leaky type (for example, a regular microstrip line) the fields are tightly bound to the guiding strip of slot. In such case the field of the transmission line on an infinite lateral extent substrate very closely approximates the field of the practical transmission line on a substrate of finite lateral extent. So, the results of the analysis with substrates of infinite lateral extent, or with the same token even with conducting side walls sufficiently far away for the guiding strip or slot, are directly useful for practical circuits.

But, this is not the case for the ideal leaky transmission line modes, clearly because in this case the field is no longer tightly bound to the guiding structure. In fact, now the field has an exponential growing behavior transversally, which can strictly never occur in practice. Thus the performance of a practical transmission line which is dominantly leaky (ideally) can be significantly different from that described by the ideal analysis discussed. It would be strongly dependent on the surroundings even far away from the guiding strip or slot. The excited characteristic wave as described by the ideal leaky analysis would be scattered off the surrounding structures and also would partly couple back to the source. To obtain a realistic picture of a practical dominantly leaky transmission line geometry one needs to analyze it as a radiating problem with finite length source distribution. The results of the ideal leaky transmission line analysis, however, would provide a qualitative picture and a good estimate of the degree of coupling to radiation or characteristic modes. A larger attenuation constant would imply a stronger coupling and a stronger effect of the surroundings on the transmission line performance. The phase progression along the transmission line, however, would be very closely described by the real part of the propagation constant obtained from the ideal leaky mode analysis.

The meaning of impedance for conventional non-leaky transmission lines needs to be reviewed for leaky transmission lines. For finite total strip current or slot voltage of the guiding strip or slot, the total power flow across the transverse cross section of a leaky mode is infinite. This is due to its growing fields in transverse direction as discussed. Thus, using power-current definition of equivalent impedance, this results in an infinite impedance of a leaky transmission line. In practice, however, the field does not grow indefinitely due to finite length and finite substrate width of practical transmission lines. Again, because of the above reasoning, the power associated with such a leaky transmission line need to be realistically characterized by solving it as a finite radiating problem (an antenna) that takes into account the effect of the surroundings. In the region of transition of a non-leaky transmission line mode (where the impedance and power are now finite, and the field profile is bound) into a leaky mode it is, however, informative to study the gradual variation of the non-leaky mode impedance (or admittance for a slotline) to an infinite value at the onset of leakage. This trend, as would be discussed, will provide practical information on the domain of physical parameters over which the tightly bound field profile of the non-leaky mode starts to spread considerably (hence, becomes more loosely bound) to finally an unbound mode at the onset of leakage.

SUPPRESSION OF DOMINANT LEAKY MODES

As discussed in [3], a stripline with two dielectric substrates (with different thickness and dielectric constants) above and below the center conductor can be dominantly leaky at all frequencies for certain relative values of the dielectric constants and thicknesses of the two substrates. It is also assertained in [3] that this dominant mode leakage can be avoided by choosing the thinner of the two substrates to be of sufficiently larger dielectric constant than that of the thicker substrate. Strictly, the dielectric constant of the thinner substrate just needs to be greater than the other substrate to avoid the leakage, but is preferably required to be sufficiently larger than the other in order to operate safely away form the onset of leakage.

Now, in this paper we present similar study on mode suppression for a conductor backed slotline [1,3] (which is found to be dominantly leaky) by loading it with a moderately high dielectric constant substrate on top. Fig.2a shows the variation of effective dielectric constant $[=(k_e/k_o)^2]$ with the cover substrate thickness, and the corresponding variation of the characteristic impedance is shown in Fig.2b. As it is clear from Fig.2a, the effective dielectric constants are larger than the dielectric constant of the parallel plate structure, which insures no leakage to the parallel plate mode [1]. As Fig.2b shows, the characteristic impedance drops sharply with decrease in the cover thickness (also, with decrease in frequency) to small values. As discussed before, this low value of characteristic impedance is indicative of onset of leakage, and transition from the bound to a unbound mode. The corresponding field spreading of a loaded conductor backed slotline as compared to that of a regular bound slotline mode (no conductor backing) is shown in Fig.3. Fig.3 clearly shows considerable field spreading of the conductor backed slotline as frequency decreases. Similar effect is also seen as the cover thickness or dielectric constant decreases, or also as the parallel plate thickness decreases. It is interesting to note that for the conductor backed slotline the coupling to parallel plate mode becomes more severe as the frequency decreases, which is unlike other layered geometries where the substrate mode coupling increases with increase in frequency.

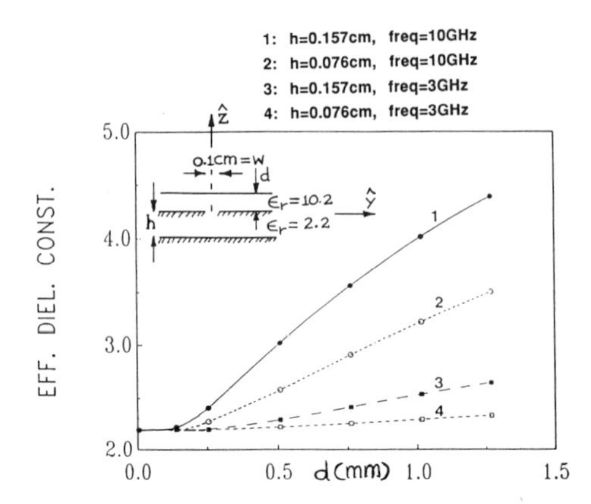

Fig.2a. Effective dielectric constants of conductor backed slotlines with a dielectric loading to avoid leakage to the parallel plate mode.

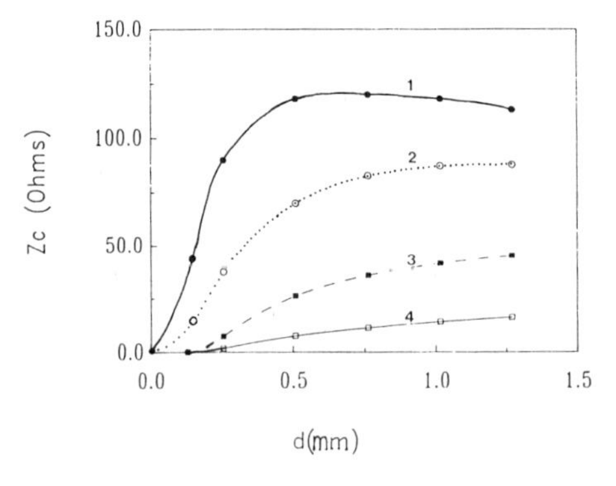

Fig.2b. Characteristic impedances, Zc, for the geometries of Fig.2a.

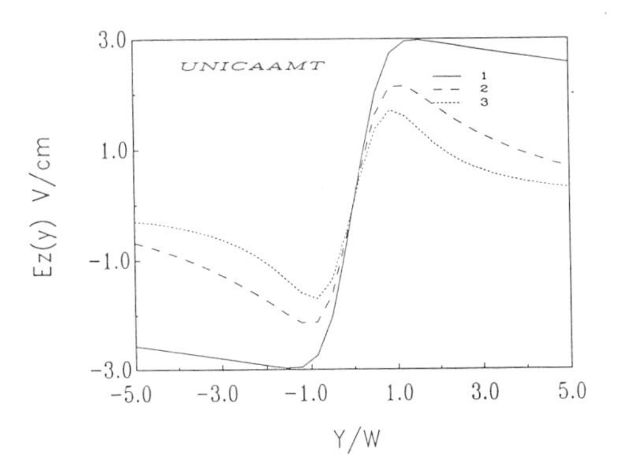

Fig.3. Comparison of field profile spreading of loaded conductor backed slotlines of Fig.2 at different frequencies (curves 1 and 2: cover: ε_r=10.2, 0.127cm, parallel plate: ε_r=2.2, h=.157cm) with that of a regular slotline (curve 3: ε_r=2.55, h=.157cm). Frequency: curves 1) 3GHz, 2)10GHz, & 3) 3GHz.

Using such studies on field spreading and/or impedance variation in the transition of onset of leakage, a suitable set of parameters for a conductor backed slotline can be chosen in order to operate safely away from the leakage effects. For example, for the parameters of curve 1 of Fig.2b, a value of the cover substrate thickness greater than about 0.5mm would be desirable. Similar studies on impedance and field spreading have also been successfully performed for the two layer stripline leakage [3].

CONCLUDING REMARKS

As discussed, due to non-local behavior of the leaky transmission line modes, practical excitation of dominant leakage modes of a printed transmission line is strongly dependent on the surrounding structure, even far away from the guiding strip or slot. How effectively these modes are excited by a finite transmission line also seem to be strongly determined by the wave launching methods used. Such studies on practical finite geometries of leaky transmission lines as compared to the results predicted by the ideal leaky transmission line analysis, and investigations on various possible methods of suppression or avoidance of such leakage effects, need to be performed to help enable proper design of multilayer integrated circuits .

REFERENCES

[1] H. Sigesawa, M. Tsuji and A. A. Oliner, "Conductor Backed Slotline and Coplanar Waveguide: Dangers and Full-Wave Analysis," IEEE MTT-S Digest, pp. 199-202, 1988.

[2] D. B. Rutledge, D. P. Neikirk and D. P. Kasilingam, "Planar Integrated Circuit Antennas," Chapter 1, Vol. 10, Millimeter Components and Techniques, Part II of Infrared and Millimeter Waves, K. J. Button, Editor.

[3] N. K. Das and D. M. Pozar, "Full-Wave Spectral-Domain Computation of Material, Radiation and Guided Wave Losses in Infinite Multilayred Printed Transmission Lines," IEEE Transactions on Microwave Theory and Techniques, January 1991, to appear.

[4] A. A. Oliner and K. S. Lee, "The Nature of the Leakage from Higher Modes on Microstrip Line," IEEE Microwave Theory and Techniques Society Symposium Digest, pp. 57-66, 1986.

[5] E. J. Denlinger, "A Frequency Dependent Solution for Microstrip Transmission Lines," IEEE Transactions on Microwave Theory and Techniques, Vol. MTT-19, pp. 30-39, January 1971.

[6] T. Itoh and R. Mittra, "Spectral Domain Approach for Calculatingthe Dispersion Characteristics of Microstrip Lines," IEEE Transactions on Microwave Theory and Techniques, Vol. MTT-21, pp. 496-499, July 1973.

[7] J. B. Davies and Mirshekar-Syahkal, "Spectral Domain Solution of Arbitrary Transmission Line With Multilayer Substrate," IEEE Transactions on Microwave Theory and Techniques, Vol. MTT-25, pp. 143-146, February 1977.

[8] N. K. Das and D. M. Pozar, "Generalized Spectral-Domain Green's Function for Multilayer Dielectric Substrates with Applications to Multilayer Transmission Lines," IEEE Transactions on Microwave Theory and Techniques, Vol. MTT-35, pp. 326-335, March 1987.

ON THE MODELING OF TRAVELING WAVE CURRENTS FOR MICROSTRIP ANALYSIS

Jeffrey S. Herd

Electromagnetic Sciences Directorate
Rome Air Development Center
Hanscom AFB, MA

ABSTRACT

This paper describes a novel technique for modeling microstrip feedlines in an infinite array of proximity coupled microstrip antennas. Details of the analysis are presented along with experimental results from an infinite array waveguide simulator.

INTRODUCTION

A traditional infinite array analysis requires that each element be confined to a unit cell which is typically one-half of a wavelength on each side. This restriction makes it difficult to model infinite arrays with microstrip feedlines, which in practice far away from the element to a power divider, feed-through, etc. It has been found that a microstrip feedline which directly contacts a microstrip patch can be effectively replaced by a vertical current ribbon at the patch edge[1,2]. Unfortunately, no such simple approximation can be used to analyze the element in figure 1 which is proximity coupled to a microstrip line.

This paper describes a technique for theoretically extending microstrip feedlines outside of the infinite array unit cell, thereby making it possible to analyze an infinite array of proximity fed elements.

THEORY

An unknown electric current density $\bar{J}$ can be determined approximately by applying a moment method procedure to the electric field integral equation. The function $\bar{J}$ is expanded in a set of linearly independent basis functions $\bar{J}_n$ with

Directions in Electromagnetic Wave Modeling
Edited by H.L. Bertoni and L.B. Felsen, Plenum Press, New York, 1991

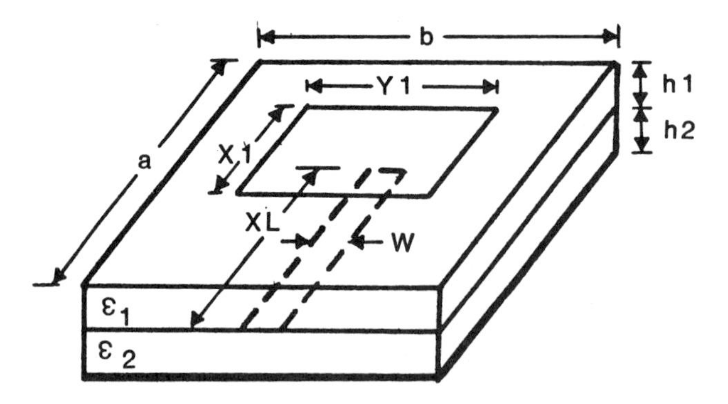

Fig. 1. Proximity coupled element geometry.

unknown coefficients, I_n. The expansion modes are chosen to closely resemble the anticipated current distributions. Figure 2 shows the three types of modes which were used in this analysis. Entire domain modes were used on the microstrip patches because of their correspondence with the cavity-like behavior of the patches. The microstrip feedline currents were modeled by incident and reflected traveling wave modes, along with overlapped piecewise sinusoidal (PWS) modes near the open end of the line to model the non-propagating currents excited by the discontinuity. As can be seen in figure 2, the traveling wave modes terminate with a quarter cycle of a sinusoid beginning at $X=X_B$ and ending at $X=X_L$. This feature replicates the expected current distribution at an open ended microstrip line, and reduces the number of PWS modes needed to capture the current behavior there.

In order to effectively replicate the fields of the incident and reflected traveling wave modes, it was necessary to theoretically extent the modes outside of the unit cell[3]. This was accomplished by recognizing that the fields of the traveling wave modes radiate very little, and therefore do not excite any substantial currents on patches or feedlines outside of the unit cell it feeds. Mathematically, this means that the reactions between the traveling wave modes and the other modes can be computed as for a single element fed by a long microstrip line. All other modes couple strongly between unit cells, and the computation of their self and mutual reactions must therefore include the infinite array contribution.

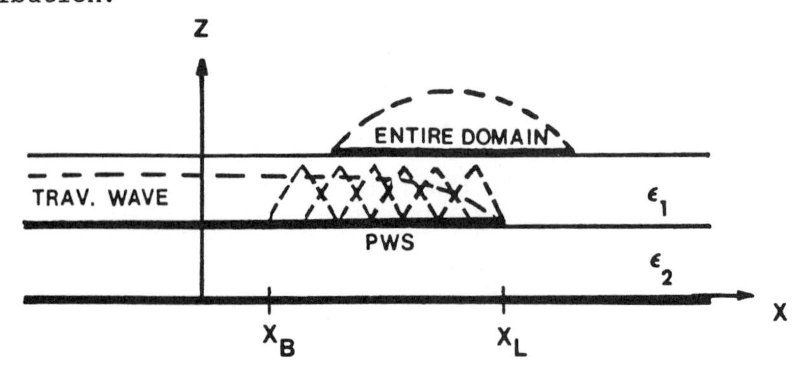

Fig. 2. Electric current expansion modes.

The resulting matrix equation has the following form;

$$\begin{bmatrix} Z_{TR} & Z_{T2} & \cdots & Z_{TN} \\ Z_{2R} & Z_{22} & \cdots & Z_{2N} \\ \vdots & \vdots & & \vdots \\ Z_{NR} & Z_{N2} & \cdots & Z_{NN} \end{bmatrix}^{-1} \begin{bmatrix} V_1 \\ V_2 \\ \vdots \\ V_N \end{bmatrix} = \begin{bmatrix} R \\ I_2 \\ \vdots \\ I_N \end{bmatrix} \tag{1}$$

where

$$V_m = \frac{-1}{4\pi^2} \int_{-\infty}^{\infty}\int_{-\infty}^{\infty} \overline{F}_m(-k_x,-k_y)\cdot\overline{\overline{Q}}(k_x,k_y)\cdot\overline{F}^{TW}_{inc}(k_x,k_y)\; dk_x dk_y \tag{2}$$

$$Z_{mR} = \frac{1}{4\pi^2} \int_{-\infty}^{\infty}\int_{-\infty}^{\infty} \overline{F}_m(-k_x,-k_y)\cdot\overline{\overline{Q}}(k_x,k_y)\cdot\overline{F}^{TW}_{ref}(k_x,k_y)\; dk_x dk_y \tag{3}$$

$$\underset{(n\neq R)}{Z_{mn}} = \frac{1}{ab} \sum_{i=-\infty}^{\infty} \sum_{j=-\infty}^{\infty} \overline{F}_m(k_x^i,k_y^j)\cdot\overline{\overline{Q}}(k_x^i,k_y^j)\cdot\overline{F}_n(-k_x^i,-k_y^j) \tag{4}$$

$$k_x^i = \frac{2\pi i}{a} + k_o u \tag{5}$$

$$k_y^j = \frac{2\pi j}{b} + k_o v \tag{6}$$

$$u = \sin\theta_o \, \cos\phi_o \tag{7}$$

$$v = \sin\theta_o \, \sin\phi_o \tag{8}$$

where R is the coefficient of the reflected traveling wave mode, k_o is the free space wavenumber, θ_o and ϕ_o are the beam pointing angles, the $\overline{F}$'s are the Fourier transforms of the current expansion modes, a and b are the unit cell dimensions in x and y, and $\overline{\overline{Q}}$ is the Fourier transformed dyadic Green's function for a two layer grounded dielectric slab[3].

Note that the voltage vector and the first column of the impedance matrix are computed from continuous integrals, while the rest consist of doubly infinite Floquet series summations. This is due to the fact that an infinite series of reactions reduces to a single reaction integral when there is negligible cell-to-cell coupling, as is the case for the traveling wave modes. This is the essential condition which makes it possible to integrate the traveling wave currents over semi-infinite lines, thereby replicating a realistic traveling wave field.

The numerical integrations in equations (2) and (3) were carried out in rectangular coordinates. In order to avoid the singularities of the spectral Green's functions at the surface wave and traveling wave poles on the real k_x, k_y plane, a path deformation technique was used[4]. By introducing an imaginary component of k_x and k_y, the integration path can be deformed sufficiently far away from the real plane to avoid numerical difficulties, as shown in figure 3. The modified integration path will give the same result as the original path as long as any poles enclosed between them are accounted for. An inspection of the incident traveling wave function shows that its Fourier transform does not exist unless k_e has a small imaginary component which is positive. Since the incident traveling wave current is the impressed current source in the analysis, the value of k_e can be chosen accordingly. As a consequence of this choice, the incident traveling wave pole moves into the third quadrant of the complex k_x plane, and is therefore enclosed between the original path and the deformed path for the k_x integration in equation (2). A value of $-2\pi j$ times the residue must be added to the deformed contour integration to account for the captured pole. Note that the reflected traveling wave pole moves down into the fourth quadrant when a small loss is introduced, and, like the surface wave poles, is not enclosed between the two paths. Values of A=-20.0, B=-1.5, C=1.5, and D=20.0 were used, and branch cuts were chosen such that the values of k_z represent outward traveling and decaying waves.

After solving equation (1), the reflection coefficient is known as a function of scan angle, and the scanning impedance can be calculated from;

$$Z_{in}(\theta,\phi) = Z_0 \frac{1 + R(\theta,\phi)}{1 - R(\theta,\phi)} \tag{9}$$

Figure 4 shows an example of the scanning reflection coefficient for an infinite array of proximity coupled elements. There is a total reflection at $\theta=57.9^\circ$ in the E-plane, and no blind spots in the H-plane.

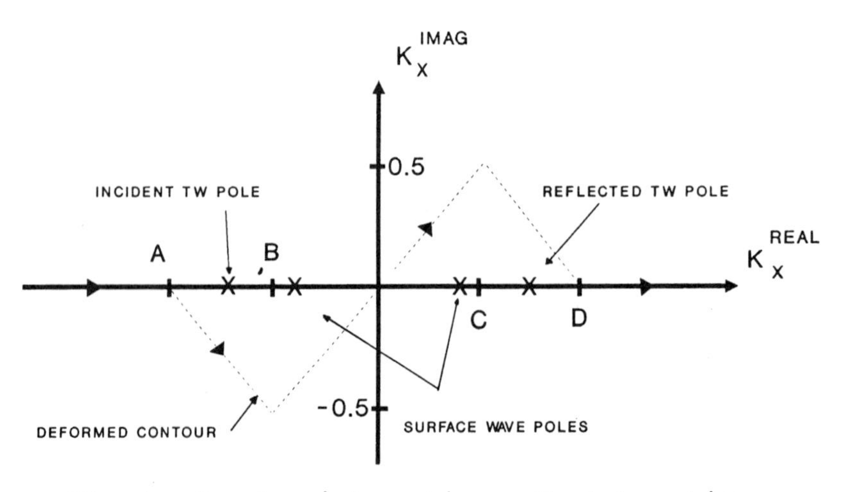

Fig. 3. Complex integration path deformation.

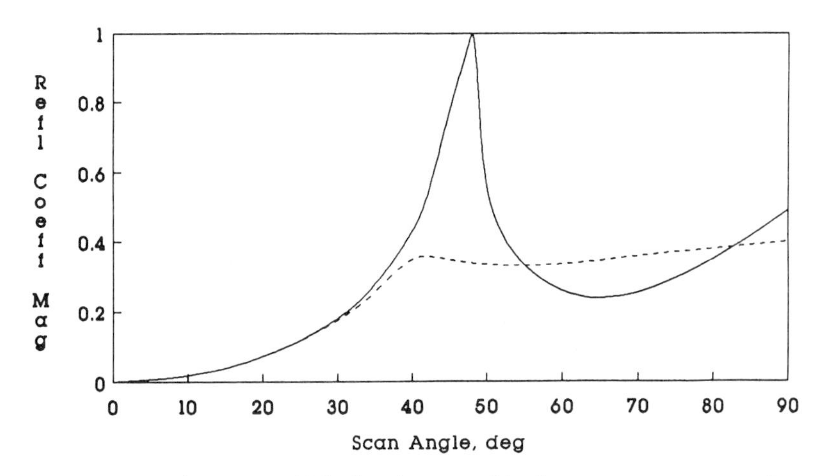

Fig. 4. Reflection coefficient magnitude vs. scan angle for element B; $h_1=h_2=0.159$cm, $X_1=Y_1=1.70$cm, $X_L=1.70$cm, $\varepsilon_1=\varepsilon_2=2.33$, a=3.35cm, b=3.60cm, W=0.40cm. E-plane= ——, H-plane= ---.

EXPERIMENT

To validate the theory, a rectangular waveguide simulator was constructed using the S-band waveguide shown in Figure 5. The waveguide is terminated in a matched load to prevent reflections back towards the radiating element. A basic requirement of the waveguide simulator is that the element must have symmetry about its center on the x and y axes. As can be seen in Figure 1, the element has a microstrip feed line on one side which violates the x-axis symmetry. However, this feature radiates very little in comparison with the microstrip patch, and therefore does not noticeably degrade the simulation.

Fig. 5. Waveguide simulator.

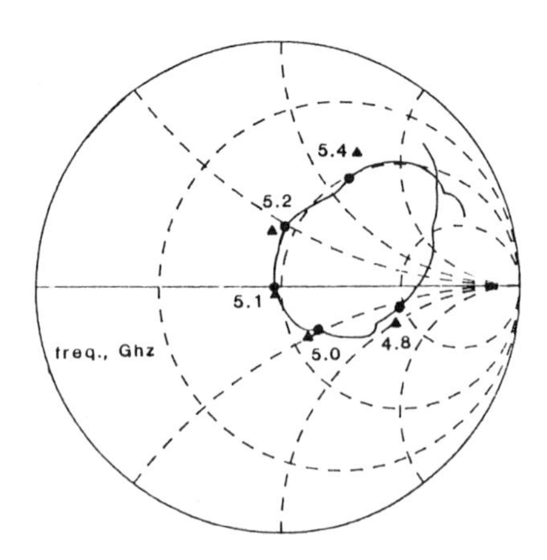

Fig. 6. Element in waveguide simulator; h_2=h_2=0.159cm, X_1=Y_1=1.80cm, X_2=Y_2=1.95cm, ε_1=ε_2=2.33, a=3.35cm, b=3.60cm, ● = measured, ▲ = predicted.

All simulator measurements were made on a calibrated Hewlett Packard 8510 network analyzer. The microstrip line was fed by a coax-to-microstrip transition through the waveguide wall, with the phase reference at the inside waveguide wall. Figure 6 shows the simulator measurements and computed data for infinite array element. Good agreement was obtained between theory and experiment.

CONCLUSION

A full wave analysis was applied to an infinite array of proximity coupled microstrip elements. The spectral dyadic Green's function was used so that all mutual coupling and surface wave effects were included. A method of moments was used to solve the electric field integral equation , and three types of expansion modes were used to efficiently model the current distribution on the feedlines and patches. By using the approximation that the microstrip feedlines couple only to the unit cell they are feeding, the incident and reflected quasi-TEM fields were found by integrating over the isolated semi-infinite line with the single element Green's function. This made it possible to model the field of a traveling wave mode with a source current which extends outside of the unit cell, while maintaining the infinite array periodicity. The theoretical results were confirmed experimentally by means of a waveguide simulator.

REFERENCES

1. Deshpande, M., and Bailey, M., "Input Impedance of Microstrip Antennas," IEEE Trans. Antennas and Prop., Vol. AP-30, No. 4, pp. 645-650, July 1982.

2. Pozar, D., "Input Impedance and Mutual Coupling of Rectangular Microstrip Antennas," IEEE Trans. Antennas and Prop., Vol AP-30, No.6, pp. 1191-1196, Nov. 1982.

3. Herd, J., "Scanning Impedance of Proximity Coupled Rectangular Microstrip Antenna Arrays," Ph.D. Diss., Univ. of Massachusetts, Sept. 1989.

4. Newman, E., and Forrai, D., "Scattering From a Microstrip Patch," IEEE Trans. Antennas and Prop., Vol. AP-35, March 1987.

S-PARAMETER CALCULATION OF ARBITRARY THREE-DIMENSIONAL LOSSY STRUCTURES BY FINITE DIFFERENCE METHOD

Detlev Hollmann*, Steffen Haffa** and Werner Wiesbeck **

* Standard Elektrik Lorenz AG
Ostendstr. 3, D-7530 Pforzheim, Germany

** Institut für Höchstfrequenztechnik und Elektronik, Universität Karlsruhe
Kaiserstr.12, D-7500 Karlsruhe, Germany

Abstract

A three-dimensional treatment of transmission line discontinuity problems by the finite difference method is presented. Maxwell's equations are solved in the frequency domain by solution of a boundary value problem. The presented method allows to compute the scattering parameters of an arbitrary structure, including coupling of higher order modes. The general formulation and the procedure of the method is described. Verification calculations are given and results for different microstrip discontinuities are included for the lossy and non-lossy case.

Introduction

Advances in the field of monolithic integrated millimeter-wave circuits have enforced the need of using exact CAD models for the circuit design. This is especially true at higher frequencies, where the influence of losses in the circuit cannot be neglected. The effect of excitation of higher order modes at discontinuities has also to be considered. A powerful field theoretical method has to be applied as a basis for valuable CAD models and as a tool for getting insight on the coupling mechanisms of discontinuities.
There have been presented several numerical methods in the literature, but most of them are only useful for specific applications or they are restricted to planar structures [1,2]. Recently, a number of authors work on field theoretical methods, which solve the problem in the time domain [3,4]. These formulations have some advantages, but there is the difficulty to derive the scattering parameters from the calculated fields. In our approach we present a more generalized version of a procedure, which has been introduced by Christ and Hartnagel [5,6] to model chip interconnections.
After a description of the general formulation, including the discretization and the derivation of the scattering parameters, verifications are given for the two-dimensional and the three-dimensional calculations. Example structures show the capabilities of the method and demonstrate the applicapability to practical problems.

Edited by H.L. Bertoni and L.B. Felsen, Plenum Press, New York, 1991

Discretization of Maxwell´s equations

A two port, connected to infinitely long transmission lines, is treated as the structure under investigation, Fig. 1. The three-dimensional structure is enclosed in a shielded rectangular box. The walls of the box are perfectly conducting except of the two planes at the ports A and B. These ports are the reference planes for the S-parameter calculation. The whole box is divided into n elementary cells of arbitrary complex permittivity and permeability including ideal and non-ideal conductors. In order to reduce the computational efforts the elementary cells are of rectangular shape in a homogeneous but non necessarily equidistant grid. As shown in Fig. 2 the space components of the electric and the magnetic fields are defined upon each elementary cell. This kind of allocation has the advantage of implicitly fulfilling the continuity conditions between two neighbored cells of different material.

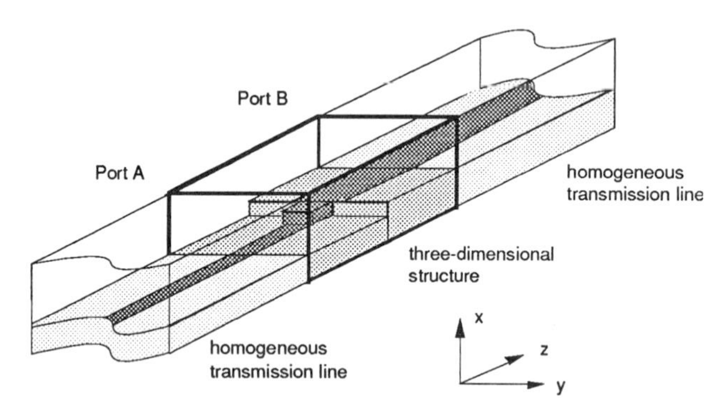

Fig.1 Structure under consideration with infinitely long transmission lines connected at ports A and B

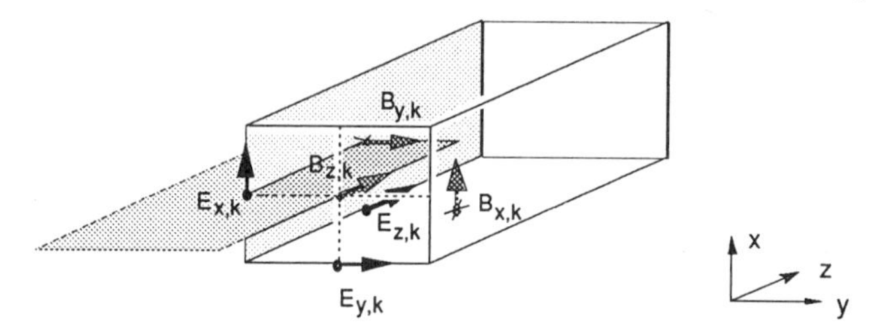

Fig. 2 Allocation of the electric and magnetic field components at the elementary cell k

In the following $\vec{E}$ and $\vec{H}$ are the electric and magnetic field strengths, $\vec{D}$ and $\vec{B}$ the flux densities and $\vec{J_e}$ the source current density. All field components and material related parameters are allowed to be complex. With a harmonic time dependence of the form exp(jωt), the first and second Maxwell's equation in integral form can be written:

$$\oint_C \vec{H} \cdot \vec{ds} = \int_A \left(j\omega\vec{D} + \kappa\vec{E} \right) \cdot \vec{dA} + \int_A \vec{J}_e \cdot \vec{dA} \tag{1a}$$

$$\oint_C \vec{E} \cdot \vec{ds} = - \int_A j\omega\vec{B} \cdot \vec{dA} \tag{1b}$$

with the relations $\vec{D} = \varepsilon \, \vec{E}$ $\quad \vec{B} = \mu \, \vec{H}$.

In this paper a source free structure is assumed ($\vec{J}_e=0$). Polarization losses ($\tan\delta$) have been taken into account by the complex permittivity ε. The finite conductivity κ is included in the frequency dependent imaginary part of the permittivity, see equation (2), which leads to a new complex dielectric constant $\varepsilon^* = \varepsilon_0 \cdot \varepsilon_k$. The values ε_k are complex in general but real for lossless structures.

$$\varepsilon^* = \varepsilon_0 \cdot \varepsilon_k = \varepsilon_0 \left[\varepsilon' - j \left(\varepsilon' \tan\delta + 120\,\pi\,\Omega \frac{\kappa}{k_0} \right) \right] \tag{2}$$

where $k_0 = \omega \cdot \sqrt{\varepsilon_0\,\mu_0}$ is the wavenumber of free space.

Applying these assumptions to eq. (1), Maxwell's equations can be discretisized for any cell k by a lowest order integration formula. The integration planes for the elementary cell k are shown in Fig. 3. The subscripts denote the space coordinates and the position of the belonging cell, relative to the actual cell k (i.e. l=left, al=above left). Combining the resulting discretisized equations leads to a relation between each space component of cell k ($E_{x,k}$, $E_{y,k}$, $E_{z,k}$) and the neighboring components. This is visualized in Fig. 4 for the $E_{x,k}$ component.
Setting up the equations for all n cells of the whole structure leads to a linear system of equations (3). This matrix equation defines a boundary value problem for the unknown electric field.

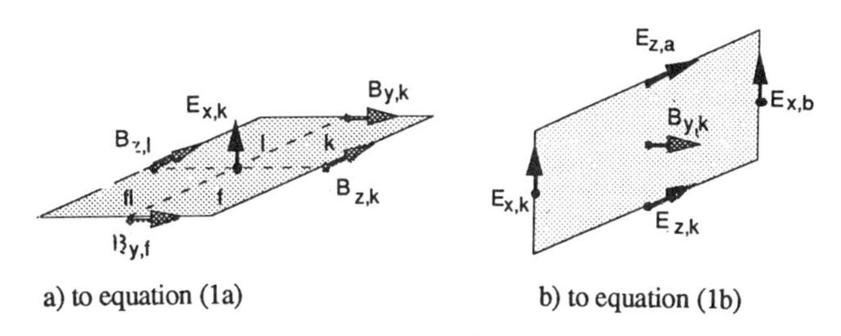

a) to equation (1a) b) to equation (1b)

Fig. 3 Integration planes for discretization of first and second Maxwell's equation

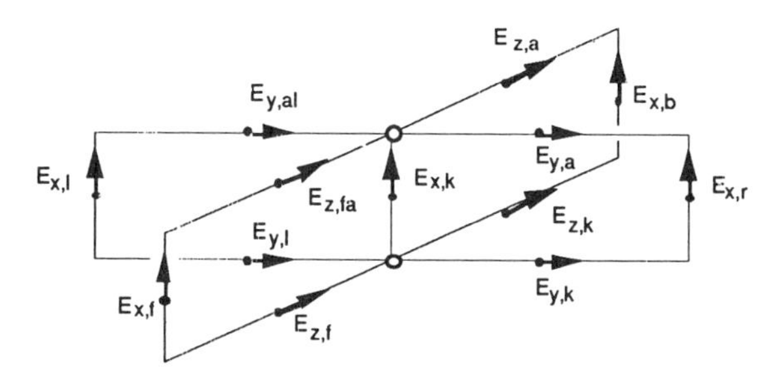

Fig. 4 $E_{x,k}$ component and dependent E components, defining a equation of the linear system of equations

$$[C]\begin{bmatrix}\vdots\\ \vec{E}_u\\ \vec{E}_k\\ \vec{E}_a\\ \vdots\end{bmatrix} = \begin{bmatrix}\vdots\\ \vec{E}_u\\ \vec{E}_k\\ \vec{E}_a\\ \vdots\end{bmatrix}_{excited} , \qquad \vec{E}_k = \begin{bmatrix}E^{(Re)}_{x,k}\\ E^{(Im)}_{x,k}\\ E^{(Re)}_{y,k}\\ E^{(Im)}_{y,k}\\ E^{(Re)}_{z,k}\\ E^{(Im)}_{z,k}\end{bmatrix} \tag{3}$$

The right hand side is given by the superposition of the modal transverse electric fields of the connecting transmission lines at port A and B. The complex values are splitted into real (Re) and imaginary (Im) parts, thus resulting a real system of equations.The coefficient matrix [C] is sparse and consists of 27 diagonals for lossy structures (μ real). The system of equations is solved iteratively by the biconjugate gradient method and leads to the electric field inside the structure.

Determination of the transverse electric modal fields at the ports

For computation of the transverse electric field at ports A and B the same procedure as in the three-dimensional case can be applied, but it simplifies because of the homogeneity of the field components in z-direction, depending on the propagation constant k_z. The z-components can be eliminated by using the continuity equation.

$$\oint\!\!\oint_A j\omega \vec{D}\, \vec{dA} = 0 \tag{4}$$

After some rearrangements this leads to the following eigenvalue problem

$$([A] - \gamma [I])\, \vec{E} = 0 \tag{5a}$$

with

$$\gamma = e^{-jk_z \Delta z} + e^{+jk_z \Delta z} , \tag{5b}$$

that is to be solved for γ. The eigenvectors related to each of the propagation constant k_z can be computed by solving (5a) and give the transverse fields of each mode. The fields are orthogonal to each other, and they are normalized, so that the modal power equals one.

Determination of the scattering matrix

For computation of the scattering matrix a linear superposition of the modal fields is applied to port A and B, and the resulting electric field inside the structure is calculated. Applying the orthogonality relation of the modal fields $\vec{E}_{ti}$, $\vec{H}_{ti}$ to the calculated field in a tangential z-plane $\vec{E}_t$ allows a modal decomposition according to the following formula.

$$\int_A \left(\vec{E}_t \times \vec{H}_{ti}\right) \vec{dA} = w_i(z) = a_i(z) + b_i(z) \tag{6}$$

Using (6) it is now possible to determine the magnitude and phase of the two waves $a_i(z)$ and $b_i(z)$ for each mode i, propagating in +z and -z-direction, by comparison of the values of the modal decomposition $w_i(z)$ at two neighboring z-planes. Combing a sufficient number of field excitations finally results in the scattering parameters. The S-matrix is generalized in the form,

Verification

The results of the two-dimensional computation have been checked against analytical solutions and results derived by other methods. The agreement depends on the division of the grid, and the error tends to zero for very small grid spacing. For practical purposes, however, only a reduced number of elementary cells can be used.
A waveguide filled with lossy dielectric material has been investigated in the frequency range below and above cut-off. The first 10 modes have been compared to the analytical solution for 3 different types of the grid, Fig. 5. The relative error of the propagation constants $\Delta\alpha$, $\Delta\beta$ is shown in Fig. 6 for the first (TE_{01}) and the seventh (TE_{21}) mode. The errors reach a maximum at cut-off, but they decrease in the region above cut-off to values below 1%. The distance of the two z-planes Δz, which is used in the calculation (see eq. (5b)), has been set to 0.01a. The influence of varying this value shows Fig. 7. The lowest errors accur for the smallest value of Δz, but there is a lower bound resulting from numerical inaccuracy.

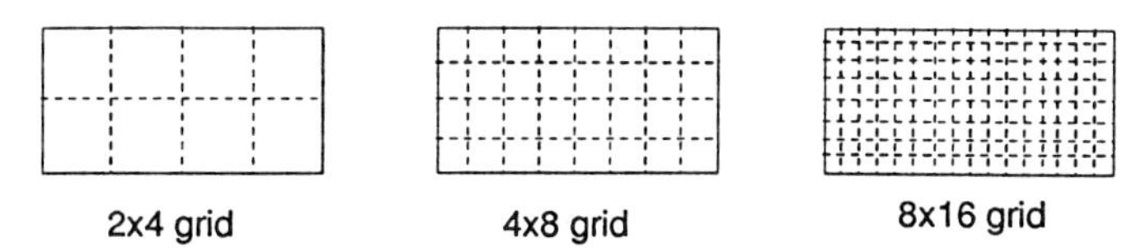

Fig. 5 Grid divisions of a waveguide filled with lossy dielectric material (ε_r=20 - j2); height h=5a, width w=10a; a: normalization constant.

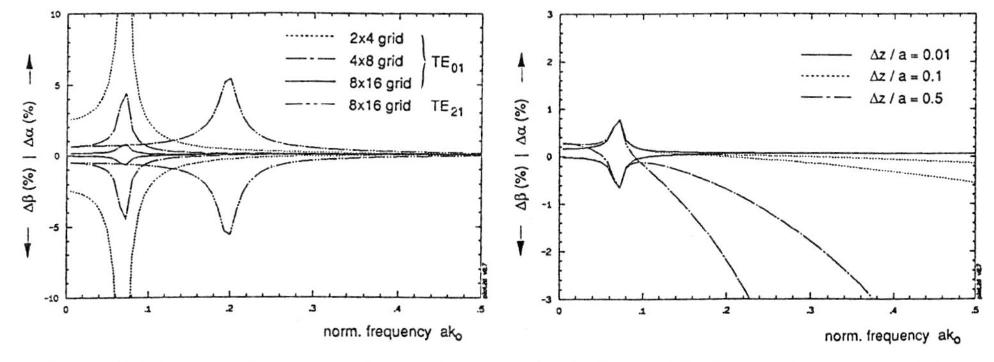

Fig. 6 Relative error of the propagation constant of the waveguide structures, Fig. 5.

Fig. 7 Errors of the 8x16 grid structure, with the z-plane spacing Δz as parameter.

A plot of the transverse electric field of a microstrip on GaAs substrate with finite metallization thickness shows Fig. 8. The influence of substrate and conductor losses on the attenuation constant demonstrates Fig. 9. The frequency dependance is ~f for substrate losses and $\sim\sqrt{f}$ for conductor losses only. At lower frequencies the attenuation constant approaches a constant value due to the increasing skin depth, which becomes comparable to the conductor thickness. This shows the feasibility of the method to deal with conductor losses. It should be mentioned however, that at least two rows of elementary cells have to be applied below the surface of the conductor to model the skin effect correctly. This increases the computational efforts.

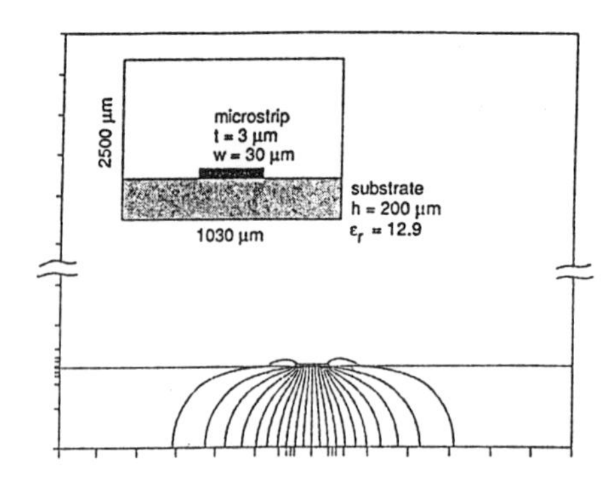

Fig. 8 Electric field line plot of a microstrip on GaAs substrate

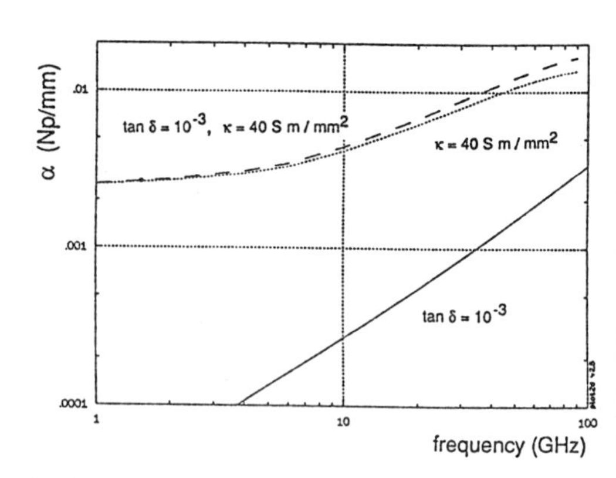

Fig. 9 Attenuation constant of the structure Fig. 8 for substrate and conductor losses and both of each.

The discretization error of the three-dimensional calculation can be checked for homogeneous structures by moving the reference planes to the middle of the structure. This is done using the propagation constant derived by the eigensystem analysis. As an example, the results of a lossy microstrip line are shown in Fig. 10 and Fig. 11. For both, the lossy and non lossy case the return loss may by estimated by the drawn envelope. The traces of return loss and transmission phase indicate the finite grid spacing in z-direction, which becomes larger with increasing frequency.

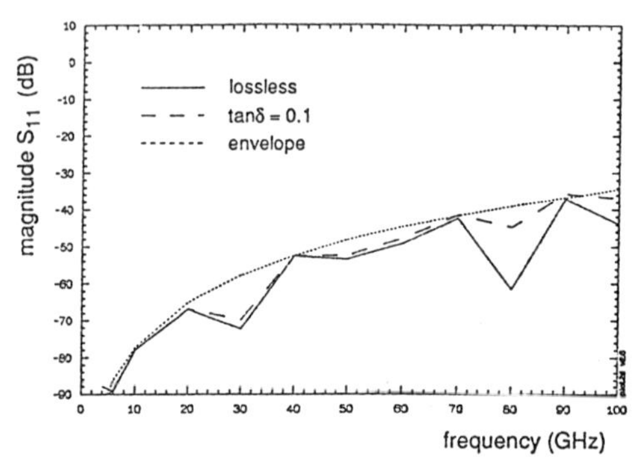

Fig. 10 Return loss of a homogeneous microstrip line (ε_r=12.9)

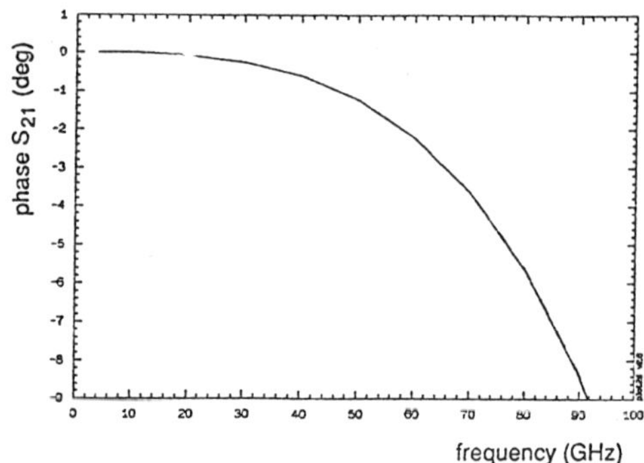

Fig. 11 Phase of S_{21} after shift of the reference plane to the middle of the structure.

A verification of an inhomogeneous structure is given in the next example. The scattering parameters of an inductive dielectric post in a rectangular waveguide have been calculated for different values of the dielectric constant of the post. The wave incident upon the post is the dominant TE_{01} mode. The results are plotted in Fig. 12 together with the data from Leviatan and Sheaffer [7]. Their approach is a moment method based on the equivalence principle, which uses filamentary currents to simulate the scattered field and the field inside the post, and a testing procedure for imposing the continuity conditions across the post surface. The agreement is very good even for high dielectric values. The small shift of the resonance frequency is due to a slightly different post diameter resulting from the discretization.

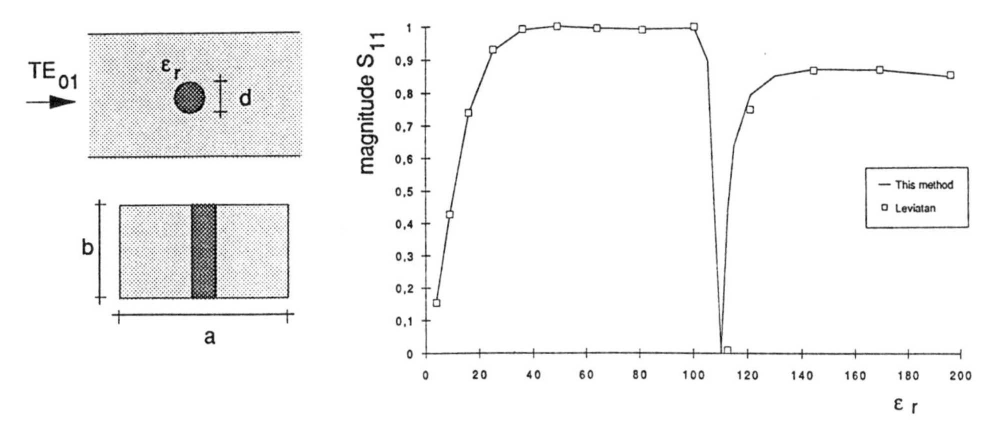

Fig. 12 Reflection coefficient magnitude $|S_{11}|$ versus ε_r for a centered dielectric post in a rectangular waveguide, compared to the results from [7]

Examples

The scattering parameters of non resonant structures are only weakly influenced by losses, if the reference plane is placed in the middle of the structure, in order to eliminate the effect of attenuation on the transmission line. However, this is not true for weak coupled resonant structures. A microstrip-resonator parallel to a microstrip line has been investigated for the lossless case and three different values of the loss tangent of the substrate. The magnitude and the phase of S_{21} are shown in Fig. 13a and 13b, respectively. For the high loss structure the region of resonance is broadened up and the magnitude decreases at higher frequencies.

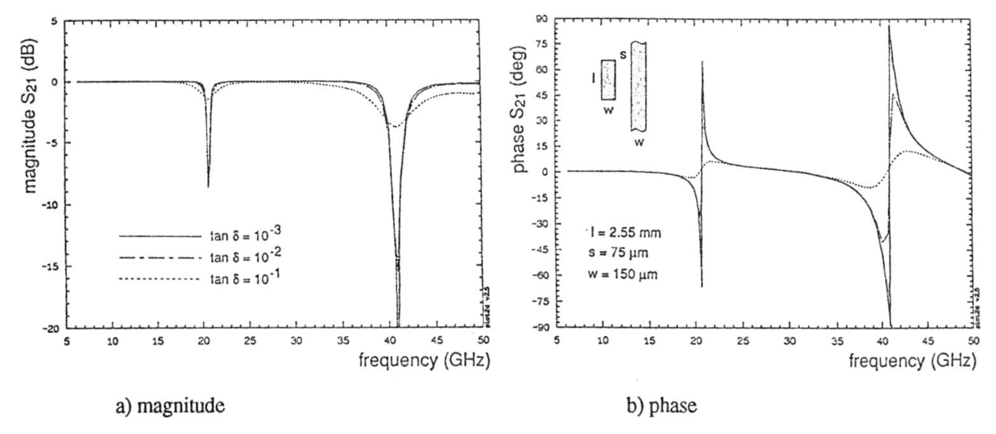

a) magnitude b) phase

Fig. 13 S_{21} of a microstrip parallel resonator with the loss tangent of the substrate as parameter ($h=150\mu m$, $\varepsilon_r=12.9$).

In many cases it is not only desirable to get the scattering parameters of a certain structure, but also to get the field distribution. Therefore the calculation has to be performed with matched ports by applying the modal fields with proper magnitude and phase. This leads to a computation with complex values even in the case of a lossless structure. To show an example, for the parallel line resonator of Fig. 13, a matched calculation has been performed at the frequency of λ-resonance. The results have been used to plot the mean value of the electrical and the magnetical energy in the substrate beneath the conductors. The incident wave propagates in z-direction.

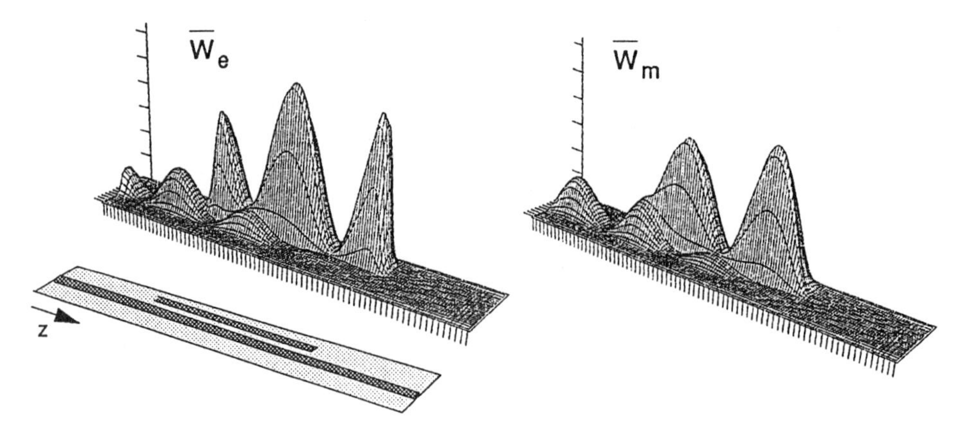

Fig. 14 Distribution of the electrical and magnetical energy in the substrate beneath the conductors. The structure is in λ-resonance.

Conclusion

The finite difference method is well suited for S-parameter computations of discontinuity problems in planar transmission lines and for arbitrary three dimensional structures. The procedure accounts for both substrate and conductor losses. The S-parameters can be computed in any required resolution in frequency. The necessity of considering losses has been proved for weak coupling resonant structures. Field plots of the computed electric and magnetic fields provide valuable insight into the coupling mechanisms inside the structure. Higher order modes exited by a discontinuity can be computed and the generalized S-parameters are used to describe the mode coupling. The achieved results are valid over a wide frequency range.

References

[1] M. Koshiba, K. Hayata, M. Suzuki, " Finite-Element Formulation in Terms of the Electric-Field Vector for Electromagnetic Waveguide Problems", *IEEE Trans on MTT*, vol. 33, pp. 900-905, June 1985

[2] S.B. Worm, R. Pregla, "Hybrid Mode Analysis of Arbitrarily Shaped Planer Microwave Structures by the Method of Lines", *IEEE Trans on MTT*, vol. 32, pp. 191-196, Feb. 1984

[3] W. Hoefer, "The Transmission-line Matrix Method - Theory and Applications", *IEEE Trans on MTT*, vol. 33, pp. 882-893, Oct. 1985

[4] X. Zhang, K. Mei, "Time-domain Finite Difference Approach to the Calculation of the Frequency-dependent Characteristics of Microstrip Discontinuities", *IEEE Trans on MTT*, vol. 36, pp. 1775-1787, Dec. 1988

[5] A. Christ, H. Hartnagel, "Three-Dimensional Finite-Difference Method for the Analysis of Microwave-Device Embedding ", *IEEE Trans on MTT*, vol. 35, pp. 688-696, Aug. 1987

[6] A. Christ, "Streumatrixberechnung mit dreidimensionalen Finite-Differenzen für Mikrowellen-Chip-Verbindungen und deren CAD-Modelle", Ph.D. Thesis, *Fortschritt-Ber., VDI Reihe 21 Nr. 3* , VDI Verlag Düsseldorf, 1988

[7] Y. Leviatan, G.S. Sheaffer, "Analysis of Inductive Dielectric Posts in Rectangular Waveguide", *IEEE Trans on MTT*, vol. 35, pp. 48-59, Jan. 1987

MODELING OF ELECTROMAGNETIC FIELDS IN LAYERED MEDIA

BY THE SIMULATED IMAGE TECHNIQUE

Y.L. Chow and J.J. Yang

Department of Electrical and Computer Engineering
University of Waterloo
Waterloo, Ontario, Canada, N2L 3G1

ABSTRACT

Integrated circuits for digital, microwave and optical purposes are fabricated on layered medium substrates. The simulated image technique is an efficient tool for modeling electromagnetic phenomena in layered medium structures. In this paper, we present the *discrete complex image theory* which finds the simulated images of a source in multilayered media. A critical comparison is made between the discrete complex image theory and other published image theories, i.e., the *exact image theory* based on the Laplace transformation of the spectral function, and the *approximate image theory* based on the Taylor expansion of the spectral function.

INTRODUCTION

The image method is a classical technique. It was originally used to solve the simple static problems of an infinite conducting plane, a conducting sphere, and an infinitely long conducting cylinder [1]. With the aid of an optimization technique, Chow et al developed a *simulated image* technique for the static field problems of arbitrarily shaped conducting bodies [2], and applied this technique to determine the capacitance of a conducting cube and a finite conducting cylinder [3].

For a static point source located on an open microstrip substrate, Silvester [4] first developed an image model by tracing the multi-reflection path. Based on this static image model, Chow [5] developed an approximate dynamic image model of the microstrip substrate. This dynamic image model is already incorporated into a MMIC simulation software EMsim™ which simulates complex MMIC circuits using the field approach [6].

For a static source located on a microstrip substrate with a metallic or a dielectric cover (i.e., a three layer medium), Farrar and Adams [39][40] performed the spectral domain analysis, and derived the static image models of doubly infinite series by taking the inverse transformation of the spectral function. In practice, a doubly infinite series converges very slowly. Moreover, for a static source in the layered medium with more than three layers, it is nearly impossible to derive the multi-fold infinite series [41].

---

* This work was supported by the Communications Research Center, Ottawa, Canada, through a contract 36001-9-3581/01-SS.

For a dynamic dipole located on a thick microstrip substrate, or embedded within a multilayered medium, or even packaged in a dielectric or metallic box, the spectral domain approach (SDA) is normally used [7]-[12]. The SDA enjoys high precision. However, the inverse transformation of the spectral function cannot be performed analytically in most cases. Due to the slow convergence of the numerical spectral integrals, the SDA may suffer from high computation cost. To calculate the spectral integrals accurately and efficiently is therefore a necessity for using the SDA for complex structures.

The techniques for calculating spectral integrals which have appeared in the literature can roughly be put into three categories. The first one is to develop the efficient numerical integration techniques, such as [8]-[12]. For calculating many spectral integrals corresponding to different source and field point locations in complex circuits, the numerical integration techniques may consume a large amount of computer time. The second category is by using an FFT algorithm, as developed in [29][30]. The third category is to develop the image theory formulations for layered medium structures [13]-[28]. This will be discussed below. The image theory solutions are normally more efficient than the numerical integration techniques. Recently, closed form asymptotic solutions were published for calculating the spectral integrals along the surface of a thick microstrip substrate [31]-[33].

From the view point of spectral domain analysis, the essence of the simulated image technique for layered medium problems is a change of the SDA integrand into a function which is inverse-transformed analytically. This is the case for both static problems and dynamic problems. Thus the major task of the simulated image technique is to rewrite the spectral function (exact or approximate), so that the spectral integral is analytically integrable.

COMPARISON OF VARIOUS SIMULATED IMAGE TECHNIQUES

The essence of the simulated image technique is to properly rewrite the spectral function into an exponential function type. Different schemes of mathematical manipulation result in different spatial simulated images. The schemes available in the literature can be classified as three types.

(a). The exact image theory developed by Lindell et al [13]-[17], is based on using the available Laplace transforms of the spectral functions. As the spectral function is represented as an integral of the exponential function, through the Laplace transformation, the resultant image source is a semi-infinite continuous line. From the view point of computing, the exact image theory converts the original integral of Sommerfeld type to an alternate integral of the image. By doing this conversion, the convergence is improved. However, the Laplace transformation of the spectral function is available only for a few of simple structures, such as the infinite half space and the two-layered microstrip substrate. Actually, the exact image theory may lose its advantage of computational efficiency for the two-layered microstrip substrate, due to the complexity of the continuous image source distribution. Moreover, for a large amount of source-to-field point distances, the exact image theory still consumes large amounts of computer time, since at least one infinite integral has to be numerically computed for each source-to-field point distance.

(b). The approximate image theory developed by Wait [18], Mahmoud et al [19]-[21], Bannister [22], is based on the proper expansion of the spectral function into a Taylor series. The resultant images are located at discrete or continuous positions. From the viewpoint of computing, this theory converts the Sommerfeld type of integral to a short series, thus it is computationally efficient. However, the approximate image theory can only give accurate spatial Green's function for short field-to-source point distances. Moreover, the Taylor series expansion of the spectral function is practical only for simple structures, such as the infinite half space problem and the two-layered medium. For structures with three layers or more, Taylor series expansion becomes impractical due to the complexity of the spectral function.

(c). The discrete complex image theory contributed by the present authors [23]-[28], is based on the numerical approximation of the spectral function. In this theory, the spectral function is sampled on a few of points. A short series of exponential functions is constructed from these sample points, through the Prony's method [36]. The accuracy of the function approximation is further improved by an optimization technique. By so doing, the Sommerfeld type of integral is converted to a short series of images *located at complex locations.* Thus the computational efficiency is high.

The discrete complex image theory is capable of handling multilayered medium structures, with little analytical effort. In addition, for the structure investigated thus far, such as the infinite half space, the open and shielded microstrip substrates and the substrate-superstrate structure, the discrete complex image theory always gives the errors of less than 1 percent. This theory has recently been incorporated into the MMIC simulation software EMsim™, which can now simulate the circuits on the thick substrate, and even with a dielectric or metallic cover (e.g., the lid of a MMIC packaging).

To compare the various image theories, we set up the following table. It is in terms of: (i) the capability (how many layers the technique can handle), (ii) the accuracy, and (iii) the simplicity for numerical computation.

Table 1. Comparison of various image theories

Type of problem	Exact image theory	Approximate image theory	Discrete complex image theory
infinite half space	C: 2 A: 2 S: 1 Ref. [13]-[16]	C: 2 A: 1 S: 2 Ref. [18]-[20]	C: 2 A: 2 S: 2 Ref. [26]
open microstrip substrate	C: 2 A: 2 S: 1 Ref. [17]	C: 2 A: 1 S: 2 Ref. [21]	C: 2 A: 2 S: 2 Ref. [25]
Substrate-superstrate structure	C: 2 A: 2 S: 1 Ref. [43]	C: 1 A: 1 S: 1	C: 2 A: 2 S: 2 Ref. [24]
More than three layers	C: 1 A: 1 S: 1	C: 0 A: 0 S: 0	C: 2 A: 2 S: 2
Transient dipole radiation	C: 2 A: 2 S: 1 Ref. [44]	C: 2 A: 1 S: 2 Ref. [45]	C: 2 A: ? S: 2 Under study

C: Capability. =2, can solve, =1, questionable, =0, cannot solve.
A: Accuracy. =2, order of 1%, =1, order of 10%, =0, cannot solve.
S: Simplicity. =2, easy to use, =1, more difficult, =0, extremely difficult and computationally time consuming

From table 1, it is seen that the rank of discrete complex image theory is higher than other image theories. The discrete complex image theory is essentially based on curve fitting of the spectral function. No matter how many layers the substrate has, this theory can always give a simple Green's function of complex images. This theory can be used for both the dynamic radiation and the static field computations in layered medium, as discussed below.

THE DISCRETE COMPLEX IMAGE THEORY FOR DYNAMIC RADIATION IN LAYERED MEDIUM

From the dipole radiation in free space, it is known that the vector potential $\vec{A}$ created by the dipole and the scalar potential Φ_q associated with one charge of the dipole have the same form of $e^{-jk_0r}/4\pi r$. For a dipole embedded in a multilayered medium, we start our formulations by using $\vec{A}$ and Φ_q, assuming that they should follow the form of free space potentials.

The spectral domain vector potential and scalar potential created by a horizontal (x-directed) electric dipole in layered medium can be generally represented as follows:

$$\tilde{G}_A^{xx} = \frac{\mu_0}{j2k_z} \cdot T_{TE} \tag{1.a}$$

$$\tilde{G}_q = \frac{1}{\varepsilon}\frac{1}{j2k_z} \cdot \left[T_{TE} + \frac{k_z^2}{k_\rho^2}\left(T_{TE} + \frac{1}{jk_z}\frac{\partial T_{TM}}{\partial z}\right)\right] \tag{1.b}$$

where the tilde " ~ " means the spectral domain quantity. In addition, ε, μ, and k_z denote the permittivity, permeability and z-directed wavenumber of the layer in which the HED is located respectively. T_{TE} and T_{TM} are plane wave transmission coefficients of the TE_z and the TM_z wave respectively, from the source plane ($z = z'$) to the field plane ($z = z$) in the layered medium. T_{TE} and T_{TM} can be derived by using the wave matrix technique [34].

An open substrate structure is examined here to show the principle of the discrete complex image theory. As shown in Fig. 1, a horizontal electric dipole of unit strength is located above the microstrip substrate ($i.e., z' \geq 0$). We are looking for the vector and scalar potential in the region $z \geq 0$. With the transmission coefficients T_{TE} and T_{TM} obtained, the vector and scalar potentials in spectral domain can be represented as follows [25]:

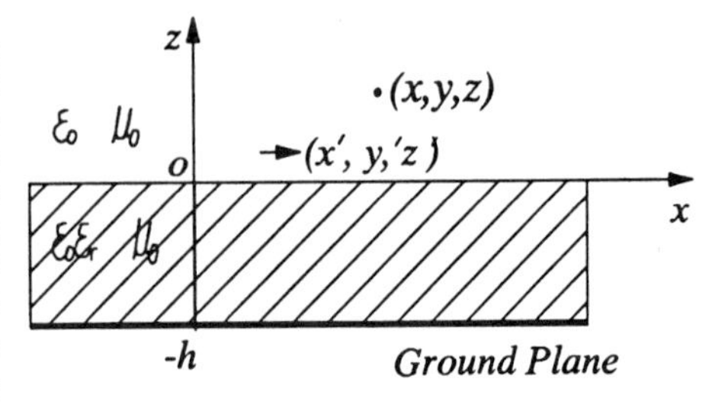

Fig. 1. Open microstrip substrate

$$\tilde{G}_A^{xx} = \frac{\mu_0}{4\pi}\frac{1}{j2k_{z0}}\left[e^{-jk_{z0}(z-z')} + R_{TE}e^{-jk_{z0}(z+z')}\right] \tag{2.a}$$

$$\tilde{G}_q = \frac{1}{4\pi\varepsilon_0}\frac{1}{j2k_{z0}}\left[e^{-jk_{z0}(z-z')} + (R_{TE}+R_q)e^{-jk_{z0}(z+z')}\right] \tag{2.b}$$

where R_{TE} and R_q are spectral functions related to T_{TE} and T_{TM} of the microstrip substrate. To get the spatial Green's functions G_A^{xx} and G_q, the following inverse-transformation should be made:

$$G = \frac{1}{4\pi}\int_{-\infty}^{+\infty} \tilde{G} \cdot H_0^{(2)}(k_\rho \rho) k_\rho dk_\rho \tag{3}$$

where G may either be G_A^{xx} or G_q. The integration path of the inverse transformation (3) may be the real axis C_0, or any other deformed contour C_1 lying in the first and third quadrants on the complex k_ρ plane, as shown in Fig. 2(a). The counterparts of C_0 and C_1 on the complex k_{z0} plane are shown in Fig. 2(b).

To gain some insight into the spectral function $R_{TE}+R_q$, its real and imaginary parts along the deformed path C_1 is plotted in Fig. 3(a). It is observed that when k_ρ is large, the real part of $(R_{TE}+R_q)$ approaches a constant. As it is known that the spectra with large k_ρ contribute mainly to the near field [35], this constant must be linked to the near field. It is also observed in Fig. 3(a) that the spectral function $(R_{TE}+R_q)$ has some extrema along the deformed path C_1. As it is known that $(R_{TE}+R_q)$ has poles on the real axis of k_ρ plane corresponding to the surface waves in the microstrip substrate, the extrema of spectral function $(R_{TE}+R_q)$ along the path C_1 must be linked to the surface waves.

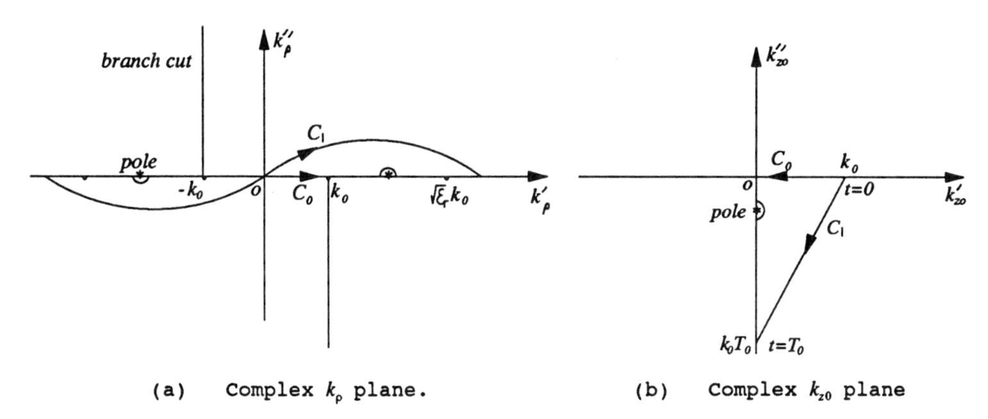

(a) Complex k_ρ plane. (b) Complex k_{z0} plane

Fig. 2 Integration paths

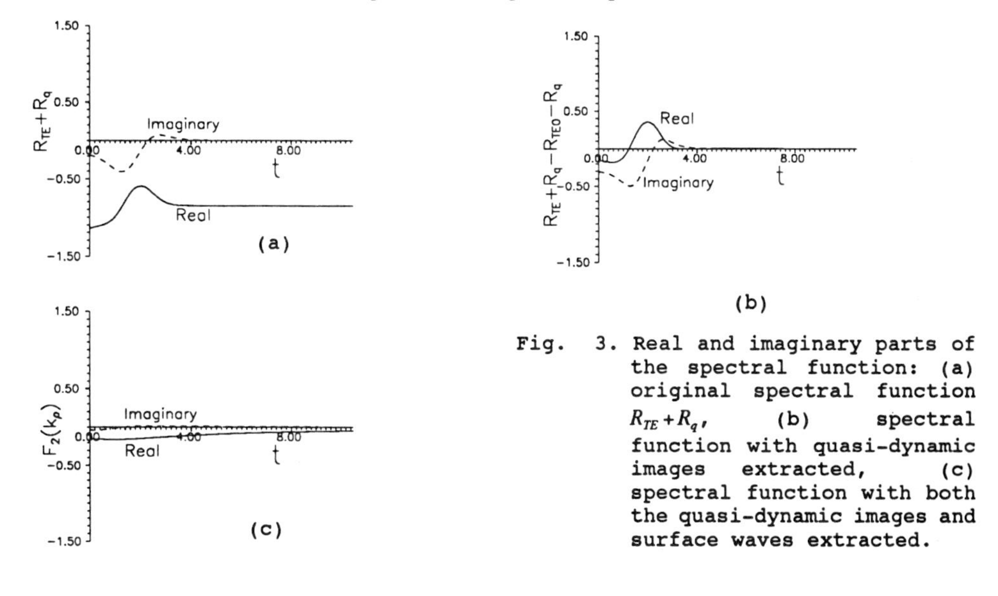

Fig. 3. Real and imaginary parts of the spectral function: (a) original spectral function $R_{TE}+R_q$, (b) spectral function with quasi-dynamic images extracted, (c) spectral function with both the quasi-dynamic images and surface waves extracted.

Fig. 3(b) shows the real and imaginary parts of the spectral function $(R_{TE}+R_q)$ with the quasi-dynamic images extracted along the path C_1. It is seen that the resultant real part then approaches zero when k_ρ is large. Fig. 3(c) shows the real and imaginary parts of spectral function $(R_{TE}+R_q)$ with both the quasi-dynamic field images and the surface wave poles extracted. It is seen that the resultant real and imaginary parts become smooth and can be easily approximated by the discrete complex images.

Based on the above observations, three steps are performed in treating the spectral function $(R_{TE}+R_q)$. That is, step 1: extraction of the quasi-dynamic images, step 2: extraction of the surface wave poles, step 3: using complex images to account for the rest of the spectral function.

With the above three extractions and approximations, the spatial Green's function G_A^{xx} and G_q can be inverse-transformed analytically, by using the Sommerfeld identity [37] and the residue theorem [38]:

$$G_A^{xx} = G_{A0}^{xx} + G_{A,ci}^{xx} + G_{A,sw}^{xx} \tag{4.a}$$

$$G_q = G_{q0} + G_{q,ci} + G_{q,sw} \tag{4.b}$$

where G_{A0}^{xx} and G_{q0} stand for the quasi-dynamic images, $G_{A,ci}^{xx}$ and $G_{q,ci}$ stand for the complex images, and $G_{A,sw}^{xx}$ and $G_{q,sw}$ stand for the surface wave residue terms. All of them can be expressed in closed form [25].

Numerical examples of G_A^{xx} and G_q calculated by the solution (4) and by the numerical integration, were given in [25]. The difference is generally less than 1 percent.

For a HED embedded in a multilayered medium with more than three layers, it is straightforward to derive the new transmission coefficients T_{TE} and T_{TM}, and get the spatial Green's function by following the same steps.

THE DISCRETE COMPLEX IMAGE THEORY FOR STATIC FIELD COMPUTATION IN MULTILAYERED MEDIUM [28].

In static case, i.e., frequency approaches zero, the scalar potential G_q associated with one point charge of a HED can be simplified as:

$$\tilde{G}_q = \frac{1}{2\varepsilon\gamma} \cdot \left(\frac{-1}{\gamma} \frac{\partial T_{TM}}{\partial z} \right) \tag{5}$$

where $\gamma = jk_z$. It is found that $\tilde{G}_q$ is exactly the same as the spectral domain scalar potential $\tilde{\Phi}$ of a point charge in layered medium, and satisfies the Possion equation and all the boundary conditions. Following the dynamic case, it is natural to speculate that the scalar potential Φ of a point charge in multilayered medium could be computed by using the discrete complex image theory. To prove this, we take an open microstrip substrate as example. This structure was studied by Silvester [4] with an infinite series solution.

As shown in Fig. 1, a point charge q_0 is located at $(x', y', 0)$. We are looking for the potential function $G_q(x, y, 0)$ at the interface. The spectral domain potential function $\tilde{G}_q$ can be easily derived as follows:

$$\tilde{G}_q = \frac{q_0}{2\varepsilon_0\gamma} \cdot \tilde{A} \quad , \quad \tilde{A} = \frac{2}{1+\varepsilon_r cth\gamma h} \tag{6}$$

Similar to the dynamic case, the spectral function $\tilde{A}$ is approximated as the summation of a constant (which is the limit of $\tilde{A}$ as $\gamma \to \infty$) and a short exponential function series, having the form:

$$\tilde{A} = \frac{2}{1+\varepsilon_r} + \sum_{i=1}^{N} a_i e^{b_i\gamma} \tag{7}$$

The approximation (7) is implemented by using Prony's method, where the approximation path is along the real axis of the γ plane. Substituting (7) into (6), and taking analytically the inverse Fourier transformation of $\tilde{G}_q$, we have:

$$G_q = \frac{1}{4\pi\varepsilon_0}\left(\frac{2}{1+\varepsilon_r}\frac{1}{r_0} + \sum_{i=1}^{N}\frac{a_i}{r_i}\right) \quad , \; N \le 5. \tag{8}$$

The first term in (8) can be considered as an *effective* homogeneous medium potential function with dielectric constant being equal to the average of two media. The short series is the complex image terms.

The complex image solution (with $N=3$) is compared with the rigorous solution of infinite series given in [4]. Table 2 lists the values of G_q computed by the complex image solution (8) and the rigorous solution. It is seen that the complex image solution can gives errors of less than 0.1 percent for all source-to-field point distances tabulated.

Table 2. Potential function $4\pi\varepsilon_0 G_q$ by the exact and by the complex image solutions (z=0, z'=0) ε_r=9.6, h=0.5m

ρ(m)	*exact image solution (1/m)*	*complex image solution (1/m)*
0.01	*18.618*	*18.618-j0.1E-4*
0.11	*1.4668*	*1.4669-j0.1E-4*
0.31	*0.3724*	*0.3724-j0.7E-5*
0.51	*0.1535*	*0.1535+j0.1E-5*
0.71	*0.0721*	*0.0721+j0.5E-5*
0.91	*0.0361*	*0.0361+j0.6E-5*

CONCLUSION

In this paper, we introduced the discrete complex image theory for modeling both static and dynamic field problems in multilayered media. This theory has the advantage of simplicity (only a few of real and complex images plus the surface wave modes) and high accuracy (less than 1% error when compared with the numerical integration of the Sommerfeld integral). Other simulated image theories are briefly reviewed and compared with the discrete complex image theory in terms of versatility, simplicity and accuracy. It is concluded that combining the moment method and the discrete complex image theory may solve rapidly a wide variety of problems such as printed antennas, microwave integrated circuits, digital circuits, as well as the integrated optical circuits. Extension of the discrete complex image theory to transient dipole radiation in layered medium is under study.

REFERENCES

[1] W.R. Smythe, "***Static and Dynamic Electricity***", McGraw-Hill, 1968, 3d edn.

[2] Y.L. Chow and C. Charalambous. *Proc. IEE*, Vol.126, 1979, pp123-125.

[3] Y.L. Chow, Y.F. Lan and D.G. Fang. *J. Appl. Phys.*, Vol.53, 1982, pp.7144-7148.

[4] P. Silvester. *Proc. IEE*, Vol.115, 1968, pp.43-48.

[5] Y.L. Chow. *IEEE Trans. on Microwave Theory and Tech.* Vol.MTT-28, 1980, pp.393-397.

[6] Y.L. Chow and G.E. Howard. to appear in *MIMICAE Journal*, Vol.1, 1991.

[7] R.H.Jansen. *IEEE Trans. on Microwave Theory and Tech.*, Vol.MTT-33, 1985, pp1043-1056.

[8] J.R. Mosig and F.E. Gardiol. *Advances in Electronics and Electron Physics*, Vol.59, 1982, pp.139-239.

[9] P.B. Katehi and N.G. Alexopolous. *Journal of Math. Phys.*, Vol.24, 1983, pp.527-533.

[10] K.A. Michalski and C.M. Bultler. *IEE Proc.*, Vol.134, pt.H, 1987, pp.93-97.

[11] D.M. Pozar. *Electromagnetics*, Vol.3, July-Dec.,1983, pp.299-209.

[12] P. Parhami, Y. Rahmat-Samii and R. Mittra. *IEEE Trans. on Antennas Propagat.*, Vol.AP-28, 1980, pp.100-104.
[13] I.V.Lindell and E.Alanen. *IEEE Trans. on Antennas and Propagat.*, Vol.AP-32, 1984, pp126-133.
[14] I.V.Lindell and E.Alanen. *IEEE Trans. on Antennas and Propagat.*, Vol.AP-32, 1984, pp.841-849.
[15] I.V.Lindell and E.Alanen. *IEEE Trans. on Antennas and Propagat.*, Vol.AP-32, 1984, pp1027-1032.
[16] I.V.Lindell, E.Alanen and H.V.Bagn. *IEEE Trans. on Antennas Propagat.*, Vol.AP-34, 1986, pp.129-137.
[17] E. Alanen and I.V. Lindell. *IEE Proc.*, Vol.133, pt.H, 1986, pp.297-304.
[18] J.R. Wait, "Characteristics of antennas over a lossy earth", in ***Antenna Theory***, part 2, editted by R.E. Collin and F.J. Zucker, Chap.23, pp.386-435, Mcgraw-Hill, New York.
[19] S.F. Mahmoud. *IEEE Trans. on Antennas and Propagat.*, Vol.AP-32, 1984, pp.679-683.
[20] S.F. Mahmoud and A. Mohsen. *IEEE Trans. on Antennas and Propagat.*, Vol.AP-33, 1985, pp.1054-1058.
[21] A.Mohsen. *IEE Proc.*, Vol.129, pt.H, 1982, pp.177-182.
[22] P.R.Bannister. *Radio Science*, Vol.21, 1986, pp.605-616.
[23] Cui Kai. *Proc. of 1985 International Symposium on Antennas and EM Theory*, Beijing, PRC, 1985, pp.318-321.
[24] D.G. Fang, J.J. Yang and G.Y. Delisle. *IEE Proc.*, Vol.135, pt.H, 1988, pp.297-303.
[25] Y.L.Chow, J.J.Yang, D.G.Fang and G.E. Howard. accepted for publication in *IEEE Trans. on Microwave Theory and Tech.*, Aug.,1990.
[26] J.J. Yang, Y.L. Chow and D.G. Fang. submitted to *IEE Proc.*, pt.H, June,1990.
[27] J.J.Yang,Y.L.Chow,D.G.Fang and G.E.Howard. submitted to *IEE Proc.*, pt.H, July,1990.
[28] Y.L. Chow, J.J.Yang and G.E.Howard. submitted to *IEEE Trans. on Microwave Theory and Tech.*, oct.,1990.
[29] L. Tsang, L.R. Brown, J.A. Kong and G. Simmons. *J. Geophys. Res.*, Vol.29, 1974, pp.2077-2080.
[30] J.J.H. Wang. *IEE Proc.*, Vol.132, pt.H, 1985, pp.58-62.
[31] S. Barkeshli and P.H. Pathak. *IEEE Trans. on Microwave Theory Tech.*, Vol.37, 1989, pp.669-679.
[32] S. Barkeshli, P.H. Pathak and M. Marin. *IEEE Trans. on Antennas Propag.*, Vol.AP-38, 1990, pp.1374-1383.
[33] S. Barkeshli and P.H. Pathak. *Radio Science*, Vol.25, 1990, pp.161-174.
[34] D.G.Fang. *J. Of East China Inst. of Tech.*, 1984, pp.55-62.
[35] J.R. Mosig and T.K. Sarkar. *IEEE Trans. on Microwave Theory Tech.*, Vol.MTT-34, 1986, pp.379-387.
[36] R.W. Hamming, "***Numerical Methods for Scientists and Engineers***", Dover Publication, Inc, New York, 1973, pp.620-622.
[37] J.A. Stratton, "***Electromagnetic Theory***", McGraw-Hill, New York, 1941, pp.576, Eq.(17).
[38] J. Duncan,"***The Elements of Complex Analysis***", John Wiley Sons, 1968.
[39] A. Farrar and A.T. Adams. *IEEE Trans. on Microwave Theory and Tech.* Vol.MTT-21, 1973, pp.494-496.
[40] A. Farrar and A.T. Adams. *IEEE Trans. on Microwave Theory and Tech.* Vol.MTT-22, 1974, pp.889-891.
[41] C. Wei, R.F. Harrington, J.R. Mautz and T.K. Sarkar. *IEEE Trans. on Microwave Theory and Tech.* Vol.MTT-32, 1984, pp.439-450.
[42] R. Crampagne, M. Ahmadpanah and J. Guiraud. *IEEE Trans. on Microwave Theory and Tech.* Vol.MTT-26, 1978, pp.82-87.
[43] E. Alanen. *1988 IEEE International Symposium on Antennas and Propagation.*, Vol.3, pp.1010-1013.
[44] K.I.Nikoskinen and I.V.Lindell. *IEEE Trans. on Antennas and Propagat.*, Vol.AP-38, 1990, pp241-250.
[45] A.D. Metwally and S.F. Mahmoud. *IEEE Trans. on Antennas Propagat.*, Vol.AP-32, 1984, pp.287-291.

MODELING INTERACTIONS WITH MATERIALS

MODELING FOR WAVES IN RANDOM MEDIA - A NEED FOR ANALYTICAL, NUMERICAL AND EXPERIMENTAL INVESTIGATIONS

Akira Ishimaru

Department of Electrical Engineering
University of Washington
Seattle, Washington 98195

ABSTRACT

This paper first presents the state of the art in the modeling of waves in random media including waves in turbulence, waves in discrete scatterers, and rough surface scattering. Secondly, we emphasize the need for interplay among analytical, numerical, and experimental investigations. Weak localization, coherent backscattering, and enhanced backscattering from rough surfaces are used as examples of these interactive investigations that are essential in constructing new analytical models and in uncovering underlying physical processes. Future directions in this field include basic theoretical studies on waves in dense media, disordered media, and nonlinear and inverse problems as well as applications in remote sensing, surface physics, geosciences, biomedical sciences, communications, and imaging.

1. INTRODUCTION

Many natural and man-made media are inhomogeneous and vary randomly in time and space, and therefore, electromagnetic waves propagating in these media also fluctuate randomly in time and space. Examples are microwave and millimeter wave scattering by ocean waves, atmospheric turbulence, rain, fog, snow, terrain, vegetation, biological media, and composite materials. These fluctuations affect communications through these media and the identification and remote-sensing of these media and of objects inside the random media.

In recent years, there has been a surging interest in wave propagation and scattering in random media [1]-[17]. This is primarily due to the discovery of new multiple scattering phenomena and an increasing awareness that a common thread underlies the work of many researchers in such diverse fields as atmospheric optics, ocean acoustics, radio physics, astrophysics, condensed matter physics, plasma physics, geophysics, bioengineering, etc. In addition, waves in random media is one of the most challenging problems to theoreticians. Thus the field of wave propagation and scattering encompasses the most practical as well as the most theoretical questions. This paper reviews some of the historical development of the field, basic theories, and applications including the most recent developments and discoveries in this field. The strong interest in this subject is reflected in the launch of a new journal, *Waves in Random Media,* by the Institute of Physics, United Kingdom in 1991; international workshops on wave propagation in random media held in Tallin, USSR in 1988, on surface and volume scattering held in Madrid in 1988, and on modern analysis of scattering phenomena held in Aix en Provence, France in 1990; and an international meeting on wave propagation in random media to be held in Seattle in 1992.

Directions in Electromagnetic Wave Modeling
Edited by H.L. Bertoni and L.B. Felsen, Plenum Press, New York, 1991

This paper first presents the state of the art in the modeling of waves in random media; and secondly, emphasizes the need for interplay among analytical, numerical, and experimental investigations in order to construct models for waves in random media. Future directions in this field are also discussed.

For problems dealing with waves in random media, exact formulations and solutions are limited, and it is necessary to devise effective approximate theories which give insight into the physical process of wave scattering as well as numerical results. Since most theories are approximate, it is possible that existing theories cannot predict some important physical phenomena. The backscattering enhancement could not have been predicted by the conventional radiative transfer theory. Here experimental discovery was the key to uncovering the underlying physical scattering process leading to a new analytical model.

In the areas of rough surface scattering, backscattering enhancement was first discovered by experiments. The development of new models can be accomplished by conducting several Monte Carlo numerical studies under various assumptions and comparing the results with the exact Monte Carlo solution. This technique has been effectively used to construct a new rough surface scattering model which includes the backscattering enhancement.

In summary, we point out that interactive investigations between experiment and modeling and numerical study and modeling are essential in constructing new analytical models and in uncovering underlying physical processes for waves in random media. Future directions in this field include basic theoretical studies on strong fluctuations, dense media, and nonlinear and inverse problems as well as interactions with the fields of condensed matter and surface physics, geosciences, and biomedical sciences.

2. WAVE PROPAGATION IN TURBULENCE AND RANDOM CONTINUUM

The propagation characteristics of coherent waves in random continuum have been investigated by many workers. Formal perturbation methods based on Feynman diagrams have been applied to this problem and the fundamental differential equation for the coherent field is called the "Dyson equation" [8]. The mass operator can be expressed by the "Bourret" approximation or the "nonlinear" approximation. The differential equations satisfied by the second- and higher-order moments are obtained [2], but the exact solutions for the moment equation have been obtained only for the second moment. The normalized variance of the intensity fluctuation, called the "scintillation index," increases with frequency and distance, but it saturates to approximately unity in the strong fluctuation region. The second-order moment formally satisfies the "Bethe-Salpeter" equation with the intensity operator [8].

Solutions to the moment equation have been obtained for the fourth moment [18] using functional integral [19] and path-integral techniques [5], [20], [21], and numerical simulation [22], [23]. The question of how the variation of the background profile affects the intensity fluctuation is also studied [24]. The correlation between the forward and backward waves contributes to the backscattering enhancement [25]. The scalar wave equation needs to be extended to include the cross polarization effects. For weak fluctuation, the probability density function (PDF) is log normal. However, for strong fluctuation, the I-K distribution is shown to be applicable [26], [27]. The enhanced backscattering from turbulence and from a mirror embedded in the turbulence has received considerable attention [12] [57].

3. TRANSPORT THEORY AND SCATTERING BY DISCRETE SCATTERERS

Let us consider propagation and scattering of microwaves through rain, fog, snow, ice, or vegetation; optical and acoustic scattering and diffusion in tissues and whole blood; and optical scattering by vegetation.

In a random distribution of discrete scatterers, waves are scattered and absorbed due to the inhomogeneities and absorption characteristics of the medium. A mathematical description of the propagation and scattering characteristics of waves can be made in two different manners: Analytical theory and transport theory. In analytical theory, we start with Maxwell's equation, take into account the statistical nature of the medium, and consider the statistical moments of the wave. In principle, this is the most fundamental approach, including all diffraction effects, and many investigations have been made using this approach. However, its drawback is the mathematical complexities involved and its limited usefulness.

Transport theory, on the other hand, does not start with Maxwell's equations. It deals directly with the transport of power through turbid media. The development of the theory is heuristic and lacks the rigor of the analytical theory. Since both the analytical and transport theories deal with the same physical problem, there should be some relation between them. In fact, many attempts have been made to derive the transport theory from Maxwell's equations with varying degrees of success. In spite of its heuristic development, however, the transport theory has been used extensively, and experimental evidence shows that the transport theory is applicable to a large number of practical problems.

The fundamental quantity in transport theory is the radiance $I(\bar{r},\hat{s})$, which is also called the specific intensity in radiative transfer theory and the brightness in radiometry. It's unit is Watt m^{-2} sr^{-1} Hz^{-1} and is the average power flux density in a given direction $\hat{s}$ within a unit solid angle within a unit frequency band. The fundamental differential equation for the radiance is the transport equation [1], [28].

Solutions of the radiative transfer equation have been obtained for a plane-parallel medium, first-order scattering, and the diffusion approximation [1], [28], [29]. The radiative transfer equation can be generalized to include all the polarization characteristics. This is done by using the Stokes parameters [I].

We can also obtain the vector radar equation including all the polarization. This approach is particularly useful for microwave remote sensing, polarimetric radars, and the polarization signature [6], [30], [31], [32].

The vector radiative transfer theory has been extended to the beam wave case and to dense media where the correlation between particles needs to be included [33], [34]. Imaging through such random media is studied using the Modulation Transfer Function (MTF) [35]. Application of the radiative transfer theory to optical diffusion in tissues is discussed using the diffusion approximation [36].

When a wave enters a medium, the radiance can be expressed as the coherent intensity and the diffuse intensity. If the medium is mostly scattering, then the diffuse intensity tends to scatter almost isotropically and the diffuse radiance has a broad angular spread. Therefore, we can expand the diffuse intensity I_d in a series of spherical harmonics. The first two terms of the expansion constitutes the diffusion theory [1]. Next consider the diffusion of a pulse in a turbid medium. First, the coherent intensity pulse $\psi_c(t)$ propagates through the medium with the velocity v of light in that medium ($v = c/n$), (n is the refractive index). This coherent pulse is scattered by the medium and generates the diffuse pulse $\psi_d(t)$. The diffuse pulse $\psi_d(t)$ satisfies an equation similar to the diffusion equation.

If the fractional volume f, which is the percent of the volume occupied by the particles, is higher than about 10%, the correlation between the particles needs to be included in the formulations. This "dense" medium theory is still a challenging problem today [6], [33], [34].

4. ROUGH SURFACE SCATTERING

Now let us consider the scattering due to a rough surface or a rough interface. If the surface height is given by $z = f(x,y)$ and its variance is given by $\sigma^2 = \langle f^2 \rangle$, the surface is considered rough only if the standard deviation σ satisfies the "Rayleigh criterion" [1]. In general, the rough surface is characterized by the height variance σ^2 and the correlation distance ℓ. If $\sigma \ll \lambda$ and $\ell \leq \lambda$, then the perturbation technique is applicable. If the radius of curvature of the surface is much greater than a wavelength, the Kirchhoff approximation is applicable. In general, the specular reflection is reduced by the roughness while the diffuse scattering increases as the roughness increases. A definitive review of rough surface scattering was written by DeSanto and Brown [11] in which the spectral methodology and the connected diagram and smoothing expansions are discussed. Scattering by randomly perturbed quasi-periodic surfaces is addressed by Yueh, Shin, and Kong [38]. The regions of validity of the Kirchhoff and perturbation solutions are studied numerically by Chen and Fung [39]. Several new techniques have been introduced in recent years which are applicable to cases beyond the range of validity of the conventional perturbation and Kirchhoff approximations. They include an equivalent surface impedance [40], the smoothing method [41], the phase perturbation technique [42], [43], and the full wave theory [44]. Recent discovery of the enhanced backscattering from very rough surfaces has spurred intensive research on rough surface scattering outside the range of validity of the conventional theories [55] [56].

5. A NEED FOR EXPERIMENTAL STUDY AND BACKSCATTERING ENHANCEMENT

There has been considerable interest in the enhanced backscattering from randomly distributed scatterers and rough surfaces. This phenomenon is important in applications to lidars, surface optics, and electron localization in disordered media.

Backscattering enhancement phenomena have been observed for many years. It has been sometimes called the "retroreflectance" or the "opposition effect" [45]. An example is "glory" which appears around the shadow of an airplane cast on a cloud underneath the airplane. The moon is brighter at full moon than at other times. Many materials such as $BaSO_4$ or $MgCO_3$ are known to cause the enhancement of scattered light in the backward direction. Some soils and vegetation are also observed to cause backscattering enhancement. In spite of many observations, the several theories proposed to explain these phenomena are inadequate. For example, the shadowing theory is based on the fact that in the scattering direction other than in the back direction, less light may be observed because of the shadowing. Also the scattering pattern of the surface or the scatterers such as Mie scattering may be often peaked in the back direction. The lens hypothesis is that the material may often act as corner reflectors or lenses resulting in peaked backscattering.

Recently, more quantitative experimental and theoretical studies of the enhancement have been reported. Watson [46] noted that the backscattered intensity is twice the multiple scattering and the first-order scattering. de Wolf showed that the backscattered intensity from turbulence is proportional to the fourth-order moment and approximately twice the multiple scattered intensity. An excellent review is given by Kravtsov and Saichev [47].

Even though the above studies had been reported, strong interest and focused efforts for research in enhanced backscattering occurred only after the experimental work of Kuga and Ishimaru in 1984 [48] who conducted an experiment which showed that the scattering from latex microspheres is enhanced in the backward direction with a sharp angular width of a fraction of a degree. It was explained theoretically by Tsang and Ishimaru that the enhanced peak is caused by the constructive interference of two waves traversing through the same particles in opposite directions [49]. This was quickly recognized as evidence of "weak Anderson localization" and that the transport of electrons in a strongly disordered material is governed by multiple scattering which leads to "coherent backscattering" [50]-[54]. It is then shown that both electron localization in disordered material and photon localization in disordered dielectrics are governed by coherent backscattering which is caused by the constructive interference of two waves traversing in opposite directions. Our experimental work in 1984 was followed by several independent optical experiments showing that the backscattering enhancement is a weak localization phenomenon. The enhanced peak value is close to 2 and the angular width is governed by the diffusion length in the medium.

This is an example of experimental work greatly helping the development of theory and the understanding of physical phenomena. It should be noted that the backscattering enhancement can not be obtained by the radiative transfer theory, because the assumption under the radiative transfer theory automatically excludes the enhanced backscattering.

6. A NEED FOR NUMERICAL STUDY AND BACKSCATTERING ENHANCEMENT

Scattering by rough surfaces also exhibits enhanced backscattering. This was first demonstrated experimentally by O'Donnell and Mendez [55] for very rough surfaces where the rms heights are of the order of a wavelength and the rms slope is close to unity. This discovery was quickly followed by numerical simulations and further study. It was then recognized that the enhancement takes place outside the range of validity of conventional perturbation and Kirchhoff approximation methods. To guide the development of the theory, numerical studies were conducted using different approximate theories and then compared with exact numerical simulations. In particular, the second-order Kirchhoff approximation with proper shadowing functions has been shown to exhibit a numerical solution close to the exact simulation. These studies lead to the development of new theories and to the understanding of physical phenomena [56].

7. CONCLUSIONS

This paper first presented the state of the art in the modeling of waves in random media. Secondly, we pointed out that interactive research between experiment and modeling and numerical study and modeling is essential in uncovering and understanding new physical phenomena for waves in random media. Future directions in this field should include basic theoretical studies as well as applications to current practical problems. These basic studies include dense media, nonlinear media, inverse problems, and disordered media. Practical applications include remote sensing of geophysical media, biological media, surface physics, composite materials, communications, and imaging.

ACKNOWLEDGMENT

This work has been supported by the National Science Foundation, the U.S. Army Research Office, the U.S. Army Engineer Waterways Experiment Station, and the Office of Naval Research. Supercomputer time was granted by the San Diego Supercomputer Center.

REFERENCES

[1] A. Ishimaru, *Wave Propagation and Scattering in Random Media,* New York, Academic Press, 1978.
[2] V. I. Tatarskii, "The Effects of the Turbulent Atmosphere on Wave Propagation," U.S. Department of Commerce, TT-68-50464, Springfield, Virginia, 1971.
[3] J. W. Strohbehn, editor, *Laser Beam Propagation in the Atmosphere,* New York, Springer Verlag, 1978.
[4] B. J. Uscinski, *The Elements of Wave Propagation in Random Media,* New York, McGraw-Hill, 1977.
[5] S. M. Flatté, *Sound Transmission Through a Fluctuating Ocean,* London, Cambridge University Press, 1979.
[6] L. Tsang, J. A. Kong, and R. T.Shin, *Theory of Microwave Remote Sensing,* Wiley-Interscience, 1985.

[7] S. M. Rytov, Yu. A. Kravtsov, and V. I. Tatarskii, *Principles of Statistical Radiophysics,* vol. I - IV, Berlin, Springer-Verlag, 1987.
[8] U. Frisch, "Wave propagation in random media," in *Probabilistic Methods in Applied Mathematics,* New York, Academic Press, 1968.
[9] F. G. Bass and I. M. Fuks, *Wave Scattering from Statistically Rough Surfaces,* Oxford, Pergamon Press, 1979.
[10] P. Beckmann and A. Spizzichino, *Scattering of Electromagnetic Waves from Rough Surfaces,* New York, Pergamon, 1963.
[11] J. A. DeSanto and G. S. Brown, "Analytical techniques for multiple scattering from rough surfaces" in *Progress in Optics XXIII,* editor, E. Wolf, Elsevier Science Publishers B.V., 1986.
[12] M. Nieto-Vesperinas and J. C. Dainty, editors, *Scattering in Volumes and Surfaces,* Amsterdam, North-Holland, 1990.
[13] V. V. Varadan and V. K. Varadan, editors, *Multiple Scattering of Waves in Random Media and Random Rough Surfaces,* Pennsylvania State University, 1987.
[14] H. C. van de Hulst, *Multiple Scattering,* New York, Academic Press, 1980.
[15] P. L. Chow, W. E. Kohler, and G. C. Papanicolaou, *Multiple Scattering and Waves in Random Media,* North-Holland, 1981.
[16] J. B. Keller, "Stochastic equations and wave propagation in random media," Proc. Symp. Applied Math, 16, pp. 145-170, 1964.
[17] A. Ishimaru, "Wave propagation and scattering in random media and rough surfaces," *Proc. IEEE* Special Issue on Electromagnetics, editor, W. K. Kahn, 1991.
[18] V. I. Tatarskii and V. U. Zavorotnyi, "Strong fluctuations in light propagation in a randomly inhomogeneous medium," in *Progress in Optics,* vol. 18, editor, E. Wolf, pp. 204-256, Amsterdam, North-Holland, 1980.
[19] K. Furutsu, "Intensity correlation functions of light waves in a turbulent medium: An exact version of the two-scale method," *Appl. Opt.,* vol. 27, pp. 2127-2144, 1988.
[20] J. L. Codona, D. P. Creamer, S. M. Flatté, R. Frehlich, and F. Henyey, "Solutions for the fourth moment of a wave propagating in random media," *Radio Sci.,* vol. 21, pp. 929-948, 1986.
[21] I. M. Besieris, "Wave-kinetic method, phase-space path integrals, and stochastic wave equations," *J. Opt. Soc. Am.,* vol. A-2, pp. 2092-2099, 1985.
[22] J. M. Martin, S. M. Flatté, "Intensity images and statistics from numerical simulation of wave propagation in three-dimensional random media," *Appl. Opt.,* vol. 27, no. 11, pp. 2111-2126, 1988.
[23] M. Spivack and B. J. Uscinski, "Accurate numerical solution of the fourth-moment equation for very large values of Γ" *J. Modern Optics,* vol. 35, no. 11, pp. 1741-1755, 1988.
[24] M. J. Beran and R. Mazar, "Intensity fluctuations in a quadratic channel," *J. Acoust. Soc. Am.,* vol. 82, no. 2, pp. 588-592, 1987.
[25] C. L. Rino, "A spectral-domain method for multiple scattering in continuous randomly irregular media," *IEEE Trans. Antennas Propag.,* vol. 36, no. 8, pp. 1114-1128, 1988.
[26] L. C. Andrews and R. L. Phillips, "Mathematical genesis of the I-K distribution for random optical fields," *J. Opt. Soc. Am.,* vol. A-3, pp. 1912-1919, 1986.
[27] E. Jakeman and P. N. Pusey, "Significance of K-distributions in scattering experiments," *Phys. Rev. Lett.,* vol. 40, pp. 546-550, 1978.
[28] S. Chandrasekhar, *Radiative Transfer,* London, Oxford University Press, 1960.
[29] G. W. Kattawar, G. N. Plass, and F. E. Catchings, "Matrix operator theory of radiative transfer," *Appl. Opt.,* vol. 12, pp. 1071-1084, 1973.
[30] W. M. Boerner, editor, *Inverse Methods in Electromagnetic Imaging,* Dordrecht, Netherlands, D. Reidel, 1985.
[31] J. J. van Zyl, H. A. Zebker, and C. Elachi, "Imaging radar polarization signatures: Theory and observations," *Radio Sci.,* vol. 22, pp. 529-543, 1987.
[32] F. T. Ulaby, R. K. Moore, and A. K. Fung, *Microwave Remote Sensing,* Reading, MA, Addison-Wesley, 1982.

[33] K.-H. Ding and L. Tsang, "Effective propagation constants of a dense nontenuous media with multi-species of particles," *J. Electromagnetic Waves and Applications,* vol. 2, no. 8, pp. 757-777, 1988.
[34] V. Twersky, "Low-frequency scattering by mixtures of correlated nonspherical particles," *J. Acoust. Soc. Am.,* vol. 84, no. 1, pp. 409-415, 1988.
[35] Y. Kuga and A. Ishimaru, "Imaging of an object behind randomly distributed particles using coherent illumination," *J. Wave-Material Interaction,* vol. 3, no. 2, pp. 105-112, 1988
[36] A. Ishimaru, "Diffusion of light in turbid material," *Appl. Opt.,* vol. 28, no. 12, pp. 2210-2215, 1989.
[37] A. Ishimaru, "Diffusion of a pulse in densely distributed scatterers," *J. Opt. Soc. Am.,* vol. 68, no. 8, pp. 1045-1050, 1978.
[38] H. A. Yueh, R. T. Shin, and J. A. Kong, "Scattering from randomly perturbed periodic and quasiperiodic surfaces," in *Progress in Electromagnetic Research,* editor, J. A. Kong, New York, Elsevier, 1989.
[39] M. F. Chen and A. K. Fung, "A numerical study of the regions of validity of the Kirchhoff and small-perturbation rough surface scattering models," *Radio Sci.,* vol. 23, no. 2, pp. 163-170, 1988.
[40] S. Ito, "Analysis of scalar wave scattering from slightly rough random surfaces: A multiple scattering theory," *Radio Sci.,* vol. 20, pp. 1-12, 1985.
[41] J. G. Watson and J. B. Keller, "Rough surface scattering via the smoothing method," *J. Acoust. Soc. Am.,* vol. 75, pp. 1705-1708, 1984.
[42] D. P. Winebrenner and A. Ishimaru, "Application of the phase-perturbation technique to randomly rough surfaces," *J. Opt. Soc. Am.,* vol. 2, no. 12, pp. 2285-2294, 1985.
[43] S. L. Broschat, L. Tsang, A. Ishimaru, and E. I. Thorsos, "A numerical comparison of the phase perturbation technique with the classical field perturbation and Kirchhoff approximations for random rough surface scattering," *J. Electromagnetic Waves and Applications,* vol. 2, no. 1, pp. 85-102, 1987.
[44] E. Bahar and M. A. Fitzwater, "Full wave theory and controlled optical experiments for enhanced scatter and depolarization by randomly rough surfaces," *Opt. Commun.,* vol. 63, pp. 355-360, 1987.
[45] W. G. Egan and T. W. Hilgeman, *Optical Properties of Inhomogeneous Materials,* New York, Academic Press, 1979.
[46] K. Watson, "Multiple scattering of electromagnetic waves in an underdense plasma," *J. Math. Phys.,* vol. 10, pp. 688-702, 1969.
[47] Yu. A. Kravtsov and A. I. Saichev, "Effects of double passage of waves in randomly inhomogeneous media," *Sov. Phys. Usp.,* vol. 25, pp. 494-508, 1982.
[48] Y. Kuga and A. Ishimaru, "Retroreflectance from a dense distribution of spherical particles," *J. Opt. Soc. Am. A,* vol. 1. no. 8, pp. 831-835, 1984.
[49] L. Tsang and A. Ishimaru, "Backscattering enhancement of random discrete scatterers," *J. Opt. Soc. Am. A,* vol. 1, no. 8, pp. 836-839, 1984.
[50] P. E. Wolf and G. Maret, "Weak localization and coherent backscattering of photons in disordered media," *Phys. Rev. Lett.,* vol. 55, pp. 2696-2699, 1985.
[51] E. Akkermans, P. E. Wolf, R. Maynard, and G. Maret, "Theoretical study of the coherent backscattering of light by disordered media," *J. Phys. France,* vol. 49, pp. 77-98, 1988.
[52] M. P. Van Albada and A. Lagendijk, "Observation of weak localization of light in a random medium," *Phys. Rev. Lett.,* vol. 55, pp. 2692-2695, 1985.
[53] P. W. Anderson, "Absence of diffusion in certain random lattices," *Phys. Rev.,* vol. 109, pp. 1492-1505, 1958.
[54] P. W. Anderson, "The questions of classical localization, A theory of white paint?" *Philos. Mag. B,* vol. 52, pp. 505-509, 1985.
[55] K. A. O'Donnell and E. R. Mendez, "Experimental study of scattering from characterized random surfaces," *J. Opt. Soc. Am. A,* vol. 4, pp. 1194-1205, 1987.
[56] A. Ishimaru and J. S. Chen, "Scattering from very rough surfaces based on the modified second-order Kirchhoff approximation with angular and propagation shadowing," *J. Acoust. Soc. Am.,* to appear 1990.
[57] A. Ishimaru, editor, Feature issue on "Wave Propagation and Scattering in Random Media," *J. Opt. Soc. Am.,* vol. 2, no. 12, pp. 2066-2348, 1985.

SPECTRAL CHANGES INDUCED BY SCATTERING FROM SPACE-TIME FLUCTUATIONS

Emil Wolf *

Department of Physics and Astronomy
University of Rochester, Rochester, NY 14627

In the majority of theoretical studies dealing with the scattering of electromagnetic waves on random media one assumes that the incident field is monochromatic and that the density fluctuations of the medium may be described by equilibrium thermodynamics. In the last few years it was found that when these restrictions are removed, a number of interesting new phenomena are revealed. In this talk a review of these developments is presented. In particular an account is given of several recent investigations[1] regarding spectral changes produced by scattering of light of any state of coherence and polarization and with arbitrary spectrum on a medium whose response is random both in space and in time, but is statistically homogeneous, isotropic and stationary, at least in the wide sense. It is shown that the spectrum of the radiation may be shifted towards the longer or towards the shorter wavelengths by the scattering process, thus inducing redshifts or blueshifts of spectral lines;[2] and moreover, that conditions may be found which ensure that these shifts imitate the Doppler effect in its essential features, even though the source of the radiation, the scatterer and the observer are all at rest relative to each other.[3,4] This possibility is of particular interest for astronomy.[5]

The limiting case of media whose response is time-independent on the macroscopic scale is also considered.[6] It is known that such media may give raise to enhanced backscattering. Our review includes a brief account of a recent analysis which showed that in enhanced backscattering redshifts of spectral lines of the kind discussed in this talk may be generated.[7]

Acknowledgement

Most of the research described in this note was supported by the Army Research Office.

REFERENCES

1. E. Wolf and J.T. Foley, "Scattering of electromagnetic fields of any state of coherence from space-time fluctuations", *Phys. Rev.* A 40:579-587 (1989).

2. J.T. Foley and E. Wolf, "Frequency shifts of spectral lines generated by scattering from space-time fluctuations", *Phys. Rev.* A. 40:588-598 (1989).

3. E. Wolf, "Correlation-induced Doppler-like frequency shifts of spectral lines", *Phys. Rev. Lett.* 63:2220-2223 (1989).

4. D.F.V. James and E. Wolf, "Doppler-like frequency shifts generated by dynamic scattering", *Phys. Letts.* A. 146:167-171 (1990).

5. D.F.V. James, M. P. Savedoff and E. Wolf, "Shifts of spectral lines caused by scattering from fluctuating random media", *Astrophys. J.* 359:67-71 (1990); see also E. Wolf, "Non-cosmological redshifts of spectral lines", *Nature* 326:363-365 (1987).

6. E. Wolf, J.T. Foley and F. Gori, "Frequency shifts of spectral lines produced by scattering from spatially random media", *J. Opt. Soc. Amer.* A. 6:1142-1149 (1989); *ibid* 7:173 (1990).

7. A. Lagendijk, "Terrestrial redshift from a diffuse light source", *Phys. Letts.* A 147:389-392 (1990).

ROUGH SURFACE SCATTERING

John A. DeSanto and Richard J. Wombell

Department of Mathematical and Computer Sciences
Colorado School of Mines
Golden, CO 80401 USA

ABSTRACT

A summary of the theoretical and computational approaches to rough surface scattering is presented. Future directions are indicated.

INTRODUCTION

There has been much recent interest in the numerical calculation of the scattering from rough surfaces. Part of this interest is due to the observation and attempts to model the enhanced backscatter phenomenon [O'Donnell and Méndez, 1987]. Because this arises for very rough surfaces and is a particularly sharp effect when the *rms* slope is large, present approximation methods are generally inadequate to describe the scattering, and the full integral equations describing the scattering must be solved numerically. The enhancement effect was not anticipated from computational solutions of theoretical models, and presently, although the models do predict an effect, there is some remaining discrepancy between these results and the experimental measurements.

We describe below three different theoretical approaches to the scattering problem and the state of their computational implementation. One method, the coordinate-coordinate (CC) approach, is preferred for its stability. This is used for our computations for the direct scattering problems we consider. For the inverse problem of surface reconstruction from incident and scattered field data the framework of a second method, the spectral-coordinate (SC) method, is used. The third approach, the spectral-spectral (SS) method, is also described. Each method requires matrix inversion for its solution, and the acronyms refer to the way the rows and columns of the specific matrix are sampled, either in coordinate-space or Fourier-transform or spectral-space. A review of the methods is available [DeSanto and Brown, 1986].

Experimental measurements are essentially averaging processes due to the finite size of any detector. For theoretical modeling, this averaging is expressed as an averaging over a statistical ensemble of surfaces. We briefly mention below how this ensemble is created. Two approaches to solving the stochastic integral equations are available. The first is to analytically average the equations and to solve numerically a resulting equation on some moment of the scattered field. This has been done in the SS-method for the first moment or the coherent specularly scattered field. The specular nature of the moment arises because

Directions in Electromagnetic Wave Modeling
Edited by H.L. Bertoni and L.B. Felsen, Plenum Press, New York, 1991

homogeneous statistics on the surface height distribution are assumed. We mention one result of this below. However, for the backscatter enhancement effect, it is necessary to calculate the incoherent intensity or second moment of the scattered field. This has as yet not been done due to the mathematical complexity of the resulting Bethe-Salpeter equation. The second method, the approach used here, is to solve the integral equation numerically for each member of the ensemble of surfaces and then compute the resulting incoherent (or coherent) intensity by averaging over the individual surface results. The intensity from an individual surface of the ensemble may or may not produce a large backscatter peak for large surface height and slope. The average, however, does.

The computations are presently limited to one-dimensional surfaces and we limit our results to the scalar case of a Dirichlet boundary condition. Electromagnetically this corresponds to the case of an electric field perpendicular to the plane described by the incident wave vector and the surface height (TE-polarization). No polarization change takes place because the surface generator is parallel to the electric field. We refer elsewhere [DeSanto and Wombell, 1991b] for extensive computational results and to a further paper [Wombell and DeSanto, 1991a] for computational results on the inverse problem. Surface realizations can be generated by means of a spectral method [Thorsos, 1988].

THEORETICAL APPROACHES AND NUMERICAL METHODS

Several approaches to theory and computations are presently being pursued.

The first is the solution of coordinate-space surface integral equations in one-dimension as used by Thorsos for the Dirichlet problem [Thorsos, 1988] and others for the more general interface problem [Maradudin, Méndez and Michel, 1989]. For the Dirichlet case a pair of equations relates the normal derivative of the field N on the surface, s, either to the incident field ψ_0 on the surface or the scattered field ψ_s evaluated at a point $\mathbf{x}$ in the field

$$\psi_0(\mathbf{x}_s) = \frac{k}{4}\int_{-\infty}^{\infty} H_0^{(1)}(k|\mathbf{x}_s - \mathbf{x}_s'|)N(x')dx' \tag{1}$$

$$\psi_s(\mathbf{x}) = -\frac{k}{4}\int_{-\infty}^{\infty} H_0^{(1)}(k|\mathbf{x} - \mathbf{x}_s'|)N(x')dx' \tag{2}$$

where $\mathbf{x}_s = (x, s(x))$ is the point on the surface and $N(x) = (ik)^{-1}\partial_n\psi(\mathbf{x}_s)$ is the normal derivative of the total field evaluated on the surface with non-unit normal and scaled with wavenumber k. By discretizing both equations, (1) can be solved for N and (2) evaluated for the scattered field. Typical published results involve repeating this process many times for a large number of (relatively short) surfaces with the same statistics and then the resulting intensities averaged in order to reduce the (large) speckle in the individual surface results. Equation (1) is solved for N using discretization and matrix inversion. Both rows and columns for the matrix result from sampling on the coordinate surface and we therefore refer to the method as the coordinate-coordinate (CC) approach. This is the major numerical method in use.

A second method, referred to as the k-space diagram or spectral-spectral (SS) method has also been formulated for scalar, electromagnetic and elastic problems for stochastic two-dimensional surfaces (and for one-dimensional surfaces with corresponding restrictions). [Zipfel and DeSanto, 1972; DeSanto, 1973, 1974; DeSanto and Brown, 1986]. The type of stochastic integral equation that arises is completely in Fourier-transform or spectral space. For a two-dimensional surface the k-space equation for the full Green's function Γ is a three-dimensional (off-shell) equation of the form

$$\Gamma(\mathbf{k}', \mathbf{k}'') = V(\mathbf{k}', \mathbf{k}'')A(\mathbf{k}' - \mathbf{k}'') + \int\int\int V(\mathbf{k}', \mathbf{k})A(\mathbf{k}' - \mathbf{k})G(k)\Gamma(\mathbf{k}, \mathbf{k}'')d\mathbf{k}\,, \tag{3}$$

with incident ($\mathbf{k}''$) and exiting ($\mathbf{k}'$) states and an integration over all intermediate ($\mathbf{k}$) states. Here G is the Fourier transform of the free-space three-dimensional Green's function (with wavenumber k_0), V is a (kinematical) vertex function of the form

$$V(\mathbf{k}',\mathbf{k}) = -\frac{2i}{(2\pi)^3}\left\{\frac{k_t'\cdot(k_t'-k_t)}{k_z'-k_z} + (k_0^2 - k_t'^2)P\frac{1}{k_z'}\right\}, \tag{4}$$

where P refers to the Cauchy principal value and A is termed the interaction function or the phase modulated amplitude spectrum. It is a standard integral in calculations of this type and has the form

$$A(\mathbf{k}) = \int\int \exp\Big[-ik_t\cdot x_t - ik_z s(x_t)\Big]\,dx_t\,, \tag{5}$$

with $x_t = (x, y)$ the transverse part of the spatial three-vector. The equation is off-shell in the sense that k_z values are not confined on the Ewald sphere in k-space. The full three-dimensional equation must be solved first and then the resulting amplitude set on-shell to find the physical scattering amplitude. To each term can be associated a Feynman-like diagram and the integral equation interpreted in propagation-interaction, etc. terminology. The three-dimensional equation cannot be solved numerically at present. It can be analytically averaged over the (Gaussian) statistical ensemble of surfaces to yield a (Dyson) equation for the mean or first moment of the scattered field. This is referred to as the coherent field and for statistically homogeneous surfaces yields a scattered field in only the specular direction. An example of the equation for the coherent scattering amplitude T which arises from this averaging process is given by

$$T(k_z', k_z'') = \sum_{j=1}^{\infty} B_i(k_z', k_z'') + \int \sum_{j=1}^{\infty} B_j(k_z', k_z) G(k) T(k_z, k_z'') dk_z\,, \tag{6}$$

again in terms of incident, exiting and intermediate states characterized by the z-component of momentum (transverse momentum is parametric). Both the Born term and the kernel are an infinite series of terms (connected diagrams). Truncation of these series yields approximations T_{ij} (i terms in the Born series, j terms in the kernel). T_{11} has been solved numerically [DeSanto and Shisha, 1974] using discrete sampling and matrix inversion. The matrix rows and columns are sampled in Fourier transform or spectral space and the method is referred to as the spectral-spectral (SS) method. It yields a higher coherent return than the single scatter result $T_{10} = B_1$ given by

$$\left|T_{10}\right|^2 = \exp(-4\Sigma^2)\,, \tag{7}$$

where $\Sigma = k\sigma\cos\theta_i$ is the Rayleigh roughness parameter. The latter is just the square of the Fourier-transform of the Gaussian *pdf* evaluated in the specular direction on-shell. The T_{11} result agreed with data for large roughness ($\Sigma \sim 2$). The results are summarized in [DeSanto and Brown, 1986]. Higher order integral equation approximations are presently computationally intractable as are higher order moment equations (e.g. the Bethe-Salpeter equation for the incoherent field which is of particular recent interest in terms of the backscatter enhancement effect). The advantage of the equations may lie in constructing approximations using state-to-state transition arguments. In order to formulate the equations the full CC equations are necessary, and the approach thus serves as a synthesis of the two methods.

The third method also involves matrix inversion for the solution. It has been published by us [DeSanto, 1985] for one-dimensional surface scattering for the full transmission problem. For the case of an interface with an arbitrary incident field, two equations relate the field on the surface, $\phi(x) = \psi(\mathbf{x}_s)$, and its scaled normal derivative, $N(x) = (ik)^{-1}\partial_n\psi(\mathbf{x}_s)$ (with a non-unit normal),

$$F^{+}(\mu) = \int_{-\infty}^{\infty} \Big[(m + \mu s'(x))\phi(x) - N(x)\Big] e^{-ik_1(\mu x - ms(x))} dx \; ; \; m = \sqrt{1-\mu^2} \; , \tag{8}$$

$$0 = \int_{-\infty}^{\infty} \Big[(\sqrt{K^2 - \mu^2} - \mu s'(x))\frac{\phi(x)}{\rho} + N(x)\Big] e^{-ik_1(\mu x + \sqrt{K^2-\mu^2} s(x))} dx \, . \tag{9}$$

Here $K = k_2/k_1$ and $\rho = \rho_2/\rho_1$, with k_1 and ρ_1 and k_2 and ρ_2 being the wavenumber and density above and below the interface, respectively. (The corresponding electromagnetic parameters describe two dielectric media with different permitivities and permeabilities.) The scattered field ψ_s and transmitted field ψ_t are then given by

$$\psi_s(x,y) = \int_{-\infty}^{\infty} A(\mu) e^{ik_1(\mu x + mz)} d\mu \; , \; \psi_t(x,y) = \int_{-\infty}^{\infty} B(p) e^{ik_2(px - qz)} dp \tag{10}$$

($q = \sqrt{1-p^2}$) where the scattering (A) and transmission (B) amplitudes are given by

$$A(\mu) = \frac{k_1}{4\pi m}\Big[-F^{-}(\mu) + \int_{-\infty}^{\infty} \Big[(m - \mu s'(x))\phi(x) + N(x)\Big] e^{-ik_1(\mu x + ms(x))} dx\Big] \; , \tag{11}$$

and

$$B(p) = \frac{k_2}{4\pi q}\int_{-\infty}^{\infty} \Big[(q + ps'(x))\frac{\phi(x)}{\rho} - \frac{1}{K}N(x)\Big] e^{-ik_2(px - qs(x))} dx \, . \tag{12}$$

The functions F^+ and F^- depend only upon the incident field ψ_0 and its z derivative, $M_0 = (ik)^{-1}\partial_z\psi_0$, at some arbitrary distance H from the surface. They are given by

$$F^{\pm}(\mu) = e^{\pm ik_1 mH} \int_{-\infty}^{\infty} \Big[\psi_0(x,H)m \mp M_0(x,H)\Big] e^{-ik\mu x} dx \, . \tag{13}$$

These equations are discretized and (8) and (9) solved for ϕ and N. Then A and B and ψ_s and ψ_t can be evaluated. The procedure again involves matrix inversion. We refer to this as the spectral-coordinate (SC) approach since the rows of the matrix formed from (8) are sampled in spectral space while the columns are sampled in coordinate space. Rino, Crystal, Koide and Ngo [1989], have solved a reduced set of similar equations for the Dirichlet problem (with limitations on the roughness) by discretization. It is found for large roughness that the method although well-posed is ill-conditioned. However, the equations are ideal for the surface inversion problem and we briefly discuss this below.

The SC method can be easily generalized in many ways. Corresponding equations for two-dimensional scalar problems are a straightforward generalization of the above. Arbitrary incident fields are a second possible generalization. The electromagnetic generalization to a dielectric interface yields a total of twelve equations ($i = 1,2,3$)

$$U_i^{\pm} = S_{1i}^{\pm} \; , \; V_{2i}^{\pm} = S_{2i}^{\pm} \; , \tag{14}$$

where the U and V integrals are expressed along the boundary $s(x_t)$ as $\mathbf{x}_s = (x,y,s)$

$$U_i^{\pm} = \int\int_{s(x_t)} G_1^{\mp}(\mathbf{x}_s)\Big\{(\mu_1/\varepsilon_1)^{1/2} J_i + \Big[\mathbf{M}^{\pm} \times (\mathbf{n} \times \mathbf{E})\Big]_i - M_i^{\pm} n_j E_j\Big\} dx_t \, , \tag{15}$$

and

$$V_i^{\pm} = \int\int_{s(x_t)} G_2^{\mp}(\mathbf{x}_s)\Big\{(\mu_2/\varepsilon_2)^{1/2} J_i + \Big[\mathbf{P}^{\pm} \times (\mathbf{n} \times \mathbf{E})\Big]_i - P_i^{\pm}\varepsilon^{-1} n_j E_j\Big\} dx_t \, , \tag{16}$$

in terms of six unknowns J_i and E_i. Here we have used the Bloch or plane wave eigenstates

$$G_1^{\mp}(\mathbf{x}) \;=\; \exp(ik_1\mathbf{M}^{\pm} \cdot \mathbf{x}) \, , \tag{17}$$

$$G_2^{\pm}(\mathbf{x}) \;=\; \exp(ik_2\mathbf{P}^{\pm}\cdot\mathbf{x}) \,, \tag{18}$$

where $k_j = (\varepsilon_j\mu_j)^{1/2}\omega/c$ in terms of the dielectric constants (ε_j) and magnetic permeabilities (μ_j) of the two media $(\varepsilon = \varepsilon_2/\varepsilon_1 \;,\;\; \mu = \mu_2/\mu_1)$, the non-unit surface normal $\mathbf{n}$ and the definitions

$$\mathbf{M}^{\pm} = (-M_t,\ \pm M_3)\ ;\ \ M_3 = (1 - M_t^2)^{1/2} \,, \tag{19}$$

and

$$\mathbf{P}^{\pm} = (-P_t,\ \pm P_3)\ ;\ \ P_3 = (1 - P_t^2)^{1/2} \,. \tag{20}$$

The functions S in (14) are evaluated at the largest positive surface excursion (s_1) and the largest negative excursion (s_2) as $(\mathbf{x}_j = (x, y, s_j))$

$$S_{1i}^{\pm} \;=\; \int\!\!\int_{s_1} G_1^{\pm}(\mathbf{x}_1)\Big\{(i/k_1)\Big[\mathbf{n}_3 \times (\vec{\nabla}\times\mathbf{E}^{(1)})\Big]_i + \Big[\mathbf{M}^{\pm}\times(\mathbf{n}_3\times\mathbf{E}^{(1)})\Big]_i - M_i^{\pm}E_3^{(1)}\Big\}dx_t \,, \tag{21}$$

and

$$S_{2i}^{\pm} \;=\; \int\!\!\int_{s_2} G_2^{\pm}(\mathbf{x}_2)\Big\{(i/k_2)\Big[\mathbf{n}_3 \times (\vec{\nabla}\times\mathbf{E}^{(2)})\Big]_i + \Big[\mathbf{P}^{\pm}\times(\mathbf{n}_3\times\mathbf{E}^{(2)})\Big]_i - P_i^{\pm}E_3^{(2)}\Big\}dx_t \,. \tag{22}$$

The spectral decompositions for the total (incident plus scattered) and transmitted electric fields on these surfaces can be represented as

$$E_i^{(1)}(\mathbf{x}) = E_i^{(0)}(\mathbf{x}) + \int\!\!\int A_i(a_t)\exp(ik_1\mathbf{a}\cdot\mathbf{x})da_t \,, \tag{23}$$

and

$$E_i^{(2)}(\mathbf{x}) = \int\!\!\int B_i(b_t)\exp(ik_2\mathbf{b}\cdot\mathbf{x})db_t \,, \tag{24}$$

with

$$\mathbf{a} = (a_t, a_3 = (1 - a_t^2)^{1/2}) \,, \tag{25}$$

and

$$\mathbf{b} = (b_t, b_3 = -(1 - b_t^2)^{1/2}) \,. \tag{26}$$

We thus have to solve six equations for the unknowns J_i and E_i on the surface, and use six equations to evaluate the scattered (A_i) and transmitted (B_i) spectral amplitudes.

COMPUTATIONAL ISSUES

The CC method is presently the best method of computation for one-dimensional problems. It is very stable even for fairly large heights and slopes. The SC method requires the evaluation of integrals with very rapidly oscillating integrands over the full range of integration. Several approaches have been tried, but for surfaces with large heights and slopes, the resulting matrix is badly conditioned. A brief discussion of why this is so is presented below. Analogous integrals also occur in the SS method which is already of higher dimensionality than either CC or SC since it is an off-shell equation. A one-dimensional rough surface yields one integration in the CC and SC methods, but is projected into an off-shell two-dimensional space for the SS method.

All the methods have been formulated in two-dimensions and for transmission problems. All are formulated for the electromagnetic problem but for polarization effects require two-dimensional integrations as well as a six-dimensional vector of boundary unknowns. Computational complexity rapidly intrudes. No full multiple scattering numerical solutions are presently available.

As an example of some of the computational problems, we consider the SC method for the Dirichlet problem. Here $\phi(x) = 0$, equations (9) and (10) are not present since we have perfect reflection, and (8) and (11) become (with examples for which $F^{-}(\mu) = 0$) for a finite surface (length L)

$$-F^{+}(\mu) = \int_{-L/2}^{L/2} N(x) e^{-ik_1(\mu x - ms(x))} dx \,. \tag{27}$$

and

$$4\pi m k_1^{-1} A(\mu) = \int_{-L/2}^{L/2} N(x) e^{-ik_1(\mu x + ms(x))} dx \,. \tag{28}$$

One procedure to solve (27) for $N(x)$ would be to expand the kernel in a Fourier series

$$e^{-ik_1(\mu x - ms(x))} = \sum_{n=-\infty}^{\infty} a_n(\mu) e^{i2\pi nx/L} \,, \tag{29}$$

and then to discretize in μ by replacing $\mu \rightarrow \mu_j = j \triangle \mu$ with $|j| \leq J$, $m(\mu) \rightarrow m_j$, $-F_1(\mu) \rightarrow F_j$, and $a_n(\mu) \rightarrow a_{jn}$. If we define the Fourier transform of N

$$\widetilde{N}_n = \int_{-L/2}^{L/2} N(x) e^{i2\pi nx/L} dx \,, \tag{30}$$

then the discretized equations become

$$F_j = \sum_n a_{jn} \widetilde{N}_n \,, \tag{31}$$

which must be solved for the n Fourier-transform samples. The matrix inversion works for $s(x)/\lambda$ small, but when this quantity gets large it becomes ill-conditioned.

Alternatively a straight-forward discretization of (27) ($x_{j'} = j' \triangle x$, $s_{j'} = s(x_{j'})$) yields the equation

$$F_j = \sum_{j'} M_{jj'} N_{j'} \qquad |j'| \leq J \,, \tag{32}$$

to solve for the $2J+1$ spatial samples of N. Here the matrix is

$$M_{jj'} = \exp\left\{\frac{i2\pi}{\lambda}(m_j s_{j'} - j \triangle \mu j' \triangle x)\right\} . \tag{33}$$

We have that $m_{-J} = m_J$, and periodicity imposes the constraint $s_{-J} = s_J$, and the requirement in the phase of (33) that $\Delta\mu \triangle x = j/(2J+1)$. We choose the sampling interval $\triangle x = \lambda/p$, $p = 2,3,4,\ldots$ where p is the number of samples per wavelength. There are $2J+1$ real m_j only if $p = 2$, but this sampling rate seriously undersamples a rapidly oscillating surface. If (27) is scaled by multiplying through by $\exp(ik_1 m \gamma s_m)$ where $\gamma > 1$ and s_m is the absolute value of the maximum surface excursion, $s(x)$ is replaced by

$$h(x) = s(x) + \gamma s_m > 0 \,, \tag{34}$$

and for $p > 2$ the rows of the matrix m for j large are purely decaying. The determinant of m is correspondingly very small. A singular value decomposition of this matrix clearly illustrates its ill-conditioned nature. The essential reason for the failure is that the problem is *not* a Fourier transform but a Weyl transform.

SURFACE INVERSION

The SC method has been very useful in developing an algorithm to reconstruct the surface from scattered data. For a shallow surface we expand the exponentials involving $s(x)$ in (27) and (28) in a two-term Taylor expansion. The results are

$$- F^{+}(\mu) \cong \int_{-L/2}^{L/2} N(x) e^{-ik_1 \mu x} \Big[1 + ik_1 m s(x)\Big] dx \,, \tag{35}$$

and

$$4\pi m k_1^{-1} A(\mu) \cong \int_{-L/2}^{L/2} N(x) e^{-ik_1 \mu x} \Big[1 - ik_1 m s(x)\Big] dx \,. \tag{36}$$

One-half the sum of (35) and (36) is

$$\delta_{+}(\mu) = \int_{-L/2}^{L/2} N(x) e^{-ik_1 \mu x} dx \,, \tag{37}$$

and one-half the difference divided by $ik_1 m$ is

$$\delta_{-}(\mu) = \int_{-L/2}^{L/2} N(x) s(x) e^{-ik_1 \mu x} dx \,. \tag{38}$$

The Fourier inverse of (37) and (38) is thus

$$F^{-1}\Big[\delta_{+}(x)\Big] = N(x) \,, \tag{39}$$

and

$$F^{-1}\Big[\delta_{-}(x)\Big] = N(x) s(x) \,. \tag{40}$$

We divide (40) by (39) to produce a reconstruction $r(x)$ of the profile $s(x)$. Examples are presented as the surface height changes and as the cone of scattered field data is narrowed around the incident direction. The incident field was a Gaussian beam and the (complex) $A(\mu)$ were computed using the CC method. The computational results can be found elsewhere [Wombell and DeSanto, 1991a and DeSanto and Wombell, 1991b].

Briefly, the method works well for shallow surfaces and for normal incidence the data aperture narrowed to $a \pm 30°$ cone about this direction before serious degradation occurs.

COHERENT AND INCOHERENT SCATTERING

The solution of the integral equation for the jth member of the ensemble of surfaces yields a complex scattered field ψ_j which is considered as a function of scattered angle and depends on the incident angle. For N_e ensemble members the ensemble averaged incoherent intensity (ic) is defined as

$$\Big\langle I_{ic} \Big\rangle = \frac{1}{E_i} \frac{1}{N_e} \sum_{j=1}^{N_e} |\psi_j|^2 \,, \tag{41}$$

normalized to the incident field energy E_i. The corresponding average coherent (c) intensity is defined in terms of the mean field $\bar{\psi}$ as

$$\begin{aligned} \Big\langle I_c \Big\rangle &= \frac{1}{E_i} |\bar{\psi}|^2 \,, \\ &= \frac{1}{E_i} \Big| \frac{1}{N_e} \sum_{j=1}^{N_e} \psi_j \Big|^2 \,, \\ &= \frac{1}{E_i N_e^2} \sum_{j=1}^{N_1} \Big|\psi_j\Big|^2 + \frac{2}{E_i N_e^2} Re \sum_{j=1}^{N_e} \sum_{p \neq j}^{N_e} \psi_j \psi_p^* \,, \end{aligned} \tag{42}$$

where *Re* represents the real part. The first term on the *rhs* of (42) is up to a factor N_e just the incoherent intensity. The second term is an interference term (IF), and (42) can be written as

$$\langle I_c \rangle = \frac{1}{N_e} \langle I_{ic} \rangle + IF \,. \tag{43}$$

The interesting feature of this equation is that if the coherent intensity is zero (as it is for example in the backscattered direction) and the incoherent intensity has a peak in the backscatter direction (as it can for large roughness and slope) then the same type of peak must occur in the interference term. For large roughness the incoherent predominates over the coherent in the specular direction [DeSanto and Wombell, 1991b].

CONCLUSIONS

The single scattering Kirchhoff approximation does not exhibit the backscattering enhancement effect although the surface normal derivative agrees surprisingly well with the Kirchhoff approximation except when the surface is tangent to the (center-angle of the Gaussian) incident field. There is evidence that the double-scatter Kirchhoff including shadowing does exhibit the effect. The exact ensemble average Fourier transform of this normal derivative on the surface resembles a diffraction pattern with some evidence of surface waves and may be directly related to the scattering properties of the surface. The coherent intensity computed using these ensemble methods appears to fall off as a Gaussian even for large roughness and this opens some questions on the analytic averaging techniques used to form moment equations as well as on the interpretation of coherent data itself. Finally some interesting surface reconstructions have also recently been done and although they are for shallow surfaces, begin to quantify the data necessary to do the reconstruction [Wombell and DeSanto, 1991a and DeSanto and Wombell, 1991b].

Finally we mention the future directions for the theoretical, computational and experimental areas for rough surface scattering. For the above three areas to interact the developments must presently be confined to one-dimensional surfaces. This may seem trivial but it is not. Firstly, the computational solutions for one-dimensional surfaces are being carried out in many places. Results in terms of energy checks, etc., on the computations are good. There is internal consistency. However, although computational backscatter enhanced results are readily available, there are many discrepancies between the computational and experimental results. We get similar effects but there is no overlay, nor is there a good physical explanation of the effect. Many things remain to be discovered in one-dimensional scattering. In particular, it should be possible to develop better scattering models and approximations, especially since the exact results are widely available to easily check.

Two-dimensional exact computations for scalar problems are a much bigger computer problem, and two-dimensional electromagnetic problems even bigger still. These are being pursued but are long and difficult problems. Better one-dimensional approximations may help this effort considerably although to be able to reach the stage presently occupied by one-dimensional problems will take several years.

Experimental surface characterization has become an issue of its own as well as for comparison to theoretical calculations. Two directions are suggested. One is to do the experiments in the microwave regime where the surface can be very accurately measured. The other is to develop interferometric techniques to better quantify optical surfaces which are presently harder to measure. In particular, we note that the algorithms used to translate experimental scattering measurements into surface profiles are often very crude single scatter models. This should be improved. Surface inversion calculations can help by providing improved algorithms.

Lastly we have noted a discrepancy for large roughness in the coherent scattering calculations done using ensemble methods and the equation averaging method. This is an area of interest for both surface scattering and random volume problems where analogous techniques are used. This is related to analytical approximation schemes mentioned above.

ACKNOWLEDGMENTS

This work was supported under contracts from the Office of Naval Research (Ocean Acoustics Division) and the Army Research Office (Geosciences Directorate).

REFERENCES

DeSanto, J.A., 1973, Scattering from a random rough surface: diagram methods for elastic media, J. Math. Phys., 14:1566.

DeSanto, J.A., 1974, Green's function for electromagnetic scattering from a random rough surface, J. Math. Phys., 15:283.

DeSanto, J.A., 1985, Exact spectral formalism for rough-surface scattering, J. Opt. Soc. Am. A2:2202.

DeSanto, J.A., and Brown, G.S., 1986, Analytical techniques for multiple scattering from rough surfaces, Progress in Optics XXIII, ed. E. Wolf, Elsevier Science Publ. B.V.

DeSanto, J.A., and Shisha O., 1974, Numerical solution of a singular integral equation in random rough surface scattering theory, J. Comput. Phys. 15:286.

DeSanto, J.A., and Wombell, R.J., 1991b, Waves in Random Media, special issue proceedings of the workshop on "Modern Analysis of Scattering Phenomena."

Maradudin, A.A., Méndez, E.R., and Michel, T., 1989, Backscattering effects in the elastic scattering of p-polarized light from a large amplitude random metallic grating, Opt. Lett. 14:151.

O'Donnell, K.A., and Méndez, E.R., 1987, Experimental study of scattering from chracterized random surfaces, J. Opt. Soc. Am., A4:1194.

Rino, C.L., Crystal, T.L., Koide, A.K., and Ngo, H.D., 1989, Numerical simulation of acoustic and electro-magnetic scattering from ocean surfaces, Vista report no. 1010, Vista Research Inc.

Thorsos, E.I., 1988, The validity of the Kirchhoff approximation for rough surface scattering using a Gaussian roughness spectrum, J. Acoust. Soc. Am., 83:78.

Wombell, R.J., and DeSanto, J.A., 1991a, The reconstruction of shallow rough-surface profiles from scattered field data, Inverse Pbs., submitted for publication.

Zipfel, G., and DeSanto, J.A., 1972, Scattering of a scalar wave from a random rough surface: a diagrammatic approach, J. Math. Phys. 13:1903.

DOMAIN OF VALIDITY OF THE WIENER-HERMITE FUNCTIONAL EXPANSION APPROACH TO ROUGH SURFACE SCATTERING

Cornel Eftimiu

St. Louis, MO

INTRODUCTION

The validation of theoretical models for the wave scattering by rough surfaces is generally hampered by the lack of systematic, statistically well-characterized experimental data. To fulfill this need, it was suggested, originally by Garcia and Stoll,[1] that Monte Carlo calculations could be profitably used instead. The purpose of the present communication is to use Monte Carlo calculations in order to validate the use of a model[2,3] based on a modified expansion of the surface currents in a series of Wiener-Hermite functionals.

Theoretical Framework

Two classical models for rough surface scattering have been long known and extensively used. One, based on perturbation theory (PT), and generally valid for slightly rough surface, was originally proposed by Rice.[4] The other, amply described in a book by Beckmann and Spizzichino,[5] uses the so-called Kirchhoff approximation (KA), and is generally valid for slowly undulating rough surfaces. A modern examination of the validity of these models, based on comparisons with Monte Carlo calculations, was presented by Thorsos.[6,7] In Fig. 1, we show the approximate domains of validity of these methods, as suggested by a study made by Chen et al.[8] In this diagram, k stands for the wavenumber, σ is the r.m.s. of the surface profile, and ℓ is the correlation length of the rough surface, assumed one-directional. While we refer to this publication for details, one should keep in mind that the validity of PT was meant to imply agreement in backscattering within 8 db with a numerical simulation over a 30-60° angular interval, whereas the validity of the KA implies agreement

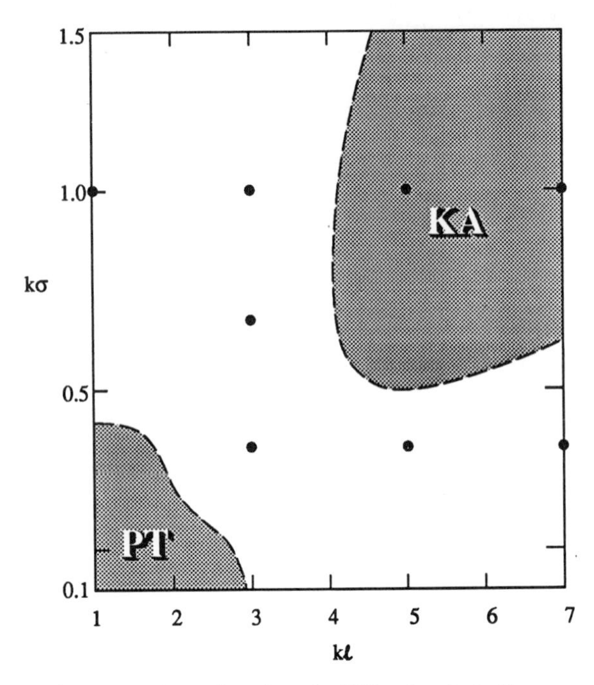

Fig. 1 Approximate domains of validity of perturbation theory (PT) and Kirchhoff approximation (KA) models.

with the numerical simulation within 2.5 db over a 0-20° angular interval. Consequently, the boundaries of the domains of validity of PT and KA, respectively, only represent estimates, which would most likely be affected if the entire angular interval were to be considered, and if bistatic instead of monostatic scattering data were used for validation purposes. Nevertheless, we believe that the qualitatively described domains of validity for PT and KA are adequate for our purposes, inasmuch as Fig. 1 clearly shows these domains to be definitely disjoint, with a gap representing a challenge for any "comprehensive" model of rough surface scattering.

Theoretical models purporting to fill this gap between the PT and KA domains of validity are not abundant. One, described as a phase-perturbation technique, was proposed by Winebrenner and Ishimaru,[9] and its validity was recently examined by comparison with Monte Carlo calculations by Broschat et al.[10] Another, based on a modified Wiener-Hermite expansion will be compared to Monte Carlo calculations below.

Numerical Comparisons

Without exception, primarily because of economy of effort considerations, the numerical Monte Carlo calculations presented here were carried out in the HH polarization case for conducting, one-directional rough surfaces, although it may be worth pointing out that, at least as far as the Wiener-Hermite model is concerned, the theoretical formulation allows both a two-directional rough surface and its being the interface of a (lossy) dielectric.

With regard to the reduction of the theoretical expressions derived[2,3] to the special case of a one-directional conducting rough surface, a few comments are in order. The theoretical study[2] of a perfectly conducting two-directional rough surface was shown to require a regularization of integrals exhibiting irreducible singularities, which was done by introducing a parameter related to the finite, albeit large, dimension of the surface. In reducing the eqs. (41)-(46) of this model to the one-directional case one has only, after regularization, to replace one factor of the Fourier transform of the correlation function by a Dirac function, expressing the fact that, in that direction, the correlation length is infinite. In this limiting process, the above-mentioned singularities disappear and the regularization becomes superfluous. Alternatively and equivalently, but perhaps mathematically more palatable, one can proceed from eq. (45) of the dielectric interface case,[3] and take the one-directional limit while keeping the imaginary part of the relative dielectric constant fixed, albeit large. After carrying out this limiting process, one can assume that the imaginary part of the dielectric constant is infinite, to recover the case of a perfectly conducting surface. Whichever the limiting procedure, the final expressions are the same and, because the manipulations are certainly straightforward and almost trivial, we shall not detail them here. The results using the theoretical, Wiener-Hermite model are indicated in Figs. 2-4 by dots. In the same figures, the continuous, irregular lines represent the Monte Carlo calculations.

The pairs of values for $k\sigma$ and $k\ell$, for which the calculations shown in Figs. 2-4 were made, were chosen to correspond to the gap between the domains of validity of PT and KA indicated in Fig. 1, and which are there highlighted by full circles. It is perhaps worth emphasizing that a comparison between the predictions of the Wiener-Hermite model and those of either PT or KA are of no direct interest at this point: such a comparison was extensively made[2,3] previously.

The comparisons made in Fig. 2 correspond to the same value of $k\sigma = 0.33$ and three values of $k\ell = 2.82, 5, 7$, respectively. The incident wave was assumed to have a wave vector contained in a plane perpendicular

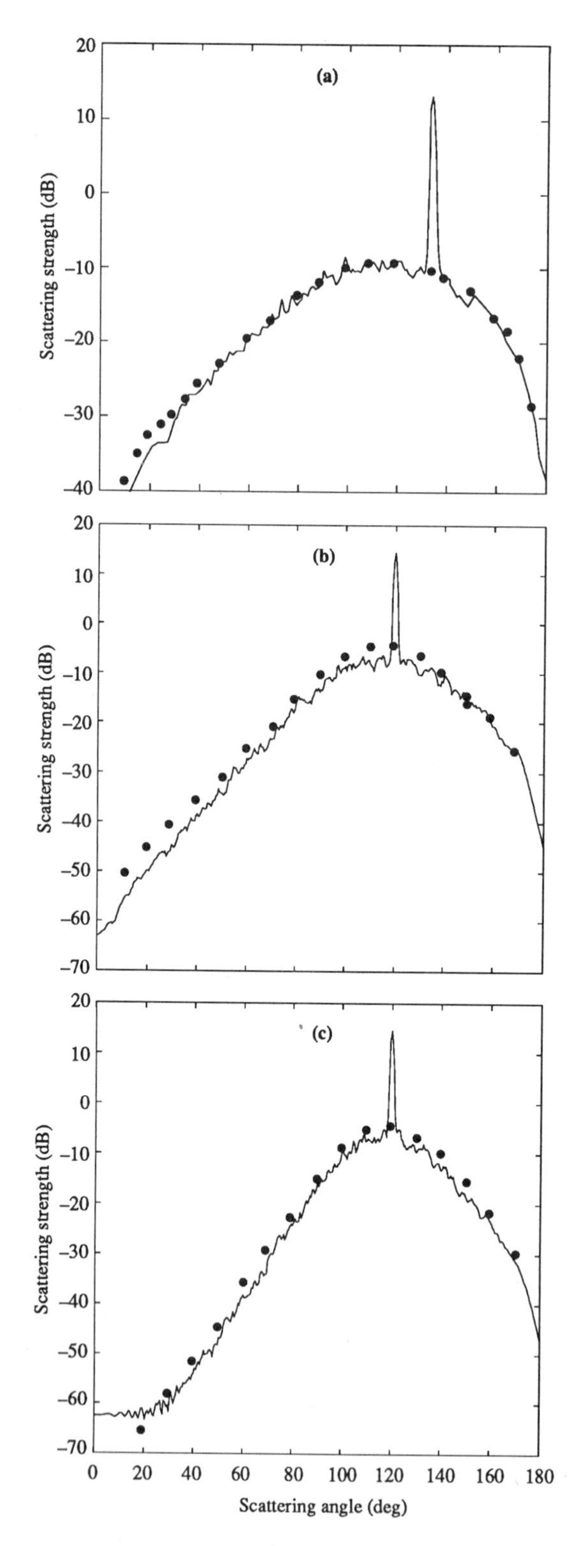

Fig. 2. Bistatic HH scattering at $k\sigma = 0.33$ and (a) $k\ell = 2.82$; (b) $k\ell = 5$; (c) $k\ell = 7$.

to the one-directional roughness of the surface, and making a 30° (in b and c) angle with the normal to the plane of the averaged surface. In Fig. 2a, which coincides with one of the cases presented by Thorsos[6] in his Fig. 8, the incident angle was 45°. The specular peak should, of course, be everywhere ignored.

The same incidence angle, but for a value $k\sigma = 0.66$, is considered in Fig. 3, which is also taken to coincide with the other case presented by Thorsos[6] in the same Fig. 8.

Finally, Fig. 4 shows the Wiener-Hermite dots as compared to the Monte Carlo calculations corresponding to a value $k\sigma = 1$ and the series of values of $k\ell = 1, 3, 5,$ and 7.

The generally good agreement noted in these figures leads to the conclusion that the domain of validity of the Wiener-Hermite model includes at least the rectangle $0 \leq k\sigma \leq 1$, $1 \leq k\ell \leq 7$. A determination of the actual boundaries of this domain requires additional systematic investigation, but there are indications that they lie well outside the established rectangle. For instance, although not shown here, we have noted excellent agreement also in a few cases, also previously considered,[6] in which $k\sigma = 2.2$ or $k\sigma = 4.4$.

Acknowledgment

This work was performed under the McDonnell Douglas Independent Research and Development program. Most of the Monte Carlo calculations shown here were done by Mr. M. Grayson, under the expert guidance of Mr. L. Bahrmasel.

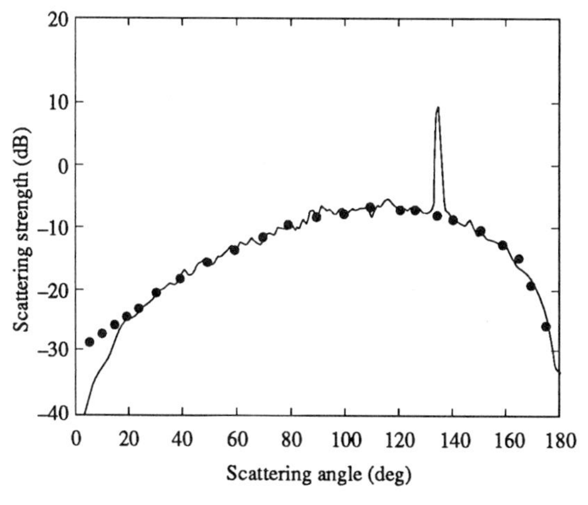

Fig. 3. Bistatic HH scattering at $k\sigma = 0.33$ and $k\ell = 2.82$.

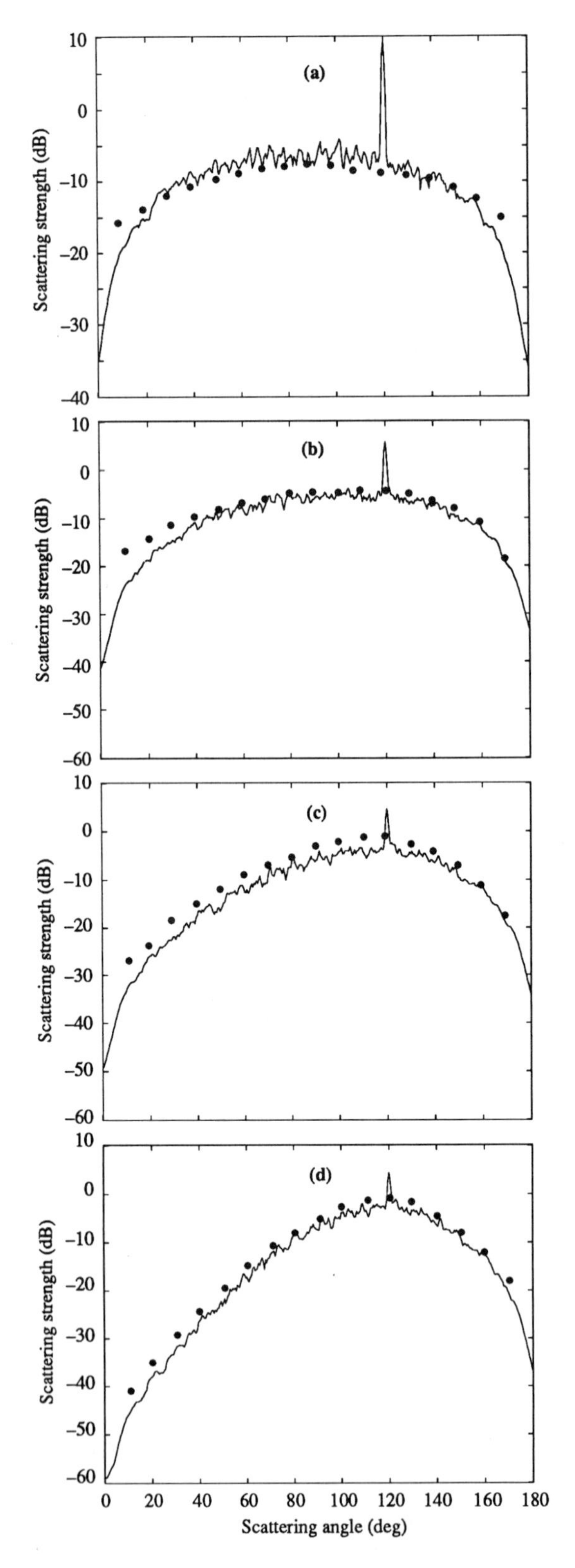

Fig. 4. Bistatic HH scattering at kσ = 1 and (a) kℓ = 1; (b) kℓ = 3; (c) kℓ = 5; (d) kℓ =7.

References

1. N. Garcia and E. Stoll, Monte Carlo Calculation for Electromagnetic-Wave Scattering from Random Rough Surfaces, Phys. Rev. Lett. 52:1798 (1984).
2. C. Eftimiu, Modified Wiener-Hermite Expansion in Rough-Surface Scattering, J. Opt. Soc. Am. A6:1584 (1989).
3. C. Eftimiu, Scattering by a Rough Dielectric Interface: A Modified Wiener-Hermite Approach, J. Opt. Soc. Am. A7:875 (1990).
4. S. O. Rice, Reflection of Electromagnetic Waves from Slightly Rough Surfaces, Comm. Pure Appl. Math. 4:351 (1951).
5. P. Beckmann and A. Spizzichino, "The Scattering of Electromagnetic Waves from Rough Surfaces," Pergammon Press, Oxford, 1963.
6. E. I. Thorsos, The Validity of the Kirchhoff Approximation for Rough Surface Scattering Using a Gaussian Roughness Spectrum, J. Acoust. Soc. Am. 83:78 (1988).
7. E. I. Thorsos and D. R. Jackson, The Validity of the Perturbation Approximation for Rough Surface Scattering Using a Gaussian Roughness Spectrum, J. Acoust. Soc. Am. 86:261 (1989).
8. M. F. Chen, S. C. Wu, and A. K. Fung, Regions of Validity of Common Scattering Models Applied to Non-Gaussian Rough Surfaces, J. Wave Mater. Interact. 2:9 (1987).
9. D. Winebrenner and A. Ishimaru, Investigation of a Surface Field Phase Perturbation Technique to Randomly Rough Surfaces, J. Opt. Soc. Am. A2:2285 (1985).
10. S. L. Broschat, E. I. Thorsos, and A. Ishimaru, Phase Perturbation technique vs. an Exact Numerical Method for Random Rough Surface Scattering, J. Electrom. Waves Appl. 3:237 (1989).

WAVE INTENSITY FLUCTUATIONS IN A ONE DIMENSIONAL DISCRETE RANDOM MEDIUM

Sasan S. Saatchi

Laboratory for Hydrospheric Processes
NASA/Goddard Space Flight Center
Greenbelt, Maryland 20771

Roger H. Lang

Dept. of Electrical Engineering and Computer Science
George Washington University
Washington, DC 20052

ABSTRACT

The propagation of electromagnetic waves in a one dimensional discrete random medium is considered. It is assumed that the medium is bounded within a slab of thickness L and the random inhomogeneities are distributed according to a Poisson impulse process of density λ. In the absence of absorption, an exact equation is obtained for the m-moments of the wave intensity using the Kolmogorov-Feller approach. In the limit of low concentration of inhomogeneities, we use a two-variable perturbation technique, valid for λ small and L large, to obtain approximate solutions for the moments of the intensity. It is shown that the wave intensity increases with the slab thickness and the peak of the higher moments occur inside the medium.

1. INTRODUCTION

The problem of propagation of scalar waves through a medium whose refractive index is a random function of a single coordinate has been widely studied in optics, radiophysics and acoustics. Theoretically, such problems are considered necessary to investigate the influence of various models of the medium on the statistical characteristics of waves. In these studies, usually, one examines the average quantities of interest such as the mean field and various moments of the wave intensity [1]-[6].

In this paper, we consider the statistical properties of the wave intensity in a layer medium with discrete random inhomogeneities. More precisely, we assume that the random inhomogeneities are distributed according to a Poisson impulse process. In our preceding articles of this series [7,8], we have analyzed the mean field and the average power reflected from the layer. There, we have proposed to use a probabilistic

Directions in Electromagnetic Wave Modeling
Edited by H.L. Bertoni and L.B. Felsen, Plenum Press, New York, 1991

approach suitable for stochastic differential equations with non-Gaussian delta correlated fluctuating coefficients. As a result, we have acquired exact equations of Kolmogorov-Feller type for the p.d.f. (probability density function) of the field and the reflection coefficient. Furthermore, approximate solutions in the limiting case of the low concentration of inhomogeneities, valid for large distance of propagation, have been obtained.

To explain succinctly, our main result comprises the limiting case where the medium is sparse and the inhomogeneities have an arbitrary scattering strength. We have been able to show that the mean field agrees with Foldy's effective field approximation and the average reflected power increases exponentially with the layer thickness. In both cases, however, the results are dependent on the scattering strength; the higher the scattering strength, the faster the wave attenuates in the medium.

As a continuation of previous work, we proceed to obtain exact equations for the m-moment of the intensity in the medium. In sections 2 and 3, the formulation of the problem and the derivation of the forward Kolmogorov-Feller equation are discussed briefly. The application of the two variable perturbation method in small density (λ) and the formulation of a compact expression for the m-moment of the wave intensity are explained in section 4. Finally, in section 5, we discuss the properties of the wave intensity and the intensity fluctuations in the medium.

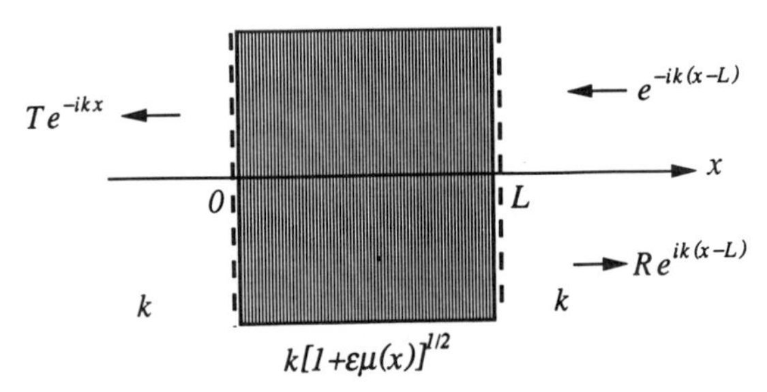

Fig. 1. Transmission and reflection from a slab with diffuse boundaries

2. FORMULATION OF THE PROBLEM

The problem is as follows. Let a one dimensional inhomogeneous medium occupy the region $0 \leq x \leq L$ and let the refractive index of the medium equal $n(x) = [1 + \varepsilon\mu(x)]^{1/2}$. Outside the slab, the medium is considered homogeneous with $n(x) = 1$. We also let $\psi(x)$ denote the complex-valued scalar wave field at a location $x\varepsilon(-\infty,\infty)$. A time dependence of $e^{-i\omega t}$ has been omitted throughout the paper. The field $\psi(x)$ then satisfies a one dimensional Helmholtz equation given by

$$\frac{d^2\psi(x)}{dx^2} + k^2[1 + \varepsilon\mu(x)]\psi(x) = 0 \qquad 0 \leq x \leq L \tag{1}$$

$$\psi(x) = \begin{cases} e^{-ik(x-L)} + Re^{-ik(x-L)} & x > L \\ Te^{-ikx} & x < 0 \end{cases} \tag{2}$$

with the two-point boundary condition

$$\frac{d\psi(0)}{dx} + ik\psi(0) = 0 \tag{3}$$

$$\frac{d\psi(L)}{dx} - ik\psi(L) = -2ik \tag{4}$$

where $\mu(x)$ is a random function of position x and R and T are the complex-valued reflection and transmission coefficients respectively (Fig. 1). The parameter ε represents the strength of the fluctuation of the permeability of the medium.

Next, we transform the two point boundary value problem into an initial value problem by considering the solution is dependent on the thickness of the slab. Assuming t = L as a new variable and using the invariant imbedding method [9], the following equations for $\psi(x) = \psi(x,t)$ and $R = R(t)$ are obtained

$$\frac{dR(t)}{dt} = 2ikR(t) + \frac{ik\varepsilon}{2}\mu(t)\,[1 + R(t)]^2\,, \quad R(0) = 0 \quad 0 \leq t \leq x \tag{5}$$

$$\begin{cases} \dfrac{d\psi(x,t)}{dt} = ik\psi(x,t) + \dfrac{ik\varepsilon}{2}\mu(t)\;[1 + R(t)]\psi(x,t) & (6) \\ \dfrac{dR(t)}{dt} = 2ikR(t) + \dfrac{ik\varepsilon}{2}\mu(t)\;[1 + R(t)]^2 \qquad x \leq t \leq L & (7) \end{cases}$$

$$\psi(x,x) = 1 + R(x), \qquad R(t)\,|_{t=x} = R(x)$$

Equation (5) has been solved to obtain the average reflected power in the medium [7], and the probability density of the reflection coefficient [8]. This result has then been used as the initial condition for equations (6) and (7) to obtain the average field in the medium [8]. Here, we follow the same method as in [8] to obtain the Kolmogorov-Feller equation for the p.d.f. of the field in the medium.

Let us first introduce the modulus and the phase of the complex functions R(t) and $\psi(x,t)$ by

$$R(t) = \sqrt{\frac{z(t) - 1}{z(t) + 1}}\,e^{i\phi(t)}\;, \quad \psi(x,t) = e^{ik(t-x)}e^{v(t)+i\theta(t)} \tag{8}$$

By making use of (8), equations (6) and (7) can then be written as a system of differential equations. Thus

$$\frac{dv}{dt} = -\frac{1}{2} k\varepsilon\mu(t)\frac{(z^2-1)^{1/2}}{z+1} \sin\phi$$

$$\frac{d\theta}{dt} = \frac{1}{2} k\varepsilon\mu(t)\left[1+ \frac{(z^2-1)^{1/2}\cos\phi}{z+1} \right]$$

$$\frac{dz}{dt} = k\varepsilon\mu(t)(z^2-1)^{1/2}\sin\phi \tag{9}$$

$$\frac{d\phi}{dt} = 2k + k\varepsilon\mu(t)\left[1 + \frac{z\cos\phi}{(z^2-1)^{1/2}}\right]$$

$$-\infty < v < \infty, \ -\pi \leq \theta \leq \pi, \ 1 \leq z < \infty, \quad -\pi \leq \phi \leq \pi$$

where the dependence on t has been omitted for simplicity. The initial conditions for eq. (9) are given by

$$v_0 = v(x) = \frac{1}{2} \ln 2\left[\frac{z+\sqrt{z^2-1}\cos\varphi}{z+1}\right] \tag{10}$$

$$\theta_0 = \theta(x) = \operatorname{arctg} \frac{\sqrt{z^2-1}\sin\varphi}{1+z+\sqrt{z^2-1}\cos\varphi} \tag{11}$$

where z and θ are also evaluated at t = x.

3. *KOLMOGOROV-FELLER METHOD*

We shall now turn our attention to the statistical description of the wave field. We assume $\mu(t)$ is a Poisson impulse process. This is a non-Gaussian delta correlated process with the following first and second order moments

$$\mu(t) = \sum_{i=1}^{\infty} \delta(t - T_i) \tag{12}$$

$$< \mu(t) > = \lambda \tag{13}$$

$$< \mu(t_1)\, \mu(t_2) > = \lambda^2 + \lambda\delta(t_1 - t_2) \tag{14}$$

where $< . >$ indicates an ensemble average and λ is the density of the Poisson points. Upon using the Markov properties of this process, an exact equation of Kolmogorov-Feller type can be written for the p.d.f. of the solution of eq. (9) as follows

$$\left(\frac{\partial}{\partial t} + 2K\frac{\partial}{\partial \varphi}\right) P(v,\theta,X,x,t) = \lambda P(V_b,\theta_b,X_b,x,t) - \lambda P(v,\theta,X,x,t) \quad (15)$$

$$\text{i.c.} \qquad P(v,\theta,X,x,x) = P(X,x)\ \delta(v-v_0)\ \delta(\theta-\theta_0) \quad (16)$$

where $X = (z,\phi)$ and v_b,θ_b and X_b correspond to the states of the variables of the stochastic differential equation (9) before an impulse occurs. The relation between the variables (v_b,θ_b,X_b) and (v,θ,X) are given in Appendix A. the function p(X,x) is the p.d.f. of the solution to equation (5) and it is given by [8]

$$P(X,x) = P(z,\varphi,x) = 1/2\pi \int_0^{\infty} s\tanh\pi s\ e^{-\lambda t[1-q(s)]}\ P_{-1/2+is}(z)\ ds \quad (17)$$

where $q(s) = P_{-1/2+is}\left(\frac{k^2\varepsilon^2}{2} + 1\right)$ is a conical Legendre function with a constant argument.

Note that the solution to equation (16) is given at a point x inside the slab for any thickness t. When this solution is known, one can readily obtain the m-moment of the wave intensity $< I^m > = < |\psi|^{2m} >$ at any point in the medium. This is done by computing the following integral

$$< I^m > = \int_{-\infty}^{\infty} dv \int_{-\pi}^{\pi} d\theta \int_1^{\infty} dz \int_{-\pi}^{\pi} d\phi\ e^{2mv}\ P(v,\theta,X,x,t) \quad (18)$$

where the function $|\psi|^{2m} = e^{2mv}$ has been obtained from (8). To acquire an analytical expression for $< I^m >$ more conveniently, we expand the function $P(v,\theta,X,t)$ in terms of the eigenfunctions $e^{in\theta+i\omega v}$ given by

$$P(v,\theta,X,x,t) = \sum_{n=-\infty}^{\infty} \int_{-\infty}^{\infty} Q_n(\omega,X,x,t)\ e^{in\theta+i\omega v}\ d\omega \quad (19)$$

By inserting (19) in (18) and integrating with respect to v,θ, and ω, we obtain

$$< I^m > = (2\pi)^2 \int_1^{\infty}\int_{-\pi}^{\pi} Q_0(2im,X,x,t)\ dzd\phi \quad (20)$$

Therefore, it is only necessary to obtain $Q_0(2im,X,x,t)$ as the generating function for the intensity moments. An equation for $Q_0(2im,X,x,t)$ can then be obtained by substituting the expansion (19) in equation (16) and invoking the orthogonality properties of the eigenfunctions and finally using $n = 0$ and $\omega = 2im$. Thus, we have

$$\frac{d}{d\tau} Q_0(2im,\hat{X},\tau) = \lambda D_0 Q_0(2im,\hat{X}_b,x,\tau) \, |1+r|^{2m} - \lambda Q_0(2im,\hat{X}_b,x,\tau) \tag{21}$$

$$\text{i.c.:} \qquad Q_0(2im,\hat{X}_b,x,x) = \frac{P(\hat{X},x)}{(2\pi)^2} \, |1+R|^{2m} \tag{22}$$

where the transformations $\tau = t$, $\alpha + 2k\tau = \phi$ and $\hat{X}(z,\alpha)$ have in turn reduced the equation to an ordinary differential form. The values of D_0, r and R are given by

$$r = \frac{k\varepsilon}{k\varepsilon - 2i} \sqrt{\frac{z-1}{z+1}} \, e^{i(\alpha+2k\tau)} \quad , \quad R = \sqrt{\frac{z-1}{z+1}} \, e^{i(\alpha+2k\tau)} \tag{23}$$

$$D_0 = \frac{1}{4^m} (k^2\varepsilon^2 + 4)^m \tag{24}$$

4. APPLICATION OF TWO-VARIABLE PERTURBATION METHOD

In this section, we seek an approximate solution to equation (21) by using small perturbation method. Here, we assume λ is small. Then, we write the solution as a two-variable perturbation expansion in powers of λ in the following form

$$Q_0(2im,\hat{X},x,\tau) = \sum_{j=0}^{\infty} Q_0^{(j)}(2im,\hat{X},x,\tau,\bar{\tau}) \, \lambda^j \tag{25}$$

where τ and $\bar{\tau} = \lambda t$ are the fast and slow variables respectively.

Next, by inserting (25) into (21) and equating equal powers of λ, a set of equations for $Q_0^{(j)}$ are obtained. The solution to the zero order equation is independent of the fast variable τ; $Q_0^{(0)} = W(\hat{X},x,\bar{\tau})$, where $W(\hat{X},x,\bar{\tau}) = W(z,\alpha,x,\bar{\tau})$ is an arbitrary function of the slow variable. This function is computed by removing the secular terms of order $\bar{\tau} = \lambda\tau$ from the higher order equations to guarantee the convergence of the expansion (25) for large τ. This yields the following equation for $W(z,\alpha,x,\bar{\tau})$

$$\left(\frac{\partial}{\partial\bar{\tau}} + 1\right) W(z,\alpha,x,\bar{\tau}) = \frac{D_0}{2\pi} \int_{-\pi}^{\pi} W(z_b,\alpha_b,x,\bar{\tau}) \, |1+r|^{2m} \, d\phi \tag{26}$$

where both z_b and α_b are periodic functions of $\phi = \alpha + 2k\tau$. By some algebraic manipulations using the form of z_b given in Appendix A, we can rewrite $|1+r|^{2m}$ as

$$|1+r|^{2m} = \frac{4^m}{(k^2\varepsilon^2 + 4)^m (z + 1)^m} (1 + z_b)^m \tag{27}$$

By using (27) in (26) and requiring that the final solution for Q_0 is periodic in ϕ, one can integrate out the ϕ dependence from (26) and the initial condition (22) to obtain

$$\left(\frac{\partial}{\partial\bar{\tau}} + 1\right) W(z,x,\bar{\tau}) = \frac{1}{2\pi(z + 1)^m} \int_{-\pi}^{\pi} W(z_b,x,\bar{\tau}) (1 + z_b)^m \, d\varphi \tag{28}$$

$$W(z,x,\lambda x) = \frac{2^m P_m(z)}{(2\pi)^3 (1 + z)^m} \int_0^{\infty} s\tanh\pi s \; e^{-\lambda x[1-q(s)]} P_{-1/2+is}(z) \, ds \tag{29}$$

where $P_m(z)$ is a Legendre polynomial of order m. Then, the m-moment of intensity is simply given by

$$< I^m > = (2\pi)^3 \int_1^{\infty} W(z,x,\lambda t) \, dz \tag{30}$$

The function $W(z,x,\lambda t)$ can be obtained by solving the integral equation (28) using the Mehler-Fock transform [10]. Here, we only show the final form for $< I^m >$ and leave out the details, since a similar procedure has been discussed in [7,8]. Therefore, we have

$$< I^m > = 2^m e^{-\lambda t} \int_0^{\infty} s\tanh\pi s \; g_m(s) \; e^{\lambda(t-x)q(s)} \int_1^{\infty} \frac{1}{(z+1)^m} P_{-1/2+is}(z) dz \tag{31}$$

where

$$g_m(s) = \int_0^{\infty} s'\tanh\pi s' \; e^{\lambda x q(s')} \, ds' \int_1^{\infty} P_m(z) \, P_{-1/2+is'}(z) \, P_{-1/2+is}(z) \, dz \tag{32}$$

By recording the following fact about the conical Legendre function [10],

$$\int_1^{\infty} \frac{1}{(z+1)^m} P_{-1/2+is}(z)dz = \frac{\pi}{\cosh \pi} K_m(s) \tag{33}$$

where

$$K_{m+1}(s) = \frac{s^2 + (m-1/2)^2}{2m} K_m(s) \quad , \qquad K_1(s) = 1 \quad , \tag{34}$$

the intensity moments $< I^m >$ can be further simplified. For $t = L$, we have

$$< I^m > = 2^m e^{-\lambda L} \int_0^{\infty} \frac{\pi s \sinh \pi s}{\cosh^2 \pi s} K_m(s)\, g_m(s)\, e^{\lambda(L-x)q(s)}\, ds \tag{35}$$

5. *AVERAGE INTENSITY AND INTENSITY FLUCTUATIONS*

From expression (35) for $< I^m >$, it follows that the average wave intensity at any point inside x inside the medium is given for $m=1$ as follows

$$<I(x,L)> = e^{-\lambda L} \int_0^{\infty} \frac{\pi s \sinh \pi s}{\cosh^2 \pi s} e^{\lambda(L-x)q(s)} \left[(s+\tfrac{i}{2}) e^{\lambda x q(s+i)} + (s-\tfrac{i}{2}) e^{\lambda x q(s-i)} \right] ds \tag{36}$$

This expression can be easily reduced to the average transmitted power, $<I(0,L)> = < |T|^2 >$ given in reference [7]. In Fig. 2, the average wave intensity is plotted versus a dimensionless quantity x/L for small and large values of λL. It is shown that the average intensity increases with the slab thickness L, but it always remains constant at the level of incident power (equal one in our case) in the middle of the slab. Figure 3 shows the average intensity for a fixed slab thickness and various values of $k\varepsilon$.

The fluctuations of the intensity in the medium is given according to the relation

$$\sigma_f = \left[< I^2(x,L) > - (< I(x,L) >)^2 \right]^{1/2} \tag{37}$$

where $< I^2(x,L) >$ is obtained by plugging $m = 2$ into (35). In Figs. 4, the distribution of the average, the second moment and the fluctuations of the wave intensity are shown.

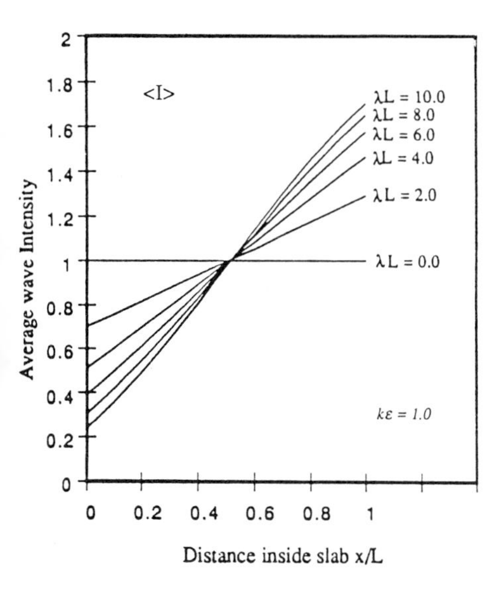

Fig. 2. Average intensity for kε = 1.0

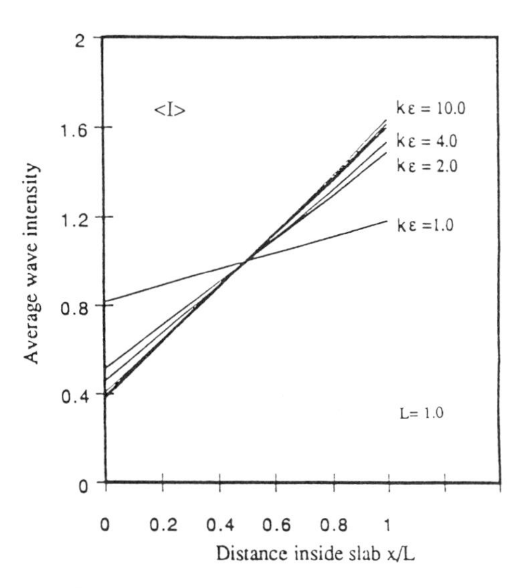

Fig. 3. Average intensity for L = 1.0

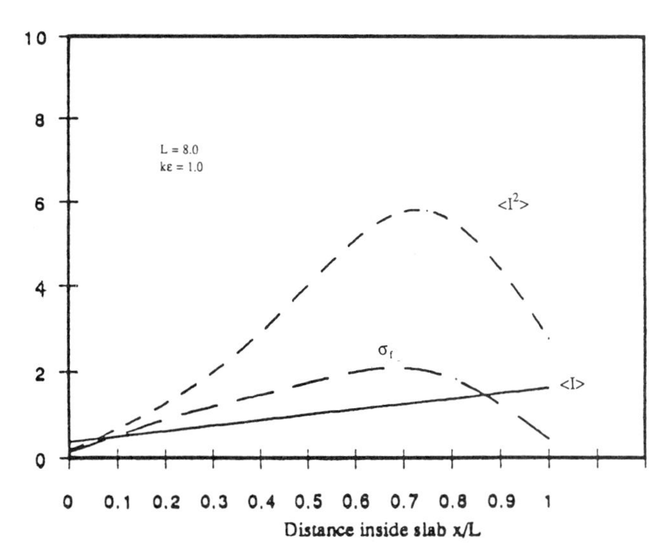

Fig. 4. First moment <I>, second moment <I²> and intensity fluctuation σ_f versus dimensionless distance x/L inside slab

References

[1] W. Kohler and G.C. Papanicolaou, "Power statistics for wave propagation in one dimension and comparison with radiative transfer theory", *J. Math. Phys. 14*, 1733-1745 (1973).

[2] W. Kohler and G.C. Papanicolaou, "Power statistics for wave propagation in one dimension and comparison with radiative transfer theory.II", *J. Math. Phys. 15*, 2186-2197 (1974).

[3] R.H. Lang,"Probability density function and moments of the field in a slab of one-dimensional random medium", *J. Math. Phys., 14*, 1921-1926 (1973).

[4] Yu.L. Gazaryan, "The one dimensional problem of propagation of waves in a medium with random inhomogeneities", *Sov. Phys. JETP 29*, 996-1003 (1969).

[5] J. Bazer, "Multiple scattering in one dimension", *J.Soc. Indust. Appl. Math. 12*, 539-579 (1964).

[6] V.I. Klyatskin, "Wave stochastic parametric resonance (wave intensity fluctuations in a one-dimensional randomly inhomogeneous medium), *Izv. Vuzov Radiofisika 22(2)*, 180-191,1979.

[7] S.S. Saatchi and R.H. Lang, "Average reflected power from a one dimensional slab of discrete scatterers", *Radio Science 25(4)*, 407-417 (1990).

[8] S.S. Saatchi and R.H. Lang,"Mean wave propagation in a slab of one-dimensional discrete random medium", submitted to *Wave Motion*, 1990.

[9] R. Bellman and R. Kalaba, "Functional equations, wave propagation, and invariant imbedding", *J. Math. Mech. 8*, 683-704 (1959).

[10] A. Erde'lyi, W. Magnus, F. Oberhettinger and F.G. Tricomi, *Higher Transcendal Functions*, McGraw-Hill Book Co., New York (1953).

APPENDIX A

The states of the variables $(v_b,\theta_b,z_b,\phi_b)$ of the system of equations (9) before an impulse occurs are related to the variables (v,θ,z,ϕ) through the following expressions [7],[8]

$$v_b = v - \gamma_1(z,\phi) \tag{A1}$$

$$\theta_b = \theta - \gamma_2(z,\phi) \tag{A2}$$

$$z_b = z - k\varepsilon\xi + \frac{k^2\varepsilon^2}{2}\eta \tag{A3}$$

$$\phi_b = \mathrm{arctg}\,\frac{\xi + k\varepsilon\eta}{z + k\varepsilon\xi + (k^2\varepsilon^2/2 - 1)\eta} \tag{A4}$$

where $\xi = \sqrt{z^2-1}\,\sin\phi$ and $\eta = z + \sqrt{z^2-1}\,\cos\phi$ and the functions γ_1 and γ_2 are defined as

$$\gamma_1(z,\phi) = \frac{1}{2}\ln\frac{1 + z + k\varepsilon\xi + k^2\varepsilon^2\eta/2}{z + 1} \tag{A5}$$

$$\gamma_2(z,\phi) = \mathrm{arctg}\,\frac{k\,\varepsilon\,(1+\eta)}{2(z+1) - k\varepsilon\xi}. \tag{A6}$$

FRACTAL ELECTRODYNAMICS AND MODELING

Dwight L. Jaggard

Complex Media Laboratory
Moore School of Electrical Engineering
University of Pennsylvania
Philadelphia, PA 19104-6390
USA

I. INTRODUCTION

Since nature has provided us with structures of almost infinite complexity and variety, it is apparent that some, and perhaps most, of these structures cannot be treated by traditional deterministic methods when modeling their interaction with electromagnetic waves. As one alternative, average properties of these interactions can be treated through statistical means and the use of averages and higher-order moments. These conventional methods are particularly successful when the scales of these structures are limited to modest ranges of variation or to the case of resonant interaction. However, certain types of wildly irregular, ramified, or variegated structures contain many scale sizes as demonstrated in the fractal surfaces of Fig. 1. It is often difficult to characterize the interaction of waves with these structures and in many cases it is not even easy to model these structures. It is just these complicated structures which are of interest to us here and which can often be modeled by the use of *fractals*. Thus, we blend fractal geometry and electromagnetics in a discipline we call *fractal electrodynamics* [1].

Fig. 1. Models of fractally corrugated surfaces. A gently undulating fractal surface (left) with small fractal dimension and a rough area-filling fractal surface (right) with larger fractal dimension.

Directions in Electromagnetic Wave Modeling
Edited by H.L. Bertoni and L.B. Felsen, Plenum Press, New York, 1991

Here we provide a tutorial summary of basic geometrical fractal concepts then note the obstacles to their use in electromagnetics followed by suggested methods to overcome these barriers. Finally we demonstrate a number of applications. These applications range from rough surface modeling and electromagnetic wave scattering to the design of robust antenna arrays using the attributes of fractal geometry. This unified viewpoint has been developed over the past half decade at the University of Pennsylvania as one approach to these problems. Here our aim is to introduce the important fundamental concepts in an intuitive way, note the salient features of the applications and so provide a gateway to other relevant problems. Analytical and computational details as well as other points of view are provided elsewhere [1].

II. BACKGROUND

Fractals provide a geometry for the description of pathological structures which are ill-defined by traditional Euclidean descriptions but which may occur naturally such as the previously shown surfaces of Fig. 1. In addition, fractals provide prescriptions for the synthesis of new electromagnetic structures and devices. Fractal structures are fundamentally characterized through their *dilation symmetry* [2] such that these structures appear identical or similar, in the mean, at all resolution scales in the regime of interest. It is this feature of *self-similarity* which is of use in the modeling of realistic problems involving the interaction of electromagnetic waves with complex natural structures. Self-similarity, mixed with the random variations of non-deterministic fractals, is the cause of long-range underlying order mixed with a short range disorder evidenced in the sample surfaces of Fig. 1. We schematically represent this balance in Fig. 2 and note that it is characteristic of many naturally occurring phenomena such as turbulence, lightening, rough coastlines, cloud formations and ocean surfaces, formed by repetitive but aperiodic forces of nature. Later we indicate methods of putting to use this tension between order and disorder in electromagnetic synthesis where the strengths of ordered systems can be combined with the strengths of their disordered counterparts.

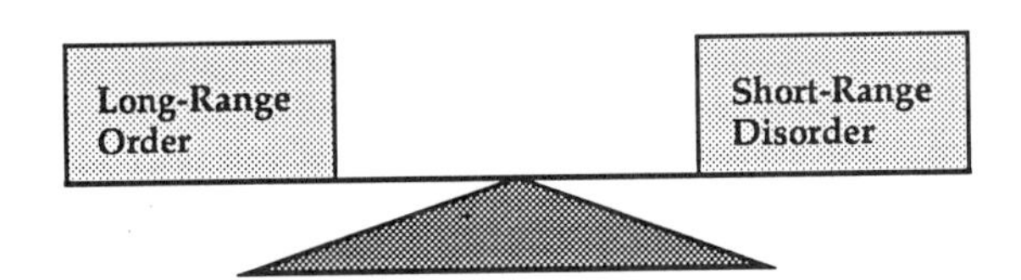

Fig. 2. One attribute of fractals is the balance between long-range order and short-range disorder. It is this attribute which make fractals ideal models for naturally occurring phenomena in which repetitive but aperiodic physical forces form natural structures (e.g., the action of water waves against a coastline). A second use of this fractal attribute can be found in electromagnetic synthesis problems in which a balance is needed between periodic lattices and random lattices (e.g., as in the tradeoffs inherent in periodic versus random antenna arrays).

Next, one requires fractal descriptors to quantify roughness. One of the canonical problems in fractal geometry has been the measurement of a rough perimeter, such as the length of a coastline, using yardsticks of various lengths. If the perimeter is self-similar, its length $L(\varepsilon)$ will increase without bound as the yardstick length ε decreases. The relation between these two quantities has been found to be $L(\varepsilon) = C\,\varepsilon^{1-D}$ where C is an appropriate constant dependent on the Euclidean dimension d and D is a number characteristic of the roughness [2]. This power-law relation between the perimeter and the yardstick size indicates that the slope of log-log plots of $L(\varepsilon)$ versus ε contains direct information regarding the fractal dimension. The fractal descriptor D approaches unity for smooth curves and two for area-filling or very rough curves. Therefore, it is appropriate to consider D as a dimension and accordingly it is denoted the *fractal dimension.*

The concept of a variable length yardstick is responsible for bringing us to the notion of a fractional fractal dimension from geometrical considerations. In Fig. 1 the two fractal surfaces are shown with dimension $2 \leq D \leq 3$. Generalizing this concept to arbitrary Euclidean dimension *d*, one can use a disks of appropriate shape and Euclidean dimension to determine the fractal dimension of a variety of objects as demonstrated in Fig. 3 for yardsticks of side ε [3]. When generalized in this manner, the method is denoted the *disk covering method.*

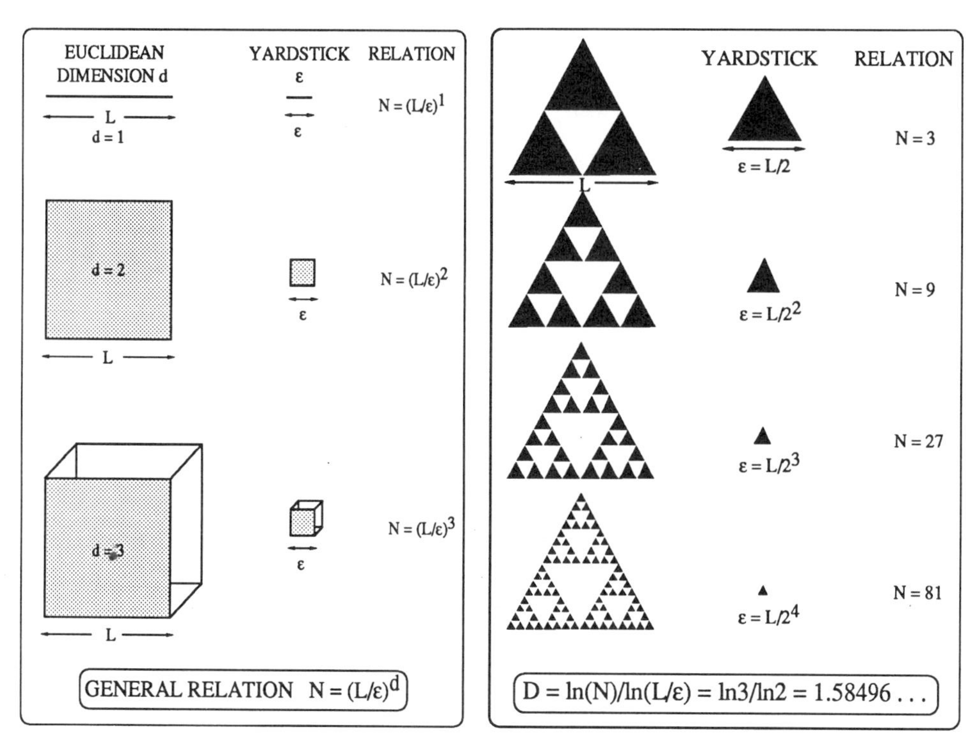

Fig. 3. Use of a physical yardstick of length ε to obtain the dimension of Euclidean objects (left) and fractal objects (right) through the *disk covering method.* Here N is the number of such yardsticks in each object of side *L*. Adapted from [1].

III. CONCEPTS OF FRACTAL ELECTRODYNAMICS

Two obstacles or barriers confront the electromagnetician who wants to attack problems involving fractal geometry. The first is to find a suitable *electromagnetic yardstick* which can act as an interrogating probe to remotely determine the roughness or fractal dimension of an object in the same way than one uses a variable length physical yardstick to measure the fractal attributes of the coastline problem. The second is to find suitable fractal functions which have bounded derivatives for modeling physical problems. We summarize, in turn, methods to overcome these two barriers.

A. Electromagnetic Yardsticks

The first problem requires the identification of an electromagnetic alternative to the physical yardstick. This alternative we denote an *electromagnetic yardstick.* Candidates include either the wavelength or the pulse width as an electromagnetic yardstick to investigate electromagnetic wave interactions in the frequency and time domains, respectively. Figure 4 indicates schematically how the physical yardstick of the physical problem is replaced by electromagnetic yardsticks for remotely interrogating rough surfaces.

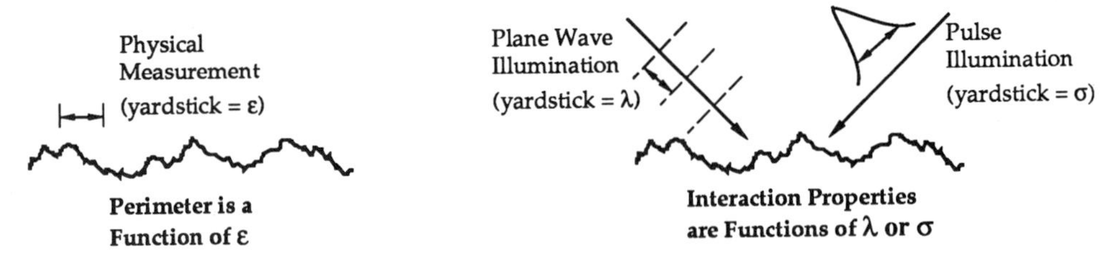

Fig. 4. Shown on the left side is the coastline problem where the perimeter is a function of yardstick length ε. Shown on the right side is the scattering problem of an interrogating wave and a fractal structure where the scattering properties are a function of the wavelength λ, for plane wave illumination) or the pulse width σ, for impulsive illumination. The parameters λ or σ act as variable length electromagnetic yardsticks which probe the fractal nature of the scatterer.

B. Bandlimited Fractals

The second problem involves the choice of suitable self-similar functions which are amenable to differentiation and provide realistic and flexible models. Strictly speaking, fractals are rough on all scales, and consequently their derivatives are unbounded. An appropriate modification is that of spatially bandlimiting otherwise fractal functions so that these functions are fractal over a regime of interest. Many such candidates occur with an appropriate family being Weierstrass functions [4] which is bandlimited [5,6]. These almost-periodic functions are composed of appropriately spaced and weighted sinusoids to insure self-similarity. A family of such bandlimited Weierstrass function is shown below in Fig. 5. This example is composed of ten sinusoids with random phases. Its line spectrum falls off with an inverse power of frequency in which the high frequency components are larger as the fractal dimension D increases thus giving rise to increasingly rough curves.

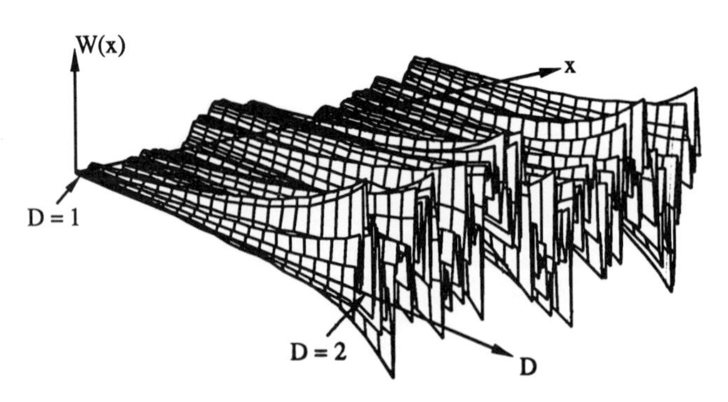

Fig. 5. Bandlimited Weierstrass function W(x) (height) as a function of coordinate x (left to right) for increasing fractal dimension (D = 1.00 back to D = 2.00 front) for ten tones. Adapted from [1].

IV. CANONICAL PROBLEMS INVOLVING SCATTERING FROM FRACTALS

There are a number of illuminating and useful examples involving the interaction of electromagnetic waves with complex objects modeled by fractals. In each case, the characteristic electromagnetic yardstick of the incident wave resonantly couples with scale lengths of comparable size in the structure and provides information concerning the roughness or fractal dimension of the object. The idea is to use the slope of log-log plots of electromagnetic parameters to indicate object roughness in a manner similar to that used to determine the fractal dimension of physical objects. Next are given several selected examples which demonstrate typical results and applications.

A. Scattering from Planar Fractal Surfaces

Consider first the interaction of electromagnetic waves with a corrugated fractal surfaces [7,8]. From first principles, it is clear that the incident wave will interact strongly with spatial frequency components of the surface so as to conserve momentum. That is, for a surface described by a fractal almost-periodic surface, the Bragg condition will be satisfied. Consider here the Kirchhoff solution for scattering from rough surfaces which takes the exact roughness profile into account [7]. This Kirchhoff solution treats the rough surface as locally flat with the assumption that the wavelength of the incident wave is small relative to the radius of curvature of the surface irregularities. Angles away from grazing incidence will be used to avoid the problem of shadowing. The geometry of interest is shown in Fig. 6 where the incident angle is taken to be 30°.

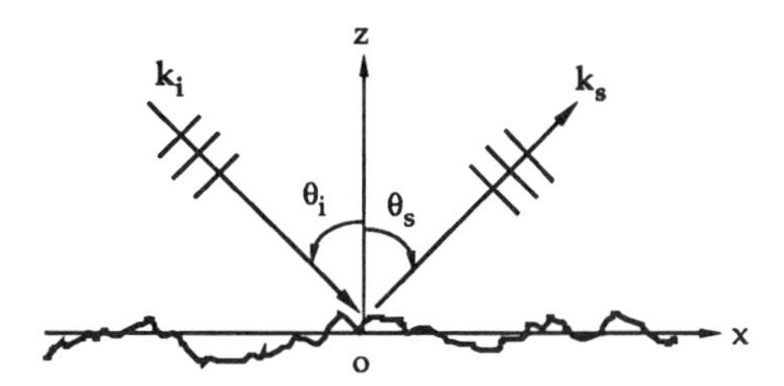

Fig. 6. The geometry of electromagnetic wave scattering in a plane from a rough fractal surface where the subscripts i and s indicate parameters associated with incident and scattered waves, respectively. Here **k** represents the wave vector and θ_i = 30°.

In order to study the relation of the fractal dimension to coupling lobe intensities, we construct log-log plots of the scattering coefficient from the specular direction to maximum angle. Plots in Fig. 7 are the scattering coefficient versus $\sin[(\theta_s - 30°)/2]$, both in dB scales. Two envelopes are indicated in the plots. The background envelope (slope ≈ -1) is due to the finite patch size and consists of the specularly reflected main beam and its sidelobes. The coupling lobe envelope (variable slope) is due to surface coupling and consists of the first-order coupling of the surface harmonics. The slope of the coupling lobe envelope is indicative of surface roughness.

The coupling lobe envelope slopes are found to be –2.50, –1.92, –1.22, and –0.72 for fractal dimensions D = 1.05, 1.30 (not shown here), 1.50 (not shown here), and 1.70, respectively. This implies that the magnitude of slope for these bandlimited fractal surfaces varies approximately as $[2.7(2 - D)]$. This result is not unlike that of scattering from fractal aggregates as discovered by others. Again, the diffracted envelope slopes provide a remote means for quantifying surface roughness. An alternative approach to fractal surface scattering using a generalized Rayleigh approach is given elsewhere [8].

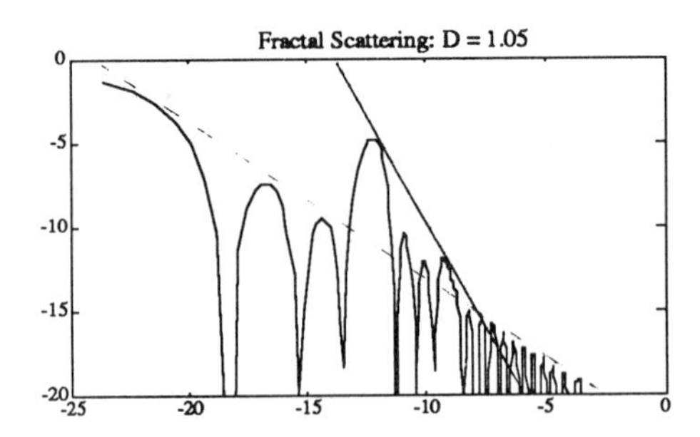

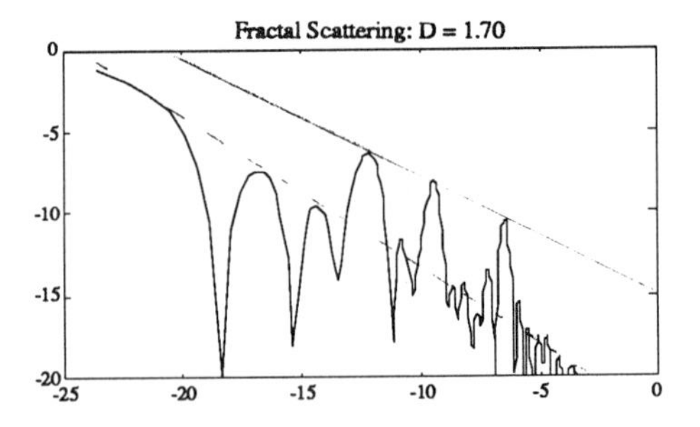

Fig. 7. The scattering coefficient versus $\sin[(\theta_s-30°)/2]$, both in dB scales, with fractal dimensions D = 1.05 and 1.70 for the fractal surface. The envelope slopes of coupling sidelobes (solid lines) vary monotonically with the fractal dimension while the background slope (dashed lines) is constant for varying D. Adapted from [7].

B. Scattering by Fractal Fibers

Scattering from fractal fibers [9,10] is considered as the bounded or curved counterpart to the rough fractal surface just examined. The geometry used for calculations involving scattering from a longitudinally corrugated fractal fiber is shown in Fig. 8 where a normally incident plane wave impinges on a azimuthally symmetric dielectric cylinder. The longitudinal surface variations are described by the bandlimited Weierstrass function described previously and the normalized differential scattering cross section NDSCS is calculated as a function of normalized size parameter Q for various fractal dimensions to recover information relating to the roughness and fractal dimension D.

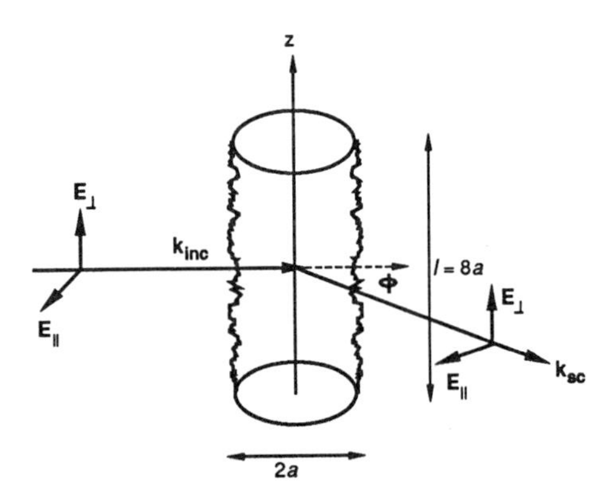

Fig. 8. Geometry for scattering of an incident plane wave by fractal fiber of length l and mean radius a. The scattering plane is defined by the incident and scattered wave vectors, $\mathbf{k}_{inc}$ and $\mathbf{k}_{sc}$.

The results analogous to the planar results given previously are shown in Fig. 9 for the fractal fiber shown above, under the Rayleigh-Gans approximation [9] for parallel polarization. Fractal and non-fractal regimes of the rough fiber are easily identified through electromagnetic interrogation. The low frequency ($Q \lesssim 1$) regime represents Rayleigh scattering and so depends only on the total volume of the fiber segment. However, high frequency scattering ($1 \lesssim Q \lesssim 50$) is due to wave interactions with the details of the scatterer and so yields fractal information concerning the fiber surface.

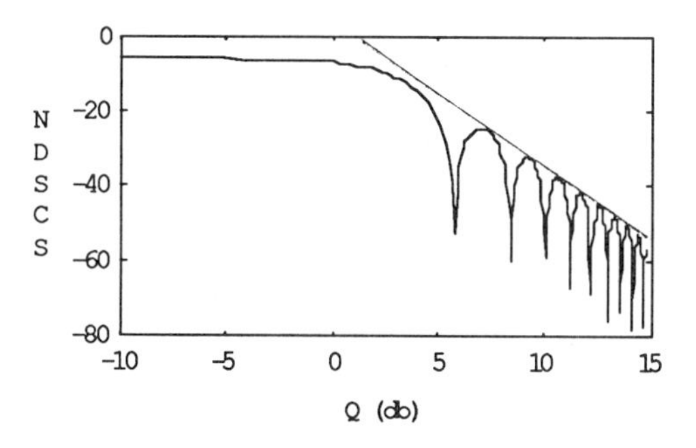

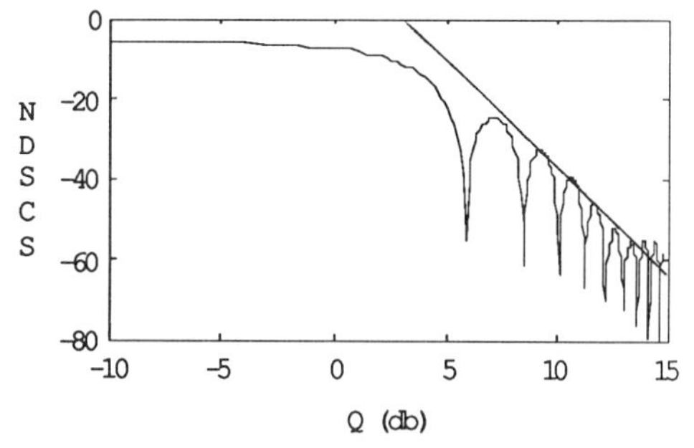

Fig. 9. Normalized differential scattering cross-sections NDSCS are plotted against the normalized size parameter Q = 2k*a* sin(ϕ/2) in logarithmic scale for the fractal fiber. For the same variance, the slopes (solid lines) of the envelopes are measured for different fractal dimension D = 1.05 (left), slope = – 3.88; and D = 1.95 (right), slope = – 4.82. Adapted from [9].

The monotonic dependence of the slope of the scattering cross-section with fractal dimension D of the surface corrugation is seen from the plots above. With increasing fractal dimensions D = 1.05 (relatively smooth corrugation), D = 1.50 (not shown here), D = 1.75 (not shown here), and D = 1.95 (relatively rough corrugation), the increasing slope values for the lobe envelopes are found to be –3.88, –4.00, –4.54, and –4.82, respectively. Thus, as for the case of scattering from a fractally corrugated surface, here the scattering characteristics shown a monotonic variation with changes in surface roughness and fractal dimension. It is apparent that increasing roughness (increasing D) leads to a faster fall off of the differential scattering cross section with frequency or normalized size parameter. The scattering of electromagnetic waves from cylinders which are fractally fluted in the azimuthal direction are considered elsewhere [10].

C. Scattering from Fractal Apertures

Diffraction by fractal phase screens and apertures has applications to the problems of wave propagation through imperfect atmospheres and diffraction by imperfect openings, respectively. The former problem leads to considerations important in modeling turbulent refractive indices and the propagation of microwaves, millimeter waves and light through multiple phase screens [6]. Here we consider the latter problem of light diffraction by fractally serrated apertures. An exact relation can be obtained between the diffracted field and a fractally serrated aperture under the physical optics approximation in the far zone. Thus, the fractal dimension of the aperture boundary can be related in closed form to the diffracted electromagnetic field [11].

An optical experiment was conducted to verify certain aspects of this calculation for a series of fractal apertures with boundaries described by the bandlimited Weierstrass function of Fig. 5. In Fig. 10 is shown the result of one such experiment in which aperture boundaries with D ranging from 1.01 (top) to 1.99 (bottom) is shown [11]. For these fractal serrations, the angular distribution of the diffracted intensity is related to the fractal dimension of the aperture perimeter in a simple manner. For small fractal dimension, the low spatial harmonics of the aperture dominate the interaction and the diffraction pattern possesses a strong angular dependence (upper right plot). However, for higher fractal dimension, all spatial harmonics of the aperture contribute almost equally and the diffraction pattern tends to lose the angular variation characteristic of its lower dimension counterpart.

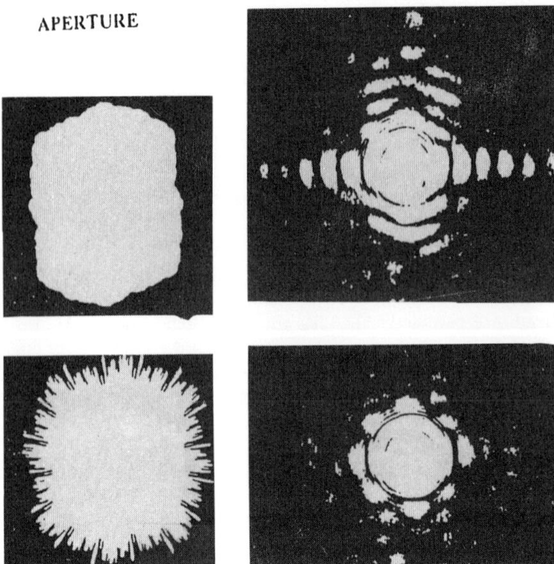

Fig. 10. Fractally serrated apertures (left) and their corresponding far zone diffracted intensity patterns (right) for two fractal dimensions. The fractal dimension of the aperture boundary varies from D = 1.01 (top) to D = 1.99 (bottom). Adapted from [11].

V. ADDITIONAL APPLICATIONS

In addition to the applications noted in the previous section where the wavelength was used as an interrogating electromagnetic yardstick, there are numerous applications involving the use of fractals and their other attributes in electromagnetic modeling and synthesis. In the following sections, several of these additional attributes are highlighted.

A. Fractal Antenna Arrays

Since fractals provide a balance between long-range order and short-range disorder, it is of engineering interest to examine cases where this attribute might be put to use. One such cases was inspired by an observation of Schelkunoff in the 1940's on the desirable properties of "arrays of arrays." This early description of self-similar antenna arrays led us to investigate more fully the properties of fractal antenna arrays in which the strengths of regular arrays, for example their moderately low sidelobes, could be combined with the strengths of random arrays such as their robustness to element failure and thinning [12]. It appears that this blending of random and regular array proprieties can be obtained through the use of random fractals.

In Fig. 11 is shown the array factors for a thinned array of traditional random design (top), a deterministic fractal array (middle) and the random fractal array (bottom) based on these fractal concepts. In the last two cases, <D> is an appropriate fractal dimension which describes the generation of the arrays from a series of subarrays. For these examples, the random fractal array provides a robust design with maximum sidelobes which are below those of typical random arrays. The top plot indicate a maximum sidelobe level of –5.7 dB for the random array versus a –12.5 dB sidelobe level for the random fractal array shown in the bottom plot. Thus, the robustness inherent in aperiodic arrays is combined with the desirable low sidelobe levels of uniform periodic arrays. This concept can, of course, be modified through the use of appropriate current tapering.

This initial exploration of fractal arrays suggests that fractal geometry, with its inherent balance between order and disorder may provide an alternative avenue of design for antenna array synthesis to those deterministic and random methods presently used. An additional virtue of this method is that the fractal prescription offers a flexible alternative to the usual periodic spacing without resorting to random arrays.

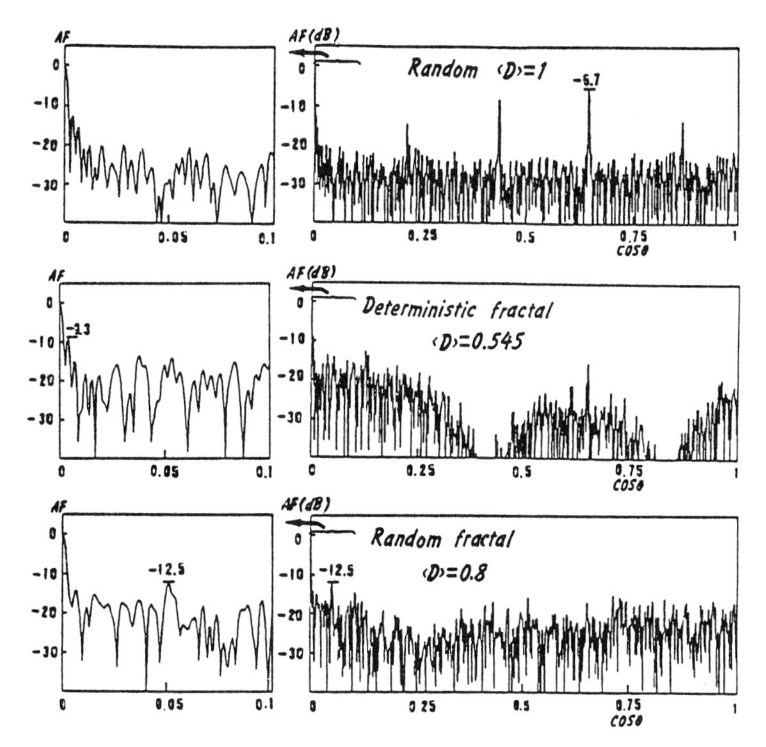

Fig. 11. Array factors AF as a function of the cosine of the array angle for a random array (top), deterministic fractal array (middle) and a random fractal array (bottom). Here <D> is the average fractal dimension of the two fractal arrays and t is the angle from the array axis. The left plots are expanded regions of the right plots near the region of the main lobe. In each case, these thinned arrays are composed of 180 elements spread over 360 wavelengths. Adapted from [12].

B. Self-Similar Methods of Calculation

Another avenue for exploration involves the the interaction of electromagnetic waves with finely divided layers such as might occur in geophysical applications or in optical devices. Fractals provide an ideal test bed for such investigations and in addition provide an avenue for investigating new methods of computation involving wave interactions with fractals.

The model used here for the variation in a dielectric fractal multi-layer is the Cantor bar model of Fig. 12 in which each state of the Cantor bar is generated in an iterative fashion by successively removing the middle third of the bar of the previous stage. This initiator/generator construction is typical of many stylized fractal constructions [1] and clearly embodies self-similarity in the structures so formed.

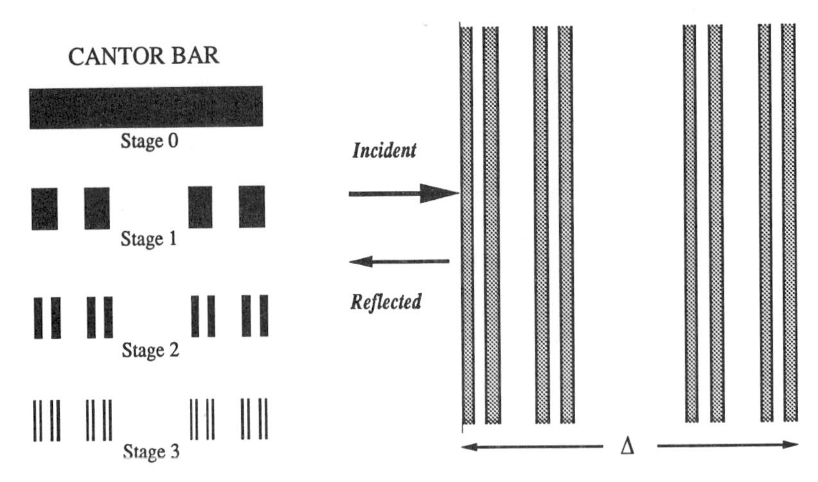

Fig. 12. The generation of a fractal Cantor bar and its stages of growth (left). A higher growth stage is generated by removing the middle third of existing bars. Growth stages 0 to 3 are shown here. Using the disk covering method, the fractal dimension of a Cantor dust is found to approach log(2)/log(3) = 0.631. . . . as the thickness of each Cantor bar approaches zero. A reflection problem (right) by a Cantor bar multi-layer involving finely divided dielectric layers with two dielectric constants determined by the cantor bar.

Here we investigate not only the interactions of waves with such fractal models but also see how self-similar structures may lead to self-similar methods of computation where the generation of the fractal object through iteration provides a model for an iterative calculation of electromagnetic waves with such models.

For the reflection and transmission problem shown in Fig. 12, there exist multiple reflections among all layers. However, a *self-similarity method of computation* can be constructed to reduces the number of calculations for many layered structures [13]. For a set of Cantor bar fractal layers of stage N, for example, only N iterations are needed. To solve the same problem with the chain matrix approach, the multiplication of $(2^{N+1} - 1)$ matrices is required.

As an example, the amplitude of the reflection coefficients as a function of this dimensionless quantity δ for Cantor bar fractal layers at different stages of growth where 1 and 1.5 are used respectively for the background and layer refractive indices as shown in Fig. 13. Here we define the quantity $\delta = \Delta/\lambda$ to be the ratio of the thickness Δ of the fractal layers to the wavelength λ of the incident wave. The amplitude of reflection coefficient is a periodic function of thickness to wavelength ratio δ. The reflection coefficient variation within a period is shown in the plots for stages 0 (left) and 3 (right).

Observing the plots in Fig. 13, the reflection coefficient variation at each higher stage is a modulated version of that associated with the previous stage. That is the reflection coefficient has self-similar properties as does the fractal multi-layer. One can extract both the stage of growth and the fractal dimension of the dielectric variation given the reflection data [13].

VI. CONCLUSIONS AND ACKNOWLEDGEMENTS

In this chapter we have explored here several problems in fractal electrodynamics in which fractal geometry and electromagnetic theory have been brought to bear on a diverse set of problems. In general we divide these problems into two categories. In the first, bandlimited fractals are used to provide realistic models of complex structures which are

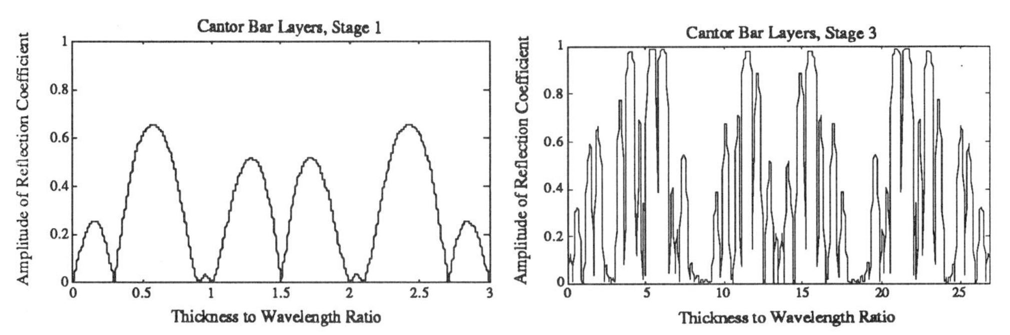

Fig. 13. Reflection coefficients calculated for Cantor bar multi-layers of the first and third stages of fractal growth as a function of thickness to wavelength ratio δ. Note that the horizontal scales are different for stages of growth so as to provide a single period of reflection coefficient variation. Adapted form [13].

amenable to analysis and calculation. We found simple relationships between the variation of scattered fields with frequency or angle and the roughness of these structures as measured by their fractal dimension. The second set of problems includes a far ranging set of additional investigations which use fundamental properties of fractals including their self-similarity and their carefully maintained balance of order and disorder to provide new tools for modeling, analysis and syntheses. This latter set of problems is almost unbounded in nature and appears fruitful for further study.

VII. ACKNOWLEDGMENT

There are mentors, colleagues and students, who are responsible for my continuing interest in the relation of geometry and electromagnetics so as to understand the interaction of waves with complex media. Professor Papas (Caltech) always desired to know the physics behind the mathematics so that the simplest picture possible emerged from a given problem. This often led to insightful relations between complicated problems in electromagnetics and their geometry. My colleagues, Dr. Y. Kim (Jet Propulsion Laboratory) and Dr. X. Sun (University of Pennsylvania), collaborated with me in the examination of the canonical problems summarized here and have been indispensable in launching the area of fractal electrodynamics.

VIII. SELECTED REFERENCES

[1] D. L. Jaggard, "On Fractal Electrodynamics," in *Recent Advances in Electromagnetic Theory*, H. N. Kritikos and D. L. Jaggard, eds., Springer-Verlag, New York (1990).

[2] Mandelbrot, B. B., *The Fractal Geometry of Nature*, W. H. Freeman and Company, San Francisco (1983).

[3] This discussion is based, in part, on a lecture of Leo P. Kadanoff, "Measuring the Properties of Fractals," presented at AT&T Bell Laboratories, Holmdel, NJ (Feb. 14, 1986); also see M. Barnsley, *Fractals Everywhere*, Academic Press, Boston (1988).

[4] Berry, M. V. and Z. V. Lewis, *Proc. R. Soc. London Ser. A 370*, 459-484 (1980).

[5] Jaggard, D. L. and Y. Kim, "Diffraction by Bandlimited Fractal Screens," *J. Opt. Soc. Am. A4*, 1055-1062 (1987).

[6] Kim, Y. and D. L. Jaggard, "A Bandlimited Fractal Model of Atmospheric Refractivity Fluctuation," *J. Opt. Soc. Am. A5*, 475-480 (1988); or Y. Kim and D. L. Jaggard, "Optical Beam Propagation in a Bandlimited Fractal Medium," *J. Opt. Soc. Am. A5*, 1419-1426 (1988).

[7] D. L. Jaggard and X. Sun, "Scattering by Fractally Corrugated Surfaces," *J. Opt. Soc. A7*, 1131 - 1139 (1990).

[8] X. Sun and D. L. Jaggard, "Wave Scattering from Non-Random Fractal Surfaces" to appear in *Opt. Comm.* (1990); or D. L. Jaggard and X. Sun, "Rough Surface Scattering: A Generalized Rayleigh Solution" to appear in *J. Appl. Phy.* (Dec. 1990).

[9] Jaggard, D. L. and X. Sun, "Backscatter Cross-Section of Bandlimited Fractal Fibers," *IEEE Trans. Ant. and Propagat. AP-37*, 1591-1597 (1989).

[10] X. Sun and D. L. Jaggard, "Scattering from Fractally Fluted Cylinders," *J. Electromagnetic Wave Appl. 4*, 599-611 (1990).

[11] Y. Kim, H. Grebel and D. L. Jaggard, "Diffraction by Fractally Serrated Apertures," to appear in *J. Opt. Soc. A.* (Nov. 1990).

[12] Kim, Y. and D. L. Jaggard, "Fractal Random Arrays," *Proc. IEEE 74,* 1278-1280, (1986).

[13] D. L. Jaggard and X. Sun, "Reflection from Fractal Multi-Layers," to appear in *Optics Lett.* (Fall 1990).

FEYNMAN-DIAGRAM APPROACH TO WAVE DIFFRACTION BY MEDIA HAVING MULTIPLE PERIODICITIES

Theodor Tamir, Kun-Yii Tu and Hyuk Lee

Department of Electrical Engineering
and Weber Research Institute
Polytechnic University, Brooklyn, NY 11201

ABSTRACT: Problems requiring the accurate determination of waves diffracted by periodically modulated media have been treated most recently by using coupled-wave techniques. By developing a field representation in terms of multiply scattered waves, we provide here an alternative approach which is particularly effective for situations involving two or more superposed volume gratings. An advantage of this approach is that the diffracted fields can be described by modified Feynman diagrams in the form of flow charts. These diagrams provide a phenomenological description of the scattering process, and they also serve as a powerful and systematic tool for the rapid evaluation of any diffracted grating orders.

INTRODUCTION

Renewed interest into the scattering of waves by periodically modulated media has recently been stimulated by applications in acousto-optics, integrated optics and optoelectronics. A corresponding basic electromagnetic problem was first explored about seventy years ago by Brillouin in the context of diffraction of light by sound,[1] and numerous analytical investigations have subsequently addressed that topic and related ones. However, pertinent review articles[2-4] reveal that those studies have dealt almost exclusively with regions containing a single periodicity in the form of a planar volume grating. Two or more overlapping periodic modulations were first considered in the context of holographic applications, which generally contain gratings with non-commensurate periodicities inclined at arbitrary angles $\Delta\phi$ with respect to each other. The boundary-value problem for such multiple gratings was treated mostly[4] by extending modal or coupled-wave methods that had previously been developed for single gratings. Those techniques can provide rigorous solutions, but they offer limited physical insight and involve very time-consuming calculations if highly accurate results are needed. Furthermore, severe numerical convergence problems appear as the angular separation $\Delta\phi$ between any two gratings becomes very small or if the number of significant diffraction orders increases appreciably.

We therefore discuss here a different approach that expresses the field diffracted by superposed gratings in terms of a physically meaningful multiple scattering process. This process can be described by Feynman diagrams which are preferably cast in the form of flow graphs or charts. In turn, the flow graphs can serve as templates for fast algorithms that are less affected by numerical convergence difficulties, and they can handle arbitrary values of $\Delta\phi$ and a large number of diffraction orders. The multiple-scattering model developed by us[5] thus provides a powerful alternative field representation, which offers conceptual and computational advantages over approaches that view the diffracted field as a collection of normal or coupled modes.

OUTLINE OF THE MULTIPLE-SCATTERING APPROACH

To describe our approach, we consider the basic configuration shown in Fig. 1, which involves two gratings inclined at an angle $\Delta\phi$ with respect to each other within a dielectric layer of thickness z_0. For simplicity, we assume isotropic media and a two-dimensional geometry with $\partial/\partial y \equiv 0$. A monochromatic plane wave with propagation vector $\mathbf{k}_0$ is incident from the left and we focus our discussion on the diffracted waves that emerge to the right. If the gratings are described by vectors $\mathbf{K}_1$ and $\mathbf{K}_2$ with magnitudes $K_1 = 2\pi/d_1$ and $K_2 = 2\pi/d_2$, respectively, the diffracted orders are identified by the wave vectors

$$\mathbf{k}_{n1,n2} = \mathbf{k}_0 + n_1\mathbf{K}_1 + n_2\mathbf{K}_2, \qquad \text{with } n_1, n_2 = 0, \pm1, \pm2, \ldots \tag{1}$$

The number pairs $n = (n_1, n_2)$ account for a finite number of propagating diffraction orders and an infinite discrete spectrum of evanescent ones, as identified by complex propagation vectors $\bar{\mathbf{k}}_{n1,n2}$ (discussed later), which are generally different from the real wave vectors $\mathbf{k}_{n1,n2}$. A few such diffracted waves are shown in Fig. 1. For practical purposes, $\mathbf{K}_1$ and $\mathbf{K}_2$ are chosen so as to suppress all undesirable (*e.g.*, $n_1 + n_2 \neq 0$ or 1) higher orders by rendering their fields evanescent along the z direction. The remaining (*e.g.*, $n_1 + n_2 = 0$ or 1) orders include the propagating (non-evanescent) fields and the balance of all other evanescent ones.

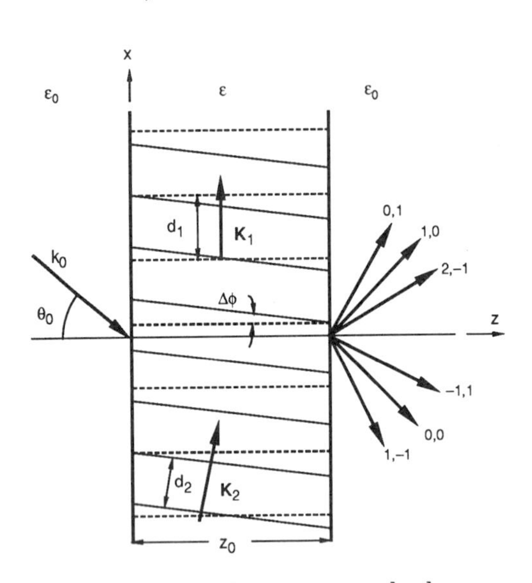

Fig. 1. Geometry of two superposed volume gratings

To neglect first-order reflection effects, we assume that the permittivity ε in the layer has an average value equal to the exterior permittivity ε_0, and choose a sinusoidal periodicity such that

$$\varepsilon = \varepsilon_0 [1 + g(\mathbf{r})], \quad \text{with } g(\mathbf{r}) = M_1 \cos(\mathbf{K}_1 \cdot \mathbf{r}) + M_2 \cos(\mathbf{K}_2 \cdot \mathbf{r}). \tag{2}$$

Here $\mathbf{r}$ is the position vector and M_1 and M_2 are modulation amplitudes, which are much smaller than unity in typical applications. If we restrict the discussion to perpendicular (TE) polarization, the electric field $E = E(\mathbf{r})$ is along y and satisfies

$$\left(\nabla^2 + k_0^2\right) E = -k_0^2\, g(\mathbf{r})\, E\,, \tag{3}$$

where a time dependence $\exp(-i\omega t)$ is implied but suppressed and $k_0 = \omega(\mu_0\varepsilon_0)^{1/2}$ is the magnitude of $\mathbf{k}_0$. The solution of Eq. (3) can be written as[6]

$$E(\mathbf{r}) = E^{(0)}(\mathbf{r}) + k_0^2 \int G(\mathbf{r},\mathbf{r}_i)\, g(\mathbf{r}_i)\, E(\mathbf{r}_i)\, d\mathbf{r}_i\,, \tag{4}$$

where $E^{(0)}$ is the homogeneous solution, the integration is over two spatial variables with respective ranges $-\infty < x_i < \infty$ and $0 < z_i < z_0$, and

$$G(\mathbf{r},\mathbf{r}_i) = \frac{1}{(2\pi)^2} \int \frac{\exp\left[i\mathbf{k}\cdot(\mathbf{r} - \mathbf{r}_i)\right]}{k^2 - k_0^2}\, d\mathbf{k}\,. \tag{5}$$

A solution can then be derived by means of the successive iteration process expressed by

$$E(\mathbf{r}) = \sum_{m=0}^{\infty} E^{(m)}(\mathbf{r})\,, \tag{6}$$

with

$$E^{(m)}(\mathbf{r}) = k_0^2 \int G(\mathbf{r},\mathbf{r}_i)\, g(\mathbf{r}_i)\, E^{(m-1)}(\mathbf{r}_i)\, d\mathbf{r}_i\,, \tag{7}$$

where m denotes the m-th iterated result and it is understood that the upper (∞) limit in Eq. (6) may actually be a finite number m_0 prescribed by a desired accuracy. By properly initiating the iteration process with

$$E^{(0)} = E_{inc} = \exp(i\mathbf{k}_0\cdot\mathbf{r}), \tag{8}$$

and by repeatedly using Eq. (7), we can derive $E^{(m)}$ up to any desired iteration level m. In particular, a first-order Born approximation is obtained if we retain only $E^{(0)}$ and $E^{(1)}$ in Eq. (6) and disregard all $E^{(m)}$ with $m > 1$.

An analytical result for any $E^{(m)}$ can be obtained by finding $E^{(1)}$, $E^{(2)}$, *etc.*, and then applying induction arguments. Because it is algebraically elaborate, this derivation is omitted here and the reader is referred elsewhere[5] to verify that

$$E^{(m)} = S_m\, \Psi_{n(m)} \exp\left[i\bar{\mathbf{k}}_{n(m)}\cdot\mathbf{r}\right], \tag{9}$$

where S_m is the sum-product operator

$$S_m = \sum_{\nu(1)} \alpha_{n(1)} \sum_{\nu(2)} \alpha_{n(2)} \sum_{\nu(3)} \alpha_{n(3)} \cdots \sum_{\nu(m)} \alpha_{n(m)}\,, \tag{10}$$

which acts on the phase correlation function

$$\Psi_{n(m)} = (-)^{m+1} \frac{1-\exp\left(-i\Delta_{n(m)}\right)}{\prod_{q=1}^{m} \Delta_{n(q)}} - \sum_{q=1}^{m-1} \frac{1-\exp\left[i\left(\Delta_{n(q)} - \Delta_{n(m)}\right)\right]}{\Delta_{n(q)} \prod_{\substack{h=1 \\ h \neq q}}^{m} \left(\Delta_{n(q)} - \Delta_{n(h)}\right)} . \tag{11}$$

In the above, the index n(q), with q = 1, 2, 3, ... , refers to a specific number pair $n(q) = n[n_1(q), n_2(q)]$, where the integers n_1 and n_2 have the same connotation as in Eq. (1). However, the functional dependence on q now implies that n(q) is the result of a sequential process given by the recursive relation

$$n(q+1) = \left[n_1(q+1), n_2(q+1)\right] = \begin{cases} \left.\begin{matrix}\left[n_1(q)+1\, , n_2(q)\right] \\ \left[n_1(q)-1\, , n_2(q)\right]\end{matrix}\right\} & \text{with } \mu(q+1) = 1 \\ \\ \left.\begin{matrix}\left[n_1(q)\, , n_2(q)+1\right] \\ \left[n_1(q)\, , n_2(q)-1\right]\end{matrix}\right\} & \text{with } \mu(q+1) = 2 \end{cases} \tag{12}$$

which is generated with initial values $n_1(0) = n_2(0) = 0$. For every q = 1, 2, 3 ... , the above sequence defines a set ν(q) containing 4^q number pairs $n(q) = n[n_1(q), n_2(q)]$. In addition, Eq. (12) indicates whether the index μ(q) should be taken as 1 or 2 in the expression

$$\alpha_{n(q)} = \frac{k_0^2 M_{\mu(q)} z}{4 w_{n(q)}} , \tag{13}$$

with

$$w_{n(q)} = [k_0{}^2 - u^2{}_{n(q)}]^{1/2} , \tag{14}$$

$$\Delta_{n(q)} = p_{n(q)} z = [w_{n(q)} - v_{n(q)}] z , \tag{15}$$

$$\bar{\mathbf{k}}_{n(q)} = \hat{\mathbf{x}} u_{n(q)} + \hat{\mathbf{z}} w_{n(q)} . \tag{16}$$

For specific $n(q) = (n_1, n_2)$, the quantities $u_{n(q)}$ and $v_{n(q)}$ are, respectively, the x and z components of the vectors $\mathbf{k}_{n1,n2}$ defined in Eq. (1), and $\hat{\mathbf{x}}$ and $\hat{\mathbf{z}}$ are unit vectors along the corresponding directions. It is important to recognize that $\bar{\mathbf{k}}_{n(q)}$ of Eq. (16) acts as a propagation vector because it determines the progression of the field components in Eq. (9). Hence, although the wave vectors $\mathbf{k}_{n1,n2}$ serve to identify diffracted waves, the actual propagation vectors are $\bar{\mathbf{k}}_{n1,n2}$ whose magnitudes satisfy the plane-wave condition that they be equal to k_0.

The formal solution for the transmitted field is thus given by Eqs. (6) and (9) through (16). As discussed below, this formulation can be systematically described by flow graphs which provide physical insight and help greatly in deriving quantitative results.

FEYNMAN DIAGRAMS AND FLOW GRAPHS

For single-grating configurations, a field representation similar to that given above was developed for acousto-optic interactions by Korpel,[7,8] who then expressed the scattering process in terms of Feynman diagrams.[9] This approach can be extended to two gratings. As an example, Fig. 2 shows some of the Feynman diagrams that must be considered when deriving the intensity of the n = (1, 0) spectral order. These diagrams can be obtained by using the formalism described above to identify those components in Eqs. (6) and (9) whose propagation factor is $\bar{\mathbf{k}}_{n1,n2}$. Because the specific n = (1, 0) order is generated by components that may occur at any iteration level m, the process that determines those diagrams must follow a systematic pattern. In particular, that process must ensure that all the requisite components are accounted for, on the one hand, and that they are arranged in some suitable quantitative order, on the other hand.

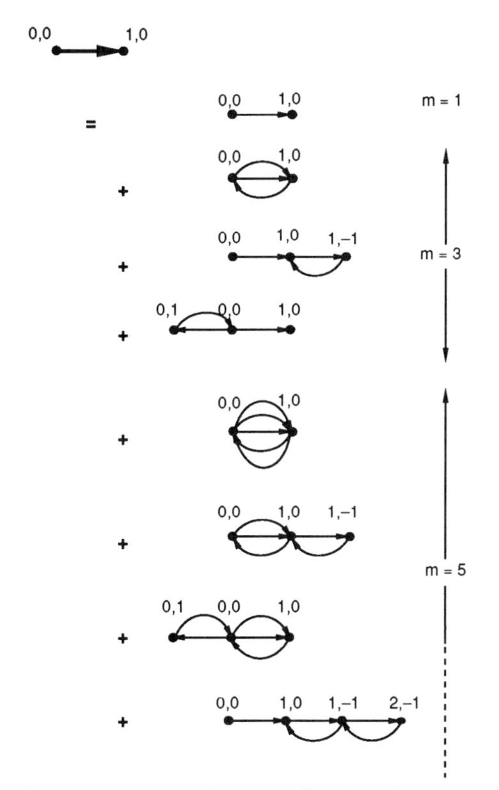

Fig. 2. Feynman diagram showing the construction of the (1, 0) diffraction order

We have therefore developed a technique that distinguishes between the effects of the two grating vectors $\mathbf{K}_1$ and $\mathbf{K}_2$ and then found that the Feynman diagrams are most conveniently cast in the form of flow charts or graphs. Such a chart is shown in Fig. 3, which accounts for all scattered-field components up to and including the third (m = 3) level.

The construction of flow charts is based on the sequence of number pairs generated by the recursive relation in Eq. (12). As defined, the sequence is initiated by the incident field $E^{(0)}$, which corresponds to the point (0, 0) at the top of Fig. 3. This field is scattered by both gratings via their $\pm\mathbf{K}_1$ or $\pm\mathbf{K}_2$ vectors, thus generating four new components indicated by the set of number pairs $\nu(1)$ = (1,0), (–1, 0), (0, 1) and (0, –1) at the first (m = 1) level in Fig. 3. The sum of these four components constitutes the field $E^{(1)}$, where the superscript identifies the pertinent scattering level, such as m = 0, 1, 2, ... in Eq. (6). Each one of these components undergoes an analogous scattering process, thus generating the set $\nu(2)$ of 16 number pairs shown at the second (m = 2) level in Fig. 3; the sum of those 16 components yields $E^{(2)}$. This scattering process continues indefinitely in a sequential fashion and generates additional $E^{(m)}$ fields at every subsequent m-th level. Hence the physical role of the iteration index m is to designate the order of the scattering process. It is then easily verified that each $E^{(m)}$ consists of 4^m field components identified by a corresponding set $\nu(m)$ of number pairs.

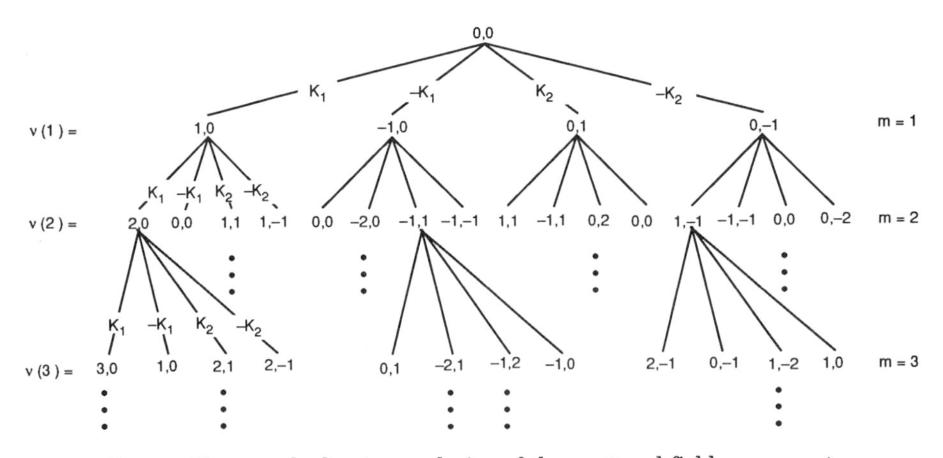

Fig. 3. Flow graph showing evolution of the scattered field components

In the flow graph, every individual component $E_{n1,n2}^{(m)}$ can be viewed as the end-product of a scattering process along a path that starts at the common initial point (0, 0) and terminates at the point (n_1, n_2) under consideration. As an example, the component $E_{-2,1}^{(3)}$ represented by the point n(m) = n(3) = (–2, 1) appears in Fig. 3 as one out of 64 number pairs in the set v(3); that particular component is accounted for by the path (0, 0) → (–1, 0) → (–1, 1) → (–2, 1). Incidentally, that path is generated by a first scattering through $-\mathbf{K}_1$ to level m = 1, followed by a scattering through $\mathbf{K}_2$ to level m = 2 and another scattering through $-\mathbf{K}_1$ to level m = 3. Hence every number pair at some level m in a flow graph is the end point of a path dictated by a particular sequence of intermediate steps n(q) with q = 1, 2 ... m; thus, for the path considered above, q = 1 yields n(1) = (–1, 0), q = 2 yields n(2) = (–1, 1), *etc.*

The field of a specific grating order n = (n_1, n_2) consists of components that may occur at several locations in a flow graph. As an example, n = (1, 0) appears in Fig. 3 at level m = 1, but not at level m = 2, and then re-appears at level m = 3. Although only two (1, 0) entries are shown in Fig. 3 at m = 3, a total of nine such components actually occur at that level because other components in the set v(2) scatter into a (1, 0) component at level m = 3. However, all those components have the same propagation factor $\bar{\mathbf{k}}_{1,0}$ so that they bear the same space-time relationship to each other. To an accuracy given by m = 3, ten separate $E_{1,0}$ components must then be combined to adequately describe the diffracted order (1, 0) of the total field E(**r**) in Eq. (3). We therefore infer for the general case that

$$E_{n1,n2} = \sum_{m=0}^{\infty} \sum_{v(m)} E^{(m)}_{n1,n2} \tag{17}$$

represents the complete diffracted field associated with a specific grating order n = (n_1, n_2) under consideration.

EVALUATION OF DIFFRACTED ORDERS

In practical situations, only a few diffracted orders $E_{n1,n2}$ are significant. It is therefore sufficient to work with reduced flow charts that, at some final scattering level m_0 being considered, retain only end points belonging to the grating orders $n = (n_1, n_2)$ that are relevant. Furthermore, some of those end points may terminate a path along which at least one branch accounts for scattering through an evanescent order; such paths can be ignored because they account for $E_{n1,n2}$ contributions whose amplitude is usually negligible.

As an example, we choose the grating vectors $\mathbf{K}_1$ and $\mathbf{K}_2$ so that the only propagating orders are those satisfying the condition that $n_1 + n_2 = 0$ or 1 in Fig. 3. If we then focus on the (1, 0) order, the corresponding reduced flow graph up to level $m_0 = 5$ is given in Fig. 4, where the components that contribute significantly to $E_{1,0}$ are shown inside rectangular frames for emphasis. Each one of those components serves as an end point that uniquely defines a scattering path. It is therefore convenient to identify those paths by their end points $n_j = n_j(n_1, n_2)$, where $j = 1, 2, \ldots j_0$, and $j_0 = j_0(n_1, n_2; m_0)$ denotes the total number of relevant (1, 0) components.

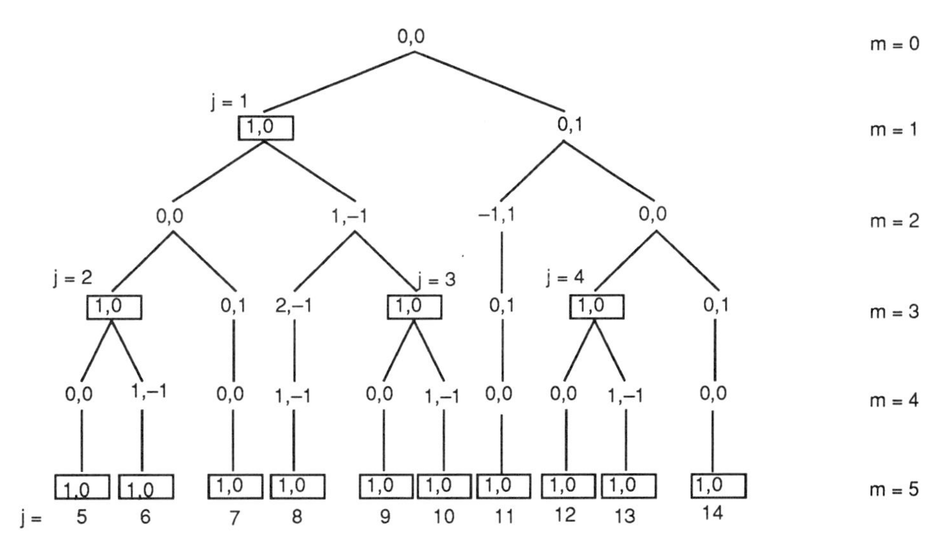

Fig. 4. Reduced flow graph for $n = (0,1)$ and m_0. The various paths are identified by $j = 1, 2, \ldots 14$ at their end points. The construction of this graph assumes that all the points satisfying $n_1 + n_2 \neq 0, 1$ can be neglected.

In the context of a reduced flow graph such as Fig. 4, we can replace Eq. (17) by the simpler expression

$$E_{n1,n2} = \sum_{j=1}^{j_0} E^{(j)}_{n1,n2}, \tag{18}$$

where the summation includes all the terms specified by the relevant paths n_j. Thus, in the example shown in Fig. 4, the diffraction order (1, 0) includes 14 terms. Using the notation of Eqs. (10) and (11), each term in the summation of

Eq. (18) can be explicitly written as

$$E^{(j)}_{n1,n2} = \left[\prod_{q=1}^{m} \alpha^{(j)}_{n(q)}\right] \Psi^{(j)}_{n(m)} \exp\left[i\overline{\mathbf{k}}^{(j)}_{n(m)} \cdot \mathbf{r}\right], \tag{19}$$

where the upper index j indicates that the specific number pairs for n(q) must be taken along the path $n_j(m)$ under consideration.

Any diffracted order of interest may then be evaluated by developing an algorithm to determine the various paths $n_j(m)$ that are needed in Eqs. (18) and (19). Such an algorithm can be simply based on Eq. (12) and therefore needs no further discussion. When applying that algorithm to evaluate Eqs. (18) and (19), the only terms that may pose difficulties are the functions $\Psi^{(j)}_{n(m)}$ given in Eq. (11). These difficulties arise because, even though every $\Psi_{n(m)}$ is finite, its computation may encounter numerical convergence problems whenever the magnitude of either a factor $\Delta_{n(q)}$ or that of a factor $\Delta_{n(q)} - \Delta_{n(h)}$ becomes very small in the denominators of Eq. (11). This correspondingly happens if a Bragg regime is encountered, *i.e.*, $\Delta_{n(q)} \to 0$, or if scatterings repeat their sequential order along the path, *i.e.*, $\Delta_{n(q)} = \Delta_{n(h)}$ for some $q \neq h$. Those difficulties are avoided by expressing Eq. (11) in the form

$$\Psi_{n(m)} = \mathcal{L}^{-1}\left\{\frac{i^m}{s\left(s + ip_{n(m)}\right)\prod_{q=1}^{m-1}\left[s + i\left(p_{n(m)} - p_{n(q)}\right)\right]}\right\}\frac{1}{z^m}, \tag{20}$$

where $\mathcal{L}^{-1}$ denotes the inverse Laplace transform and $p_{n(m)}$ was defined in Eq. (15). A particularly interesting situation occurs if $n_j(m)$ is a path along which all the scattering processes satisfy Bragg conditions exactly. In that case, we have phase synchronism, *i.e.*, all $\Delta_{n(q)} = 0$ along that entire path. We can then verify that Eqs. (11) or (20) reduce to $\Psi_{n(m)} = (i)^m/m!$. This much simpler expression was used to show[5] that our formulation yields results that are identical to those obtained for special cases in the past.

From all of the above considerations, it follows that flow graphs delineate the scattering process and thus describe the evolution of the various components in a systematic manner. A major advantage of flow graphs is that their pattern provides by itself the algorithm needed in calculating any of the relevant diffraction orders. In addition, the hierachical structure of flow graphs helps greatly in evaluating the highest level m_0 needed to achieve numerical results within any desirable accuracy. Such an evaluation can readily include a systematic error analysis[5] which estimates the error incurred if convenient approximations are made, *e.g.*, levels higher than m_0 are neglected.

EXTENSION OF THE MULTIPLE SCATTERING FORMALISM

The representation of the scattered field was restricted here to two gratings having a pure sinusoidal variation in order to simplify the discussion.

However, the formal solution and its corresponding flow graphs can easily be extended to configurations having more than two gratings, and to periodicities having general (non-sinusoidal) variations.

The extension to a larger number N of gratings is implemented by assuming that, instead of using only the integers 1 and 2, we take $\mu = 1, 2, 3, \ldots N$ in all expressions involving the integers n_μ, such as $\mathbf{k}_{n1,\, n2,\, \ldots\, nN} = \mathbf{k}_0 + n_1\mathbf{K}_1 + n_2\mathbf{K}_2 + \ldots + n_N\mathbf{K}_N$, $E_{n1,n2,\ldots nN}$, *etc.* Of course, the sequence of number pairs n(q) denoting flow paths for two gratings will then be replaced accordingly by sequences of number triplets, quadruplets, *etc.*, for N = 3, 4, *etc.*

As an example, for three (N = 3) superposed gratings, Eq. (12) is replaced by

$$n(q+1) = [n_1(q+1), n_2(q+1), n_3(q+1)]$$

$$= \begin{cases} \left.\begin{array}{l} [n_1(q)+1, n_2(q), n_3(q)] \\ [n_1(q)-1, n_2(q), n_3(q)] \end{array}\right\} \begin{array}{l} \text{with} \\ \mu(q+1) = 1 \end{array} \\ \left.\begin{array}{l} [n_1(q), n_2(q)+1, n_3(q)] \\ [n_1(q), n_2(q)-1, n_3(q)] \end{array}\right\} \begin{array}{l} \text{with} \\ \mu(q+1) = 2 \end{array} \\ \left.\begin{array}{l} [n_1(q), n_2(q), n_3(q)+1] \\ [n_1(q), n_2(q), n_3(q)-1] \end{array}\right\} \begin{array}{l} \text{with} \\ \mu(q+1) = 3 \end{array} \end{cases} \tag{21}$$

In this case, each wave component is scattered into 6 new waves at every scattering level, so that 6^m scattered components appear at every m-th level. By induction, we may easily verify for N superposed gratings that each scattering generates 2N new waves and the m-th scattering level includes $(2N)^m$ components. The number of components increases therefore very rapidly with N, but the progression of the scattered waves remains systematic. Hence the evaluation of all diffracted orders can be adequately handled by a computer program that expands upon the scheme adopted for the simpler N = 2 case.

To extend the formalism to non-sinusoidal gratings, we can assume that the periodic modulation for each one of them is given by the Fourier series

$$g_\mu(\mathbf{r}) = M_\mu\, [a_{\mu 1} \cos(\mathbf{K}_\mu \cdot \mathbf{r}) + a_{\mu 2} \cos(2\mathbf{K}_\mu \cdot \mathbf{r}) + a_{\mu 3} \cos(3\mathbf{K}_\mu \cdot \mathbf{r}) + \ldots\,], \tag{22}$$

where $\mu = 1, 2, 3, \ldots N$ refers to specific gratings, as before. In terms of flow graphs, Eq. (22) implies that every grating scatters waves through vectors given by $\pm\mathbf{K}_\mu$, $\pm 2\mathbf{K}_\mu$, $\pm 3\,\mathbf{K}_\mu$, *etc.*, rather through only $\pm\mathbf{K}_\mu$. The additional scattered components can then be accounted for by suitably augmenting the flow graphs and the formulation of Eqs. (10), (11) and (12) or (21).

The above extensions are thus straightforward and they can generalize the present results to a large class of problems involving incident plane waves. Other extensions are being contemplated, such as those involving bounded beams instead of plane waves, or pulsed signals rather than monochromatic waves. These are more interesting but considerably more complex situations, which require extending the present formulation into the spectral (**k**) and/or the temporal (t) domains.

CONCLUDING REMARKS

We have discussed a new approach to the diffraction of waves by periodically modulated media, which can easily accommodate the presence of several gratings having arbitrary periodic variations. This approach is based on characterizing the diffracted field in terms of a multiple scattering process that can be phrased in terms of flow graphs. The pattern of such flow graphs serves as the basis for algorithms that provide a convenient procedure to accurately evaluate the intensities of the scattered spectral orders. The multiple scattering approach can thus serve as a powerful tool in the analysis of problems dealing with plane-wave diffraction by superposed gratings, and it has promising capabilities to address analogous situations involving incident fields that are bounded in either space or time, or both.

ACKNOWLEDGEMENT

The material presented here is based upon work supported by the National Science Foundation and by the New York State Science and Technology Foundation.

REFERENCES

1. M. Born and E. Wolf, *Principles of Optics* (Pergamon, New York, 1980), Chap. 19, p. 593.
2. C. Elachi, "Waves in active and passive periodic structures: A review," Proc. IEEE **64**, 1666-1976 (1976).
3. T. K. Gaylord and M. T. Moharam, "Analysis and applications of optical diffraction by gratings," Proc. IEEE **73**, 894-937 (1985).
4. E. N. Glytsis and T. K. Gaylord, "Rigorous 3-D coupled wave diffraction analysis of multiple superposed gratings in anisotropic media," Appl. Opt. **28**, 2401-2421 (1989).
5. K.-Y. Tu, T. Tamir and H. Lee, "Multiple-scattering theory of wave diffraction by superposed volume gratings," J. Opt. Soc. Am. A **7**, 1421-1436 (1990).
6. K. Fujiwara, "Application of higher order Born approximation to multiple elastic scattering of electrons by crystals," J. Phys. Soc. Japan **14**, 1513-1524 (1959).
7. A. Korpel, "Two-dimensional plane wave theory of strong acousto-optic interaction in isotropic media," J. Opt. Soc. Am. **69**, 678-683 (1979).
8. A. Korpel and T.-C. Poon, "Explicit formalism for acousto-optic multiple plane wave scattering," J. Opt. Soc. Am. **70**, 817-820 (1980).
9. R. D. Mattuck, *A guide to Feynman diagrams in the many-body problem* (McGraw-Hill, New York, NY, 1967).

ELECTROMAGNETIC BANDGAP ENGINEERING IN THREE-DIMENSIONAL PERIODIC DIELECTRIC STRUCTURES

K. M. Leung

Department of Physics
Polytechnic University
Brooklyn, NY 11201

ABSTRACT

Recent theoretical calculations of photonic band structures using plane-wave expansion and the Korringa-Kohn-Rostoker method in three-dimensional periodic dielectric media are reported. Results are given for scalar waves and vector waves. The merits and drawbacks of the two different methods will be discussed. Comparisons with the available experimental results will also be made. The inadequacy of scalar wave calculations for photonic bands will be pointed out.

INTRODUCTION

It is well-known that practically all of the interactions in condensed matter physics, as well as atomic, molecular and chemical physics have at least a portion whose origin can be traced to electromagnetism. It is therefore possible to alter these interactions by appropriately changing the nature of the electromagnetic modes that are allowed in the material. In particular, if the material has a spatially periodic refractive index in three-dimension, the electromagnetic spectrum will consist of allowed and forbidden bands. And if the refractive index modulation is sufficiently high for strong multiple scattering to occur, and if the Brillouin zone associated with the periodic structure is close to spherical, then the forbidden bands in different directions of the zone may completely overlap so as to form a frequency gap in the photon density of states (the number of electromagnetic modes per unit frequency range). This means that photons with frequencies within the gap are totally absent within the material. This can lead to a number of fascinating phenomena such as severe inhibition of spontaneous emission and drastic enhancement of exciton lifetimes.[1] [2] The ability to inhibit spontaneous emission has some extremely important consequences since spontaneous emission plays a fundamental role in limiting the performance of many optical and electronic devices such as semiconductor lasers, heterojunction bipolar transistors, and solar cells. Other potential applications of such a forbidden frequency gap include the modification of basic properties of atomic, molecular and chemical systems[3] [4] and the possibility of studying mobility edges and Anderson localization of photons within the gap with the introduction of some randomness into the system. [5] [6]

Based on rather general arguments, it is clear that in order to have a true photonic band gap, there are three basic properties that the material must possess. (1) Since the size of the forbidden gap increases with the scattering strength, the dielectric modulation must be very strong so that multiple scattering occurs. (2) The length scale of the modulation must also be comparable to the wavelength of interest. (3) In order for the gaps in different directions of the Brillouin zone to align to form a true gap, the basic lattice structure must have a Brillouin zone which is as close to a sphere as possible. Of all common crystal structures, the face-centered

Directions in Electromagnetic Wave Modeling
Edited by H.L. Bertoni and L.B. Felsen, Plenum Press, New York, 1991

cubic (fcc) structure has a most spherical lowest order Brillouin zone, and therefore has been chosen as the optimal lattice structure.[1] [5]

The basic idea of fabricating three-dimensional periodically modulated dielectric materials for the purpose of altering the appropriate electromagnetic modes to strongly suppress spontaneous emission was due to Dr. Eli Yablonovitch at Bellcore.[1] To totally suppress the one-photon spontaneous emission rate, requires a material with a true bandgap. The first experimental attempt to fabricate a crystal with such a bandgap was carried out by Yablonovitch and co-worker[7] [8] [9] with microwaves in order that the required periodic structures can readily be fabricated. The materials were chosen so that their microwave refractive indices are close to those of semiconducting materials in the visible or near infrared, in order that the experimental results can be scaled down in size to those wavelengths where applications will be most interesting. The crystals that were made consist of either dielectric spheres or spherical voids located on a fcc lattice. The dielectric spheres were made of Al_2O_3 and have a microwave refractive index of 3.06. They were embedded in thermal-compression-molded dielectric foam of refractive index 1.01. The volume-filling fraction of spheres was varied by adjusting the distance between adjacent spheres. Crystals consisting of spherical voids, on the other hand, were fabricated by drilling a series of hemispheres on the face of dielectric plates which are then stacked together so that the hemispheres face one another, forming spherical voids. The volume fraction of spheres is varied by changing the hemispherical diameter. The plates were made of commercial low-loss dielectric with a microwave refractive index of 3.6. Out of the twenty-one samples that were made, twenty showed "semi-metallic" behavior, i.e. photonic bands in different directions of the Brillouin zone partially overlap, and only the one with a 86% volume fraction of spherical voids was reported to have a forbidden frequency gap that opens up in all directions.

It should be pointed out that in atomic physics, the phenomena of inhibited spontaneous emission has in fact been demonstrated with Rydberg atoms in a microwave cavity.[10] [11] [12] More recently, weak photonic band effects have also been observed in the visible in a flourescent lifetime measurement of dye molecules in a fcc colloidal crystal of polystyrene spheres.[13]

In this work, we will discuss the theoretical calculations of photonic band structures. First we will consider calculations that are based on the use of the scalar wave equation. Results based on the use of plane-wave expansion and the Korringa-Kohn-Rostoker (KKR) method will be presented. The merits as well as drawbacks of these two different methods will also be discussed. The inadequacy of photonic band structure calculations based on the scalar wave equation will be pointed out. Next, results of the *first* calculation based on full vector waves will be presented. Detailed comparisons with the available experimental data will also be made. Good agreements between theory and experiment are obtained for the effective long wavelength refractive index of the medium, and for the gap sizes at the L- and X-points in the Brillouin zone as a function of the volume fraction of spheres. However, in contrast with the experimental finding, we do not find a true gap for the fcc structure with either dielectric spheres or spherical voids. The problem is due to the underlying symmetry of the structure which causes the gap to collapse at the W-point in the Brillouin zone. Ways to achieve a true photonic gap will be discussed.

SCALAR WAVE BAND STRUCTURES

Thus far, there are a total of four independent calulations of the photonic band structures of fcc arrays of spheres using the scalar wave equation.[14] [15] [16] [17] Our results, which we obtained by employing plane-wave expansion and the KKR method, are basically the same as those reported by Satpathy et al.,[17] but are rather different than those in the other two studies.[14] [15] We will discuss these two rather differentmethods separately in the following sections.

Plane Wave Expansion for Scalar Wave

We start with the scalar wave equation

$$-\nabla^2\psi + V\psi = k_b^2\,\psi, \tag{1}$$

where ψ is the electric field, and the "potential" V is given by

$$V = k_b^2\left(1 - \frac{\varepsilon}{\varepsilon_b}\right) \tag{2}$$

with $\varepsilon = \varepsilon_a$ inside the spheres, $\varepsilon = \varepsilon_a$ inside the host, and $k_b = \sqrt{\mu\varepsilon_b}\,\omega/c$. In terms of the Fourier coefficients

$$c_{\mathbf{k}} = \int d\mathbf{r}\, e^{-i\mathbf{k}\cdot\mathbf{r}}\psi(\mathbf{r}), \tag{3}$$

and

$$V_{\mathbf{G}} = \frac{1}{\Omega}\int_{cell} d\mathbf{r}\, e^{-i\mathbf{G}\cdot\mathbf{r}}\, V(\mathbf{r}), \tag{4}$$

where $\mathbf{G}$ is a reciprocal lattice vector of the fcc lattice, Ω is the volume of the unit cell, and the integral in Eq. (4) is over the volume of the unit cell, the scalar wave equation becomes

$$(k_b^2 - |\mathbf{k} - \mathbf{G}|^2)\, c_{\mathbf{k}\text{-}\mathbf{G}} - k_b^2 \sum_{\mathbf{G}} U_{\mathbf{G}\text{-}\mathbf{G}'} c_{\mathbf{k}\text{-}\mathbf{G}'} = 0, \tag{5}$$

where we have defined $U = V/k_b^2$. This equation can be recast into the standard eigenvalue form if we divide it by $k_b^2\,|\mathbf{k}\text{-}\mathbf{G}|$, let $d_{\mathbf{k}\text{-}\mathbf{G}} = |\mathbf{k}\text{-}\mathbf{G}|\, c_{\mathbf{k}\text{-}\mathbf{G}}$, and treat k_b^{-2} as our new eigenvalue. The result is

$$\left(\frac{1-U_0}{|\mathbf{k}\text{-}\mathbf{G}|^2} - \frac{1}{k_b^2}\right) d_{\mathbf{k}\text{-}\mathbf{G}} - \frac{1}{|\mathbf{k}\text{-}\mathbf{G}|}{\sum_{\mathbf{G}'}}' \frac{U_{\mathbf{G}\text{-}\mathbf{G}'}}{|\mathbf{k}\text{-}\mathbf{G}'|} d_{\mathbf{k}\text{-}\mathbf{G}'} = 0, \tag{6}$$

where the prime over the summation sign signifies that the $\mathbf{G}' = \mathbf{G}$ term is to be omitted. The photonic band structure is then obtained by finding the eigenvalues of this equation for each value of $\mathbf{k}$.

Before we give the results of our calculations that are specific to the present problem, we want to make a few general remarks. First, it is easy to see that in the empty lattice limit, i.e. $V\rightarrow 0$, the eigenvalues are given by $k_b^2 = |\mathbf{k}+\mathbf{G}|^2$ and are at least doubly degenerate, because the photon can have different states of polarization. The band structure can be found in most solid state textbooks, except that one must remember that the bands are usually given for electrons and therefore represent basically ω^2 versus $\mathbf{k}$, rather than ω versus $\mathbf{k}$. It is also true that in this limit, most of the levels are highly degenerate, especially at high symmetry points in the Brillouin zone. And for $\mathbf{k}$ varying from the Γ-point to the edge of the Brillouin zone, the dispersion curves are straight lines given by $k_b = k$. A detailed plot of the free-photon bands for the fcc lattice can be found in a recent paper, [17] except that the degeneracy factor for each level should be multiplied by a factor of two.

For $V \neq 0$, depending on the symmetry of V, some of these degeneracies are lifted, and the dispersion curves originating from the Γ-point are linear only near the Γ-point, where k is small compared with the magnitude of the smallest nonzero reciprocal lattice vector of the lattice. If we plot k_b versus k for the photon bands, then the slope of the straight portion is no longer unity, but should be given by (n_b/n_{eff}) where n_{eff} is the effective long wavelength refractive index of the entire medium.

For fixed values of the relative refractive index and volume- filling fraction, the lowest lying frequency gap is expected to have the the largest width. We find that this is true in all the cases that we have studied, and therefore we shall only report results for the lowest few bands. Moreover, it is important to note that for scalar waves, because the two different states of polarization are ignored, the lowest gap is expected to lie between the first and second allowed bands. This situation is very different from the actual vector wave case where the lowest gap is expected to lie between the second and third allowed bands. As we will see later on, this is the primary reason why scalar wave results are too optimistic in predicting a gap to open up for a minimum threshold dielectric ratio of only 1.7.

For the case of dielectric spheres considered here, we have

$$\frac{\varepsilon_{\mathbf{G}}}{\varepsilon_b} = \delta(\mathbf{G}) - 3f(1\text{-}r)g(Ga), \tag{7}$$

where the function $g(x) = (\sin x - x\cos x)/x^3$, and a is the radius of the sphere. This result applies as long as the spheres do not touch. When the volume filling fraction is larger than the close-packed value of $f_c = 0.74$ the spheres actually overlap, and we see that there are three separated cases that have to be considered. These three cases correspond to $c/\sqrt{8} < a < c/\sqrt{6}$, $c/\sqrt{6} < a < c\sqrt{3/16}$, and $c\sqrt{3/16} < a < c/2$, where c is the length of the side of the conventional unit cube for the fcc lattice. In the first case, both the spheres and the host material form an infinite multiply-connected domain. In the last two cases, the host material breaks up into disconnected star-shape islands while the spheres form an infinite multiply-connected domain. We find that for $a = c/\sqrt{6}$, the volume filling fraction is 0.964. Therefore the sample which was found experimentally to have a gap in the photon density of states and has a volume filling fraction $f = 0.86$ actually belongs to the first case. It can be shown that for this case

$$\frac{\varepsilon_{\mathbf{G}}}{\varepsilon_b} = \delta(\mathbf{G}) - (1\text{-}r)\left[3fg(Ga) - \frac{2}{\Omega}\sum_{\mathbf{Q}} I(\mathbf{Q})\right] \tag{8}$$

where

$$I(\mathbf{Q}) = \frac{2\pi}{Q_\rho}\int_{a_c}^{a} dz\,\cos(Q_z z)\sqrt{a^2-z^2}\,J_1(Q_\rho\sqrt{a^2-z^2}). \tag{9}$$

In the above equations, J_1 is the first order Bessel function, a_c is defined as $c/\sqrt{8}$, and two of the six $\mathbf{Q}$ vectors are given by $Q_\rho = \sqrt{K_x^2+(K_y \pm K_z)^2/2}$ and $Q_z = (K_y \mp K_z)/\sqrt{2}$ in cylindrical coordinates. The remaining four vectors are given by cyclic permutations of x, y, and z. Although the integrals can be expressed in terms of Lommel's functions of two variables and various schemes for computing them are available[18] , we find it more convenient in our work to simply compute them numerically.

The band structure for a dielectric ratio $\varepsilon_a/\varepsilon_b = 12.25$ and a volume-filling fraction of spheres $f = 0.375$ is shown as solid curves in Fig. 1. The optimal value of f for the creation of a gap in the photon density of states has also been calculated as a function of the dielectric ratio.[16] The minimum threshold dielectric ratio for the gap is found to be about 1.7.

Korringa-Kohn-Rostoker Method

In the KKR method,[19] [20] one starts by defining the "free-particle" Green's function $\Gamma(\mathbf{r},\mathbf{r}')$ by the equation

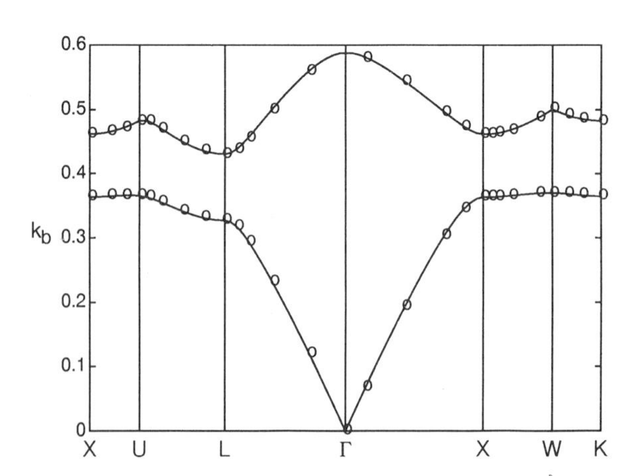

Fig. 1 - The solid curves give the band structure calculated with plane-wave expansion. The circles are our results calculated with the KKR method. The volume-filling fraction of spheres is 0.375 and the dielectric ratio is 12.25.

$$(-\nabla^2 - k_b^2)\,\Gamma(\mathbf{r},\mathbf{r}') = -\delta(\mathbf{r}-\mathbf{r}'). \tag{10}$$

Using the conventional integral equation method one easily obtain the secular equation

$$\int_{r'<r-\delta} d\hat{\mathbf{r}}' \cdot [G(\mathbf{r},\mathbf{r}')\nabla_{\mathbf{r}'}\psi(\mathbf{r}') - \psi(\mathbf{r}')\nabla_{\mathbf{r}'}G(\mathbf{r},\mathbf{r}')] = 0, \tag{11}$$

where $G(\mathbf{r},\mathbf{r}')$ is the lattice Green's function which is defined in terms of the usual Green's function by the equation

$$G(\mathbf{r},\mathbf{r}') = \sum_{\mathbf{R}} \Gamma(\mathbf{r},\mathbf{r}'+\mathbf{R})\, e^{i\mathbf{k}\cdot\mathbf{R}}, \tag{12}$$

and δ is an infinitesimal positive quantity. In Eq. (12), the sum is over all the real space lattice vectors, $\mathbf{R}$. Using spherical wave expansions for ψ and G, one can show that the eigenvalues k_b^2 are determined by the equation

$$\det \Lambda_{l,m;l',m'} = 0, \tag{13}$$

where the matrix Λ is given by

$$\Lambda_{l,m;l',m'} = A_{l,m;l',m'}\left[j_{l'}(k_a a)\frac{dj_{l'}(k_b a)}{da} - j_{l'}(k_b a)\frac{dj_{l'}(k_a a)}{da}\right]$$
$$+ k_b \delta_{l,l'}\delta_{m,m'}\left[j_l(k_a a)\frac{dn_l(k_b a)}{da} - n_l(k_b a)\frac{dj_l(k_a a)}{da}\right], \tag{14}$$

with

$$A_{l,m;l',m'} = -\frac{4\pi^2}{\Omega}\frac{1}{j_l(k_b r)\, j_{l'}(k_b r')}\sum_{\mathbf{G}}\frac{j_l(|\mathbf{k}+\mathbf{G}|r)\, j_{l'}(|\mathbf{k}+\mathbf{G}|r')}{|\mathbf{k}+\mathbf{G}|^2 - k_b^2} Y^*_{lm}(\mathbf{k}+\mathbf{G})\, Y_{l'm'}(\mathbf{k}+\mathbf{G})$$
$$- k_b^2\, \delta_{l,l'}\delta_{m,m'}\frac{n_l(k_b r')}{j_l(k_b r)}, \tag{15}$$

for $r<r'<a$. In Eqs. (14) and (15), j_l and n_l are respectively the l-th order cylindrical Bessel function of the first and second kind. It turns out that A depends not separately on l,m and l',m', but on a smaller number of constants $D_{L,M}$ through the relation

$$A_{l,m;l',m'} = 4\pi \sum_L D_{L,m-m'} C_{L,m-m';l,m;l',m'}, \tag{16}$$

where L is to be summed from $|l-l'|$ to $|l+l'|$ in unit steps. The C coefficients defined by

$$C_{L,M;l,m;l',m'} = \int d\hat{\mathbf{r}}\, Y_{LM}(\mathbf{r})\, Y^*_{lm}(\mathbf{r})\, Y_{l'm'}(\mathbf{r}), \tag{17}$$

are proportional to the Clebsch Gordon coefficients.[21] The constants D_{LM} are given by

$$D_{LM} = -\frac{4\pi}{\Omega}\frac{1}{k_b^L}\sum_{\mathbf{G}}\frac{|\mathbf{k}+\mathbf{G}|^L}{|\mathbf{k}+\mathbf{G}|^2 - k_b^2} Y^*_{LM}(\mathbf{k}+\mathbf{G}). \tag{18}$$

Direct evaluation of D_{LM} using this equation involves a lattice sum whose convergence is notoriously slow. However, the sum can be computed rather easily if it is first split up into three separate sums using the Ewald's lattice sum method. The result yields

$$D_{LM} = D^{(1)}_{LM} + D^{(2)}_{LM} + D^{(3)}_{LM}, \tag{19}$$

where

$$D^{(1)}_{LM} = -\frac{4\pi}{\Omega}\frac{1}{k_b^L}\sum_{\mathbf{G}}\frac{|\mathbf{k}+\mathbf{G}|^L}{|\mathbf{k}+\mathbf{G}|^2 - k_b^2} Y^*_{LM}(\mathbf{k}+\mathbf{G})\exp\left[(k_b^2 - |\mathbf{k}+\mathbf{G}|^2)/\eta\right] \tag{20}$$

$$D^{(2)}_{LM} = \frac{i^L(-2)^{L+1}}{\sqrt{\pi}\, k_b^L}\sum_{\mathbf{R}\neq 0} R^L e^{i\mathbf{k}\cdot\mathbf{R}}\, Y^*_{LM}(\mathbf{R}) \int_{\sqrt{\eta}/2}^{\infty} d\xi\, \xi^{2L}\exp\left(-\xi^2 R^2 + \frac{k_b^2}{4\xi^2}\right), \tag{21}$$

and

$$D^{(3)}_{LM} = -\delta_{L,0}\,\delta_{M,0}\frac{\sqrt{\eta}}{2\pi}\sum_{n=0}^{\infty}\frac{(k_b^2/\eta)^n}{(2n-1)n!}. \tag{22}$$

In Eq. (21), $\mathbf{R}$ is to summed over all the real space lattice vectors, except $\mathbf{R}=0$. And in Eqs.

(20)-(22), η is an arbitrary non-negative number, whose value is to be adjusted so that the sums involved with $D_{LM}^{(1)}$ and $D_{LM}^{(2)}$ are both rapidly converging. We find that $\eta=0.6$ is close to optimal in our calculations.

After calculating the matrix elements in Λ we have to solve the infinite order determinantal equation in EQ. (13). Note that l takes on integer values from 0 to ∞, while m varies from $-l$ to l in unit steps. We find that the results for the photonic bands converge already with a maximum l value of 3. However, convergence is not met if one uses a $l_{\max}=2$. [14] The results of our calculation using the KKR method are shown by the open circles in Fig. 1. They are in excellent agreement with our plane-wave calculations.

It is interesting to compare the plane-wave expansion and KKR method for photonic band structure calculations. We find that the plane-wave method is very easy to implement because of the simplicity of the method, and can yield very accurate results only using 100-200 plane-waves (for a dielectric ratio between 1/16 and 16). It can also readily handle dielectric modulations not only of the form of dielectric spheres and spherical voids, but practically any other shapes as well. This advantage turns out to be especially important, as we will see below that in order to obtain a true photonic bandgap for the fcc structure, the unit cell must be filled with a nonspherical basis. In contrast, the KKR method is rather difficult to implement because of the complexcity of the method. Moreover, the determinantal equation in Eq. (13) does not have the normal eigenvalue form (the eigenvalues are not located only along the diagonal, but appear in all the matrix elements in a very complicated fashion), therefore ones has to resort to a root finding scheme to locate the eigenvalues. Of course one must be extremely careful not to miss any eigenvalues. This procedure can become particularly difficult to implement, especially when some of the eigenvalues are degenerate. The KKR method, however, has a merit in that accurate results for the lowest few bands are obtained with only a 16×16 Λ matrix, using $l_{\max}=3$. This rapid convergence is partly due to the use of the integral equation approach which reduces the dimensionality of the problem by one, and partly because we use the natural eigenfunctions, rather than plane-waves, to expand the field and the Green's function.

VECTOR WAVE BAND STRUCTURES

We now describe the full vector wave calculations of photonic band structures.[22] We start with Maxwell's equations and eliminate the electric field in favor of the magnetic field **H** to obtain, for monochromatic waves of frequency ω, the equation

$$\nabla\times\left(\frac{\varepsilon_b}{\varepsilon(\mathbf{r})}\nabla\times\mathbf{H}\right) = k_b^2\,\mathbf{H}, \tag{23}$$

where $k_b = \omega\sqrt{\mu\varepsilon_b}/c_0$, and $\varepsilon = \varepsilon_a$ inside the spheres and $\varepsilon = \varepsilon_b$ inside the host. It turns out that it is much better to work with the magnetic field[23] because (1) both it's tangential and normal components at the surface of the sphere are continuous, and therefore it can be represented in terms of a smaller number of plane waves and thus leads to more rapidly convergent numerical results, and (2) we can take advantage of it's transverse nature (i.e. $\nabla\times\mathbf{H} = 0$) and reduce the size of the relevant matrices by one third, thus reducing memory cost and execution time.

Because of it's transverse nature, the magnetic field can be expanded in terms of plane waves in the form:

$$\mathbf{H}(\mathbf{r}) = \sum_{\mathbf{G}}\sum_{\lambda=1}^{2} H_{\mathbf{G},\lambda}\,\hat{\mathbf{e}}_\lambda\, e^{i(\mathbf{k}+\mathbf{G})\cdot\mathbf{r}}, \tag{24}$$

where **k** is a wavevector in the Brillouin zone of the lattice, **G** is a reciprocal lattice vector, and $\hat{\mathbf{e}}_1$, $\hat{\mathbf{e}}_2$ are unit vectors perpendicular to $\mathbf{k}+\mathbf{G}$. With this form of the magnetic field, Eq. (23) can be written in Fourier space in the form:

$$\sum_{\mathbf{G}'}\sum_{\lambda'=1}^{2} M_{\mathbf{G},\mathbf{G}'}^{\lambda,\lambda'} H_{\mathbf{G}',\lambda'} = k_b^2\, H_{\mathbf{G},\lambda}, \tag{25}$$

where the matrix M is given by

$$M_{\mathbf{G},\mathbf{G}'} = |\mathbf{k}+\mathbf{G}|\ |\mathbf{k}+\mathbf{G}'|\ \varepsilon^{-1}_{\mathbf{G},\mathbf{G}'} \begin{pmatrix} \hat{\mathbf{e}}_2\cdot\hat{\mathbf{e}}_2' & -\hat{\mathbf{e}}_2\cdot\hat{\mathbf{e}}_1' \\ -\hat{\mathbf{e}}_1\cdot\hat{\mathbf{e}}_2' & \hat{\mathbf{e}}_1\cdot\hat{\mathbf{e}}_1' \end{pmatrix} \tag{26}$$

and $\varepsilon_{\mathbf{G},\mathbf{G}'} = \varepsilon_{\mathbf{G}-\mathbf{G}'}$ is the Fourier transform of $\varepsilon(\mathbf{r})$, and is defined by

$$\varepsilon_{\mathbf{G}} = \frac{1}{\Omega}\int d\mathbf{r}\, e^{-i\mathbf{G}\cdot\mathbf{r}}\, \frac{\varepsilon(\mathbf{r})}{\varepsilon_b}, \tag{27}$$

with Ω as the volume of the fcc primitive cell. This equation gives an infinite order determinantal equation that normally can only be solved numerically by truncation. If N reciprocal lattice points are included, then a $2N\times 2N$ matrix equation has to be solved. The photon band structure is then obtained by finding the eigenvalues k_b^2 of the resulting matrix for each value of $\mathbf{k}$.

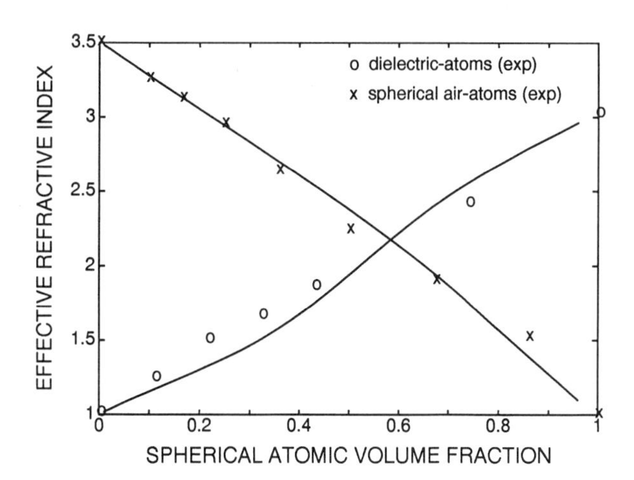

Fig. 2 - The effective long wavelength refractive index for two basic crystal structures as a function of the volume-filling fraction. The solid lines are our computed results. The experimental values are shown by the (o) and (x) points respectively for dielectric spheres and spherical voids.

With the plane-wave expansion method, we have calculated the photon band structure for various values of volume-filling fraction f and relative dielectric constant, $r = \varepsilon_a/\varepsilon_b$. We find that the results for the lowest few bands converge reasonably fast. To within an accuracy of about a few tenths of one percent, we find that the use of 400 $\mathbf{G}$ vectors in the plane-wave expansion are sufficient for f ranging from 0 to 0.96 and for r ranging from 1/12.25 to 9.179. First we show the results for the effective long wavelength refractive index in Fig. 2 for both the dielectric spheres and spherical voids as a function of the volume-filling fraction. The results are seen to be in excellent agreement with the experimental results of Yablonovitch and Gmitter.[7]

Next we present, in Fig. 3, the computed results for the eigenvalues for the second and third bands at the L- and X-points in the Brillouin zone as a function of the volume-filling fraction for the case of spherical voids. These results are normalized to the center frequency of the lowest gap at the X-point. The agreement with the experimental results [7] is fairly good. In particular we find that the X-gap goes to zero for $f\approx 0.66$. This is very close to the experimental value of 0.68. The physical origin of this null gap has been fully discussed by Yablonovitch and Gmitter, and accordingly we plot the gap width at the X-point as a negative quantity for $f>0.66$. For $f=0.86$ our results for the gap sizes at L and X are both slightly smaller than those observed in the experiment.

We have also calculated the entire photonic band structure for $\mathbf{k}$ along the symmetry directions in the Brillouin zone. Results are obtained for r varying from 1/16 to 1 for spherical voids and from 1 to 16 for dielectric spheres. For each value of r, the volume-filling fraction is

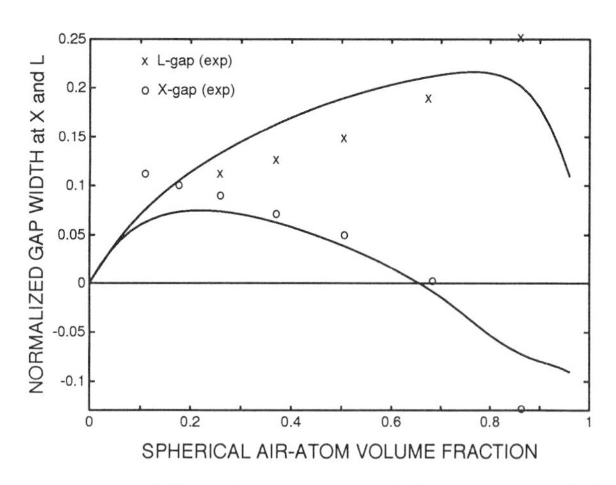

Fig. 3 - The computed forbidden frequency gap width at the L- and X-points as a function of the volume-filling fraction of spherical voids embedded in a dielectric material with a dielectric constant of 3.5. The experimental values at the L- and X-points are labeled respectively by (x) and (o). These results are all normalized to the center frequency of the lowest gap at the X-point.

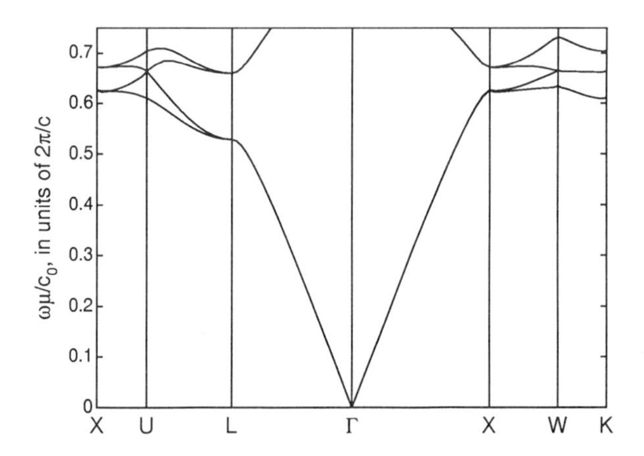

Fig. 4 - The computed photonic band structure for a 86% volume-filling fraction of spherical voids embedded in a dielectric material with a dielectric constant of 3.5. The voids are closer than close-packed, and are overlapping.

varied from 0 to 0.96 for spherical voids and from 0 to 0.74 for dielectric spheres. Fig. 4 shows the results for a 86% volume filling fraction of spherical voids embedded in a dielectric material with a refractive index of 3.5. These parameters correspond to the case in which a common gap was found experimentally. We see that although the overall band structure agrees reasonably well with the experiment, our computed band structure does not have a common gap. This is due to the fact that the second and third bands appear to be degenerate at the W-point. Fig. 4 also suggests that the second and third bands are degenerate along the W to K direction. (Note that the K- and U-points are connected by a reciprocal lattice vector, and are therefore equivalent points in the Brillouin zone.) However, further investigation shows that this degeneracy away from the W point in the K direction is purely accidental and it is absent in general atother values of r and f. The same is also true for the near degenaracy of the second and third bands at the U point.

In trying to obtain a true photonic bandgap, we find that the basic problem is at the W point where the degeneracy is found to persist for this dielectric ratio of $r=1/12.25$ for f ranging from 0 to 0.96. Further calculation indicates that this remains true at least down to a value of $r=1/16$ for all values of f, and that this degeneracy is symmetry related. In the experiment, [7] the band structures were obtained by measuring the phase and intensity of microwaves transmitted across the samples. The incident waves are either TE or TM polarized. For wavevectors within the X-U-L and X-W-K planes, there is a clear separation of the electromagnetic modes into two groups, one couple only to the TE waves, and the other only to the TM waves. This comes about because of the reflection symmetry of the crystal structure. From Γ to X, both of these modes are degenerate. However from X to W, these mode are not longer degenerate, the frequencies of the TE modes both below and above the bandgap increase while the frequencies of the TM modes remain practicallly constant. And at the W-point, the frequency of the TE mode at the bottom of the band coincides with the frequency of the TM mode at the top of the bandgap, thus causing the bandgap to completely collapse there.

In the case of dielectric-atoms, we find that the gap at the W-point opens up for $r>7.8$. This comes about because of a different mode frequency ordering. Unfortunately there is no overlap in the gaps at the L- and X-points for these values of r for a true gap to develop. We also find that for r around 16 the second and third bands at X appear to be degenerate and therefore a common gap does not exist either. We have not gone to values of r too much larger than 16 because of suitable optical materials with such a large dielectric constant and little disscipation are currently not available, and because in our calculation the size of the matrix required for accurate band results becomes prohibitively too large.

CONCLUSION

It is clear from our work that in order to obtain a true photonic bandgap it is important to remove the degeneracy at the W-point. This can be accomplished in many different ways. For example one can replace the repeating unit with a nonspherical basis. One can expect that an 86% volume fraction of voids, as found by Yablonovitch and Gmitter, [7] should not be far from the optimal value. This is confirmed in a very recent work [23] on the diamond lattice. In the case of spherical voids in a host with a dielectric constant of $\varepsilon_b = 12.25$, the fractional size of the true gap can be as large as 28.8%. Moreover, the minimum threshold value of the dielectric ratio for the creation of a gap is only 2. This will certainly make the choice of materials much easier when such photonic bandgap structures will be fabricated for applications in the visible and near-infrared regions.

Acknowledgements - We are grateful to Dr. Eli Yablonovitch for introducing to us the idea of photonic band strucutres, and for many illuminating discussions. This work is supported by the U.S. ONR under contract N00014-88-0050, by the Naval Research Laboratory under contract N00014-89-J-2078, and by the Joint Services Electronics Program.

REFERENCES

1. E. Yablonovitch, Phys. Rev. Lett. **58**, 2059 (1987).
2. E. Yablonovitch, T.J. Gmitter, and R. Blat, Phys. Rev. Lett. **61**, 2546 (1988).
3. G. Kurizki and A. Z. Genack, Phys. Rev. Lett. **61**, 2269 (1988).
4. S. John and J. Wang, Phys. Rev. Lett. **64**, 2418, (1990).
5. S. John, Phys. Rev. Lett. **58**, 2486 (1987).
6. S. John, Comm. Cond. Matt. Phys. **14**, 193 (1988).

7. E. Yablonovitch and T. J. Gmitter, Phys. Rev. Lett. **61**, 1950 (1989).

8. E. Yablonovitch, J. Opt. Soc. Am. A**7**, 1792 (1990).

9. E. Yablonovitch, in *Analogies in Optics and Micro Electronics*, eds. W. van Haeringen and D. Lenstra, Kluwer Academic Publishers, Netherlands (1990).

10. R.G. Hulet, E.S. Hifer, and D. Kleppner, Phys. Rev. Lett. **55**, 2137 (1985).

11. W. Jhe, A. Anderson, E.A. Hinds, D. Meschede, L. Moi, and S. Haroche, Phys. Rev. Lett. **58**, 1320 (1987).

12. D.J. Heinzen and M.S. Feld, Phys. Rev. Lett. **59**, 2623 (1987).

13. J. Martorell and N. Lawandy, Phys. Rev. Lett. **65**, 1877 (1990).

14. S. John and R. Rangarajan, Phys. Rev. B, **38**, 10101 (1988).

15. E. N. Economou and A Zdetsis, Phys. Rev. B, **40**, 1334 (1989).

16. K. M. Leung and Y. F. Liu, Phys. Rev. B, **41**, 10188, (1990).

17. S. Satpathy, Z. Zhang, and M. R. Salehpour, Phys. Rev. Lett. **64**, 1239 (1990).

18. G. N. Watson, *A Treatise on the Theory of Bessel Function*, Ch. 16, p. 522 (Cambridge University Press, 1966).

19. J. Korringa, Physica (Amsterdam) **13**, 392 (1947).

20. W. Kohn and N. Rostoker, Phys. Rev. **94**, 1111 (1954).

21. M.E. Rose, *Elementary Theory of Angular Momentum*, John Wiley and Sons, New York (1957).

22. K.M. Leung and Y.F. Liu, Phys. Rev. Lett. **65**, (1990).

23. K.M. Ho, C.T. Chan, and C.M. Soukoulis, Phys. Rev. Lett. **65**, (1990).

MODELING OF SUPERCONDUCTIVITY FOR EM BOUNDARY VALUE PROBLEMS

R. Pous, G. C. Liang, and K. K. Mei

Department of Electrical Engineering and Computer Science
and the Electronics Research Laboratory
University of California, Berkeley, CA 94720

1 INTRODUCTION

High-temperature superconductivity has spawned a wide search for electronics applications. Various experiments on microwave circuits, including resonators, filters, phase shifters, and small antennas have been performed with promising results. The design of these components will necessarily involve numerical simulation, which requires the solution of superconductive EM boundary value problems. In this paper, several approaches to this solution are presented. In the first approach, superconductors are treated as negative dielectric materials. Secondly, the superconductor surface impedance condition is used at the boundaries. Finally, the problem is solved using perfectly conducting boundaries, and perturbation is used to approximate the desired parameters. The above methods are used to solve the scattering by a superconducting cylinder, propagation in a superconducting parallel-plate waveguide, and radiation by a superconducting short dipole. Very good agreement is found between the negative dielectric model and the surface impedance method. The perfect conductor approximation also gives very good estimates of the fields, but fails to predict some important effects, such as the change in resonant frequency of the dipole with temperature.

2 SUPERCONDUCTORS AS NEGATIVE DIELECTRICS

In earlier work we have shown how a superconductor operating below its critical temperature can be modeled as a negative dielectric material [1], [2], and [3]. This approach makes analyzing these materials much easier for antenna and microwave engineers. In this section, the derivation of the dielectric constant of a superconductor based on the two-fluid model is summarized [4], [5].

*This work was supported by the California State MICRO Program, and industrial sponsor HUGHES AIRCRAFT COMPANY.

The two-fluid model of a superconductor postulates that the conduction electrons in the material are divided into two groups: normal or unpaired electrons, and superconducting or paired electrons. The unpaired electrons behave like electrons in a normal metal, colliding as they travel pulled by the electric field. The paired electrons, on the other hand, travel in what are called 'Cooper pairs', and macroscopically behave like travelling in a collisionless path.

The local equations relating the currents and the electric field are[1]

$$m\frac{<\vec{v}_n>}{\tau} = -e\vec{E} \qquad\qquad m\frac{\partial \vec{v}_s}{\partial t} = -e\vec{E} \tag{1}$$

$$\vec{J}_n = -n_n e <\vec{v}_n> \qquad\qquad \vec{J}_s = -n_s e\vec{v}_s \tag{2}$$

where n_n, n_s, $<\vec{v}_n>$, and $\vec{v}_s$ are the volumetric densities and velocities of the normal and superconducting electrons, respectively (in the case of the normal electrons we use the average velocity), and τ is the average momentum relaxation time of the normal electrons. The total electron density can be written as $n = n_s + n_n$. The empirical formula for the dependence of n_n and n_s on temperature is

$$\frac{n_n}{n} = \overline{T}^4 \qquad\qquad \frac{n_s}{n} = 1 - \overline{T}^4 \tag{3}$$

where T_c and $\overline{T} = T/T_c$ are the critical and normalized temperatures of the superconductive material, respectively.

The generalized dielectric constant of the material, is defined by

$$\nabla \times \vec{H} = j\omega\epsilon_o\vec{E} + \vec{J}_n + \vec{J}_s = j\omega\epsilon\vec{E} \tag{4}$$

Substituting (1) and (2) in (4) yields

$$\epsilon_r = \frac{\epsilon}{\epsilon_o} = \left(1 - \frac{\omega_s^2}{\omega^2}\right) - j\frac{\omega_n^2}{\omega\omega_r} = \frac{\epsilon'}{\epsilon_o} - j\frac{\epsilon''}{\epsilon_o} \tag{5}$$

where $\omega_o^2 = e^2 n/m\epsilon_o$, $\omega_r = 1/\tau$, $\omega_n^2 = \omega_o^2\overline{T}^4$, and $\omega_s^2 = \omega_o^2 - \omega_n^2$. It is noted that only three parameters are needed to characterize the material, namely, the fluid resonant frequency ω_o, the relaxation frequency ω_r, and the normalized temperature $\overline{T}$ (or, alternatively, ω_n, ω_s, and ω_r). Typically, ϵ' is negative, with $|\epsilon'| \gg \epsilon_o \gg |\epsilon''|$, which is the reason superconductors can be modeled as negative dielectric materials. In (5) it can be seen that the generalized dielectric constant is dispersive, with a plasma-like frequency dependence. It is also important to clarify that the above discussion only applies for frequencies far below the gap frequency of the material.

[1]In fact, the first equation in (1) should include the term $m\partial <\vec{v}_n> /\partial t$ but, for all frequencies of interest, this term is negligible.

Although electrons in a superconductor react differently to the applied electric field than those in a normal conductor, and the surface impedances of both materials lie on different lines in the complex plane, both approach the perfect conductor limit ($Zs = 0$) as $\omega \rightarrow 0$ and $\sigma \rightarrow \infty$, respectively (Fig. 1). We can use perturbational approximations for problems involving superconductors whenever $\omega \ll \omega_s$, in the same way that we use them for good, but not perfect conductors.

3 APPROXIMATE SOLUTIONS TO SUPERCONDUCTIVE BOUNDARY VALUE PROBLEMS

Even though negative dielectric materials can be treated in a similar way as conventional dielectrics in the solution of boundary value problems, much effort can be spared if an approximate solution is obtained. The most common approximation is to use the surface impedance boundary condition at the interface

$$\vec{E} \cdot \hat{n} = Zs\vec{H} \cdot \hat{n} \tag{6}$$

where Zs is the surface impedance of the material, and $\hat{n}$ is the unit vector normal to the boundary. This approximation is only valid when the radius of curvature of the interface is much larger than a wavelength. Moreover, for frequencies low enough so that $\epsilon_r \gg 1$, and for $T \ll T_c$, we may obtain an approximate solution by perturbing the solution found using perfectly conducting boundaries. In the following sections, we will present several examples using both rigorous and approximate approaches.

3.1 Scattering by an Infinitely Long Superconducting Circular Cylinder

The scattering by a superconductive cylinder has an exact closed-form solution that we can use to test the two approximations. In order to find the exact solution, we divide the fields into three components: incident field, scattered field, and the field within the cylinder. The incident field is a plane wave propagating in the $-\hat{x}$ direction. TMz and TEz cases are studied separately.

A) NEGATIVE DIELECTRIC MODEL

If we choose the axis of the cylinder to be the $\hat{z}$ coordinate axis, then the fields for the TMz case can be expressed in cylindrical coordinates as

$$E_z^i = E_o \sum_{n=-\infty}^{\infty} j^n J_n(k_o\rho)e^{jn\phi} \tag{7}$$

$$E_z^s = E_o \sum_{n=-\infty}^{\infty} a_n j^n H_n^{(2)}(k_o\rho)e^{jn\phi} \tag{8}$$

$$E_z^d = E_o \sum_{n=-\infty}^{\infty} b_n j^n I_n(\alpha\rho)e^{jn\phi} \tag{9}$$

$$H_\phi^i = \frac{-jE_o}{\eta_o} \sum_{n=-\infty}^{\infty} j^n J_n'(k_o\rho) e^{jn\phi} \tag{10}$$

$$H_\phi^s = \frac{-jE_o}{\eta_o} \sum_{n=-\infty}^{\infty} a_n j^n H_n^{(2)\prime}(k_o\rho) e^{jn\phi} \tag{11}$$

$$H_\phi^d = \frac{E_o}{\eta_o \overline{Z}s} \sum_{n=-\infty}^{\infty} b_n j^n I_n'(\alpha\rho) e^{jn\phi} \tag{12}$$

where we define $\eta_o = \sqrt{\mu_o/\epsilon_o}$, $k_o = \omega\sqrt{\mu_o\epsilon_o}$, $\alpha = k_o\sqrt{-\epsilon_r}$, and $\overline{Z}s = Zs/\eta_o = 1/\sqrt{\epsilon_r}$. Enforcing continuity of the tangential fields E_z and H_ϕ at $\rho = a$ leads to the following expression for the tangential magnetic field on the surface of the cylinder,

$$\begin{aligned} H_\phi(\phi) = \frac{2E_o}{\pi\eta_o k_o a} \Bigg[& \frac{1}{H_0^{(2)}(k_o a) - j\frac{\overline{Z}s}{c_o} H_1^{(2)}(k_o a)} \\ & + 2\sum_{n=1}^{\infty} j^n \frac{\cos n\phi}{\left(1 - j\frac{\overline{Z}s}{c_n}\frac{n}{k_o a}\right) H_n^{(2)}(k_o a) + j\frac{\overline{Z}s}{c_n} H_{n-1}^{(2)}(k_o a)} \Bigg] \end{aligned} \tag{13}$$

where

$$c_o = \frac{I_1(\alpha a)}{I_o(\alpha a)} \qquad c_n = \frac{I_{n-1}(\alpha a)}{I_n(\alpha a)} - \frac{n}{\alpha a} \tag{14}$$

are constants that are approximately 1 for large values of αa. Similarly, for TEz incidence we find the tangential magnetic field on the surface of the cylinder

$$\begin{aligned} H_z(\phi) = \frac{2H_o}{\pi k_o a} \Bigg[& \frac{1}{\overline{Z}s c_o H_0^{(2)}(k_o a) - jH_1^{(2)}(k_o a)} \\ & + 2\sum_{n=1}^{\infty} j^n \frac{\cos n\phi}{\left(\overline{Z}s c_n - j\frac{n}{k_o a}\right) H_n^{(2)}(k_o a) + jH_{n-1}^{(2)}(k_o a)} \Bigg] \end{aligned} \tag{15}$$

B) Surface impedance approximation

To find the solution for the surface impedance approximation, we simply have to enforce (6) on the surface of the cylinder. Only the incident and scattered fields are needed in this case (the field inside the cylinder is assumed to decay exponentially). It is straight forward to show that the expressions for the tangential magnetic fields are similar to (13) and (15), but with $c_o = c_n = 1$. Since $\lim_{x\to\infty} I_p(x) = e^x/\sqrt{2\pi x}$ this approximation will be valid for

$$\alpha a = 2\pi\sqrt{-\epsilon_r} \cdot \frac{a}{\lambda} \gg 1 \Rightarrow \frac{a}{\lambda} \gg \frac{1}{2\pi\sqrt{-\epsilon_r}} \tag{16}$$

C) Perfect conductor approximation

The perfect conductor approximation is easily found by setting $\overline{Z}s = 0$ in (13) and (15). This is a more crude approximation, but it is reasonable in low frequency regions. A better accuracy is expected for quantities depending on the integration of the field, such as dissipated power or radar cross section.

D) NUMERICAL RESULTS

The following results have been computed for a typical YBCO high-T_c superconductor ($T_c = 88$ K). We assume a surface resistance of $0.3\,m\Omega$ at 77 K and 10 GHz, and a penetration depth $\lambda(0) = 140\,nm$. The surface impedance can then be expressed as

$$Zs = Rs_{10GHz}\left(\frac{f\,(GHz)}{10}\right)^2 + j\omega\mu_o\frac{\lambda(0)}{\sqrt{1-\overline{T}^4}} \tag{17}$$

which at 77 K results in

$$Zs(f) \simeq 3\cdot 10^{-6} f^2\,(GHz) + j1.1\cdot 10^{-3} f\,(GHz) \quad (\Omega) \tag{18}$$

From the this expression it is easy to find the alternative material parameters mentioned earlier to be $f_o \simeq 550\,THz$, $f_r \simeq 250\,GHz$, and $\overline{T} = 0.875$. These values are used for the following simulations, unless otherwise indicated. Figure 2 shows the ϕ component of the TMz surface magnetic field for the exact simulation and the two approximate methods. It is observed that the fields obtained with the three different methods cannot be distinguished even for cylinders with a small radius. Figure 3 shows the error in the computation of the dissipated power for the perfect conductor approximation when compared to the negative dielectric model. The error in the surface impedance case is of the order of the computer roundoff error.

3.2 Parallel-Plate Waveguide Analysis

The propagation of waves inside a superconducting parallel-plate waveguide also has an analytical solution. The mathematical analysis closely parallels that of the propagation in a dielectric slab [3], [6]. For a waveguide having width b and a separation a between the plates ($b \gg a$), it is straightforward to find an equation for the propagation constant γ,

$$\frac{ua}{2}\tanh\frac{ua}{2} = \frac{1}{\epsilon_r}\frac{va}{2} \tag{19}$$

where $u^2 = -\gamma^2 - k_o^2$, $v^2 = -\gamma^2 + \alpha^2$. However, (19) is transcendental, and has to be solved numerically. The surface impedance approximation is easily found through the transmission line parameters

$$Y' = j\omega\epsilon_o\frac{b}{a}, \qquad Z' = j\omega\mu_o\frac{a}{b} + 2\frac{Zs}{b}, \qquad \gamma = \sqrt{Z'Y'} \tag{20}$$

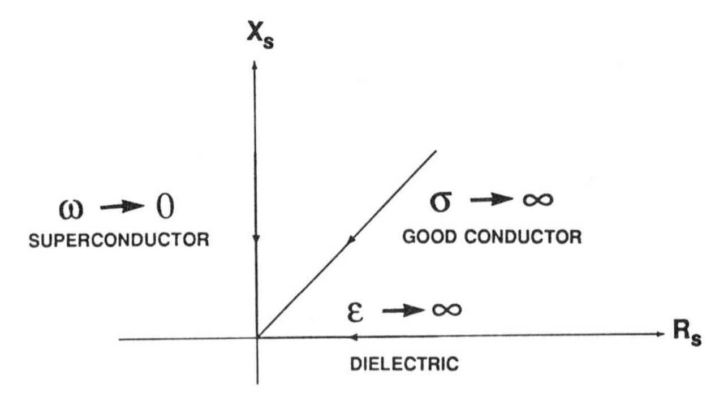

Fig. 1. Surface impedance of a superconductor (at T = 0), a good conductor, and a conventional dielectric material, as they approach the perfect conductor limit ($Z_s = 0$).

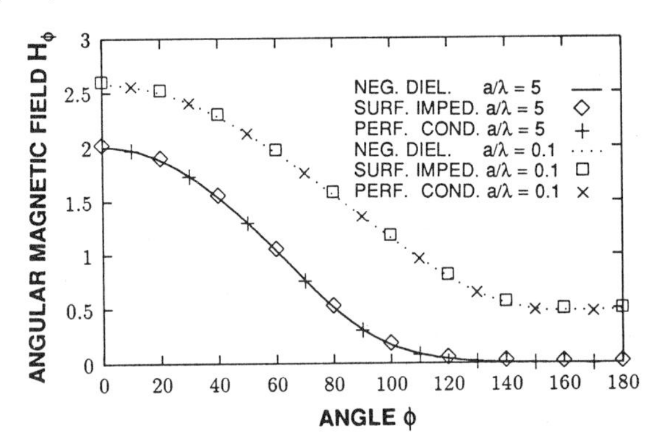

Fig. 2. Angular component of the surface magnetic field H_ϕ for two cylinder radii, obtained by three different methods: (1) Negative dielectric model, (2) Surface impedance condition, and (3) Perfect conductor approximation (a/λ = 5, and a/λ = 0.1, f = 50 GHz, TMz incidence).

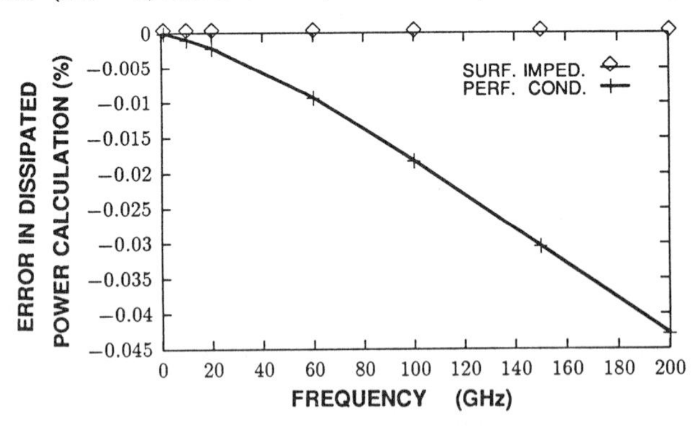

Fig. 3. Percentage error in the calculation of the dissipated power density versus frequency. The values found with the surface impedance and perfect conductor approximations are compared to those found using the negative dielectric model (a/λ = 1, TMz incidence).

Figure 4 shows a comparison of the exact and approximate attenuation constants ($\alpha = Re(\gamma)$), computed with the above expressions.

3.3 Superconducting Dipole

In this section we will present the numerical results for a superconducting dipole using the method of moments. Since we are considering the superconductor to be a negative dielectric material, the radiation of the dipole can be solved as the scattering by a dielectric cylinder, where the incident field is the field due to a magnetic frill generator located at the center of the dipole. We use the same approach as in [7] to find the solution. Figure 5 shows the computed dipole currents for the negative dielectric model and surface impedance condition. Again, in this case, the currents found using the surface impedance condition are very accurate.

One of the most interesting features of a superconducting dipole is the dependence of the resonant frequency on temperature. At higher temperatures the surface impedance becomes more and more inductive, making the dipole electrically longer, and the resonant frequency smaller. Figure 6 shows the resonant frequency shift with respect to temperature. This information can be very important for narrowband antennas and circuits, where the precise knowledge of the operating frequency is critical. It is also interesting to compare the input resistance of a superconducting dipole with that of a copper dipole. Figure 7 shows the input resistance of a superconducting dipole. Since the radiation resistance is dominant, the input resistance is very similar to that of a conventional dipole. Figure 8 shows the change in the input resistance if a YBCO dipole at 77 K is replaced by a copper dipole at 77 K. Finally, Fig. 9 shows the efficiency of a superconducting dipole, compared to that of a copper dipole. We see that very short superconducting dipoles would be efficient radiators at frequencies at which the loss resistance would dominate over the radiation resistance in a copper dipole.

4 CONCLUSION

A simple characterization of superconductors based on the classical two-fluid model has been presented, which allows to solve EM boundary value problems involving superconductors by treating them as negative dielectrics. First, these methods are tested in the scattering by a superconductive cylinder. The solutions found using the surface impedance is almost identical to that found using the negative dielectric model. Also, a perturbational solution for the dissipated power is computed, with very small error. Secondly, two methods are used to find the attenuation constant of a parallel-plate waveguide, also showing very good agreement. Finally, a short dipole is analyzed. Both the negative dielectric model and the surface impedance approximation succeed in predicting the most important features of superconductor dipoles, such as the variation of resonant frequency with temperature, and their high efficiency at small frequencies.

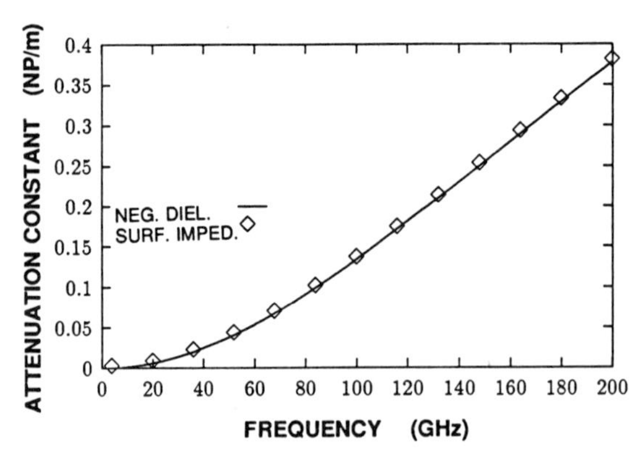

Fig. 4. Comparison of the attenuation constants found using the negative dielectric model, and the surface impedance condition for a parallel-plate waveguide (waveguide width b=1cm, space between the two plates a = 0.5 mm).

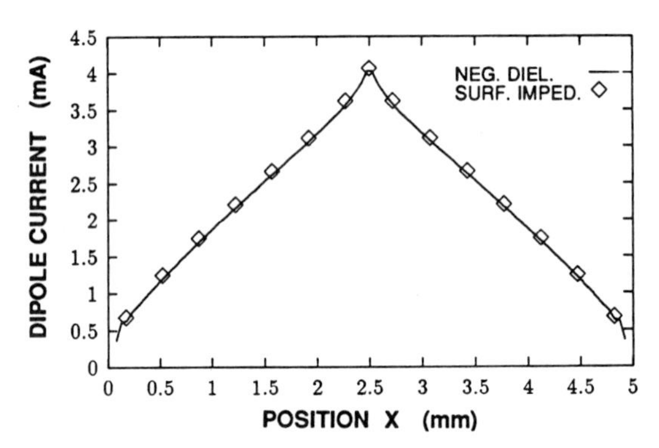

Fig. 5. Current distribution of a 5-mm YBCO dipole with a radius a = 0.1 mm, at f = 15 GHz (T = 77 K, V_{in} = 1V).

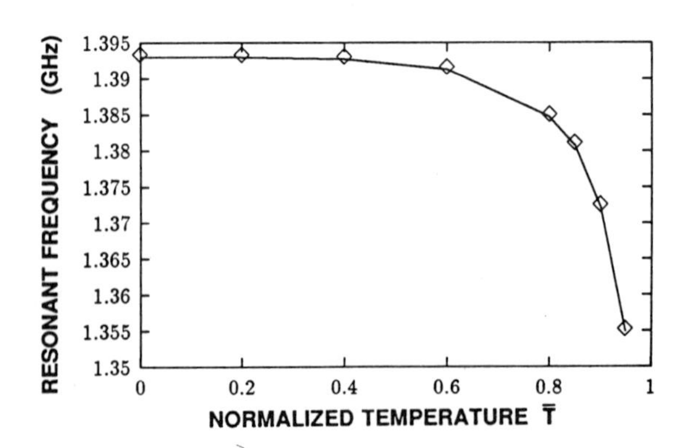

Fig. 6. Resonant frequency of a 10-cm YBCO dipole of radius a = 1 mm versus normalized temperature.

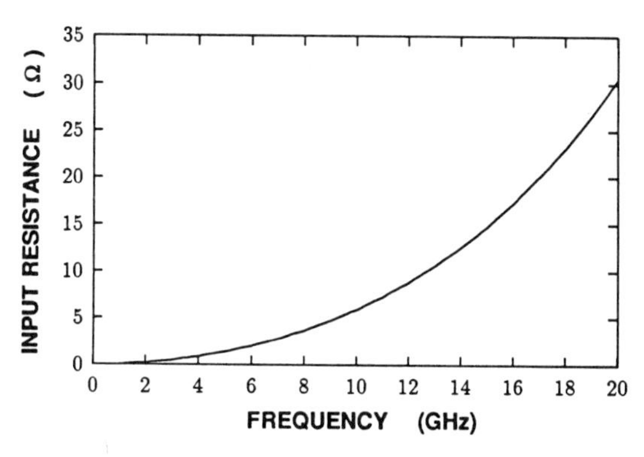

Fig. 7. Input resistance of a 5-mm YBCO dipole with radius a = 0.1 mm versus frequency (T = 77 K).

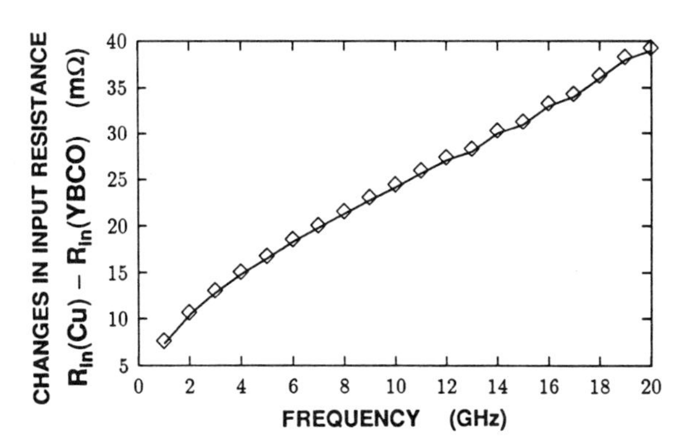

Fig. 8. Change of input resistance when a YBCO dipole is replaced by a copper dipole (both at 77 K).

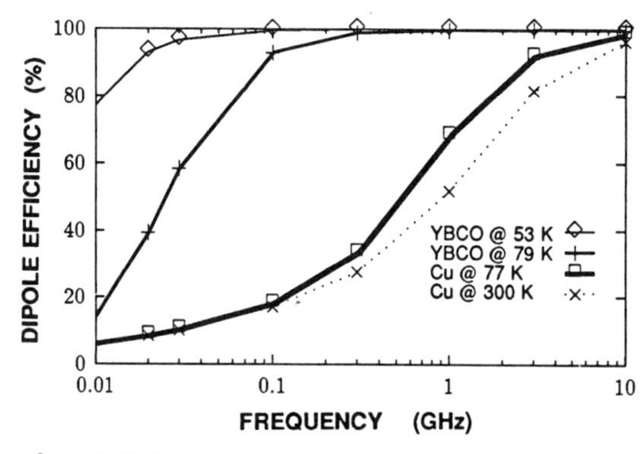

Fig. 9. Efficiency of a superconductor and a copper dipole versus frequency for different temperatures (the dipole length L = 5 mm, and the radius a = 0.05 mm).

REFERENCES

[1] K. K. Mei, G. C. Liang, and T. Van Duzer, "Electrodynamics of Superconductors," presented at *1989 IEEE AP-S Intl. Symp.*, San Jose, June 1989.

[2] K. K. Mei, G. C. Liang, "Electrodynamics of Superconductors," submitted to the *IEEE Trans. on Microwave Theory and Tech.*, Special Issue on Superconductivity, Sept. 1991.

[3] G. C. Liang, Y. W. Liu, and K. K. Mei, "Propagation Properties of a Superconductive Stripline," *IEEE AP-S Int. Symp.*, Dallas, pp. 728-731, May 1990.

[4] T. Van Duzer and C. W. Turner, *Principles of Superconductive Devices and Circuits,* New York, NY: Elsevier, 1981.

[5] J. I. Gittleman, and B. Rosenblum, "Microwave Propertics of Superconductors," *Proc. of the IEEE*, vol. 52, pp. 1138-1147, Oct. 1964.

[6] R. E. Collin, *Field Theory of Guided Waves,* New York, NY: McGraw-Hill, 1960.

[7] J. H. Richmond, "Scattering by a Dielectric Cylinder of Arbitrary Cross Section Shape," *IEEE Trans. Antennas and Propagat.*, vol. AP-13, pp. 334-341, May 1965.

DERIVATION AND APPLICATION OF

APPROXIMATE BOUNDARY CONDITIONS

Thomas B.A. Senior

Department of Electrical Engineering and Computer Science
University of Michigan
Ann Arbor, MI 48109-2122

I. INTRODUCION

Approximate boundary conditions (or abc's) can be very helpful in simplifying the analytical or numerical solution of scattering problems involving complex structures, and are becoming more important as we seek to model some of the complicated materials in use today. Some versions have been around for a long time, for example, the classical condition $E_{tan} = 0$ at the surface of a metal. Although often regarded as exact, it is in fact an approximation for all metals even at microwave frequencies.

The purpose of an abc is to simulate the material and geometric properties of the surface involved. Consider, for example, an opaque body illuminated by an electromagnetic field. Knowing the material properties it is, in principle, possible to find the scattered field outside by taking into account the fields which penetrate the body, but the task would be greatly simplified if the properties could be simulated by a boundary condition imposed at the outer surface and involving only the external fields, thereby converting a multi-media problem into a single medium one. It is not even necessary that we take the surface where the boundary condition is imposed to coincide with the actual surface of the body. The only requirement is that in the region of interest the field obtained using the postulated condition approximate the exact field to an adequate degree of accuracy.

If the exact solution of the original problem were known, we could always construct a posterior an equivalent boundary condition which, if imposed at the surface, would reproduce the field throughout the exterior region. But such a condition would be unique to this particular situation, and to be useful an abc must be applicable to a class of incident field and body configurations over and beyond the particular one for which it was derived. It goes without saying that it must not be incident field dependent.

2. STANDARD IMPEDANCE CONDITION

The most common abc is an impedance boundary condition which can be derived by considering the simple problem of a plane wave incident on a material half space. If the interface is z = 0, we have

$$E_x = -\eta H_y , \quad E_y = \eta H_x$$

where η is a function of the material properties, and these can be written as

Directions in Electromagnetic Wave Modeling
Edited by H.L. Bertoni and L.B. Felsen, Plenum Press, New York, 1991

$$\hat{n} \times (\hat{n} \times \bar{E}) = -\eta\, \hat{n} \times \bar{H} \tag{1}$$

where $\hat{n}$ is the unit vector outward normal. This last provides the natural extension to a curved surface. We can also admit the possibility of anisotropy, in which case the boundary conditions becomes

$$\hat{n} \times (\hat{n} \times \bar{E}) = -\bar{\bar{\eta}} \bullet \hat{n} \times \bar{H}. \tag{2}$$

Since η is simply the impedance looking into the half space, the boundary condition is physically meaningful and attractive in its simplicity. If the material occupying the half space is homogeneous and isotropic with $|N| \gg 1$ where N is the complex refractive index, then

$$\eta = Z \tag{3}$$

where Z is the intrinsic impedance of the material. In this form the boundary condition is attributable to Leontovich [1948] and was widely used in Russian work on ground wave propagation during World War II. For a curved surface whose principal radii of curvature are ρ_1 and ρ_2, the same condition applies provided

$$|\text{Im.}N|\, k_o\, \rho_{1,2} \gg 1.$$

To this order, the boundary condition is unaffected by any slow variations of the material properties, either laterally or in depth.

Reverting to the case of a planar surface, if the properties vary in depth, a more accurate expression for η obtained using an analysis similar to that developed by Rytov [1940] is

$$\eta = Z\left\{1 - \frac{1}{2ik_oN}\frac{\partial}{\partial z}(\ln Z) + O(N^{-2})\right\}, \tag{4}$$

and the correction term is of the same order as the effect of any curvature that the surface may have. In contrast, the effect of any lateral variation of the surface properties is smaller by a factor N. A particular example of a surface whose properties vary in depth is a homogeneous metal-backed layer of thickness τ. A widely used approximation is then

$$\eta = -iZ \tan(Nk_o\tau), \tag{5}$$

but this is effective only for very thin layers and even then its accuracy diminishes as the incidence becomes oblique.

3. GENERALIZED IMPEDANCE CONDITIONS

What we would like is to improve the accuracy afforded by the standard impedance boundary condition, and one way to do this is to include additional field components and/or derivatives. To see how to do it, consider again the case of a planar surface $z = 0$. By differentiating the standard IBC with respect to x and y we obtain the alternative forms

$$\left(\frac{\partial}{\partial z} + ik_o\frac{\eta}{Z_o}\right)E_z = 0, \qquad \left(\frac{\partial}{\partial z} + ik_o\frac{Z_o}{\eta}\right)H_z = 0, \tag{6}$$

and a logical extension of these is

$$\prod_{m=1}^{M}\left(\frac{\partial}{\partial z}+ik_o\,\Gamma_m\right)E_z=0,\qquad \prod_{m=1}^{M'}\left(\frac{\partial}{\partial z}+ik_o\,\Gamma'_m\right)H_z=0. \tag{7}$$

These provide additional degrees of freedom which can be used to improve the simulation of the surface properties, and have been referred to as generalized impedance boundary conditions (GIBCs) whose order is determined by the highest derivative present. A standard impedance boundary condition (SIBC) is then a GIBC of order unity. An alternative representation of a GIBC is

$$\sum_{m=0}^{M}\frac{a_m}{(ik_o)^m}\frac{\partial^m}{\partial z^m}E_z=0,\qquad \sum_{m=0}^{M'}\frac{a'_m}{(ik_o)^m}\frac{\partial^m}{\partial z^m}H_z=0, \tag{8}$$

and it is a trivial matter to relate the a_m and a'_m to Γ_m and Γ_m respectively.

Boundary conditions of this type were first introduced by Karp and Karal [1967] in a study of the surface waves that can be supported by a dielectric coating. The idea was revived a decade later [Engquist and Majda, 1977] in connection with the use of fictitious boundaries to limit the region of computation in a finite element solution of an exterior scattering problem. What is needed here is a highly absorbing surface. The same objective can be achieved by considering the one-way wave equation [Trefethen and Halpern, 1986; Keller and Givoli, 1989], or by seeking a form of radiation condition that can be applied at a finite (and possibly small) distance from a scattering body and still guarantee that the scattered field is outgoing [Morgan, 1990]. But in spite of the similarity to the development of approximate boundary conditions, our objective is somewhat different in that we are seeking to model an actual surface using a boundary condition typically applied at the surface itself.

Just as we obtained (6) from the SIBC (1) by differentiation with respect to x and y, so can we recover (1) from (6) by tangential integration. In the same way we can express a GIBC in terms of tangential field components, and provided M = M' and duality is enforced, there is a standard way for doing so. As an example, in the third order case [Senior and Volakis, 1989]

$$\hat{z}\times\left(\hat{z}\times\left\{\bar{E}+\frac{1}{ik_o\,(a_3+a_1)}\nabla\left[a_2\,\hat{z}\cdot\bar{E}-\frac{a_3}{ik_o}\nabla_s\cdot\bar{E}\right]\right\}\right)$$

$$=-\frac{a_2+a_o}{a_3+a_1}Z_o\,\hat{z}\times\left\{\bar{H}+\frac{1}{ik_o\,(a'_3+a'_1)}\nabla\left[a'_2\,\hat{z}\cdot\bar{H}-\frac{a'_3}{ik_o}\nabla_s\cdot\bar{H}\right]\right\} \tag{9}$$

provided

$$\frac{a'_2+a'_o}{a'_3+a'_1}=\frac{a_3+a_1}{a_2+a_o} \tag{10}$$

where $\nabla_s\cdot(\,)$ is the surface divergence. Since only the ratios of the constants occur, there are actually three parameters at our disposal to simulate the surface, and we observe that one effect of going to a higher order is to make the boundary condition less local in character through the inclusion of tangential derivatives of the fields. But in contrast to a first order condition, the extension to a curved surface is not trivial, and in general it is <u>not</u> enough to simply replace z by the outward normal n without any change in the values of the constants.

4. SPECIFYING THE CONSTANTS

Several techniques have been proposed for choosing the constants, and some of the more effective ones are as follows. In the case of a planar surface whose reflection coefficients are known either analytically or numerically, the constants can be determined by matching the reflection coefficients to those implied by a GIBC. For an incident plane wave, the reflection coefficient for E_z is

$$R(\phi) = -\sum_{m=1}^{M} \frac{\Gamma_m - \sin\phi}{\Gamma_m + \sin\phi} \tag{11}$$

with a similar expression for H_z, and this can be used to match the known values. Results have been developed for single and multi-layer coatings backed by a metal. In general, increasing the order of the condition leads to an improvement in accuracy for a given thickness of coating, or to an increase in the allowed thickness for a given accuracy, and this is illustrated in Fig. 1 showing the maximum coating or layer/thickness for a 1% amplitude or 1 degree phase error. By going to a fourth order condition, the increase in thickness is substantial.

The constants can also be found by expanding the fields in the dielectric layer, either in a Tayler series in τ when the coating is a low contrast one, or asymptotically for large $|N|$ in the case of a high contrast coating, and these same techniques are also applicable to a curved layer [Senior and Volakis, 1990]. Thus, for a metal-backed homogeneous dielectric layer whose outer surface is γ = constant where α, β, γ are orthogonal curvilinear coordinates with metric coefficients h_α, h_β, h_γ, the second order boundary condition obtained by Taylor expansion in powers of τ is

$$\hat{\gamma} \times \left\{ \hat{\gamma} \times \left[\bar{E} - \frac{\tau}{\varepsilon} \nabla (h_\gamma E_\gamma) \right] \right\} = -Z \frac{1 - ik_o \tau h_\gamma \mu/\eta_m}{1 - ik_o \tau h_\gamma \varepsilon \eta_m} \hat{\gamma} \times \left[\bar{H} - \frac{\tau}{\mu} \nabla (h_\gamma H_\gamma) \right]. \tag{12}$$

Similarly, for a high contrast homogeneous dielectric body,

$$\hat{\gamma} \times \left\{ \hat{\gamma} \times \left[\bar{E} - \frac{i}{2k_o h_\gamma} \frac{1}{N\varepsilon} \nabla (h_\gamma E_\gamma) \right] \right\} = -Z \bar{\bar{\Gamma}} \bullet \hat{\gamma} \times \left[\bar{H} - \frac{i}{2k_o h_\gamma} \frac{1}{N\varepsilon} \nabla (h_\gamma H_\gamma) \right]$$

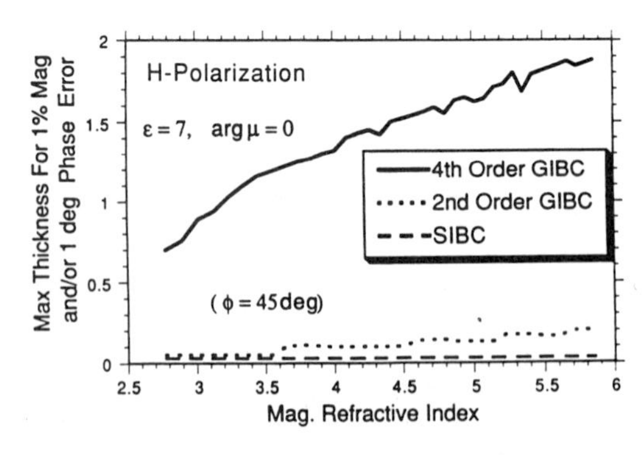

Fig. 1. Maximum thickness of a homogeneous metal-backed slab for a 1 percent/1 degree error when illuminated at 45 degrees.

where

$$\bar{\bar{\Gamma}} = \Gamma \hat{\alpha}\hat{\alpha} + \frac{1}{\Gamma}\hat{\beta}\hat{\beta}$$

with

$$\Gamma = 1 - \frac{i}{2k_oNh_\gamma}\frac{\partial}{\partial\gamma}\left(\ln\frac{h_\alpha}{h_\beta}\right) - \frac{1}{4(k_oN)^2h_\gamma}\left[\frac{1}{h_\alpha}\frac{\partial}{\partial\gamma}\left\{\frac{h_\alpha}{h_\gamma}\frac{\partial}{\partial\gamma}\left(\ln\frac{h_\alpha}{h_\beta}\right)\right\}\right.$$

$$\left. - \frac{1}{2h_\gamma}\frac{\partial}{\partial\gamma}\left(\ln\frac{h_\alpha}{h_\beta}\right)\frac{\partial}{\partial\gamma}(\ln h_\alpha h_\beta)\right], \qquad (13)$$

and we note that the main effect of the curvature is to modify the coefficient multiplying the right hand side. If the radii of curvature are sufficiently large, we can use the boundary condition developed for a planar surface.

Even a second order GIBC can provide a substantial improvement over a first order one (or SIBC), particularly when surface wave effects are important. This is shown in Fig. 2 for the case of a metal cylinder with a uniform dielectric coating [Syed and Volakis, 1990]. The field is computed 0.05 wavelengths from the surface: the solid line is the exact Mie series solution, and the two dashed lines are the eigenfunction solutions obtained using first and second order boundary conditions Within the shadow the SIBC results are in error by as much as 6 dB, whereas the data based on the second order GIBC are virtually exact.

Since a GIBC is well suited to the simulation of a coated surface, there have been numerous analyses of half planes and wedges subject to these boundary conditions [for example, Bernard, 1987; Senior, 1989; Rojas and Al-hekail, 1989]. The standard Wiener-Hopf and Maliuzhinets techniques are applicable just as they are for first order boundary conditions, and since a GIBC is a product of first order derivative factors, the key parts of the solutions are just products of split functions or Maliuzhinets functions. But as we are now realizing, things are not quite as simple as they seem [Senior, 1990]. There are subtle difficulties associated with GIBCs that are not shared by first order conditions, and because of these, many of the solutions in the literature are incomplete or incorrect.

5. THEORETICAL CONSIDERATIONS

With a first order condition a boundary value problem is well-posed provided the standard edge condition is enforced at any edges that are present, and the resulting solution is unique. Since the problem is also self-adjoint, the Green's function is symmetric, implying that the reciprocity condition concerning the interchange of transmitter and receiver is satisfied automatically. In contrast, with a boundary condition of higher order than the first, the problem is not in general self-adjoint. This is certainly true when there are edges or material junctions present, and the simple specification of an edge condition is no longer sufficient to ensure uniqueness. Since reciprocity is an essential feature of a physically-meaningful result, it now must be enforced explicitly, and while the arbitrariness inherent in the solution allows this to be done, additional constraints are necessary to guarantee a unique solution [Senior, 1990].

None of this is surprising if a GIBC is viewed as a derivative operation applied to a lower order condition, and for a half plane subject to an Nth order GIBC a rigorous application of the Wiener-Hopf method requires that the solution be carried out for the Nth or (N-1)th tangential integral of the original field components depending on whether the boundary condition was expressed in terms of the normal or tangential fields respectively. The constants thus introduced are associated with the solution of the source-free problem, and even the edge condition and the requirement for reciprocity are insufficient to specify all of the constants. The same is true with Maliuzhinets' method where we are now faced with the solution of an inhomogeneous difference equation, and for a second order boundary condition, the single constant that remains can be related to the value of a surface field component at the edge [Senior and Ricoy, 1990]. Such information must be developed from a knowledge of the particular structure modelled by the boundary condition.

Even for a GIBC of arbitrary order, progress is now being made in understanding the additional constraints that must be imposed to ensure a unique solution [Ricoy and Volakis, 1990], and with information of this type, GIBCs could become an effective tool in the analytical and numerical solution of scattering problems.

Though attention has been confined to boundary conditions, it should be noted that the same comments apply to higher order transition conditions used to simulate a partially transparent layer. A first order simulation is simply a resistive or conductive sheet; a second order condition was proposed as long ago as 1969 by Weinstein, and numerous papers documenting the potential advantages of higher order transition conditions have appeared in the last few years.

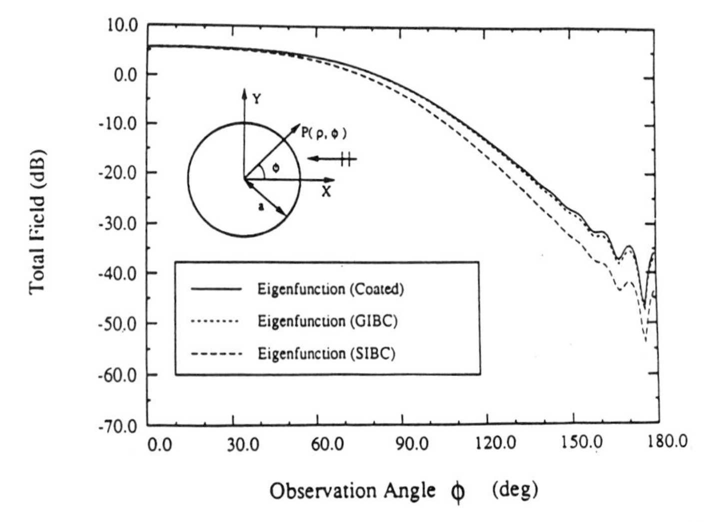

Fig. 2. Total field computed 0.05λ from the surface of a coated metal cylinder using first (----) and second (••••) order boundary conditions, compared with the exact (—) solution.

REFERENCES

Bernard, J.-M.L., 1987, Diffraction by a metallic wedge covered with a dielectric material, Wave Motion, 9:543.

Engquist, B., and Majda, A., 1977, Absorbing boundary conditions for the numerical simulation of waves, Math. Comput., 31:629.

Karp, S.N., and Karal, F.C., Jr., 1967, Generalized impedance boundary conditions with application to surface wave structures, in Electromagnetic Wave Theory, Part I, J. Brown, ed., Pergamon, London.

Keller, J.B., and Givoli, D., 1989, Exact non-reflecting boundary conditions, J. Comp. Phys., 82:172.

Leontovich, M.A., 1948, Investigations on Radiowave Propagation, Part II, Printing House of USSR Academy of Sciences, Moscow.

Morgan, M.A., 1990, Finite Element and Finite Difference Methods in Electromagnetic Scattering, Elsevier, New York.
Ricoy, M.A., and J.L. Volakis, 1990, Derivation of generalized transition/boundary conditions for planar multiple layer structures, Radio Sci. (to be published).
Rojas, R.G., and Al-hekail, Z. 1989, Generalized impedance/resistive boundary conditions for electromagnetic scattering problems, Radio Sci., 24:1.
Rytov, S.M., 1940, Computation of the skin effect by the perturbation method, J. Exp. Theor. Phys., 10:180.
Senior, T.B.A., 1989, Diffraction by a right-angled second order impedance wedge, Electromag., 9:113.
Senior, T.B.A., 1990, Diffraction by a generalized impedance half plane, Radio Sci. (to be published).
Senior, T.B.A., and Ricoy, M.A., 1989, On the use of generalized impedance boundary conditions, IEEE AP-S and URSI Symposia, Dallas, Texas.
Senior, T.B.A., and Volakis, J.L., 1990, Generalized impedance boundary conditions in scattering, Proc. IEEE (to be published).
Syed, H.H., and Volakis, J.L., 1990, Diffraction by coated cylinders using higher order impedance boundary conditions, submitted to IEEE Trans. Antennas Propagat.
Trefethen, L.N., and Halpern, L., 1986, Well-posedness of one-way wave equations and absorbing boundary conditions, Math. Comput. 47:421.
Weinstein, A.L., 1969, The Theory of Diffraction and the Factorization Method, Golem Press, Boulder, Colorado.

CHIRALITY IN ELECTRODYNAMICS: MODELING AND APPLICATIONS

Dwight L. Jaggard and Nader Engheta

Complex Media Laboratory
Moore School of Electrical Engineering
University of Pennsylvania
Philadelphia, PA 19104-6390
USA

I. INTRODUCTION

Recently, attention has been focused on the area of wave propagation, radiation and guidance in chiral media due to their diverse applications and availability. The role of *chirality* or handedness and its importance in a variety of fields such as geometry, chemistry, optics, particle physics and life sciences has been studied since the early part of nineteenth century. *Chiral electrodynamics*, or *electromagnetic chirality*, addresses the effects of handedness in electrodynamics, and these effects can be observed in chiral materials, which, in the optical regime are also known as optically active materials. Owing to the handed nature of their constituents, these materials themselves possess an intrinsic handedness.

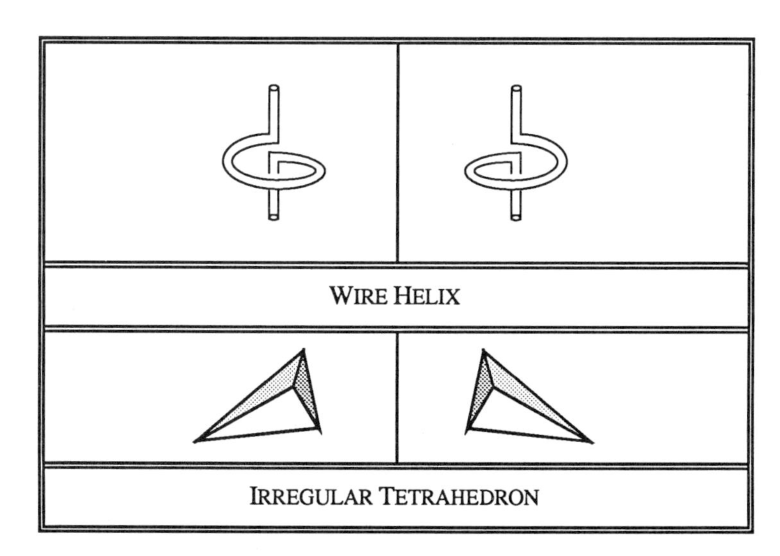

Fig. 1. Two samples of chiral objects, a wire helix and an irregular tetrahedron and their enantiomorphs.

Directions in Electromagnetic Wave Modeling
Edited by H.L. Bertoni and L.B. Felsen, Plenum Press, New York, 1991

The concept of chiral material is derived from the geometric notion of chirality. A *chiral* object is one which cannot be brought into congruence with its mirror image. Such objects lack bilateral symmetry. The mirror image of a chiral object is denoted its *enantiomorph*. Chiral objects and their enantiomorphs form a right-handed and left-handed pairs as shown in Fig. 1. Objects which are not chiral, are said to be *achiral*. A chiral medium can be modeled by a collection of chiral objects, of same handedness, embedded in a host matrix. If the objects are randomly oriented in the host, the material is isotropic.

Concerning wave interaction with chiral media, it has been shown that, for time-harmonic electromagnetic fields, a homogeneous, isotropic chiral medium can be described by the following constitutive relations [1]:

$$\mathbf{D} = \varepsilon_c \mathbf{E} + i\,\xi_c \mathbf{B} \qquad \text{and} \qquad \mathbf{H} = i\,\xi_c \mathbf{E} + (1/\mu_c)\,\mathbf{B} \tag{1}$$

where **E**, **B**, **D** and **H** are electromagnetic field vectors and ε_c, μ_c, ξ_c are, in general, complex and represent the dielectric constant, permeability and chirality admittance of the chiral medium, respectively. The chirality admittance provides the cross linkages between inducing electric field and their magnetic effects and inducing magnetic fields and their electric effects. This admittance is introduced to take into account the handedness properties of the material as indicated by its sign. Its absolute value is the measure of material chirality. From an operational point of view, it is this chirality admittance which allows one to taylor the properties of such materials and introduces a host of new phenomena and novel devices and components. From a physical point of view it is the handedness of the medium which causes it to interact with RCP and LCP eigenmodes differently. Such media exhibits *electromagnetic chirality* which embraces both optical activity and circular dichroism. Optical activity refers to the rotation of the plane of polarization of optical waves by chiral media. Likewise, circular dichroism indicates a change in the polarization ellipticity of optical waves by such media. These phenomena are due to the presence of two unequal characteristic wavenumbers corresponding to two circularly polarized eigenmodes with opposite handedness.

Using the above constitutive relations and the source-free Maxwell equations, we have obtained the following chiral Helmholtz equation

$$\nabla \times \nabla \times \mathbf{C} - 2\omega\mu_c\xi_c\, \nabla \times \mathbf{C} - k^2\, \mathbf{C} = 0 \tag{2}$$

where **C** is any one of the electromagnetic field vectors **E**, **H**, **B** and **D**, with $k = \omega\sqrt{\mu_c\varepsilon_c}$ where ω is the radian frequency of the time-harmonic fields [2]. We have shown that there exist two eigenmodes of propagation, a right-handed and a left-handed circularly polarized (RCP and LCP) plane wave with a pair of wavenumbers given by

$$k_{\pm} = \pm\omega\mu_c\xi_c + \sqrt{k^2 + (\omega\mu_c\xi_c)^2}\,. \tag{3}$$

These are the wavenumbers noted previously and give rise to both optical activity and circular dichorism characteristic of chiral media. Table 1 summarizes these results for chiral media and compare them to simple media.

During the past few years, there have been new developments at the University of Pennsylvania in the area of chiral materials and their diverse applications. We have shown that in chiral materials, the extra degrees of freedom afforded by the chirality admittance ξ_c introduce several novel and notable features in electromagnetic and optical wave propagation, radiation, guidance, scattering and absorption. In the following, we discuss selected applications of chiral materials to scattering, absorption, guiding and radiation.

Table 1. Comparison of characteristics of a chiral medium and a simple isotropic medium

Characteristics	Simple Medium	Chiral Medium
Constitutive Relations	$\mathbf{D} = \varepsilon \mathbf{E}$ $\mathbf{H} = (1/\mu) \mathbf{B}$	$\mathbf{D} = \varepsilon_c \mathbf{E} + i \xi_c \mathbf{B}$ $\mathbf{H} = i \xi_c \mathbf{E} + (1/\mu_c) \mathbf{B}$
Wave Equation	$\nabla \times \nabla \times \mathbf{C} - k^2 \mathbf{C} = 0$ $(\mathbf{C} = \mathbf{E}, \mathbf{H}, \mathbf{D} \text{ or } \mathbf{B})$	$\nabla \times \nabla \times \mathbf{C} - k^2 \mathbf{C} - 2\omega\mu_c\xi_c \nabla \times \mathbf{C} = 0$
Wavenumbers	$k = \omega\sqrt{\mu\varepsilon}$	$k_{\pm} = \pm\, \omega\mu_c\xi_c + \sqrt{k^2 + (\omega\mu_c\xi_c)^2}$
Eigenmodes	Linear Polarization Allowed	RCP and LCP Only

II. CHIROSORB™ AND CHIROSHIELD™

With the present interest in radar cross-section management and control, there is an increasing emphasis on methods to reduce the reflection properties of conducting and composite bodies through the application of coatings. One traditional method is the use of Salisbury and Dallenbach screens [3], which are placed on targets to significantly absorb incident radiation at specified frequencies. By introducing the concept of chiral impedance matching, denoted *chirosorb*™ [4] and noting the increased absorption afforded by the use of chiral material, we introduce a Salisbury and Dallenbach shield alternative made from chiral materials which we call *chiroshield*™ [5]. By incorporating electromagnetic chirality to these shields they offer unique advantages, such as increased absorption in thin layers for a relatively wide range of frequencies, over conventional designs.

The center and right panels of Fig. 2 shows the reflection coefficient amplitude for the magnetic (upper) and electric (lower) *chiroshield*™ and their achiral counterparts, respectively. The vertical axis is a logarithmic measure of the reflection while the frequency axis is a measure of $(\omega - \omega_o)/\omega_o$ where ω is the radian frequency of the incident wave. The chiral material thickness is $\Delta = 0.2\lambda_0$ where λ_0 is the free-space wavelength corresponding to ω_0. The loss axis is a measure of the loss tangent μ^i/μ^r for the two cases under consideration where the superscripts i and r indicate imaginary and real parts, respectively. Here the degree of chirality is specified to be $\eta\xi_c = 1$ where η is the free-space impedance. The desirable low plateau of relatively constant reflection amplitude is characteristic of these two chiral screens and is found over a broad range of frequency and loss values. The reduction in the level of reflection achieved by the *chiroshield*™ as compared to its conventional counterparts is considerable, even for thin screens. For example, a chiral magnetic screen of this design with fractional wavelength thickness provides an additional reduction in reflection of approximately 20 dB as compared to its achiral counterpart for a loss tangent of approximately 4. Such reduction is enhanced with increasing loss due to the advantageous effect of chirality. The advantages offered by *chiroshield*™ include a ~20 dB to ~30 dB additional reduction in reflection over its conventional counterpart and a significant increase in bandwidth for the cases examined to date.

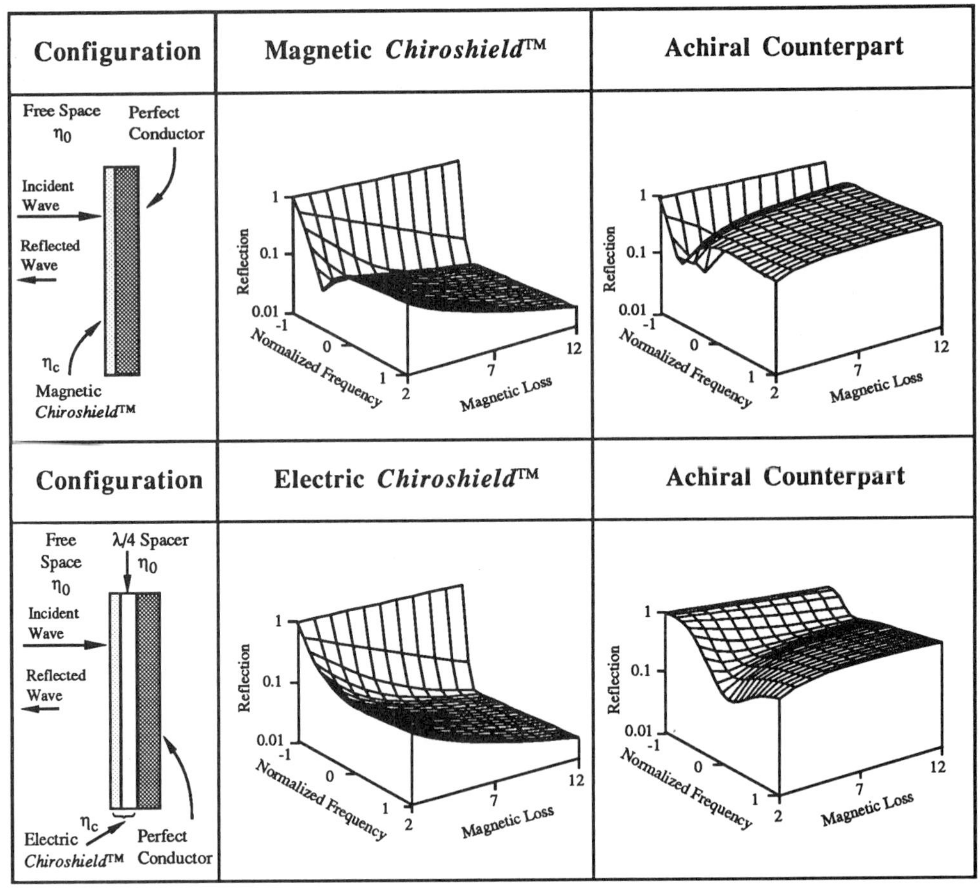

Fig. 2. In the left panel is shown the magnetic *chiroshield*™ (upper) and the electric *chiroshield*™ (lower). In the center and right panels are shown the reflection coefficient amplitude, on logarithmic scale, as a function of normalized frequency $(\omega - \omega_0)/\omega_0$ and magnetic loss tangent μ^i/μ^r. The chiral material thickness is $\Delta = 0.2\lambda_0$ where λ_0 is the free-space wavelength corresponding to ω_0 while the spacer thickness (lower) varies with frequency so that the shield is always a quarter wavelength from the perfect conductor at all frequencies. Adapted from [5].

III. CHIROWAVEGUIDE™

Guided-wave structures filled with chiral materials, which we call *chirowaveguides*™ exhibit novel and unique features which can be used in device construction. In such waveguides we have obtained the Helmholtz equations governing the longitudinal components of **E** and **H** which are the guided wave counterparts to (2) as

$$\begin{cases} \nabla_t^2 E_z + \left[\dfrac{k_+^2 + k_-^2}{2} - h^2 \right] E_z + (2i\,\omega^2\mu^2\xi_c)\, H_z = 0 \\ \nabla_t^2 H_z + \left[\dfrac{k_+^2 + k_-^2}{2} - h^2 \right] H_z - (2i\,\omega^2\mu^2\xi_c/\eta_c^2) E_z = 0 \end{cases} \tag{4}$$

where h is the waveguide propagation constant and η_c is the chiral impedance [6]. The above equations reveal the fact that the longitudinal components of **E** and **H** and their corresponding transverse components are coupled through coefficients which are proportional to the chirality admittance ξ_c. Therefore, in such waveguides, the propagating modes are all hybrid and individual TE, TM, or TEM mode cannot be supported. Furthermore, in such waveguides, these hybrid modes are bifurcated [6]. Figure 3 shows a parallel-plate chirowaveguide filled with isotropic homogeneous chiral materials and its dispersion diagram. The diagram clearly presents the mode bifurcation in such waveguides, which is one of the novel features of electromagnetic chirality in guided-wave structures.

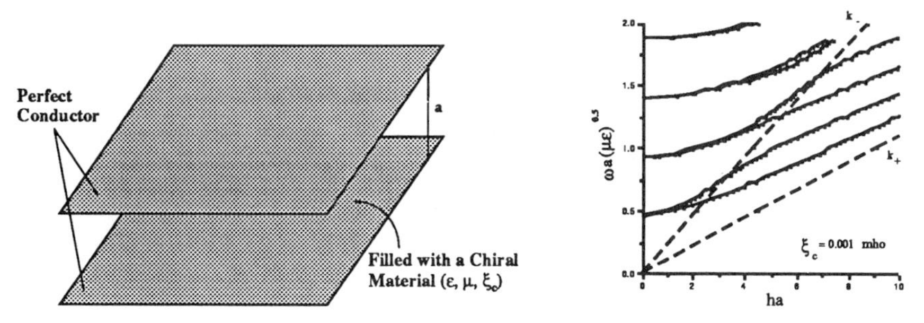

Fig. 3. A parallel-plate chirowaveguide™ filled with an isotropic homogeneous chiral material (left) and its dispersion diagram (right). Adapted from [6].

These structures have potential applications to integrated optical and microwave and millimeter-wave devices such as directional couplers, TE↔TM converter [7], polarization selective photonic switches, and polarization filters are also discussed.

IV. CHIROARRAYS™

Since a chiral medium exhibits polarization birefringence and therefore the wavelengths of polarizations with opposite handedness are different, electromagnetic sources producing various states of polarization in a non-chiral medium, function differently when they are embedded in a chiral medium. An important example of this sort of radiator is a distributed source composed of a linear array of dipoles embedded in a chiral medium [8]. Since there is an inherent geometrical spacing which defines the array, it intuitively appears that the two eigenmodes of the medium will see an array of differing effective geometry.

Using the chiral dyadic Green's function [2], Fig. 4 presents the far-field radiation pattern of an array of fifteen elements, spaced a half-wavelength apart. As the phase shift α is varied from nominally broadside to increasing values beam splitting occurs at a critical phase shift α_c. The evolution of the beam splitting is clearly shown below for six values of the phase shift α with positive chiral admittance ξ_c [8]. First is shown the broadside case ($\alpha = 0$) in part a) followed by increasing phase shift until beam splitting occurs ($\alpha > \alpha_c$) in part b). As the phase shift is increased further, a grating lobe appears in the beam for the larger wavenumber k_+ as first noted in part c). As the phase shift increases still further so that $\alpha > k_- d$, the visible range begins to exclude the main lobe and the negative eigenmode beam decreases in size as shown in part e). In the limit as $\alpha \to \pi$, almost all of the beam energy in the negative eigenmode vanishes and is converted to the positive eigenmode beam as noted in part f). Therefore, we demonstrate the ability of chiral media to allow linear arrays to radiate circularly polarized modes.

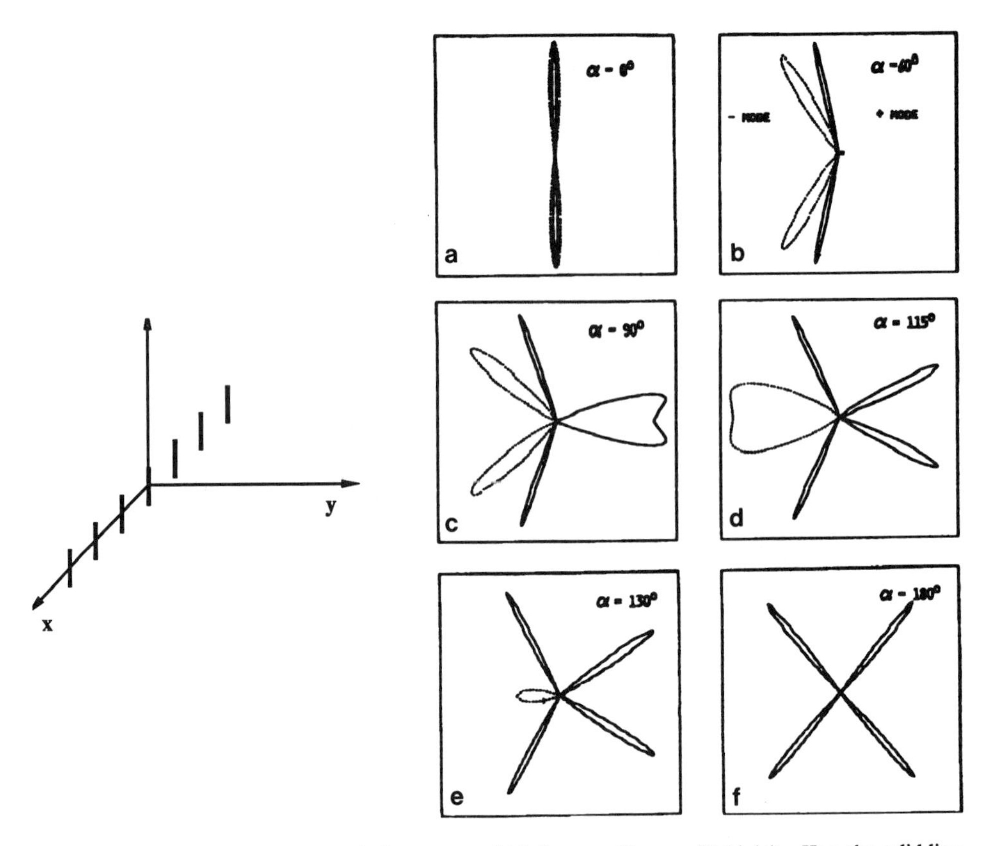

Fig. 4. Array geometry (left) and radiation pattern of 15-element chiroarray™ (right). Here the solid line represents the positive mode while the dashed line represents the negative mode. We use (a) $\alpha = 0°$, (b) $\alpha = 60°$ (beam splitting begins), (c) $\alpha = 90°$, (d) $\alpha = 115°$, (e) $\alpha = 130°$ and (f) $\alpha = 180°$ (complete negative eigenmode suppression). Adapted from [8].

V. CHIROSTRIP™ ANTENNAS

The increasing use of microwave and millimeter-wave integrated-circuit antennas and components in radars, telecommunications, and sensing and surveillance systems has focused attention on various problems of wave propagation with centimeter and millimeter wavelengths in the complex materials used in such devices. Loss of electromagnetic energy in dielectric layers used in microwave and millimeter-wave elements is a major constraint in the design of such elements. In addition, generation of surface wave in conventional dielectric substrates and superstrates in microwave integrated-circuit devices in general, and in microstrip antennas in particular, is an undesirable phenomenon in the performance of such devices. In fact, the presence of surface waves decreases the radiation efficiency of printed-circuit antennas, generates unwanted sidelobes, and reduces the bandwidth of such antennas.

A considerable amount of research from both the theoretical and the experimental point of view, has been directed towards the development of new methods, the synthesis of new materials with low loss and complex properties, and the analysis of electromagnetic properties of such materials in order to have broadband and more efficient microwave and millimeter-wave antennas. New synthetic materials are needed to reduce the energy loss, and more importantly to decrease or eliminate surface-wave propagation in substrates and superstrates used in microwave and millimeter-wave elements. Chiral materials due to their double-mode characteristics are potential candidates for such applications.

We have introduced the idea of a new class of printed-circuit antennas, which we name *chirostrip™ antennas* [9]. These radiators consist of microstrip lines printed on a layer of chiral substrate as shown in Fig. 5. Preliminary theoretical investigation has revealed some novel and unique radiation properties associated with chirostrip™ antennas. For example, it is found that the surface wave power can be reduced [10] and the radiation efficiency and radiation resistance of microstrip antennas can be improved when chiral materials are used as antenna substrates [11]. It is also shown that the state of polarization of radiation fields of chirostrip™ antennas is different from that of conventional microstrip antennas. We anticipate that the polarization flexibility of radiated fields and higher efficiency offered by chirostrip™ antennas can be used in radar and surveillance systems.

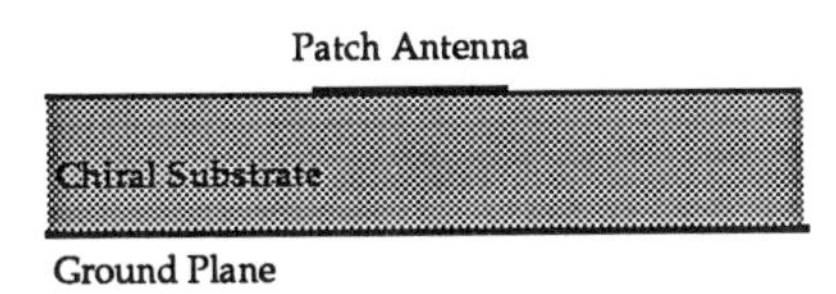

Fig. 5. Chirostrip™ antenna which consists of a microstrip radiator over chiral substrate. Adapted from [10]

VI. CHIRODOME™

The antenna radomes made from chiral materials, which we name *chirodomes™*, are of particular interest since the polarization birefringence present in chiral materials may give rise to new radiation characteristics. We have investigated the interaction of radiation emitted by an electromagnetic source with a sphere made from an isotropic lossless chiral material [12]. As an illustrative example, a short electric dipole found at the center of the chiral sphere, as shown in Fig. 6, has been studied in detail, and the radiated fields and radiation resistance have been examined [12]. It is found that, in this case, the dipole's radiated fields are, in general, elliptically polarized. Furthermore, by choosing the sphere's size and material parameters properly, circular polarization may be obtained. It is also demonstrated that the radiation resistance of the dipole depends on the sphere's size and increases monotonically with chirality parameter ξ_c. Figure 6 presents the radiation resistance of a short electric dipole covered with a sphere of chirodome™ as a function of sphere normalized radius $\rho = (k_+ + k_-)a$ for several values of normalized chirality $\Omega = \xi_c\eta$. Here the vertical axis presents the radiation resistance normalized to the radiation resistance of the same dipole without the chirodome™.

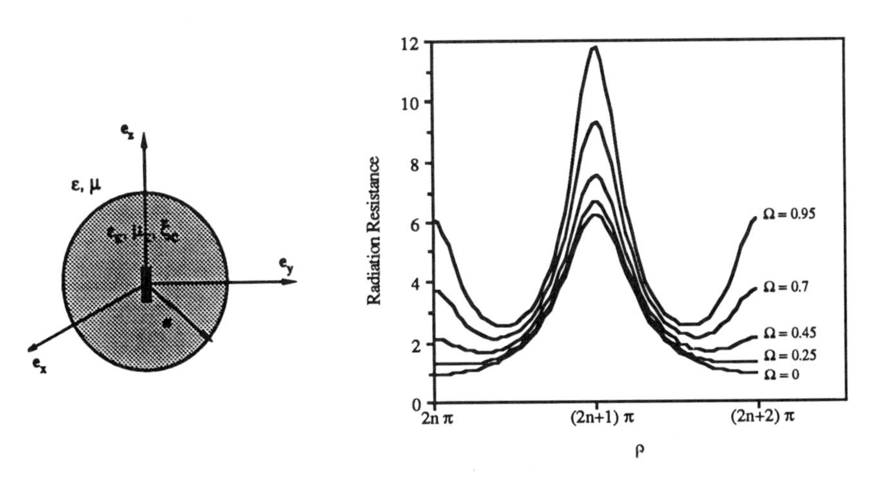

Fig. 6. Antenna immersed in chiral radome (left) and radiation resistance (right) normalized to the achiral case for the given configuration as a function of normalized chirality admittance $\eta\xi_c = \Omega$ and normalized frequency $r = (k_+ + k_-)\,a$. Adapted from [12].

Results of this study have potential applications to novel radome designs in the microwave and millimeter-wave regimes, and to multipolarized antennas with polarization control.

VII. Scatterers in Chiral Media

Scatterers in chiral media have characteristic scattering patterns which depend on the interaction between the degree of chirality and its characteristic electrical length. The former is quantified by $\eta\xi_c$ while the latter is given by kL where k is the free-space wavelength of the incident field and L is the characteristic dimension of the scatterer.

As an example, we consider the scattering of plane waves from wire scatterers immersed in an unbounded chiral medium characterized by the constitutive relations (1). Chirality introduces characteristic bow-tie shaped induced currents found by solving the appropriate integral equation [13] with the chiral dyadic Green's function [2]. The scattered fields are found from the usual methods. In Fig. 7, these induced currents (left) and differential scattering cross-section (right) are shown for a wire scatter in an infinite chiral medium. The current plots are given for the case $kL = 24\pi$ while the scattering cross-sections are given when $kL = 12\pi$. The two plots correspond to normalized chirality admittance values of $\eta\xi_c = 0.0$ and 1.0, respectively, for the top and bottom plots. As the chirality increases, the main lobe becomes narrower, implying increased backscatter and decreased beamwidth. In addition to the main backscatter lobe, two side-lobes emerge. These side-lobes delineate a region where no scattering occurs as seen in the lower right plot. This region we denote the *forbidden zone*. Physically, the forbidden zone appears as a result of Cerenkov-like effects within the medium due to traveling waves of current along the wire.

The results shown here and elsewhere [13] can be grouped into three regions of interest. They are termed the *subchiral*, *chiral*, and *superchiral* regions and refer to the cases $\eta\xi_c kL \lesssim 1$, $\eta\xi_c kL \approx 10\pi$ and $\eta\xi_c kL \gtrsim 48\pi$, respectively. In the subchiral region, the results are similar to that of the achiral case while in the superchiral region the results are independent of wire length. In the intermediate chiral region, detailed calculations are needed to view the tradeoff between electrical length and degree of chirality.

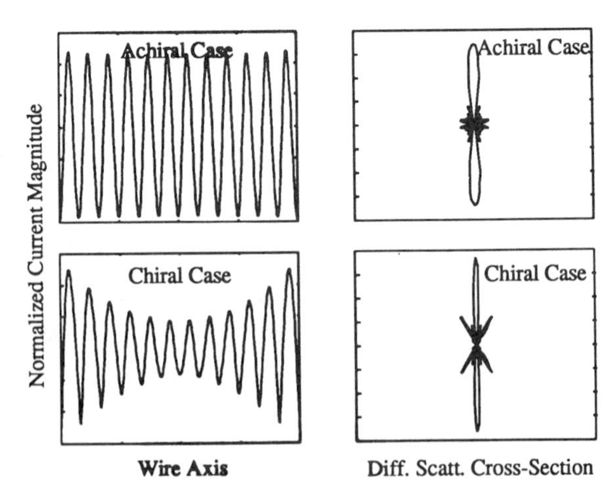

Fig. 7. Induced currents (left) for a wire with $kL = 24\ \pi$ immersed in an achiral (top) and chiral media. Associated differential scattering cross-sections (right) for the achiral (top) and chiral (bottom) cases where $kL = 12\ \pi$. Adapted from [13].

VIII. CONCLUSIONS

In this chapter, we have reviewed the concept of *electromagnetic chirality* which addresses the roles of chirality in electrodynamics, and have shown that chiral materials exhibit handedness in their interaction with electromagnetic waves. We have analyzed and modeled, both physically and mathematically, propagation, radiation and guidance of electromagnetic waves in such media. It has been demonstrated that the extra degrees of freedom introduced by chirality in modeling of such materials have led to a wide variety of novel applications in the areas of shielding and radar cross-section management, microwave and optical devices and components, and antenna and radome designs.

IX. SELECTED REFERENCES

[1] D. L. Jaggard, D. L., Mickelson, A. R., and C. H. Papas, "On Electromagnetic Waves in Chiral Media," *Appl. Phys. 18*, pp. 211-216 (1979).

[2] S. Bassiri, N. Engheta, and C. H. Papas, "Dyadic Green's Function and Dipole Radiation in Chiral Media," *Alta Frequenza LV-2*, 83-88 (1986).

[3] R. L. Fante and M. T. McCormack, "Reflection Properties of the Salisbury Screen," *IEEE Trans. AP-36*, 1443-1454 (1988).

[4] D. L. Jaggard and N. Engheta, "*Chirosorb*™ as an Invisible Medium," *Electron. Lett. 25*, 173-174 (1989).

[5] D. L. Jaggard and N. Engheta, "*Chiroshield*™: A Salisbury/Dallenbach Shield Alternative," *Electron. Lett., 26*, 173-174, 1990. For spherical example, see D. L. Jaggard, J. C. Liu and X. Sun "Spherical *Chiroshield*™," *Electron. Lett. 27*, 77-79 (1991).

[6] N. Engheta and P. Pelet, "Modes in Chirowaveguides," *Optics. Letters. 25*, 173-174, (1989).

[7] P. Pelet and N. Engheta, "Coupled-Mode Theory in Chirowaveguides," *J. of Appl. Phys. 67*, 2742-2745 (1990).

[8] D. L. Jaggard, X. Sun and N. Engheta, "Canonical Sources and Duality in Chiral Media," *IEEE Trans. Ant. and Propagat. AP-36*, 1007-1013 (1988).

[9] N. Engheta, "The Theory of Chirostrip Antennas," *Proceedings of the 1988 URSI International Radio Science Symposium*, 213, Syracuse, New York, (June 1988).

[10] N. Engheta and P. Pelet, "Reduction of Surface Wave in *Chirostrip*™ Antenna," *Electronics Letters 27*, 5-7 (1991).

[11] P. Pelet and N. Engheta, "Chirostrip Antenna: Line Source Problem," to appear in *Journal of Electromagnetic Waves and Applications* (1991).

[12] N. Engheta and M. W. Kowarz, "Antenna Radiation in The Presence of A Chiral Sphere," *J. of Appl. Phys. 67*, No. 2, 639-647 (1990).

[13] D. L. Jaggard, J. C. Liu, A. C. Grot and P. Pelet, "Wire Scatterers in Chiral Media," to appear in *Optics Lett.* (1991).

ELECTROMAGNETIC WAVE MODELING FOR REMOTE SENSING

S. V. Nghiem and J. A. Kong

Department of Electrical Engineering and Computer Science
and Research Laboratory of Electronics
Massachusetts Institute of Technology
Cambridge, Massachusetts 02139, USA

T. Le Toan

Centre d'Etude Spatiale des Rayonnements
CNRS, 31400 Toulouse, France

1. Introduction

Presented in this paper is a layer random medium model for fully polarimetric remote sensing of geophysical media. The strong permittivity fluctuation theory is used to calculate effective permittivities and the distorted Born approximation is applied to obtain polarimetric scattering coefficients. In scattering layers, the embedded scatterers are generally modeled with a non-spherical correlation function with orientation described by a probability density function [1]. The model accounts for multiple interactions due to the medium interfaces, coherent effects of wave propagation, first-order cross-polarized return, and multiple scattering to some extent. The paper is composed of five sections. After this introduction, section 2 reviews polarimetric scattering descriptions under consideration in terms of scattering coefficients, covariance matrix, and Mueller matrix. Relations between the matrix elements and the scattering coefficients will be shown. Section 3 presents the theoretical model formulated from Maxwell's equations to derive the polarimetric scattering coefficients. Section 4 shows results for some geophysical media such as snow, sea ice, and soybean. Physical insights provided by the theoretical model are used to explain the behaviors of the corresponding covariance matrix and the polarization signatures calculated with Mueller matrix. Finally, section 5 summarizes this paper.

2. Polarimetric Descriptions

In this section, polarimetric descriptions characterizing scattering properties of the random media are considered. For polarimetric backscattering from reciprocal media, the scattering coefficients are defined by [2]

$$\sigma_{\mu\tau\nu\kappa} = \lim_{\substack{r\to\infty \\ A\to\infty}} \frac{4\pi r^2}{A} \frac{\left\langle E_{\mu s} E_{\nu s}^* \right\rangle}{E_{\tau i} E_{\kappa i}^*} \tag{1}$$

where subscripts μ, ν, τ, and κ can be h or v polarization and the subscripts i and s stand for incident and scattered waves, respectively. The components of the scattered field in (1) are obtained by measuring the h and the v returns while the incident field is transmitted exclusively with h or v polarization.

Directions in Electromagnetic Wave Modeling
Edited by H.L. Bertoni and L.B. Felsen, Plenum Press, New York, 1991

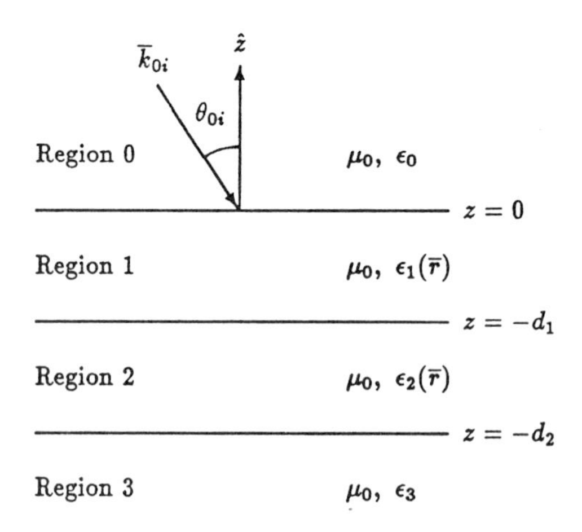

Figure 3.1 Scattering configuration

In terms of the scattering coefficients (1), the polarimetric covariance matrix $\overline{\overline{C}}$ and its normalized form are, respectively

$$\overline{\overline{C}} = \begin{bmatrix} \sigma_{hhhh} & \sigma_{hhhv} & \sigma_{hhvv} \\ \sigma^*_{hhhv} & \sigma_{hvhv} & \sigma_{hvvv} \\ \sigma^*_{hhvv} & \sigma^*_{hvvv} & \sigma_{vvvv} \end{bmatrix} = \sigma \begin{bmatrix} 1 & \beta\sqrt{e} & \rho\sqrt{\gamma} \\ \beta^*\sqrt{e} & e & \xi\sqrt{\gamma e} \\ \rho^*\sqrt{\gamma} & \xi^*\sqrt{\gamma e} & \gamma \end{bmatrix} \tag{2}$$

in which diagonal element σ_{hhhh}, σ_{hvhv}, and σ_{vvvv} are conventional backscattering coefficient σ_{hh}, σ_{hv}, and σ_{vv}, respectively. In the normalized $\overline{\overline{C}}$, $\sigma = \sigma_{hhhh}$, intensity ratio γ and e and correlation coefficient ρ, β, and ξ are

$$\gamma = \frac{\sigma_{vvvv}}{\sigma}, \quad e = \frac{\sigma_{hvhv}}{\sigma} \tag{3}$$

$$\rho = \frac{\sigma_{hhvv}}{\sigma\sqrt{\gamma}}, \quad \beta = \frac{\sigma_{hhhv}}{\sigma\sqrt{e}}, \quad \xi = \frac{\sigma_{hvvv}}{\sigma\sqrt{\gamma e}} \tag{4}$$

The elements of the Mueller matrix $\overline{\overline{M}}$, which is a 4×4 matrix, can also be expressed in terms of the scattering coefficients [3]; however, it contains the same amount of information as the covariance matrix does since both matrices are fully expressible with the complete set of polarimetric backscattering coefficients with 9 independent parameters.

3. Random Medium Model

The scattering configuration is depicted in Fig. 3.1. Region 0 is air with real permittivity ϵ_0. Region 1 is a scattering medium with isotropic scatterers randomly embedded whose electrical property can be characterized by inhomogeneous permittivity $\epsilon_1(\overline{r})$. Region 2 contains scatterers with non-spherical shape constituting an anisotropic random medium which has spatially dependent permittivity $\epsilon_2(\overline{r})$. Depending on the shape and orientation of the non-spherical scatterers, region 2 can be effectively anisotropic. If the random medium is azimuthally symmetric, region 2 becomes effectively uniaxial with vertical optic axis and further reduces to isotropic effectively when the scatterers are randomly oriented. In the calculation of the scattering coefficients, the effective permittivity of region 2 is kept vertically uniaxial due to the azimuthal symmetry of the media under consideration. Region 3 is the underlying half space with homogeneous permittivity ϵ_3. The three regions are assumed to have

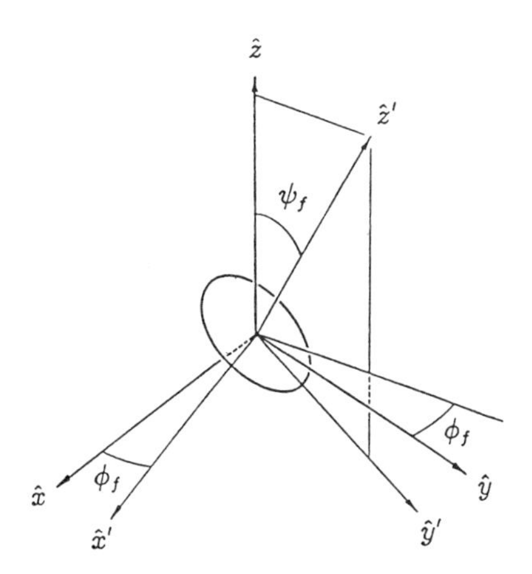

Figure 3.2 Geometry of a scatterer in region 2

identical permeability μ_0. The interfaces are at $z = 0$, $z = -d_1$, and $z = -d_2$ in the global Cartesian coordinate system $(\hat{x}, \hat{y}, \hat{z})$ shown in Fig. 3.1. The local orientation of a scatterer in region 2 is illustrated in Fig. 3.2.

In the phasor notation defined with $e^{-i\omega t}$ [4], time-harmonic total field $\overline{E}_0(\overline{r})$, $\overline{E}_1(\overline{r})$, and $\overline{E}_2(\overline{r})$ respectively in region 0, 1, and 2 satisfy the following wave equations

$$\nabla \times \nabla \times \overline{E}_0(\overline{r}) - k_0^2 \overline{E}_0(\overline{r}) = 0 \tag{5}$$

$$\nabla \times \nabla \times \overline{E}_1(\overline{r}) - k_0^2 \frac{\epsilon_1(\overline{r})}{\epsilon_0} \overline{E}_1(\overline{r}) = 0 \tag{6}$$

$$\nabla \times \nabla \times \overline{E}_2(\overline{r}) - k_0^2 \frac{\epsilon_2(\overline{r})}{\epsilon_0} \overline{E}_2(\overline{r}) = 0 \tag{7}$$

where the wave number is $k_0 = \omega\sqrt{\mu_0 \epsilon_0}$ and ω is the angular frequency. The strong fluctuation theory [1,5] will be used in the calculations of the random-medium effective permittivities. Deterministic permittivity $\overline{\overline{\epsilon}}_{g1} = \epsilon_{g1}\overline{\overline{I}}$, where $\overline{\overline{I}}$ is the unit dyadic, and $\overline{\overline{\epsilon}}_{g2}$ are respectively introduced in both sides of (6) and (7). By defining the following vectors for $m = 1$ and $m = 2$

$$k_0^2 \overline{\overline{Q}}_m(\overline{r}) \cdot \overline{E}_m(\overline{r}) = k_0^2 \left[\frac{\epsilon_m(\overline{r})\overline{\overline{I}} - \overline{\overline{\epsilon}}_{gm}}{\epsilon_0} \right] \cdot \overline{E}_m(\overline{r}) \tag{8}$$

as effective sources, wave equation (6) and (7) for the scattering random media can be rewritten as

$$\nabla \times \nabla \times \overline{E}_m(\overline{r}) - k_0^2 \frac{\overline{\overline{\epsilon}}_{gm}}{\epsilon_0} \cdot \overline{E}_m(\overline{r}) = k_0^2 \overline{\overline{Q}}_m(\overline{r}) \cdot \overline{E}_m(\overline{r}) \tag{9}$$

By the elimination of secular terms, $\overline{\overline{\epsilon}}_{g1}$ and $\overline{\overline{\epsilon}}_{g2}$ are determined. Physically, $\overline{\overline{\epsilon}}_{g1}$ and $\overline{\overline{\epsilon}}_{g2}$ are the effective permittivity tensors in the very low frequency limit where the scattering loss is negligible compared to the absorption loss [6].

In integral equation form, the total field in region $m = 0, 1, 2$ is the superposition of the mean field and the scattered field; explicitly,

$$\overline{E}_m(\overline{r}) = \overline{E}^{(0)}_m(\overline{r}) + k_0^2 \sum_{n=1}^{2} \int_{V_n} d\overline{r}_n \, \overline{\overline{G}}_{mn}(\overline{r}, \overline{r}_n) \cdot \overline{\overline{Q}}_n(\overline{r}_n) \cdot \overline{E}_n(\overline{r}_n) \tag{10}$$

Mean field $\overline{E}^{(0)}_m(\overline{r})$ is the solution to the homogeneous wave equations where the effective sources vanish in the absence of the scatterers. As a particular solution to the inhomogeneous wave equations in the presence of the scatterers in region $n = 1, 2$ occupying volume V_n, the scattered field in (10) is the integrals of the products between the effective source and dyadic Green's functions $\overline{\overline{G}}_{mn}(\overline{r}, \overline{r}_n)$ defined by

$$\nabla \times \nabla \times \overline{\overline{G}}_{mn}(\overline{r}, \overline{r}_n) - k_0^2 \frac{\overline{\overline{\epsilon}}_{gm}}{\epsilon_0} \cdot \overline{\overline{G}}_{mn}(\overline{r}, \overline{r}_n) = \delta(\overline{r} - \overline{r}_n) \overline{\overline{I}} \tag{11}$$

where first subscript m in $\overline{\overline{G}}_{mn}(\overline{r}, \overline{r}_n)$ denotes the observation region containing observation point $\overline{r}$, second subscript n stands for source region $n = 1, 2$ containing source point $\overline{r}_n$, and $\delta(\overline{r} - \overline{r}_n)$ is the Dirac delta function. It is seen from (11) that an observation point in a scattering region can coincide with a source point in the same region ($m = n = 1, 2$) giving rise to the singularity of dyadic Green's function $\overline{\overline{G}}_{nn}(\overline{r}, \overline{r}_n)$ which can be decomposed into

$$\overline{\overline{G}}_{nn}(\overline{r}, \overline{r}_n) = PV \overline{\overline{G}}_{nn}(\overline{r}, \overline{r}_n) - \delta(\overline{r} - \overline{r}_n) k_0^{-2} \overline{\overline{S}}_n \,, \quad n = 1, 2 \tag{12}$$

where PV is the principal value and dyadic coefficient $\overline{\overline{S}}_n$ is conformed with the shape of the source exclusion volume and determined by the condition of secular-term elimination [5]. By using the decomposed Green's function (12), the singular part in the integrand on the right-hand side of (10) for $m = n$ can be extracted and then combined with total field $\overline{E}_n(\overline{r})$ on the left-hand side to form external field $\overline{F}_n(\overline{r})$ in scattering region $n = 1, 2$

$$\overline{F}_n(\overline{r}) = \left[\overline{\overline{I}} + \overline{\overline{S}}_n \cdot \overline{\overline{Q}}_n(\overline{r})\right] \cdot \overline{E}_n(\overline{r}) \tag{13}$$

In terms of external field $\overline{F}_n(\overline{r})$, the vector source (8) can be redefined by introducing scatterer $\overline{\overline{\xi}}_n(\overline{r})$ such that

$$k_0^2 \overline{\overline{\xi}}_n(\overline{r}) \cdot \overline{F}_n(\overline{r}) = k_0^2 \overline{\overline{Q}}_n(\overline{r}) \cdot \overline{E}_n(\overline{r}) \tag{14}$$

in which the scatterers $\overline{\overline{\xi}}_n(\overline{r})$ for the isotropic ($n = 1$) and the anisotropic ($n = 2$) random media are

$$\overline{\overline{\xi}}_n(\overline{r}) = \overline{\overline{Q}}_n(\overline{r}) \cdot \left[\overline{\overline{I}} + \overline{\overline{S}}_n \cdot \overline{\overline{Q}}_n(\overline{r})\right]^{-1} \tag{15}$$

By applying the distorted Born approximation [5,7] to (10) with the new definition of the sources in (15), the total field observed in region 0 is

$$\overline{E}_0(\overline{r}) = \overline{E}^{(0)}_0(\overline{r}) + k_0^2 \sum_{n=1}^{2} \int_{V_n} d\overline{r}_n \left\langle \overline{\overline{G}}_{0n}(\overline{r}, \overline{r}_n) \right\rangle \cdot \overline{\overline{\xi}}_n(\overline{r}_n) \cdot \left\langle \overline{F}_n(\overline{r}_n) \right\rangle \tag{16}$$

where effective permittivity $\overline{\overline{\epsilon}}_{effn}$ is used to calculate the mean dyadic Green's functions and the mean fields. The polarimetric scattering coefficients can then be obtained from (1) with the following ensemble average of the scattered field in (16)

$$\left\langle \overline{E}_{0s}(\overline{r}) \cdot \overline{E}^*_{0s}(\overline{r}) \right\rangle = \sum_{i,j,k,l,m}^{x,y,z} k_0^4 \int_{V_1} d\overline{r}_1 \int_{V_1} d\overline{r}^o_1 \, C_{\xi 1jklm}(\overline{r}_1, \overline{r}^o_1)$$

$$\cdot \left[\left\langle G_{01ij}(\overline{r}, \overline{r}_1) \right\rangle \left\langle F_{1k}(\overline{r}_1) \right\rangle \right] \cdot \left[\left\langle G_{01il}(\overline{r}, \overline{r}^o_1) \right\rangle \left\langle F_{1m}(\overline{r}^o_1) \right\rangle \right]^*$$

$$+ \sum_{i,j,k,l,m}^{x,y,z} k_0^4 \int_0^{\pi} d\psi_f \int_0^{2\pi} d\phi_f P(\psi_f, \phi_f) \int_{V_2} d\overline{r}_2 \int_{V_2} d\overline{r}^o_2 \, C_{\xi 2jklm}(\overline{r}_2, \overline{r}^o_2; \psi_f, \phi_f)$$

$$\cdot \left[\left\langle G_{02ij}(\overline{r}, \overline{r}_2) \right\rangle \left\langle F_{2k}(\overline{r}_2) \right\rangle \right] \cdot \left[\left\langle G_{02il}(\overline{r}, \overline{r}^o_2) \right\rangle \left\langle F_{2m}(\overline{r}^o_2) \right\rangle \right]^* \tag{17}$$

where $P(\psi_f, \phi_f)$ is the probability density function of the orientation. For random media $n = 1$ and $n = 2$, $C_{\xi njklm}$ in (17) is the $jklm$ element of fourth-rank correlation tensor $\overline{\overline{C}}_{\xi n}$ defined as

$$C_{\xi 1jklm}(\overline{r}_1, \overline{r}^o_1) = \left\langle \xi_{1jk}(\overline{r}_1) \xi^*_{1lm}(\overline{r}^o_1) \right\rangle \tag{18a}$$

$$C_{\xi 2jklm}(\overline{r}_2, \overline{r}^o_2, \psi_f, \phi_f) = \left\langle \xi_{2jk}(\overline{r}_2) \xi^*_{2lm}(\overline{r}^o_2) \middle| \psi_f(\overline{r}_2), \phi_f(\overline{r}_2) \right\rangle \tag{18b}$$

To calculate the effective permittivities and the polarimetric scattering coefficients, the correlation functions need to be considered. In region 1, the scatterers are described with isotropic correlation function having the normalized form

$$R_{\xi 1}(\overline{r}) = \exp\left(-\frac{r}{\ell_1}\right) \tag{19}$$

where ℓ_1, is the correlation length. In region 2, the scatterers are modeled with non-spherical correlation function in the local coordinates as illustrated in Fig 3.2. The normalized form of the local correlation function is

$$R_{\xi 2}(\overline{r}') = \exp\left(-\sqrt{\frac{x'^2}{\ell^2_{2x'}} + \frac{y'^2}{\ell^2_{2y'}} + \frac{z'^2}{\ell^2_{2z'}}}\right) \tag{20}$$

with correlation lengths $\ell_{2x'}$, $\ell_{2y'}$ and $\ell_{2z'}$. The orientation of the non-spherical scatterers is described with probability density function $P(\psi_f, \phi_f)$ where ψ_f and ϕ_f are polar and azimuthal orientation angles. For ellipsoidal scatterers such as brine inclusions in sea ice, the orientation is preferably vertical and azimuthally random; i.e., $P(\psi_f, \phi_f) = \delta(\psi_f)/(2\pi)$. For randomly oriented spheroidal scatterers such as disk-like or needle-like leaves, $P(\psi_f, \phi_f) = \sin\psi_f/(4\pi)$ and $\ell_{2x'} = \ell_{2y'} \equiv \ell_{2\rho'}$. For the effective permittivities, the results from strong fluctuation theory have the form

$$\overline{\overline{\epsilon}}_{effn} = \overline{\overline{\epsilon}}_{gn} + \epsilon_0 \overline{\overline{\xi}}_{effn} \tag{21}$$

where $\overline{\overline{\xi}}_{effn}$ is the effective dyadic scatterer for region $n = 1, 2$ [1].

The ensemble average of the scattered field can now be found from (17) with the dyadic Green's functions of the layer medium, the means fields, and the correlation functions. Then, the scattering coefficients can be obtained from (1) and expressed as

$$\sigma_{\mu\tau\nu\kappa} = \pi k_0^4 \sum_{a,b,c,d}^{-1,1} \Psi^{ab}_{1\mu\tau} \Psi^{cd*}_{1\nu\kappa} \mathcal{I}^{abcd}_1 + \pi k_0^4 \int_0^{\pi} d\psi_f \int_0^{2\pi} d\phi_f P(\psi_f, \phi_f) \sum_{p,q,r,s}^{\substack{ou,od\\eu,ed}} \sum_{j,k,l,m}^{x,y,z} \Psi^{pq}_{2\mu\tau,jk} \Psi^{rs*}_{2\nu\kappa,lm} \mathcal{I}^{pqrs}_{2jklm} \tag{22}$$

where analytical expressions for coefficient Ψ's and $\mathcal{I}$'s can be found from [3]. For a configuration without the top scattering layer, the layer random medium model is applied by setting the thickness and variances of region 1 to zero. If the boundaries are rough, the effect can be estimated by incoherently adding the contribution from the rough surface scattering taking into account the total propagation loss to calculate

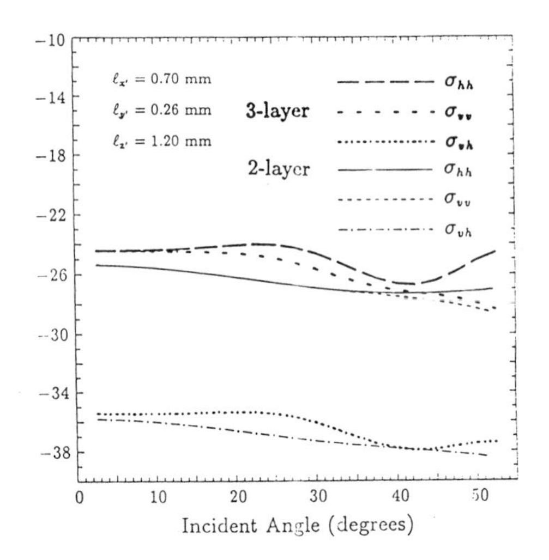

Figure 4.1 Backscattering coefficients of bare and snow covered sea ice

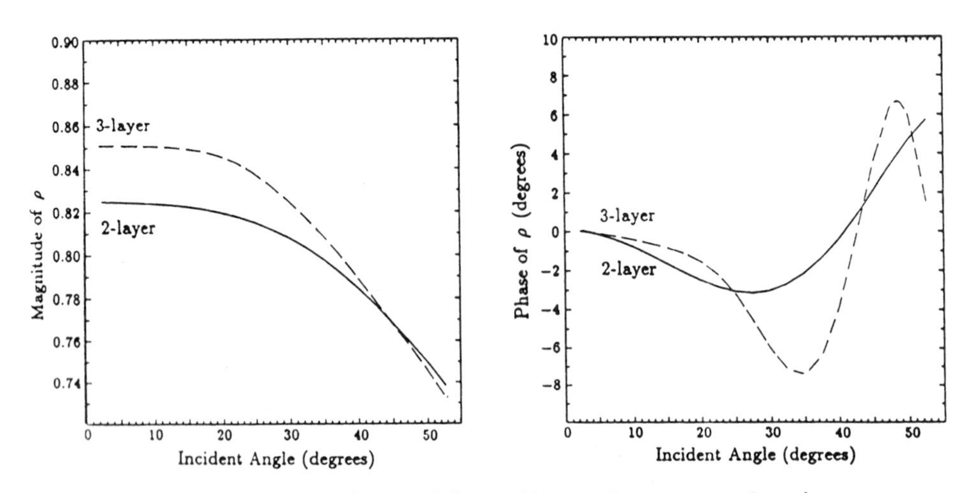

Figure 4.2 Correlation coefficients of bare and snow covered sea ice

the conventional scattering coefficient; formulas for rough surface scattering are in [6]. For an azimuthally symmetric random medium, cross term σ_{hhhv} and σ_{hvvv} are zero. Nevertheless, σ_{hv} is non-zero under the first-order distorted Born approximation due to the non-spherical shape of the scatterers; higher-order contribution to σ_{hv} is, however, neglected.

In this section, the random medium model is formulated and the polarimetric backscattering coefficients are obtained under the distorted Born approximation with the strong permittivity fluctuation theory. In the next section, the model is applied to study the polarimetric scattering properties from vegetation. Polarization signatures and their relations to the corresponding covariance matrices are also investigated.

4. Results and Discussions

a. Application to Remote Sensing of Snow and Sea Ice

Consider first a layer of bare sea ice over sea water. The sea ice is 6.5 cm thick and composed of composed of an ice background with permittivity $\epsilon_{2b} = (3.15 + i0.002)\epsilon_0$ and a 4.2%-volume fraction of brine inclusions having a permittivity $\epsilon_{2s} = (37.5 + i43.1)\epsilon_0$ with correlation length of $\ell_{2x'} = 0.70$ mm, $\ell_{2y'} = 0.26$ mm, and $\ell_{2z'} = 1.20$ mm. These parameters are chosen to simulate young sea ice. The underlying sea water has permittivity $\epsilon_3 = (45 + i40)\epsilon_0$. From the strong fluctuation theory, the sea ice is shown to be effectively anisotropic with optic axis in the vertical direction. The conventional backscattering coefficients at frequency 5.0 GHz are plotted as a function of incident angle in Fig. 4.1 and the correlation coefficient ρ is shown in Fig. 4.2.

To identify the effects of the snow cover, dry snow composed of 20.0% ice with permittivity $\epsilon_{1s} = (3.15 + i0.002)\epsilon_0$ and correlation length $\ell_1 = 0.3$ mm is put over the sea ice. For the purpose of comparison, the conventional backscattering coefficients are also plotted in Fig 4.1 and ρ in Fig. 4.2. It is seen that scattering coefficients are enhanced due to the dry snow cover. Furthermore, the boundary effect is manifested in form of the oscillation on σ_{hh} and σ_{vv}. The oscillation is also observed in the phase of ρ. The magnitude of ρ, however, does not exhibit the oscillation while clearly retaining the similar characteristics as observed directly from the bare sea ice except that $|\rho|$ is higher in the snow cover case due to the isotropy of the snow layer. Thus, the correlation coefficient ρ can carry information from both the covering snow layer the lower anisotropic sea ice layer in a rather distinctive manner.

b. Application to Remote Sensing of Vegetation

The model is applied to a soybean canopy. The data are provided by *Centre d'Etude Spatiale des Rayonnements.* Backscattering coefficients σ_{hh}, σ_{vv}, and σ_{hv} of wave frequency 5.3 GHz and 0-50° incident angles together with ground truth were measured at various growth stages of the soybean from 17 July to 30 September 1986. A comparison for conventional backscattering coefficients of the soybean is shown in Fig. 4.3.1 over the range of incident angles from 0° to 50°. In this case, the configuration is composed of a soybean canopy with height=44 cm, $f_{2s} = 0.29\%$, $\epsilon_{eff2} = (1.05 + i0.0202)\epsilon_0$, $\ell_{2\rho'} = 6.5$ mm, and $\ell_{2z'} = 0.2$ mm and the underlying medium is soil with $\epsilon_3 = (11.9 + i2.34)\epsilon_0$, rms roughness height $\sigma_s = 1.6$ cm, and surface correlation length $\ell_s = 14.0$ cm. For copolarized returns σ_{hh} and σ_{vv}, the theoretical and experimental results are in good agreement. By comparing the contributions from the scattering due to the soybean canopy and the scattering due to the rough soil surface, the trend observed in σ_{hh} and σ_{vv} can be interpreted physically. At small incident angles, the rough surface scattering contribution is important when the total attenuation of the soybean, especially at low vegetation fractional volume, allows the soil surface to be seen by the wave. The rough surface contribution, however, rapidly diminishes as the incident angle increases. At larger incident angles, the volume scattering due to the soybean canopy becomes dominant and the copolarized returns slowly decreases as the incident angle increases. These scattering mechanisms explain why the returns are high near normal incident angle, decrease rather quickly up to incident angle about 20°, and slowly decrease as the incident angle increases to 50°. For cross polarized return σ_{hv}, the theoretical values are lower than the experimental data. One reason is that higher-order contribution to the cross term are neglected in the calculation of volume, surface, and volume-surface scattering. A comparison between theoretical and experimental results as a function of time is also shown in Fig. 4.3.2.

An application of the theoretical model is to simulate scattering coefficients and study their sensitivity to interested biophysical parameters such as vegetation fractional volume and soil moisture for inversion consideration. Biophysical parameters are interrelated; therefore, their trends observed in the growth of the vegetation will be imposed on the simulation to obtain physically meaningful results. Followed these trends, a set of physical parameters is prepared to input in the model to calculate the scattering coefficients. First, the simulation is done at a low incident angle of 10° where the wave can sense the soil moisture better due to a shorter path length than at higher incident angle. The results are plotted in Fig. 4.4 for young soybean (from early to fully grown) which indicates that σ_{hh} and σ_{hh} are rather sensitive to both vegetation fractional volume (f_s) and soil moisture (m_s) for $f_s < 0.3\%$. Thus, if fractional volume is determined, then the soil moisture can be found. Another simulation at a high incident angle of 40° (Fig. 4.5) shows that, while σ_{hh} is still dependent on soil moisture, σ_{vv} is insensitive to soil moisture but sensitive to fractional volume in the range of interest ($f_s < 0.3\%$). This information can thus give useful suggestions for applications in inversion.

To investigate the polarimetric scattering properties, the physical parameters of the soybean are used as inputs to the model to calculate the polarization signatures of the soybean with the Mueller matrix [3]. As an example, the copolarized signature (case of Fig. 4.3.1) at the incident angle of 40° is shown in Fig. 4.6. The signature exhibits local maxima at h and v polarization and the symmetry about the v polarization manifesting the azimuthal symmetry of the soybean. The signature also shows the existence of the pedestal, contributed by the cross term e and the imbalance between $(\gamma + 1)$ and $2\sqrt{\gamma}\mathrm{Re}\rho$ [3], which is physically due to the non-spherical shape of the scatterers. For other growth stages, the corresponding copolarized signatures are displayed in Fig. 4.6 showing deeper saddles at early and late stages. This is caused by the soil surface effect which is more pronounced for early and late soybean when the vegetation fractional volume is low.

5. Summary

In this paper, the fully polarimetric backscattering coefficients have been obtained with the layer random medium model. The top layer is modelled as an isotropic random medium, the middle layer contains non-spherical scatterers, and the underlying layer is a homogeneous medium. The strong fluctuation theory is used to calculated the effective permittivities of the scattering layers and the distorted Born approximation is applied to derive the scattered fields. The model is applied to geophysical media such as snow, sea ice, and vegetation. Cross polarized return is non-zero under the

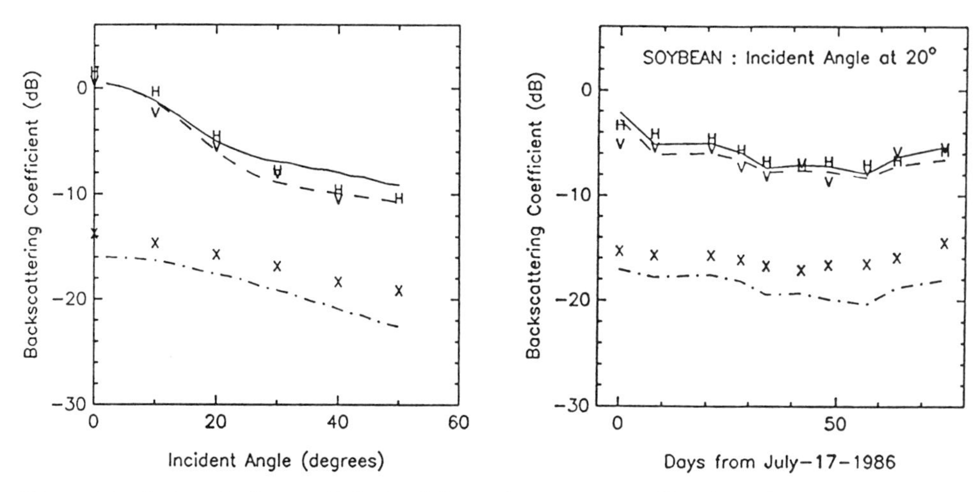

Figure 4.3 Backscattering coefficients of the soybean canopy : H and continuous curve for σ_{hh}, V and dash curve for σ_{vv}, and X and dash-dot curve for σ_{hv}; letters are for experimental data and curves for theoretical results

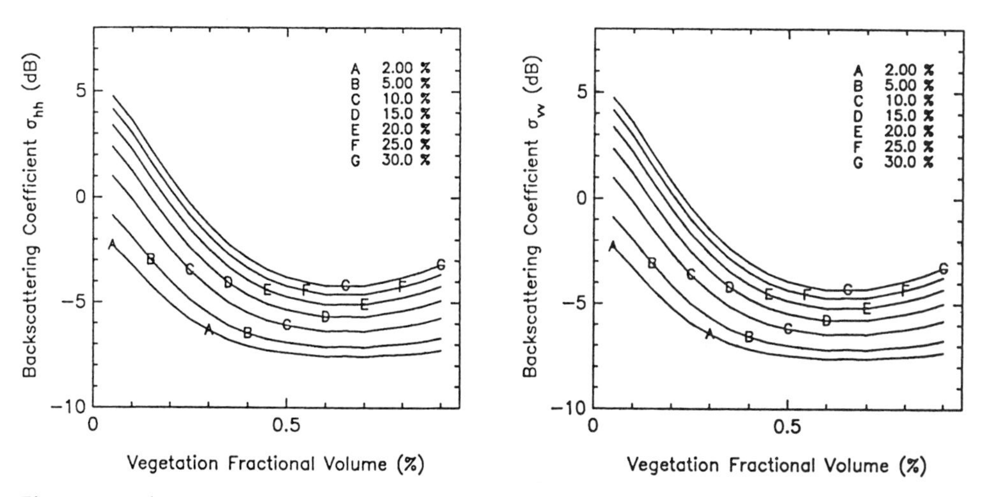

Figure 4.4 Simulation of σ_{hh} and σ_{vv} for young soybean at 10° incident angle as a function of vegetation fractional volume (0.05 to 0.90%) and volumetric soil moisture content (2.0 to 30.0%)

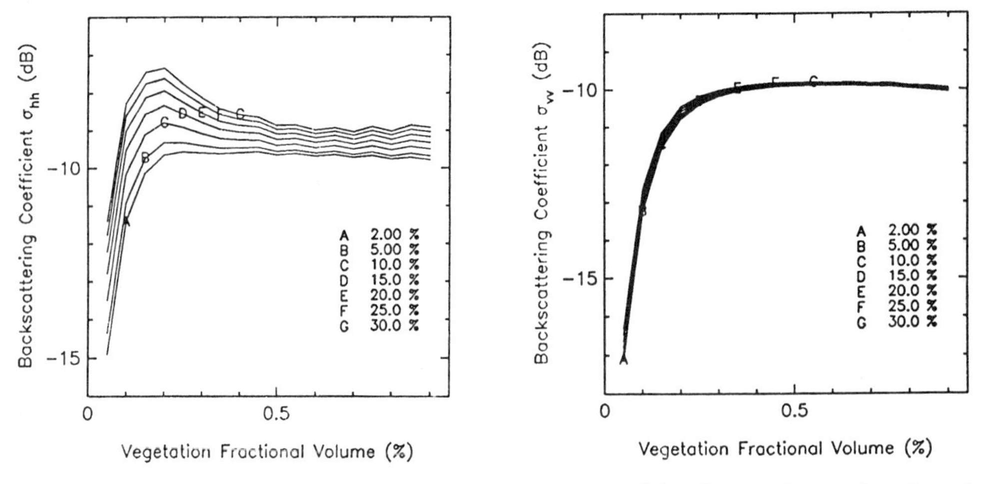

Figure 4.5 Simulation of σ_{hh} and σ_{vv} for young soybean at 40° incident angle as a function of vegetation fractional volume (0.05 to 0.90%) and volumetric soil moisture content (2.0 to 30.0%)

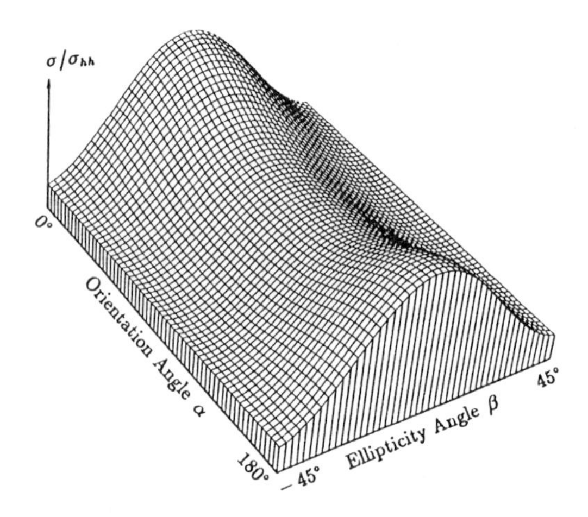

Figure 4.6 Normalized copolarization signature of the soybean at 40° incident angle

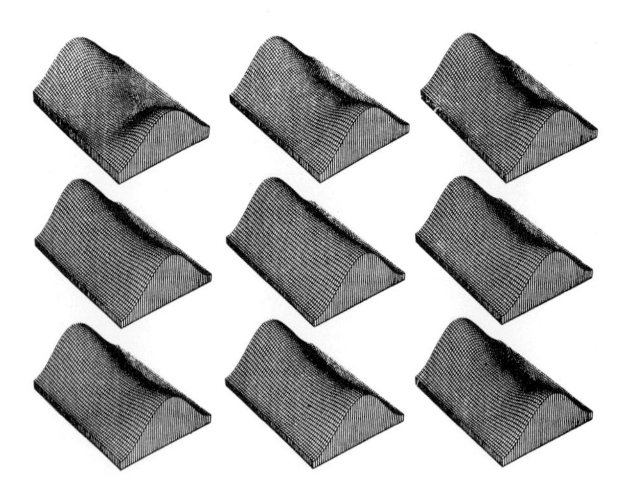

Figure 4.7 Normalized copolarization signatures of the soybean at 40° incident angle from early to late stage (time increases for signatures from left to right and top to bottom)

first-order distorted Born approximation due to the non-spherical shape of the scatterers. The complete set of polarimetric scattering coefficients contain more information about the remotely sensed vegetation as compared to the conventional scattering coefficients and therefore improve the identification and classification of terrain type. Furthermore with simulation and sensitivity analysis, the model can provide useful suggestions to applications in inversion.

Further comparisons with calibrated polarimetric scattering data together with detail ground truth will provide more affirmative validation and/or improvement of the existing model. For geophysical media with more complicate physical and structural properties, further model development is necessary to incorporate the complexity. Experimental data are therefore necessitated to validate the new theoretical developments. Sensitivity study using validated model will provide directions for application in inversion of interested parameters in future research.

Acknowledgments

This work was supported by the NASA Contract 958461, the ARO Contract DAAL03-88-J-0057, the ARMY Corp of Engineers Contract DACA39-87-K-0022, the NASA Contract NAGW-1617, the ONR Contract N00014-89-J-1107, and the European Space Agency.

References

[1] Yueh, H. A., R. T. Shin, and J. A. Kong, "Scattering from randomly oriented scatterers with strong permittivity fluctuations," accepted for publication in *Journal of Electromagnetic Waves and Applications,* 1990.
[2] Borgeaud, M., S. V. Nghiem, R. T. Shin, and J. A. Kong, "Theoretical Models for Polarimetric Microwave Remote Sensing of Earth Terrain," *Journal of Electromagnetic Waves and Applications,* **3**, 61–81, 1989.
[3] Nghiem, S. V., M. Borgeaud, J. A. Kong, and R. T. Shin, "Polarimetric remote sensing of geophysical media with layer random medium model," *Progress in Electromagnetics Research,* edited by J. A. Kong, **3**, Chapter 1, 1-73, 1990.
[4] Kong, J. A., *Electromagnetic Wave Theory,* Wiley-Interscience, New York, 1986.
[5] Tsang, L., and J. A. Kong, "Scattering of Electromagnetic Waves from Random Media with Strong Permittivity Fluctuations," *Radio Science*, **16**, 303–320, 1981.
[6] Tsang, L., J. A. Kong, and R. T. Shin, *Theory of Microwave Remote Sensing,* Wiley-Interscience, New York, 1985.
[7] Tsang, L., and J. A. Kong, "Application of Strong Fluctuation Random Medium Theory to Scattering from Vegetation-like Half Space," *IEEE Transactions on Geoscience and Remote Sensing,* **GE–19**, 62–69, 1981.

APPROXIMATE SCATTERING MODELS IN INVERSE SCATTERING: PAST, PRESENT, AND FUTURE

Anthony J. Devaney†

Department of Electrical and Computer Engineering
Northeastern University
Boston, Mass. 02115

INTRODUCTION

There exist a number of applications in industry and the military that involve what is generally known as *inverse scattering*. Examples include radar target detection and estimation, radar imaging, nondestructive evaluation (NDE), biological and geophysical imaging, and structure determination using X-rays and electron beams. Although the applications appear very different they all involve estimating (reconstructing) the properties of a scatterer from scattered field data and have a common underlying mathematical structure.

The general approaches to inverse scattering can be roughly separated into two categories: (i) exact approaches which employ no simplifying approximations to the forward scattering model, and (ii) approximate approaches that employ simplified, approximate forward scattering models. Examples of research falling within category (i) is the Gelfand Levitan [1] and related treatments of the inverse Sturm Liouville problem and the various generalizations of these procedures to multidimensions [2]. Examples of category (ii) research include the classical theory of X-ray crystallography [3] and its various generalizations [4-6] and diffraction tomography [7-13].

As a general rule, category (ii) research is aimed more at practical applications and, as such, is less concerned with questions of mathematical rigor than it is with problems associated with actually implementing and testing approximate reconstruction algorithms. The research reported here falls into category (ii) and has as its primary goal the review of approximate forward scattering models and their implementation in reconstruction algorithms that have potential use in various industrial and military applications.

INVERSE SCATTERING THEORY

The canonical inverse scattering problem that I deal with in this paper can be stated as follows. Let the scalar field $\psi(\mathbf{r}, k)$ be related to a scattering potential via

†Also with A.J. Devaney Associates, 355 Boylston St., Boston, MA 02116

Directions in Electromagnetic Wave Modeling
Edited by H.L. Bertoni and L.B. Felsen, Plenum Press, New York, 1991

the reduced wave equation

$$[\nabla^2 + k^2 - V(\mathbf{r}, k)]\psi(\mathbf{r}, k) = 0 \tag{1}$$

where $k = \frac{2\pi}{\lambda}$ is a real or complex parameter (with $\Im k \geq 0$), $\nabla = \hat{\mathbf{x}}\frac{\partial}{\partial x} + \hat{\mathbf{y}}\frac{\partial}{\partial y} + \hat{\mathbf{z}}\frac{\partial}{\partial z}$ is the gradient operator in three-dimensions and $V(\mathbf{r}, k)$ is a (generally complex) scattering potential. Further, let ψ obey the asymptotic condition that it equals the sum of a known incident wave ψ_0 plus an outgoing scattered wave ψ^s that obeys the Sommerfeld radiation condition [14]. The inverse problem consists of estimating (reconstructing) the scattering potential V as a function of position $\mathbf{r}$ for one or more values of the parameter k from limited observations of the scattered wave ψ^s obtained in some finite set of scattering experiments.

The above scattering model includes, as special cases, acoustic scattering in constant density medium having a variable bulk modulus (velocity), certain applications of electromagnetic scattering (e.g., optical scattering within the scalar wave approximation), and quantum mechanical potential scattering. With minor modifications, the model applies to the general acoustic scattering problem [5,15] and the general electromagnetic scattering problem [5]. In both the acoustic and optical scattering applications the parameter $k = \frac{2\pi}{\lambda}$ is the wavenumber of a background medium (e.g., free space in optics) and the scattering potential is linearly related to the square of the complex index of refraction of a scatterer embedded in the background.

The canonical inverse scattering problem defined above presents enormous difficulties to the researcher due to the nonlinear and nonlocal mapping that exists between the scattering potential and the scattered wavefield (the "data"). This mapping is best expressed in terms of the integral equation of scattering (the Lippmann Schwinger or "L.S."equation) that results from incorporating the boundary conditions into the general solution of the reduced wave equation (1) [16]

$$\psi(\mathbf{r}, k) = \psi_0(\mathbf{r}, k) + \int d^3r' V(\mathbf{r}', k)\psi(\mathbf{r}', k)G_0(|\mathbf{r} - \mathbf{r}'|, k) \tag{2}$$

where $G_0(R, k) = \frac{-1}{4\pi}e^{ikR}/R$ is the outgoing wave Green function associated with the wavenumber k (i.e., in the background medium).

It is clear from Eq.(2) that the mapping $V \rightarrow \psi$ is nonlocal. That it is also nonlinear follows at once from employing a Liouville Neumann expansion for the scattered wave component of ψ [16]. The first term in this expansion is the so-called Born approximation

$$\psi_B^s(\mathbf{r}, k) = \int d^3r' V(\mathbf{r}', k)\psi_0(\mathbf{r}', k)G_0(|\mathbf{r} - \mathbf{r}'|, k) \tag{3}$$

which has been employed extensively in the inverse scattering problem [3-13,17,18]. As is well known in regard to the direct problem (the problem of computing ψ given V) the use of higher order terms in the Liouville Neumann series (sometimes called the Born series) to improve the accuracy of the first order Born solution (3) is not a good approach due to the very slow convergence rate of the series. Thus, it is not a trivial matter to go beyond the Born approximation (3) in inverse scattering theory and a good deal of current and past research has been devoted to precisely this problem.

RYTOV APPROXIMATION

The first order Born approximation requires that the magnitude of the incident wave be much larger than the magnitude of the scattered wave component throughout the support volume of the scatterer. This, in turn, restricts *both* the magnitude of the scattering potential and the radius of the support volume of this quantity. This latter restriction can be partially alleviated if the so-called ***Rytov approximation*** [5-13,19-21] is employed instead of the Born approximation. The Rytov approximation is the first term in a perturbation expansion of the complex phase of the total wavefield ψ about the complex phase ϕ_0 of the (known) incident wavefield. One finds that the complex phase ϕ of the total wavefield satisfies a nonlinear equation of the Ricatti type and that the Rytov approximation corresponds to linearizing this equation with respect to the phase perturbation $\delta\phi = \phi - \phi_0$. The Rytov approximation $\delta\phi_R$ to the phase perturbation is found to be functionally related to the Born approximation to the scattered field ψ_B^s via the simple equation [19]

$$\delta\phi_R = \psi_B^s / \psi_0 \tag{4}$$

Although the Born and Rytov approximations are functionally related via Eq.(4) their domains of validity are quite different (c.f., Ref. 20) and, in particular, the Rytov approximation avoids the saturation problem that occurs with the Born approximation when applied to scatterers having large support volumes. For this reason the Rytov approximation is usually employed in applications where the propagation paths are very long such as in ultrasound and geophysical tomography [5-13] (so-called diffraction tomography†). Although the Rytov approximation extends the domain of validity of the Born approximation to large, weakly inhomogeneous scatterers it still requires that the magnitude of the scattering potential be small in comparison with $|k|^2$ and, hence, has limited applicability.

It is possible to employ more accurate scattering models than the simple Born and Rytov approximations and still obtain a relatively simple inverse scattering problem. The possible extensions can be roughly divided into the following two categories:

1. Perturbational methods such as the distorted wave Born and Rytov approximations.
2. Non-perturbational approaches such as the "Heitler approximation."

Each of these approximation schemes and its relevance to inverse scattering will now be described.

DISTORTED WAVE APPROXIMATIONS

In the distorted wave approximations it is assumed that the scattering potential can be decomposed into the sum of a known "background" component V_0 and an unknown component δV that is pointwise small in magnitude compared with the

†When the wavefield data for the inverse problem consists of near field measurements in the forward scattering direction the inverse scattering problem formally resembles the inverse problem in computed tomography (CT) and for this reason is generally referred to as *diffraction tomography*.

sum $k^2 + V_0$. A generalized form of the Lippmann Schwinger equation (2) applies as well as a generalized form of the linearized Ricatti equation for the complex phase perturbation relative to the phase in the background medium [21]. The distorted wave procedure is called the *distorted wave Born approximation* (DWBA) when it is based on a linearization of the wavefield and the *distorted wave Rytov approximation* (DWRA) when it is based on a linearization of the complex phase perturbation.

One possible approach to the inverse scattering problem within the distorted wave Born approximation (DWBA) is based on an the integral equation derived in [22] that relates the perturbation of the scattering potential δV to an "image" of this quantity generated directly from the scattered wavefield data. This integral equation is given by (see Eqs.(13)-(15) in Ref. [22]):

$$\mathcal{I}(\mathbf{r}, k) = \int d^3r' \delta V(\mathbf{r}', k) K(\mathbf{r}, \mathbf{r}', k). \tag{5}$$

Here, $\mathcal{I}$ is an image of δV generated directly from the scattered wavefield data (see Eq.(13) in [22]) and

$$K(\mathbf{r}, \mathbf{r}', k) = \frac{8}{k^2} [\Im G_0(\mathbf{r}, \mathbf{r}', k)]^2 \tag{6}$$

where $\Im G_0$ is the imaginary part of the outgoing wave Green function *in the background medium.* The mathematical operation of forming the image $\mathcal{I}$ can be interpreted in terms of a *backpropagation* (time reversal) of the scattered field data into the support region of the scatterer [22] (see also discussion below).

In [22] it was shown that the "solution" of the inverse scattering problem within the distorted wave Born approximation (DWBA) reduces to solving Eq.(5) for δV. A general procedure was outlined which was based on using Mercer's expansion [23] to represent the kernel K in terms of its eigenfunctions and thereby generate a projection of δV onto the set of these eigenfunctions having non-zero eigenvalues. It was shown that in the case of the first order Born approximation these eigenfunctions are the plane waves and that this approach reduced to the method employed in [18] for solving the inverse scattering problem in the first order Born approximation. This latter method led ultimately to the *filtered backpropagation algorithm* [10] of diffraction tomography. The difficulty that arises in the general case of non-uniform backgrounds is the computation of these eigenfunctions which is generally non-trivial except for the simplest type of background.

A special case of great importance in certain applications occurs when the space dependence of the background potential V_0 varies slowly relative to both the wavenumber $\lambda = \Re \frac{2\pi}{k}$ and the perturbation δV. This case is important in geophysical applications where the "slow" background of a geological formation is known either from surface seismic surveys or from borehole data and in NDE applications where one is interested in imaging a defect in an otherwise known homogeneous background medium. In this case the scattering volume $\mathcal{V}$ can be partitioned into a number of "isoplanatic" regions $\mathcal{V}_j$, $j = 1, \ldots, N$ within each of which the background Green function G_0 can be taken to be that of a homogeneous medium having effective parameter $k_j^2 = k^2 + < V_0 >$ where $< V_0 >$ is a local average of the background potential within the j'th isoplanatic area. The integral equation can then be approximately solved locally within each isoplanatic area. This approach has been applied by the author and one of his graduate students in cross-well diffraction tomography in geophysical applications where the background is assumed to consist of a number of plane parallel homogeneous layers [24].

Some of the questions to be addressed in connection with the DWBA are: (i) How is this approach implemented in practice? (ii) What error bounds apply to the solutions generated by the approach? (iii) What are the practical limitations of the approach? (iv) Can the method be employed in an iterative fashion where the background potential V_0 is updated by the estimated δV. An interesting avenue of research in connection with the distorted wave approximations is the use of alternative approaches in solving the linearized Lippmann Schwinger equation for the unknown scattering potential. Such approaches might include, for example, iterative methods of solution analogous to the algebraic reconstruction techniques employed in computed tomography (CT) [25-27] and the use of the *singular value decomposition* which has been employed in connection with the inverse scattering problem within the usual constant background Born approximation [28].

NON-PERTURBATIONAL METHODS

The Born and Rytov approximations and their distorted wave generalizations are *perturbational* in that they result from a (perturbation) expansion of the field (or complex phase) in a power series in a strength parameter associated with the scattering potential $V(\mathbf{r}, k)$. An alternative approach to obtaining an approximate scattering model is to simply "approximate" the field within the integral term in the L.S. equation (or its complex phase counterpart) by some physically based model and base the inverse scattering procedure on the integral equation that results. An example of such a procedure is provided by the use of the *Heitler Approximation* by the author and A. Weglein [30]. The Heitler approximation, which was first employed by W. Heitler in his treatment of radiation damping, is best described in the context of the equation satisfied by the *Transition operator* [30]. However, in order to avoid a prolonged discussion regarding the Transition operator I will discuss the approximation and the resulting inverse scattering approach directly in terms of the Lippmann Schwinger equation satisfied by the field. The interpretation of the Heitler approximation presented below is new and I believe that it provides a solid justification for its use in a number of inverse scattering applications.

The Heitler approximation to the wavefield ψ is obtained by neglecting the real part of the uniform background Green function G_0 in the Lippmann Schwinger equation (2); i.e.,

$$\psi_H(\mathbf{r}, k) = \psi_0(\mathbf{r}, k) + \int d^3r' V(\mathbf{r}', k)\psi_H(\mathbf{r}', k)\Im G_0(|\mathbf{r} - \mathbf{r}'|, k) \quad (7)$$

where ψ_H is the field ψ within the Heitler approximation with $\Im G_0$ being the imaginary part of the uniform background Green function. The inverse scattering problem within the Heitler approximation is formulated by using the Heitler field ψ_H as the internal field in the computation of the scattered wave to obtain

$$\psi_H^s(\mathbf{r}, k) = \int d^3r' V(\mathbf{r}', k)\psi_H(\mathbf{r}', k)G_0(|\mathbf{r} - \mathbf{r}'|, k), \quad (8)$$

where ψ_H^s is the Heitler approximation to the scattered wave. The inverse scattering problem then reduces to jointly solving Eqs.(7) and (8) for the scattering potential in terms of observations of the scattered wavefield ψ_H^s.

A justification for employing the Heitler approximation is based on the fact that the imaginary part of the Green function, which is sometimes referred to as the "Schwinger function", is responsible for all virtual interactions within the scatterer and, as such, is responsible for long-range interactions while the real part of the Green function is responsible for short-range "self" interactions. The approximation is thus a renormalization of the Liouville Neumann series solution of (2) that includes all virtual processes. It can be expected that the Heitler approximation will be excellent for extended systems of discrete scattering centers such as occur in crystallography and in certain NDE applications.

A second justification for using the Heitler approximation follows from an interpretation of the Heitler field ψ_H as being the *backpropagation of the total field into the interior of the scatterer volume.* Thus, for the inverse scattering problem the Heitler approximation corresponds to replacing the internal field by the total backpropagated field. It is interesting to note in this connection that the first order Born approximation employs the backpropagation of the incident wave as the internal field. The Heitler approximation thus improves on the Born approximation by including the contribution of the backpropagated scattered field component to the internal field.

In the following discussion of the Heitler approximation I will consider only the case where the scattering amplitude is known for all scattering directions and some set of incident wave directions. For this case Eq.(8) yields the following expression for the scattering amplitude within the Heitler approximation

$$f_H(\mathbf{s}, \mathbf{s}_0) = \int d^3r V(\mathbf{r}, k)\psi_H(\mathbf{r}, k)e^{-ik\mathbf{s}\cdot\mathbf{r}}, \tag{9}$$

where $f_H(\mathbf{s}, \mathbf{s}_0)$ is the Heitler approximation to the scattering amplitude. Here, $\mathbf{s}_0$ is the unit propagation vector of the incident plane wave and $\mathbf{s}$ is the unit vector in the direction of the scattered field measurement. The inverse scattering problem then requires that Eqs.(9) and (7) be jointly solved for V in terms of $f_H(\mathbf{s}, \mathbf{s}_0)$ specified for all scattering directions $\mathbf{s}$ and some set of incident wave directions $\mathbf{s}_0$.

Equations (7) and (9) can be combined into a single integral equation by making use of the expansion

$$\Im G_0(\mathbf{r} - \mathbf{r}'; k) = \frac{1}{4\pi} \int d\Omega_s e^{ik\mathbf{s}\cdot(\mathbf{r}-\mathbf{r}')}. \tag{10}$$

In this expansion, $d\Omega_s$ is the differential solid angle associated with the unit vector $\mathbf{s}$ and the integration is over 4π steradians. On substituting Eq.(10) into Eq.(7) and making use of Eq.(9) one finds that

$$\psi_H(\mathbf{r}, k) = e^{ik\mathbf{s}_0\cdot\mathbf{r}} + \frac{1}{4\pi} \int d\Omega_s f_H(\mathbf{s}, \mathbf{s}_0)e^{ik\mathbf{s}\cdot\mathbf{r}}. \tag{11}$$

The quantity on the right-hand side of Eq.(11) can be shown to be precisely the backpropagation of the total field into the interior of the region of space occupied by the scatterer. Thus, as mentioned above, the Heitler approximation amounts to using this backpropagated field as the interior field in the expression for the scattering amplitude.

The final step is simply to substitute Eq.(11) into Eq.(9) to obtain

$$f_H(\mathbf{s}, \mathbf{s}_0) = \tilde{V}[k(\mathbf{s} - \mathbf{s}_0)] + \frac{1}{4\pi} \int d\Omega_{s'} f_H(\mathbf{s}', \mathbf{s}_0)\tilde{V}[k(\mathbf{s} - \mathbf{s}')] \tag{12}$$

where

$$\tilde{V}(\mathbf{K}) = \int d^3r V(\mathbf{r}, k)e^{-i\mathbf{K}\cdot\mathbf{r}} \tag{13}$$

is the three-dimensional Fourier transform of the scattering potential. Eq.(12) is recognized as being an integral equation for the transform of the scattering potential in terms of the (measured) scattering amplitude. The inverse scattering problem then reduces to looking for solutions to this equation given $f_H(\mathbf{s}, \mathbf{s}_0)$ for some set of incident wave directions $\mathbf{s}_0$. It is clear that the lowest order solution to the equation is simply the Born approximation

$$f_H(\mathbf{s}, \mathbf{s}_0) = \tilde{V}[k(\mathbf{s} - \mathbf{s}_0)]. \tag{14}$$

Once the integral equation (13) is solved for the transform of the scattering potential the corresponding scattering potential is readily obtained by employing the integral transform described in [18]; i.e.,

$$\hat{V}(\mathbf{r}) = \frac{k^2}{2\pi^2} \int d\Omega_s \int d\Omega_{s_0} |\mathbf{s} - \mathbf{s}_0| \tilde{V}[k(\mathbf{s} - \mathbf{s}_0)] e^{ik(\mathbf{s}-\mathbf{s}_0)\cdot\mathbf{r}} \tag{15}$$

where $\hat{V}$ is the approximate solution of the inverse scattering problem that has minimum L^2 norm (see discussion in [18]).

SUMMARY

I have in this paper reviewed a number of approximate scattering models that are of use in inverse scattering applications. As of this writing I believe that it is fair to say that the constant background Born and Rytov approximations appear to be still the most popular and most used although the distorted wave approximations are being used increasingly in various studies [31,32]. Unfortunately, the separation between the constant background Born and Rytov approximations and more accurate models, such as the distorted wave approximations is vast, and it is questionable whether any general, easily implemented, inverse scattering procedures can be developed for these more accurate scattering models.

REFERENCES

1. P.C. Sabatier, "Basic concepts and methods of inverse problems," in *Tomography and Inverse Problems*, ed. P.C. Sabatier. Philadelphia: Adam Hilger, 1987.

2. R.G. Newton, "The Marchenko and Gelfand-Levitand methods in the inverse scattering problem in one and three dimensions," in *Conference on Inverse Scattering: Theory and Application* Philadelphia: SIAM Press, 1983.

3. J.M. Cowley, *Diffraction Physics*. New York: North-Holland, 1984.

4. E. Wolf, "Three-dimensional structure determination of semi-transparent objects from holographic data," *Opt. Commun.* **1**, p. 153, 1969.

5. K.J. Langenberg, "Applied inverse problems for acoustic, electromagnetic and elastic wave scattering," in *Tomography and Inverse Problems*, ed. P.C. Sabatier. Philadelphia: Adam Hilger, 1987.

6. A.J. Devaney, "Inverse source and scattering problems in ultrasonics," *IEEE Trans. Sonics and Ultra.* **SU-30**, pp. 355-364, 1983.

7. A.C. Kak, "Computerized tomography with x-ray emission and ultrasound sources," *Proc. IEEE*, Vol. 67, pp. 1245-1272, 1979.

8. A.J. Devaney, *Mathematical Topics in Diffraction Tomography*, Final Project Report NSF Grant # *DMS-8460595*, 1985. Available through the U.S. Goverment Printing Office.

9. A.J. Devaney, "Reconstructive tomography with diffracting wavefields," *Inverse Problems*, Vol. 2, pp.161-183, 1986.

10. A.J. Devaney, "A filtered backpropagation algorithm for diffraction tomography," *Ultrasonic Imaging*, Vol. 4, pp. 336-350, 1982.

11. S.X. Pan and A.C. Kak, "A computational study of reconstruction algorithms for diffraction tomography," *IEEE Trans. Acoustics, Speech and Signal Processing*, Vol. ASSP-31, pp. 1262-1275, 1982.

12. R.K. Mueller, M. Kaveh, and G. Wade, "Reconstructive tomography and applications to ultrasonics," *Proc. IEEE*, Vol. 67, pp. 567-587, 1979.

13. A.J. Devaney, "A computer simulation study of diffraction tomography," *IEEE Trans. on Biomed. Eng.*, Vol. BME-30, pp. 377-386, 1982.

14. A. Sommerfeld, *Partial Differential Equations in Physics*. New York: Academic Press, p. 189, 1967.

15. A.J. Devaney, "Variable density diffraction tomography," *J. Acous. Soc. Am.* **78**, pp. 120-130, 1985.

16. J.R. Taylor, *Scattering Theory*. New York: Wiley, 1972.

17. A.J. Devaney, "Nonuniqueness in the inverse scattering problem," *J. Math. Phys.* **19**, pp. 1526-1532, 1978.

18. A.J. Devaney, "Inversion formula for inverse scattering within the Born approximation," *Opt. Commun.* **7**, p. 111, 1982.

19. A.J. Devaney, "Inverse scattering theory within the Rytov approximation," *Opt. Lett.* **6**, pp. 374-376, 1981.

20. M. Oristaglio, "Accuracy of the Born and Rytov approximations for reflection and refraction at a plane interface," *J. Opt. Soc. Am.* **2**, pp. 1987-1989, 1985.

21. G. Beylkin and M. Oristaglio, "Distorted wave Born and Rytov approximations," *Opt. Commun.* **53**, pp. 213-216, 1985.

22. A.J. Devaney and M. Oristaglio, "Inversion procedure for inverse scattering within the distorted wave Born approximation," *Phys. Rev. Letts.* **51**, p.237, 1983.

23. P.M. Morse and H. Feshbach, *Methods of Theoretical Physics*. New York, McGraw-Hill, Part I, Chap. 8, 1953.

24. A.J. Devaney and D.H. Zhang, "Geophysical diffraction tomography in a layered background," (submitted to *Wave Motion*).

25. R. Gordon and G.T. Herman, "Three-dimensional reconstruction from projections: A review of algorithms," *Int. Rev. Cytol.*, Vol. 38, pp. 111-151, 1974.

26. R. Gordon, "A tutorial on ART," *IEEE Trans. Nuclear Science*, Vol. NS-21, pp. 78-93, 1974.

27. F. Natterer, *The Mathematics of Computerized Tomography.* New York, John Wiley, 1986.

28. O. Brander and B. DeFacio, "The role of filters and the singular-value decomposition for the inverse Born approximation," *Inverse Problems* **2**, pp. 375-393, 1986.

29. A.J. Devaney, "The limited view problem in diffraction tomography," *Inverse Problems*, Vol.5, p. 501, 1989.

30. A.J. Devaney and A.B. Weglein, "Inverse scattering within the Heitler approximation," *Inverse Problems*, Vol. 5, p. L49, 1989.

31. R.J. Wombell and M. A. Fiddy, "Inverse scattering within the distorted wave Born approximation," *Inverse Problems*, Vol. 4, p. L23, 1988.

32. W.C. Chew and Y.M. Wang, "Reconstruction of two-dimensional permittivity distribution using the distorted Born iterative method," *IEEE Trans. Med. Imag.*, Vol. 9, p. 218, 1990.

WAVE MODELING FOR INVERSE PROBLEMS
WITH ACOUSTIC, ELECTROMAGNETIC, AND ELASTIC WAVES

Karl J. Langenberg

Dept. Electrical Engineering
University of Kassel
3500 Kassel, Germany

INTRODUCTION

Applied inverse problems comprise radar remote sensing, geophysical exploration, medical diagnostics, nondestructive testing a.s.o., and as such, acoustic, electromagnetic, and elastic waves are under concern. Therefore, appropriate models have to be found to solve the inverse scattering problem for these types of waves algorithmically. Essentially, the linearization of the direct as well as the inverse scattering problem is most often required, and the underlying model is either the weak scattering (Born) approximation, or the physical optics (Kirchhoff) approximation. This allows a unified treatment of the scalar — acoustic — as well as the vector inverse scattering problem for electromagnetic and elastic waves, thus yielding full polarimetric backpropagation inversion schemes. In order to check the validity of the linearization and the influence of insufficient experimental data due to aperture or frequency bandwidth limitations, simulations are required utilizing appropriate numerical codes. Here, we essentially present results for acoustic and elastic wave scattering obtained with our AFIT and EFIT Finite Difference codes.

ACOUSTIC WAVE SCATTERING AND ULTRASONIC DATA FIELDS

Let us denote the scalar pressure in a medium with zero shear and acoustic wave speed c_0 by $\Phi(\underline{\mathbf{R}}, t)$, $\underline{\mathbf{R}}$ being the vector of position and t the time; $\Phi(\underline{\mathbf{R}}, t)$ has to satisfy a wave equation

$$\Delta\Phi(\underline{\mathbf{R}}, t) - \frac{1}{c_0^2}\frac{\partial^2\Phi(\underline{\mathbf{R}}, t)}{\partial t^2} = -q(\underline{\mathbf{R}}, t) - q_c(\underline{\mathbf{R}}, t) \quad , \tag{1}$$

where $q(\underline{\mathbf{R}}, t)$ denotes the source of the incident field $\Phi_i(\underline{\mathbf{R}}, t)$ being confined to a spatial region of compact support. The equivalent volume source $q_c(\underline{\mathbf{R}}, t)$ represents the scatterer — equally of compact support — with surface S_c and volume V_c, which is embedded in the otherwise homogeneous and lossless medium. It produces a scattered field $\Phi_s(\underline{\mathbf{R}}, t)$ given by

$$\Phi_s(\underline{\mathbf{R}}, \omega) = \int_{-\infty}^{+\infty}\int_{-\infty}^{+\infty}\int_{-\infty}^{+\infty} q_c(\underline{\mathbf{R}}', \omega)G(\underline{\mathbf{R}} - \underline{\mathbf{R}}', \omega)\,\mathrm{d}^3\underline{\mathbf{R}}' \quad , \tag{2}$$

where, via application of a Fourier transform with respect to time according to

$$\Phi_s(\underline{\mathbf{R}}, \omega) = \int_{-\infty}^{\infty} \Phi(\underline{\mathbf{R}}, t)\mathrm{e}^{j\omega t}\,\mathrm{d}t \quad , \tag{3}$$

the scalar free space time harmonic Green function

$$G(\underline{\mathbf{R}} - \underline{\mathbf{R}}', \omega) = \frac{e^{jk_0|\underline{\mathbf{R}} - \underline{\mathbf{R}}'|}}{4\pi|\underline{\mathbf{R}} - \underline{\mathbf{R}}'|} \tag{4}$$

comes in; $k_0 = \omega/c_0$ is the wave number. Three different types of scatterers can be distinguished through proper selection of the equivalent volume sources. Choosing a (lossless) penetrable scatterer with interior wave speed $c(\underline{\mathbf{R}})$, we have [1]

$$q_c^{\text{pen}}(\underline{\mathbf{R}}, \omega) = -k_0^2 O(\underline{\mathbf{R}}) \Phi(\underline{\mathbf{R}}, \omega) \tag{5}$$

with the object function $O(\underline{\mathbf{R}})$ defined by

$$O(\underline{\mathbf{R}}) = \left[1 - \frac{k^2(\underline{\mathbf{R}})}{k_0^2}\right] \Gamma(\underline{\mathbf{R}}) \tag{6}$$

and the characteristic function $\Gamma(\underline{\mathbf{R}})$ of the volume V_c

$$\Gamma(\underline{\mathbf{R}}) = \begin{cases} 0 & \text{for } \underline{\mathbf{R}} \notin V_c \\ 1 & \text{for } \underline{\mathbf{R}} \in V_c \end{cases} \quad . \tag{7}$$

If we consider a perfect scatterer with an either acoustically soft (s) or rigid (r) surface, Dirichlet or Neumann boundary conditions apply yielding

$$q_c^{\text{s}}(\underline{\mathbf{R}}, \omega) = -\gamma(\underline{\mathbf{R}}) \underline{\mathbf{n}} \cdot \nabla \Phi(\underline{\mathbf{R}}, \omega) \tag{8}$$

$$q_c^{\text{r}}(\underline{\mathbf{R}}, \omega) = -\Phi(\underline{\mathbf{R}}, \omega) \nabla \cdot \gamma(\underline{\mathbf{R}}) \underline{\mathbf{n}} \tag{9}$$

with the singular function $\gamma(\underline{\mathbf{R}})$ of the surface S_c with outward unit normal $\underline{\mathbf{n}}$ according to

$$\gamma(\underline{\mathbf{R}}) = -\underline{\mathbf{n}} \cdot \nabla \Gamma(\underline{\mathbf{R}}) \quad . \tag{10}$$

Obviously, all q_c's are dependent upon the *total* field inside or on the surface of the scatterer being responsible for the nonlinearity of the direct as well as the inverse scattering problem. An appropriate linearization reduces this dependence to $\Phi_i(\underline{\mathbf{R}}, \omega)$ putting the *scattered* field to zero in (5), (8) and (9). This requires some specification of the incident field, and in Fig. 1 (top) we consider an example.

Even though nondestructive testing of materials deals with elastic waves in solids, a "scalarization" of elastic wave scattering is often tolerable; an intuitive argument will be given in a subsequent section of this paper. Therefore, our example for an incident field is in terms of a piezoelectric transducer model. Fig. 1 shows a (twodimensional) "specimen" of finite size with a Dirichlet boundary condition, which models a steel block within a scalar approximation. The black bar on the surface represents a finite radiating aperture, where $q(\underline{\mathbf{R}}, t)$ is prescribed in terms of a finite bandwidth pulse, which is also displayed in Fig. 1 (top). The resulting time domain wavefront snapshot has been computed with the so-called AFIT code (Acoustic Finite Integration Technique) [2], which is an acoustic version of the Finite Integration Technique to solve Maxwell's equations numerically [4]. We observe two circular cylindrical wavefronts emanating from the edges of the aperture, and a spatially confined plane wave front as the "geometric-optical" wavefront of the aperture being tangential to the edge wavefronts. By the way, the superposition of these wavefronts as a function of the duration of the exciting pulse determines the Fresnel region of the radiating aperture [5]. Here, we notice that a plane wave approximation of the aperture radiation field might be appropriate to model an incident field, especially for inverse scattering.

In order to get an impression of the data field, which is available for inverse scattering purposes, Fig. 1 (bottom) displays a time domain wavefront snapshot after the aperture radiation field has reached a (twodimensional) strip-like scatterer with Dirichlet boundary condition modeling a crack in a solid. Suppose now, that the transmitting aperture is switched to a receiving mode with the capability of recording the scattered field "pointwise" — say, as an

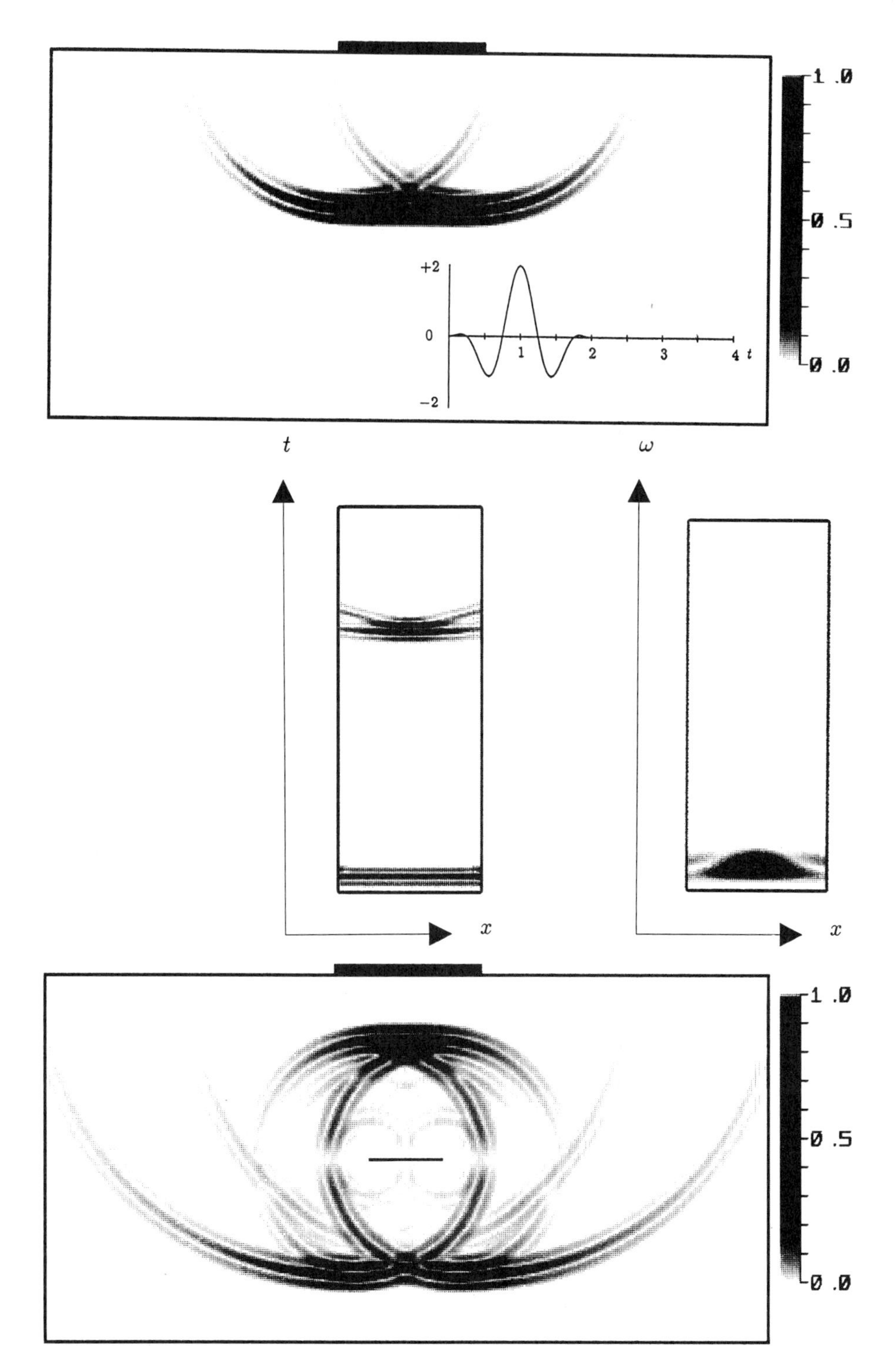

Figure 1. Top: Acoustic radiation from a finite twodimensional aperture located on a surface with a Dirichlet boundary condition. Bottom: Scattering of the above wavefronts by a twodimensional strip with Dirichlet boundary condition. Middle: Corresponding xt-data field as "recorded" within the receiving aperture and pertinent $x\omega$-data (magnitude) as obtained via Fourier transform.

array of receivers — along the linear x-coordinate on the surface of the specimen. The result is the xt-data field displayed in the middle of Fig. 1. This finite aperture finite bandwidth data field is a typical model of the data being available for ultrasonic nondestructive testing. A *linear* recording aperture yields *twodimensional* data allowing for twodimensional images in terms of slices — tomograms — of the specimen being orthogonal to its surface. Typically, 128 aperture samples ("array elements") are recorded versus 256 time samples, which results in 128 Kbytes of data. To obtain *threedimensional* images, *twodimensional* xy-apertures have to be scanned yielding about 16 Mbytes of data for a corresponding experiment. Therefore, limited on-site storage and processing capabilities led to inverse scattering concepts applying "holographic" data reduction.

SINGLE FREQUENCY HOLOGRAPHY

The argument behind holography is as follows. Take Fourier transforms of the xt-data with respect to t for every x to obtain $x\omega$-data, which, for the impulse used in Fig. 1, are centered around a carrier frequency ω_0 (compare Fig. 1, middle). Since the information of *every* t-sample of the original data field for a certain prescribed x enters the (complex) ω_0-sample, one might hope that selecting — or recording — single frequency data only — say, $x\omega_0$-data — produces twodimensional images; correspondingly, $xy\omega_0$-data should produce threedimensional images.

Holographic processing of twodimensional $xy\omega_0$-data is then in terms of the backpropagation version of Huygens' principle. The latter one states, that *outside* a closed measurement surface S_M surrounding the scatterer completely, where the scattered field as well as its normal derivative is available, we can compute $\Phi_s(\underline{\mathbf{R}},\omega)$ in terms of the integral

$$\Phi_s(\underline{\mathbf{R}},\omega) = \int\!\!\int_{S_M} \left[\Phi_s(\underline{\mathbf{R}}',\omega)\frac{\partial G}{\partial n'} - G(\underline{\mathbf{R}}-\underline{\mathbf{R}}',\omega)\frac{\partial \Phi_s}{\partial n'}\right] \mathrm{d}S' \ . \tag{11}$$

Inside S_M the application of (11) yields only $\Phi_s(\underline{\mathbf{R}},\omega) = -\Phi_i(\underline{\mathbf{R}},\omega)$, which says nothing about the scatterer. A remedy consists in the definition of a generalized holographic field $\Theta_H(\underline{\mathbf{R}},\omega)$ for $\underline{\mathbf{R}}$ inside S_M via the backpropagation ansatz [7, 8]

$$\Theta_H(\underline{\mathbf{R}},\omega) = -\int\!\!\int_{S_M} \left[\Phi_s(\underline{\mathbf{R}}',\omega)\frac{\partial G^*}{\partial n'} - G^*(\underline{\mathbf{R}}-\underline{\mathbf{R}}',\omega)\frac{\partial \Phi_s}{\partial n'}\right] \mathrm{d}S' \tag{12}$$

utilizing the complex conjugate G^* of Green's function. For planar measurement surfaces given by, say, the plane $z = 0$ of a cartesian coordinate system, the backpropagation integral (12) is of convolutional type with respect to the aperture variables x and y, and, hence, processing into the half-space $z \leq 0$, where the scatterer resides, is then possible in terms of twodimensional Fourier transforms for every prescribed value of z. A single frequency threedimensional "image" $\Theta_H(x,y,z \leq 0,\omega_0)$ is obtained that way.

An example with synthetic data is given in Fig. 2. Obviously, if — either by chance or by a priori knowledge — we process into *that* plane, where the scatterer resides, a correct image is obtained with lateral resolution of $\pi c_0/\omega_0$ [9]. If this plane is not known a priori, only a blurred image results from single frequency holography; this is also true, if we choose a plane even deeper in the material. Single frequency holography has only lateral, but no axial resolution, even if it is generalized to a closed measurement surface [6]. Therefore, it is not appropriate to image arbitrarily shaped scatterers.

SYNTHETIC APERTURE FOCUSING TECHNIQUE (SAFT)

The investigation of twodimensional single frequency holography told us, that we have to cope with the complete threedimensional xyt-data field in order to solve the threedimensional

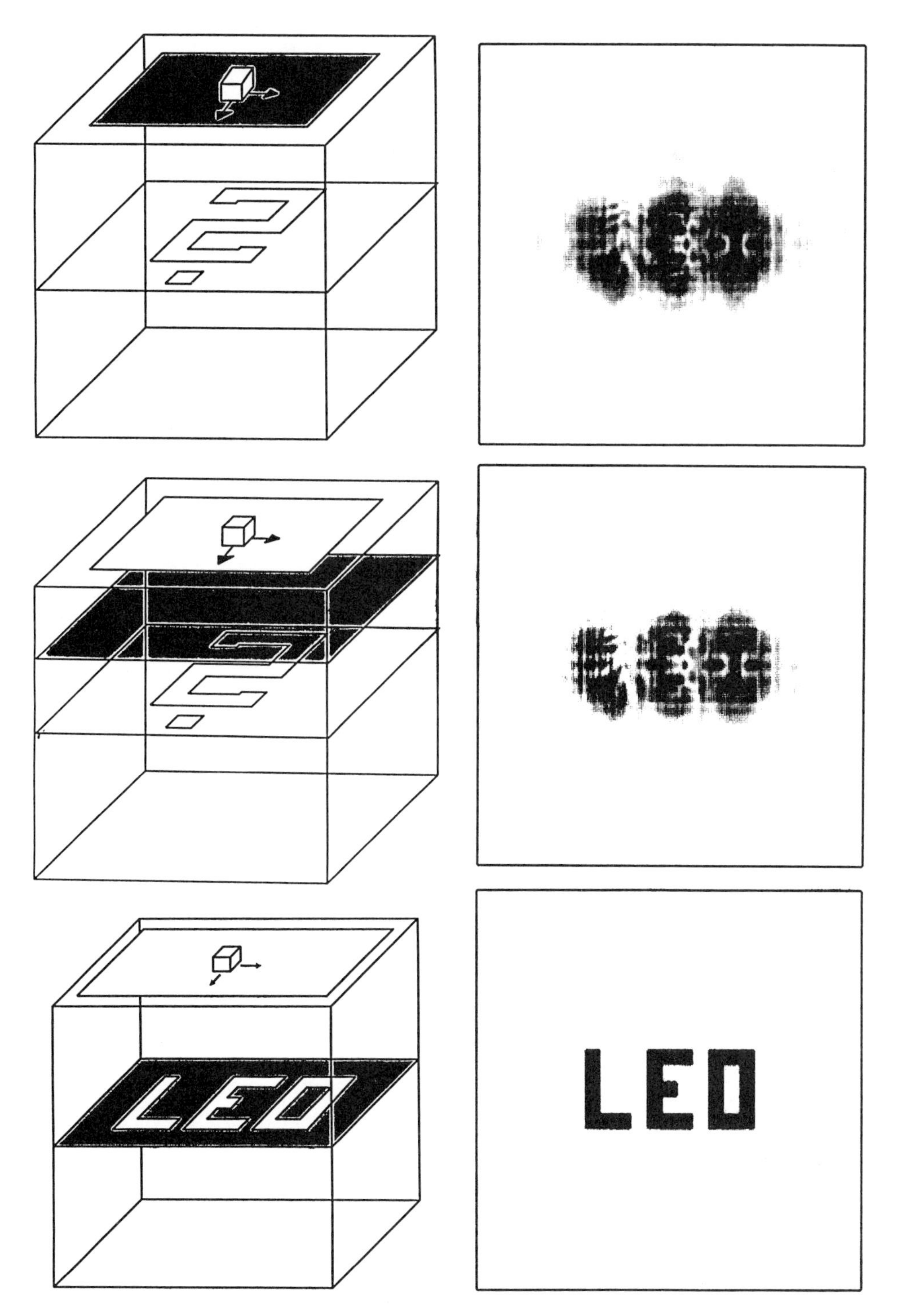

Figure 2. Holographic processing of twodimensional single frequency synthetic data obtained by plane wave scattering from the letters "LEO", which have been chosen as zero-thickness scatterers in honor of Professor Leo B. Felsen. Top: Recorded magnitude of the complex valued hologram from the still unknown scatterer. Middle: Holographic backpropagation into a plane below the surface, which does not contain the scatterer. Bottom: Holographic backpropagation into *that* plane, where the scatterer resides.

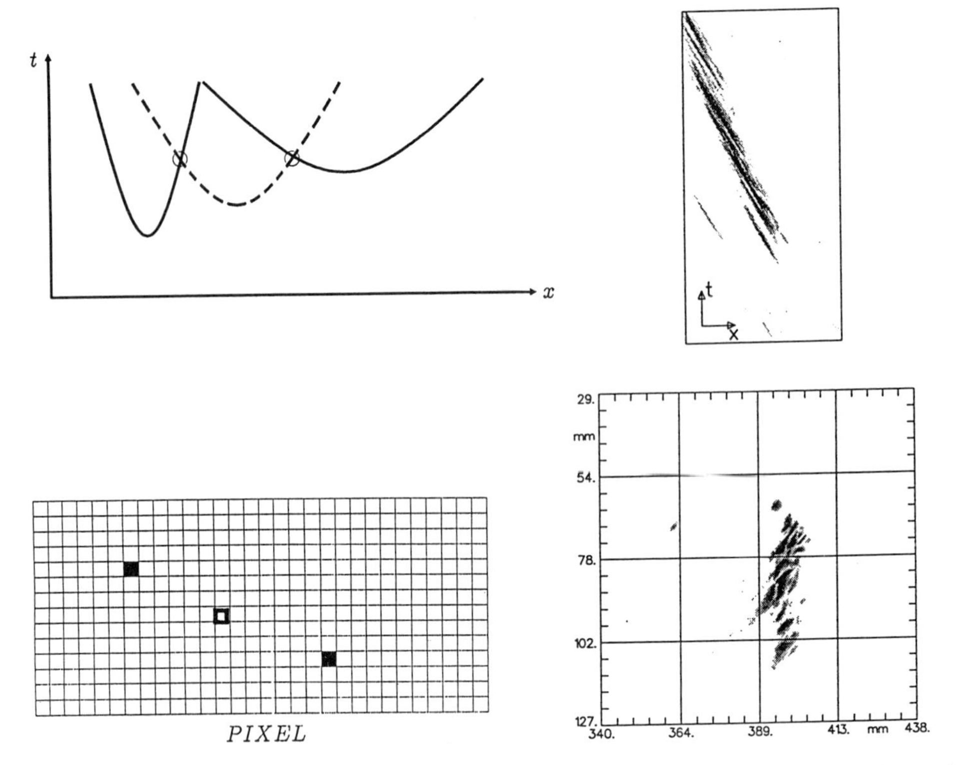

Figure 3. Synthetic Aperture Focusing Technique. Left: Principle. Right: Experimental Result.

inverse scattering problem. The considerable amount of data to be processed stimulated the intuitive proposition of an algorithm named L-SAFT for Synthetic Aperture Focusing Technique in an L-version [10] for "linear", i.e. a linear aperture scan along x yielding xt-data is used as input into a heuristic focusing technique with tomograms as output, *even if the scatterer is threedimensional.* This should at least provide — apart from the axial resolution — lateral resolution in one direction, namely in the plane of the tomogram.

The simplest threedimensional scattering geometry is a point scatterer at some spatial point $\underline{\mathbf{R}}_0$ within a solid; let $\underline{\mathbf{R}}_0$ be given by its cartesian coordinates $x_1 < x_0 < x_2, y_0, z_0 < 0$. Considering the pertinent xt-data field obtained by a scan along x between x_1 and x_2 for $z = 0$ and $y = y_0$ provides us with a phenomenological, but very intuitive algorithm. Therefore, we postulate an equivalent volume source in terms of a delta-function

$$q_c(\underline{\mathbf{R}}, t) = \delta(\underline{\mathbf{R}} - \underline{\mathbf{R}}_0)\delta(t) \quad , \tag{13}$$

which is switched on at $t = 0$ as a delta-pulse $\delta(t)$. This source yields the following "scattered" field in the time domain

$$\Phi_s(t) = \frac{\delta\left(t - \frac{1}{c_0}|\underline{\mathbf{R}} - \underline{\mathbf{R}}_0|\right)}{4\pi|\underline{\mathbf{R}} - \underline{\mathbf{R}}_0|} \quad . \tag{14}$$

Within the linear synthetic aperture, due to the δ-function behavior of $\Phi_s(\underline{\mathbf{R}}, t)$, nonzero xt-data are only obtained for

$$c_0 t = |\underline{\mathbf{R}} - \underline{\mathbf{R}}_0| \quad . \tag{15}$$

This is a hyperbola in data space.

Fig. 3 (left) exhibits the data space for two "physically real" point scatterers, indicated by the solid squares. Dividing the spatial region to be imaged into pixels, we can define a hypothetic pixel scatterer — given by the empty square — with a pertinent hypothetic dashed "data" hyperbola. Phenomenological algorithmic imaging exploits the following idea: imagine, the hypothetic pixel is a physically real pixel; we would then have to focus all the data on the pertinent dashed hyperbola into that pixel , but there are no data — the pixel being only a hypothetic scatterer — except at the two points indicated by the circles, where the hypothetic hyperbola intersects the real data hyperbolas. The same is true for *all* possible hypothetic pixels, which can be algorithmically checked one after each other with a computer. Only when the moving "computer pixel" meets physically real scatterers, data summation is over real data hyperbolas, yielding high amplitudes at those image space pixels versus neglectable amplitudes at all hypothetic pixels. Thus, an "image" is obtained via a synthetic aperture focusing technique. Fig. 3 (right) demonstrates that this procedure really works; experimental xt-data have been recorded from a crack oriented vertically with respect to the linear synthetic aperture — requiring 45°-degree shear wave excitation — yielding a SAFT-image with a lateral resolution depending mainly upon the center frequency of the pulse, the size of the transducer, and the size of the aperture. Even though numerous experiments have demonstrated the feasability of SAFT for practical applications [10], the fundamental question remains as to what degree and under what assumptions the image of Fig. 3 is "quantitative", i.e. reveals the "real" defect.

SAFT VERSUS DIFFRACTION TOMOGRAPHY

The generalized holographic field as defined in (12) provides an equation to compute a field quantitiy *inside* the measurement surface; the question is now, how it relates to the equivalent volume sources representing the scatterer. The answer is given in terms of an integral equation associated with the names of Porter and Bojarski [9], which can be derived applying Green's theorem to (12) and recognizing (2). It is of convolutional type and involves the imaginary part G_I of the free space Green function as kernel

$$\Theta_H(\underline{\mathbf{R}},\omega) = 2j \int_{-\infty}^{+\infty}\int_{-\infty}^{+\infty}\int_{-\infty}^{+\infty} q_c(\underline{\mathbf{R}}',\omega) G_I(\underline{\mathbf{R}}-\underline{\mathbf{R}}',\omega)\, \mathrm{d}^3\underline{\mathbf{R}}' \ . \tag{16}$$

The advantage of (16) is the accessability of Θ_H throughout *all* space, the disadvantage is, that it cannot simply be solved by deconvolution, because of the distributional properties of the multidimensional spatial Fourier transform of $G_I(\underline{\mathbf{R}},\omega)$, denoted by $\tilde{G}_I(\underline{\mathbf{K}},\omega)$, the vector $\underline{\mathbf{K}}$ being the Fourier vector variable corresponding to $\underline{\mathbf{R}}$ according to

$$\tilde{G}_I(\underline{\mathbf{K}},\omega) = \int_{-\infty}^{+\infty}\int_{-\infty}^{+\infty}\int_{-\infty}^{+\infty} G_I(\underline{\mathbf{R}},\omega) \mathrm{e}^{-j\underline{\mathbf{K}}\cdot\underline{\mathbf{R}}}\, \mathrm{d}^3\underline{\mathbf{R}} \ . \tag{17}$$

As a matter of fact, we obtain

$$\tilde{\Theta}_H(\underline{\mathbf{K}},\omega) = \frac{j\pi}{k_0}\tilde{q}_c(\underline{\mathbf{K}},\omega)\delta(K-k_0) \tag{18}$$

where $K = |\underline{\mathbf{K}}|$; (18) reveals, that, for a single frequency ω, $\tilde{\Theta}_H(\underline{\mathbf{K}},\omega)$ is nonzero only on a sphere of radius k_0 in spatial Fourier space, the so-called Ewald-sphere. Fortunately, data outside the Ewald-sphere *are* available on behalf of our time domain experiment, which ensures ω to vary within the bandwidth of the exciting ultrasound. Hence, sweeping frequency in (18) should enable us to invert $\tilde{\Theta}_H(\underline{\mathbf{K}},\omega)$ from a finite support in $\underline{\mathbf{K}}$-space to yield the equivalent volume sources explicitly. Unfortunately, as we already stated earlier, all the q_c's are dependent upon the *total* field $\Phi(\underline{\mathbf{R}},\omega)$ localized in or on the scatterer. Hence, their variation with frequency is unknown; the complete spatial structure of $\Phi(\underline{\mathbf{R}},\omega)$ — as a function of $\underline{\mathbf{R}}$ and, therefore, as a function of $\underline{\mathbf{K}}$ for its Fourier transform — is changed if ω is changed. This tells us that frequency sweeping yields an uncontrollable variation of $q_c(\underline{\mathbf{R}},\omega)$, or $\tilde{q}_c(\underline{\mathbf{K}},\omega)$; the inverse problem is nonlinear with respect to the equivalent sources. Physically, this is due to the fact, that local scattering centers on the defect are not independent of each other but interact via radiation.

Until today, to end up with fast algorithms, the only remedy to the nonlinear inversion problem is its *linearization*. Here, essentially the first order Born approximation for the penetrable scatterer, or the Kirchhoff or physical optics approximation for the perfect scatterer come in. For simplicity, let us discuss the case of weak scattering by a penetrable scatterer being associated with the name of Born. Weak scattering applies intuitively if the scatterer is present in the real world to produce a scattered field *outside*, but simultaneously, it is absent as not to disturb the incident field while propagating through it, i.e. the scattered field is zero *inside*. Mathematically, the Born approximated scattered field is the first term in its Neumann series expansion. Therefore, instead of (5) we put

$$q_c(\underline{\mathbf{R}},\omega) \simeq -k_0^2 O(\underline{\mathbf{R}})\Phi_i(\underline{\mathbf{R}},\omega) \quad , \tag{19}$$

which obviously requires some specification concerning the incident field now in order to be inserted into (18). The previous discussion of ultrasonic transducer models has revealed that plane wave representations of incident fields are quite appropriate, i.e. we assume — $\hat{\underline{\mathbf{k}}}$ being the unit-vector of propagation —

$$\Phi_i(\underline{\mathbf{R}},\omega) = \Phi_0(\omega)e^{jk_0\hat{\underline{\mathbf{k}}}\cdot\underline{\mathbf{R}}} \quad , \tag{20}$$

where $\Phi_0(\omega)$ is the frequency spectrum of the excitation (compare Fig. 1, top). After some mathematical manipulation [9], insertion of (20) into (18) finally yields

$$O(\underline{\mathbf{R}}) = -\frac{2}{\pi}\Re\int_0^\infty \frac{1}{k_0^2\Phi_0(\omega)}\hat{\underline{\mathbf{k}}}\cdot\nabla\left[\Theta_H(\underline{\mathbf{R}},\omega)\mathrm{e}^{-jk_0\hat{\underline{\mathbf{k}}}\cdot\underline{\mathbf{R}}}\right]\,\mathrm{d}k_0 \quad , \tag{21}$$

where the k_0-integration accounts for the frequency sweeping in our experiment. The above equation is an algorithmic recipe of data processing to obtain an image of the scatterer in terms of its object function. This recipe is quantitative in the sense, that any approximations and assumptions, which have been made to formulate it, are well defined. These comprise:

- "Scalarization" of the elastic wave field in a solid handling it as a truly acoustic wave field
- Linearization of the inverse scattering problem treating the scatterer as only weak
- Plane wave illumination
- Deconvolution with the frequency spectrum $\Phi_0(\omega)$ as it appears in the denominator of (21)
- Closed measurement surface.

By the way, the corresponding linearization for the perfectly soft (rigid) scatterer, simulating a void or a crack, is in terms of the physical optics (Kirchhoff) approximation, which says, that the equivalent surface sources are proportional to the incident field on the illuminated side being zero on the dark side, as it is *strictly* true for plane surfaces only [11].

Of course, any violation of the above-mentined items will cause degradations and blurring of the image, but with equation (21), in contrast to phenomenological algorithmic SAFT-imaging, one is able to study these influences quantitatively.

Surprisingly enough, the backpropagation ansatz of (12) comprises the exact mathematical formulation of the phenomenological SAFT algorithm. This can be seen considering the k_0-integral in (21) as an inverse Fourier integral with regard to frequency for $t = 0$; the result is an *exact* time domain backpropagation algorithm [9], where the introduction of certain additional approximations yields the algorithmic procedure as described in terms of Fig. 3.

Hence, quantitative algorithmic imaging based on inverse scattering theory has by now two advantages: it provides quantitative algorithms, and it contains phenomenological imaging as a special well defined case. Furthermore, and this might be the most important feature, (21) can be very effectively computed, at least when planar or circular cylindrical measurement surfaces are involved, which is often the case. This is readily seen, if a particular representation of the generalized holographic field in terms of the far-field scattering amplitude $H(\underline{\hat{\mathbf{R}}},\omega)$ is introduced. The latter one is defined by (2) for $R \gg R'$ and $k_0 R \gg 1$ through

$$\Phi_s^{\text{far}}(\underline{\mathbf{R}},\omega) = H(\underline{\hat{\mathbf{R}}},\omega)\frac{e^{jk_0R}}{R} \tag{22}$$

with

$$H(\underline{\hat{\mathbf{R}}},\omega) = \frac{1}{4\pi}\int_{-\infty}^{+\infty}\int_{-\infty}^{+\infty}\int_{-\infty}^{+\infty} q_c(\underline{\mathbf{R}}',\omega)\mathrm{e}^{-jk_0\underline{\hat{\mathbf{R}}}\cdot\underline{\mathbf{R}}'}\,\mathrm{d}^3\underline{\mathbf{R}}' \ , \tag{23}$$

where $\underline{\hat{\mathbf{R}}} = \underline{\mathbf{R}}/R$. Notice, (23) says that the scattering amplitude is related to the spatial Fourier transform of the equivalent sources on the Ewald-sphere, i.e. (23) can be rewritten as

$$H(\underline{\hat{\mathbf{R}}},\omega) = \frac{1}{4\pi}\tilde{q}_c(\underline{\mathbf{K}} = k_0\underline{\hat{\mathbf{R}}},\omega) \ , \tag{24}$$

because $\underline{\mathbf{K}}$ is restricted to a sphere with radius k_0 via $\underline{\mathbf{K}} = k_0\underline{\hat{\mathbf{R}}}$.

From (12) we can compute the following representation of the generalized holographic field

$$\Theta_H(\underline{\mathbf{R}},\omega) = \frac{jk_0}{2\pi}\int\int_{S^2} H(\underline{\hat{\mathbf{R}}}',\omega)\mathrm{e}^{jk_0\underline{\hat{\mathbf{R}}}'\cdot\underline{\mathbf{R}}}\,\mathrm{d}^2\underline{\hat{\mathbf{R}}}' \tag{25}$$

for $\underline{\mathbf{R}}$ taken in the far-field [1]; here, S^2 denotes the unit-sphere. Insertion into the imaging recipe (21) yields the triple integral

$$\int\int_{S^2}\int_0^{\infty} H(\underline{\hat{\mathbf{R}}}',\omega)\underline{\hat{\mathbf{k}}}\cdot(\underline{\hat{\mathbf{R}}}' - \underline{\hat{\mathbf{k}}})\mathrm{e}^{jk_0(\underline{\hat{\mathbf{R}}}'-\underline{\hat{\mathbf{k}}})\cdot\underline{\mathbf{R}}}\,\mathrm{d}k_0\mathrm{d}^2\underline{\hat{\mathbf{R}}}' \ , \tag{26}$$

which, due to the exponentials, looks very much like a spatial inverse Fourier integral, provided the Fourier vector $\underline{\mathbf{K}}$ is defined by

$$\underline{\mathbf{K}} = k_0(\underline{\hat{\mathbf{R}}}' - \underline{\hat{\mathbf{k}}}) \ . \tag{27}$$

Computation of the required Jacobian results in

$$\mathrm{d}^3\underline{\mathbf{K}} = -k_0^2\underline{\hat{\mathbf{k}}}\cdot(\underline{\hat{\mathbf{R}}}' - \underline{\hat{\mathbf{k}}})\,\mathrm{d}k_0\mathrm{d}^2\underline{\hat{\mathbf{R}}}' \ , \tag{28}$$

making it possible to process (26) with threedimensional FFT-techniques. Hence, $O(\underline{\mathbf{R}})$ can be effectively computed as soon as $H(\underline{\hat{\mathbf{R}}}',\omega)$, i.e. far-field data, is available. This might a priori not be the case; then, from (11) we find

$$H(\underline{\hat{\mathbf{R}}},\omega) = -\frac{1}{4\pi}\int\int_{S_M}\left[\frac{\partial\Phi_s}{\partial n'} + jk_0\underline{\mathbf{n}}'\cdot\underline{\hat{\mathbf{R}}}\Phi_s(\underline{\mathbf{R}}',\omega)\right]\mathrm{e}^{-jk_0\underline{\hat{\mathbf{R}}}\cdot\underline{\hat{\mathbf{R}}}'}\,\mathrm{d}S' \ , \tag{29}$$

which is a computational scheme to obtain the scattering amplitude from measurements on an arbitrary surface S_M. As a matter of fact, for planar and circular cylindrical surfaces, FFT-techniques can be used as well to evaluate (29) numerically [11, 3]. First of all, this results in extremely fast algorithmic schemes making threedimensional imaging possible, and, as a byproduct, the normal derivative of the scattered field on the measurement surface is not required in terms of explicit measurements as it can be computed from the field itself [12]. From a conceptual viewpoint, the mapping of the "data" $H(\underline{\hat{\mathbf{R}}}',\omega)$ into $\underline{\mathbf{K}}$-space via (27) is a procedure, which closely resembles algorithms applied for X-ray computer tomography, except that diffraction effects of the acoustic wave field are properly accounted for. Therefore, quantitative algorithmic imaging exploiting Fourier transform relationships explicitly has been named diffraction tomography; here, for obvious reasons, we call it F(ourier)T(ransform)-SAFT.

Fig. 4 (left) shows an example for threedimensional FT-SAFT imaging. A T-shaped tube in a solid is displayed via geometry rendering, and, simultaneously, an image obtained from an xyt-data field taken on the planar top surface of the pertinent specimen is added as an isocontour surface. A clear indication of the top surface of the tube as it is "visible" from the transducer is obvious; additional crack fields close to the tube have been imaged, making it possible to evaluate their size, shape and orientation nondestructively in a quantitative manner.

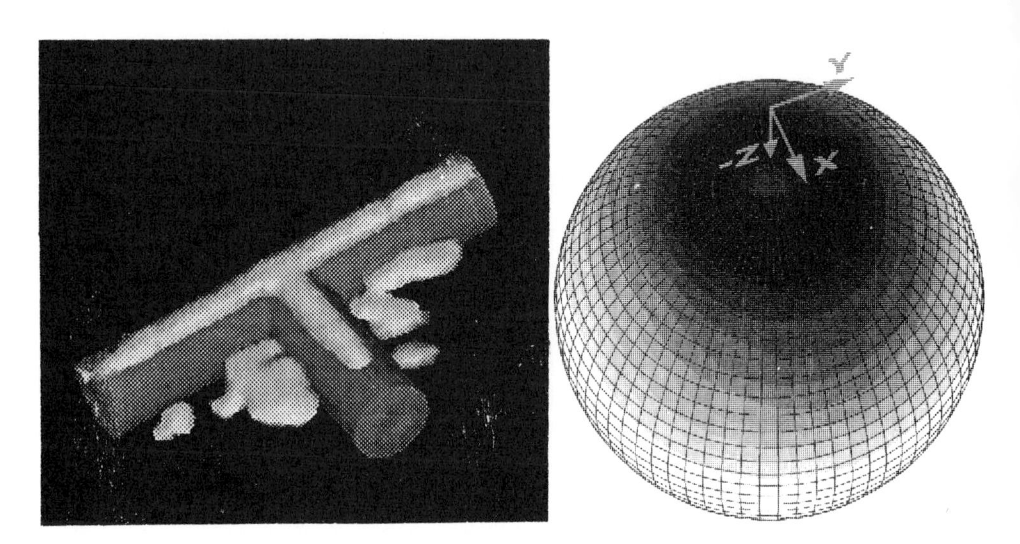

Figure 4. Left: Experimentally obtained FT-SAFT image of cracks close to a T-shaped tube in a solid. Right: Surface current reconstruction on a perfectly conducting sphere as obtained from polarimetric synthetic data.

ULTRASONIC DATA FIELDS FOR ELASTIC WAVES

Let us repeat the model computation of Fig. 1, but this time for a real solid with nonzero shear. Our numerical EFIT code [13] provides us with the results of Fig. 5. Typically, the pressure waves of Fig. 1 remain nearly unchanged; instead, they are now supplemented by circular cylindrical shear waves emanating from the edges of the aperture with about one half of the pressure wave speed, together with plane so-called head wavefronts to fulfill the boundary condition on the surface of the specimen. In addition, Rayleigh surface wave pulses being somewhat slower than the shear waves emanate from both aperture edges. Fig. 5 (bottom) also shows, how these waves are scattered by a crack in a solid, and obviously, the mess of wavefronts is still increased, since every pressure as well as every shear wave is not only scattered but also mode-converted into each other. As a consequence, the xt-data field, even though rather "crowded", contains a lot of information beyond simple scalar acoustic scattering; this is all the more true, if more complicated scattering geometries are under concern. Interesting enough, a full elastodynamic diffraction tomography can be formulated along the same ideas as in the previous section [14], which allows to use all the information of all the wavefronts properly.

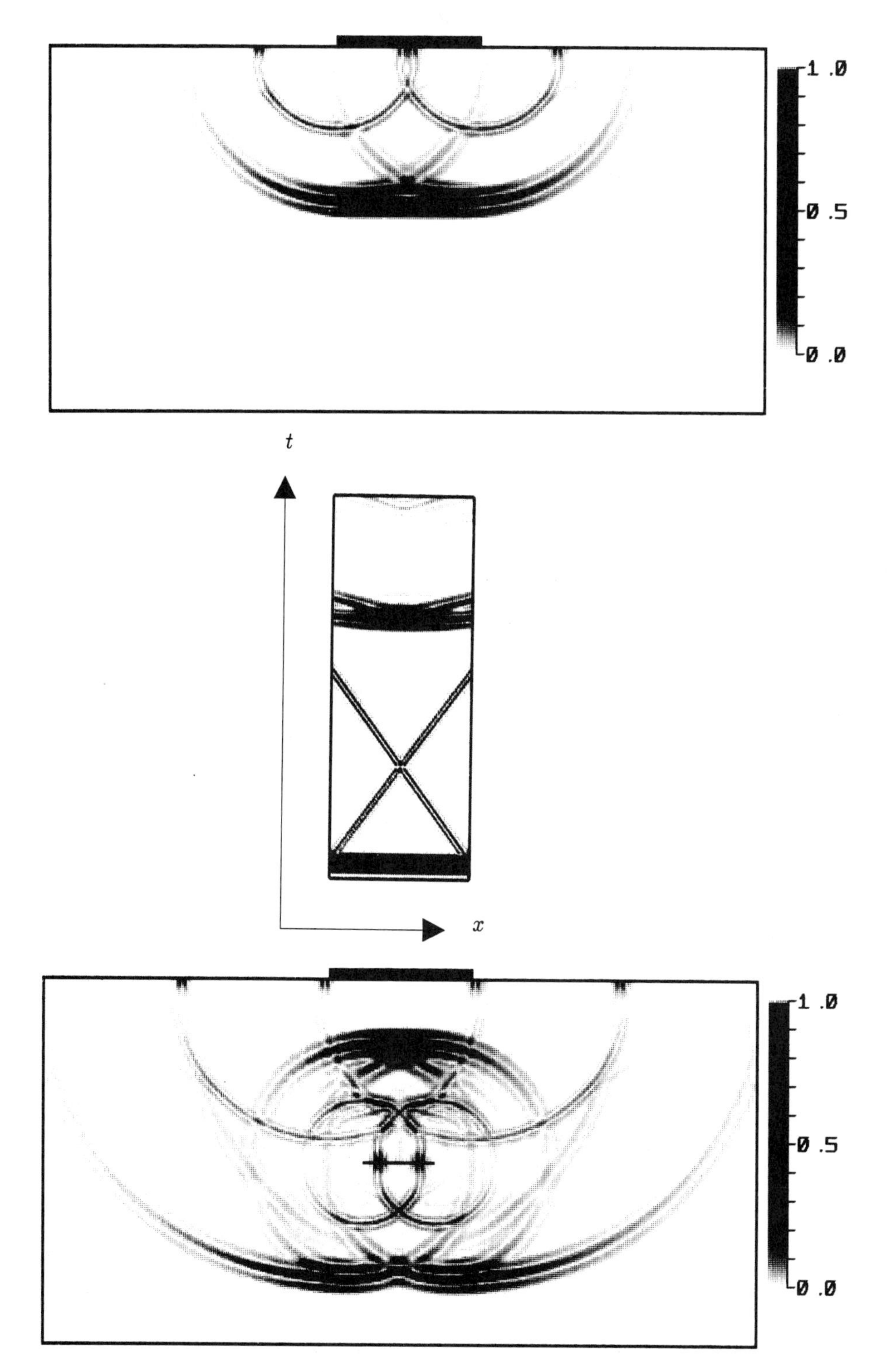

Figure 5. Top: Elastic waves radiated from a finite twodimensional aperture located on the surface of a solid with stress-free boundary. Bottom: Scattering of the above wavefronts by a twodimensional strip with stress-free faces. Middle: Corresponding xt-data field as "recorded" within the receiving aperture.

ELECTROMAGNETIC INVERSE SCATTERING

Electromagnetic wave scattering results in depolarization; this information can be properly processed in terms of a dyadic backpropagation scheme based on the vector version of the Porter-Bojarski integral equation. The result is an explicit equation for the surface current, for instance, on a perfectly conducting target, where the underlying model is again the physical optics approximation [15]; an example is given in Fig. 4, right.

REFERENCES

[1] K.J. Langenberg: Introduction to the Special Issue on Inverse Problems, Wave Motion **11** (1989) 99-112

[2] R. Marklein, P. Fellinger: Mathematisch-numerische Modellierung der Ausbreitung und Beugung akustischer Wellen, Seminar "Modelle und Theorien für die Ultraschallprüfung" der Deutschen Gesellschaft für zerstörungsfreie Prüfung , Berlin 1990

[3] T. Kreutter, S. Klaholz, A. Brüll, J. Sahm, A.Hecht: Optimierung und Anwendung eines schnellen Abbildungsalgorithmus für die Schmiedewellenprüfung, ibid.

[4] T. Weiland: On the Numerical Solution of Maxwell's Equations and Applications in the Field of Accelerator Physics, Particle Accelerators **15** (1984)

[5] U. Aulenbacher, K.J. Langenberg: Analytical Representation of Transient Ultrasonic Phsed-Array Near- and Far-Fields, J. Nondestr. Eval. **1** (1980) 53

[6] K.J. Langenberg, M. Fischer, M. Berger, G. Weinfurter: Imaging Performance of Generalized Holography, J. Opt. Soc. Am. **3** (1986) 329

[7] R.P. Porter: Diffraction-Limited Scalar Image Formation with Holograms of Arbitrary Shape, J. Opt. Soc. Am. **60** (1970) 1051

[8] N.N. Bojarski: Exact Inverse Scattering Theory, Radio Science **16** (1981) 1025

[9] G.T. Herman, H.K. Tuy, K.J. Langenberg. P. Sabatier: *Basic Methods of Tomography and Inverse Problems*, Adam Hilger, Bristol 1987

[10] V. Schmitz, W. Müller, G. Schäfer: Practical Experiences with L-SAFT, in: *Review of Progress in Quantitative Nondestructive Evaluation*, Eds.: D.O. Thompson, D.E. Chimenti, Plenum Press, New York 1986

[11] K. Mayer, R. Marklein, K.J. Langenberg, T. Kreutter: Threedimensional Imaging System based on Fourier Transform Synthetic Aperture Focussing Technique, Ultrasonics **28** (1990) 241-255

[12] A.J. Devaney, G. Beylkin: Diffraction Tomography Using Arbitrary Transmitter and Receiver Surfaces, Ultrasonic Imaging **6** (1984) 181

[13] P. Fellinger, K.J. Langenberg: Numerical Techniques for Elastic Wave Propagagtion and Scattering, in: *Elastic Waves and Ultrasonic Nondestructive Evaluation*, Eds.: S.K. Datta, J.D. Achenbach, Y.S. Rajapakse, North-Holland, Amsterdam 1990

[14] K.J. Langenberg, T. Kreutter, K. Mayer, P. Fellinger: Inverse Scattering and Imaging, ibid.

[15] K.J. Langenberg, M. Brandfaß, P. Fellinger, T. Gurke, T. Kreutter: A Unified Theory of Multidimensional Electromagnetic Vector Inverse Scattering within the Kirchhoff or Born Approximation, in: Vector Inverse Methods in Radar Target Imaging, Ed.: H. Überall, Springer, Berlin 1991 (to appear)

SELF-FOCUSING REVISITED: SPATIAL SOLITONS, LIGHT BULLETS AND OPTICAL PULSE COLLAPSE

Y. Silberberg, J. S. Aitchison, A. M. Weiner, D. E. Leaird,
M. K. Oliver, J. L. Jackel, and P.W.E. Smith

Bellcore
331 Newman Springs Road
Red Bank, New Jersey 07701-7040

Self-focusing and self-trapping is a subject which is almost as old as the field of nonlinear optics. It is well-known that self-trapping in an optical Kerr medium is unstable in three dimensions, but it is stable when diffraction is limited to one transverse dimension, such as in a planar waveguide. Recently we have looked, both experimentally and theoretically, into several aspects of self-action of intense optical fields in nonlinear media. In this talk we will stress some of the open questions that pose a challenge for numerical works in the field.

First, we will report on recent experiments that demonstrate the formation of self-trapped beams in planar glass waveguides. We have demonstrated that a beam that diffracts at low powers is self-trapped at high powers. The self-trapped beam can be described as a spatial soliton, a beam that retains its spatial profile as it propagates along the waveguide. We have also observed interactions between two bright spatial solitons. When the power levels are increased above those required to form fundamental spatial solitons, we have observed the break-up of the beam into a triply-peaked profile. This phenomenon is explained as the result of weak two-photon absorption in the waveguide. We will show through numerical modeling how two-photon absorption leads to the breakup of spatial and temporal solitons.

The second part of the talk will examine self-focusing with ultrashort pulses when temporal reshaping effects due to dispersion are comparable to the spatial focusing effects. We show theoretically that in the presence of dispersion a properly constructed pulse could undergo spatial-temporal collapse to generate extremely short and intense pulses of light. We will discuss the condition for such a process, and the effects that are expected to limit it. Light bullets - pulses that propagate without changing their spatial or temporal shape - are possible if saturation of the nonlinear processes can be obtained. There are many unanswered questions which are related to these processes. Most of them can be answered only by numerical modeling of this complex spatial-temporal process.

PARAMETRIC EXCITATION OF WHISTLER WAVES BY CIRCULARLY POLARIZED ELECTROMAGNETIC PUMPS IN A NONUNIFORM MAGNETOPLASMA: MODELING AND ANALYSIS

S.P. Kuo

Weber Research Institute
Polytechnic University
Farmingdale, New York 11735

Robert J. Barker

Air Force Office of Scientific Research
Bolling Air Force Base
Washington, DC 20332-6448

ABSTRACT

Possible generation of whistler waves by Tromsϕ HF heater in an overdense plasma having a density profile increasing linearly with altitude is investigated. Two coupled mode equations are derived and combined, after Laplace-transformed in time, into an equation in the form of the Schrödinger equation. Thus, the threshold field and growth rate of the instability can be determined by the eigenvalues of the eigenvalue equation defined by the Bohr-Sommerfeld quantization condition. It is shown that the HF heater wave of both R-H and L-H polarizations can parametrically decay into whistler wave and a Langmuir sideband. The Langmuir sideband is an anti-Stokes (i.e., frequency upshifted) line in the case of the R-H (o-mode) pump, while it is a Stokes (i.e., frequency downshifted) line when an L-H (x-mode) pump is employed. Since whistler waves have a broad range of frequency, the simultaneously excited Langmuir waves can have a much broader frequency bandwidth than those excited by the parametric decay instability and can be either frequency-upshifted or downshifted.

INTRODUCTION

In the Arecibo HF heating experiments, the backscatter spectrum of the 430 MHz UHF radar is usually enhanced at frequencies near 430 MHZ$\pm f_p$, [Showen and Kim, 1978] where f_p is the frequency of heater-excited plasma wave width which is slightly less than the frequency f_{HF} of the HF heater by the frequency of ion acoustic wave and its multiple. The radar returns at these two sidebands are often referred to as HF-enhanced plasma lines (HFPLs). Similar spectral feature (HFPLs) is also detected by the EISCAT 224 MHz VHF radar during the ionospheric HF-modification experiments carried out near Tromsϕ, Norway [Hagfors et al., 1983]. However, such a spectral feature displaced from the radar frequency by slightly less than the heater frequency is still undetectable by the EISCAT 933 MHz UHF radar. It is believed that the excessive Landau damping impedes the parametric excitation of plasma waves and ion acoustic waves at twice the wavenumber of 933 MHz radar signal.

On the other hand, a new spectral feature in the backscatter spectrum of EISCAT 933 MHz radar was observed throughout most of the observing period of the heating experiments performed with o-mode heater transmitting near Tromsϕ, Norway on Aug. 16-18, 1986. The radar returns were enhanced at frequencies offset from the radar frequency by a frequency a few hundred KHz more than the heater frequency of 4.04 MHz [Isham et al., 1989]. Moreover, running alternately with the radar in the chirped and in the unchirped mode it was shown that the enhanced plasma lines seemed to emanate from very localized regions. Similar phenomenon was also observed in Gorky's heating experiments [Frolov, 1990].

In addition to the several well-studied parametric processes including the decay of HF heater into a Langmuir wave and an ion acoustic wave [Kuo and Cheo, 1978], the filamentation instability of HF heater [Kuo and Schmidt, 1983; Kuo and Lee, 1983], and Langmuir sideband [Kuo et al., 1983] and parametric excitation of plasma modes at upper hybrid resonance [Kuo and Lee, 1982], the parametric decay of HF heater wave into a Langmuir wave and a whistler wave has also been shown to be a viable process occurring in EISCAT's heating experiments [Kuo and Lee, 1989]. In that work, a left-handed (L-H) circularly polarized heater wave (x-mode) propagating upward along a downward geomagnetic field in an overdense ionosphere has been considered and the excited Langmuir sideband has a frequency less than that of the heater by the frequency of the concomitantly excited whistler wave. The purpose of that work is to explore a viable mechanism of VLF/LF (whistler) wave generation in the ionosphere by HF heater for potential telecommunication applications.

In this paper, a unified analysis of two similar parametric instability processes which excite the Langmuir wave together with whistler waves by either L-H (x-mode) or R-H (o-mode) HF heater [Kuo and Barker, 1990] is presented. In the case of R-H heater, the pump (heater) and the decay mode (whistler wave) are polarized in opposite sense, and hence, the frequency of the excited Langmuir wave will be shown to be at the sum (instead of the difference) of the heater frequency and the concomitantly excited whistler frequency, i.e., Langmuir sideband is an anti-Stokes line. Since the frequency of whistler wave is in the range of a few hundred KHz, the excited Langmuir wave (anti-Stokes line) having a frequency larger than the heater frequency by a few hundred KHz as observed in the Tromsø's heating experiments is expected.

THEORY

The Tromsø HF transmitter is steerable to launch heater waves propagating along the earth's magnetic field in either o-mode (R-H) or x-mode (L-H) polarization. In view of these facts, we will present our theory as follows in a simple geometry of propagation, which is appropriate for describing the parametric excitation of whistler waves by HF heater at Tromsø, Norway.

Consider a circularly polarized heater (L-H or R-H) propagating upward along a downward geomagnetic field in an overdense ionosphere. The ionospheric density (n) shown in Fig. 1 is assumed to have a linear profile, increasing with altitudes (z) as $n=n_0(1+z/L)$ where L is the scale size of the background ionospheric inhomogeneity. The heater wave is reflected at a critical height before reaching the ionospheric F peak and becomes a standing wave, whose energy follows an Airy function distribution illustrated in Fig. 2.

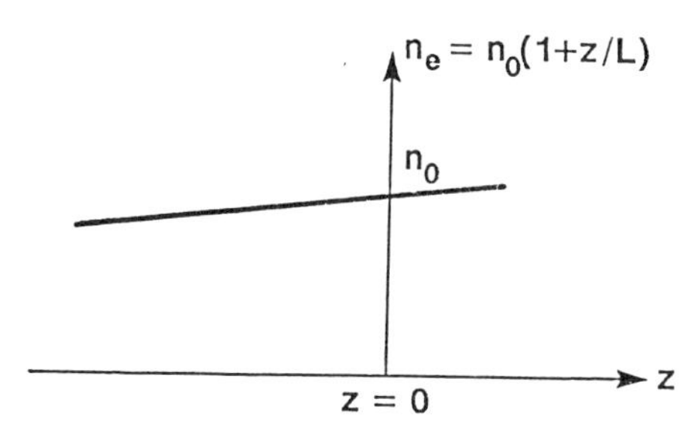

Fig. 1 The ionospheric electron density below F-peak is modelled by a linearly increasing function of altitude. z represents the altitude.

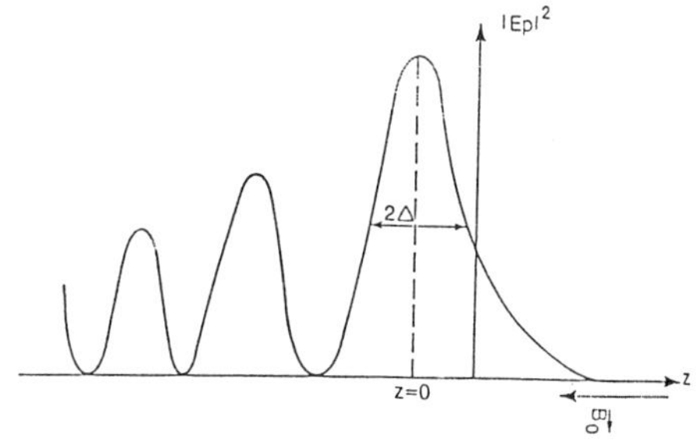

Fig. 2 Standing wave pattern of heater wave near the reflection height.

The first peak of the Airy function is located at z= 0 and its width is denoted by 2Δ. The heater wave field in this region is then modeled by $\vec{E}_{p\pm}=(\hat{x}\pm i\hat{y})\epsilon_p$+c.c., where $\epsilon_p=[\epsilon_0/(1-iz/\Delta)]\ \exp(-i\omega_0 t)$ whose phasor, $\epsilon_0/(1-iz/\Delta)$, is proportional to a complex Lorentzian function, the ± signs refer to the R-H and L-H circular polarization, respectively. The velocity responses of electrons to the heater fields are found to be

$$\vec{V}_{p\pm} = -i(\hat{x}\pm i\hat{y})\, e\epsilon_p/m[(\omega_0\pm\Omega_e) + i\nu_e] + \text{c.c.} = \vec{v}_{p\pm} e^{-i\omega_0 t} + \text{c.c.} \qquad (1)$$

where ν_e is the electron collision frequency.

Through the nonlinearity of the plasma, a beating current driven by the heater wave field in the electron density fluctuations of the Langmuir sidebands $(\omega_\ell, \vec{k}_\ell)$ is generated. This current consists of components of R-H and L-H circular polarizations (with respect to the z-axis) and of sum and difference of the frequencies of the heater and the sideband. The component of the beating current with L-H circular polarization and having the difference frequency ω acts as the source of the whistler wave $(\omega,\vec{k})$. Let $\delta n_{\ell}=\delta\tilde{n}_{\ell}\exp(-i\omega_{\ell}t)+c.c.$ be the electron density fluctuation associated with the Langmuir sideband, this source current density of the whistler wave can be singled out from the total beating current density $\delta\vec{J}_{B\pm}=-e\delta n_{\ell}\vec{V}_{p\pm}$ to be $\delta\vec{J}^{NL}_{\omega+}=-e\delta\tilde{n}_{\ell}\,\vec{V}^{*}_{p+}e^{-i\omega t}+c.c.$ and $\delta\vec{J}^{NL}_{\omega-}=-e\delta\tilde{n}^{*}_{\ell}\,\vec{V}_{p-}e^{-i\omega t}+c.c.$ for R-H and L-H heater waves respectively, where the frequency and wave vector matching relations, $\omega=\pm(\omega_{\ell}-\omega_{o})$ and $\vec{k}(o)=\pm\vec{k}_{\ell}(o)=\hat{z}k$, are satisfied for parametric coupling process. It should be noted that in order to match the polarization of this source current with that of the whistler wave, the frequency matching condition for the R-H pump case requires the frequency of the Langmuir sideband to be upshifted from the heater frequency by the whistler frequency. This is in contrast to the L-H pump case [Kuo and Lee, 1989] that only frequency downshifted Langmuir sideband is excited. Including this source current density, the wave equation of the whistler wave becomes

$$\left(\frac{\partial^2}{\partial t^2}-c^2\frac{\partial^2}{\partial z^2}\right)\delta\vec{E}_{\omega\pm}-4\pi n_o e\frac{\partial}{\partial t}\delta\vec{V}_{\omega\pm}=-4\pi\frac{\partial}{\partial t}\delta\vec{J}^{NL}_{\omega\pm} \tag{2}$$

where $\delta\vec{E}_{\omega\pm}$ are the electric field of the whistler wave and $\delta\vec{V}_{\omega\pm}$ are the linear velocity responses of electrons to the fields.

For the whistler wave, we express $\delta\vec{E}_{\omega\pm}=(\hat{x}-i\hat{y})\ \epsilon_{\omega\pm}\exp(-i\omega t)+c.c.$ Thus, $\delta\vec{V}_{\omega\pm}=-i(\hat{x}-i\hat{y})\ [(e\epsilon_{\omega\pm}/m)/(\omega-\Omega_e+i\nu_e)]\ \exp(-i\omega t)+c.c.$ and Eq. (2) reduces to

$$\left\{\frac{\partial^2}{\partial t^2}+i\,[\omega_p^2(z)/(\omega-\Omega_e+i\nu_e)]\frac{\partial}{\partial t}-c^2\frac{\partial^2}{\partial z^2}\right\}\epsilon_{\omega\pm}$$

$$=\mp i[e/m(\omega_0\pm\Omega_e\mp i\nu_e)]\frac{\partial}{\partial t}\left(\epsilon_{p\mp}\frac{\partial}{\partial z}\epsilon_{\ell\pm}\right) \tag{3}$$

where $\omega_p^2(z)=4\pi e^2 n(z)/m$; $\epsilon_{p+}=\epsilon_p=\epsilon^*_{p-}$ and $\epsilon_{\ell+}=\epsilon_{\ell}=\epsilon^*_{\ell-}$.

Equation (3) shows that the whistler wave fields are produced by coupling the sideband wave fields to the pump fields through the nonlinearity of plasma. This is the coupled mode equation of whistler decay mode.

We next derive the coupled mode equation for the Langmuir wave. The driving force of the plasma oscillation is the Lorentz force experienced by the oscillating electrons in the wave magnetic fields. Responding to the electric fields of the heater wave and whistler wave, electrons oscillate, respectively, with velocities $\vec{V}_{p\pm}$ and $\delta\vec{V}_{\omega\pm}$ transverse to the background magnetic field. These cross-field motions interact with the wave magnetic fields, $\delta\vec{B}_{\omega\pm}$ and $\delta\vec{B}_{p\pm}$, induce a longitudinal Lorentz force, $-(e/c)(\delta\vec{V}_{\omega}\times\vec{B}_{p\pm}+\vec{V}_{p\pm}\times\delta\vec{B}_{\omega})$ which is responsible for the longitudinal plasma oscillation, where the notations $\delta\vec{V}_{\omega}=\delta\vec{V}_{\omega+}$ and $\delta\vec{B}_{\omega}=\delta\vec{B}_{\omega+}$ for the R-H case and $\delta\vec{V}_{\omega}=\delta\vec{V}^{*}_{\omega-}$ and $\delta\vec{B}_{\omega}=\delta\vec{B}^{*}_{\omega-}$ for the L-H pump case are used. Including this force in the electron momentum equation and combining this equation with the electron continuity equation we obtain the following coupled mode equation for the Langmuir wave

$$[\frac{\partial^2}{\partial t^2} + \nu_e \frac{\partial}{\partial t} + (\omega_p^2 - v_t^2 \frac{\partial^2}{\partial z^2})] \epsilon_{\ell\pm}$$

$$= \pm 2(e\omega_p^2/m) [\epsilon^*_{p\pm} \frac{\partial}{\partial z} \epsilon^*_{\omega\pm} /\omega^*(\omega_0 \pm \Omega_e \mp i\nu_e) \tag{4}$$

$$+ \epsilon^*_{\omega\pm} \frac{\partial}{\partial z} \epsilon^*_{p\pm} /\omega_0(\omega^* - \Omega_e - i\nu_e)]^*$$

where v_t is the thermal velocity of electrons.

It is clear from the two coupled mode equations (3) and (4), that the whistler wave and Langmuir wave are coupled through the HF heater wave, which acts as the pump of the concerned parametric instability. To proceed the instability analysis further, it is desirable to reduce these two equations into first-order differential equations. This can be done by introducing the transformations [Rosenbluth, 1972; White et al., 1973],

$$\epsilon_\ell = a_1 \exp \{i[k_1 z - \omega_1 t + \int_0^z \Delta k_1(z')dz']\} \tag{5}$$

and

$$\epsilon_{\omega\pm} = a_{2\pm} \exp \{i[k_2 z - \omega_2 t + \int_0^z \Delta k_2(z')dz']\} \tag{6}$$

where $k_1 = k_\ell(0) = \pm k = \pm k_2(0)$, $\omega_1^2 = \omega_\ell^2 = \omega_p^2(0) + k_1^2 v_t^2$ and $\omega_2 = \omega = k_2^2 c^2 \Omega_e / \omega_p^2(0)$, $\Delta k_1 = [\omega_p^2(0) - \omega_p^2(z)]/2k_1 v_t^2$ and $\Delta k_2 = (\omega/2k_2\Omega_e c^2)[\omega_p^2(z) - \omega_p^2(0)]$. Note that $\omega_1 \mp \omega_2 - \omega_0 = \delta$ for a mismatch frequency and $\Delta k = \Delta k_1 \mp \Delta k_2 = \mp k'z$, where $k' = (\omega_p(0)/2kL)(\omega/\Omega_e c^2 + 1/v_t^2)$.

Substituting (5) and (6) into (3) and (4) and assuming that $|\partial a_1/\partial t| << \omega_\ell |a_1|$, $|\partial a_{2\pm}/\partial t| << \omega |a_{2\pm}|$, $|\partial a_1/\partial z| << |k_1| |a_1|$, and $|\partial a_{2\pm}/\partial z| << |k_2| |a_{2\pm}|$, we derive the following first-order differential equations for a_1 and $a_{2\pm}$

$$\left(\frac{\partial}{\partial t} \pm V_1 \frac{\partial}{\partial z} + \nu_1\right) a_{1\pm} = -(e/m)[\omega_p^2(z)/\omega_1] [k/\omega(\omega_0 \pm \Omega_e) \mp 1/\omega_0 \Omega_e \Delta]$$
$$\cdot a_{p\pm} a_{2\pm} \exp[+ik'z^2/2] \exp[i\delta t] \tag{7}$$

and

$$\left(\frac{\partial}{\partial t} + V_2 \frac{\partial}{\partial z} + \nu_2\right) a_{2\pm} = \pm(e/m)[\Omega_e \omega_2 k/\omega_p^2(z)(\omega_0 \pm \Omega_e)]$$
$$\cdot a_{p\mp} a_{1\pm} \exp[-ik'z^2/2] \exp[-i\delta t] \tag{8}$$

where the notations $a_{1+}=a_1=a_{1-}^*$ are used; $a_p=\epsilon_p$; $V_1=kv_t^2/\omega_1$ and $V_2=2kc^2\Omega_e/\omega_p^2(0)$ are the group velocities of Langmuir wave and whistler wave, respectively; $\nu_1=\nu_e/2$ and $\nu_2=\nu_e\omega/\Omega_e$. The simplified coupled mode equations, (7) and (8), will be analyzed in the next section for the parametric excitation of Langmuir wave and whistler wave by a nonuniform HF heater wave in a nonuniform ionospheric plasma.

ANALYSIS

After being Laplace-transformed in time, equations (7) and (8) can be written as

$$\left(p+\nu_1\pm V_1\frac{\partial}{\partial z}\right)A_{1+} = -(e/m)[\omega_p^2(z)/\omega_1][k/\omega(\omega_o\pm\Omega_e)\mp 1/\omega_o\Omega_e\,\Delta] \cdot a_{p\pm}A_{2\pm}\exp(+ik'z^2/2) + a_{10}(z) \tag{9}$$

$$\left(p+\nu_2-i\delta+V_2\frac{\partial}{\partial z}\right)A_{2\pm} = \pm(e/m)[\Omega_e(\omega_2-\delta)k/\omega_p^2(z)(\omega_0\pm\Omega_e)] \cdot a_{p\mp}A_{1\pm}\exp(-ik'z^2/2) + a_{20}(z) \tag{10}$$

where p is the Laplace transform variable; $A_{1\pm}=\mathbf{L}[a_{1\pm}]$ and $A_{2\pm}=\mathbf{L}[a_{2\pm}]$, and $\mathbf{L}[\,]=\int_0^\infty[\,]\exp(-pt)dt$ represents the Laplace transform; $a_{10}(z)$ and $a_{20}(z)$ are the initial values of $a_{1\pm}$ and $a_{2\pm}$, respectively. In the following eigenmode analysis, $a_{10}(z)$ and $a_{20}(z)$ can be set to be zeros.

Combining (9) and (10) leads to a second-order differential equation for $A_{1\pm}$

$$\left[\frac{d^2}{dz^2}+f_{1\pm}(z)\frac{d}{dz}+f_{2\pm}(z)\right]A_{1\pm}=0 \tag{11}$$

where

$$f_{1\pm}(z)=\pm(p+\nu_1)/V_1+(p+\nu_2-i\delta)/V_2-[ik'z+1/(L+z)\pm i/(\Delta\mp iz)] \tag{12}$$

and

$$f_{2\pm}(z) = (e/m)^2[k\Omega_e(\omega_2-\delta)/\omega_1(\omega_0\pm\Omega_e)][k/\omega(\omega_0\pm\Omega_e)\mp 1/\omega_0\Omega_e\,\Delta] \cdot \epsilon_o^2/V_1V_2(1+z^2/\Delta^2)\mp[(p+\nu_1)/V_1][ik'z+1/(L+z)\pm i/(\Delta\mp iz)] \pm[(p+\nu_1)/V_1][(p+\nu_2-i\delta)/V_2] \tag{13}$$

Let $A_{1\pm}(z)=\Psi_{\pm}(z)\exp[-1/2\int_0^z f_{1\pm}(z')dz']$, equation (11) can be transformed into

$$\left[\frac{d^2}{dz^2}+f_{\pm}(z)\right]\Psi_{\pm}(z)=0 \tag{14}$$

where $f_{\pm}(z)=f_{2\pm}(z)-(1/2)\,[(1/2)f_{1\pm}^2(z)+(d/dz)f_{1\pm}(z)]$.

Both the effects of the background density inhomogeneity (i.e., $L\neq\infty$) and finite instability zone (i.e., $\Delta\neq\infty$) are taken into account and included in $f_{\pm}(z)$. However, since $L>>\Delta$, the effect of background density inhomogeneity is negligible in comparison with that of finite instability zone. Therefore, $f_{\pm}(z)$ can be approximately expressed as

$$\begin{aligned} f_{\pm}(z)\simeq\alpha_{\pm}^2/(1+z^2/\Delta^2)\;-\;(1/4)[\mp(p+\nu_1)/V_1+(p+\nu_2-i\delta)/V_2 \\ -ik'z\mp i/(\Delta\mp iz)]^2+(1/2)[ik'-1/(\Delta\mp iz)^2] \end{aligned} \tag{15}$$

where $\alpha_{\pm}^2=(e/m)^2\,[k\Omega_e\,\omega_2/\omega_1(\omega_0\pm\Omega_e)]\,[k/\omega(\omega_0\pm\Omega_e)\mp 1/\omega_0\Omega_e\,\Delta]\,\epsilon_0^2/V_1V_2$.

The eigenvalue of (14) needs to be determined for finding the threshold fields of the heater wave and the growth rates of the instability. The requirement of threshold fields is, in general, imposed by two additive effects. They are the collisional loss of excited modes and the convective loss across the boundaries of finite instability zone. These two effects are evaluate separately as follows.

(a) Collisional loss

We first consider the damping of excited modes due to collisions by letting $\Delta\rightarrow\infty$. It is equivalent to the study of instability in a uniform, unbounded ionospheric plasma. Then, A_1 and $f_2(z)$ become spatially independent functions; equation (13) reduces to

$$f_{2\pm}=\alpha_{\pm}^2\pm[(p+\nu_1)/V_1]\,[(p+\nu_2)/V_2] \tag{16}$$

Note that $\delta=0=k'$ because the neglect of background density inhomogeneity (i.e., $L\rightarrow\infty$) and $1/\Delta$ in $\alpha_{\pm}$ should be replaced by k_0, the wavenumber of the pump (heater).

The dispersion relation of the instability, determined from equation (11), is obtained by setting $f_{2\pm}=0$ which leads to growth rates (p) of the instability to be

$$p=(1/2)\left\{-(\nu_1+\nu_2)+[(\nu_1-\nu_2)^2\mp 4\alpha_{\pm}^2V_1V_2]^{1/2}\right\} \tag{17}$$

Taking p= 0 in (17), we derive the threshold condition for exciting the instability

$$\left[\frac{e\epsilon_o}{(m(\omega_0 \pm \Omega_e)}\right]_{th} = (\nu_1\nu_2\omega_1/\Omega_e\, k^2)^{1/2}/[\omega(\omega_o \pm \Omega_e)k_o/k\omega_o\Omega_e \mp 1]^{1/2} \quad (18)$$

The results show that, in the case of R-H (o-mode) pump, instability exists only in the region $\omega(\omega_o+\Omega_e)k_o > k\omega_o\Omega_e$.

(b) Convective loss

We next examine the effect of finite instability zone, namely, the convective loss of excited modes which is assumed to dominate over the collisional damping loss in determining the thresholds of the instability. In this case, equation (15) is simplified to be

$$f_\pm(z) = \alpha_\pm^2/(1+z^2/\Delta^2) - (1/4)[p(V_1 \mp V_2)/V_1V_2]^2 - (1/2\Delta^2) \quad (19)$$

In obtaining (19), we have also assumed that $\delta \simeq \mp V_2/\Delta$.

The zeros of $f_\pm(z)$ are then found to be

$$z_{1,2} = \pm\Delta\left\{4\alpha_\pm^2 V_1^2V_2^2/[p^2(V_1 \mp V_2)^2 + 2(V_1V_2/\Delta)^2] - 1\right\}^{1/2}$$

$$= \pm z_t \quad (20)$$

These are the positions of two real turning points for equation (14), which has the following solutions

$$\Psi^I(z) = f_\pm^{-1/4}(z)\exp\left[i\int_{z_1}^{z} f_\pm^{1/2}(z')dz'\right] \qquad \text{for } z > z_1$$

$$\Psi^{II}(z) = f_\pm^{-1/4}(z)\exp\left[-i\int_{z_2}^{z} f_\pm^{1/2}(z')dz'\right] \qquad \text{for } z < z_2$$

well behaved at $z \to \pm\infty$. Thus, spatially localized but temporally growing modes exist provided that a positive eigenvalue p can be found from the eigenvalue equation defined by the Bohr-Sommerfeld quantization condition [Landau and Lifshitz, 1965]

$$\left(m + \frac{1}{2}\right)\pi = \int_{z_2}^{z_1} f_\pm^{1/2}(z)dz \quad (21)$$

where m=0,1,2,... stands for the number of modes of the eigenfunction $\Psi(z)$ between the two turning points z_1 and z_2.

The integral in (21) can be expressed in terms of complete elliptic integral for numerical analysis. However, this integral can be much simplified for

$z_t/\Delta << 1$, and a simple approximate form of (21) results as

$$z_t/\Delta \simeq [(2m+1)/\alpha_{\pm}\Delta]^{1/2} \tag{22}$$

Substituting (20) into (22) to eliminate z_t and considering the $m=0$ case for the requirement of lowest thresholds, we obtain the growth rate of the instability. It is

$$p = [2V_1V_2/\Delta(V_1 \mp V_2)]\ [(\alpha_{\pm}\Delta)^3/(1+\alpha_{\pm}\Delta) - 1/2]^{1/2} \tag{23}$$

Setting $p=0$ in (23) leads to the following expression for the threshold fields of heater wave.

$$\left[\frac{e\epsilon_o}{m(\omega_o \pm \Omega_e)}\right]_{th} \cong \frac{\sqrt{2}\,v_t c}{\omega_{po}\Delta} \tag{24}$$

where $\omega_{po}=(4\pi e^2 n_0/m)^{1/2}$, and $k/\omega(\omega_0 \pm \Omega_e) >> 1/\omega_0\Omega_e\,\Delta$ is assumed.

DISCUSSION

We have formulated a mechanism whereby whistler wave and Langmuir wave can be parametrically excited by an HF heater in the overdense ionospheric heating experiments at Tromsϕ, Norway. The required threshold fields of heater wave are derived and examined separately for both R-H and L-H pumps and for two cases, wherein either the collisional loss or convective loss is considered to be the dominant damping process of the proposed instabilities. A quantitative analysis of the instability is carried out here with the following parameters of Tromsϕ's heating experiments [Stubbe et al., 1984]: $\omega_0/2\pi=4.04$ MHz, $\Omega_e/2\pi=1.4$ MHz, $v_t=1.3\times10^5$ m/sec, $\Delta\sim150$ m, $\nu_e\sim500$ Hz, and $\omega_{po}\simeq\omega_0$.

It is found that both calculations of the threshold fields from (18) and (24) have nearly the same magnitude, i.e., 1 v/m for both R-H and L-H pumps. This indicates that the collisional loss and convective loss play equally important roles as the linear damping mechanisms of the proposed instabilities. The effective threshold fields imposed by both damping mechanisms is, therefore, $\sqrt{2}$ v/m. The growth rates of the instabilities with the effective threshold fields can be calculated from (23) with modified width (Δ') of the first peak of the heater wave energy distribution. The modified width is given by $\Delta'=\Delta[(1-x)/(1+x)]^{1/2}$ where $x=\epsilon_{th}^2/\epsilon_0^2$ and ϵ_{th} is the threshold field (given in (18)) imposed by the collisional loss as the dominant damping mechanism. Conceptually, the width modification results from the reduction of heater wave energy by collisions.

The effective threshold fields ($\sim$1.4 v/m) of the instabilities can be exceeded by the peak field intensity of pump waves transmitted from the Tromsϕ HF heater, especially, after the swelling effect on the wave field has been taken into account. If the effective field intensity of heater wave near the reflection height is assumed to be 3 v/m, whistler waves with frequencies ranging from a few tens of kHz up to, say, several hundred kHZ can be excited in a few seconds according to (23). The theory predicts that Langmuir waves excited concomitantly with whistler waves at Tromsø may have a much broader frequency bandwidth than those excited by the parametric decay instability. The frequency bandwidth and the nature of the frequency shift (upshift or downshift) of excited Langmuir waves are determined by the wave frequency of excited whistler waves and the polarization (R-H or L-H) of the pump respectively. Stimulated electromagnetic emissions with a broad frequency band were indeed observed in the Tromsø heating experiments [Stubbe et al., 1984]. HF-enhanced plasma lines having an upshifted broad frequency band were also observed recently when R-H (o-mode) pump was transmitted [Isham et al., 1990].

ACKNOWLEDGMENT

This work was supported by NSF under Grant No. ATM-8816467.

REFERENCES

Frolov, V.L., 1990 URSI General Assembly, private communication.

Hagfors, T., Kofman, W., Kopka, H., Stubbe, P. and Aijanen, T., 1983, "Observations of Enhanced Plasma Lines by EISCAT during Heating Experiments., Radio Sci., 18:861.

Isham, B., Kofman, W., Hagfors, T., Nordland, J., Thide, B., Lattoz, C. and Stubbe, P., 1990, "New Phenomena Observed by EISCAT during an RF Ionospheric Modification Experiment," Radio Sci., 25:251.

Kuo, S.P. and Cheo, B.R., 1978, "Parametric Excitation of Coupled Plasma Waves," Phys. Fluids, 21:1753.

Kuo, S.P. and Lee, M.C., 1982, "On the Parametric Excitation of Plasma Modes at Upper Hybrid Resonance," Phys. Lett., 91A:444.

Kuo, S.P. and Schmidt, G., 1983, "Filamentation Instability in Magnetoplasmas," Phys. Fluids, 26:2529.

Kuo, S.P., Cheo, B.R. and Lee, M.C., 1983, "The Role of Parametric Decay Instabilities in Generating Ionsopheric Irregularities," J. Geophys. Res., 88:417.

Kuo, S.P. and Lee, M.C., 1983, "Earth's Magnetic Field Fluctuations Produced by Filamentation Instabilities of Electromagnetic Heater Waves," J. Geophys. Res., 10:979.

Kuo, S.P. and Lee, M.C., 1989, "Parametric Excitation of Whistler Waves by HF Heater," J. Atoms. Terr. Phys., 51:727.

Kuo, S.P. and Barker, R.J., 1990 "Parametric Excitation of Whistler Wave and Frequency Upshifted Langmuir Sideband by L-H Circularly Polarized HF Heater," XXIII General Assembly of URSI, Prague, Czechoslovakia, Aug. 27-Sept. 5, 1990.

Landau, L.D. and Lifshitz, E.M., 1965, Quantum Mechanics, Nonrelativistic Theory, Pergamon Press, Oxford.

Rosenbluth, M.N., 1972, "Parametric Instabilities in Inhomogeneous Media," Phys. Rev. Lett., 29:565.

Showen, R.L. and Kim, D.M., 1978, "Time Variations of HF-Induced Plasma Waves," J. Geophys. Res., 83:623.

Stubbe, P., Kopka, H., Thide, B. and Derblom, H., 1984, "Stimulated Electromagnetic Emission: A New Technique to Study the Parametric Decay Instability in the Ionosphere," J. Geophys. Res., 89:7523.

White, R.B., Liu, C.S. and Rosenbluth, M.N., 1973, "Parametric Decay of Obliquely Incident Radiation," Phys. Rev. Lett., 31:520.

NONLINEAR MODULATIONAL INSTABILITY OF AN ELECTROMAGNETIC PULSE IN A NEUTRAL PLASMA

D.J. Kaup

Clarkson University
Department of Mathematics and Computer Science and Physics
and the Institute for Nonlinear Studies
Potsdam, NY 13699-5815

ABSTRACT

Using the propagation of an intense electromagnetic pulse through a neutral plasma as a model, we demonstrate how one can obtain information about the nonlinear behavior of modulated waves. Assuming no resonant instabilities, the envelope can be shown to evolve over long time scales according to a vector form of the well-known cubic nonlinear Schroedinger (NLS) equation. In the weakly relativistic regime, three distinct nonlinear effects contribute terms cubic in the amplitude and thus can be of comparable magnitude: ponderomotive forces, relativistic corrections, and harmonic generation. Modulational stability of any given system is shown to depend on polarization, frequency, composition, and these dependences are given. In the special case of a cold positron-electron plasma, the model is strictly modulationally stable for both linear and circular polarization. However the presence of an ambient magnetic field can make a decisive difference. Now modulational instability can arise within a broad range of frequencies and values of B_0, in particular for a pure positron-electron plasma. For the case of intensely propagating (relativistic) EM plane plasma waves, we find that the circularly-polarized waves are modulationally unstable under a range of conditions. We use a unified approach which illustrates how the modulational instability changes as one moves from the weakly relativistic case to the fully relativistic case. And we are able to give the first self-consistent calculation of the modulational instability properties of a slowly-modulated, fully relativistic envelope. Modulated, intensely propagating EM waves couple (in general) to longitudinal motions via the ponderomotive force. The effect of longitudinal motions is comparable to that of relativistic nonlinearities. A correct and proper expansion procedure requires solution of all the field equations up to the appropriate order, including the longitudinal equations. In particular, in the extreme relativistic limit of either an electron-positron or ion-electron plasma, waves with frequencies below twice the relativistic plasma frequency ω_p, where $\omega_p^2 \equiv 4\pi e^2 n_0(1/(m\gamma_0) + 1/(M\Gamma_0))$, are unstable. Growth occurs on a very short timescale comparable to the time for a wave packet to move past a fixed point. (This timescale is much shorter than that for spreading of the wave packet due to linear dispersion.)

Directions in Electromagnetic Wave Modeling
Edited by H.L. Bertoni and L.B. Felsen, Plenum Press, New York, 1991

INTRODUCTION

The propagation of a relativistically-strong, modulated electromagnetic pulse in a plasma is more complicated than appears at first glance. In the case of uniform propagation (i.e., no modulation), some exact solutions (for circular polarization) are indeed known[1] which are described by a nonlinear dispersion relation. Chian and Kennel[2] pointed out the possible application to pulsar micropulses and obtained a scalar NLS (NonLinear Schroedinger) equation. Others have studied the problem in various forms[3-5], but many of the results are incorrect and even violate the force equations in leading order. The problem evidently contains subtleties and requires the careful application of singular perturbation methods.

Taking pieces of the problem one at a time, Kates and Kaup[6] first studied the weakly relativistic limit in the absence of an ambient magnetic field. In this limit, the equations can be reduced to a set of coupled NLS equations, with one coefficient of dispersion and three nonlinear coefficients. They presented an accurate evaluation of these nonlinear coefficients and demonstrated that these coefficients were crucial in determining the qualitative behavior of the envelope, and in particular whether or not modulational stability and soliton type solutions will occur. The special case of an electron-positron plasma exhibit several peculiarities, such as accidental (near) cancellations in certain limits. One could obtain the correct nonlinear coefficients only by considering all three sources of nonlinearity right from the beginning. When all terms were included, they found that in the nonrelativistic limit, the claims of modulational instability for electromagnetic pulses propagating in a neutral positron-electron plasma were actually reversed. A cold positron-electron plasma is modulationally stable for both circular and linear polarization.

However, there is strong recent observational evidence[7-9] that pulsar microstructure may result from modulations of a coherent pulse. If nonlinear modulational instability is the cause of micropulses, then some additional element must enter the picture: e.g., finite temperature, ambient magnetic fields, possible presence of ions, or even fully relativistic amplitudes. As shown in Kates[6], finite temperature can indeed induce modulational instability for both circular and linear polarization, but the effect is confined to frequencies near the plasma frequency. And, if some positive ions are present, then modulational instability is also possible. In a following paper, Kates and Kaup[10] showed that in the presence of an ambient magnetic field, a circularly polarized weakly relativistic pulse propagating parallel to the magnetic field in a cold plasma could also go modulationally unstable under a range of conditions depending on composition and the frequency regime.

Here I shall also report some of our more recent results for the fully relativistic case, where the pulse intensity is so intense that the particles achieve relativistic velocities. This problem has its own subtleties, which include a coupling to the longitudinal components of order unity. Now we find that intensely propagating modulated pulses are subject to modulational instabilities even without ambient magnetic fields. Whereas the instability in the weakly relativistic limit requires an interplay between nonlinearity and dispersion, for intense propagation the effects of linear dispersion are evidently small compared to the dominant terms, which arise from nonlinearity. As we shall see, this system has instabilities on timescales comparable to the time for a wave-packet to propagate past a fixed point. This timescale is much shorter than that of the weakly relativistic case, which is comparable to the time for spreading of the wave packet.

In any singular perturbation method, one seeks uniformly valid asymptotic expansions[11] with respect to a small parameter, which we will call ϵ (in this problem ϵ is related to the typical amplitude of the vector potential). Here, the domain of uniform validity is of order $1/\epsilon$ cycles in space and $1/\epsilon^2$ cycles in time. The singular perturbation method which we will apply, known as the method of multiple time scales, has been described in several textbooks (see for example Van Dyke[12] or Nayfeh[13]) and has been applied to a wide variety of problems. Since we are interested in the evolution of a pulse with arbitrary modulation, it will not be possible to assume dependence on only a single independent variable (although it is possible to verify consistency of our results for particular cases with the work of Kozlov, et. al.[14], in which fully nonlinear solutions depending on one independent variable were found). Instead, we allow the functions in this problem to depend, not only on a "fast" (phase) variable, but also on "slow" time and space variables. These will be chosen to make the effects of nonlinearity enter the calculation at the same order as the effects of linear spreading of the wave packet.

Consider an envelope of length L of waves with wavelength near λ. If $\epsilon \equiv \frac{e}{mc\omega}|E| \sim \frac{\lambda}{L}$, then nonlinear effects will typically act on a timescale $T_0 = \lambda/(\epsilon^2 c)$ comparable to the timescale for the spreading of packet due to linear dispersion. By routine dimensional analysis it is possible to see that ponderomotive forces, relativistic corrections, and harmonic generation might have comparable effects on this timescale. Our calculation shows that, in general, all three effects indeed play a role. The ponderomotive force depends quadratically on the amplitude and leads to a slowly varying longitudinal field, corresponding physically to radiation pressure, which in the weakly relativistic case leads only to slow longitudinal motions and modifies the background density. At cubic order, the modification of the background couples back to the fundamental. Independently of the composition, temperature, or frequency, the effects of harmonic generation are identically zero for circular polarization – but in general only for circular polarization.

Our model comprises a neutral, fully-ionized plasma, consisting of singly-ionized atoms (of charge $+e$ and mass M) and electrons (of charge $-e$ and mass m); the plasma is modeled as a cold fluid. The fluid is characterized by the electron density n, the electron velocity $\vec{v}$, the ion density N, and the ion velocity $\vec{V}$. (Although we will refer to the positive component as "the ions," our model of course also describes a positron-electron plasma if $m = M$.) The plasma is further characterized by a vector potential $\vec{A}$ and an electrostatic potential ϕ. The functions n, $\vec{v}$ N, $\vec{V}$, $\vec{A}$, and ϕ are assumed to depend on t and z, but not on x and y. It is convenient to write the continuity and Euler equations for the electrons in the form

$$\partial_t n + \vec{\nabla} \cdot (n\vec{v}) = 0 \tag{1a}$$

$$d(m\gamma\vec{v} - \frac{e}{c}\vec{A}) - e\vec{\nabla}\phi + \frac{e}{c}\Sigma_{\ell=1}^{3} v_\ell \vec{\nabla} A_\ell = 0 \tag{1b}$$

where m is the electron mass, and

$$d = \partial_t + \vec{v} \cdot \vec{\nabla} \tag{1c}$$

$$\gamma = (1 - v^2/c^2)^{-1/2} \tag{1d}$$

For the ions, the respective equations are

$$\partial_t N + \vec{\nabla} \cdot (N\vec{V}) = 0 \tag{2a}$$

$$D(M\Gamma\vec{V} + \frac{e}{c}\vec{A}) + e\vec{\nabla}\phi - \frac{e}{c}\Sigma_{\ell=1}^{3} V_\ell \vec{\nabla} A_\ell = 0 \tag{2b}$$

where M is the ion mass and

$$D = \partial_t + \vec{V} \cdot \vec{\nabla} \tag{2c}$$

$$\Gamma = (1 - V^2/c^2)^{-1/2} \tag{2d}$$

Of course Maxwell's equations must hold:

$$\vec{\nabla} \times \vec{B} = \frac{4\pi e}{c}(n\vec{v} - N\vec{V}) + \frac{1}{c}\partial_t \vec{E} \tag{3a}$$

$$\vec{\nabla} \cdot \vec{E} = 4\pi e(N - n) \tag{3b}$$

with

$$\vec{B} = \vec{\nabla} \times \vec{A} \tag{3c}$$

$$\vec{E} = -\vec{\nabla}\phi - \frac{1}{c}\partial_t \vec{A} \tag{3d}$$

and where we impose the radiation gauge

$$\vec{\nabla} \cdot \vec{A} = 0 \quad . \tag{3e}$$

This is a nonlinear system of equations is to be solved for the evolution of some initial data given on the interval $-\infty < z < \infty$ for the functions $(n, \vec{v}, N, \vec{V}, \vec{A}, \phi)$, satisfying the usual initial-value constraints. (The constraints will be satisfied automatically at later times).

ZERO MAGNETIC FIELD, WEAKLY RELATIVISTIC CASE

First we will consider only the weakly relativistic case, where we are interested in a particular class of solutions, namely those which correspond to the physical idea of a slowly modulated, weak, nearly sinusoidal disturbance about a uniform quiescent medium (which is an obvious exact solution of Eqs. (1-3)). We assume that well-posed initial data can be given which evolve in this way.

To make these assumptions precise, consider first the electron density and the vector potential, which are assumed to have asymptotic expansions of the form

$$n = n_0 + \epsilon n_1 + \epsilon^2 n_2 + \epsilon^3 n_3 + \ldots \tag{4a}$$

$$\vec{A} = \quad \epsilon \vec{A}_1 + \epsilon^2 \vec{A}_2 + \epsilon^3 \vec{A}_3 + \ldots \tag{4b}$$

with respect to the positive dimensionless parameter $\epsilon \ll 1$, where the first term n_0 in (4a) is a constant. Since the velocities will also scale with ϵ, it is evident that for sufficiently small ϵ the motion is only weakly relativistic and thus we will be able to expand the relativistic factors γ and Γ about unity.

Similarly, the other variables are expanded as follows:

$$N = N_0 + \epsilon N_1 + \epsilon^2 N_2 + \epsilon^3 N_3 + \ldots \tag{4c}$$

$$\vec{v} = \quad \epsilon \vec{v_1} + \epsilon^2 \vec{v_2} + \epsilon^3 \vec{v_3} + \ldots \tag{4d}$$

$$\vec{V} = \quad \epsilon \vec{V_1} + \epsilon^2 \vec{V_2} + \epsilon^3 \vec{V_3} + \ldots \tag{4e}$$

$$\phi = \quad \epsilon \phi_1 + \epsilon^2 \phi_2 + \epsilon^3 \phi_3 + \ldots \tag{4f}$$

Note that $n_0 = \text{const}$, $N_0 = n_0$, and all other zeroth-order quantities vanish. (Except where otherwise stated, numerical subscripts indicate the formal order in ϵ.)

Now, it is routine to work out the <u>linearized</u> solutions of (1)-(3) corresponding to a sum over Fourier components of transverse electromagnetic EM waves[1]. (By "linearized" solutions, we mean solutions obtained by simply truncating (1)-(3) at order ϵ.) A typical Fourier component of the vector potential perturbation behaves in this linearized theory like $\vec{a}(\omega)e^{i(kz-\omega t)}$, where k and ω are related by a dispersion relation to be given below. Here we are interested in the <u>nonlinear</u> theory in which different Fourier components are coupled. In order to study the evolution of a slowly modulated beam, we assume that the first-order perturbation of the vector potential takes the form

$$\vec{A}_1 = \vec{a} e^{i(kz-\omega t)} + c.c. \tag{5}$$

where "c.c." means the complex conjugate, k is the wave number, ω is the frequency. The functions a_x and a_y (where the subscripts refer to the component) are not strictly constant, but depend on the slow variables

$$Z \equiv \epsilon z, \quad T \equiv \epsilon t, \quad \tau \equiv \epsilon^2 t \tag{6}$$

With the introduction of slow coordinates, we employ the following standard procedure: We collect terms in all equations according to their formal order of magnitude. ("Order" always means with respect to ϵ.) We then solve order by order. Every derivative of a slowly varying function is down by one or two orders. For this reason, the transverse equations at leading order are the same as those of the truncated linearized theory. (Not so for the longitudinal equations, as we shall shortly see.) Slow derivatives of the lower-order terms will appear at higher orders. According to standard singular-perturbation procedure[13], we seek solutions of the higher-order equations which satisfy the requirement of uniform validity, not only over a cycle, but also over the slow scales. (In general, not all of the higher-order equations need to be solved explicitly, but instead provide integrability conditions which affect the lower orders.) In this way, the evolution of the slowly varying functions will be determined.

With this strategy in mind, we now solve the first-order equations. The longitudinal components are identically zero or can be chosen to vanish without loss of generality. The transverse components can be treated as a coupled, linear, homogeneous, self-adjoint system of the form

$$L(\psi_1) = 0 \quad . \tag{7}$$

For appropriate initial data, the first-order solution can be chosen to represent an electromagnetic mode, and the solution thus takes the form

$$\vec{v}_1 = \frac{e}{mc} \vec{A}_1 \tag{8a}$$

$$\vec{V}_1 = \frac{-e}{Mc} \vec{A}_1 \tag{8b}$$

$$\phi_1 = 0 = n_1 = N_1 = A_{1z} = v_{1z} = V_{1z} \tag{8c}$$

Note that for this mode the velocities and the vector potential are purely transverse in first order. The first-order part of (3b) guarantees quasineutrality at first order. The z-component of (3a) is then satisfied at first order automatically.

The dispersion relation is

$$\omega^2 = \omega_p^2 + c^2 k^2 \tag{9}$$

where

$$\omega_p^2 = 4\pi e^2 n_0 \left(\frac{1}{m} + \frac{1}{M} \right) \tag{10}$$

The slow dependence of $\vec{a}$ is of course as yet undetermined.

As we have seen, the longitudinal components of the velocities vanish in first order. However, for a system with slowly modulated waves, the third-order part of the longitudinal component of the force equations (4b) and (5b) leads to a nonvanishing, slowly varying longitudinal component of the velocities at second order. This happens even in a positron-electron plasma, despite the fact that the third-order longitudinal electric field (second-order electrostatic potential) vanishes.

At second order there are three possible types of terms with different phase dependence, which we will discuss in turn. Decomposing the second-order electron density n_2 as

$$n_2 = n_2^{(0)} + \left[n_2^{(1)} e^{i(kz-\omega t)} + c.c.\right] + \left[n_2^{(2)} e^{2i(kz-\omega t)} + c.c.\right] \quad , \tag{11}$$

we note that the first term $n_2(0)$ corresponds to a second-order correction to the background density. A similar decomposition will be made for the other variables. From here on superscripts (0), (1), (2), etc. refer to coefficients of terms with corresponding phase dependence. Terms such as $n_2^{(0)}$ will be refered to as "DC" terms. Terms such as $n_2^{(1)}$ correspond to a correction to the fundamental. Terms such as $n_2^{(2)}$ correspond to the harmonic, which in general arises due to nonlinearity, although in specific cases it may vanish identically.

At all orders $p > 1$, the fundamental mode is described by a linear, inhomogeneous system of the general form

$$L(\psi_p) = N_p(\psi_1, \ldots \psi_{p-1}) \quad , \tag{12}$$

where L is the linear, self-adjoint operator of Eq. (7), and where N_p is a (possibly) nonlinear functional of the lower-order terms. Taking $\theta \equiv kx - \omega t$ and defining the scalar product

$$\langle f, g \rangle \equiv \int_0^{2\pi} d\theta \; f^* g \tag{13}$$

of two functions f and g over a cycle, one finds using the self-adjointness of L that N_p must be orthogonal (with respect to (13)) to the general homogeneous solution ψ_H of (7). Otherwise, secular terms, e.g., terms of the form $\theta e^{i\theta}$ would arise. Secular terms would obviously violate the uniformity condition stated above, because the pth order would eventually become comparable in size to the terms of order $p-1$.

Without giving the details, we shall summarize the results[6] for second order. For the DC terms, Although the electrostatic potential could vanish (as it does for a positron-electron plasma), there is always at least a longitudinal velocity shift due to the radiation pressure. This is the source of the ponderomotive nonlinearity. For the fundamental, the requirement of orthogonality (13) implies simply

$$(\partial_T + v_g \partial_Z) \vec{a} = 0 \tag{14}$$

where

$$v_g = \frac{d\omega}{dk} = \frac{c^2 k}{\omega} \tag{15}$$

Note that Eq. (14) is the same as the group velocity condition for a wave packet in the truncated linear theory. (Nonlinearities will play a role in the fundamental at third order.)

The nonlinear terms in the Euler equations (1b) and (2b) also induce harmonic terms in the solution. A special feature of the harmonic terms are that they all are proportional to $\vec{a} \cdot \vec{a}$. Thus circular polarization (vanishing $\vec{a} \cdot \vec{a}$) is a special case (for any mass ratio), because harmonic generation is absent. In this case, charge quasi-neutrality is maintained at second order. However, in general the second-order density perturbations of electrons and ions are not equal: charge separation over the fast scales is already present at second order! It is these terms which will have serious consequences for the relativistic case. The electron-positron case is again special because the harmonic, second-order potential perturbation is zero. Otherwise, a second-order electric field varying as the harmonic will be present in general (except for circular polarization).

In third order, it is again possible to derive the evolution of the slow functions ($\vec{a}$, etc.) as a consequence of the requirement that the third-order equations contain no secular terms, as explained above. Since DC and harmonic terms are of course orthogonal to the operator L of Eq. (12), it suffices to consider the fundamental. Here we have nonlinear contributions from the relativistic corrections, ponderomotive terms, and the harmonic terms. The final result is

$$2i\omega\partial_\tau\vec{a}+\frac{\omega_p^2c^2}{\omega^2}\partial_Z^2\vec{a}+\vec{a}^*(\vec{a}\cdot\vec{a})(C_H+C_R)$$

$$+(C_P+2C_R)\vec{a}(\vec{a}^*\cdot\vec{a})=0 \tag{16}$$

In the above, the subscripts H (harmonic), P (ponderomotive), and R (relativistic) indicate the origin of the corresponding nonlinear coefficient. These nonlinear coefficients are given by

$$C_H=\frac{-e^2k^2\omega_p^2}{2\omega^2c^2(4\omega^2-\omega_p^2)}\left[4\omega^2\frac{\frac{1}{m^3}+\frac{1}{M^3}}{\frac{1}{m}+\frac{1}{M}}-\frac{\omega_p^2}{mM}\right] \tag{17a}$$

$$C_P=-\frac{e^2\omega_p^2}{Mm\;c^2v_g^2} \tag{17b}$$

$$C_R=\frac{e^2\omega_p^2}{2c^4}\,\frac{\frac{1}{m^3}+\frac{1}{M^3}}{\frac{1}{m}+\frac{1}{M}} \tag{17c}$$

In the cases of linear and circular polarization, the vector NLS reduces to a scalar NLS. For circular polarization, we obtain

$$2i\omega\partial_\tau a+\frac{\omega_p^2c^2}{\omega^2}\partial_Z^2a+2(C_P+2C_R)a^*a^2=0 \tag{18}$$

where $\vec{a}=a\cdot(\hat{x}+i\hat{y})$. For linear polarization, we find

$$2i\omega\partial_\tau a+\frac{\omega_p^2c^2}{\omega^2}\partial_Z^2a+(C_H+3C_R+C_p)a^*a^2=0 \tag{19}$$

where $\vec{a}=a\cdot\hat{x}$. Since the coefficient of dispersion in (18) and (19) is positive definite, then the wave is modulationally unstable if the corresponding sum of the nonlinear coefficients is positive.

We now consider two special cases:

ELECTRON-POSITIVE ION PLASMA: If the plasma is composed of positive ions and electrons, then $m/M \ll 1$ and we find

$$C_H+C_R\doteq\frac{3e^2\omega_p^4}{2m^2c^4(4\omega^2-\omega_p^2)}>0 \tag{20a}$$

$$C_P+2C_R\doteq\frac{e^2\omega_p^2}{c^4m^2}\left[1-\frac{m\omega^2}{Mc^2k^2}\right] \tag{20b}$$

Except for frequencies very near the plasma frequency, both polarizations are evidently modulationally unstable. The contribution of the ponderomotive coefficient is down by $O(m/M)$ compared to the others, except near the plasma frequency. The positive (destabilizing) contribution of the relativistic coefficient outweighs the negative (stabilizing) contribution of the harmonic coefficient.

ELECTRON-POSITRON PLASMA: Next, let us consider an electron-positron plasma. Setting $M=m$ in Eqs. (17a-c), we find

$$C_H+C_R=\frac{e^2\omega_p^4}{2m^2c^4\omega^2}>0 \tag{21a}$$

$$C_P+2C_R=\frac{-e^2\omega_p^4}{m^2c^6k^2}<0 \tag{21b}$$

$$C_H+3C_R+C_p=-\frac{e^2\omega_p^2}{2m^2c^6k^2\omega^2}(\omega^2+\omega_p^2)<0 \tag{21c}$$

Since $C_H + 3C_R + C_P < 0$ and $C_P + 2C_R < 0$, both polarizations are evidently modulationally stable, contrary to the claim of Chian and Kennel[2]. Their solution violates the z-component of the force equation at leading order and thus ignores ponderomotive effects, which (as we saw) are stabilizing for both polarizations. For linear polarization, three is the additional stabilizing effect of mode-mode interactions.

AMBIENT MAGNETIC FIELD CASE

Let us now consider extending the above to the case when an ambient magnetic field is present[10]. Now we have two additional frequencies present. Define the electron-cyclotron frequency

$$\omega_- = e|B_0|/(mc) \tag{22a}$$

and the ion-cyclotron frequency

$$\omega_+ = e|B_0|/(Mc) \tag{22b}$$

where B_0 is the ambient magnetic field, assumed to be aligned along the z-axis and parallel to the direction of propagation. (The two cyclotron frequencies obviously coincide in the case of an electron-positron plasma.)

It is instructive to review briefly properties of the solutions of the linearized system obtained by simply truncating Eqs. (1-3) at $O(\epsilon)$. (The linearized theory of circularly-polarized electromagnetic-wave propagation parallel to an ambient magnetic field can be found in[15,16].) A typical Fourier component of the vector potential perturbation behaves in this linearized theory like $\vec{a}(\omega)e^{i(kz-\omega t)}$, where

$$\vec{a} = a \cdot (\hat{x} + i\hat{y}). \tag{23}$$

The dispersion relation for right-hand circular polarization is

$$\omega^2 = 4\pi e^2 n_0/(m\nu) + 4\pi e^2 n_0/(M\mu) + k^2c^2 \tag{24}$$

where

$$\nu \equiv 1 - eB_0/(mc\omega) = 1 - \omega_-/\omega \tag{25a}$$

$$\mu \equiv 1 + eB_0/(Mc\omega) = 1 + \omega_+/\omega \tag{25b}$$

The dispersion relation for left-hand circular polarization can be obtained from (25) by substituting for ν and μ expressions in which B_0 is replaced by $-B_0$.

If the masses are unequal, then four electromagnetic branches can be distinguished. If the masses are equal (electron-positron plasma), then two branches are present. The requirements $k^2 > 0$ and $\omega^2 > 0$ will in general place restrictions on the allowable frequencies of a propagating electromagnetic wave. Unlike the unmagnetized case, propagation can occur below the plasma frequency. In the linear theory it is of course also possible to consider the evolution of initial data corresponding to "linear" polarization; evidently Faraday rotation of the polarization will occur.

In addition to the electromagnetic modes, there are also longitudinal Langmuir oscillations at the plasma frequency. In the linearized theory, these are of course decoupled from electromagnetic waves. However, in the modulated nonlinear theory we again have that the ponderomotive force induces coupling between transverse and longitudinal modes.

The ponderomotive term is given simply by

$$C_P = -\left(\frac{e^2}{c^4 Mm}\right)\frac{\omega}{k\omega_p^2 v_g F}(2k\omega v_g - \omega^2 - c^2k^2)^2 \tag{26a}$$

The relativistic coefficient is given explicitly by

$$C_R = \left(\frac{e^2}{c^4 Mm}\right) \frac{\omega v_g M^2 m^2 \omega_p^2}{2c^2 k(M+m)} \left[1/(m^3\nu^4) + 1/(M^3\mu^4)\right] \tag{26b}$$

We note that the coefficients C_P and C_R reduce to the corresponding formulas (17b-c) if B_0 vanishes.

The condition for modulational instability is now

$$\omega\omega_{kk}(C_P + 2C_R) < 0 \tag{27}$$

In the limit of large Ω^2/ω_p^2, which is certainly a good approximation for the pulsar environment, one finds that $\omega_{kk}\omega < 0$. In the limit of $B_0 \to 0$, we have the opposite result of $\omega_{kk}\omega > 0$. In the absence of an ambient magnetic field, it was positive nonlinear coefficients which led to modulational instability; here it will be negative nonlinear coefficients which will give rise to modulational instability.

Including thermal effects and as long as the sound speeds are much smaller than the group velocity of the electromagnetic wave, (which is usually the case for nonrelativistic gas temperatures) one obtains[10]

$$C_P + 2C_R = \left(\frac{e^2}{c^2 m^2}\right) \frac{{\omega_p}^2 \omega^4}{2\Omega^4} \left[\frac{4\omega^2 - \omega_p^2}{\Omega^2} - \frac{c_s^2 + C_s^2}{2c^2} + \cdots\right] \tag{28}$$

Now we can have modulational instability, see (27), only if

$$\omega^2 < \frac{1}{4}\omega_p^2 + \Omega^2 \frac{c_s^2 + C_s^2}{2c^2} \tag{29}$$

For large Ω^2, we see that there is a very wide frequency range for even moderate temperatures.

According to pulsar models[17,18], the pulsar magnetosphere is composed of secondary electrons and positrons, and strong magnetic fields (10^{12} Gauss) are present, corresponding to an electron plasma frequency of the order 10^{19} Hz, as compared to a plasma frequency of a few Megahertz. If we apply the statistical model of Rickett[19], as described[2], a modulationally unstable pulse could be responsible for the micropulse structure observed[20].

INTENSE PROPAGATION

Now let us take up the case of intense propagation. (By "intense propagation," we mean that either or both components may attain fully relativistic velocities.) In this case, we will be interested in the conditions under which plane, circularly polarized waves in a cold, fully ionized plasma are subject to modulational instabilities. It is crucial for a proper physical understanding of modulational instabilities in intense propagation to realize that when a circularly-polarized system is modulated, momentum is exchanged between the electromagnetic field and the charged particles due to ponderomotive forces. (The exchange occurs on the timescale of the modulation.) The consequences of this coupling between longitudinal and transverse modes manifest themselves even in the weakly relativistic case[6,10] as a contribution to the nonlinear coefficients of the nonlinear Schroedinger equation (NLS). In the fully relativistic case, the effect is even stronger. Either or both components of the plasma may acquire relativistic longitudinal velocities, notwithstanding the fact that the unmodulated solution is purely transverse.

An intensely propagating plasma wave corresponds to a fully nonlinear solution of the Maxwell-Euler equations describing the plasma. For the case of a circularly-polarized, unmodulated electromagnetic wave, a class of exact solutions depending on an amplitude parameter is known[1]. These particular solutions satisfy a "nonlinear dispersion relation," that is, one that depends on the amplitude parameter.

Now consider the problem of the evolution of modulated waves. This problem has to be treated with singular perturbation methods. It is tempting to apply the formalism of Karpman and Krushkal[21], which involves essentially inspection of the (known) "nonlinear dispersion relation" satisfied by unmodulated waves. However this procedure leds to incorrect and inconsistent result unless the assumption of[21] – that the "nonlinear dispersion relation" of modulated waves should have the same functional dependence on amplitude as that of unmodulated waves – is satisfied, which is not known *a priori.* Indeed, it is violated in the case of plasma electromagnetic waves, due to the coupling between transverse and longitudinal modes mentioned above. As a result of ponderomotive forces, both the densities and the relativistic factors in the dispersion relation will vary strongly along the propagation direction on the scale of the modulations. Thus, the dispersion relation depends not only on the amplitude, but also on its derivatives, whereas the dispersion relation of unmodulated waves (by definition) does not include derivatives of the amplitude.

To study modulations, one replaces the amplitude parameter a_0 with a function a depending on the auxiliary variables Z and T. However, this generalization alone does not lead to an approximate solution of the field equations, even for $\epsilon << 1$, even for the electron-positron case, and even for circular polarization. As it turns out, the ponderomotive force induces slowly varying z-components of velocity $V \equiv V_z$, $v \equiv v_z$ (unless the gradient of a happens to vanish, which is the trivial case), which as we will see cannot be neglected. In general, if a changes by order unity, then v and V can even be relativistic.

These longitudinal velocities couple via the continuity equation and the other field equations to the densities and all other slowly-varying quantities. A longitudinal electric field also arises in general. However, we note that longitudinal motions play an important role even when the longitudinal electrostatic field vanishes or is of higher order.

The simplest allowable circular-polarization Ansatz thus consists of introducing – in addition to the amplitude $a(T,Z)$ – variable densities $n(T,Z)$ and $N(T,Z)$, z-components of velocity $v(T,Z)$ and $V(T,Z)$, electrostatic potential $\Phi(T,Z)$, and phase $\Theta(T,Z)$. Finally, we note that the relativistic factors γ and Γ will now of course depend on the z-component of the velocities, v and V, which vanish in the steady solution.

We thus write our modulated circular-polarization Ansatz in the form

$$n = n(T,Z) \tag{30a}$$

$$N = N(T,Z) \tag{30b}$$

$$\vec{A} \equiv (a(T,Z)e^{i\theta}, ia(T,Z)e^{i\theta}, 0) + c.c. \tag{30c}$$

$$\vec{v} \equiv \frac{e}{mc\gamma(T,Z)}\vec{A} + v(T,Z)\hat{z} \tag{30d}$$

$$\vec{V} \equiv \frac{-e}{Mc\Gamma(T,Z)}\vec{A} + V(T,Z)\hat{z} \tag{30e}$$

$$\Phi = \Phi(T,Z) \tag{30f}$$

where

$$\theta \equiv \Theta(T,Z)/\epsilon \tag{30g}$$

$$k \equiv \partial\theta/\partial z = \partial\Theta/\partial Z \tag{30h}$$

$$\omega \equiv -\partial\theta/\partial t = -\partial\Theta/\partial T \tag{30i}$$

$$\gamma^2 = \left[1 + 4\frac{e^2}{m^2c^2}a^2\right]/(1 - v^2/c^2) \tag{30j}$$

$$\Gamma^2 = \left[1 + 4\frac{e^2}{M^2c^2}a^2\right]/(1 - V^2/c^2) \quad (30k)$$

and where $a(T, Z)$ is considered to be a real quantity. The phase information is contained in Θ. Note again that v and V refer to the z-components of the respective vectors and not to the magnitudes. This notation will be convenient in what follows.

We now substitute our Ansatz (30) into the field equations (1-3). From the continuity and Euler equations (1-2), we obtain

$$\partial_T n + \partial_Z(nv) = 0 \quad (31a)$$

$$\partial_T N + \partial_Z(NV) = 0 \quad (31b)$$

$$-m(\partial_T + v\partial_Z)(\gamma v) = -e\partial_Z\Phi + 2\frac{e^2}{mc^2\gamma}\partial_Z(a^2) \quad (31c)$$

$$-M(\partial_T + V\partial_Z)(\Gamma V) = +e\partial_Z\Phi + 2\frac{e^2}{Mc^2\Gamma}\partial_Z(a^2). \quad (31d)$$

Equation (3b) yields

$$\epsilon^2\frac{\partial^2\Phi}{\partial Z^2} = 4\pi e(n - N). \quad (31e)$$

The z-component of (3a) is satisfied automatically by virtue of (31a-e). The transverse components of (3a) imply

$$\partial_T(\omega a^2) + c^2\partial_Z(ka^2) = 0 \quad (31f)$$

$$\left[-\omega^2 + c^2k^2 + 4\pi e^2 n/(m\gamma) + 4\pi e^2 n/(M\Gamma)\right] a = \epsilon^2(c^2 a_{ZZ} - a_{TT}) \quad (31g)$$

Together with the definitions Eqs. (30h-k), Eqs. (31) comprise a system of seven coupled partial differential equations with respect to the independent variables T and Z in seven real unknowns a, n, N, v, V, Φ, and Θ. (In what follows, these will be refered to as the "envelope variables" and Eqs. (31) as the "envelope equations.") An exact solution of Eqs. (31) together with the circular-polarization Ansatz (30) is equivalent to an exact solution of the original system (1-3). Note that as yet no approximation has been made: The envelope eqs. (31) hold for any positive constant ϵ. There is no restriction on the size of a.

Our main interest is to determine how an intense pulse will evolve if it is close to a constant steady solution. So let us expand about this constant steady solution. Take

$$a = a_0 \quad (32a)$$

$$n = N = n_0 \quad (32b)$$

$$v = V = \Phi = 0 \quad (32c)$$

$$\omega = \omega_0 \quad (32d)$$

$$k = k_0 \quad (32e)$$

$$\gamma = \gamma_0 \quad (32f)$$

$$\Gamma = \Gamma_0 \quad (32g)$$

where the subscript zero indicate constant quantities and

$${\omega_0}^2 = {k_0}^2c^2 + {\omega_p}^2 \quad (33a)$$

$${\omega_p}^2 = 4\pi e^2 n_0\left[1/(m\gamma_0) + 1/(M\Gamma_0)\right] \quad (33b)$$

$${\gamma_0}^2 = 1 + 4\frac{e^2}{m^2c^2}{a_0}^2 \quad (34a)$$

$$\Gamma_0{}^2 = 1 + 4\frac{e^2}{M^2c^2}a_0{}^2 \tag{34b}$$

We linearize about (32) by taking

$$a = a_0 + \epsilon' a_1 + \ldots \tag{35a}$$

$$n = n_0 + \epsilon' n_1 + \ldots \tag{35b}$$

$$N = n_0 + \epsilon' N_1 + \ldots \tag{35c}$$

$$v = \epsilon' v_1 + \ldots \tag{35d}$$

$$V = \epsilon' V_1 + \ldots \tag{35e}$$

$$\Phi = \epsilon' \phi_1 + \ldots \tag{35f}$$

$$\omega = \omega_0 - \epsilon'\left(i\nu + \frac{\partial\Lambda}{\partial T}\right) + \ldots \tag{35g}$$

$$k = k_0 + \epsilon'\frac{\partial\Lambda}{\partial Z} + \ldots \tag{35h}$$

where ϵ' is the amplitude expansion variable. Note that we have linearized in amplitude only. The linearization of k and ω (35g,h) is a consequence of Eqs. (31). The phase Λ is itself a quantity of $O(1)$. We assume that $0 < \epsilon' << 1$ and keep only linear terms in ϵ'. However, all terms in ϵ are kept for now.

We may solve for Λ and the variables with subscript 1 by Fourier transform. Consider a Fourier component $e^{i(KZ-\Omega T)}$. Defining

$$1/R_n \equiv 1/(m\gamma_0)^n - (-)^n/(M\Gamma_0)^n \tag{36a}$$

$$\omega_p{}^2 \equiv 4\pi e^2 n_0/R_1 \tag{36b}$$

$$w \equiv \Omega/cK, \tag{36c}$$

we obtain the dispersion relation

$$\frac{(\omega_0 w - ck_0)^2}{w^2-1} = (w^2-1)\frac{R_1}{R_3}\frac{a_0^2e^2\omega_p^2}{c^2} + \frac{1}{4}\Omega^2\epsilon^2(w^2-1)$$
$$+\frac{R_1^2}{R_2^2}\frac{\omega_p^2a_0^2e^2}{w^2c^2(1-\epsilon^2\Omega^2/\omega_p^2)}. \tag{37}$$

The relative importance of the terms containing ϵ differs sharply from the weakly to the fully relativistic case. In the weakly relativistic case, one can show that that the dispersion relation (37) gives rise to growing solutions (imaginary Ω) under precisely the same conditions which give rise to modulational stability of the NLS.

In order to apply (37) to the weakly relativistic case, we treat a_0 as a quantity of $O(\epsilon)$,

$$a_0 = \epsilon a_1 \tag{38}$$

The leading order of (37) then implies that the left-hand side of (37) is of $O(\epsilon)$, so we take

$$w = ck_0/\omega_0 + \epsilon w_1 + \ldots \tag{39}$$

and examine the next order, $O(\epsilon^2)$, in (37). Solving for w_1^2, we find

$$w_1^2 = \frac{1}{4}\frac{K^2}{k_0^2}\frac{\omega_p^2}{\omega_0^2} + \frac{a_1^2 e^2}{c^2}\frac{R_1}{R_3}\frac{\omega_p^6}{c^4 k_0^4}$$

$$-\frac{a_1^2 e^2}{c^2}\frac{R_1^2}{R_2^2}\frac{\omega_p^4 \omega_0^2}{c^4 k_0^4} \tag{40}$$

Negative values of w_1^2 correspond to instability. Since the terms containing K^2 are positive definite (stabilizing), it suffices to set $K^2 = 0$ to find the condition for instability. It is

$$\omega_p^2 < \frac{R_1 R_3}{R_2^2}\omega_0^2 \tag{41}$$

In the nonrelativistic limit, one can show

$$\frac{1}{mM} = \frac{R_1}{R_3} - \frac{R_1^2}{R_2^2} \tag{42}$$

with which one can show that, as expected, negative w_1^2 is possible over exactly the same frequency range that (18) is modulational unstable over.

A glance at Eqs. (34) shows that at least one of the components will attain relativistic velocities if a_0 is comparable to mc/e. In this case, the terms containing ϵ in the dispersion relation (37) can generally be neglected. The dispersion relation can then be expressed as a fourth-order polynomial in w:

$$w^2(\omega_0 w - ck_0)^2 = \omega_p^2 \alpha^2 (w^2 - \beta^2)(w^2 - 1) \tag{43}$$

where

$$\alpha^2 \equiv \frac{a_0^2 e^2}{c^2}\frac{R_1}{R_3} \tag{44a}$$

$$\beta^2 \equiv 1 - \frac{R_1 R_3}{R_2^2} = \frac{R_3}{R_1 M \Gamma m \gamma} \tag{44b}$$

Let us consider first the particular special case of an electron-positron plasma. Eq. (43) then factors into perfect squares on both sides since $1/R_2 = 0$. One thus obtains the pair of quadratic equations

$$w^2(\omega_0 - \alpha\omega_p) - wck_0 + \alpha\omega_p = 0 \tag{45a}$$

$$w^2(\omega_0 + \alpha\omega_p) - wck_0 - \alpha\omega_p = 0 \tag{45b}$$

The solution of (45a) is

$$2w = \frac{1}{\omega_0 - \alpha\omega_p}\left(ck_0 \pm ((\omega_0 - 2\alpha\omega_p)^2 - \omega_p^2)^{1/2}\right) \tag{46}$$

Clearly w will have a complex root if the radical is imaginary. This will occur whenever

$$\omega_0^2 < (2\alpha + 1)^2 \omega_p^2 \tag{47}$$

In the extreme relativistic limit, α approaches 1/2 and the range of instability is from $\omega_0 = \omega_p$ up to $2\omega_p$.

A second special case of interest is the extreme relativistic limit of an ion-electron plasma. A glance at the values of Γ_0 and γ_0 and evaluation of Eq. (44b) shows that in the limit $a_0 >> Mc/e$, we have $\beta \to 1$, and therefore instability occurs if (47) is valid.

A third special case of interest is the case $m\gamma_0 << M\Gamma_0$. This limit corresponds to the case of a nonrelativistic or moderately relativistic plasma in which the ions are much heavier than the electrons. In this limit,

$$\alpha \approx e a_0 m \gamma_0 / c \tag{48a}$$

$$\beta \approx m\gamma_0/(M\Gamma_0) << 1 \tag{48b}$$

and dividing out the trivial double root $w = 0$, we see that the dispersion relation reduces to

$$(\omega_0 w - ck_0)^2 = \alpha^2 \omega_p^2 (w^2 - 1) \tag{49}$$

The condition for instability is now

$$\omega_0^2 < \omega_p^2 \frac{1+\alpha^2}{2} \tag{50}$$

SUMMARY

We now briefly discuss some of the astrophysical implications of our results. Recent observational data[7-9] strongly support the hypothesis that pulsar micropulses are a "temporal" phenomenon and can be interpreted within the amplitude-modulated noise model[19].

As discussed[2-5], nonlinear modulational instability would provide a natural mechanism for amplitude modulations. Now, according to pulsar models[17,18], the pulsar magnetosphere is composed of secondary electrons and positrons. However, as we have seen here, in the absense of an ambient magnetic field, intense propagation, and postive ions, a cold electron-positron plasma does not exhibit modulational instability for either linear or circular polarization. (Finite temperature affects this result only just above the plasma frequency.) On the other hand, we have that a cold electron positive-ion plasma is modulationally unstable for either polarization. So if pulsar micropulses really are caused by modulational instability, then at least one new feature must enter the problem.

Of the possible new features, there are three which seem reasonable to consider. From the above result on the modulational instability of an electron-positive ion plasma, one would expect that the inclusion of some positive ions could drive the electron-positron plasma modulationally unstable. In a three-component neutral plasma consisting of electrons, positrons, and positive ions, we expect that a critical ratio (positive-ion density)/(positron density) will exist – depending perhaps on parameters such as the frequency – since the limiting cases (electron-positron, electron-ion) give opposing results.

The second feature would be the presence of ambient magnetic fields, which could strongly affect the modulational instability of any pulse in a plasma, in particular an electron-positron plasma. We find that indeed such an ambient magnetic field could drive an electromagnetic pulse modulationally unstable, [see eq. (29)], and if the field was extremely strong as expected, then a very wide frequency band of instability could be open.

Third, in pulsar magnetospheres, relativistic electron velocities are fully expected. Under these circumstances, since higher-order corrections to the nonrelativistic nonlinear coefficients could be important, some other scheme for analyzing the stability of an intense pulse had to be devised.

We have presented here such a scheme for studying the problem of modulational instability of a relativistic electromagnetic pulse propagating in a two-component neutral plasma. The general condition for modulational instability is given for imaginary solutions for ω_0 in (43). This condition reduces to the earlier conditions found in the nonrelativistic cases[6,10] in the appropriate limit. In the extreme relativistic case or in the electron-positron case, (50) is the general result. In the extreme relativistic limit, the range of instability for ω_0 is from ω_p up to $2\omega_p$. However remember that for relativistic velocities, the plasma frequency is reduced by a factor of $\gamma^{1/2}$ from what it would be if the particles in the plasma were nonrelativistic. In other words, as the pulse intensity is increased, the general range of instability is reduced because of the decrease in the value of ω_p. But there is always a range of ω_0 just above the plasma frequency which is unstable.

It is interesting to note how the modulational instability depends on the linear dispersion. In the nonrelativistic case, it is as important as the nonlinear terms. As the pulse intensity is increases, the dispersion term becomes less and less important, finally producing only an $O(\epsilon^2)$ correction to the regions of modulational instability.

ACKNOWLEDGEMENTS

This research was supported in part by the ONR through grant N00014-88-K-0153 and the AFOSR thru grant AFOSR-89-0510.

REFERENCES

1. Max, C.: Phys. Fluids 16, 1277; ibid. 1480 (1973).
2. Chian, A. and Kennel, C: Astrophys. & Sp. Sci. 97, 9 (1983).
3. Mofiz, U.A., DeAngelis, U. & Forlani, A. 1984 *Plasma Physics and Controlled Fusion* 26, 1099.
4. Mofiz, U.A., DeAngelis, U. & Forlani, A. 1985 Phys. Rev. A31, 951.
5. Mofiz, U.A. and Podder, J., 1987 Phys. Rev. A36, 1811.
6. Kates, R. and Kaup, D.J., 1989 J. Plasma Physics (GB) 41, 507.
7. Gil, J: 1986, Astroph. J. 308, 691.
8. Smirnova, T.V., Soglasnov, V.A., Popov, M.V. & Novikov, A.Yu: 1986 Sov. Astron. 30, 51.
9. Smirnova, 1988 Sov. Astron. Lett. 14, 20.
10. Kates, R. and Kaup, D.J., 1989 J. Plasma Physics (GB) 41, 521.
11. Kates, R., Ann. Phys. 1981 132, 1.
12. Van Dyke, M. 1964 *Perturbations in Fluid Mechanics*, Academic Press, N.Y.
13. Nayfeh, A. 1981 *Introduction to Perturbation Techniques*, Wiley.
14. Kozlov, V., Litvak, A., and Suvarov, E., 1979 Sov. Phys. JETP 49 75.
15. Akhiezer, A.L., et el.: 1975, Plasma Electrodynamics, Pergammon, Oxford, Chapter 5.
16. Luenow, W.:1968, Plasma Phys. 10, 973.
17. Ruderman M. and Sutherland, P.: 1975, Astrophys. J. 196, 51.
18. Arons J. and Scharlemann, E.:1979, Astrophys, J. 231, 854.
19. Rickett, B: 1975 Astrophys. J. 197, 185.
20. Cordes, J.: 1979, Space Sci Rev. 24 567.
21. Karpman, V. and Krushkal, E. :1969, Soviet Phys. JETP 28, 277.
22. Luenow, W.:1968, Plasma Phys. 10, 973.

INDEX

Accelerator designs, 229
Acoustic finite integration technique, 518
Acoustic wave, 146,532
Acousto-optic interactions, 451
Active circuit, 299
Adaptive mesh generation, 239
Adaptive refinement, 240
Adiabatic modes, 16
Algorithmic imaging, 523
Analytical theory, 399
Anderson localization, 457
Anisotropic media, 114
Annular waveguide, 134
Ansatz, 124
Antenna radiation, 126
Approximate boundary conditions, 477
Astigmatic beam, 80
Asymptotics, 198
  diffraction theory, 23
  evaluation, 92
  expansions, 542
  ray theory, 101,113,118
  representations, 329
  techniques, 324

Backpropagation, 510, 517
Backscatter, 417
  cross sections, 128,142
  enhancement, 398,405,407
  spectrum, 531
Banded-matrix solvers, 187
Bandlimited fractals, 438
Basis functions, 251, 324
Beams, 12
Bethe-Salpeter equation, 409
Biconjugate gradient method, 382
Biological media, 79
Bit operations, 186
Bohr-Sommerfeld quantization, 538
Born approximation, 498,508,524
Born scattering, 114,120
Born series, 508
Boundary elements, 101
Boundaries
  absorbing, 171,181,294
  dispersive, 287
Bragg conditions, 439,454
Brillouin zone, 457

Cagniard method, 145
Cantor bar model, 443
Chebyshev polynomials, 126
Chiral electrodynamics, 485
Chirality, 485
Chiroarrays™, 489
Chiroshield™, 487
Chirosorb™, 487
Chirostrip™ Antennas, 490
Chirowaveguides™, 488
Circular dichroism, 486
Coherent intensity, 413
Collisional loss, 537
Complex dielectric constant, 381
Complex rays, 13,71
Complex source, 14,79,87,88
Computation bandwidth, 186,192
Computational complexity, 185
Computational electromagnetics, 197
Computer packages, 273
Computer performance, 185
Conductor losses, 310
Conjugate gradient method, 193,215,260,339
Constitutive relatons, 486
Convective loss, 538
Cooper pairs, 468
Coordinate-coordinate approach, 407
Coplanar stripline, 365
Coplanar waveguide, 314
Corner functions, 276
Coupled cavity linear accelerators, 232
Cray, 185,198
Current-based hybrid analysis, 124

Curvilinear coordinate, 179
Curvilinear space cell, 203
Cyclotron frequency, 548

Debye relaxations, 205
Delaunay-Voronoi compl. grids, 172
Density fluctuation, 534
Density of states, 460
Dielectric radomes, 153
Dielectric slabs and cones, 153
Diffraction orders, 448
Diffraction tomography, 510,523
Dilation symmetry, 436
Dispersive materials, 205
Double-stub filter, 258
Drift-tube linear accelerator, 231
Dynamic ray equations, 117
Dyson equation, 398,409

Edge condition, 324,481
Edge elements, 171,253
Edge-diffracted field, 50
Eikonal equation, 116
Elastic waves, 518,526
Electromagnetic yardstick, 437
Ellipsoidal reflector, 49
Enantiomorph, 486
Equivalence principle, 104
Equivalent edge currents, 47
Ewald's lattice sum, 461
Ewald-sphere, 523
Exciton lifetimes, 457

Faraday rotation, 548
FET, 300
Feynman diagrams, 409,448
Field transient, 136
Finite difference, 101,113,171,197,300
Finite element, 101,171,194,239,251,355,479
FLOP count, 185
Flow graphs, 451
Fluctuation theory, 495
Fourier-Bessel transforms, 43,256
Fractal antenna arrays, 442
Fractal apertures, 441
Fractal dimension, 436
Fractal electrodynamics, 435
Fractal fibers, 440
Fractal multi-layer, 445
Fractal surfaces, 435
Fractals, 435
Frequency domain solution, 173
Frequency-domain, 198
Fresnel approximation, 24
Fresnel length, 88

Gabor expansion, 71
Gabor lattice, 13
Gabor representation, 23,28
Galerkin method, 125, 250, 309,320,340,357
Gaussian beams, 12,47,51,70,79,92,101,113,413
Gaussian synthesis, 23
Generalized dielectric constant, 468
Generalized quadratic operator, 33
Generalized ray expansion, 72
Geometrical optics, 70
Geometrical ray theory, 118
Geometrical theory of diffraction,12,47,198
Gridding methods, 202
Group velocity, 116
Guided modes, 15
Gunn diodes, 302

Half-plane, 482
Heitler approximation, 511
HF heating, 531
Hilbert transform, 88
Holographic data reduction, 520
Huygens principle, 520
Hybrid methods, 72,101,123,138
Hybrid ray-mode, 106
Hyperthermia, 79
Hysteresis, 242

Image theory, 388
Impedance boundary condition, 477
Incoherent intensity, 413
Inhomogeneous medium, 161
Integral equation,124 148,154,166,198,216,250, 307,347,408,508,523
Intensity moments, 429
Intrinsic mode, 17
Inverse scattering, 507,517
Ionosphere, 161

Jacobi-Bessel, 43
Johns matrix technique, 287
Jungle gym structure, 230

Kerr law, 241
Kerr medium, 529
Kirchhoff approximation, 71,400,417
Kirchhoff solution, 439
Klein-Gorden equation, 33
Kolmogorov-Feller equation, 426
Korringa-Kohn-Rostoker Method, 460

Lagrange manifold, 27
Landau-Pollak Theorem, 23
Langmuir sideband, 531
Langmuir wave, 532

Layered dielectric waveguide, 314
Leaky wave antenna, 305
Leaky waves, 319,363
Lippmann Schwinger equation, 511
Lorentz force, 534

MAFIA codes, 230
Maliuzhinets method, 482
Marching-in-time, 201
Material slab Green's function, 330
Meander line, 259,343
Mehler-Fock transform, 431
Mesh current, 253
Mesh reflectors, 41
Method of moments, 101,106,198,218,275,
  308,319, 329, 364,473
Microstrip, 250,281,305,310,319,329,347,355
  363,383,387,491
  antennas, 265,351,490
  feedlines, 371
  junctions, 258
  patches, 359,371
  resonator, 385
Mie series, 481
MMICs, 249
Mobile satellite system, 53
Modal propagator matrix, 135
Model complexity, 186
Model-based parameter estimation, 194
Modeling
  frequency domain, 187
  time domain, 188
Monolithic circuits, 307
Monte Carlo solution, 398, 417
Mueller matrix, 502
Multiple gratings, 447

Negative dielectric materials, 468
Nondestructive testing, 518
Nonlinear characteristics, 305
Nonlinear modulational instability, 542
Nonlinear optical waveguides, 239
Nonlinear Schroedinger equation, 542
Nonlocal mapping, 508
Nyquist theorem, 341

Ohmic losses, 341
One-way wave equation, 479
Optical activity, 486

Pseudo-mesh, 249
Parabolic approximation, 31
Parabolic equation, 103
Parabolic reflector, 52
Parallel computers, 185
Parallel plate modes, 321
Parallel-plate waveguide, 367,471
Parametric excitation, 532
Patch antennas, 305
Periodic structure, 302,457
Periodically modulated dielectric, 458
Personal access satellite system, 53
Perturbation theory, 417,430
Phase-perturbation technique, 418
Phase-space, 70
Photonic band gap, 457
Physical optics, 47,524
Physical theory of diffraction, 47
Piecewise sinusoidal modes, 372
Planar dipole antenna, 217
Planar volume grating, 447
Plane interface, 97
Plasma, 531,541
Plasma oscillation, 534
Pockington's equation, 308
Poisson impulse rpcoess, 428
Polarimetric backscattering, 495
Polarization currents, 275
Predictive dynamic range, 204
Processing speed, 207
Prony's method, 393
Pulsed Beam, 87

Quasi-TEM structures, 277

Radar cross section, 190,197,471
Radiance, 399
Radiative transfer theory, 399
Radio-frequency quadrupole, 230
Radomes, 491
Radon transform, 93
Random medium, 397,425,496
Ray and beam tracking methods, 70
Ray tube, 15
Rayleigh criterion, 400
Rayleigh surface wave, 526
Rayleigh-Gans approximation, 440
Rays, 12
Reflector antennas, 39
Relativistic regime, 541
Remote sensing, 495
Resonant mode expansion, 137
Richmond's formulation, 154
Rooftop functions, 218,275,339
Rytov approximation, 509

Salisbury and Dallenbach screens, 487
Saturable nonlinearity, 241
Scattering, 67,133
  cross-sections, 492
  matrix, 382
  rough surface, 398,407,417,502

Sea ice, 501
Sea water, 501
Seismic waves, 113
Self-focusing, 244,529
Self-similarity, 436,444
Serrated aperture, 441
Shear waves, 526
Sheet impedance, 275
Simulated image technique, 388
Singular perturbation method, 542
Singular value decomposition, 511
Slit-coupled cavity, 133
Slot antenna, 129
Slotline, 322
Snow, 501
Soil, 501
Soliton, 529
Soybean canopy, 501
Space-grid time-domain codes, 201
Space-time fluctuations, 405
Spatial harmonics, 441
Spatial influence zone, 23
Spatial sampling, 186,194
Spectral changes, 405
Spectral domain, 339,364
Spectral influence zone, 25
Spectral-coordinate method, 407
Spectral-spectral method, 407
Spontaneous emission, 457
Statistical ensemble, 407
Stubs, 310,343
Subsectional functions, 251
Superconductivity, 467
  cylinder, 469
  dipole, 473
Surface distortion compensation, 53
Surface impedances, 203,400,467
Surface-wave, 125
Synthetic aperture focusing, 520
System matrix, 187,188

T-matrix, 101
Target electrical size, 202
Target Q factor, 202
Temporal sampling, 186
Time domain, 133,176,234
Time domain diakoptics, 288
Time-stepping, 201
Transducer model, 518
Transport equation, 117
Transport theory, 399
Transverse field transmission matrix, 266
Transverse switching, 243
Traveling wave transistor, 300
Truncation planes, 201
Two-fluid model, 468
Two-Photon absorption, 529

Underwater acoustics, 111
Uniform asymptotic theory, 47

Variable grid size, 177
Vector radiative transfer theory, 399

Waveguide iris, 295
Waveguide simulator, 375
Weak Anderson localization, 400
Weierstrass functions, 438
Weyl composition equation, 32
Weyl pseudo-differential operator, 35
Whistler waves, 531
Wiener-Hermite expansion, 418
Wiener-Hopf method, 482
Wraparound antennas, 265
Wraparound current, 43